University of
Hertfordshire

College Lane, Hatfield, Herts. AL10 9AB

Learning and Information Services
College Lane Campus Learning Resources Centre, Hatfield

For renewal of Standard and One Week Loans,
please visit the web site **http://www.voyager.herts.ac.uk**

This item must be returned or the loan renewed by the due date.
The University reserves the right to recall items from loan at any time.
A fine will be charged for the late return of items.

PRENTICE HALL SIGNAL PROCESSING SERIES

Alan V. Oppenheim, *Series Editor*

Discrete-Time Speech
Signal Processing

ISBN 0-13-242942-X

90000

Discrete-Time Speech Signal Processing

Principles and Practice

Thomas F. Quatieri

Massachusetts Institute of Technology
Lincoln Laboratory

Prentice Hall PTR
Upper Saddle River, NJ 07458
www.phptr.com

Library of Congress Cataloging-in-Publication Data

Quatieri, T. F. (Thomas F.)
    Discrete-time speech processing: principles and practice / Thomas F.
Quatieri.
            p.   cm. -- (Prentice-Hall signal processing series)
Includes bibliographical references and index.
    ISBN 0-13-242942-X
    1. Speech processing systems. 2. Discrete-time systems   I. Title.
II. Series.
    TK7882.S65   Q38   2001
    006.5--dc21

2001021821

Editorial/production supervision: *Faye Gemmellaro*
Production assistant: *Jodi Shorr*
Acquisitions editor: *Bernard Goodwin*
Editorial assistant: *Michelle Vincenti*
Marketing manager: *Dan DePasquale*
Manufacturing manager: *Alexis Heydt*
Cover design director: *Jerry Votta*
Cover designers: *Talar Agasyan, Nina Scuderi*
Composition: *PreTEX, Inc.*

PH
PTR

©2002 Prentice Hall PTR
Prentice-Hall, Inc.
Upper Saddle River, NJ 07458

Prentice Hall books are widely used by corporations and government agencies for training, marketing, and resale.

The publisher offers discounts on this book when ordered in bulk quantities.

For more information, contact:
Corporate Sales Department
Phone: 800-382-3419   Fax: 201-236-7141
Email: corpsales@prenhall.com

Or write:
Prentice Hall PTR
Corporate Sales Department
One Lake Street
Upper Saddle River, NJ 07458

MATLAB is a registered trademark of The MathWorks, Inc.

All product names mentioned herein are the trademarks of their respective owners.

Printed in the United States of America

10 9 8 7 6 5 4 3 2 1

ISBN 0-13-242942-X

Pearson Education LTD.
Pearson Education Australia PTY, Limited
Pearson Education Singapore, Pte. Ltd.
Pearson Education North Asia Ltd.
Pearson Education Canada, Ltd.
Pearson Education de Mexico, S.A. de C.V.
Pearson Education—Japan
Pearson Education Malaysia, Pte. Ltd.

This book is dedicated to my wife Linda

and to our parents and family.

Contents

Foreword

Speech and hearing, man's most used means of communication, have been the objects of intense study for more than 150 years—from the time of von Kempelen's speaking machine to the present day. With the advent of the telephone and the explosive growth of its dissemination and use, the engineering and design of evermore bandwidth-efficient and higher-quality transmission systems has been the objective and providence of both engineers and scientists for more than seventy years. This work and investigations have been largely driven by these real-world applications which now have broadened to include not only speech synthesizers but also automatic speech recognition systems, speaker verification systems, speech enhancement systems, efficient speech coding systems, and speech and voice modification systems. The objectives of the engineers have been to design and build real workable and economically affordable systems that can be used over the broad range of existing and newly installed communication channels.

Following the development of the integrated circuit in the 1960s, the communication channels and the end speech signal processing systems changed from analog to purely digital systems. The early laboratories involved in this major shift in implementation technology included Bell Telephone Laboratories, MIT Lincoln Laboratory, IBM Thomas Watson Research Laboratories, the BB&N Speech Group, and the Texas Instruments Company, along with numerous excellent university research groups. The introduction by Texas Instruments in the 1970s of its Speak-and-Spell product, which employed extensive digital integrated circuit technology, caused the entire technical, business, and marketing communities to awaken to the endless system and product possibilities becoming viable through application of the rapidly developing integrated circuit technologies.

As more powerful integrated circuits became available, the engineers would take their existing working systems and try to improve them. This meant going back and studying their existing models of speech production and analysis in order to gain a more complete understanding of the physical processes involved. It also meant devising and bringing to bear more powerful mathematical tools and algorithms to handle the added complexity of the more detailed analysis. Certain methodologies became widely used partly because of their initial success, their viability, and their ease of analysis and implementation. It then became increasingly difficult to change an individual part of the system without affecting the other parts of the system. This logical design procedure was complicated and compromised by the ever-present reducing cost and increasing power of the digital integrated circuits used.

In the midst of all this activity lay Lincoln Laboratory with its many and broad projects in the speech area. The author of this timely book has been very actively involved in both the engineering and the scientific aspects of many of those projects and has been a major contributor to their success. In addition, he has developed over the course of many years the graduate course in speech analysis and processing at MIT, the outgrowth of which is this text on the subject.

In this book you will gain a thorough understanding of the basic scientific principles of speech production and hearing and the basic mathematical tools needed for speech signal representation, analysis, and manipulation. Then, through a plethora of applications, the author illustrates the design considerations, the system performance, and the careful analysis and critique of the results. You will view these many systems through the eyes of one who has been there, and one with vision and keen insight into figuring out why the systems behave the way they do and where the limitations still exist.

Read carefully, think continually, question always, try out the ideas, listen to the results, and check out the extensive references. Enjoy the magic and fascination of this broad area of the application of digital technology to voice communication through the experiences of an active researcher in the field. You will be richly rewarded.

James F. Kaiser
Visiting Professor, Department of Electrical and Computer Engineering
Duke University
Durham, NC

Preface

This text is in part an outgrowth of my MIT graduate course *Digital Speech Signal Processing*, which I have taught since the Fall of 1990, and in part a result of my research at MIT Lincoln Laboratory. As such, principles are never too distant from practice; theory is often followed by applications, both past and present. This text is also an outgrowth of my childhood wonder in the blending of signal and symbol processing, sound, and technology. I first felt this fascination in communicating with two cans coupled by twine, in playing with a toy Morse code, and in adventuring through old ham radio equipment in my family's basement. My goals in this book are to provide an intensive tutorial on the principles of discrete-time speech signal processing, to describe the state-of-the-art in speech signal processing research and its applications, and to pass on to the reader my continued wonder for this rapidly evolving field.

The text consists of fourteen chapters that are outlined in detail in Chapter 1. The "theory" component of the book falls within Chapters 2–11, while Chapters 12–14 consist primarily of the application areas of speech coding and enhancement, and speaker recognition. Other applications are introduced throughout Chapters 2–11, such as speech modification, noise reduction, signal restoration, and dynamic range compression. A broader range of topics that include speech and language recognition is not covered; to do so would result in a survey book that does not fill the current need in this field. The style of the text is to show not only when speech modeling and processing methods succeed, but also to describe limitations of the methods. This style makes the reader question established ideas and reveals where advancement is needed. An important tenet in this book is that anomaly in observation is crucial for advancement; as reflected by the late philospher Thomas Kuhn: "Discovery commences with the awareness of anomaly, i.e., with the recognition that nature has somehow violated the paradigm-induced expectations that govern normal science."[1]

The text body is strongly supplemented with examples and exercises. Each exercise set contains a number of MATLAB problems that provide hands-on experience with speech signals and processing methods. Scripts, workspaces, and signals, required for the MATLAB exercises, are located on the Prentice Hall companion website (**http://www.phptr.com/quatieri/**) Also on this website are audio demonstrations that illustrate a variety of principles and applications

---

[1] T. Kuhn, *The Structure of Scientific Revolution*, University of Chicago Press, 1970.

from each chapter, including time-scale modification of the phrase "as time goes by" shown on the front cover of this book. The book is structured so that application areas that are not covered as separate topics are either presented as examples or exercises, e.g., speaker separation by sinusoidal modeling and restoration of old acoustic recordings by homomorphic processing. In my MIT speech processing course, I found this approach to be very effective, especially since such examples and exercises are fascinating demonstrations of the theory and can provide a glimpse of state-of-the-art applications.

The book is also structured so that topics can be covered on different levels of depth and breadth. For example, a one-semester course on discrete-time speech signal processing could be taught with an emphasis on fundamentals using Chapters 2–9. To focus on the speech coding application, one can include Chapter 12, but also other applications as examples and exercises. In a two-semester course, greater depth could be given to fundamentals in the first semester, using Chapters 2–9. In the second semester, a focus could then be given to advanced theories and applications of Chapters 10–14, with supplementary material on speech recognition.

I wish to express my thanks to the many colleagues, friends, and students who provided review of different chapters of this manuscript, as well as discussions on various chapter topics and style. These include Walt Andrews, Carlos Avendano, Joe Campbell, Mark Clements, Jody and Michael Crocetta, Ron Danisewicz, Bob Dunn, Carol Epsy-Wilson, Allen Gersho, Terry Gleason, Ben Gold, Mike Goodwin, Siddhartan Govindasamy, Charles Jankowski, Mark Kahrs, Jim Kemerling, Gernot Kubin, Petros Maragos, Rich McGowen, Michael Padilla, Jim Pitton, Mike Plumpe, Larry Rabiner, Doug Reynolds, Dan Sinder, Elliot Singer, Doug Sturim, Charlie Therrien, and Lisa Yanguas. In addition, I thank my MIT course students for the many constructive comments on my speech processing notes, and my teaching assistants: Babak Azifar, Ibrahim Hajjahmad, Tim Hazen, Hanfeng Yuan, and Xiaochun Yang for help in developing class exercise solutions and for feedback on my course notes. Also, in memory of Gary Kopec and Tom Hanna, who were both colleagues and friends, I acknowledge their inspiration and influence that live on in the pages of this book.

A particular thanks goes to Jim Kaiser, who reviewed nearly the entire book in his characteristic meticulous and uncompromising detail and has provided continued motivation throughout the writing of this text, as well as throughout my career, by his model of excellence and creativity. I also acknowledge Bob McAulay for the many fruitful and highly motivational years we have worked together; our collaborative effort provides the basis for Chapters 9, 10, and parts of Chapter 12 on sinusoidal analysis/synthesis and its applications. Likewise, I thank Hamid Nawab for our productive work together in the early 1980s that helped shape Chapter 7, and Rob Baxter for our stimulating discussions that helped to develop the time-frequency distribution tutorials for Chapter 11. In addition, I thank the following MIT Lincoln Laboratory management for flexibility given me to both lecture at MIT and perform research at Lincoln Laboratory, and for providing a stimulating and open research environment: Cliff Weinstein, Marc Zissman, Jerry O'Leary, Al McLaughlin, and Peter Blankenship. I have also been very fortunate to have the support of Al Oppenheim, who opened the door for me to teach in the MIT Electrical Engineering and Computer Science Department, planted the seed for writing this book, and provided the initial and continued inspiration for my career in digital signal processing. Thanks also goes to Faye Gemmellaro, production editor; Bernard Goodwin, publisher; and others at Prentice Hall for their great care and dedication that helped determine the quality of the finished book product.

Finally, I express my deepest gratitude to my wife Linda, who provided the love, support, and encouragement that was essential in a project of this magnitude and who has made it all meaningful. Linda's voice example on the front cover of this book symbolizes my gratitude now and "as time goes by."

Thomas F. Quatieri
MIT Lincoln Laboratory[2]

[2] This work was sponsored by the Department of Defense under Air Force contract F19628–00-C–0002. Opinions, interpretations, conclusions, and recommendations are those of the author and not necessarily endorsed by the United States Air Force.

Introduction

1.1 Discrete-Time Speech Signal Processing

Speech has evolved as a primary form of communication between humans. Nevertheless, there often occur conditions under which we measure and then transform the speech signal to another form in order to enhance our ability to communicate. One early case of this is the transduction by a telephone handset of the continuously-varying speech pressure signal at the lips output to a continuously-varying (analog) electric voltage signal. The resulting signal can be transmitted and processed electrically with analog circuitry and then transduced back by the receiving handset to a speech pressure signal. With the advent of the wonders of digital technology, the analog-to-digital (A/D) converter has entered as further "transduction" that samples the electrical speech samples, e.g., 8000 samples per second for telephone speech, so that the speech signal can be digitally transmitted and processed. Digital processors with their fast speed, low cost and power, and tremendous versatility have replaced a large part of analog-based technology.

The topic of this text, *discrete-time speech signal processing*, can be loosely defined as the manipulation of sampled speech signals by a digital processor to obtain a new signal with some desired properties. Consider, for example, changing a speaker's rate of articulation with the use of a digital computer. In the modification of articulation rate, sometimes referred to as *time-scale modification* of speech, the objective is a new speech waveform that corresponds to a person talking faster or slower than the original rate, but that maintains the character of the speaker's voice, i.e., there should be little change in the pitch (or rate of vocal cord vibration) and spectrum of the original utterance. This operation may be useful, for example, in fast scanning of a long recording in a message playback system or slowing down difficult-to-understand speech. In this

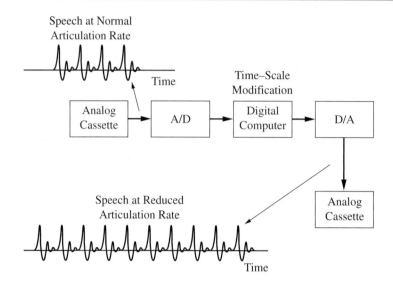

Figure 1.1 Time-scale modification as an example of discrete-time speech signal processing.

application, we might begin with an analog recording of a speech utterance (Figure 1.1). This *continuous-time* waveform is passed through an A/D waveform converter to obtain a sequence of numbers, referred to as a *discrete-time signal*, which is entered into the digital computer. Discrete-time signal processing is then applied to obtain the required speech modification that is performed based on a model of speech production and a model of how articulation rate change occurs. These speech-generation models may themselves be designed as analog models that are transformed into discrete time. The modified discrete-time signal is converted back to analog form with a digital-to-analog (D/A) converter, and then finally perhaps stored as an analog waveform or played directly through an amplifier and speakers. Although the signal processing required for a high-quality modification could conceivably be performed by analog circuitry built into a redesigned tape recorder,[1] current digital processors allow far greater design flexibility. Time-scale modification is one of many applications of discrete-time speech signal processing that we explore throughout the text.

1.2 The Speech Communication Pathway

In the processing of speech signals, it is important to understand the pathway of communication from speaker to listener [2]. At the linguistic level of communication, an idea is first formed in the mind of the speaker. The idea is then transformed to words, phrases, and sentences according to the grammatical rules of the language. At the physiological level of communication, the

---

[1] Observe that time-scale modification cannot be performed simply by changing the speed of a tape recorder because this changes the pitch and spectrum of the speech.

brain creates electric signals that move along the motor nerves; these electric signals activate muscles in the vocal tract and vocal cords. This vocal tract and vocal cord movement results in pressure changes within the vocal tract, and, in particular, at the lips, initiating a sound wave that propagates in space. The sound wave propagates through space as a chain reaction among the air particles, resulting in a pressure change at the ear canal and thus vibrating the ear drum. The pressure change at the lip, the sound propagation, and the resulting pressure change at the ear drum of the listener are considered the acoustic level in the speech communication pathway. The vibration at the ear drum induces electric signals that move along the sensory nerves to the brain; we are now back to the physiological level. Finally, at the linguistic level of the listener, the brain performs speech recognition and understanding.

The linguistic and physiological activity of the speaker and listener can be thought of as the "transmitter" and "receiver," respectively, in the speech communication pathway. The transmitter and receiver of the system, however, have other functions besides basic communications. In the transmitter there is feedback through the ear which allows monitoring and correction of one's own speech (the importance of this feedback has been seen in studies of the speech of the deaf). Examples of the use of this feedback are in controlling articulation rate and in the adaptation of speech production to mimic voices. The receiver also has additional functions. It performs voice recognition and it is robust in noise and other interferences; in a room of multiple speakers, for example, the listener can focus on a single low-volume speaker in spite of louder interfering speakers. Although we have made great strides in reproducing parts of this communication system by synthetic means, we are far from emulating the human communication system.

1.3 Analysis/Synthesis Based on Speech Production and Perception

In this text, we do not cover the entire speech communication pathway. We break into the pathway and make an analog-to-digital measurement of the acoustic waveform. From these measurements and our understanding of speech production, we build engineering models of how the vocal tract and vocal cords produce sound waves, beginning with analog representations which are then transformed to discrete-time. We also consider the receiver, i.e., the signal processing of the ear and higher auditory levels, although to a lesser extent than the transmitter, because it is imperative to account for the effect of speech signal processing on perception.

To preview the building of a speech model, consider Figure 1.2 which shows a model of vowel production. In vowel production, air is forced from the lungs by contraction of the muscles around the lung cavity. Air then flows past the vocal cords, which are two masses of flesh, causing periodic vibration of the cords whose rate gives the pitch of the sound; the resulting periodic puffs of air act as an excitation input, or source, to the vocal tract. The vocal tract is the cavity between the vocal cords and the lips, and acts as a resonator that spectrally shapes the periodic input, much like the cavity of a musical wind instrument. From this basic understanding of the speech production mechanism, we can build a simple engineering model, referred to as the *source/filter* model. Specifically, if we assume that the vocal tract is a linear time-invariant system, or filter, with a periodic impulse-like input, then the pressure output at the lips is the

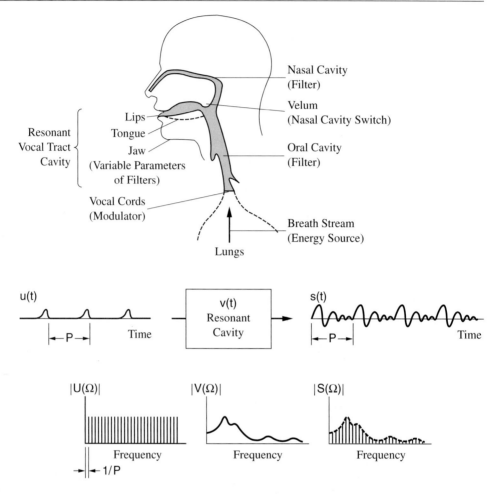

Figure 1.2 Speech production mechanism and model of a steady-state vowel. The acoustic waveform is modeled as the output of a linear time-invariant system with a periodic impulse-like input. In the frequency domain, the vocal tract system function spectrally shapes the harmonic input.

convolution of the impulse-like train with the vocal tract impulse response, and therefore is itself periodic. This is a simple model of a steady-state vowel. A particular vowel, as, for example, "a" in the word "f_ather," is one of many basic sounds of a language that are called *phonemes* and for which we build different production models. A typical speech utterance consists of a string of vowel and consonant phonemes whose temporal and spectral characteristics change with time, corresponding to a *changing* excitation source and vocal tract system. In addition, the time-varying source and system can also *nonlinearly* interact in a complex way. Therefore, although our simple model for a steady vowel seems plausible, the sounds of speech are not always well represented by linear time-invariant systems.

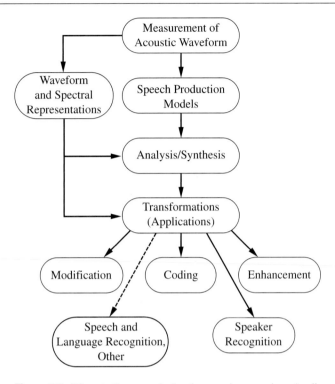

Figure 1.3   Discrete-time speech signal processing overview. Applications within the text include speech modification, coding, enhancement, and speaker recognition.

Based on discrete-time models of speech production, we embark on the design of speech *analysis/synthesis* systems (Figure 1.3). In analysis, we take apart the speech waveform to extract underlying parameters of the time-varying model. The analysis is performed with temporal and spectral resolution that is adequate for the measurement of the speech model parameters. In synthesis, based on these parameter estimates and models, we then put the waveform back together. An objective in this development is to achieve an identity system for which the output equals the input when no speech manipulation is performed. We also investigate waveform and spectral representations that do not involve models, but rather various useful mathematical representations in time or in frequency from which other analysis/synthesis methods can be derived. These analysis/synthesis methods are the backbone for applications that transform the speech waveform into some desirable form.

1.4   Applications

This text deals with applications of discrete-time speech analysis/synthesis primarily in the following areas: (1) speech modification, (2) speech coding, (3) speech enhancement, and (4) speaker recognition (Figure 1.3). Other important application areas for discrete-time signal processing, including speech recognition, language recognition, and speech synthesis from text, are not given; to do so would require a deeper study of statistical discrete-time signal processing and linguistics than can be satisfactorily covered within the boundaries of this text. Tutorials in these areas can be found in [1],[3],[4],[5],[6],[7].

Modification — The goal in speech modification is to alter the speech signal to have some desired property. Modifications of interest include time-scale, pitch, and spectral changes. Applications of time-scale modification are fitting radio and TV commercials into an allocated time slot and the synchronization of audio and video presentations. In addition, speeding up speech has use in message playback, voice mail, and reading machines and books for the blind, while slowing down speech has application to learning a foreign language. Voice transformations using pitch and spectral modification have application in voice disguise, entertainment, and concatenative speech synthesis. The spectral change of frequency compression and expansion may be useful in transforming speech as an aid to the partially deaf. Many of the techniques we develop also have applicability to music and special effects. In music modification, a goal is to create new and exotic sounds and enhance electronic musical instruments. Cross synthesis, used for special effects, combines different source and system components of sounds, such as blending the human excitation with the resonances of a musical instrument. We will see that separation of the source and system components of a sound is also important in a variety of other speech application areas.

Coding — In the application of speech coding, the goal is to reduce the information rate, measured in bits per second, while maintaining the quality of the original speech waveform.[2] We study three broad classes of speech coders. Waveform coders, which represent the speech waveform directly and do not rely on a speech production model, operate in the high range of 16–64 kbps (bps, denoting bits per second). Vocoders are largely speech model-based and rely on a small set of model parameters; they operate at the low bit rate range of 1.2–4.8 kbps, and tend to be of lower quality than waveform coders. Hybrid coders are partly waveform-based and partly speech model-based and operate in the 4.8–16 kbps range with a quality between waveform coders and vocoders. Applications of speech coders include digital telephony over constrained-bandwidth channels, such as cellular, satellite, and Internet communications. Other applications are video phones where bits are traded off between speech and image data, secure speech links for government and military communications, and voice storage as with computer voice mail where storage capacity is limited. This last application can also benefit from time-scale compression where both information reduction and voice speed-up are desirable.

Enhancement — In the third application—speech enhancement—the goal is to improve the quality of degraded speech. One approach is to preprocess the speech waveform before it is degraded. Another is postprocessing enhancement after signal degradation. Applications of preprocessing include increasing the broadcast range of transmitters constrained by a peak power transmission limit, as, for example, in AM radio and TV transmission. Applications of postprocessing include reduction of additive noise in digital telephony and vehicle and aircraft

---

[2] The term *quality* refers to speech attributes such as naturalness, intelligibility, and speaker recognizability.

communications, reduction of interfering backgrounds and speakers for the hearing-impaired, removal of unwanted convolutional channel distortion and reverberation, and restoration of old phonograph recordings degraded, for example, by acoustic horns and impulse-like scratches from age and wear.

Speaker Recognition — This area of speech signal processing exploits the variability of speech model parameters across speakers. Applications include verifying a person's identity for entrance to a secure facility or personal account, and voice identification in forensic investigation. An understanding of the speech model features that cue a person's identity is also important in speech modification where we can transform model parameters for the study of specific voice characteristics; thus, speech modification and speaker recognition can be developed synergistically.

1.5 Outline of Book

The goal of this book is to provide an understanding of discrete-time speech signal processing techniques that are motivated by speech model building, as well as by the above applications. We will see how signal processing algorithms are driven by both time- and frequency-domain representations of speech production, as well as by aspects of speech perception. In addition, we investigate the capability of these algorithms to analyze the speech signal with appropriate time-frequency resolution, as well as the capability to synthesize a desired waveform.

Chapter 2 reviews the foundation of discrete-time signal processing which serves as the framework for the remainder of the text. We investigate some essential discrete-time tools and touch upon limitations of these techniques, as manifested through the uncertainty principle and the theory of time-varying linear systems that arise in a speech signal processing context. Chapter 3 describes qualitatively the main functions of the speech production mechanism and the associated anatomy. Acoustic and articulatory descriptors of speech sounds are given, some simple linear and time-invariant models are proposed, and, based on these features and models, the study of phonetics is introduced. Implications of sound production mechanisms for signal processing algorithms are discussed. In Chapter 4, we develop a more quantitative description of the acoustics of speech production, showing how the heuristics of Chapter 3 are approximately supported with linear and time-invariant mathematical models, as well as predicting other effects not seen by a qualitative perspective, such as a nonlinear acoustic coupling between the source and system functions.

Based on the acoustic models of Chapters 3 and 4, in Chapter 5 we investigate pole-zero transfer function representations of the three broad speech sound classes of periodic (e.g., vowels), noise-like (e.g., fricative consonants), and impulsive (e.g., plosive consonants), loosely categorized as "deterministic," i.e., with a periodic or impulsive source, and "stochastic," i.e., with a noise source. There also exist many speech sounds having a combination of these sound elements. In this chapter, methodologies are developed for estimating all-pole system parameters for each sound class, an approach referred to as *linear prediction* analysis. Extension of these methods is made to pole-zero system models. For both all-pole and pole-zero analysis, corresponding synthesis methods are developed. Linear prediction analysis first extracts the system component and then, by inverse filtering, extracts the source component. We can think of the source extraction as a method of deconvolution. Focus is given to estimating the source function during periodic sounds, particularly a "pitch synchronous" technique, based on the closed phase of the glottis, i.e., the slit between the vocal cords. This method of glottal flow waveform estimation reveals a nonlinear coupling between the source and the system. Chapter 6 describes an alternate means of deconvolution of the source and system components, referred to as *homomorphic filtering*. In this approach, convolutionally combined signals are mapped to additively combined signals on which linear filtering is applied for signal separation. Unlike linear prediction, which is a "parametric" (all-pole) approach to deconvolution, homomorphic filtering is "nonparametric" in that a specific model need not be imposed on the system transfer function in analysis. Corresponding synthesis methods are also developed, and special attention is given to the importance of phase in speech synthesis.

In Chapter 7, we introduce the *short-time Fourier transform* (STFT) and its magnitude for analyzing the spectral evolution of time-varying speech waveforms. Synthesis techniques are developed from both the STFT and the STFT magnitude. Time-frequency resolution properties of the STFT are studied and application to time-scale modification is made. In this chapter, the STFT is viewed in terms of a filter-bank analysis of speech which leads to an extension to constant-Q analysis and the *wavelet transform* described in Chapter 8. The wavelet transform represents one approach to addressing time-frequency resolution limitations of the STFT as revealed through the *uncertainty principle*. The filter-bank perspective of the STFT also leads to an analysis/synthesis method in Chapter 8 referred to as the *phase vocoder*, as well as other filter-bank structures. Also in Chapter 8, certain principles of auditory processing are introduced, beginning with a filter-bank representation of the auditory front-end. These principles, as well as others described as needed in later chapters, are used throughout the text to help motivate various signal processing techniques, as, for example, signal phase preservation. The analysis stage of the phase vocoder views the output of a bank of bandpass filters as sinewave signal components. Rather than relying on a filter bank to extract the underlying sinewave components and their parameters, an alternate approach is to explicitly model and estimate time-varying parameters of sinewave components by way of spectral peaks in the short-time Fourier transform. The resulting *sinewave analysis/synthesis* scheme, described in Chapter 9, resolves many of the problems encountered by the phase vocoder, e.g., a characteristic phase distortion problem, and provides a useful framework for a large range of speech applications, including speech modification, coding, and speech enhancement by speaker separation.

Pitch and a voicing decision, i.e., whether the vocal tract source is periodic or noisy, play a major role in the application of speech analysis/synthesis to speech modification, coding, and enhancement. Time-domain methods of pitch and voicing estimation follow from specific analysis techniques developed throughout the text, e.g., linear prediction or homomorphic analysis. The purpose of Chapter 10 is to describe pitch and voicing estimation, on the other hand, from a frequency-domain perspective, based primarily on the sinewave modeling approach of Chapter 9.

Chapter 11 then deviates from the main trend of the text and investigates advanced topics in nonlinear estimation and modeling techniques. Here we first go beyond the STFT and wavelet transforms of the previous chapters to time-frequency analysis methods including the *Wigner distribution* and its variations referred to as *bilinear time-frequency distributions*. These distributions, aimed at undermining the uncertainty principle, attempt to estimate important fine-structure speech events not revealed by the STFT and wavelet transform, such as events that occur *within* a glottal cycle. In the latter half of this chapter, we introduce a second approach to analysis of fine structure whose original development was motivated by *nonlinear aeroacoustic models* for spatially distributed sound sources and modulations induced by nonacoustic fluid motion. For example, a "vortex ring," generated by a fast-moving air jet from the glottis and traveling along the vocal tract, can be converted to a secondary acoustic sound source when interacting with vocal tract boundaries such as the epiglottis (false vocal folds), teeth, or inclusions in the vocal tract. During periodic sounds, in this model secondary sources occur within a glottal cycle and can exist simultaneously with the primary glottal source. Such aeroacoustic models follow from complex nonlinear behavior of fluid flow, quite different from the small compression and rarefraction perturbations associated with acoustic sound waves in the vocal tract that are given in Chapter 4. This aeroacoustic modeling approach provides the impetus for the high-resolution *Teager energy operator* developed in the final section of Chapter 11. This operator is characterized by a time resolution that can track rapid signal energy changes within a glottal cycle.

Based on the foundational Chapters 2–11, Chapters 12, 13, and 14 then address the three application areas of speech coding, speech enhancement, and speaker recognition, respectively. We do not devote a separate chapter to the speech modification application, but rather use this application to illustrate principles throughout the text. Certain other applications not covered in Chapters 12, 13, and 14 are addressed sporadically for this same purpose, including restoration of old acoustic recordings, and dynamic range compression and signal separation for signal enhancement.

1.6 Summary

In this chapter, we first defined discrete-time speech signal processing as the manipulation of sampled speech signals by a digital processor to obtain a new signal with some desired properties. The application of time-scale modification, where a speaker's articulation rate is altered, was used to illustrate this definition and to indicate the design flexibility of discrete-time processing. We saw that the goal of this book is to provide an understanding of discrete-time speech signal processing techniques driven by both time- and frequency-domain models of speech production, as well as by aspects of speech perception. The speech signal processing algorithms are also motivated by applications that include speech modification, coding, and enhancement and speaker recognition. Finally, we gave a brief outline of the text.

BIBLIOGRAPHY

[1] J.R. Deller, J.G. Proakis, and J.H.L. Hansen, *Discrete-Time Processing of Speech*, Macmillan Publishing Co., New York, NY, 1993.

[2] P.B. Denes and E.N. Pinson, *The Speech Chain: The Physics and Biology of Spoken Language*, Anchor Press-Doubleday, Garden City, NY, 1973.

[3] F. Jelinek, *Statistical Methods for Speech Recognition*, The MIT Press, Cambridge, MA, 1998.

[4] W.B. Kleijn and K.K. Paliwal, eds., *Speech Coding and Synthesis*, Elsevier, 1995.

[5] D. O'Shaughnessy, *Speech Communication: Human and Machine*, Addison-Wesley, Reading, MA, 1987.

[6] L.R. Rabiner and B.H. Juang, *Fundamentals of Speech Recognition*, Prentice Hall, Englewood Cliffs, NJ, 1993.

[7] M.A. Zissman, "Comparison of Four Approaches to Automatic Language Identification of Telephone Speech," *IEEE Trans. on Speech and Audio Processing*, vol. 4, no. 1, pp. 31–44, Jan. 1996.

A Discrete-Time Signal Processing Framework

2.1 Introduction

In this chapter we review the foundation of discrete-time signal processing which serves as a framework for the discrete-time speech processing approaches in the remainder of the text. We investigate some essential discrete-time tools and touch upon the limitations of these techniques, as manifested through the time-frequency uncertainty principle and the theory of time-varying linear systems, that will arise in a speech processing context. We do not cover all relevant background material in detail, given that we assume the familiarity of the reader with the basics of discrete-time signal processing and given that certain topics are appropriately cited, reviewed, or extended throughout the text as they are needed.

2.2 Discrete-Time Signals

The first component of a speech processing system is the measurement of the speech signal, which is a continuously varying acoustic pressure wave. We can transduce the pressure wave from an acoustic to an electrical signal with a microphone, amplify the microphone voltage, and view the voltage on an oscilloscope. The resulting analog waveform is denoted by $x_a(t)$, which is referred to as a *continuous-time signal* or, alternately, as an *analog waveform*. To process the changing voltage $x_a(t)$ with a digital computer, we sample $x_a(t)$ at uniformly spaced time instants, an operation that is represented in Figure 2.1 by a sampler with T being the sampling interval. The sampler is sometimes called a *continuous-to-discrete (C/D) converter* [7] and its output is a series of numbers $x_a(nT)$ whose representation is simplified as

$$x[n] = x_a(nT).$$

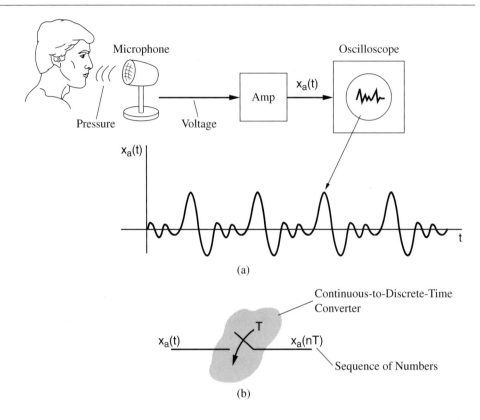

Figure 2.1 Measurement (a) and sampling (b) of an analog speech waveform.

The series $x[n]$ is referred to as a *discrete-time signal* or *sequence*. Unless otherwise specified, when working with samples from an analog signal, we henceforth use the terms *discrete-time signal*, *sequence*, and (for simplicity) *signal* interchangeably. We have assumed that the analog signal $x_a(t)$ is sampled "fast enough" to be recoverable from the sequence $x[n]$, a condition that is called the *Nyquist criterion* to which we return at the end of this chapter.

The C/D converter that generates the discrete-time signal is characterized by infinite amplitude precision. Therefore, although the signal $x[n]$ is discrete in time, it is continuous in amplitude. In practice, however, a physical device does not achieve this infinite precision. As an approximation to a C/D converter, an *analog-to-digital (A/D) converter* quantizes each amplitude to a finite set of values closest to the actual analog signal amplitude [7]. The resulting *digital signal* is thus discrete in time and in amplitude. Associated with discrete-time signals are discrete-time systems whose input and output are sequences. Likewise, digital systems are characterized by digital inputs and outputs. The principal focus of this text is discrete-time signals and systems, except where deliberate amplitude quantization is imparted, as in speech coding for bit rate and bandwidth reduction, which is introduced in Chapter 12.

Some special sequences serve as the building blocks for a general class of discrete-time signals [7]. The unit sample or "impulse" is denoted by

$$\delta[n] = 1, \qquad n = 0$$
$$= 0, \qquad n \neq 0.$$

The unit step is given by

$$u[n] = 1, \qquad n \geq 0$$
$$= 0, \qquad n < 0$$

and can be obtained by summing the unit sample: $u[n] = \sum_{k=-\infty}^{n} \delta[k]$. Likewise, the unit sample can be obtained by differencing the unit step with itself shifted one sample to the right, i.e., forming the first backward difference: $\delta[n] = u[n] - u[n-1]$. The exponential sequence is given by

$$x[n] = A\alpha^n$$

where if A and α are real, then $x[n]$ is real. Moreover, if $0 \leq \alpha \leq 1$ and $A > 0$, then the sequence $x[n]$ is positive and decreasing with increasing n. If $-1 \leq \alpha \leq 0$, then the sequence values alternate in sign. The sinusoidal sequence is given by

$$x[n] = A\cos(\omega n + \phi)$$

with frequency ω, amplitude A, and phase offset ϕ. Observe that the discrete-time sinusoidal signal is periodic[1] in the time variable n with period N only if $N = $ integer $= 2\pi k/\omega$. The complex exponential sequence with complex gain $A = |A|e^{j\phi}$ is written as

$$x[n] = Ae^{j\omega n}$$
$$= |A|e^{j\phi}e^{j\omega n}$$
$$= |A|\cos(\omega n + \phi) + j|A|\sin(\omega n + \phi).$$

An interesting property, which is a consequence of being discrete, is that the complex exponential sequence is periodic in the frequency variable ω with period 2π, i.e., $Ae^{j(\omega+2\pi)n} = Ae^{j\omega n}$. This periodicity in ω also holds for the sinusoidal sequence. Therefore, in discrete time we need to consider frequencies only in the range $0 \leq \omega < 2\pi$. The complex exponential and the above four real sequences serve as building blocks to discrete-time speech signals throughout the text.

---

[1] This is in contrast to its continuous-time counterpart $x_a(t) = A\cos(\Omega t + \phi)$ that is always periodic with period $= 2\pi/\Omega$. Here the uppercase frequency variable Ω is used for continuous time rather than the lower case ω for discrete time. This notation will be used throughout the text.

2.3 Discrete-Time Systems

Discrete-time signals are often associated with discrete-time systems. A discrete-time system can be thought of as a transformation $T[x]$ of an input sequence to an output sequence, i.e.,

$$y[n] = T(x[n]).$$

When we restrict this transformation to have the properties of linearity and time invariance, we form a class of *linear time-invariant* (LTI) systems. Let $x_1[n]$ and $x_2[n]$ be the inputs to a discrete-time system. Then for arbitrary constants a and b, the system is linear if and only if

$$T(ax_1[n] + bx_2[n]) = aT(x_1[n]) + bT(x_2[n])$$

which is sometimes referred to as the "principle of superposition" [7]. A system is time-invariant if a time shift in the input by n_o samples gives a shift in the output by n_o samples, i.e.,

$$\text{if } y[n] = T(x[n]),$$

$$\text{then } y[n - n_o] = T(x[n - n_o]).$$

An important property of an LTI system is that it is completely characterized by its impulse response, which is defined as the system's response to a unit sample (or impulse). Given an LTI system, the output $y[n]$ for an input $x[n]$ is given by a sum of weighted and delayed impulse responses, i.e.,

$$y[n] = \sum_{k=-\infty}^{\infty} x[k]h[n - k]$$

$$= x[n] * h[n] \tag{2.1}$$

which is referred to as the convolution of $x[n]$ with $h[n]$, where "$*$" denotes the convolution operator. We can visualize convolution either by the weighted sum in Equation (2.1) or by flipping $h[n]$ in time and shifting the flipped $h[n]$ past $x[n]$. The length of the resulting sequence $y[n] = x[n] * h[n]$ is $M + L - 1$. Since convolution is commutative, i.e, $x[n] * h[n] = h[n] * x[n]$ (Exercise 2.1), we can also flip $x[n]$ and run it past the response $h[n]$. The convolution operation with $h[n]$ is sometimes referred to as *filtering* the input $x[n]$ by the system $h[n]$, an operation useful in our modeling of speech production and in almost all speech processing systems.

Two useful properties of LTI systems are *stability* and *causality*, which give a more restricted class of systems (Example 2.1). In a stable system, every bounded input produces a bounded output, i.e., if $|x[n]| < \infty$, then $|y[n]| < \infty$ for all n. A necessary and sufficient condition for stability is that $h[n]$ be absolutely summable, i.e.,

$$\sum_{n=-\infty}^{\infty} |h[n]| < \infty.$$

A causal system is one where for any time instant n_o, $y[n_o]$ does not depend on $x[n]$ for $n > n_o$, i.e., the output does not depend on the future of the input. A necessary and sufficient condition for causality is that $h[n] = 0$, for $n < 0$. One can argue this necessary and sufficient condition by exploring the signal-flip interpretation of convolution (Exercise 2.2). A consequence of causality is that if $x_1[n] = x_2[n]$ for $n < n_o$, then $y_1[n] = y_2[n]$ for $n < n_o$.

EXAMPLE 2.1   The two properties of stability and causality are illustrated with an LTI system having an exponentially decaying impulse response $h[n] = Aa^n$ for $n \geq 0$ and zero otherwise. The system is causal because $h[n] = 0$ for $n < 0$ and, when $|a| < 1$, the system is stable because the impulse response is absolutely summable:

$$\sum_{n=-\infty}^{\infty} |h[n]| = A \sum_{n=0}^{\infty} |a|^n$$

$$= \frac{A}{1 - |a|}$$

where we have used the geometric series relation $\sum_{n=0}^{\infty} b^n = \frac{1}{1-b}$ for $|b| < 1$. If, on the other hand, $|a| \geq 1$, then the geometric series does not converge, the response is not absolutely summable, and the system is unstable. Note that, according to this condition, a system whose impulse response is the unit step function, i.e., $h[n] = u[n]$, is unstable.   ▲

The terminology formulated in this section for systems is also used for sequences, although its physical meaning for sequences is lacking. A *stable sequence* is defined as an absolutely summable sequence, and a *causal sequence* is zero for $n < 0$. A causal sequence will also be referred to as a *right-sided sequence*.

2.4 Discrete-Time Fourier Transform

The previous section focused on time-domain representations of signals and systems. Frequency-domain representations, the topic of this section, are useful for the analysis of signals and the design of systems for processing signals. We begin with a review of the Fourier transform.

A large class of sequences can be represented as a linear combination of complex exponentials whose frequencies lie in the range[2] $[-\pi, \pi]$. Specifically, we write the following pair of equations:

$$x[n] = \frac{1}{2\pi} \int_{-\pi}^{\pi} X(\omega)e^{j\omega n} d\omega$$

$$X(\omega) = \sum_{n=-\infty}^{\infty} x[n]e^{-j\omega n}. \tag{2.2}$$

This pair of equations is known as the *discrete-time Fourier transform* pair representation of a sequence. For convenience, the phrase *Fourier transform* will often be used in place of *discrete-*

---

[2] Recall that $e^{j(\omega+2\pi)n} = e^{j\omega n}$.

time Fourier transform. Equation (2.2) represents $x[n]$ as a superposition of infinitesimally small complex exponentials $d\omega X(\omega)e^{j\omega n}$, where $X(\omega)$ determines the relative weight of each exponential. $X(\omega)$ is the Fourier transform of the sequence $x[n]$, and is also referred to as the "analysis equation" because it analyzes $x[n]$ to determine its relative weights. The first equation in the pair is the *inverse Fourier transform,* also referred to as the "synthesis equation" because it puts the signal back together again from its (complex exponential) components. In this text, we often use the terminology "analysis" and "synthesis" of a signal. We have not yet explicitly shown for what class of sequences such a Fourier transform pair exists. Existence means that (1) $X(\omega)$ does not diverge, i.e., the Fourier transform sum converges, and (2) $x[n]$ can be obtained from $X(\omega)$. It can be shown that a sufficient condition for the existence of the pair is that $x[n]$ be absolutely summable, i.e., that $x[n]$ is stable [7]. Therefore, all stable sequences and stable system impulse responses have Fourier transforms.

Some useful properties of the Fourier transform are as follows (Exercise 2.3):

P1: Since the Fourier transform is complex, it can be written in polar form as

$$X(\omega) = X_r(\omega) + jX_i(\omega)$$
$$= |X(\omega)|e^{j\angle X(\omega)}$$

where the subscripts r and i denote real and imaginary parts, respectively.

P2: The Fourier transform is periodic with period 2π:

$$X(\omega + 2\pi) = X(\omega)$$

which is consistent with the statement that the frequency range $[-\pi, \pi]$ is sufficient for representing a discrete-time signal.

P3: For a real-valued sequence $x[n]$, the Fourier transform is conjugate-symmetric:

$$X(\omega) = X^*(-\omega)$$

where $*$ denotes complex conjugate. Conjugate symmetry implies that the magnitude and real part of $X(\omega)$ are even, i.e., $|X(\omega)| = |X(-\omega)|$ and $X_r(\omega) = X_r(-\omega)$, while its phase and imaginary parts are odd, i.e., $\angle X(\omega) = -\angle X(-\omega)$ and $X_i(\omega) = -X_i(-\omega)$. It follows that if a sequence is not conjugate-symmetric, then it must be a complex-valued sequence.

P4: The energy of a signal can be expressed by Parseval's Theorem as

$$\sum_{n=-\infty}^{\infty} |x[n]|^2 = \frac{1}{2\pi}\int_{-\pi}^{\pi} |X(\omega)|^2 d\omega \tag{2.3}$$

which states that the total energy of a signal can be given in either the time or frequency domain. The functions $|x[n]|^2$ and $|X(\omega)|^2$ are thought of as *energy densities,* i.e., the energy per unit time and the energy per unit frequency, because they describe the distribution of energy in time and frequency, respectively. Energy density is also referred to as *power* at a particular time or frequency.

EXAMPLE 2.2   Consider the shifted unit sample

$$x[n] = \delta[n - n_o].$$

The Fourier transform of $x[n]$ is given by

$$X(\omega) = \sum_{n=-\infty}^{\infty} \delta[n - n_o]e^{-j\omega n}$$

$$= e^{-j\omega n_o}$$

since $x[n]$ is nonzero for only $n = n_o$. This complex function has unity magnitude and a linear phase of slope $-n_o$. In time, the energy in this sequence is unity and concentrated at $n = n_o$, but in frequency the energy is uniformly distributed over the interval $[-\pi, \pi]$ and, as seen from Parseval's Theorem, averages to unity.   ▲

More generally, it can be shown that the Fourier transform of a displaced sequence $x[n - n_o]$ is given by $X(\omega)e^{-j\omega n_o}$. Likewise, it can be shown, consistent with the similar forms of the Fourier transform and its inverse, that the Fourier transform of $e^{j\omega_o n}x[n]$ is given by $X(\omega - \omega_o)$. This later property is exploited in the following example:

EXAMPLE 2.3   Consider the decaying exponential sequence multiplied by the unit step:

$$x[n] = a^n u[n]$$

with a generally complex. Then the Fourier transform of $x[n]$ is given by

$$X(\omega) = \sum_{n=0}^{\infty} a^n e^{-j\omega n}$$

$$= \sum_{n=0}^{\infty} (ae^{-j\omega})^n$$

$$= \frac{1}{1 - ae^{-j\omega}}, \quad |ae^{j\omega}| = |a| < 1$$

so that the convergence condition on a becomes $|a| < 1$. If we multiply the sequence by the complex exponential $e^{j\omega_o n}$, then we have the following Fourier transform pair:

$$e^{j\omega_o n} a^n u[n] \leftrightarrow \frac{1}{1 - ae^{-j(\omega - \omega_o)}}, \quad |a| < 1.$$

An example of this later transform pair is shown in Figure 2.2a,b where it is seen that in frequency the energy is concentrated around $\omega = \omega_o = \frac{\pi}{2}$. The two different values of a show a broadening of the Fourier transform magnitude with decreasing a corresponding to a faster decay of the exponential. From the linearity of the Fourier transform, and using the above relation, we can write the Fourier transform pair for a real decaying sinewave as

$$2a^n \cos(\omega_o n)u[n] \leftrightarrow \frac{1}{1 - ae^{-j(\omega - \omega_o)}} + \frac{1}{1 - ae^{-j(\omega + \omega_o)}}, \quad |a| < 1$$

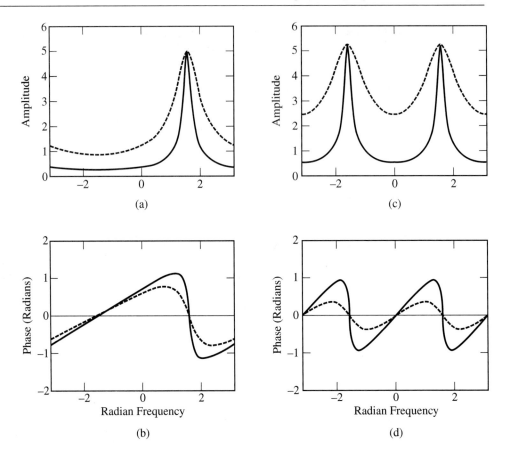

Figure 2.2 Frequency response of decaying complex and real exponentials of Example 2.3: (a) magnitude and (b) phase for complex exponential; (c) magnitude and (d) phase for decaying sinewave (solid for slow decay [$a = 0.9$] and dashed for fast decay [$a = 0.7$]). Frequency $\omega_o = \pi/2$.

where we have used the identity $\cos(\alpha) = \frac{1}{2}(e^{j\alpha} + e^{-j\alpha})$. Figure 2.2c,d illustrates the implications of conjugate symmetry on the Fourier transform magnitude and phase of this real sequence, i.e., the magnitude function is even, while the phase function is odd. In this case, decreasing the value of a broadens the positive and negative frequency components of the signal around the frequencies ω_o and $-\omega_o$, respectively.   ▲

Example 2.3 illustrates a fundamental property of the Fourier transform pair representation: A signal cannot be arbitrarily narrow in time and in frequency. We return to this property in the following section.

The next example derives the Fourier transform of the complex exponential, requiring in frequency the unit impulse which is also called the Dirac delta function.

EXAMPLE 2.4   In this case we begin in the frequency domain and perform the inverse Fourier transform. Consider a train of scaled unit impulses in frequency:

$$X(\omega) = \sum_{r=-\infty}^{\infty} A2\pi\delta(\omega - \omega_o + r2\pi)$$

where 2π periodicity is enforced by adding delta function replicas at multiples of 2π (Figure 2.3a). The inverse Fourier transform is given by[3]

$$x[n] = \frac{1}{2\pi} \int_{-\pi}^{\pi} A2\pi\delta(\omega - \omega_o)e^{j\omega n}d\omega$$

$$= Ae^{j\omega_o n}$$

which is our familiar complex exponential. Observe that this Fourier transform pair represents the time-frequency dual of the shifted unit sample $\delta[n - n_o]$ and its transform $e^{jn_o\omega}$. More generally, a shifted Fourier transform $X(\omega - \omega_o)$ corresponds to the sequence $x[n]e^{j\omega_o n}$, a property alluded to earlier.   ▲

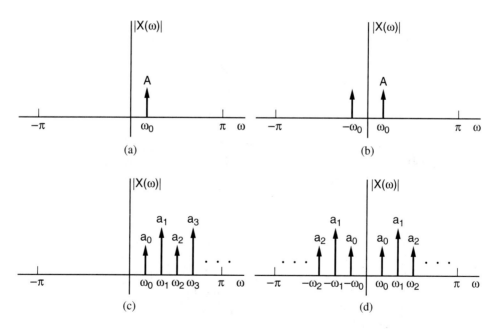

Figure 2.3   Dirac delta Fourier transforms of (a) complex exponential sequence, (b) sinusoidal sequence, (c) sum of complex exponentials, and (d) sum of sinusoids. For simplicity, π and 2π factors are not shown in the amplitudes.

---

[3] Although this sequence is not absolutely summable, use of the Fourier transform pair can rigorously be justified using the theory of generalized functions [7].

Using the linearity of the Fourier transform, we can generalize the previous result to a sinusoidal sequence as well as to multiple complex exponentials and sines. Figure 2.3b–d illustrates the Fourier transforms of the following three classes of sequences:

Sinusoidal sequence

$$A\cos(\omega_o n + \phi) \leftrightarrow \pi A e^{j\phi}\delta(\omega - \omega_o) + \pi A e^{-j\phi}\delta(\omega + \omega_o)$$

Multiple complex exponentials

$$\sum_{k=0}^{N} a_k e^{j\omega_k n + \phi_k} \leftrightarrow \sum_{k=0}^{N} 2\pi a_k e^{j\phi_k}\delta(\omega - \omega_k)$$

Multiple sinusoidals

$$\sum_{k=0}^{N} a_k \cos(\omega_k n + \phi_k) \leftrightarrow \sum_{k=0}^{N} \pi a_k e^{j\phi_k}\delta(\omega - \omega_k) + \pi a_k e^{-j\phi_k}\delta(\omega + \omega_k)$$

For simplicity, each transform is represented over only one period; for generality, phase offsets are included.

2.5 Uncertainty Principle

We saw in Example 2.3 a fundamental property of the Fourier transform pair: A signal cannot be arbitrarily narrow in time and in frequency. We saw in Figure 2.2 that the Fourier transform increased in spread as the time sequence decreased in width. This property can be stated more precisely in the *uncertainty principle*. To do so requires a formal definition of the width of the signal and its Fourier transform. We refer to these signal characteristics as *duration*, denoted by $D(x)$ and *bandwidth*,[4] denoted by $B(x)$, and define them respectively as

$$D(x) = \sum_{n=-\infty}^{\infty} (n - \bar{n})^2 |x[n]|^2$$

$$B(x) = \int_{-\pi}^{\pi} (\omega - \bar{\omega})^2 |X(\omega)|^2 d\omega \qquad (2.4)$$

where $\bar{n}$ is the *average time* of the signal, i.e., $\bar{n} = \sum_{n=-\infty}^{\infty} n|x[n]|^2$, and $\bar{\omega}$ is its *average frequency*, i.e, $\bar{\omega} = \int_{-\infty}^{\infty} \omega |X(\omega)|^2 d\omega$ [2]. In these definitions, in order that the time and frequency averages be meaningful, we assume that the signal energy is unity, i.e., $\sum_{n=-\infty}^{\infty} |x[n]|^2 = \frac{1}{2\pi}\int_{-\pi}^{\pi} |X(\omega)|^2 d\omega = 1$ or that the signal has been normalized by its energy. These duration and bandwidth values give us a sense of the concentration of a signal, or of its Fourier transform, about its average location. The definitions of *signal* or *transform width* are motivated by the definition of the variance, or "spread," of a random variable. In fact,

---

[4] A more traditional definition of *bandwidth*, not necessarily giving the same value as that in Equation (2.4), is the distance between the 3 dB attenuation points around the average frequency.

$|x[n]|^2$ and $|X(\omega)|^2$ viewed as energy densities, we will see later in the text, are analogous to the probability density function used in defining the variance of a random variable [2]. It follows that normalizing the magnitude functions in Equation (2.4) by the total signal energy ensures probability density-like functions that integrate to unity.

The uncertainty principle states that the product of signal duration and bandwidth cannot be less than a fixed limit, i.e.,

$$D(x)B(x) \ \geq \ \frac{1}{4}. \tag{2.5}$$

Proof of the uncertainty principle is given by first applying Parseval's Theorem to obtain

$$D(x)B(x) \ \geq \ \int_{-\pi}^{\pi} \omega X^*(\omega)\frac{dX(\omega)}{d\omega}d\omega \tag{2.6}$$

from which Equation (2.5) follows. The reader is led through the derivation in Exercise 2.5. The principle implies that a wide signal gives a narrow Fourier transform, and a narrow signal gives a wide Fourier transform.[5] The uncertainty principle will play a major role in spectrographic, and, more generally, time-frequency analysis of speech, especially when the speech waveform consists of dynamically changing events or closely-spaced time or frequency components.

It is important to look more carefully at our definition of bandwidth in Equation (2.4). Observe that for a real sequence, from the conjugate-symmetry property, the Fourier transform magnitude is even. Thus the average frequency is zero and the bandwidth is determined by the distance between the spectral energy concentrations in positive and negative frequency. The resulting bandwidth, therefore, is not necessarily indicative of the distribution of energy of physically meaningful quantities such as system resonances (Exercise 2.4). The bandwidth of the signals with discrete-time Fourier transform magnitudes in Figure 2.2c is such a case. As a consequence, complex sequences, as those corresponding to the transform magnitudes in Figure 2.2a, or only the positive frequencies of a real sequence, are used in computing bandwidth. In the latter case, we form a sequence $s[n]$ with Fourier transform

$$S(\omega) \ = \ X(\omega), \quad 0 \leq \omega \leq \pi$$
$$= \ 0, \qquad -\pi \ < \ \omega \ < \ 0$$

where we implicitly assume that $S(\omega)$ is periodic with period 2π. The sequence $s[n]$ must be complex because conjugate symmetry is not satisfied, and so can be represented by real and

---

[5] The nomenclature "uncertainty" is somewhat misleading because there is no uncertainty in the measurement of the signal or its Fourier transform. This terminology evolved from Heisenberg's uncertainty principle in the probabilistic context of quantum mechanics where it was discovered that the position and the momentum of a particle cannot be measured at a particular time simultaneously with absolute certainty. There is a mathematical similarity, and *not* a similarity in the physical interpretation, in Heisenberg's uncertainty principle with Equation (2.5) because the position and momentum functions are related through the Fourier transform. The squared magnitude of the Heisenberg functions represent the *probability* of measuring a particle with a certain position and momentum, respectively, unlike the deterministic magnitude of a signal and its Fourier transform [2].

imaginary components

$$s[n] = s_r[n] + js_i[n]. \tag{2.7}$$

Equation (2.7), the inverse Fourier transform of $S(\omega)$, is called the *analytic signal* representation of $x[n]$ and is used occasionally later in the text. It can be shown that the real component $s_r[n] = \frac{x[n]}{2}$ and that the imaginary part $s_i[n]$ can be obtained from $s_r[n]$ through the frequency-domain operation (Exercise 2.6):

$$S_i(\omega) = H(\omega)S_r(\omega) \tag{2.8}$$

where $H(\omega)$ is called the *Hilbert transformer* and is given by

$$\begin{aligned} H(\omega) &= -j, \quad 0 \le \omega < \pi \\ &= j, \quad -\pi \le \omega < 0 \end{aligned} \tag{2.9}$$

with inverse Fourier transform (left to the reader to show)

$$\begin{aligned} h[n] &= \frac{2}{\pi} \frac{\sin^2(\pi n/2)}{n}, \quad n \ne 0 \\ &= 0, \qquad\qquad\qquad n = 0. \end{aligned} \tag{2.10}$$

Because $s[n]$ is complex, it can be expressed in polar form as

$$s[n] = A[n]e^{j\theta[n]} \tag{2.11}$$

where $A[n] = |s[n]|$ and $\theta[n] = \angle s[n]$. $A[n]$ is often called the *envelope* of the underlying signal $x[n]$; it follows that the analytic signal $s[n]$ is sometimes referred to as its *complex envelope*. It is useful to express the phase $\theta[n]$ in the form

$$\theta[n] = \int_{-\infty}^{nT} \omega(\tau)d\tau \tag{2.12}$$

where $\omega(t)$ is referred to as the *instantaneous frequency* of the underlying analog signal, i.e., $\theta[n]$ is obtained by sampling the phase of the continuous-signal counterpart $s(t) = A(t)e^{j\theta(t)}$. The discrete-time instantaneous frequency then is the derivative of the phase $\theta(t)$ evaluated at time samples $t = nT$ with T the sampling interval, i.e., $\omega[n] = \frac{d\theta(t)}{dt}|_{t=nT}$. We will see later in the text that the instantaneous frequency plays an important role in many speech signal representations.

The analytic signal is close in form to a more familiar alternate complex signal representation. Specifically, suppose that we are given a real sequence $x[n] = a[n]\cos(\phi[n])$. Then the complex *quadrature* rendition of this signal is of the form

$$s_q[n] = a[n]e^{j\phi[n]}. \tag{2.13}$$

The condition under which the envelope and phase of the quadrature (Equation (2.13)) and analytic (Equation (2.11)) signal representations are equivalent is developed in Exercise 2.7. The analytic and quadrature signal representations will play an important role within certain speech analysis/synthesis systems, especially those based on filter-bank, sinewave, and resonant representations that we will see throughout the text.

   In closing this section, we observe that although the uncertainty principle imposes a constraint between the duration and bandwidth of a signal, the bandwidth is not determined solely by the signal duration. Cohen [2] has shown that the bandwidth of a signal, as defined in Equation (2.4), can be expressed in continuous time for a signal of the form $s(t) = a(t)e^{\phi(t)}$ as (Exercise 2.8)

$$B = \int_{-\infty}^{\infty} \left(\frac{da(t)}{dt} \right)^2 dt + \int_{-\infty}^{\infty} \left(\frac{d\phi(t)}{dt} - \bar{\Omega} \right)^2 a^2(t) dt \qquad (2.14)$$

where Ω denotes frequency for continuous time. The first term in Equation (2.14) is the average rate of change of the signal envelope $a(t)$ and corresponds *in part* to signal duration; a shorter signal gives a more rapidly changing envelope. The second term is the average deviation of the instantaneous frequency $\frac{d\phi(t)}{dt}$ around the average frequency $\bar{\Omega}$ and thus is a measure of frequency spread. The Fourier transform pair provides an "average" signal characterization that does not necessarily reveal such finer aspects of the signal. The importance of these bandwidth components in the speech context, and the more general importance of a finer signal representation than that provided by the Fourier transform, will become clear throughout the text when we describe the rapidly varying nature of certain speech sounds and when we introduce alternate tools for analysis of fine detail in the signal.

2.6 *z*-Transform

A limitation of the Fourier transform is that it cannot represent most unstable sequences, i.e., sequences that are not absolutely summable. The *z*-transform is a generalization of the Fourier transform that allows for a much larger class of sequences and is given by

$$X(z) = \sum_{n=-\infty}^{\infty} x[n]z^{-n} \qquad (2.15)$$

which is an infinite geometric series in the complex variable $z = re^{j\omega}$. For $r = 1$, the *z*-transform reduces to the Fourier transform

$$X(\omega) = X(z)|_{z=e^{j\omega}}$$

where for convenience we have abused notation by replacing $e^{j\omega}$ with ω. Therefore, we can think of the *z*-transform as a generalization of the Fourier transform that permits a broader class of signals as linear combinations of complex exponentials whose magnitude may or may not be unity; sequences that are not absolutely summable may be made to converge. We can see this property by rewriting Equation (2.15) as

$$X(re^{j\omega}) = \sum_{n=-\infty}^{\infty} (x[n]r^{-n})e^{-j\omega n} \qquad (2.16)$$

where the sequence $x[n]r^{-n}$ may be absolutely summable when $x[n]$ is not.

Associated with the z-transform is a *region of convergence* (ROC) defined as all values of r such that

$$\sum_{n=-\infty}^{\infty} |x[n]r^{-n}| < \infty.$$

This condition of absolute summability is a sufficient condition for the convergence of $X(z)$. Since $|z| = |re^{j\omega}| = r$, we can view the ROC in the z-plane. In particular, the ROC is generally an annular (ring) region in the z-plane. For $r = 1$, the z-transform becomes the Fourier transform and so the Fourier transform exists if the ROC includes the unit circle. As we have deduced from Equation (2.16), however, it is possible for the z-transform to converge when the Fourier transform does not.

As with the Fourier transform, there exists an inverse z-transform, also referred to as the z-transform "synthesis equation." The z-transform pair is written as

$$x[n] = \frac{1}{2\pi j} \oint_C X(z)z^{n-1}dz$$

$$X(z) = \sum_{n=-\infty}^{\infty} x[n]z^{-n}$$

where the integral in the inverse transform is taken over C, a counterclockwise closed contour in the region of convergence of $X(z)$ encircling the origin in the z-plane, and which can be formally evaluated using residue theory [7]. Often we are interested in the specific class of *rational* functions of the form

$$X(z) = \frac{P(z)}{Q(z)}$$

where the roots of the polynomials $P(z)$ and $Q(z)$ define the zeros, i.e., values of z such that $P(z) = 0$, and poles, i.e., values of z such that $Q(z) = 0$, respectively. When the z-transform is known to be rational, a partial fraction expansion in terms of single and multiple poles (order two or higher) can be used [7]. We now look at a few simple sequences whose rational z-transforms form the building blocks of a more general class of rational functions. The ROC for these examples is illustrated in Figure 2.4.

EXAMPLE 2.5   Consider the shifted impulse

$$x[n] = \delta[n - n_o].$$

Then the z-transform is given by

$$X(z) = \sum_{n=-\infty}^{\infty} \delta[n - n_o]z^{-n}$$

$$= z^{-n_o}$$

where the ROC includes the entire z-plane, not including $z = 0$ for $n_o > 0$.

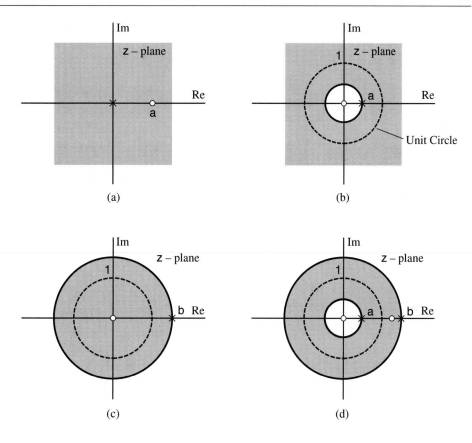

Figure 2.4 Region of convergence (ROC) of the *z*-transform for Examples 2.5–2.7: (a) $x[n] = \delta[n] - a\delta[n - 1]$; (b) $x[n] = a^n u[n]$; (c) $x[n] = -b^n u[-n - 1]$; (d) $x[n] = a^n u[n] - b^n u[-n - 1]$. Poles are indicated by small crosses and zeros by small circles, and the ROC is indicated by shaded regions.

Consider now the difference of unit samples

$$x[n] = \delta[n] - a\delta[n - 1]$$

with a generally complex. From linearity of the *z*-transform operator and the previous result, the *z*-transform is given by

$$X(z) = 1 - az^{-1}$$

where the ROC again includes the entire *z*-plane, not including $z = 0$. For this case, we say that $X(z)$ has a pole at $z = 0$, i.e., the transform goes to infinity, and a zero at $z = a$, i.e., the transform takes on value zero.   ▲

EXAMPLE 2.6   Consider the decaying exponential

$$x[n] = a^n u[n]$$

where a is generally complex. From the convergence property of a geometric series (Example 2.1), the z-transform is given by

$$X(z) = \frac{1}{1 - az^{-1}}, \quad |z| > |a|$$

where the ROC includes the z-plane outside of the radius a, i.e., $|z| > |a|$. For this case, $X(z)$ has a pole at $z = a$ and a zero at $z = 0$. For $|a| < 1$ the ROC includes the unit circle.

A similar case is the time-reversed version of the decaying exponential

$$x[n] = -b^n u[-n - 1].$$

Again, from the convergence property of a geometric series, the z-transform is given by

$$X(z) = \frac{1}{1 - bz^{-1}}, \quad |z| < |b|$$

but now the ROC includes the z-plane *inside* of the radius b, i.e., $|z| < |b|$. For this case, $X(z)$ has a pole at $z = b$ and a zero at $z = 0$. For $|b| > 1$, the ROC includes the unit circle.   ▲

From the above examples, we see that both the z-transform and an ROC are required to specify a sequence. When a sequence consists of a sum of components, such as the sequences of the previous examples, then, in general, the ROC is the intersection of the ROC for individual terms, as we show in the following example:

EXAMPLE 2.7   We now combine the elemental sequences of the previous example. Let $x_1[n]$ and $x_2[n]$ denote the decaying exponential and its time-reversal, respectively. Then the sum of the two sequences is given by

$$x[n] = x_1[n] + x_2[n] = a^n u[n] - b^n u[-n - 1].$$

From the linearity of the z-transform, $X(z)$ is given by

$$
\begin{aligned}
X(z) &= X_1(z) + X_2(z) \\
&= \frac{1}{1 - az^{-1}} + \frac{1}{1 - bz^{-1}}, \quad |a| < |z| < |b|
\end{aligned}
$$

where, because both sequences must converge to have $x[n]$ converge, the ROC is an *annulus* in the z-plane defined by $|a| < |z| < |b|$. Rewriting $X(z)$ as

$$X(z) = \frac{2\left(1 - \left[\frac{a+b}{2}\right] z^{-1}\right)}{(1 - az^{-1})(1 - bz^{-1})},$$

we see that $X(z)$ is characterized by two poles at $z = a, b$, one zero at $z = \frac{a+b}{2}$, and a zero at $z = 0$.   ▲

The z-transforms of the previous examples can be generalized to complex a and b; the poles and zeros are not constrained to the real axis, but can lie anywhere in the z-plane. For a general rational $X(z)$, factoring $P(z)$ and $Q(z)$ gives the z-transform in the form

$$X(z) = Az^{-r} \frac{\prod_{k=1}^{M_i}(1 - a_k z^{-1}) \prod_{k=1}^{M_o}(1 - b_k z)}{\prod_{k=1}^{N_i}(1 - c_k z^{-1}) \prod_{k=1}^{N_o}(1 - d_k z)} \tag{2.17}$$

where $(1 - a_k z^{-1})$ and $(1 - c_k z^{-1})$ correspond to zeros and poles inside the unit circle and $(1 - b_k z)$ and $(1 - d_k z)$ to zeros and poles outside the unit circle with $|a_k|$, $|b_k|$, $|c_k|$, $|d_k| < 1$. If equality holds in any of these relations, then a zero or pole lies on the unit circle. The term z^{-r} represents a shift of the sequence with respect to the time origin (Example 2.5). For a real sequence $x[n]$, the poles and zeros will occur in complex conjugate pairs. The ROC is generally an annulus bounded by the poles, the specific shape depending on whether the signal is right-sided ($x[n] = 0$ for $n < 0$) or left-sided ($x[n] = 0$ for $n > 0$), or two-sided (neither right-sided nor left-sided). The ROC may or may not include the unit circle. For a right-sided sequence, the ROC extends outside the outermost pole, while for a left-sided sequence, the ROC extends inside the innermost pole. For a two-sided sequence, the ROC is an annulus in the z-plane bounded by poles and, given convergence, not containing any poles. These different configurations are illustrated in Figure 2.5.

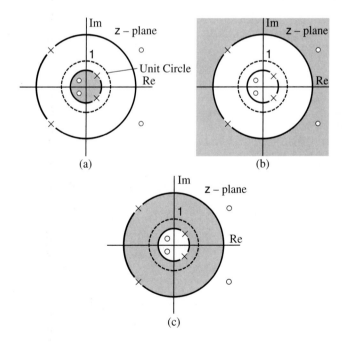

Figure 2.5 Poles and zeros, and region of convergence (ROC) of rational z-transforms: (a) left-sided, (b) right-sided, and (c) two-sided. The ROC is indicated by shaded regions.

A specific case, important in the modeling of speech production, is when $x[n]$ is stable and causal; in this case, because the unit circle must be included in the ROC, all poles are inside the unit circle and the ROC is outside the outermost pole. This configuration is shown later in Figure 2.8 and will be further described in Section 2.8.

It was stated earlier that a rational z-transform can be represented by a partial fraction expansion. The details of this partial fraction expansion are given in [7]; here it suffices to describe a particular signal class. Consider the case where the number of poles is greater than the number of zeros. Then, for the case of no poles outside the unit circle, we can write Equation (2.17) as

$$X(z) = \sum_{k=1}^{N_i} \frac{A_k}{(1 - c_k z^{-1})}. \tag{2.18}$$

This *additive* decomposition, in contrast to the earlier *multiplicative* representation of component poles, will be useful in modeling of speech signals, as in the representation of resonances of a vocal tract system impulse response.

2.7 LTI Systems in the Frequency Domain

We have completed our brief review of frequency-domain representations of sequences; we now look at similar representations for systems. Consider the complex exponential $x[n] = e^{j\omega n}$ as the input to an LTI system. From the convolutional property

$$y[n] = x[n] * h[n]$$

$$= \sum_{k=-\infty}^{\infty} h[k] e^{j\omega(n-k)}$$

$$= e^{j\omega n} \sum_{k=-\infty}^{\infty} h[k] e^{-j\omega k}. \tag{2.19}$$

The second term in Equation (2.19) is the Fourier transform of the system impulse response which we denote by $H(\omega)$

$$H(\omega) = \sum_{k=-\infty}^{\infty} h[k] e^{-j\omega k} \tag{2.20}$$

so that

$$y[n] = H(\omega) e^{j\omega n}.$$

Therefore, a complex exponential input to an LTI system results in the same complex exponential at the output, but modified by $H(\omega)$. It follows that the complex exponential is an eigenfunction of an LTI system, and $H(\omega)$ is the associated eigenvalue [7].[6] $H(\omega)$ is often referred to as the

---

[6] This eigenfunction/eigenvalue terminology is also often used in mathematics. In linear algebra, for example, for a matrix A and a vector $\underline{x}$, $A\underline{x} = \lambda \underline{x}$ where $\underline{x}$ is the eigenfunction and λ the eigenvalue.

system *frequency response* because it describes the change in $e^{j\omega n}$ with frequency; its z-domain generalization, $H(z)$, is referred to as the *system function* or sometimes the *transfer function*.

The following example exploits the eigenfunction property of the complex exponential:

EXAMPLE 2.8   A sinusoidal sequence can be expressed as

$$x[n] = A\cos(\omega_o n + \phi)$$

$$= \frac{A}{2}e^{j\phi}e^{j\omega_o n} + \frac{A}{2}e^{-j\phi}e^{-j\omega_o n}.$$

Then, by superposition, the output to an LTI system $H(\omega)$ to the input $x[n]$ is given by

$$y[n] = H(\omega_o)\frac{A}{2}e^{j\phi}e^{j\omega_o n} + H(-\omega_o)\frac{A}{2}e^{-j\phi}e^{-j\omega_o n}$$

$$= \frac{A}{2}[H(\omega_o)e^{j\phi}e^{j\omega_o n} + H^*(\omega_o)e^{-j\phi}e^{-j\omega_o n}]$$

where we have used the conjugate symmetry property of the Fourier transform, $H^*(-\omega_o) = H(\omega_o)$. Using the relation $a + a^* = 2\text{Re}[a] = 2|a|\cos(\theta)$ where $a = |a|e^{j\theta}$, the output can be expressed as

$$y[n] = A|H(\omega_o)|\cos[\omega_o n + \phi + \angle H(\omega_o)]$$

where we have invoked the polar form $H(\omega) = |H(\omega)|e^{j\angle H(\omega)}$. Because the system is linear, this result can be generalized to a sum of sinewaves, i.e., for an input of the form

$$x[n] = \sum_{k=0}^{N} A_k\cos(\omega_k n + \phi_k)$$

the output is given by

$$y[n] = \sum_{k=0}^{N} A_k|H(\omega_k)|\cos[\omega_k n + \phi_k + \angle H(\omega_k)].$$

A similar expression is obtained for an input consisting of a sum of complex exponentials.   ▲

Two important consequences of the eigenfunction/eigenvalue property of complex exponentials for LTI systems to be proven in Exercise 2.12 are stated below.

Convolution Theorem — This theorem states that convolution of sequences corresponds to multiplication of their corresponding Fourier transforms. Specifically, if

$$x[n] \leftrightarrow X(\omega)$$

$$h[n] \leftrightarrow H(\omega)$$

and if

$$y[n] = x[n] * h[n]$$

then

$$Y(\omega) = X(\omega)H(\omega).$$

Windowing (Modulation) Theorem — The following theorem is the dual of the Convolution Theorem. Let

$$x[n] \leftrightarrow X(\omega)$$
$$w[n] \leftrightarrow W(\omega)$$

and if

$$y[n] = x[n]w[n]$$

then

$$Y(\omega) = \frac{1}{2\pi} \int_{-\pi}^{\pi} X(\Theta)W(\omega - \Theta)d\Theta$$

$$= \frac{1}{2\pi}X(\omega) \circledast W(\omega)$$

where $\circledast$ denotes *circular convolution*, corresponding to one function being circularly shifted relative to the other with period 2π. We can also think of each function being defined only in the interval $[-\pi, \pi]$ and being shifted modulo 2π in the convolution. The "windowing" terminology comes about because when the duration of $w[n]$ is short relative to that of $x[n]$, we think of $w[n]$ as viewing (extracting) a short piece of the sequence $x[n]$.

EXAMPLE 2.9   Consider a sequence consisting of a periodic train of unit samples

$$x[n] = \sum_{k=-\infty}^{\infty} \delta[n - kP]$$

with Fourier transform (Exercise 2.13)

$$X(\omega) = \sum_{k=-\infty}^{\infty} \frac{2\pi}{P}\delta(\omega - \frac{2\pi}{P}k).$$

Suppose that $x[n]$ is the input to an LTI system with impulse response given by

$$h[n] = a^n u[n]$$

with Fourier transform

$$H(\omega) = \frac{1}{1 - ae^{-j\omega}}, \qquad |a| < 1.$$

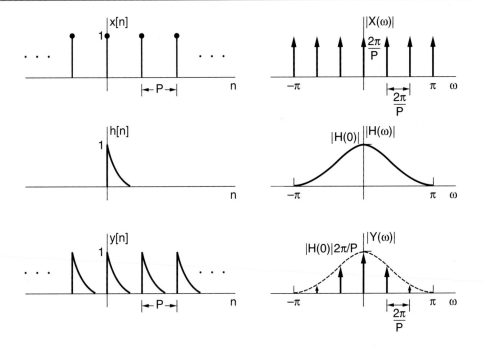

Figure 2.6 Illustration of the Convolution Theorem in Example 2.9.

Then from the Convolution Theorem

$$Y(\omega) = H(\omega)X(\omega)$$

$$= \frac{1}{1 - ae^{-j\omega}} \sum_{k=-\infty}^{\infty} \frac{2\pi}{P} \delta\left(\omega - \frac{2\pi}{P}k\right)$$

$$= \frac{2\pi}{P} \sum_{k=-\infty}^{\infty} \frac{1}{1 - ae^{-j\frac{2\pi}{P}k}} \delta\left(\omega - \frac{2\pi}{P}k\right).$$

We think of the magnitude of the system function as the *spectral envelope* to the train of Dirac delta pulses (Figure 2.6). This representation will be particularly important in modeling voiced speech sounds such as vowels. ▲

The importance of Example 2.10, the dual of Example 2.9, will become apparent later in the text when we perform short-time analysis of speech.

EXAMPLE 2.10   Consider again the sequence $x[n]$ of Example 2.9, and suppose that the sequence is multiplied by a Hamming window of the form [7]

$$w[n] = 0.54 - 0.46 \cos\left[\frac{2\pi n}{N_w - 1}\right], \quad 0 \leq n \leq N_w - 1$$

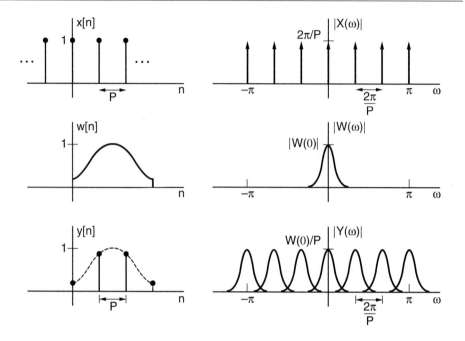

Figure 2.7 Illustration of the Windowing Theorem in Example 2.10.

with Fourier transform denoted by $W(\omega)$ (the reader should sketch the Fourier transform of $w[n]$). Then from the Windowing Theorem

$$Y(\omega) = \frac{1}{2\pi} W(\omega) \circledast X(\omega)$$

$$= \frac{1}{2\pi} W(\omega) \circledast \sum_{k=-\infty}^{\infty} \frac{2\pi}{P} \delta\left(\omega - \frac{2\pi}{P}k\right)$$

$$= \frac{1}{P} \sum_{k=-\infty}^{\infty} W(\omega) \circledast \delta\left(\omega - \frac{2\pi}{P}k\right)$$

$$= \frac{1}{P} \sum_{k=-\infty}^{\infty} W\left(\omega - \frac{2\pi}{P}k\right).$$

Therefore, the window function is replicated at the uniformly spaced frequencies of the periodic pulse train (Figure 2.7).   ▲

Both the Convolution and Windowing Theorems can be generalized with the z-transform [7]. For example, if $y[n] = x[n] * h[n]$, then $Y(z) = X(z)H(z)$, where the ROC of $Y(z)$ is the intersection of the ROCs of $X(z)$ and $H(z)$. Extension of the Windowing Theorem is left to the reader to ponder.

2.8 Properties of LTI Systems

We saw in Equation (2.17) that an important class of LTI systems is represented by a rational z-transform. In this section, we constrain this class to satisfy properties of stability and causality that we cited earlier. LTI systems with rational z-transforms that satisfy these two properties are referred to as *digital filters*.[7]

2.8.1 Difference Equation Realization

The output of a digital filter is related to the input by an Nth-order *difference equation* of the form

$$y[n] = \sum_{k=1}^{N} \alpha_k y[n-k] + \sum_{k=0}^{M} \beta_k x[n-k] \tag{2.21}$$

which corresponds to a rational system function. The starting condition of the difference equation is that of initial rest because linearity and time-invariance require that the output be zero for all time if the input is zero for all time [7]. We can evaluate the impulse response by a recursive computation of the difference equation for input $x[n] = \delta[n]$ (assuming that the response is causal) or by means of the z-transform. In the latter case, we apply the z-transform to both sides of the difference equation, and use linearity of the z-transform operator and the delay property (i.e., $x[n-n_o]$ has z-transform $X(z)z^{-n_o}$) to obtain

$$Y(z) = \sum_{k=1}^{N} \alpha_k Y(z) z^{-k} + \sum_{k=0}^{M} \beta_k X(z) z^{-k}$$

or

$$H(z) = \frac{Y(z)}{X(z)} = \frac{\sum_{k=0}^{M} \beta_k z^{-k}}{1 - \sum_{k=1}^{N} \alpha_k z^{-k}}$$

which is a rational function with numerator and denominator that are polynomials in variable z^{-1}. The transfer function can be expressed in the factored form of Equation (2.17), but since we assume that the system is causal and stable, poles are restricted to lie inside the unit circle. This constraint is understood by recalling that causality implies rightsidedness, which implies that the ROC is outside the outermost pole, and that stability implies that the ROC includes the unit circle so that all poles must fall inside the unit circle. The zeros, on the other hand, can fall inside or outside the unit circle (Figure 2.8). The factored transfer function is then reduced to

$$H(z) = A z^{-r} \frac{\prod_{k=1}^{M_i}(1 - a_k z^{-1}) \prod_{k=1}^{M_o}(1 - b_k z)}{\prod_{k=1}^{N_i}(1 - c_k z^{-1})} \tag{2.22}$$

with $|a_k|, |b_k|, |c_k| < 1$, with $M_i + M_o = M$ and $N_i = N$, and where the terms $(1 - a_k z^{-1})$ and $(1 - c_k z^{-1})$ represent, respectively, zeros and poles inside the unit circle and $(1 - b_k z)$ represents zeros outside the unit circle.

---

[7] The term *digital filters* more generally includes nonlinear filters where superposition does not hold. Also note that the causality constraint is artificial. We will see later in the text that noncausal filters are viable, as well as quite useful, in digital computer processing of speech.

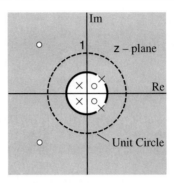

Figure 2.8 Pole-zero configuration
for a causal and stable discrete-time
system.

2.8.2 Magnitude-Phase Relationships

When all zeros as well as poles of a rational transfer function $H(z)$ are inside the unit circle, the transfer function is said to be *minimum phase*, and the corresponding impulse response is referred to as a *minimum-phase sequence* [7]. Such zeros and poles will sometimes be called "minimum-phase components" of the transfer function. An arbitrary causal and stable digital filter, however, will have its poles, but not necessarily its zeros, inside the unit circle. Zeros (or poles) outside the unit circle are referred to as *maximum-phase* components. Therefore, $H(z)$ is generally *mixed-phase*, consisting of a minimum-phase and maximum-phase component, i.e.,

$$H(z) \; = \; H_{min}(z) H_{max}(z)$$

where $H_{min}(z)$ consists of all poles and zeros inside the unit circle, and $H_{max}(z)$ consists of all zeros outside the unit circle. Although not strictly following our definitions, we also include in these terms zeros on the unit circle. The terminology "minimum-phase" and "maximum-phase" is also applied to discrete-time signals, as well as to systems and their impulse responses.

Alternatively, any digital filter can be represented by the cascade of a minimum-phase "reference" system $H_{rmp}(z)$ and an all-pass system $A_{all}(z)$:

$$H(z) \; = \; H_{rmp}(z) A_{all}(z)$$

where the all-pass system is characterized by a frequency response with unity magnitude for all ω. In particular, an arbitrary rational all-pass $A_{all}(z)$ can be shown to consist of a cascade of factors of the form $[\frac{1-a^*z}{1-az^{-1}}]^{\pm 1}$ where $|a| < 1$ (Exercise 2.10). Consequently, such all-pass systems have the property that their poles and zeros occur at conjugate reciprocal locations; by this, we mean that a zero at $z = \frac{1}{a^*}$ implies a pole at $z = a$. It follows that if $H(z)$ contains Q zeros then, because $|H(z)| = |H_{rmp}(z)|$, there exists a maximum of 2^Q different phase functions (excluding linear phase), and thus 2^Q different sequences, for the given magnitude function.[8] These phase functions can be obtained by reflecting zeros about the unit circle to their conjugate reciprocal locations by multiplying $H(z)$ by $[\frac{1-a^*z}{1-az^{-1}}]^{\pm 1}$. Because the all-pass function with unity power ($+1$) has negative phase (Exercise 2.10), if we flip a zero of a minimum-phase function outside the unit circle, then the resulting phase is more negative than the original. The term *minimum phase-lag* would thus be more precise than the commonly used *minimum phase* [7].

There are a number of important properties of minimum-phase sequences that have important consequences for speech modeling and processing. A minimum-phase sequence is uniquely specified by the magnitude of its Fourier transform. This result will be proven formally in Chapter 10 in the context of the complex cepstrum representation of a sequence, but can be seen intuitively from the example of a stable rational z-transform. In this case, for a given Fourier-transform magnitude, there is only one sequence with all zeros (or poles) inside the unit circle. Likewise, the Fourier transform phase of a minimum-phase sequence $H(\omega)$ uniquely specifies the sequence (to within a scale factor).

Another useful property relates to the energy concentration of a minimum-phase sequence. From Parseval's theorem, all sequences with the same Fourier transform magnitude have the same energy, but when we flip zeros (or poles) inside and outside the unit circle to their conjugate reciprocal locations, this energy gets distributed along the time axis in different ways. It can be shown that a finite-length (all-zero) minimum-phase sequence has energy most concentrated near (and to the right of) the time origin, relative to all other finite-length causal sequences with the same Fourier transform magnitude, and thus tends to be characterized by an abrupt onset or what is sometimes referred to as a fast "attack" of the sequence.[9] This property can be formally expressed as [7]

$$\sum_{n=0}^{m} |h_{rmp}[n]|^2 \leq \sum_{n=0}^{m} |h[n]|^2, \quad m \geq 0 \tag{2.23}$$

where $h[n]$ is a causal sequence with the Fourier transform magnitude equal to that of the reference minimum-phase sequence $h_{rmp}[n]$. As zeros are flipped outside the unit circle, the energy of the sequence is delayed in time, the maximum-phase counterpart having maximum energy delay (or phase lag) [7]. Similar energy localization properties are found with respect to poles. However, because causality strictly cannot be made to hold when a z-transform contains maximum-phase poles, it is more useful to investigate how the energy of the sequence shifts with respect to the time origin. As illustrated in Example 2.11, flipping poles from inside to outside the unit circle to their conjugate reciprocal location moves energy to the left of the time origin, transforming the fast attack of the minimum-phase sequence to a more gradual onset. We will see throughout the text that numerous speech analysis schemes result in a minimum-phase vocal tract impulse response estimate. Because the vocal tract is not necessarily minimum phase, synthesized speech may be characterized in these cases by an unnaturally abrupt vocal tract impulse response.

---

[8] Because we assume causality and stability, the poles lie inside the unit circle. Different phase functions, for a specified magnitude, therefore are not contributed by the poles.

[9] It is of interest to note that a sufficient but not necessary condition for a causal sequence to be minimum phase is that $|h[0]| > \sum_{n=1}^{\infty} |h[n]|$ [9].

EXAMPLE 2.11   An example comparing a mixed-phase impulse response $h[n]$, having poles inside and outside the unit circle, with its minimum-phase reference $h_{rmp}[n]$ is given in Figure 2.9. The minimum-phase sequence has pole pairs at $0.95e^{\pm j0.1}$ and $0.95e^{\pm j0.3}$. The mixed-phase sequence has pole pairs at $0.95e^{\pm j0.1}$ and $\frac{1}{0.95}e^{\pm j0.3}$. The minimum-phase sequence (a) is concentrated to the right of the origin and in this case is less "dispersed" than its non-minimum-phase counterpart (c). Panels (b) and (d) show that the frequency response magnitudes of the two sequences are identical. As we will see later in the text, there are perceptual differences in speech synthesis between the fast and gradual "attack" of the minimum-phase and mixed-phase sequences, respectively.   ▲

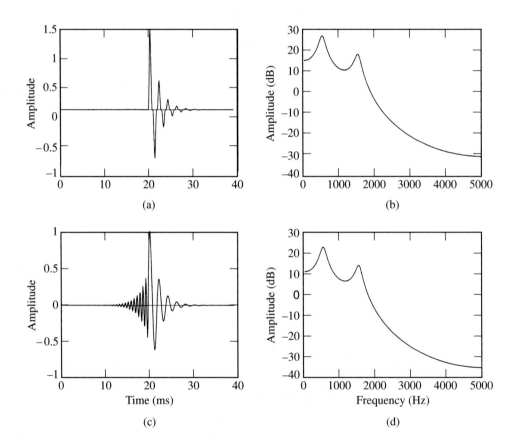

Figure 2.9 Comparison in Example 2.11 of (a) a minimum-phase sequence $h_{rmp}[n]$ with (c) a mixed-phase sequence $h[n]$ obtained by flipping one pole pair of $h_{rmp}[n]$ outside the unit circle to its conjugate reciprocal location. Panels (b) and (d) show the frequency response magnitudes of the minimum- and mixed-phase sequences, respectively.

2.8.3 FIR Filters

There are two classes of digital filters: *finite impulse response* (FIR) and *infinite impulse response* (IIR) filters [7],[10]. The impulse response of an FIR filter has finite duration and corresponds to having no denominator in the rational function $H(z)$, i.e, there is no feedback in the difference Equation (2.21). This results in the reduced form

$$y[n] = \sum_{r=0}^{M} \beta_r x[n - r].$$ (2.24)

Implementing such a filter thus requires simply a train of delay, multiply, and add operations. By applying the unit sample input and interpreting the output as the sum of weighted delayed unit samples, we obtain the impulse response given by

$$h[n] = \beta_n, \quad 0 \le n \le M$$
$$= 0, \quad \text{otherwise}$$

Because $h[n]$ is bounded over the duration $0 \le n \le M$, it is causal and stable. The corresponding rational transfer function in Equation (2.22) reduces to the form

$$X(z) = A z^{-r} \prod_{k=1}^{M_i} (1 - a_k z^{-1}) \prod_{k=1}^{M_o} (1 - b_k z)$$

with $M_i + M_o = M$ and with zeros inside and outside the unit circle; the ROC is the entire z-plane except at the only possible poles $z = 0$ or $z = \infty$.

FIR filters can be designed to have perfect linear phase. For example, if we impose on the impulse response symmetry of the form $h[n] = h[M - n]$, then under the simplifying assumption that M is even (Exercise 2.14), $H(\omega) = A(\omega)e^{-j\omega(M/2)}$ where $A(\omega)$ is purely real, implying that phase distortion will not occur due to filtering [10], an important property in speech processing.[10]

2.8.4 IIR Filters

IIR filters include the denominator term in $H(z)$ and thus have feedback in the difference equation representation of Equation (2.21). Because symmetry is required for linear phase, most[11] IIR filters will not have linear phase since they are right-sided and infinite in duration. Generally, IIR filters have both poles and zeros. As we noted earlier for the special case where the number of zeros is less than the number of poles, the system function $H(z)$ can be expressed in a partial fraction expansion as in Equation (2.18). Under this condition, for causal systems, the impulse response can be written in the form

$$h[n] = \sum_{k=1}^{N_i} A_k c_k^n u[n]$$

---

[10] This does not mean that $M(\omega)$ is positive. However, if the filter $h[n]$ has most of its spectral energy where $M(\omega) > 0$, then little speech phase distortion will occur.

[11] A class of linear phase IIR filters has been shown to exist [1]. The transfer function for this filter class, however, is not rational and thus does not have an associated difference equation.

where c_k is generally complex so that the impulse response is a sum of decaying complex exponentials. Equivalently, because $h[n]$ is real, it can be written by combining complex conjugate pairs as a set of decaying sinusoids of the form

$$h[n] = \sum_{k=1}^{N_i/2} B_k |c_k|^n \cos(\omega_k n + \phi_k) u[n]$$

where we have assumed no real poles and thus N_i is even. Given a desired spectral magnitude and phase response, there exist numerous IIR filter design methods [7],[10].

In the implementation of IIR filters, there exists more flexibility than with FIR filters. A "direct-form" method is seen in the recursive difference equation itself [7],[10]. The partial fraction expansion of Equation (2.18) gives another implementation that, as we will see later in the text, is particularly useful in a parallel resonance realization of a vocal tract transfer function. Suppose, for example, that the number of poles in $H(z)$ is even and that all poles occur in complex conjugate pairs. Then we can alter the partial fraction expansion in Equation (2.18) to take the form

$$\begin{aligned}
X(z) &= \sum_{k=1}^{N_i/2} \frac{A_k(1 - p_k z^{-1})}{(1 - c_k z^{-1})(1 - c_k^* z^{-1})} \\
&= \sum_{k=1}^{N_i/2} \frac{A_k(1 - p_k z^{-1})}{1 - u_k z^{-1} + v_k z^{-2}}
\end{aligned} \tag{2.25}$$

which represents $\frac{N_i}{2}$ second-order IIR filters in parallel. Other digital filter implementation structures are introduced as needed in speech analysis/synthesis schemes throughout the text.

2.9 Time-Varying Systems

Up to now we have studied linear systems that are time-invariant, i.e., if $x[n]$ results in $y[n]$, then a shifted input $x[n - n_o]$ results in a shifted output $y[n - n_o]$. In the speech production mechanism, however, we often encounter *time-varying linear systems*. Although superposition holds in such systems, time-invariance does not. A simple illustrative example is a "system" which multiplies the input by a sequence $h[n]$. The system is linear because $(\alpha x_1[n] + \beta x_2[n])h[n] = \alpha x_1[n]h[n] + \beta x_2[n]h[n]$, but it is not time-invariant because, in general, $x[n - n_o]h[n] \neq x[n - n_o]h[n - n_o]$.

A time-varying linear system is characterized by an impulse response that changes for each time m. This system can be represented by a two-dimensional function $g[n, m]$, which is the impulse response at time n to a unit sample applied at time m. The response, for example, to a unit sample at time $m = 0$ is $g[n, 0]$, while the response to $\delta[n - n_o]$ is $g[n, n_o]$. The two-dimensional function $g[n, m]$ is sometimes referred to as *Green's function* [8]. Because the system is linear and because an input $x[n]$ consists of a sum of weighted and delayed unit samples, i.e., $x[n] = \sum_{m=-\infty}^{\infty} x[m]\delta[n - m]$, then the output to the input $x[n]$ is given by

$$y[n] = \sum_{m=-\infty}^{\infty} g[n, m]x[m] \tag{2.26}$$

which is a superposition sum, but *not* a convolution. We can see how Equation (2.26) differs from a convolution by invoking an alternate two-dimensional representation which is the response of the system at time n to a unit sample applied m samples earlier at time $[n - m]$. This new function, called the *time-varying unit sample response* and denoted by $h[n, m]$, is related to Green's function by $h[n, m] = g[n, n - m]$ or, equivalently, $h[n, n - m] = g[n, m]$. We can then write the time-varying system output as (Exercise 2.16)

$$y[n] = \sum_{m=-\infty}^{\infty} h[n, n - m]x[m]$$

$$= \sum_{m=-\infty}^{\infty} h[n, m]x[n - m] \tag{2.27}$$

where we have invoked a change in variables and where the weight on each impulse response corresponds to the input sequence m samples in the past. When the system is time-invariant, it follows that Equation (2.27) is expressed as (Exercise 2.16)

$$y[n] = \sum_{m=-\infty}^{\infty} h[m]x[n - m] \tag{2.28}$$

which is the convolution of the input with the impulse response of the resulting linear time-invariant system.

It is of interest to determine whether we can devise Fourier and z-transform pairs, i.e., a frequency response and transfer function, for linear time-varying systems, as we had done with linear time-invariant systems. Let us return to the familiar complex exponential as an input to the linear time-varying system with impulse response $h[n, m]$. Then, from Equation (2.27), the output $y[n]$ is expressed as

$$y[n] = \sum_{m=-\infty}^{\infty} h[n, m]e^{j\omega(n-m)}$$

$$= e^{j\omega n} \sum_{m=-\infty}^{\infty} h[n, m]e^{-j\omega m}$$

$$= e^{j\omega n} H(n, \omega) \tag{2.29}$$

where

$$H(n, \omega) = \sum_{m=-\infty}^{\infty} h[n, m]e^{-j\omega m}$$

which is the Fourier transform of $h[n, m]$ at time n evaluated with respect to the variable m and referred to as the *time-varying frequency response*. Equivalently, we can write the time-varying frequency response in terms of Green's function as (Exercise 2.17)

$$H(n, \omega) = e^{j\omega n} \sum_{m=-\infty}^{\infty} g[n, m]e^{-j\omega m} \tag{2.30}$$

which, except for the linear phase factor $e^{j\omega n}$, is the Fourier transform of Green's function at time n.

Because the system of interest is linear, its output to an arbitrary input $x[n]$ is given by the following superposition [8] (Exercise 2.15):

$$\begin{aligned} y[n] &= \sum_{m=-\infty}^{\infty} h[n, m]x[n - m] \\ &= \frac{1}{2\pi} \int_{-\pi}^{\pi} H(n, \omega)X(\omega)e^{j\omega n} d\omega \end{aligned} \tag{2.31}$$

so that the output $y[n]$ of $h[n, m]$ at time n is the inverse Fourier transform of the *product* of $X(\omega)$ and $H(n, \omega)$, $X(\omega)H(n, \omega)$ which can be thought of as a generalization of the Convolution Theorem for linear time-invariant systems. This generalization, however, can be taken only so far. For example the elements of a cascade of two time-varying linear systems, i.e., $H_1(n, \omega)$ followed by $H_2(n, \omega)$, do not generally combine in the frequency domain by multiplication and the elements cannot generally be interchanged, as illustrated in the following example. Consequently, care must be taken interchanging the order of time-varying systems in the context of speech modeling and processing.

EXAMPLE 2.12   Consider the linear time-varying multiplier operation

$$y[n] = x[n]e^{j\omega_o n}$$

cascaded with a linear time-invariant ideal low-pass filter $h[n]$, as illustrated in Figure 2.10. Then, in general, $(x[n]e^{j\omega_o n}) * h[n] \neq (x[n] * h[n])e^{j\omega_o n}$. For example, let $x[n] = e^{j\frac{\pi}{3}n}$, $\omega_o = \frac{\pi}{3}$, and $h[n]$ have lowpass cutoff frequency at $\frac{\pi}{2}$. When the lowpass filter follows the multiplier, the output is zero; when the order is interchanged, the output is nonzero.   ▲

We will see in following chapters that under certain "slowly varying" conditions, linear time-varying systems can be approximated by linear time-invariant systems. The accuracy of this approximation will depend on the time-duration over which we view the system and its input, as well as the rate at which the system changes. More formal conditions have been derived by Matz and Hlawatsch [5] under which a "transfer function calculus" is allowed for time-varying systems.

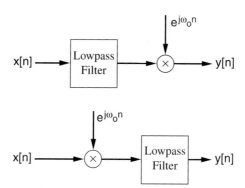

Figure 2.10 Cascade configurations of linear time-invariant and time-varying systems.

2.10 Discrete Fourier Transform

The Fourier transform of a discrete-time sequence is a *continuous* function of frequency [Equation (2.2)]. Because, in practice, when using digital computers we cannot work with continuous frequency, we need to sample the Fourier transform, and, in particular, we want to sample finely enough to be able to recover the sequence. For sequences of finite length N, sampling yields a new transform referred to as the *discrete Fourier transform* or DFT. The DFT pair representation of $x[n]$ is given by

$$x[n] = \frac{1}{N} \sum_{k=0}^{N-1} X(k) e^{j \frac{2\pi}{N} kn}, \quad 0 \leq n \leq N - 1$$

$$X(k) = \sum_{n=0}^{N-1} x[n] e^{-j \frac{2\pi}{N} kn}, \quad 0 \leq k \leq N - 1. \tag{2.32}$$

The sequence $x[n]$ and transform $X(k)$, although specified only in the interval $[0, N-1]$, take on an *implicit* periodicity with period N in operations involving the DFT [7]. Observe that if we sample the discrete-time Fourier transform (continuous in frequency) at uniformly spaced samples $\omega = \frac{2\pi}{N} k$, we obtain values of the discrete Fourier transform. The two equations of the DFT pair are sometimes referred to as the analysis (DFT) and synthesis (inverse DFT) equations for the DFT.

The properties of the DFT are similar to those of the discrete-time Fourier transform. These include the DFT counterparts to the four properties given in Section 2.3. For example, Parseval's theorem for the DFT can be shown to be given by

$$\sum_{n=0}^{N-1} |x[n]|^2 = \frac{1}{N} \sum_{k=0}^{N-1} |X(k)|^2.$$

The functions $|x[n]|^2$ and $|X(k)|^2$ are still thought of as "energy densities," i.e., the energy per unit time and the energy per unit frequency, because they describe the distribution of energy in time and frequency, respectively.

There are, however, some important distinctions from the discrete-time Fourier transform. Because in the DFT representation we think of $x[n]$ as one period of a periodic sequence with period N, the argument of $x[n]$ is computed modulo N, i.e., $x[n] = x[n \ modulo \ N]$, which is also denoted by the subscript notation $x[n] = x[(n)_N]$. Consequently, many operations are performed modulo N. Consider, for example, a shifted unit sample.

EXAMPLE 2.13   For a shifted unit sample, the delay is computed modulo N:

$$x[n] = \delta[(n - n_o)_N]$$

which we can think of as shifting the periodic construction of the unit sample (with period N), sometimes referred to as a *circular shift* or *rotation*, and then extracting the resulting sequence over $0 \leq n \leq N - 1$. The DFT is given by

$$X(k) = \sum_{n=0}^{N-1} \delta[(n - n_o)_N] e^{-j\frac{2\pi}{N}kn}$$

$$= e^{-j\frac{2\pi}{N}kn_o}$$

because $e^{-j\frac{2\pi}{N}kn}$ is periodic in n with period N and because $x[n]$ is nonzero for only $n = n_o$.   ▲

Consider now the Convolution Theorem introduced in Section 2.6. We saw that convolution of two sequences, sometimes referred to as a *linear* convolution, corresponds to multiplication of their discrete Fourier transforms. Multiplication of DFTs of two sequences, on the other hand, corresponds to a *circular* convolution of the (implied) periodic sequences [7]. Specifically, let $x[n]$ have N-point DFT $X(k)$ and $h[n]$ have N-point DFT $H(k)$. Then the inverse DFT of $X(k)H(k) = Y(k)$ is not the linear convolution, i.e., $y[n] = x[n] * h[n]$, but is rather the circular convolution $y[n] = h[n] \circledast x[n]$, where one function is circularly shifted relative to the other with period N. We can also think of each sequence being defined only in the interval $[0, N)$ and being shifted modulo N in the convolution. Suppose now that $x[n]$ is nonzero over the interval $0 \leq n \leq M - 1$ and $h[n]$ is nonzero over the interval $0 \leq n \leq L - 1$. The linear and circular convolutions are equivalent only when the sum of the sequence durations (minus one) is less than the DFT length, i.e.,

$$x[n] \circledast h[n] = x[n] * h[n], \quad n = 0, 1, 2, \ldots N - 1$$

only if $M + L - 1 \leq N$. The implication is that, because circular convolution is performed modulo N, zero padding of the respective sequences is required to obtain a linear convolution. Similar considerations must be made in frequency for the DFT realization of the Windowing Theorem. A more extensive description of the DFT realization of the Convolution and Windowing Theorems can be found in [7].

Finally, there exists an implementation of the DFT through the fast Fourier transform (FFT). The straightforward DFT requires on the order N^2 operations, i.e., multiples and additions, while the FFT requires order $N \log(N)$ [7]. In doing convolution with the DFT, it is often faster to perform two forward FFTs and one inverse FFT rather than the direct shift, multiply, and accumulate operations.

2.11 Conversion of Continuous Signals and Systems to Discrete Time

2.11.1 Sampling Theorem

We began this review by sampling a continuous-time speech waveform to generate a discrete-time sequence, i.e., $x[n] = x_a(t)|_{t=nT}$, where T is the sampling interval. An implicit assumption was that the sampling is fast enough so that we can recover $x_a(t)$ from $x[n]$. The condition under which recovery is possible is called the *Sampling Theorem*.

> **Sampling Theorem:**   Suppose that $x_a(t)$ is sampled at a rate of $F_s = \frac{1}{T}$ samples per second and suppose $x_a(t)$ is a bandlimited signal, i.e., its continuous-time Fourier transform $X_a(\Omega)$ is such that $X_a(\Omega) = 0$ for $|\Omega| \geq \Omega_N = 2\pi F_N$. Then $x_a(t)$ can be uniquely determined from its uniformly spaced samples $x[n] = x_a(nT)$ if the sampling frequency F_s is greater than twice the largest frequency of the signal, i.e., $F_s > 2F_N$. The largest frequency in the signal F_N is called the *Nyquist frequency*, and $2F_N$, which must be attained in sampling for reconstruction, is called the *Nyquist rate*.

For example, in speech we might assume a 5000 Hz bandwidth. Therefore, for signal recovery we must sample at $\frac{1}{T} = 10000$ samples/s corresponding to a $T = 100\ \mu s$ sampling interval.

The basis for the Sampling Theorem is that sampling $x_a(t)$ at a rate of $\frac{1}{T}$ results in spectral duplicates spaced by $\frac{2}{T}$, so that sampling at the Nyquist rate avoids aliasing, thus preserving the spectral integrity of the signal. The sampling can be performed with a periodic impulse train with spacing T and unity weights, i.e., $p(t) = \sum_{k=-\infty}^{\infty} \delta(t - kT)$. The impulse train resulting from multiplication with the signal $x_a(t)$, denoted by $x_p(t)$, has weights equal to the signal values evaluated at the sampling rate, i.e.,

$$x_p(t) = x_a(t)p(t)$$

$$= \sum_{k=-\infty}^{\infty} x_a(kT)\delta(t - kT). \tag{2.33}$$

The impulse weights are values of the discrete-time signal, i.e., $x[n] = x_a(nT)$, and therefore, as illustrated in Figure 2.11, the cascade of sampling with the impulse train $p(t)$ followed by conversion of the resulting impulse weights to a sequence is thought of as an ideal A/D converter (or C/D converter). In the frequency domain, the impulse train $p(t)$ maps to another impulse train with spacing $2\pi F_s$, i.e., the Fourier transform of $p(t)$ is $P(\Omega) = \frac{2\pi}{T} \sum_{k=-\infty}^{\infty} \delta(\Omega - k\Omega_s)$ where $\Omega_s = 2\pi F_s$. Using the continuous-time version of the Windowing Theorem, it fol-

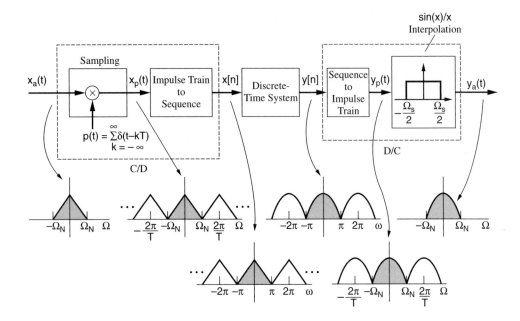

Figure 2.11   Path from sampling to reconstruction $(2\Omega_N = \Omega_s)$.

lows that $P(\Omega)$ convolves with the Fourier transform of the signal $x_a(t)$, thus resulting in a continuous-time Fourier transform with spectral duplicates

$$X_p(\Omega) = \frac{1}{T} \sum_{k=-\infty}^{\infty} X_a(\Omega - k\Omega_s) \tag{2.34}$$

where $\Omega_s = 2\pi F_s$. Therefore, the original continuous-time signal $x_a(t)$ can be recovered by applying a lowpass analog filter, unity in the passband $\left[-\frac{\Omega_s}{2}, \frac{\Omega_s}{2}\right]$, and zero outside this band.

This perspective also leads to a reconstruction formula which *interpolates* the signal samples with a sin function. Using the continuous-time version of the Convolution Theorem, application of an ideal lowpass filter of width Ω_s corresponds to the convolution of the filter impulse response with the signal-weighted impulse train $x_p(t)$. Thus, we have a reconstruction formula given by

$$x_a(t) = \sum_{n=-\infty}^{\infty} x_a(nT) \frac{\sin(\pi(t - nT)/T)}{\pi(t - nT)/T}$$

because the function $\frac{\sin(\pi(t-nT)/T)}{\pi(t-nT)/T}$ is the inverse Fourier transform of the ideal lowpass filter. As illustrated in Figure 2.11, the cascade of the conversion of the sequence $y[n] = x[n]$ to a continuous-time impulse train (with weights $x_a(nT)$) followed by lowpass filtering is thought of as a *discrete-to-continuous (D/C) converter*. In practice, however, a *digital-to-analog converter* is used. Unlike the D/C converter, because of quantization error and other forms of distortion, D/A converters do not achieve perfect reconstruction.

The relation between the Fourier transform of $x_a(t)$, $X_a(\Omega)$, and the discrete-time Fourier transform of $x[n] = x_a(nT)$, $X(\omega)$, can now be deduced. When the Sampling Theorem holds, over the frequency interval $[-\pi, \pi]$ $X(\omega)$ is a frequency-scaled (or frequency-normalized) version of $X_a(\Omega)$. Specifically, over the interval $[-\pi, \pi]$ we have

$$X(\omega) = \frac{1}{T} X_a \left(\frac{\omega}{T}\right), \qquad |\omega| \leq \pi.$$

This relation can be obtained by first observing that $X_p(\Omega)$ can be written as

$$X_p(\Omega) = \sum_{n=-\infty}^{\infty} x_a(nT) e^{-j\Omega Tn} \tag{2.35}$$

and then by applying the continuous-time Fourier transform to Equation (2.33), and comparing this result with the expression for the discrete-time Fourier transform in Equation (2.2) [7]. We see that if the sampling is performed exactly at the Nyquist rate, then the normalized frequency π corresponds to the highest frequency in the signal. For example, when $F_N = 5000$ Hz, then π corresponds to 5000 Hz.

The entire path, including sampling, application of a discrete-time system, and reconstruction, as well as the frequency relation between signals, is illustrated in Figure 2.11. In this illustration, the sampling frequency equals the Nyquist rate, i.e., $2\Omega_N = \Omega_s$. The discrete-time system shown in the figure may be, for example, a digital filter which has been designed in discrete time or derived from sampling an analog filter with some desired properties.

A topic related to the Sampling Theorem is the decrease and increase of the sampling rate, referred to as *decimation* and *interpolation* or, alternatively, as *downsampling* and *upsampling*, respectively. Changing the sampling rate can be important in speech processing where one parameter may be deemed to be slowly-varying relative to another; for example, the state of the vocal tract may vary more slowly, and thus have a smaller bandwidth, than the state of the vocal cords. Therefore, different sampling rates may be applied in their estimation, requiring a change in sampling rate in waveform reconstruction. For example, although the speech waveform is sampled at, say, 10000 samples/s, the vocal cord parameters may be sampled at 100 times/s, while the vocal tract parameters are sampled at 50 times/s. The reader should briefly review one of the numerous tutorials on decimation and interpolation [3],[7],[10].

2.11.2 Sampling a System Response

In the previous section, we sampled a continuous-time waveform to obtain discrete-time samples for processing by a digital computer or other discrete-time-based system. We will also have occasion to transform analog systems to discrete-time systems, as, for example, in sampling a continuous-time representation of the vocal tract impulse response, or in the replication of the spectral shape of an analog filter. One approach to this transformation is to simply sample the continuous-time impulse response of the analog system; i.e., we perform the continuous-to-discrete-time mapping

$$h[n] = h_a(nT)$$

where $h_a(t)$ is the analog system impulse response and T is the sampling interval. This method of discrete-time filter design is referred to as the *impulse invariance* method [7].

Similar to sampling of continuous-time waveforms, the discrete-time Fourier transform of the sequence $h[n]$, $H(\omega)$, is related to the continuous-time Fourier transform of $h_a(t)$, $H_a(\Omega)$, by the relation

$$H(\omega) = \frac{1}{T} H_a \left(\frac{\omega}{T} \right), \qquad |\omega| \leq \pi$$

where we assume $h_a(t)$ is bandlimited and the sampling rate is such to satisfy the Nyquist criterion [7]. The frequency response of the analog signal is therefore preserved. It is also of interest to determine how poles and zeros are transformed in going from the continuous- to the discrete-time filter domains as, for example, in transforming a continuous-time vocal tract impulse response. To obtain a flavor for this style of conversion, consider the continuous-time rendition of the IIR filter in Equation (2.25), i.e.,

$$h_a(t) = \sum_{k=1}^{N} A_k e^{s_k t} u(t)$$

whose Laplace transform is given in partial fraction expansion form (the continuous counterpart to Equation (2.18)) [7]

$$H_a(s) = \sum_{k=1}^{N} \frac{A_k}{s - s_k}.$$

Then the impulse invariance method results in the discrete-time impulse response

$$h[n] = h_a[nT] = \sum_{k=1}^{N} A_k e^{(s_k T)n} u[n]$$

whose z-transform is given by

$$H(z) = \sum_{k=1}^{N} \frac{A_k}{1 - e^{(s_k T)} z^{-1}}$$

with poles at $z = e^{(s_k T)}$ inside the unity circle in the z-plane ($|e^{(s_k T)}| = e^{(\mathrm{Re}[s_k]T)} < 1$ when $\mathrm{Re}[s_k] < 0$) is mapped from poles in the s-plane from $s = s_k$ located to the left of the $j\Omega$ axis. Poles being to the left of the $j\Omega$ axis is a stability condition for causal continuous systems. Although the poles are mapped inside the unit circle, the mapping of the zeros depends on both the resulting poles and the coefficients A_k in the partial fraction expansion. It is conceivable, therefore, that a minimum-phase response may be mapped to a mixed-phase response with zeros outside the unit circle, a consideration that can be particularly important in modeling the vocal tract impulse response.

Other continuous-to-discrete-time conversion methods, e.g., the bilinear transformation [7], will be described as needed throughout the text.

2.11.3 Numerical Simulation of Differential Equations

In a loose sense, the impulse invariance method can be thought of as a *numerical simulation* of a continuous-time system by a discrete-time system. Suppose that the continuous-time system is represented by a differential equation. Then a discrete-time simulation of this analog system could alternatively be obtained by approximating the derivatives by finite differences, e.g., $\frac{d}{dt}x(t)$ at $t = nT$ is approximated by $\frac{x(nT)-x((n-1)T)}{T}$. In mapping the frequency response of the continuous-time system to the unit circle, however, such an approach has been shown to be undesirable due to the need for an exceedingly fast sampling rate, as well as due to the restriction on the nature of the frequency response [6]. Nevertheless, in this text we will have occasion to revisit this approach in a number of contexts, such as in realizing analog models of speech production or analog signal processing tools in discrete time. This will become especially important when considering differential equations that are not necessarily time-invariant, are possibly coupled, and/or which may contain a nonlinear element. In these scenarios, approximating derivatives by differences is one solution option, where digital signal processing techniques are applied synergistically with more conventional numerical analysis methods. Other solution options exist, such as the use of a *wave digital filter* methodology to solve coupled, time-varying, nonlinear differential equations [4].

2.12 Summary

In this chapter we have reviewed the foundation of discrete-time signal processing which will serve as a framework for the remainder of the text. We reviewed discrete-time signals and systems and their Fourier and z-transform representations. A fundamental property of the Fourier transform is the uncertainty principle that imposes a constraint between the duration and bandwidth of a sequence and limits our ability to simultaneously resolve both in time and frequency dynamically changing events or events closely spaced in time and frequency. We will investigate this limitation in a speech context more fully in Chapters 7 and 11.

In this chapter, we introduced the concepts of minimum- and mixed-phase sequences and looked at important relationships between the magnitude and phase of their Fourier transforms. A property of these sequences, which we will see influences the perception of synthesized speech, is that a minimum-phase sequence is often characterized by a sharper "attack" than that of a mixed-phase counterpart with the same Fourier transform magnitude. Also in this chapter, we briefly reviewed some considerations in obtaining a discrete-time sequence by sampling a continuous-time signal, and also reviewed constraints for representing a sequence from samples of its discrete-time Fourier transform, i.e., from its DFT. Finally, we introduced the notion of time-varying linear systems whose output is not represented by a convolution sum, but rather by a more general superposition of delayed and weighted input values. An important property of time-varying systems is that they do not commute, implying that care must be taken when interchanging their order. The importance of time-varying systems in a speech processing context will become evident as we proceed deeper into the text.

EXERCISES

2.1 Prove the commutativity of discrete-time convolution, i.e., $x[n] * h[n] = h[n] * x[n]$.

2.2 Argue, through the "flip and shift" interpretation of discrete-time convolution, that a necessary and sufficient condition for causality of an LTI system is that the system impulse response be zero for $n < 0$.

2.3 In this exercise, you are asked to consider a number of properties of the Fourier transform.

    (a) Prove the 2π periodicity of the Fourier transform.

    (b) Prove the conjugate symmetry of the Fourier transform for a real sequence $x[n]$. Then, using this property, show the resulting following properties: (1) The magnitude and real part of $X(\omega)$ are even, i.e., $|X(\omega)| = |X(-\omega)|$ and $X_r(\omega) = X_r(-\omega)$; (2) Its phase and imaginary parts are odd, i.e., $\angle X(\omega) = -\angle X(-\omega)$ and $X_i(\omega) = -X_i(-\omega)$.

    (c) Prove Parseval's Theorem. *Hint:* In Equation (2.3), write $|X(\omega)|^2 = X(\omega)X^*(\omega)$ and substitute the Fourier transform expression for $X(\omega)$.

2.4 In this exercise, you explore the notion of duration and bandwidth of a sequence. Assume that the signal energy is unity.

    (a) Show that the average frequency location $\bar{\omega} = \int_{-\pi}^{\pi} \omega |X(\omega)|^2 d\omega$ of a real sequence $x[n]$ is zero. Compute the average time location $\bar{n} = \sum_{n=-\infty}^{\infty} n|x[n]|^2$ of an even complex sequence.

    (b) Determine the bandwidth $\int_{-\pi}^{\pi} (\omega - \bar{\omega})^2 |X(\omega)|^2 d\omega$ of the sequences with Fourier transforms in Figure 2.12.

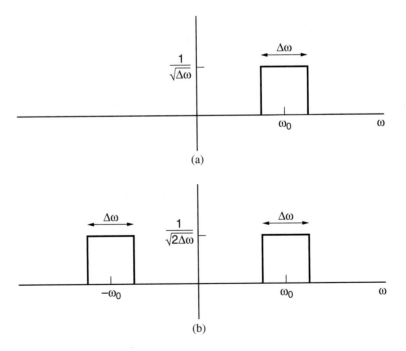

Figure 2.12 Fourier transforms with different bandwidths: (a) complex sequence; (b) real sequence.

2.5 In this exercise, you are asked to prove the uncertainty principle in Equation (2.5).

(a) Show that if $x[n]$ has Fourier transform $X(\omega)$, then differentiating in frequency results in the pair

$$nx[n] \leftrightarrow j\frac{d}{d\omega}X(\omega).$$

(b) Using your result from part (a), derive Equation (2.6) using the Schwartz inequality: $|\int_a^b x(u)y(u)du|^2 \le \int_a^b |x(u)|^2 du \int_a^b |y(u)|^2 du$.

(c) Complete the derivation of the uncertainty principle. *Hint:* Write the Fourier transform $X(\omega)$ in terms of its magnitude and phase, integrate, and then determine terms that are always positive.

2.6 In this exercise, you are asked to investigate the analytic signal representation.

(a) Show that the real part of the analytic signal representation of $x[n]$ in Equation (2.7) is given by $s_r[n] = x[n]/2$. Show that the imaginary part $s_i[n]$ is obtained by applying the Hilbert transformer

$$H(\omega) = \begin{cases} -j & 0 \le \omega < \pi \\ +j & -\pi \le \omega < 0 \end{cases}$$

to $s_r[n] = x[n]/2$ as in Equation (2.8).

(b) Find the output of the Hilbert transformer when the input sequence is given by

$$x[n] = \cos(\omega_0 n)$$

where $0 < \omega_0 < \pi$.

(c) Find the magnitude and phase of the analytic signal

$$s[n] = s_r[n] + js_i[n]$$

for the sequence $x[n]$ of part (b).

(d) Repeat parts (b) and (c) for the input sequence

$$x[n] = \cos(\omega_0 n)\left[\frac{\sin(\omega_1 n)}{\pi n}\right]$$

where $0 < \omega_0 - \omega_1 < \pi$ and $0 < \omega_0 + \omega_1 < \pi$.

2.7 In this exercise, you are asked to find the condition under which the analytic and quadrature representations of the real signal $x[n] = a[n]\cos(\phi[n])$ are equivalent. We write the quadrature and analytic signal representations, respectively, as

$$s_q[n] = a[n]e^{j\phi[n]}$$

and

$$s_a[n] = \frac{1}{2\pi}\int_0^\pi X(\omega)e^{j\omega n}d\omega.$$

(a) Show that

$$S_q(\omega) = X(\omega) + j\sum_{n=-\infty}^{\infty} a[n]\sin(\phi[n])e^{-j\omega n}$$

and therefore

$$2X(\omega) = S_q(\omega) + S_q^*(-\omega).$$

(b) Let $S_q(\omega)$ and $S_a(\omega)$ denote the Fourier transforms of the quadrature and analytic signals, respectively. Show that

$$2S_a(\omega) = 0, \quad \omega < 0$$
$$= S_q(\omega) + S_q^*(-\omega), \quad \omega \geq 0.$$

(c) Using the following mean-squared difference as a criterion of closeness between the quadrature and analytic signals:

$$C = \sum_{n=-\infty}^{\infty} |2s_a[n] - s_q[n]|^2$$

(the factor of 2 is needed to account for removing negative frequencies in obtaining the analytic signal), and using Parseval's Theorem and your result from part (b), show that

$$C = \frac{2}{\pi} \int_{-\pi}^{0} |S_q(\omega)|^2 d\omega$$

which is twice the energy of the quadrature signal for negative frequencies.

(d) Based on the closeness criterion of part (c), state whether the analytic and quadrature representations are equivalent for each of the following two signals:

$$x_1[n] = \cos(\omega n)$$
$$x_2[n] = a^n \cos(\omega n)$$

where $0 \leq \omega < \pi$. Explain your reasoning.

2.8 Consider a complex continuous-time signal of the form $s(t) = a(t)e^{j\phi(t)}$. Show that the bandwidth of $s(t)$, as defined in Equation (2.4), can be expressed as

$$B = \int_{-\infty}^{\infty} \left(\frac{da(t)}{dt}\right)^2 dt + \int_{-\infty}^{\infty} \left(\frac{d\phi(t)}{dt} - \bar{\Omega}\right)^2 a^2(t)dt \qquad (2.36)$$

where $\bar{\Omega}$ is the signal's average frequency, i.e., $\bar{\Omega} = \int_{-\infty}^{\infty} \Omega |S(\Omega)|^2 d\Omega$. *Hint:* Use the relation $\int_{-\infty}^{\infty} \Omega^2 |S(\Omega)|^2 d\Omega = \int_{-\infty}^{\infty} |\frac{d}{dt}s(t)|^2 dt$.

2.9 The autocorrelation function for a real-valued stable sequence $x[n]$ is defined as

$$c_{xx}[n] = \sum_{k=-\infty}^{\infty} x[k]x[n+k].$$

(a) Show that the z-transform of $c_{xx}[n]$ is

$$C_{xx}(z) = X(z)X(z^{-1}).$$

Determine the region of convergence for $C_{xx}(z)$.

(b) Suppose that $x[n] = a^n u[n]$. Sketch the pole-zero plot for $C_{xx}(z)$, including the region of convergence. Also find $c_{xx}[n]$ by evaluating the inverse z-transform of $C_{xx}(z)$.

(c) Specify another sequence, $x_1[n]$, that is not equal to $x[n]$ in part (b) but that has the same autocorrelation function, $c_{xx}[n]$, as $x[n]$ in part (b).

2.10 Show that the magnitude of the all-pass transfer function $\frac{1-a^*z}{1-az^{-1}}$ is unity on the unit circle. Show that the phase of this all-pass function is negative on the unit circle.

2.11 We saw in the text that replacing a zero (or pole) by its conjugate reciprocal counterpart does not alter the magnitude of its Fourier transform. In this exercise, we explore the change in Fourier transform phase with such operations.

(a) Consider a minimum-phase zero corresponding to the component $Q(z) = \frac{1}{(1-qz^{-1})}$ ($|q| < 1$) of the rational z-transform $X(z)$. Denote the phase of $Q(z)$ along the unit circle by $\theta_1(\omega)$ and write the polar representation of $X(z)$ along the unit circle as $X(\omega) = A(\omega)e^{j\theta(\omega)}$. Give an expression for the phase of the new z-transform when the zero is flipped to its conjugate reciprocal location.

(b) Suppose now that the zero in part (a) is flipped to its *conjugate reciprocal pole location*, i.e., $\frac{1}{(1-qz^{-1})}$ becomes $(1 - q^*z)$. Show that the phase of the resulting z-transform is unaltered.

2.12 Use the eigenfunction property and superposition principle of linear systems to derive the Convolution Theorem. Then argue informally for the Windowing Theorem using the fact that the discrete-time Fourier transform and its inverse have similar structures. Specifically, the discrete-time Fourier transform differs from its inverse by a scale factor, a sign change in the exponent of the complex exponential, and by a sum over all time rather than a circular integration over frequency.

2.13 Suppose

$$x[n] = \sum_{k=-\infty}^{\infty} \delta[n - kP].$$

Show that the Fourier transform of $x[n]$ is given by

$$X(\omega) = \sum_{k=-\infty}^{\infty} \frac{2\pi}{P} \delta\left(\omega - \frac{2\pi}{P}k\right).$$

Hint: Work backward and observe that the inverse Fourier transform of $X(\omega)$ can be expressed as $\frac{1}{P} \sum_{k=0}^{P-1} e^{j\frac{2\pi}{P}kn}$.

2.14 By imposing on an FIR impulse response "symmetry" of the form

$$h[n] = h[M - n]$$

show under the simplifying assumption that M is even that

$$H(\omega) = A(\omega)e^{-j\omega(M/2)}$$

where $A(\omega)$ is real and therefore its frequency response has *linear phase*.

2.15 Show that the output $y[n]$ of a linear time-varying system with response $h[n, m]$ is the inverse Fourier transform of the *product* of $X(\omega)$ and $H(n, \omega)$, as in Equation (2.31).

2.16 In this exercise, you show that the output expression Equation (2.26) for a linear time-varying system reduces to a convolution when the system is time-invariant.

(a) Argue that Green's function $g[n, m]$ and the time-varying unit sample response $h[n, m]$ are related by $h[n, m] = g[n, n - m]$ or, equivalently, $h[n, n - m] = g[n, m]$.

(b) From part (a), derive Equation (2.27).

(c) Argue that if the system is time-invariant, then Green's function $g[n, m]$ depends only on the difference $[n - m]$, which is the number of samples between the application of the unit sample input at time m and the output at time n, i.e., $g[n, m] = g[n - m]$.

(d) From your results in parts (a)–(c), show that when the system is time-invariant $h[n, m] = h[m]$, and thus the output of the linear time-varying system in Equation (2.27) reduces to

$$y[n] = \sum_{m=-\infty}^{\infty} h[m]x[n - m].$$

2.17 Derive Equation (2.30), i.e., show that the frequency response of a linear time-varying system can be written in terms of Green's function as

$$H(n, \omega) = e^{j\omega n} \sum_{m=-\infty}^{\infty} g[n, m]e^{-j\omega m}.$$

2.18 Consider an input sequence $x[n]$ which is processed by an LTI system with impulse response $h[n]$ to yield an output sequence $y[n]$. Let $X(\omega)$, $H(\omega)$, and $Y(\omega)$ denote the Fourier transforms of the input sequence, system impulse response, and output sequence, respectively. That is,

$$Y(\omega) = X(\omega)H(\omega)$$

where the system frequency response is given by

$$H(\omega) = \frac{A(\omega)}{1 - C(\omega)}.$$

(a) We can estimate $Y(\omega)$ by formulating an iteration of the form

$$Y_k(\omega) = A(\omega)X(\omega) + C(\omega)Y_{k-1}(\omega)$$

where

$$Y_{-1}(\omega) = 0.$$

Derive a closed-form expression for $Y_k(\omega)$.

(b) Show that if

$$|C(\omega)| < 1,$$

then $Y_k(\omega)$ converges to $Y(\omega)$ for each ω, as k increases. That is, show

$$\lim_{k \to \infty} Y_k(\omega) = Y(\omega).$$

Hint: Use a geometric series.

(c) Propose a similar iterative scheme for recovering $X(\omega)$ from $Y(\omega)$. Discuss briefly why such iterative methods are useful in overcoming practical problems in recovering convolutionally distorted signals.

2.19 In Exercise 2.6, you found the analytic signal representation of the sequence

$$x[n] = \left[\frac{\sin(\omega_1 n)}{\pi n} \right] \cos(\omega_0 n)$$

which allowed you to compute the magnitude (which can be thought of as the "envelope" of $x[n]$) $a[n] = \sin(\omega_1 n)/\pi n$ and also the phase $\phi[n] = \omega_0 n$. In this exercise, you are asked to stretch out $x[n]$ along the time axis without changing the sinewave frequency ω_0. The approach of this exercise is similar in style to that used later in the text in modifying the articulation rate of a speaker, i.e., in making a speaker sound as though he/she is talking faster or slower. This operation is also referred to as *time-scale modification.*

(a) Design an interpolation scheme to stretch out the envelope of $x[n]$ along the time axis by a factor of two. The interpolation will first require putting zeros between every other sample of $a[n]$.

(b) Using the phase function from Exercise 2.6, design an interpolation scheme to stretch the phase by two in time while maintaining the underlying sinewave frequency. This will involve a conventional interpolator followed by multiplicative scaling.

(c) Use the results of parts (a) and (b) to obtain the time-scaled signal

$$\tilde{x}[n] = \left[\frac{\sin(\omega_1 n/2)}{\pi(n/2)} \right] \cos(\omega_0 n).$$

2.20 (MATLAB) In this MATLAB exercise, you investigate some properties of a windowed speech waveform. On the companion website, you will find in the directory Chap_exercises/chapter2 a workspace *ex2M1.mat*. Load this workspace and plot the speech waveform labeled *speech1_10k*. This speech segment was taken from a vowel sound that is approximately periodic (sometimes referred to as "quasi-periodic"), is 25 ms in duration, and was sampled at 10000 samples/s.

(a) Plot the log-magnitude of the Fourier transform of the signal over the interval $[0, \pi]$, using a 1024-point FFT. The signal should be windowed with a Hamming window of two different durations, 25 ms and 10 ms, with the window placed, in each case, at the signal's center. Show the log-magnitude plot for each duration. In doing this exercise, use MATLAB functions *fft.m* and *hamming.m*.

(b) From the speech waveform, estimate the period in seconds of the quasi-periodic waveform. As we will see in Chapter 3, the *pitch* is the reciprocal of the period and equals the rate of vocal cord vibration. In discrete time, the pitch of the waveform is given (in radians) by $\frac{2\pi}{P}$, where P is the period in samples. From Example 2.10, we see that for a discrete-time periodic impulse train, the pitch is the spacing between window transform replicas in frequency. Using your result from part (a) and Example 2.10, argue which spectral estimate from part (a), i.e., from a 25 ms or from a 10 ms window, is more suitable to pitch estimation. Propose a method of pitch estimation using the log-magnitude of the Fourier transform.

2.21 (MATLAB) In this MATLAB exercise, you will listen to various versions of a bandpass-filtered speech waveform. On the companion website, you will find in the directory Chap_exercises/chapter2 a workspace *ex2M2.mat*. Load this workspace and plot the speech waveforms labeled *speech2_10k* and *speech3_10k*. These speech passages were sampled at 10000 samples/s.

(a) Use the MATLAB function *remez.m* to design a lowpass filter, i.e.,

$$b_l[n] = remez(102, [0 .5 .6 1], [1 1 0 0]).$$

Plot the log-magnitude of its Fourier transform over the interval $[0, \pi]$, using a 1024-point FFT. In doing this exercise, use MATLAB function *fft.m*.

(b) Normalize $b_l[n]$ by the sum of its values and then design a highpass filter as

$$b_h[n] = \delta[n - 52] - b_l[n].$$

Plot the log-magnitude of the Fourier transform of $b_h[n]$ over the interval $[0, \pi]$, using a 1024-point FFT. In doing this exercise, use MATLAB function *fft.m* and the script *high_low.m* provided in the directory Chap_exercises/chapter2.

(c) Using the MATLAB function *conv.m*, filter the speech waveforms *speech2_10k* and *speech3_10k* with the lowpass and highpass filters you designed in parts (a) and (b).

(d) It is sometimes said that speech possesses redundant information in frequency, i.e., the same information may appear in different frequency bands. Using the MATLAB function *sound.m*, listen to the lowpass- and highpass-filtered speech from part (c) and describe its intelligibility. Comment on the notion of spectral redundancy.

BIBLIOGRAPHY

[1] M.A. Clements and J.W. Pease, "On Causal Linear Phase IIR Digital Filters," *IEEE Trans. Acoustics, Speech, and Signal Processing*, vol. 37, no. 4, pp. 479–485, April 1989.

[2] L. Cohen, *Time-Frequency Analysis*, Prentice Hall, Englewood Cliffs, NJ, 1995.

[3] R.E. Crochiere and L.R. Rabiner, *Multirate Digital Signal Processing*, Prentice Hall, Englewood Cliffs, NJ, 1983.

[4] A. Fettweis, "Wave Digital Filters: Theory and Practice," *Proc. IEEE*, vol. 74, no. 2, pp. 270–327, 1986.

[5] G. Matz and F. Hlawatsch, "Time-Frequency Transfer Function Calculus (Symbolic Calculus) of Linear Time-Varying Systems (Linear Operators) Based on a Generalized Underspread Theory," *J. of Mathematical Physics*, vol. 39. no. 8, pp. 4041–4070, Aug. 1998.

[6] A.V. Oppenheim and R.W. Schafer, *Digital Signal Processing*, Prentice Hall, Englewood Cliffs, NJ, 1975.

[7] A.V. Oppenheim and R.W. Schafer, *Discrete-Time Signal Processing*, Prentice Hall, Englewood Cliffs, NJ, 1989.

[8] M.R. Portnoff, "Time-Frequency Representation of Digital Signals and Systems Based on Short-Time Fourier Analysis," *IEEE Trans. Acoustics, Speech, and Signal Processing*, vol. ASSP–28, no. 1, pp. 55–69, Feb. 1980.

[9] T.F. Quatieri and A.V. Oppenheim, "Iterative Techniques for Minimum Phase Signal Reconstruction from Phase or Magnitude," *IEEE Trans. Acoustics, Speech, and Signal Processing*, vol. ASSP–29, no. 6, pp. 1187–1193, Dec. 1981.

[10] L.R. Rabiner and R.W. Schafer, *Digital Processing of Speech Signals*, Prentice Hall, Englewood Cliffs, NJ, 1978.

Production and Classification of Speech Sounds

3.1 Introduction

A simplified view of speech production is given in Figure 3.1, where the speech organs are divided into three main groups: the lungs, larynx, and vocal tract. The lungs act as a power supply and provide airflow to the larynx stage of the speech production mechanism. The larynx modulates airflow from the lungs and provides either a periodic puff-like or a noisy airflow source to the third organ group, the vocal tract. The vocal tract consists of oral, nasal, and pharynx cavities, giving the modulated airflow its "color" by spectrally shaping the source. Sound sources can also be generated by constrictions and boundaries, not shown in Figure 3.1, that are made within the vocal tract itself, yielding in addition to noisy and periodic sources, an impulsive airflow source. We have here idealized the sources in the sense that the anatomy and physiology of the speech production mechanism does not generate a perfect periodic, impulsive, or noise source.[1] Following the spectral coloring of the source by the vocal tract, the variation of air pressure at the lips results in a traveling sound wave that the listener perceives as speech.

There are then three general categories of the source for speech sounds: *periodic*, *noisy*, and *impulsive*, although combinations of these sources are often present. Examples of speech sounds generated with each of these source categories are seen in the word "shop," where the "sh," "o," and "p" are generated from a noisy, periodic, and impulsive source, respectively. The reader should speak the word "shop" slowly and determine where each sound source is occurring, i.e., at the larynx or at a constriction within the vocal tract.

---

[1] This idealization also assumes a flat (white) noise spectrum. *Noise* and its white subclass are defined formally in a stochastic signal framework in Chapter 5.

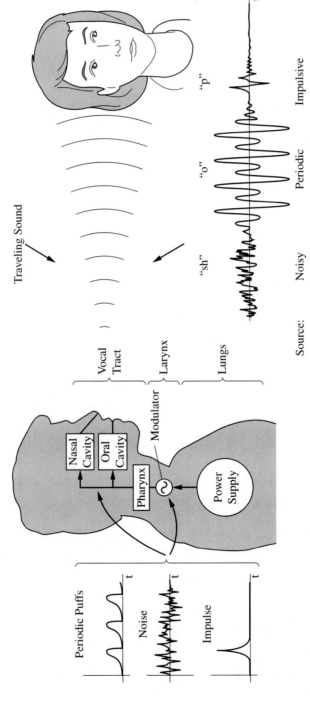

Figure 3.1 Simple view of speech production. The sound sources are idealized as periodic, impulsive, or (white) noise and can occur in the larynx or vocal tract.

Such distinguishable speech sounds are determined not only by the source, but by different vocal tract configurations, and how these shapes combine with periodic, noisy, and impulsive sources. These more refined speech sound classes are referred to as *phonemes*, the study of which is called *phonemics*. A specific phoneme class provides a certain meaning in a word, but within a phoneme class, as we will see in a moment, there exist many sound variations that provide the same meaning. The study of these sound variations is called *phonetics*. Phonemes, the basic building blocks of a language, are concatenated, more or less, as discrete elements into words, according to certain phonemic and grammatical rules. This chapter provides a qualitative description of the speech production mechanism and the resulting variety of phonetic sound patterns, and, to a lesser extent, how these sound patterns differ among different speakers. Implications for the design of digital signal processing algorithms will be illustrated. In Chapter 4, we refine this qualitative description with more quantitative mathematical models.

In Section 3.2, we first describe the anatomy and physiology of the different organ groups and show how these organ groups result in source inputs and vocal tract configurations that contribute generally to making different speech sounds. Time- and frequency-domain properties of the source and its spectral shaping by the vocal tract are illustrated, and these result in a number of important definitions, such as the pitch and harmonics of a periodic source and the formants of the vocal tract. In this section, we also elaborate on sound categorization based on source only: periodic, noisy, and impulsive sound sources. In Section 3.3, we deviate and develop the spectrogram, which is a means to illustrate the spectral evolution of a sound; in Chapter 7, the spectrogram will be studied more formally. Having four tools in hand—the time-waveform, spectrogram, source classification, and vocal tract configurations—we then embark in Section 3.4 on the study of phonetics. In Section 3.5, we take a wider temporal view of the speech waveform, i.e., across phonetic boundaries of individual speech sounds, and study the *prosodics* of speech, which is the rhythm (timing of the phonemes) and intonation (changing pitch of the source) over phrases and sentences. In Section 3.6, we give a flavor for the perceptual aspect of phonetics, i.e., how the auditory system might perceive a speech sound, and how various properties of sound production are important in the distinguishing of different speech phonemes. We will see in later chapters how characteristics of speech production, used as perceptual cues, can drive the development and selection of signal processing algorithms.

3.2 Anatomy and Physiology of Speech Production

Figure 3.2 shows a more realistic view of the anatomy of speech production than was shown in Figure 3.1. We now look in detail at this anatomy, as well as at the associated physiology and its importance in speech production.

3.2.1 Lungs

One purpose of the lungs is the inhalation and exhalation of air. When we inhale, we enlarge the chest cavity by expanding the rib cage surrounding the lungs and by lowering the diaphragm that sits at the bottom of the lungs and separates the lungs from the abdomen; this action lowers the air pressure in the lungs, thus causing air to rush in through the vocal tract and down the trachea

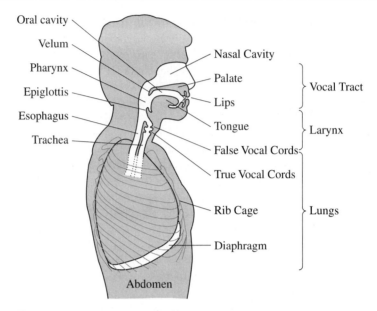

Figure 3.2 Cross-sectional view of the anatomy of speech production.

into the lungs. The trachea, sometimes referred to as the "windpipe," is about a 12-cm-long and 1.5–2-cm-diameter pipe which goes from the lungs to the epiglottis. The epiglottis is a small mass, or "switch," which, during swallowing and eating, deflects food away from entering the trachea. When we eat, the epiglottis falls, allowing food to pass through a tube called the esophagus and into the stomach. When we exhale, we reduce the volume of the chest cavity by contracting the muscles in the rib cage, thus increasing the lung air pressure. This increase in pressure then causes air to flow through the trachea into the larynx. In breathing, we rhythmically inhale to take in oxygen, and exhale to release carbon dioxide.

During speaking, on the other hand, we take in short spurts of air and release them steadily by controlling the muscles around the rib cage. We override our rhythmic breathing by making the duration of exhaling roughly equal to the length of a sentence or phrase. During this timed exhalation, the lung air pressure is maintained at approximately a constant level, slightly above atmospheric pressure, by steady slow contraction of the rib cage, although the air pressure varies around this level due to the time-varying properties of the larynx and vocal tract.

3.2.2 Larynx

The larynx is a complicated system of cartilages, muscles, and ligaments[2] whose primary purpose, in the context of speech production, is to control the *vocal cords* or *vocal*

---

[2] Some definitions useful throughout this chapter are: (1) muscles are tissue that contract when stimulated; (2) cartilage is rigid, yet elastic bony tissue, but not as hard as bone, helping to support organs in place; (3) ligaments are tough bands of tissue that connect bones to bones and also support organs in place.

folds[3] [10]. The vocal folds are two masses of flesh, ligament, and muscle, which stretch between the front and back of the larynx, as illustrated in Figure 3.3. The folds are about 15 mm long in men and 13 mm long in women. The *glottis* is the slit-like orifice between the two folds. The folds are fixed at the front of the larynx where they are attached to the stationary thyroid cartilage. The thyroid cartilage is located at the front (or Adam's apple) and sides of the larynx. The folds are free to move at the back and sides of the larynx; they are attached to the two arytenoid cartilages that move in a sliding motion at the back of the larynx along with the cricoid cartilage. The size of the glottis is controlled in part by the arytenoid cartilages, and in part by muscles within the folds. Another important property of the vocal folds, in addition to the size of the glottis, is their tension. The tension is controlled primarily by muscle within the folds, as well as the cartilage around the folds. The vocal folds, as well as the epiglottis, close during eating, thus providing a second protection mechanism. The false vocal folds, above the vocal folds (Figure 3.2), provide a third protection. They also extend from the Adam's apple to the arytenoids. They can be closed and they can vibrate, but they are likely open during speech production [4]. We see then that a triple barrier is provided across the windpipe through the action of the epiglottis, the false vocal folds, and the true vocal folds. All three are closed during swallowing and wide open during breathing.

There are three primary states of the vocal folds: breathing, voiced, and unvoiced. In the *breathing state*, the arytenoid cartilages are held outward (Figure 3.3b), maintaining a wide glottis, and the muscles within the vocal folds are relaxed. In this state, the air from the lungs flows freely through the glottis with negligible hindrance by the vocal folds. In speech production, on the other hand, an obstruction of airflow is provided by the folds. In the *voicing state*, as, for example, during a vowel, the arytenoid cartilages move toward one another (Figure 3.3a). The vocal folds tense up and are brought close together. This partial closing of the glottis and increased fold tension cause self-sustained oscillations of the folds. We can describe how this oscillation comes about in three steps [10] (Figure 3.4a).

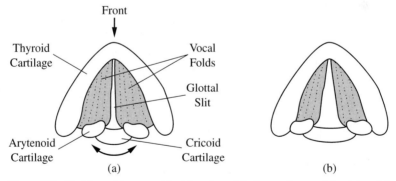

Figure 3.3 Sketches of downward-looking view of the human larynx: (a) voicing; (b) breathing.

SOURCE: K.N. Stevens, *Acoustic Phonetics*, The MIT Press [33]. ©1998, Massachusetts Institute of Technology. Used by permission.

---

[3] The more accurate term is "vocal folds," since the masses are actually not cords. The term "vocal cords" originated with an early erroneous anatomical study [30]. Although we use the term "vocal folds" more often, we apply the two terms interchangeably throughout the text.

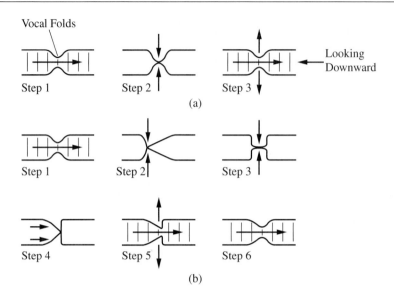

Figure 3.4 Bernoulli's Principle in the glottis: (a) basic horizontal open/close voicing cycle; (b) refinement of (a) with vertical vocal fold motion. Vertical lines represent airflow in the direction of the arrows.

Suppose the vocal folds begin in a loose and open state. The contraction of the lungs first results in air flowing through the glottis. According to a fluid dynamic property called Bernoulli's Principle, as the airflow velocity (i.e., the velocity of air particles) increases, local pressure in the region at the glottis decreases. At the same time, tension in the vocal folds increases. This increase in tension of the folds, together with the decrease in pressure at the glottis, causes the vocal folds to close shut abruptly. Air pressure then builds behind the vocal folds as the lungs continue to contract, forcing the folds to open. The entire process then repeats and the result is periodic "puffs" of air that enter the vocal tract.

Thus far, we have illustrated the vocal folds as vibrating horizontally, perpendicular to the tracheal wall. The vocal fold movement, however, is generally not so simple. For example, both horizontal and vertical movement of the folds may occur simultaneously, as illustrated in Figure 3.4b. During the time when the glottis is open, because the lower parts of the fleshy folds are more flexible than the upper parts, there is a time delay between the closing of the two regions, as seen in Steps 1–3 of Figure 3.4b. Additional vertical movement then occurs because there is also a time delay between the opening of the two regions. When the air pressure below the glottis increases during the time when the glottis closes, the lower region of the folds is first pushed up, followed by the upper region, as seen in Steps 4–6. Such complexity has led to a nonlinear two-mass model [11] (Figure 3.5), as well as more elaborate nonlinear multi-component models describing various modes of vibration along the folds themselves [39]. The masses m_k, nonlinear spring constants s_k, and damping constants τ_k in such mechanical models correspond, respectively, to the masses, tensions, and resistances within the vocal folds and the surrounding cartilage.

According to our description of the airflow velocity in the glottis, if we were to measure the airflow velocity at the glottis as a function of time, we would obtain a waveform approximately

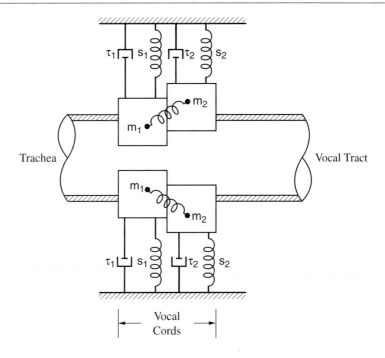

Figure 3.5 Two-mass mechanical model of Flanagan and Ishizaka with masses m_1 and m_2, resistances τ_1 and τ_2, and spring constants s_1 and s_2.

SOURCE: J.L. Flanagan and K. Ishizaka, "Computer Model to Characterize the Air Volume Displaced by the Vibrating Vocal Cords" [11]. ©1978, Acoustical Society of America. Used by permission.

similar to that illustrated in Figure 3.6 that roughly follows the time-varying area of the glottis. Typically, with the folds in a closed position, the flow begins slowly, builds up to a maximum, and then quickly decreases to zero when the vocal folds abruptly shut. The time interval during which the vocal folds are closed, and no flow occurs, is referred to as the glottal *closed phase*; the time interval over which there is nonzero flow and up to the maximum of the airflow velocity is referred to as the glottal *open phase*, and the time interval from the airflow maximum to the time of glottal closure is referred to as the *return phase*. The specific flow shape can change with the speaker, the speaking style, and the specific speech sound. In some cases, the folds do not even close completely, so that a closed phase does not exist. For simplicity throughout this text, we will often refer to the glottal airflow velocity as simply the *glottal flow*.

The time duration of one glottal cycle is referred to as the *pitch period* and the reciprocal of the pitch period is the corresponding *pitch*, also referred to as the *fundamental frequency*. The term "pitch" might lead to some confusion because the term is often used to describe the subjectively perceived "height" of a complex musical sound even when no single fundamental frequency exists. In this text, however, we use the term in the above strict sense, i.e., pitch is synonomous with fundamental frequency. In conversational speech, during vowel sounds, we

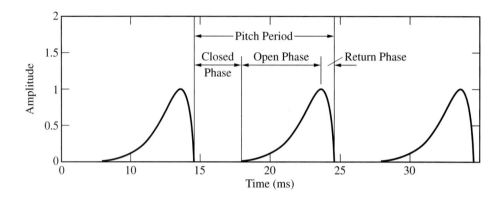

Figure 3.6   Illustration of periodic glottal airflow velocity.

might see typically one to four pitch periods over the duration of the sound, although, as we will see in the discussion of prosodics, the number of pitch periods changes with numerous factors such as stress and speaking rate. The rate at which the vocal folds oscillate through a closed, open, and return cycle is influenced by many factors. These include vocal fold muscle tension (as the tension increases, so does the pitch), the vocal fold mass (as the mass increases, the pitch decreases because the folds are more sluggish), and the air pressure behind the glottis in the lungs and trachea, which might increase in a stressed sound or in a more excited state of speaking (as the pressure below the glottis increases, so does the pitch). The pitch range is about 60 Hz to 400 Hz. Typically, males have lower pitch than females because their vocal folds are longer and more massive.

    A simple mathematical model of the glottal flow is given by the convolution of a periodic impulse train with the glottal flow over one cycle. The following example shows glottal flow waveforms with different shapes and pitch periods, as well as how the simple convolutional model lends insight into the spectral nature of the glottal airflow.

EXAMPLE 3.1   Consider a glottal flow waveform model of the form

$$u[n] = g[n] * p[n] \tag{3.1}$$

where $g[n]$ is the glottal flow waveform over a single cycle and $p[n] = \sum_{k=-\infty}^{\infty} \delta[n - kP]$ is an impulse train with spacing P. Because the waveform is infinitely long, we extract a segment by multiplying $x[n]$ by a short sequence called an *analysis window* or simply a *window*. The window, denoted by $w[n, \tau]$, is centered at time τ, as illustrated in Figure 3.7, and the resulting waveform segment is written as

$$u[n, \tau] = w[n, \tau](g[n] * p[n]).$$

Using the Multiplication and Convolution Theorems of Chapter 2, we obtain in the frequency domain

$$U(\omega, \tau) = \frac{1}{P} W(\omega, \tau) \circledast \left[\sum_{k=-\infty}^{\infty} G(\omega)\delta(\omega - \omega_k) \right]$$

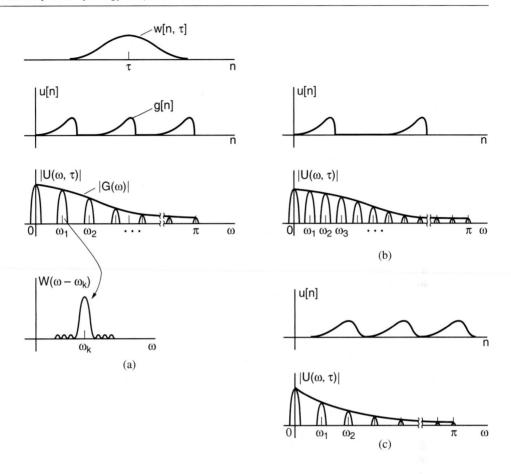

Figure 3.7   Illustration of periodic glottal flow in Example 3.1: (a) typical glottal flow and its spectrum; (b) same as (a) with lower pitch; and (c) same as (a) with "softer" or more "relaxed" glottal flow.

$$= \frac{1}{P} \sum_{k=-\infty}^{\infty} G(\omega_k) W(\omega - \omega_k, \tau)$$

where $W(\omega, \tau)$ is the Fourier transform of $w[n, \tau]$, where $G(\omega)$ is the Fourier transform of $g[n]$, where $\omega_k = \frac{2\pi}{P}k$, and where $\frac{2\pi}{P}$ is the fundamental frequency or pitch. As illustrated in Figure 3.7, the Fourier transform of the window sequence is characterized by a narrow main lobe centered at $\omega = 0$ with lower surrounding sidelobes. The window is typically selected to trade off the width of the mainlobe and attenuation of the sidelobes. Figure 3.7 illustrates how the Fourier transform magnitude of the waveform segment changes with pitch and with characteristics of the glottal flow. As the pitch period decreases, the spacing between the frequencies $\omega_k = \frac{2\pi}{P}k$, which are referred to as the *harmonics* of the glottal waveform, increases, as can be seen by comparing Figures 3.7a and 3.7b. The first harmonic is also the fundamental frequency, and the other harmonics occur at integer multiples of the fundamental frequency. Located at each harmonic is a translated window Fourier transform $W(\omega - \omega_k)$ weighted by $G(\omega_k)$; as the pitch changes, the harmonics can be thought of as sliding under $G(\omega)$. As the glottal flow over a cycle becomes more smooth, i.e., a gradual rather than an abrupt closing, then the "spectral shaping" by $G(\omega)$ of the harmonically-spaced window Fourier transforms becomes more lowpass, as seen by comparing Figures 3.7a and 3.7c. We can see, based on these sliding and spectral shaping properties, why the magnitude of the spectral shaping function, in this case $|G(\omega)|$, is sometimes referred to as a *spectral envelope* of the harmonics.   ▲

We saw in the previous example that the Fourier transform of the periodic glottal waveform is characterized by harmonics. Typically, the spectral envelope of the harmonics, governed by the glottal flow over one cycle, has, on the average, a -12 dB/octave rolloff, although this changes with the specific nature of the airflow and the speaker characteristics. With more forceful speaking, for example, the glottal closure may be more abrupt (e.g., Figure 3.7a, b) with perhaps an average -9 dB/octave slope being more typical [29]. In more "relaxed" voicing, the vocal folds do not close as abruptly, and the glottal waveform has more rounded corners (e.g., Figure 3.7c), with an average -15 dB/octave rolloff, typically. Exercise 3.18 explores some specific cases. The model in Example 3.1 is ideal in the sense that, even for sustained voicing—i.e., a vowel uttered by a speaker trying to hold steady pitch and vocal tract shape—a fixed pitch period is almost never maintained in time but can randomly vary over successive periods, a characteristic referred to as pitch "jitter." In addition, the amplitude of the airflow velocity within a glottal cycle may differ across consecutive pitch periods, even in a sustained vowel, a characteristic called amplitude "shimmer." These variations are due, perhaps, to time-varying characteristics of the vocal tract and vocal folds. Pitch jitter and shimmer, however, have also been speculated to be due to nonlinear behavior in the speech anatomy whereby successive cyclic variations may alternate on each glottal cycle [38] or may *appear* random while being the result of an underlying deterministic (chaotic) system [15]. The jitter and shimmer over successive pitch periods help give the vowel its naturalness, in contrast to a monotone pitch and fixed amplitude that can result in a machine-like sound. In addition to naturalness, however, the extent and form of jitter and shimmer can contribute to voice character. A high degree of jitter, for example, results in a voice with a *hoarse* quality which can be characteristic of a particular speaker or can be created under specific speaking conditions such as with stress or fear. The time- and frequency-domain properties of this condition are further studied in Exercise 3.2.

We have described two states of the vocal folds: breathing and voicing. The last state of the vocal folds is *unvoicing*. This state is similar to the breathing state in there being no vocal fold vibration. In the unvoiced state, however, the folds are closer together and more tense than in the breathing state, thus allowing for turbulence to be generated at the folds themselves. Turbulence at the vocal folds is called *aspiration*. Aspiration occurs in normal speech as with "h" in the word "he." Such sounds are sometimes called "whispered" sounds because turbulence is also created at the vocal folds when we whisper. Whispering is not simply a reduction in volume, because when we whisper the vocal folds do not oscillate. In certain voice types, aspiration occurs normally simultaneously with voicing, resulting in the *breathy* voice, by maintaining part of the vocal folds nearly fixed and somewhat open to produce turbulence and part of the vocal folds in oscillation. Nevertheless, aspiration occurs to some extent in all speakers and the amount of aspiration may serve as a distinguishing feature. The physiological change, then, in creating the breathy voice is distinctly different from that of the hoarse voice which, as we saw earlier, is associated with pitch jitter. Figure 3.8 shows a comparison of vocal fold configurations for aspiration (whispering), voicing, and aspirated voicing.

There are also other forms of vocal fold movement that do not fall clearly into any of the three states of breathing, voicing, or unvoicing. We point out these different voice types because, as we will see, they can pose particularly large challenges in speech signal processing and, contrary to being "idiosyncratic," they occur quite often. One such state of the vocal folds is the *creaky* voice where the vocal folds are very tense, with only a short portion of the folds in oscillation, resulting in a harsh-sounding voice with a high and irregular pitch. (Look ahead to Figure 10.15b.) In *vocal fry*, on the other hand, the folds are massy and relaxed with an abnormally low and irregular pitch [27],[40], which is characterized by secondary glottal pulses close to and overlapping the primary glottal pulse within the open phase, as illustrated in Figure 3.9a. We use the term "glottal pulse" loosely in this chapter to mean a glottal airflow velocity waveform over a single glottal cycle. In vocal fry, the true vocal folds may couple with the false vocal folds, producing the secondary glottal pulses. Vocal fry occurs even in the normal voice at the end of a phrase or word where the muscles of the larynx relax and the lung pressure is decreasing. Another atypical voice type is the *diplophonic* voice where again secondary glottal pulses occur between the primary pulses but within the closed phase, away from the primary pulse [18], as illustrated in Figure 3.9b. Diplophonia often occurs as extra flaps in low-pitch speakers and, as with vocal fry, in normal voices at the end of a phrase or word.[4] An example of a low-pitch diplophonic voice is provided later, in Figure 3.16. In the diplophonic and vocal fry voice types, a simple model in discrete time for the occurrence of a secondary glottal pulse is given by the modified glottal flow waveform $\tilde{g}[n] = g[n] + \alpha g[n - n_o]$, where $g[n]$ is the primary glottal pulse, where n_o is the spacing[5] between the primary and secondary glottal pulses, and α is an attenuation factor on the secondary pulse. We assume here the same shape of the

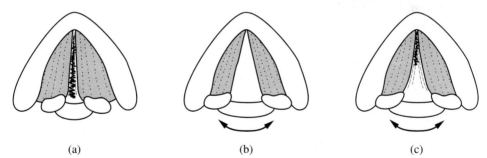

      (a)   (b)   (c)

Figure 3.8 Sketches of various vocal fold configurations: (a) aspiration (whispering), (b) voicing, and (c) aspirated voicing. Arrows indicate vocal fold vibration, while ragged lines indicate turbulence.

---

[4] Pitch period doubling also often occurs at the end of a phrase or word where the vocal cords are relaxed. In Figure 3.13, the "o" in "to" shows this phenomenon.

[5] We assume that the spacing in continuous time is $t_o = n_o T$ (with T being the sampling interval) so that in discrete time the spacing is represented by the integer n_o.

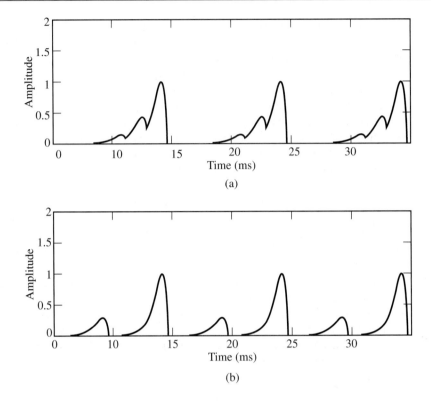

Figure 3.9 Illustration of secondary-pulse glottal flow: (a) vocal fry; (b) diplophonia.

secondary and primary glottal pulses, but generally they may differ. The presence of $\alpha g[n - n_o]$ introduces zeros into the z-transform of the glottal waveform (Exercise 3.3). Other abnormal voice types include the falsetto voice where there occurs extreme swings in vocal fold tension and relaxation allowing for abnormally large pitch fluctuations. Some speakers, especially in the singing voice, may regularly induce a rapid pitch modulation, referred to as pitch *vibrato*, over a smaller range to give the utterance more flavor or musicality. We will see examples of some of these voice types throughout the text.

3.2.3 Vocal Tract

The vocal tract is comprised of the oral cavity from the larynx to the lips and the nasal passage that is coupled to the oral tract by way of the velum. The oral tract takes on many different lengths and cross-sections by moving the tongue, teeth, lips, and jaw and has an average length of 17 cm in a typical adult male and shorter for females, and a spatially-varying cross section of up to 20 cm^2. If we were to listen to the pressure wave at the output of the vocal folds during voicing, we would hear simply a time-varying buzz-like sound which is not very interesting. One purpose of the vocal tract is to spectrally "color" the source, which is important for making perceptually distinct speech sounds. A second purpose is to generate new sources for sound production.

Spectral Shaping — Under certain conditions, the relation between a glottal airflow velocity input and vocal tract airflow velocity output can be approximated by a linear filter with resonances, much like resonances of organ pipes and wind instruments. The resonance frequencies of the vocal tract are, in a speech science context, called *formant frequencies* or simply *formants*. The word "formant" also refers to the entire spectral contribution of a resonance so we often use the phrases "formant bandwidth" and "formant amplitude" (at the formant frequency). Formants change with different vocal tract configurations. With different vowels, for example, the jaw, teeth, lips, and tongue, are generally in different positions. Panel (a) of Figure 3.10 shows the tongue hump high in the front and back of the palate (upper wall of mouth), each position corresponding to different resonant cavities and thus different vowels.

The peaks of the spectrum of the vocal tract response correspond approximately to its formants. More specifically, when the vocal tract is modeled as a time-invariant all-pole linear system then, as we will see in Chapter 4, a pole at $z_o = r_o e^{j\omega_o}$ corresponds approximately to a vocal tract formant. The frequency of the formant is at $\omega = \omega_o$ and the bandwidth of the formant is determined by the distance of the pole from the unit circle (r_o). Because the poles of a real sequence typically occur in complex conjugate pairs (except for the case of a pole falling on the real axis), only the positive frequencies are used in defining the formant frequencies, and the formant bandwidth is computed over positive frequencies using, for example, the definitions of bandwidth in Chapter 2. Under the linear time-invariant all-pole assumption, each vocal tract shape is characterized by a collection of formants. Because the vocal tract is assumed stable with poles inside the unit circle, the vocal tract transfer function can be expressed either in product or partial fraction expansion form:

$$H(z) = \frac{A}{\prod_{k=1}^{N_i}(1 - c_k z^{-1})(1 - c_k^* z^{-1})}$$

$$= \sum_{k=1}^{N_i} \frac{\tilde{A}}{(1 - c_k z^{-1})(1 - c_k^* z^{-1})} \tag{3.2}$$

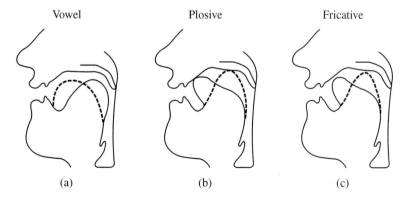

| Vowel | Plosive | Fricative |
| (a) | (b) | (c) |

Figure 3.10 Illustration of changing vocal tract shapes for (a) vowels (having a periodic source), (b) plosives (having an impulsive source), and (c) fricatives (having a noise source).

where $(1 - c_k z^{-1})$ and $(1 - c_k^* z^{-1})$ are complex conjugate poles inside the unit circle with $|c_k| < 1$. The formants of the vocal tract are numbered from the low to high formants according to their location; the first formant is denoted by F_1, the second formant by F_2, and so on up to the highest formant. Generally, the frequencies of the formants decrease as the vocal tract length increases; as a consequence, a male speaker tends to have lower formants than a female, and a female has lower formants than a child. Under a vocal tract linearity and time-invariance assumption, and when the sound source occurs at the glottis, the speech waveform, i.e., the airflow velocity at the vocal tract output, can be expressed as the convolution of the glottal flow input and vocal tract impulse response, as illustrated in the following example:

EXAMPLE 3.2   Consider a periodic glottal flow source of the form

$$u[n] = g[n] * p[n]$$

where $g[n]$ is the airflow over one glottal cycle and $p[n]$ is the unit sample train with spacing P. When the sequence $u[n]$ is passed through a linear time-invariant vocal tract with impulse response $h[n]$, the vocal tract output is given by

$$x[n] = h[n] * (g[n] * p[n]).$$

A window centered at time τ, $w[n, \tau]$, is applied to the vocal tract output to obtain the speech segment

$$x[n, \tau] = w[n, \tau]\{h[n] * (g[n] * p[n])\}.$$

Using the Multiplication and Convolution Theorems of Chapter 2, we obtain in the frequency domain the Fourier transform of the speech segment

$$X(\omega, \tau) = \frac{1}{P} W(\omega, \tau) \circledast \left[H(\omega)G(\omega) \sum_{k=-\infty}^{\infty} \delta(\omega - \omega_k) \right]$$

$$= \frac{1}{P} \sum_{k=-\infty}^{\infty} H(\omega_k)G(\omega_k)W(\omega - \omega_k, \tau)$$

where $W(\omega, \tau)$ is the Fourier transform of $w[n, \tau]$, where $\omega_k = \frac{2\pi}{P}k$, and where $\frac{2\pi}{P}$ is the fundamental frequency or pitch. Figure 3.11 illustrates that the spectral shaping of the window transforms at the harmonics ω_1, ω_2, ... ω_N is determined by the spectral envelope $|H(\omega)G(\omega)|$ consisting of a glottal and vocal tract contribution, unlike in Example 3.1, where only the glottal contribution occurred. The peaks in the spectral envelope correspond to vocal-tract formant frequencies, F_1, F_2, ... F_M. The general upward or downward slope of the spectral envelope, sometimes called the *spectral tilt*, is influenced by the nature of the glottal flow waveform over a cycle, e.g., a gradual or abrupt closing, and by the manner in which formant tails add. We also see in Figure 3.11 that the formant locations are not always clear from the short-time Fourier transform magnitude $|X(\omega, \tau)|$ because of sparse sampling of the spectral envelope $|H(\omega)G(\omega)|$ by the source harmonics, especially for high pitch.   ▲

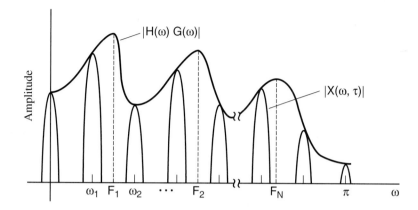

Figure 3.11   Illustration of relation of glottal source harmonics $\omega_1, \omega_2, \ldots \omega_N$, vocal tract formants $F_1, F_2, \ldots F_M$, and the spectral envelope $|H(\omega)G(\omega)|$.

This example illustrates the important difference between a *formant*, or resonance, frequency and a *harmonic* frequency. A formant corresponds to the vocal tract poles, while the harmonics arise from the periodicity of the glottal source. The spectrum of the vocal tract, for a perfectly periodic source, is, in essence, *sampled* at the harmonic frequencies; with this idealized perfect periodicity, there is spectral information only at the harmonics. In the development of signal processing algorithms that require formants, this sparcity of spectral information can perhaps be a detriment to formant estimation. In some situations, on the other hand, the spectral sampling at harmonics can be exploited to enhance perception of a sound, as in the singing voice.

EXAMPLE 3.3   A soprano singer often sings a tone whose first harmonic (fundamental frequency ω_1) is much higher than the first formant frequency (F_1) of the vowel being sung [37]. As shown in Figure 3.12, when the nulls of the vocal tract spectrum are sampled at the harmonics, the resulting sound is weak, especially in the face of competing instrumentals. To enhance the sound, the singer creates a vocal tract configuration with a widened jaw which increases the first formant frequency (Exercise 3.4), and can match the frequency of the first harmonic, thus generating a louder sound[6] [37] (Figure 3.12). In training, the singer is asked to "Hear the next tone within yourself before you start to sing it" because a widening of the jaw requires some articulatory anticipation [37].   ▲

We have seen that the nasal and oral components of the vocal tract are coupled by the velum. When the vocal tract velum is lowered, introducing an opening into the nasal passage, and the oral tract is shut off by the tongue or lips, sound propagates through the nasal passage and out through the nose. The resulting *nasal* sounds, e.g., "m" as in "<u>m</u>eet," have a spectrum that is dominated by low-frequency formants of the large volume of the nasal cavity. Because the nasal cavity, unlike the oral tract, is essentially constant, characteristics of nasal sounds may

---

[6] A singer will also lower his/her larynx in such a way as to introduce a new high-frequency formant between 2500–3000 Hz, a frequency region where the background instrumental is low, to help further enhance the sound [37].

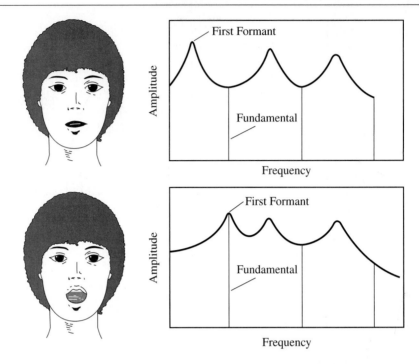

Figure 3.12 Illustration of formant movement to enhance the singing voice of a soprano: (a) first harmonic higher than first formant frequency; (b) first formant matched to first harmonic frequency.

Source: J. Sundberg, "The Acoustics of the Singing Voice" [37]. ©1977, Laszlo Kubinyi and Gabor Kiss. Used by permission.

be particularly useful in speaker identifiability. The velum can be lowered even when the oral tract is open. When this coupling occurs, we obtain a *nasalized vowel*. One effect of the nasal passage is that the formant bandwidths of the oral tract become broader because of loss of energy through the nasal passage. A second effect is the introduction of *anti-resonances*, i.e., zeros, in the vocal tract transfer function due to the absorption of energy at the resonances of the nasal passage [29].

The previous discussion has assumed a linear time-invariant vocal tract. Formants, however, are time-varying because the vocal tract changes in time. Although the vocal tract is almost time-invariant for steady-state sounds, as with a sustained vowel, in normal conversational speech the vocal tract is continuously and significantly changing. This time-variation will influence signal analysis techniques. We will return to this speech production characteristic in the discussion of transitional speech sounds.

Source Generation — We have seen that different vocal tract shapes correspond to different resonant cavities; different vocal tract shapes can also result in different sound sources. The panel (b) of Figure 3.10 shows a complete closure of the tract, the tongue pressing against the palate, required in making an impulsive sound source. There is a build-up of pressure behind the closure and then an abrupt release of pressure. Panel (c) shows another sound source created with the tongue close to the palate, but not completely impeded, for the generation of turbulence and thus a noise source. As with a periodic glottal sound source, a spectral shaping similar to that described in Example 3.2 also occurs for either type of input, i.e., an impulsive or noise source; this spectral shaping is performed by a resonant vocal tract cavity whose formants change with different vocal tract configurations, such as those illustrated in panels (b) and (c) of Figure 3.10. There is not, however, harmonic structure in the impulsive or noise source spectrum, but rather the source spectrum is shaped at all frequencies by $|H(\omega)|$. Keep in mind that we have idealized the impulsive and noise sources to have flat spectra; in practice, these sources will themselves have a non-flat spectral shape.

There is yet one other source type that is generated within the vocal tract, but is less understood than noisy and impulsive sources occurring at oral tract constrictions. This source arises from the interaction of *vortices* with vocal tract boundaries such as the false vocal folds, teeth, or occlusions in the oral tract [1],[38]. The reader may have intuition about the nature of a vortex; for the moment, let's think of a vortex in the oral tract as a tiny rotational airflow. For voiced speech, the vortices move possibly as a train from the glottis to the lips along the oral tract and are predicted to initiate from the air jet emanating from the glottis during vocal fold vibration [1],[38]. Vortices can also arise during fricative sounds with resulting sources distributed along the oral tract [19]. There is evidence that sources due to vortices influence the temporal and spectral, and perhaps perceptual, characteristics of speech sounds [1],[19],[38]. We delay further discussion of these vortical sound sources until Chapter 11.

3.2.4 Categorization of Sound by Source

There are various ways to categorize speech sounds. For example, we can categorize speech sounds based on different sources to the vocal tract; we have seen that different sources are due to the vocal fold state, but are also formed at various constrictions in the oral tract. Speech sounds generated with a periodic glottal source are termed *voiced*; likewise, sounds not so generated are called *unvoiced*. There are a variety of unvoiced sounds, including those created with a noise source at an oral tract constriction. Because the noise of such sounds comes from the friction of the moving air against the constriction, these sounds are sometimes referred to as *fricatives* (Figure 3.10c). An example of frication is in the sound "th" in the word "thin" where turbulence is generated between the tongue and the upper teeth. The reader should hold the "th" sound and feel the turbulence. A second unvoiced sound class is *plosives* created with an impulsive source within the oral tract (Figure 3.10b). An example of a plosive is the "t" in the word "top." The location of the closed or partial constriction corresponds to different plosive or fricative sounds, respectively. We noted earlier that a barrier can also be made at the vocal folds by partially closing the vocal folds, but without oscillation, as in the sound "h" in "he." These are *whispered* unvoiced speech sounds. These voiced and unvoiced sound categories, however, do not relate exclusively to the source state because a combination of these states can also be made whereby vocal fold vibration occurs simultaneously with impulsive or noisy sources. For example, with "z" in the word "zebra," the vocal folds are vibrating and, at the same time, noise is created at a vocal tract constriction behind the teeth against the palate. Such sounds are referred to as *voiced fricatives* in contrast to *unvoiced fricatives* where the vocal folds do not vibrate simultaneously with frication. There also exist *voiced plosives* as counterparts to *unvoiced plosives* as with the "b" in the word "boat." Examples of some of these sound classes are shown in Figure 3.13 in the sentence, "Which tea party did Baker go to?"

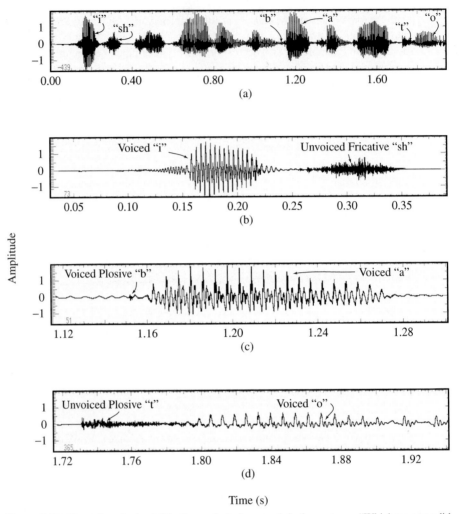

Figure 3.13 Examples of voiced, fricative, and plosive sounds in the sentence, "Which tea party did Baker go to?": (a) speech waveform; (b)–(d) magnified voiced, fricative, and plosive sounds from (a). (Note the "sh" is a component of an affricate to be studied in Section 3.4.6.)

     This loose classification provides a stepping stone to Section 3.4 where distinctive features of these sound classes will be further studied and where we will combine this source categorization with different vocal tract configurations to form the more complete classification of elements of a language.

3.3 Spectrographic Analysis of Speech

We have seen that a speech waveform consists of a sequence of different events. This time-variation corresponds to highly fluctuating spectral characteristics over time. For example, in the word "to," the plosive "t" is characterized by high-frequency energy corresponding to a vocal tract configured as a short cavity at the front of the oral tract. The "t" is followed by the vowel "o," which is characterized by low-frequency energy corresponding to a vocal tract configured as a long cavity along the oral tract. We will show quantitatively in Chapter 4 how such spectral energy shifts occur with different cavity lengths and cross sections. A single Fourier transform of the entire acoustic signal of the word "to" cannot capture this time-varying frequency content. In contrast, the *short-time Fourier transform* (STFT) consists of a separate Fourier transform of pieces of the waveform under a sliding window. We have already introduced this sliding window in Examples 3.1 and 3.2 and denoted it by $w[n, \tau]$, where τ is the position of the window center. The window is typically tapered at its end (Figure 3.14) to avoid unnatural discontinuities in the speech segment and distortion in its underlying spectrum. The Hamming window, for example, is given by the sequence $w[n, \tau] = 0.54 - 0.4 \cos[\frac{2\pi(n-\tau)}{N_w - 1}]$ for $0 \leq n \leq N_w - 1$ and zero otherwise, with N_w as the window duration. As we mentioned earlier, the window is typically selected to trade off the width of its mainlobe and attenuation of its sidelobes. The effect of specific window shapes will be further discussed in Chapter 7. In practice, the window does not necessarily move one sample at a time, but rather moves at some *frame interval* consistent with the temporal structure one wants to reveal.

The Fourier transform of the windowed speech waveform, i.e., the STFT, is given by

$$X(\omega, \tau) = \sum_{n=-\infty}^{\infty} x[n, \tau] \exp[-j\omega n] \qquad (3.3)$$

where

$$x[n, \tau] = w[n, \tau] x[n]$$

represents the windowed speech segments as a function of the window center at time τ. The *spectrogram* is a graphical display of the magnitude of the time-varying spectral characteristics and is given by

$$S(\omega, \tau) = |X(\omega, \tau)|^2$$

which can be thought of as a two-dimensional (2-D) "energy density," i.e., a generalization of the one-dimensional (1-D) energy density associated with the Fourier transform, describing the relative energy content in frequency at different time locations, i.e., in the neighborhood of (ω, τ), as we move, for example, from plosive to voiced to fricative sounds. We will have more to say about $S(\omega, \tau)$ as a 2-D energy density in following chapters.[7] We could plot $S(\omega, \tau)$ for each window position τ to represent the spectral time variations, but we would soon run out of space. A more compact time-frequency display of the spectrogram places the spectral magnitude measurements vertically in a three-dimensional mesh or two-dimensionally with intensity coming out of the page. This later display is illustrated in Figure 3.14 where the Fourier transform magnitudes of the segments $x[n, \tau]$ are shown laid out on the 2-D time-frequency grid. The figure also indicates two kinds of spectrograms: *narrowband*, which gives good spectral resolution, e.g., a good view of the frequency content of sinewaves with closely

---

[7] The notion of $S(\omega, \tau)$ as a 2-D energy density follows from the relation $\sum_{n=-\infty}^{\infty} |x[n]|^2 = \frac{1}{2\pi} \int_{-\pi}^{\pi} \sum_{\tau=-\infty}^{\infty} |S(\omega, \tau)|^2 d\omega$ that holds under certain conditions on the window $w[n, \tau]$.

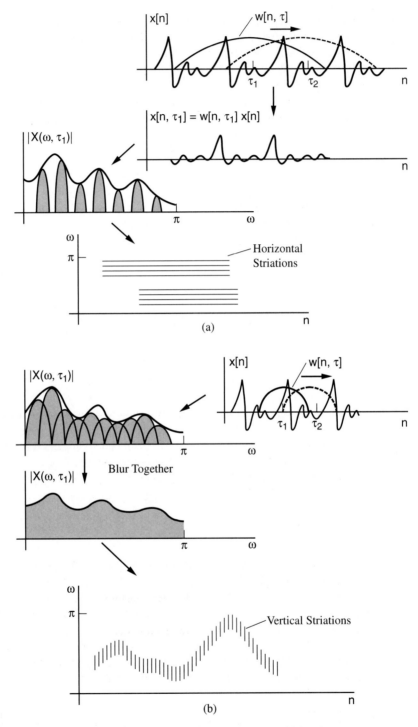

Figure 3.14 Formation of (a) the narrowband and (b) the wideband spectrograms.

spaced frequencies, and *wideband*, which gives good temporal resolution, e.g., a good view of the temporal content of impulses closely spaced in time.[8] We introduce the two classes of spectrograms using speech sounds with a voiced source as an example.

For voiced speech, we have approximated the speech waveform as the output of a linear time-invariant system with impulse response $h[n]$ and with a glottal flow input given by the convolution of the glottal flow over one cycle, $g[n]$, with the impulse train $p[n] = \sum_{k=-\infty}^{\infty} \delta[n - kP]$. This results in the windowed speech waveform expressed as

$$x[n, \tau] = w[n, \tau]\{(p[n] * g[n]) * h[n]\}$$
$$= w[n, \tau](p[n] * \tilde{h}[n])$$

where we have written the glottal waveform over a cycle and vocal tract impulse response as lumped into $\tilde{h}[n] = g[n] * h[n]$. Using the result of Example 3.2, the spectrogram of $x[n]$ can therefore be expressed as

$$S(\omega, \tau) = \frac{1}{P^2} | \sum_{k=-\infty}^{\infty} \tilde{H}(\omega_k) W(\omega - \omega_k, \tau)|^2 \tag{3.4}$$

where

$$\tilde{H}(\omega) = H(\omega)G(\omega)$$

and where $\omega_k = \frac{2\pi}{P} k$ and $\frac{2\pi}{P}$ is the fundamental frequency.

Narrowband Spectrogram — The difference between the narrowband and wideband spectrogram is the length of the window $w[n, \tau]$. For the narrowband spectrogram, we use a "long" window with a duration of typically at least two pitch periods. Under the condition that the main lobes of shifted window Fourier transforms are non-overlapping and that corresponding transform sidelobes are negligible, Equation (3.4) leads to the approximation

$$S(\omega, \tau) \approx \frac{1}{P^2} \sum_{k=-\infty}^{\infty} |\tilde{H}(\omega_k)|^2 |W(\omega - \omega_k, \tau)|^2. \tag{3.5}$$

This approximation is left as an exercise (Exercise 3.8). We see then that using a long window gives a short-time Fourier transform of voiced speech that consists of a set of narrow "harmonic lines," whose width is determined by the Fourier transform of the window, which are shaped by the magnitude of the product of the glottal flow Fourier transform and vocal tract transfer function. The narrowband spectrogram gives good frequency resolution because the harmonic lines are "resolved"; these harmonic lines are seen as horizontal striations in the time-frequency plane of the spectrogram. The long window, however, covers several pitch periods and thus is unable to reveal fine periodicity changes over time; it also smears closely spaced temporal

---

[8] More precise definitions of temporal and spectral resolution are given later in the text. For our purpose here an intuition for the concepts is sufficient.

events and thus gives poor time resolution, as with a plosive that is closely spaced to a succeeding voiced sound.

Wideband Spectrogram — For the *wideband* spectrogram, we choose a "short" window with a duration of less than a single pitch period (Figure 3.14); shortening the window widens its Fourier transform (recall the uncertainty principle). The wide Fourier transform of the window, when translated to harmonics, will overlap and add with its neighboring window transforms and smear out the harmonic line structure, roughly tracing out the spectral envelope $|\tilde{H}(\omega)|$ due to the vocal tract and glottal flow contributions. In an alternative temporal perspective, since the window length is less than a pitch period, as the window slides in time it "sees" essentially pieces of the periodically occurring sequence $\tilde{h}[n]$ (assuming tails of previous responses have died away). For the steady-state voiced sound, we can therefore express the wideband spectrogram (very) roughly (Exercise 3.9 asks the reader to complete the argument) as

$$S(\omega, \tau) \approx \beta |\tilde{H}(\omega)|^2 E[\tau] \qquad (3.6)$$

where β is a constant scale factor and where $E[n]$ is the energy in the waveform under the sliding window, i.e., $E[\tau] = \sum_{n=-\infty}^{\infty} |x[n, \tau]|^2$, that rises and falls as the window slides across the waveform. In this case, where the window $w[n, \tau]$ is short, and less than a pitch period, the spectrogram shows the formants of the vocal tract in frequency, but also gives vertical striations in time every pitch period, rather than the harmonic horizontal striations as in the narrowband spectrogram. These vertical striations arise because the short window is sliding through fluctuating energy regions of the speech waveform.

In our description of the narrowband and wideband spectrograms, we have used the example of voiced speech. Similar reasoning can be made for fricative and plosive sounds. With regard to fricatives, the squared STFT magnitude of noise sounds is often referred to as the *periodogram*, which is characterized by random wiggles around the underlying function $|\tilde{H}(\omega)|^2$. The periodogram is developed formally in a stochastic process framework later in the text. For plosives, the spectrogram reveals the general spectral structure of the sound as the window $w[n, \tau]$ slides across the signal. For these sound classes, both the narrowband and wideband spectrograms show greater intensity at formants of the vocal tract; neither, however, typically shows horizontal or vertical pitch-related striations because periodicity is not present except when the vocal folds are vibrating simultaneously with these noise or impulsive sounds. With plosive sounds, the wideband spectrogram is often preferred because it gives better temporal resolution of the sound's components, especially when the plosive is closely surrounded by vowels.

Figure 3.15 compares the narrowband (20-ms Hamming window) and wideband (4-ms Hamming window) spectrograms for a particular utterance. The spectrograms were computed with a 512-point FFT. For the narrowband spectrogram, the 20-ms Hamming window was shifted at a 5-ms frame interval, and for the wideband spectrogram, the 4-ms Hamming window was shifted at a 1-ms frame interval. Both spectrograms reveal the speech spectral envelope $|\tilde{H}(\omega)| = |H(\omega)G(\omega)|$ consisting of the vocal tract formant and glottal contributions. Notice, however, the distinctive horizontal and vertical striations in the narrowband and wideband spectrograms, respectively. Observe, however, that occasionally the vertical striations are barely visible in the wideband spectrogram when the pitch is very high. Observe also a difference in

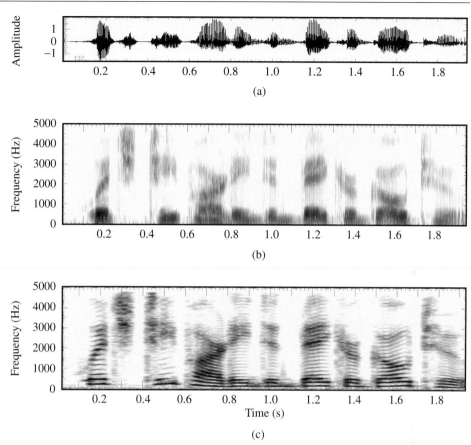

Figure 3.15 Comparison of measured spectrograms for the utterance, "Which tea party did Baker go to?": (a) speech waveform; (b) wideband spectrogram; (c) narrowband spectrogram.

time and frequency resolution between the two spectrograms; for example, the short-time spectrum of the short-duration speech sound "t" in the words "<u>t</u>ea" and "<u>t</u>o," across time, is blurry in the narrowband spectrogram while sharp in the wideband spectrogram. Figure 3.16 gives a similar comparison for an utterance that transitions from normal voicing into diplophonic voicing as the pitch becomes very low. In this case, the pitch is so low that horizontal striations are barely visible in the narrowband spectrogram, in spite of an increased window length of 40-ms to improve resolution of harmonic lines. In the wideband spectrogram, one clearly sees vertical striations corresponding to both the primary glottal pulses and secondary diplophonic pulses.

3.4 Categorization of Speech Sounds

In Section 3.2, we described the anatomy of speech production, the vocal folds and vocal tract being the two primary components, and described the mechanism of speech production, i.e., how we generate sounds with our speech anatomy and physiology. We saw that a sound source can be created with either the vocal folds or with a constriction in the vocal tract, and, based on

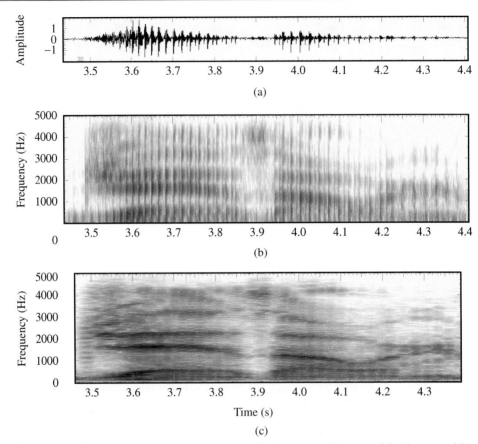

Figure 3.16 Comparison of measured spectrograms for the utterance "Jazz hour" that has a transition into diplophonia: (a) speech waveform; (b) wideband spectrogram; (c) narrowband spectrogram.

the various sound sources, we proposed a general categorization of speech sounds. Section 3.3 then deviated from the flow of this chapter to describe spectrographic analysis for the study of time-varying spectral characteristics of speech. We are now in a position to study and classify speech sounds from the following different perspectives:

1. The nature of the source: periodic, noisy, or impulsive, and combinations of the three;

2. The shape of the vocal tract. The shape is described primarily with respect to the place of the tongue hump along the oral tract and the degree of the constriction of the hump, sometimes referred to as the *place and manner-of-articulation*, respectively. The shape of the vocal tract is also determined by possible connection to the nasal passage by way of the velum;

3. The time-domain waveform which gives the pressure change with time at the lips output;

4. The time-varying spectral characteristics revealed through the spectrogram.

With these four speech descriptors, we embark on a brief study of the classification of speech sounds. We focus on the English language, but from time to time discuss characteristics of other languages.

3.4.1 Elements of a Language

A fundamental distinctive unit of a language is the phoneme; the phoneme is distinctive in the sense that it is a speech sound class that differentiates words of a language [29]. For example, the words "cat," "bat," and "hat" consist of three speech sounds, the first of which gives each word its distinctive meaning, being from different phoneme classes. We saw earlier, and we will discuss further below, that many sounds provide this distinctive meaning, and such sounds represent a particular phoneme. To emphasize the distinction between the concept of a phoneme and sounds that convey a phoneme, the speech scientist uses the term *phone* to mean a particular instantiation of a phoneme. As we discussed in this chapter's introduction, this distinction is also seen in the different studies of phonemics and phonetics.

Different languages contain different phoneme sets. Syllables contain one or more phonemes, while words are formed with one or more syllables, concatenated to form phrases and sentences. Linguistics is the study of the arrangement of speech sounds, i.e., phonemes and the larger speech units built from phonemes, according to the rules of a language. Phonemes can differ across languages, but certain properties of the grammatical rules combining phonemes and larger units of a language may be common and instinctual [30]. There are various ways to study speech sounds that make up phoneme classes; the use of the above first two descriptors in this study is sometimes referred to as *articulatory phonetics*, while using the last two is referred to as *acoustic phonetics*. One broad phoneme classification for English is in terms of vowels, consonants, diphthongs, affricates, and semi-vowels. Figure 3.17 shows this classification, along with various subgroups, where each phoneme symbol is written within slashes according to both the International Phonetic Alphabet and an orthographic (alphabetic spelling) representation. An insightful history of the various phoneme symbol representations is described in [6]. In the remainder of this text, we use the orthographic symbols.

Phonemes arise from a combination of vocal fold and vocal tract articulatory *features*. Articulatory features, corresponding to the first two descriptors above, include the vocal fold state, i.e., whether the vocal folds are vibrating or open; the tongue position and height, i.e., whether it is in the front, central, or back along the palate and whether its constriction is partial or complete; and the velum state, i.e., whether a sound is nasal or not. It has been hypothesized that the first step in the production of a phone is to conceive in the brain the set of articulatory features that correspond to a phoneme. A particular set of speech muscles is responsible for "activating" each feature with certain relative timing. It is these features that we may store in our brain for the representation of a phoneme. In English, the combinations of features are such to give 40 phonemes, while in other languages the features can yield a smaller—e.g., 11 in Polynesian, or a larger, e.g., 141 in the "click" language of Khosian[9]—phoneme set [30]. The rules of a language string together its phonemes in a particular order; for example, in Italian,

---

[9] A click used in the Khosian language is made by the lips and tongue body and with air drawn into the oral tract. The positions of the lips and tongue are features of the language that combine with other features, such as whether the vocal folds are vibrating or not, to form the Khosian phoneme set.

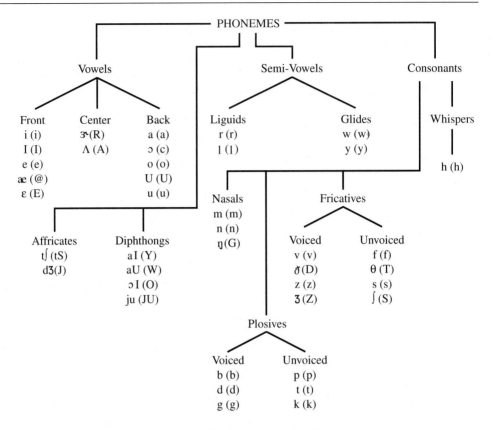

Figure 3.17 Phonemes in American English [6],[32]. Orthographic symbols are given in parentheses to the left of the International Phonetic Alphabet symbols.

consonants are not normally allowed at the end of words. The ordering of the phonemes is also determined in part by the underlying articulatory features of the phones; for example, vibration of the vocal folds or a particular vocal tract shape can constrain or influence the following sound.

A phoneme is not strictly defined by the precise adjustment of articulators; for example, the tongue hump forming a 0.1-mm constriction with the palate, 3 cm along the oral tract, will likely correspond to the same phoneme when these specifications are changed by a few percent. The articulatory properties are influenced by adjacent phonemes, rate and emphasis in speaking, and the time-varying nature of the articulators. The variants of sounds, or phones, that convey the same phoneme are called the *allophones* of the phoneme [29]. Consider, for example, the words "butter," "but," and "to," where the /t/ in each word is somewhat different with respect to articulation, being influenced by its position within the word. Therefore, although the allophones of a phoneme do have consistent articulatory features, the fine details of these features vary in different conditions. In this sense, then, the concept of a phoneme as a distinctive unit of a language is abstract.

In speech production, the articulatory features ultimately lead to the speech waveform and its acoustic temporal and spectral features, corresponding to the above third and fourth descriptors, such as the time delay of a plosive before a voiced sound and vocal tract formants. In the *motor theory of perception* [3], such acoustic properties are measured by the auditory system and ultimately are mapped in the brain to the set of articulatory features that define the phoneme, i.e., in perceiving the phoneme the listener reconstructs the set of articulatory features for that phoneme. Later in this chapter, we return to this paradigm, as well as to a different view where articulatory features are not the end perceptual representation. We now begin a short study of the classification of speech sounds, using both articulatory and acoustic characterizations. For each phoneme class, we describe source and system (vocal tract) articulators, and the resulting spectral and waveform characteristics that give a phoneme its distinction.

3.4.2 Vowels

The largest phoneme group is that of vowels. Vowels contain three subgroups defined by the tongue hump being along the front, central, or back part of the palate.

Source: The source is quasi-periodic puffs of airflow through the vocal folds vibrating at a certain fundamental frequency. We use the term "quasi" because perfect periodicity is never achieved; henceforth, the term "periodic" will be used in this sense. A simple model of the source waveform and spectrum and its modification by the vocal tract was given in Examples 3.1 and 3.2. In English, the pitch of the periodic source does not distinguish phonemes as in some languages such as Chinese.

System: Each vowel phoneme corresponds to a different vocal tract configuration. The vocal tract shape is a function of the tongue, the jaw, the lips, and the velum which is closed in non-nasalized vowels, i.e., the nasal passage is not coupled to the oral tract. In addition to their degree of openness, the lips can contribute to the vocal tract configuration by being rounded, which can increase the effective vocal tract length. Recite the phoneme /u/ in the word "boot" and you will feel the lips become rounded and protruded. The tongue, which is the primary determinant of vocal tract shape, has three general places of articulation: front, center, or back of the oral cavity. The degree of constriction by the tongue is another shape determinant. A comparative example is given with the vowel /a/ as in "f<u>a</u>ther" and with the vowel /i/ as in "<u>e</u>ve" [32]. For the vowel /a/ the vocal tract is open at the front, the tongue is raised at the back, and there is a low degree of constriction by the tongue against the palate. For the vowel /i/ the vocal tract is open at the back, the tongue is raised at the front, and there is a high degree of constriction of the tongue against the palate. These examples are included in Figure 3.18, which illustrates the vocal tract profiles for all English vowels in terms of tongue position and degree of constriction [31]. Keep in mind that Figure 3.18 shows the oral cavity and does not include the pharynx, the region just above the glottis, which can also influence formant locations. X-ray studies of the complete vocal tract for different phonemes are found in the early work of Fant [8], as well as in more recent magnetic resonance imaging studies [35].

Spectrogram: The particular shape of the vocal tract determines its resonances. Qualitative rules based on physical principles have been developed by Stevens [33] for mapping changes in vocal tract shape to formant movement. Perturbations in cross-section at various points of a uniform reference tube (approximately modeling the vowel /Λ/), by narrowing of the front,

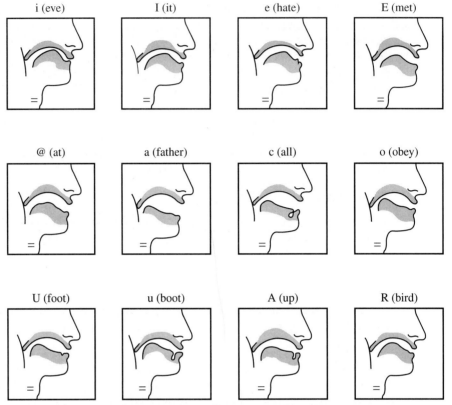

Figure 3.18 Vocal tract profiles for vowels in American English. The two horizontal lines denote voicing.

SOURCE: R.K. Potter, G.A. Kopp, and H.G. Kopp, *Visible Speech* [31]. ©1966, Dover Publications, Inc. Used by permission.

central, or back of the oral cavity by the tongue and jaws, are mapped to certain changes in formant location. In Chapter 4, we will study quantitatively the relation between vocal tract shape and formants using a concatenated acoustic tube model. The wideband spectrograms and spectral slices of the narrowband spectrograms of the two vowels /a/ and /i/ are shown in Figure 3.19. The first formant of /a/ is dominant and falls at roughly 800 Hz, while the second and third weaker formants are at roughly 1200 Hz and 2300 Hz, respectively. For the vowel /i/, the first formant is at about 400 Hz and the second and third formants are at about 2000 Hz and 3000 Hz, respectively, with the third being greater in amplitude than its counterpart in /a/. The wideband spectrograms in these and following examples are obtained with a 4-ms window and a 1-ms frame interval. The narrowband spectral slices are obtained with a 20-ms and 30-ms window for the /a/ and /i/, respectively, and a 5-ms frame interval.

Waveform: Certain vowel properties seen in the spectrogram are also seen in the speech waveform within a pitch period. As illustrated in Figure 3.19, for the vowel /a/ the dominant first

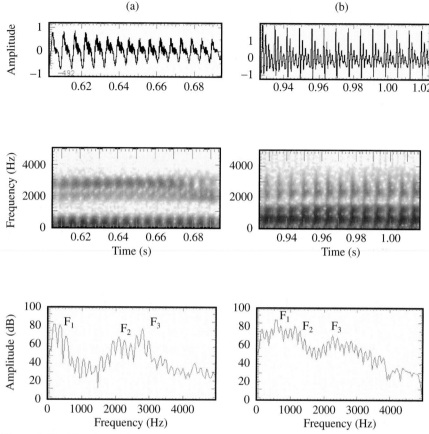

Figure 3.19 Waveform, wideband spectrogram, and spectral slice of narrowband spectrogram for two vowels: (a) /i/ as in "eve"; (b) /a/ as in "father." The first three formants F_1, F_2, and F_3 are marked on the spectral slices.

formant gives a low-frequency damped oscillation while the second and third weaker formants give no visible high-frequency energy. In contrast, for the vowel /i/, the first formant gives a very low-frequency damped oscillation and the third strong formant gives a visible high-frequency oscillation superimposed on the low-frequency formant.

In spite of the specific properties of different vowels, there is much variability of vowel characteristics among speakers. We noted earlier that articulatory differences in speakers is one cause for allophonic variations. The place and degree of constriction of the tongue hump and cross-section and length of the vocal tract, and therefore the vocal tract formants, will vary with the speaker. Peterson and Barney [28],[32] measured the first (F_1) and second (F_2) formants from a spectrogram for a large range of speakers. Vowels deemed to be "perceptually equivalent" were used. A plot of F_1 and F_2 on a 2-D grid reveals approximate elliptical clusters corresponding to the different vowels and shows a large range of variation in F_1 and F_2 for each vowel group. This variability presents a challenge to speech recognition algorithms that rely on invariance of vowel spectral properties across speaker, but aids in speaker recognition where spectral variability with speaker is required.

3.4.3 Nasals

The second large phoneme grouping is that of consonants. The consonants contain a number of subgroups: nasals, fricatives, plosives, whispers, and affricates. We begin with the nasals since they are closest to the vowels.

Source: As with vowels, the source is quasi-periodic airflow puffs from the vibrating vocal folds.

System: The velum is lowered and the air flows mainly through the nasal cavity, the oral tract being constricted; thus sound is radiated at the nostrils. The nasal consonants are distinguished by the place along the oral tract at which the tongue makes a constriction (Figure 3.20). The two nasals that we compare are /m/ as in "<u>m</u>o" and /n/ as "<u>n</u>o." For /m/, the oral tract constriction occurs at the lips and for /n/ the constriction is with the tongue to the gum ridge.

Spectrogram: The spectrum of a nasal is dominated by the low resonance of the large volume of the nasal cavity. The resonances of the nasal cavity have a large bandwidth because viscous losses are high as air flows along its complexly configured surface, quickly damping its impulse response. The closed oral cavity acts as a side branch with its own resonances that change with the place of constriction of the tongue; these resonances absorb acoustic energy and thus are anti-resonances (zeros) of the vocal tract. The anti-resonances of the oral tract tend to lie beyond the low-frequency resonances of the nasal tract; a result of this is that for nasals there is little high-frequency energy passed by the vocal tract transfer function. For the /m/ in Figure 3.21b, there is a low F_1 at about 250 Hz with little energy above this frequency. A similar pattern is seen for the /n/ in Figure 3.21a. Observe that at the release of the constriction of the nasal there is an abrupt change in the spectrogram when the sound is radiated from the mouth. The formant transitions that follow the release are quite different for the nasals /m/ and /n/; these transitions, which reflect the manner in which the oral cavity transitions into its steady vowel position, are an important perceptually distinguishing characteristic of the two nasals [33].

m (me)   n (no)   G (sing)

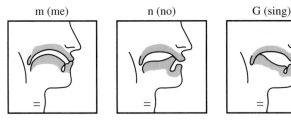

Figure 3.20   Vocal tract configurations for nasal consonants. Oral tract constrictions occur at the lips for /m/, with the tongue tip to the gum ridge for /n/, and with the tongue body against the palate near the velum for /ng/. Horizontal lines denote voicing.

SOURCE: R.K. Potter, G.A. Kopp, and H.G. Kopp, *Visible Speech* [31]. ©1966, Dover Publications, Inc. Used by permission.

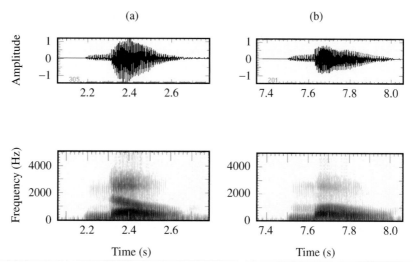

Figure 3.21 Wideband spectrograms of nasal consonants (a) /n/ in "<u>n</u>o" and (b) /m/ in "<u>m</u>o."

Waveform: The waveforms for both the /m/ and /n/ are dominated by the low, wide-bandwidth F_1 formant; within each glottal cycle, there is seen a rapidly damped oscillation. Other resonances are not high enough in energy to be seen.

A phenomenon we referred to earlier as nasalization of vowels is related to the generation of nasals in that the velum is partially open. The nasal cavity couples with the oral cavity and introduces anti-resonances (zeros) into the vocal tract system function. The open nasal cavity acts as a side chamber that introduces the anti-resonances by absorbing energy at certain frequencies, thus serving the same role as the oral tract for a nasal consonant. There is also some broadening of the bandwidth of the resonances of the oral cavity due to the oral-nasal tract coupling. In vowel nasalization, the speech sound is primarily due to the sound at the lips and not the sound at the nose output, which is very weak. Vowels adjacent to nasal consonants tend to be nasalized. Certain speakers characteristically nasalize their vowels by keeping their velum partially open. In English, unlike some languages such as French, Polish, and Portuguese, vowel nasalization is not used to differentiate phonemes [30].

3.4.4 Fricatives

Fricative consonants are specified in two classes: voiced and unvoiced fricatives.

Source: In unvoiced fricatives, the vocal folds are relaxed and not vibrating. Noise is generated by turbulent airflow at some point of constriction along the oral tract, a constriction that is narrower than with vowels. The degree of the constriction somewhat colors the spectral character of the noise source, although this is a secondary effect, the vocal tract spectral coloring being primary.

System: The location of the constriction by the tongue at the back, center, or front of the oral tract, as well as at the teeth or lips, influences which fricative sound is produced. The constriction separates the oral tract into front and back cavities with the sound radiated from the front cavity. Although the front cavity dominates the spectral shaping of the sound, the back cavity introduces anti-resonances in the transfer function, absorbing energy at approximately its own resonances. Because the front cavity is shorter than the full oral cavity and because anti-resonances of the back cavity tend to be lower in frequency than the resonances of the front cavity, the resulting transfer function consists primarily of high-frequency resonances which change with the location of the constriction.

Voiced fricatives have a similar noise source and system characteristic to unvoiced fricatives; for voiced fricatives, however, the vocal folds often vibrate *simultaneously* with noise generation at the constriction and a periodicity of the noisy airflow is seen. Recite the voiced fricative /z/, as in "<u>z</u>ebra," and you will feel the vocal folds vibrating while noise is generated. Generally, fricatives occur in voiced/unvoiced pairs. We compare the unvoiced fricative /f/ as in "<u>f</u>or" and the matching voiced fricative /v/ as in "<u>v</u>ote." In /f/, the vocal folds are not vibrating and the constriction occurs by the teeth against the lips. In contrast, for /v/ the vocal folds are vibrating and again the constriction is formed by the teeth against the lips (Figure 3.22).

When the vocal folds vibrate in a voiced fricative, the periodic airflow from the glottis passes through the back oral cavity to the constriction. At the constriction, frication takes place only when the airflow velocity of the periodic puffs is "high enough." According to fluid dynamical principles, the airflow velocity must exceed a constant called the Reynolds number, which is a function of the density and viscosity of the air medium as well as the geometry of the constriction [20]. This implies that frication is approximately synchronized with airflow velocity. The glottal waveform shape therefore can be thought of as modulating a noise source. This leads to a simplified model of voiced frication given in the following example:

EXAMPLE 3.4   A voiced fricative is generated with both a periodic and noise source. The periodic glottal flow component can be expressed as

$$u[n] = g[n] * p[n]$$

where $g[n]$ is the glottal flow over one cycle and $p[n]$ is an impulse train with pitch period P. In a simplified model of a voiced fricative, the periodic signal component $u[n]$ is passed through a linear time-invariant vocal tract with impulse response $h[n]$. The output at the lips due to the periodic glottal source is given by

$$x_g[n] = h[n] * (g[n] * p[n]).$$

In the model of the noise source component of the voiced fricative, the vocal tract is constricted along the oral tract and air flows through the constriction, resulting in a turbulent airflow velocity source at the constriction that we denote by $q[n]$. In this simplified model, the glottal flow $u[n]$ modulates this noise function $q[n]$ (assumed white noise). The modulated noise then excites the front oral cavity that has impulse response $h_f[n]$. The output flow at the lips due to the noise source is expressed as

$$x_q[n] = h_f[n] * (q[n]u[n]).$$

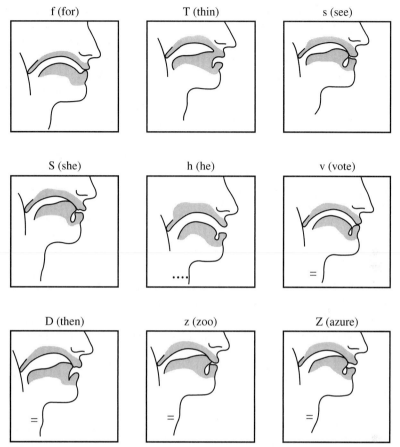

Figure 3.22 Vocal tract configurations for pairs of voiced and unvoiced fricatives. Horizontal lines denote voicing and dots denote aspiration.

SOURCE: R.K. Potter, G.A. Kopp, and H.G. Kopp, *Visible Speech* [31]. ©1966, Dover Publications, Inc. Used by permission.

We assume in our simple model that the results of the two airflow sources add, so that the complete output of the lips is given by

$$x[n] = x_g[n] + x_q[n]$$
$$= h[n] * u[n] + h_f[n] * (q[n]u[n]).$$

The spectral characteristics of $x[n]$ are studied in Exercise 3.10.

In this simple model, we have ignored that the modulating function $u[n]$ is modified by the oral cavity and that the noise response $x_q[n]$ can be influenced by the back cavity. We have also not accounted for sources from nonlinear effects other than the modulation process, one possibility being distributed sources due to traveling vortices.   ▲

In voiced fricatives, however, voicing does not always occur simultaneously with noise generation. Simultaneous voicing may occur only early on or not at all during the frication. A voiced fricative can also be distinguished from its unvoiced counterpart by a shorter duration of frication prior to the onset of voicing in a following vowel. The timing of the onset of voicing after frication thus provides a cue in the distinction of these sounds. The formant transitions from the frication into the following vowel also serve to distinguish between voiced and unvoiced fricative counterparts; for voiced fricatives, the voicing occurs sooner into the transition, thus accentuating the transition relative to the weaker noise excitation of its unvoiced counterpart during the initial part of the transition.[10] Generally then, there are multiple cues that help in these distinctions.

Spectrogram: Unvoiced fricatives are characterized by a "noisy" spectrum while voiced fricatives often show both noise and harmonics. The spectral nature of the sound is determined by the location of the tongue constriction. For example, with an /S/ the frication occurs at the palate, and with an /f/ at the lips. The /S/ has a highpass spectrum corresponding to a short upper oral cavity. For the /f/ there is little front cavity, so its spectrum is almost flat with a mild upward trend. A comparison of the unvoiced fricative /f/ with the voiced fricative /v/ is given in Figure 3.23. The noise component of each fricative has a wide spectrum focused in the high-frequency region (1000–5000 Hz). The voiced fricative, however, is characterized by the additional harmonic structure due to the oscillating vocal folds, as revealed in the spectral slices as well as the spectrograms of Figure 3.23. The influence of the surrounding vowels on the formant transitions to and from the fricatives can also be seen in the spectrograms.

Waveform: For the unvoiced/voiced fricative pair, the waveform of the unvoiced fricative contains noise, while that of the voiced fricative contains noise superimposed on periodicity during the fricative region, as seen in Figure 3.23.

Whisper: Although the whisper is a consonant similar in formation to the unvoiced fricative, we place the whisper in its own consonantal class. We saw earlier that with a whisper the glottis is open and there is no vocal fold vibration. Turbulent flow is produced, however, at the glottis, rather than at a vocal tract constriction. The spectral characteristics of the whisper depend on the size of the glottis, which influences the spectrum of the noise source, and the resonant cavity at the onset of the vowel. An example is /h/, the sole whisper in English, as in "he." Other whispers exist outside the English language.

3.4.5 Plosives

As with fricatives, plosives are both unvoiced and voiced.

Source and System: With unvoiced plosives, a "burst" is generated at the release of the buildup of pressure behind a total constriction in the oral tract. We have idealized this burst as an impulsive source, although in practice there is a time spread and turbulent component to this source. The constriction can occur at the front, center, or back of the oral tract (Figure 3.24). There is no vibration of the folds. The sequence of events is: (1) complete closure of the oral tract and buildup of air pressure behind closure; during this time of closure, no sound is radiated from the lips; (2) release of air pressure and generation of turbulence over a very short-time duration,

---

[10] Within an unvoiced/voiced consonant pair, the formant transitions are similar but differ in their excitation. Across consonant pairs, on the other hand, differences in formant transitions from a consonant to a following vowel help in the distinction of consonants.

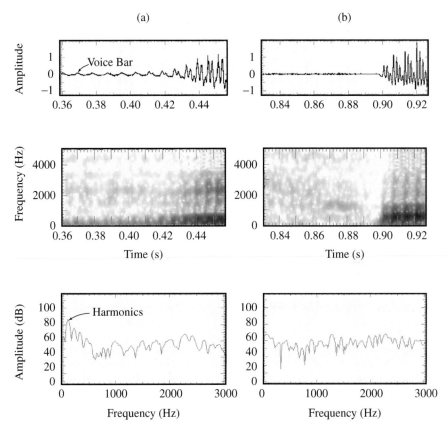

Figure 3.23 Waveform, wideband spectrogram, and narrowband spectral slice of voiced and unvoiced fricative pair: (a) /v/ as in "vote"; (b) /f/ as in "for." Spectral slices taken in fricative regions over a 20-ms window.

i.e., the burst ("impulsive") source, which excites the oral cavity in front of the constriction; (3) generation of aspiration due to turbulence at the open vocal folds (before onset of vibration) as air rushes through the open oral cavity after the burst; and (4) onset of the following vowel about 40–50 ms after the burst. The *voiced onset time* is the difference between the time of the burst and the onset of voicing in the following vowel. The length of the voice onset time and the place of constriction vary with the plosive consonant.

With voiced plosives, as with unvoiced plosives, there is a buildup of pressure behind an oral tract constriction, but the vocal folds can also vibrate. When this vibration occurs, although the oral tract is closed, we hear a low-frequency vibration due to its propagation through the walls of the throat. This activity is referred to as a "voice bar."[11] After the release of the burst, unlike the unvoiced plosive, there is little or no aspiration, and the vocal folds continue to vibrate

---

[11] Voice bars are exploited by other species besides humans. Male frogs emit a mating call by forcing air from their lungs through vocal folds into the mouth and nostrils. During this maneuver the mouth and nostrils are closed tightly and a thin wall sac at the base of the mouth is blown up like a balloon. The vibrating vocal folds propagate sound to the sac, which radiates the mating call into the external air [26].

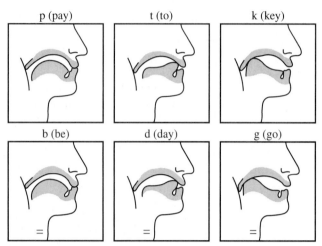

Figure 3.24 Vocal tract configurations for unvoiced and voiced plosive pairs. Horizontal lines denote voicing.

SOURCE: R.K. Potter, G.A. Kopp, and H.G. Kopp, *Visible Speech* [31]. ©1966, Dover Publications, Inc. Used by permission.

into the following vowel; there is much shorter delay between the burst and the voicing of the vowel onset. Figure 3.25 compares an abstraction of the unvoiced plosive and a voiced plosive with the presence of a voice bar. In Figure 3.26, we compare an actual voiced/unvoiced plosive pair: /g/ as in "go" and /k/ as in "baker." The phoneme /k/ is unvoiced with a constriction at the velum; the phoneme /g/ is characterized by the same constriction, but with vibrating folds.

Vocal fold vibration, and thus a voice bar, does not always occur during the burst of a voiced plosive, and therefore is not the only distinguishing feature between the two sound classes. The length of the voice onset time can also provide the distinction. Perceptual experiments indicate that if the release of the burst and onset of voicing are within 20 ms of each other, the consonant is considered voiced; otherwise, it is judged as unvoiced. The muscles controlling the vocal fold positions and tension are very precisely activated in time to generate the different voice onset times [10]. The formant transitions from the onset of the burst into the following vowel also help in distinguishing between voiced and unvoiced plosive counterparts; as with voiced fricatives, the voicing in voiced plosives occurs sooner into the transition, thus accentuating the transition relative to the weaker aspiration excitation of its unvoiced counterpart.

Spectrogram and Waveform: For the unvoiced plosive /k/, we observe (Figure 3.26) in both the waveform and the spectrogram a gap of near silence, followed by an abrupt burst, and then aspiration noise. The spectrogram at the time of the burst is governed by the shape of the oral cavity in front of the constriction (which is excited by the burst) and the spectral character of the burst itself. The aspiration then acts as an excitation to the vocal tract as it transitions from its constricted state. Finally, when the vocal folds vibrate, the resulting periodic puffs excite the vocal tract as it enters into its steady vowel state. The formant trajectories through aspiration and into voicing reflect this changing vocal tract shape. For the counterpart voiced plosive /g/, a low-frequency voice bar is seen in the spectrogram prior to the burst. Observe also that the voice onset

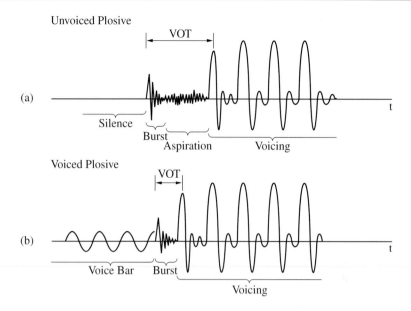

Figure 3.25 A schematic representation of (a) unvoiced and (b) voiced plosives. The voiced onset time is denoted by VOT.

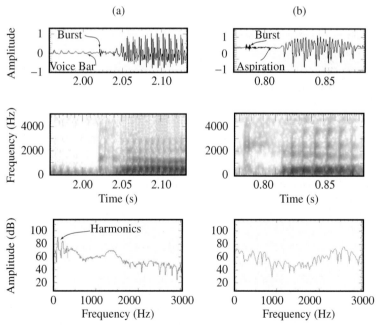

Figure 3.26 Waveform, wideband spectrogram, and narrowband spectral slice of voiced and unvoiced plosive pair: (a) /g/ as in "go"; (b) /k/ as in "key." Spectral slices are taken in burst regions over a 40-ms window in (a) and a 25-ms window in (b).

time is negligible for the /g/ in contrast to that for the /k/. Given a negligible aspiration stage, the formant trajectories sometimes appear more visible, being excited throughout by voicing. As with voiced fricatives, vocal fold vibration need not be present for a plosive to be "voiced"; the voice onset time and formant transitions can be sufficient to make the distinction.

In the following example, we explore a time-varying linear system model for the voiced plosive.

EXAMPLE 3.5   We have seen that a voiced plosive is generated with a burst source and can also have present a periodic source throughout the burst and into the following vowel. Assuming the burst occurs at time $n = 0$, we idealize the burst source as an impulse $\delta[n]$. The glottal flow velocity model for the periodic source component is given by

$$u[n] = g[n] * p[n]$$

where $g[n]$ is the glottal flow over one cycle and $p[n]$ is an impulse train with spacing P. Assume that the vocal tract is linear but time-varying, due to the changing vocal tract shape during its transition from the burst to a following steady vowel. The vocal tract output cannot, therefore, be obtained by the convolutional operator. Rather, the vocal tract output can be expressed using the time-varying impulse response concept introduced in Chapter 2. In our simple model, the periodic glottal flow excites a time-varying vocal tract, with impulse response denoted by $h[n, m]$, while the burst excites a time-varying front cavity beyond a constriction, denoted by $h_f[n, m]$. The sequences $h[n, m]$ and $h_f[n, m]$, as we described in Chapter 2, represent time-varying impulse responses at time n to a unit sample applied m samples earlier at time $n - m$. We can then write the output with a generalization of the convolution operator as

$$x[n] = \sum_{m=-\infty}^{\infty} h[n, m]u[n - m] + \sum_{m=-\infty}^{\infty} h_f[n, m]\delta[n - m]$$

where we assume the two outputs can be linearly combined. In this model, we have not accounted for aspiration that might occur before the onset of voicing, as well as other effects described in Example 3.4. ▲

3.4.6 Transitional Speech Sounds

A number of the phone examples we have presented thus far were illustrated as "stationary," such as the sounds for the phonemes /a/ and /i/ in Figure 3.19, in the sense that their underlying articulators hold an almost fixed configuration and the resulting formants appear nearly constant. We also illustrated the movement of one phone to the next with articulators transiting from one configuration to another, for example, as manifested in the spectral change in going from the /g/ to /o/ in "go" of Figure 3.26. Such speech sounds are "nonstationary" and in some cases the rapid transition across two articulatory states *defines* the sound. This time-varying nature of articulators is more the norm than the exception. The articulators are almost always in transition between states, often never reaching a desired state in typically spoken speech. Nonstationarity is further imparted through a phenomenon known as *coarticulation*, which involves the anticipation of the following sound, sometimes many phones in the future, and therefore a blending of the articulatory states. Such nonstationarity poses interesting challenges to speech signal processing algorithms that typically assume (and often require) stationarity over intervals of 10–20 ms.

Phonemes Defined by Transition — Diphthongs: Diphthongs have a vowel-like nature with vibrating vocal folds. Diphthongs, however, cannot be generated with the vocal tract in a steady configuration; they are produced by varying in time the vocal tract smoothly between two vowel configurations and are characterized by a *movement* from one vowel *target* to another. The term target, as implied, refers to seeking a particular vocal tract configuration for a phoneme, but not necessarily achieving that configuration. Four diphthongs in the American English language are: /Y/ as in "hide," /W/ as in "out," /O/ as in "boy," and /JU/ as in "new." As an example, Figure 3.27 shows the spectrogram of the diphthong /O/ in "boy," where we see a rapid movement of the formants (especially F_2) as the vocal tract changes from its configuration for the /u/ to that for the /i/ vowel sound. Such formant transitions characterize the diphthong.

Semi-Vowels: This class is also vowel-like in nature with vibrating folds. There are two categories of semi-vowels: glides (/w/ as in "we" and /y/ as in "you") and liquids (/r/ as in "read" and /l/ as in "let"). The glides are dynamic and transitional and often occur before a vowel or are surrounded by vowels, passing from a preceding vowel and moving toward the following vowel, thus in this latter case being similar to diphthongs. Glides differ from diphthongs in two ways. In glides, the constriction of the oral tract is greater during the transition and the speed of the oral tract movement is faster than for diphthongs. These articulations result in weaker, but faster formant transitions [29]. An example is /w/ in the phrase "away." At the onset of the /w/, the tongue hump is high and the lips are highly constricted. The formants of the /w/ "glide" rapidly from the preceding /a/ to the following /e/. The liquids /l/ and /r/ are similar to the glides in being voiced, in the speed of movement, and in the degree of oral constriction. The liquids differ from the glides in the formation of the constriction by the tongue; the tongue

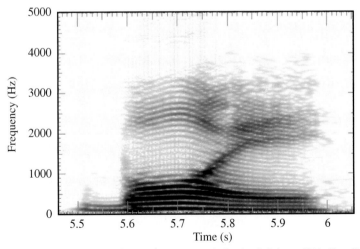

Figure 3.27 Narrowband spectrogram example for the diphthong /O/ in "boy."

is shaped in such a way as to form side branches, different from all other oral tract formations thus far [33]. The presence of these side branches can introduce anti-resonances. With /l/ the tongue tip is pressed against the palate, and is shaped to form two openings, one on each side of the contact point with air passing on each side. For the liquid /r/, two different tongue formations are known to exist, one with a tongue tip raised at the palate as shown in Figure 3.28 and creating a space under it, and the other with the tongue tip lowered and the tongue hump bunched near the palate forming a side branch to the primary oral cavity [7],[33].[12] In addition, the liquids are typically characterized by an effective constriction of the pharynx, quite unusual for American English consonants; magnetic resonance image vocal tract profiles have shown such constrictions [7],[24]. The formant transitions differ for all four semi-vowels; nevertheless, these transitions are smooth except for the liquid /l/, whose spectral evolution is discontinuous because the tongue tip holds and releases its contact with the palate.

Affricates: This sound is a counterpart of diphthongs, consisting of consonant plosive-fricative combinations, rapidly transiting from plosives to fricatives. The articulation of affricates is similar to that of fricatives. The difference is that for affricates, the fricative is preceded by a complete constriction of the oral cavity, formed at the same place as for the plosive. An example is the /tS/ in the word "chew," which is the plosive /t/ followed by the fricative /S/. The voiced counterpart to /tS/ is the affricate /J/ as in "just," which is the voiced plosive /d/ followed by the voiced fricative /Z/.

Coarticulation — Although our vocal fold/vocal tract muscles are programmed to seek a target state or shape, often the target is never reached. Our speech anatomy cannot move to a desired position instantaneously and thus past positions influence the present. Furthermore, to make anatomical movement easy and graceful, the brain anticipates the future, and so the articulators at any time instant are influenced by where they have been and where they are going. *Coarticulation* refers to the influence of the articulation of one sound on the articulation of another sound in the same utterance and can occur on different temporal levels. In "local" coarticulation, articulation of a phoneme is influenced by its adjacent neighbors or by neighbors close in time. Consider,

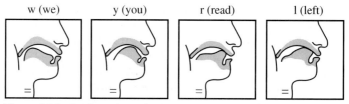

Figure 3.28   Configurations of glides and liquids. Horizontal lines denote voicing.

Source: R.K. Potter, G.A. Kopp, and H.G. Kopp, *Visible Speech* [31]. ©1966, Dover Publications, Inc. Used by permission.

---

[12] It is interesting that these two very different vocal tract formations can result in similar acoustic characteristics [13]. We discuss this observation further in Section 3.6.2.

for example, the words "horse" and "horseshoe." In the two cases, the /s/ is spoken differently, the /s/ in "horseshoe" *anticipating* the /S/. (Here we see two allophones of the phoneme /s/.) Another example of local coarticulation is in the fricative /s/ anticipating the following phomeme in the word "sweep." Here /w/ influences the formant near 3500 Hz in /s/ to move downward by about 500 Hz, in contrast to the same formant /s/ in "seep" [29]. This occurs because the extending and narrowing of the lips in forming /w/ lowers the resonances of the mouth cavity in front of the frication of /s/. This configuration of the lips occurs *during* the /s/ frication in "sweep," thus influencing its spectral structure. On the other hand, during the frication of the /s/ in "seep," the lips are more open. There also exists "global" coarticulation, where the articulators are influenced by phonemes that occur some time in the future beyond the succeeding or nearby phonemes; this can be even further generalized to a higher level where entire phrase structures are anticipated. Each phoneme then contains parts of other surrounding phonemes, giving each sound an added redundancy which may aid in effective communication.[13]

The anticipation of phonemes, words, and phrase groups may be tied closely to the duration of exhalation from the lungs. That is, words, phrases, and sentences are formed into *breath groups*. There is evidence that consonant and vowel durations are shortened or lengthened depending on the length of the exhalation, and this duration change requires anticipation of what is to come; if early events were not timed in anticipation of later events, then the later events would have to be compressed in time and spoken too rapidly to be audible. For example, the number of pitch periods for each vowel in an utterance may be set so that no vowel in the utterance is too short. The *short-term memory* theory of the programming of vocal fold/tract muscles hypothesizes that the human can store a limited set of instructions for muscle movement [30]; as new instructions enter the storage buffer, old instructions leave, compressing as many instructions as possible without sacrificing the integrity of the sound.

3.5 Prosody: The Melody of Speech

Long-time variations, i.e., changes extending over more than one phoneme, in pitch (intonation), amplitude (loudness), and timing (articulation rate or rhythm) follow what are referred to as the rules of *prosody* of a language. These rules are followed to convey different meaning, stress, and emotion. For example, the average pitch contour of a declarative statement tends to increase at the onset of an utterance, decline slowly, then make a rapid drop at the end of the utterance. For a question, on the other hand, the pitch contour increases over time relative to that of the same statement made in declarative form. In addition, a speaker also adds his/her own speaker-dependent prosodic refinements such as a characteristic articulation rate and pitch fluctuation.

Consider changes made in the production mechanism and prosody during the stress of a speech sound. The subglottal pressure, i.e., the pressure just below the glottis, during a normally spoken utterance is typically about constant, but falling near the termination of the

---

[13] Short pieces of a phrase can sometimes be replaced by another sound without a change in intelligibility by the listener as, for example, when replacing certain short speech segments by noise. This result may reflect phonological redundancy by the speaker, but also expectation by the listener. It is also interesting to note that, in addition to temporal redundancy, there is evidence of frequency redundancy in that certain spectral bands can be eliminated with little effect on intelligibility of a sound [22].

utterance where there is also often a corresponding decline in pitch. We can impart, however, variations on this average subglottal pressure. When adding stress to a speech sound, we increase pressure within the lungs. The resulting increased subglottal pressure causes an increase in pitch and loudness. This increased pressure will also cause a more abrupt glottal closure by accentuating the Bernoulli effect on airflow through the glottis. A more abrupt closure of the glottis, as seen in Example 3.1, corresponds to more energy in high frequencies of the glottal spectrum, and thus a "harder" voice quality which can also characterize a stressed sound. An increase in pitch is also imparted due to an increase in vocal fold tension. A rise in subglottal pressure and in vocal fold tension is also responsible for a rising pitch over the duration of a question. We can impart these variations globally, as in going from a soft to a shouting voice in argument, or more locally in time to stress or change intonation on a particular syllable, word, or phrase.

EXAMPLE 3.6   Figure 3.29 shows a comparison of the utterance, "Please do this today," which in the first case is spoken normally, while in the second case, the word "today" is stressed. The pitch, spectral content, and loudness are all seen to be generally higher in the stressed case for the word "today" (Exercise 3.13). The reader should speak the utterance in the two voice styles. Feel your lungs contract during the stressing of the word "today." The increase in lung contraction causes an increase in subglottal pressure. Another change in stressing a sound is its increase in duration as seen in the waveforms and spectrograms of the word "today" in Figure 3.29. Although the speaker planned to stress only the word "today," the entire sentence is longer, likely due to the anticipation of stressing the last word of the sentence. Stressing a sound is generally accompanied by an increase in the duration of vowels and consonants. For a vowel, the number of pitch periods increases, as we can see in Figure 3.29. For consonants, the time of frication or time prior to closure of a constriction, thus building up greater pressure, will increase. For example, for the /t/ of the stressed word "today," the preceding silence gap is longer and the burst louder than in its normally spoken counterpart.   ▲

We saw in Example 3.6 that an increase in duration is also characteristic of a stressed sound. One theory proposed by Lieberman [21] suggests that subglottal pressure, vocal fold tension, and phoneme duration are set within a time period reference which is the duration of an exhalation. We earlier called the collection of word and phrase groups within a single exhalation a breath group, and saw that the duration of a breath group affects the coarticulation of phonemes. Lieberman's hypothesis states that, in addition, the durational and timing patterns of phonemes are also determined by their degree of stress within the breath group. The muscles of the larynx that control vocal fold tension and the muscles around the lungs that control subglottal pressure are timed in their contraction relative to the breath group duration and the sounds marked for stress and pitch intonation within the breath group.

We have seen thus far two factors affecting the duration of vowels and consonants: coarticulation and stress within a breath group. There are other factors as well. Vowels are typically longer than consonants. One hypothesis for this difference is that vowels, in addition to relaying intelligibility of the message, by being typically longer carry prosodic components such as rhythm and intonation, while consonants, by being shorter, carry more of the information load.

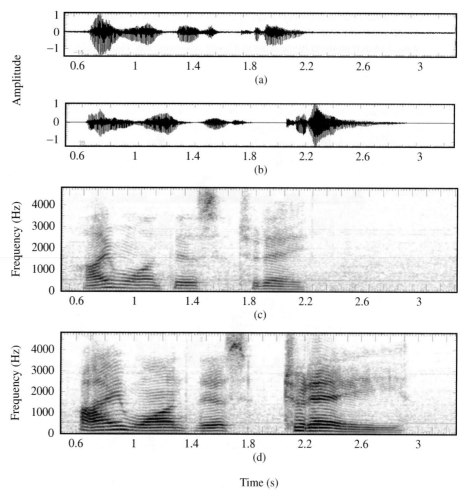

Figure 3.29 Comparison of "Please do this today," where "today" is spoken in a normal and stressed style: (a) waveform of normal; (b) waveform of stressed; (c)–(d) spectrograms of (a)–(b).

Although consonants are often shorter than vowels, time is still required to process them by the perceptual system. This processing time may be provided when a vowel follows a consonant. Duration is also affected by other factors, such as position and importance of a word within a breath group or the relative position of vowels and consonants; e.g., a vowel is shorter when followed by a voiceless consonant than a voiced consonant. An understanding of these timing patterns is important because they indicate how speech is temporally organized in the brain and motor system. A better understanding of timing also has the practical benefit of improved speech synthesis techniques.

Related to the timing of speech events is the rate of articulation of a speaker, a topic we will revisit throughout the text when we describe signal processing systems that modify or measure

[37] J. Sundberg, "The Acoustics of the Singing Voice," *Scientific American*, pp. 82–91, March 1977.

[38] H.M. Teager and S.M. Teager, "Evidence for Nonlinear Sound Production Mechanisms in the Vocal Tract," chapter in *Speech Production and Speech Modeling*, W.J. Hardcastle and A. Marchal, eds., NATO Adv. Study Inst. Series D, vol. 55, Bonas, France; Kluwer Academic Publishers, Boston, MA, pp. 241–261, 1990.

[39] B. Story and I.R. Titze, "Voice Simulation with a Body Cover Model of the Vocal Folds," *J. Acoustical Society of America*, vol. 97, pp. 1249–1260, 1995.

[40] R.L. Whitehead, D.E. Metz, and B.H. Whitehead, "Vibratory Patterns of the Vocal Folds during Pulse Register Phonation," *J. Acoustical Society of America*, vol. 75, no. 4, pp. 1293–1297, April 1984.

CHAPTER **4**

Acoustics of Speech Production

4.1 Introduction

In the previous chapter, we used the term "sound" without a formal definition. In this chapter, we give a more quantitative description of the phenomenon of sound through the *wave equation*. The wave equation, derived in Section 4.2, is a linear partial differential equation whose solution describes pressure and velocity of air particles in a sound wave as a function of time and space, and provides a means to approximately describe the acoustics of speech production. In arriving at the wave equation, we remove nonlinear contributions to the relationship between particle pressure and velocity. This simplification, although valid under a small particle velocity assumption, prohibits rotational flows, such as vortices, alluded to in Chapter 3 and further described in Chapter 11.

In Section 4.3, we gain intuition for sound in the vocal tract by using a simple lossless uniform tube model. This model is then extended to a nonuniform tube with energy loss, including vocal tract wall vibration, viscosity, and thermal conduction of air particles, and radiation loss at the lips. These losses are represented by partial differential equations coupled to the wave equation, and the full equation set is solved by numerical simulation to obtain airflow velocity and pressure at the lips output. Although solution by numerical simulation gives insight into properties of the speech waveform, it does not result in a closed-form transfer function from the glottis to the lips. In Section 4.4, an alternative to numerical simulation is presented in which the vocal tract is approximated by a concatenation of acoustic tubes, each of small length and of uniform cross-section. With this concatenated acoustic tube model and the wave equation, we derive a transfer function of air particle velocity from the glottis to the lips. This transfer function is first derived in continuous time and is then converted to a discrete-time form using the impulse invariance method of Chapter 2.

The notion of a vocal tract transfer function assumes that the vocal tract is linear and time-invariant. More accurate models invoke a time-varying vocal tract and nonlinear coupling between the vocal tract input, i.e., the glottal airflow velocity, and the pressure in the vocal tract resonant cavity. In Section 4.5, a simplified model of the nonlinear interaction between the glottal source and a time-varying vocal tract is described. Numerical simulation of the model shows certain time- and frequency-domain properties of speech not seen with a linear model, an important example being the "truncation" effect in speech production where a rapid bandwidth increase in vocal tract formants during glottal opening corresponds to an abrupt decay of the speech waveform within a glottal cycle.

4.2 Physics of Sound

4.2.1 Basics

The generation of sound involves the vibration of particles in a medium. In speech production, air particles are perturbed near the lips, and this perturbation moves as a chain reaction through free space to the listener; the perturbation of air molecules ends in the listener's ear canal, vibrating the ear drum and initiating a series of transductions of this mechanical vibration to neural firing patterns that are ultimately perceived by the brain. The mechanism of the chain reaction in sound generation can be illustrated by a simple analogy. Consider a set of pool balls configured in a row on a pool table. The striking of the cue ball sets up a chain reaction whereby each ball in the series is struck until the last ball reaches the intended pool-table pocket. Each ball has a "particle velocity" (m/s) and there is a "pressure" (newtons/m^2) felt by each ball due to the interaction between balls. After the first ball is struck, a pressure increase is felt successively by each ball, as well as a change in velocity. As the balls move closer together, we say they are in a *compression* state which occurs locally in space, i.e., in the vicinity of each ball. On the other hand, if the pool stick were to move fast enough away from the first ball, we can stretch our imagination to envision a vacuum set up behind the first ball, thus moving each ball successively in the other direction, creating a local decrease in pressure. For this case, the balls move farther apart which is referred to as a *rarefaction* state near each ball. If we move the pool stick sinusoidally, and fast enough, we can imagine creating a *traveling wave* of local compression and rarefaction fluctuations.

A deficiency in this pool-ball analogy is that there is no springiness between the balls as there is between air particles. Air particles stick together and forces exist between them. To refine our analogy, we thus connect the pool balls with springs. Again we have two states: (1) compression (pushing in the springs between balls) and (2) rarefaction (pulling out the springs between balls). Also, as before, a traveling wave of compression and rarefaction fluctuations is created by pushing and pulling the springs. Observe that the pressure and velocity changes are not permanent; i.e., the balls are not permanently displaced, and this is why we refer to the changes in pressure and velocity as *local* fluctuations in compression and rarefaction.

In describing sound propagation, we replace the pool balls with air molecules and the springs with forces between the air molecules; in Figure 4.1 we depict these forces with little springs. As with the pool balls with connecting springs, air molecules are locally displaced, they

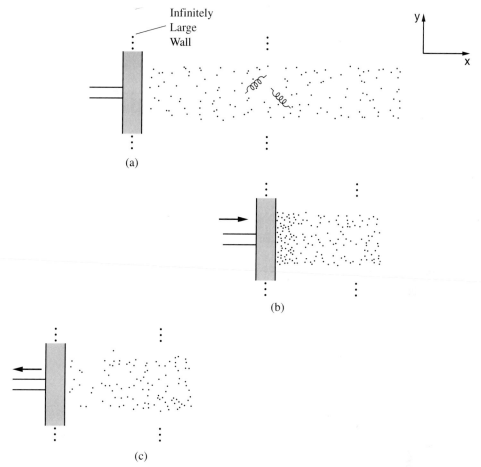

Figure 4.1 Compression and rarefaction of air particles in front of an infinitely large wall: (a) illustration of springiness among air particles; (b) compression; (c) rarefaction.

have a particle velocity, and there is pressure built up among the molecules due to the springiness and collisions that occur with themselves.[1]

Consider now placing a moving wall in front of the air molecules, and suppose that the wall is infinitely large in the y, z direction (z coming out of the page). As with the balls attached by springs, by moving the wall to the right or to the left, we create a chain reaction of compression or rarefaction, respectively, and, with a sinusoidal movement of the wall, we create a traveling wave of compression and rarefaction fluctuations. When the wall is pushed to the right, the molecules in front of the wall are compressed and collide with one another and build up pressure. The pressure increase in front of the wall moves as a chain reaction to the right. When pulled back, the piston creates a region of rarefaction in front of the wall. The local decrease in pressure travels as a chain reaction to the left. When moved sinusoidally, the wall generates

---

[1] Pressure = force/unit area = time rate of change of momentum/unit area [2].

a traveling wave of rarefaction and compression fluctuations characterized by local sinusoidal changes in particle pressure. In addition, local changes in particle velocity and displacement accompany pressure changes. Here the sound travels as a *plane wave* longitudinally along the x axis of Figure 4.1 [2]. This is in contrast to perturbations of air particles in other forms, i.e., other than the longitudinal flow wave of the compression-rarefaction type, such as rotational or jet flow [14].

We are now in a position to give a formal definition of a sound wave.

A *sound wave* is the propagation of a disturbance of particles through an air medium, or, more generally, any medium,[2] without the permanent displacement of the particles themselves. Associated with the disturbance are local changes in particle pressure, displacement, and velocity. The term *acoustic* is used in describing properties or quantities associated with sound waves, e.g., the "acoustic medium."

A pressure disturbance will reside about some ambient (atmospheric) pressure. The pressure variation at the ear causes the ear's diaphragm to vibrate, thus allowing the listener to "perceive" the sound wave. For a sinusoidal disturbance, with the sinusoidally-varying wall, there are a number of important properties that characterize the sound wave (Figure 4.2).

The *wavelength* is the distance between two consecutive peak compressions (or rarefactions) in space (not in time), and is denoted by λ. If you take a snapshot at a particular time instant, you will see a sinusoidally-varying pressure or velocity pattern whose crests are separated by the wavelength. Alternately, the wavelength is the distance the wave travels in one cycle of the vibration of air particles. The *frequency* of the sound wave, denoted by f, is the number of cycles of compression (or rarefaction) of air particle vibration per second. If we were to place a probe in the air medium, we would see a change in pressure or in particle velocity of f cycles in one second. Therefore, the wave travels a distance of f wavelengths in one second. The velocity of sound, denoted by c, is therefore given by $c = f\lambda$. Because the speed of sound $c = f\lambda$ and observing that the radian frequency $\Omega = 2\pi f$, then $\Omega/c = 2\pi/\lambda$. We denote $2\pi/\lambda$ by k, which we call the *wavenumber*. At sea level and 70°F, $c = 344$ m/s; this value is approximate because the velocity of sound varies with temperature. The velocity of sound is distinctly different from particle velocity; the former describes the movement of the traveling compression and rarefaction through the medium, and the latter describes the local movement of the particles which, on the average, go nowhere (as we saw earlier in the pool-ball analogy).

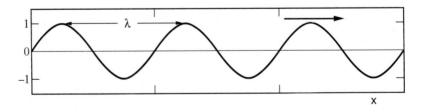

Figure 4.2   Traveling sinusoidal wave characterized by wavelength λ and frequency f.

---

[2] Sound waves can also propagate in, for example, solid and liquid media.

EXAMPLE 4.1   Suppose the frequency of a sound wave $f = 50$ cycles/s (or Hertz, denoted by Hz) and assume that the velocity of sound at sea level is $c = 344$ m/s. Then the wavelength of the sound wave $\lambda = 6.88$ m. For $f = 1000$ Hz, $\lambda = 0.344$ m, while for $f = 10000$ cycles/s, $\lambda = 0.0344$ m. This wide range of wavelengths occurs in the propagation of speech sounds.   ▲

In describing sound propagation, it is important to distinguish pressure change due to variations that occur rapidly or slowly [2]. An *isothermal* process is a slow variation that stays at constant temperature. Isothermal compression of a gas results in an increase in pressure because a given number of molecules are forced into a smaller volume and will necessarily collide with each other (and the boundaries of an enclosure) more often, thus increasing the time rate of change of momentum. However, because the variation is slow, there is time for the heat generated by collisions to flow to other parts of the air medium; thus the temperature remains constant. On the other hand, an *adiabatic* process is a fast variation in which there is no time for heat to flow away and thus there occurs a temperature change in the medium. The heat generated causes an even greater number of collisions. The molecules get hotter faster and collide with each other (or the boundaries of an enclosure) more frequently. They have greater momentum (they are moving faster) and thus transfer more momentum to each other during collisions. It follows that an adiabatic gas is "stiffer" (than an isothermal gas) since it takes more force to expand or compress it.[3] For most of the audible range, a sound wave is an adiabatic process,[4] a property that we use in deriving the wave equation.

4.2.2 The Wave Equation

Imagine a small cube of air particles sitting in space with volume $\Delta x \Delta y \Delta z$ and which is characterized by mass, pressure, and velocity. As before, a vibrating wall lies to the left of the cube and is infinite in extent (Figure 4.3), so all quantities are one-dimensional (1-D), i.e., propagating waves are planar with no change in the y or z direction. The pressure within the cube is a function of both time and space and is denoted by $p(x, t)$, and fluctuates about an ambient or average (atmospheric) pressure P_o. The pressure is the amount of force acting over a unit area of surface and is measured in newtons/m^2 and increases with the number of particle collisions and change in particle momentum. Therefore, $p(x, t)$ is an incremental pressure, the total pressure being expressed as $P_o + p(x, t)$. Henceforth, the average pressure P_o will be dropped because the wave equation requires only incremental pressure. The atmospheric pressure is typically $P_o = 10^5$ newtons/m^2, while the threshold of hearing, i.e., the minimum perceivable pressure above atmospheric pressure, is about $p(x, t) = 2(10^{-5})$ newtons/m^2 at 1000 Hz (while the threshold of pain is about 20 newtons/m^2) [23]. The ear is thus extremely

---

[3] Think about a bicycle pump. If you move the pump handle very quickly, the temperature rises and the local heat energy increases, not having time to distribute itself within the pump. As you pump faster, the hot air becomes stiffer, i.e., the gas is adiabatic. On the other hand, if you move the handle slower, then the instantaneous temperature rise has time to distribute itself and there is negligible average temperature increase; the pumping thus becomes easier. This is an isothermal process.

[4] This result is determined by measuring the relative speed of a thermal and acoustic wave. At very low frequency, the adiabatic condition may break down for sound waves.

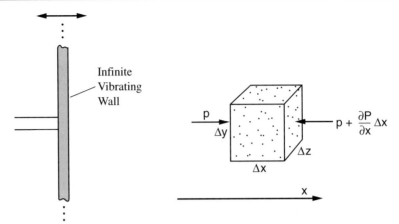

Figure 4.3 Cube of air in front of an infinite vibrating wall. Cube configuration is shown with first-order pressure change across the cube. Arrows indicate forces on two vertical surfaces.

sensitive to pressure changes. The particle velocity is the rate of change in the location of an air particle and fluctuates about zero average velocity. The particle velocity is measured in m/s and is denoted by $v(x, t)$. The density of air particles, denoted by $\rho(x, t)$, is the mass per unit volume and is measured in kg/m^3 around an average density ρ_o, the total density being $\rho_o + \rho(x, t)$.

There are three laws of physics used to obtain the desired relations between pressure and velocity of the air particles within the cube [2]. Newton's Second Law of Motion states that the total force on the cube is the mass times the acceleration of the cube and is written as $F = ma$. This law predicts that a constant applied force produces a constant acceleration of the cube. The Gas Law, from thermodynamics, relates pressure, volume, and temperature and, under the adiabatic condition—the case of interest for speech sound propagation—reduces to the relation $PV^\gamma = C$, where P is the total pressure on the cube of volume V, C is a constant, and $\gamma = 1.4$ is the ratio of the specific heat of air at constant pressure to the specific heat of air at constant volume [2]. Given that the cube is assumed deformable, this formula predicts that if the pressure increases, then the volume of the cube decreases. The third law is the Conservation of Mass: The total mass of gas inside the deformable cube must remain fixed. This law states that a pressure differential can deform the cube, but the total number of particles in the deformable cube remains constant. The moving boundaries of the cube are defined by the original particles. The local density may change, but the total mass does not change. Using these three laws, we can derive the wave equation. We do not give the complete derivation of the wave equation, but rather give a flavor of one approach to its derivation.

The first step in deriving the wave equation relies on Newton's Second Law of Motion. Three assumptions are made that result in equations that are linear and computationally tractable [2],[27]. First, we assume there is negligible friction of air particles in the cube with those outside the cube, i.e., there is no shearing pressure due to horizontal movement of the air. We refer to this shearing pressure as *viscosity*. Therefore, the pressure on the cube is due to only forces on the two vertical faces of the cube, as illustrated in Figure 4.3. Our second assumption is that the cube of air is small enough so that the pressure change across the cube in the horizontal dimension (Δx) is of "first order," corresponding to sounds of not extremely large intensity.

This means that the second- and higher-order terms in a Taylor series expansion of the pressure function with respect to the x argument can be neglected, resulting in

$$p(x + \Delta x, t) \approx p(x, t) + \frac{\partial p}{\partial x} \Delta x$$

where $\frac{\partial p}{\partial x}$ is the rate at which the pressure increases in the x direction (left to right). Henceforth, we drop the (x, t) notation unless explicitly needed. The third assumption is that the density of air particles is constant in the cube and equal to the average atmospheric density, i.e., $\rho(x, t) + \rho_o = \rho_o$, which we will denote by simply ρ.

Because the pressure is force divided by the surface area of the vertical face of the cube, the net force in the x direction on the cube is the pressure difference across the cube multiplied by the surface area:

$$F = - \left(\frac{\partial p}{\partial x} \Delta x \right) \Delta y \Delta z.$$

We have assumed the density in the cube to be constant. Therefore, the mass m of the cube is given by

$$m = \rho \Delta x \Delta y \Delta z.$$

The acceleration of the cube of air is given by $a = \frac{dv}{dt}$, where $v(x, t)$ denotes the velocity of the particles in the cube.[5] Finally, from Newton's Second Law of Motion, $F = ma$, we have the net force acting on the cube of air as

$$-\left(\frac{\partial p}{\partial x} \Delta x \right) \Delta y \Delta z = \rho \Delta x \Delta y \Delta z \left(\frac{dv}{dt} \right)$$

and cancelling terms we have

$$-\frac{\partial p}{\partial x} = \rho \frac{\partial v}{\partial t} \tag{4.1}$$

which is accurate "in the limit," i.e., as Δx gets very small the approximations become more accurate because the differential pressure becomes "1st order" and the density constant.

The reader might have observed a slight-of-hand in arriving at the final form in Equation (4.1); the total derivative $\frac{dv}{dt}$ in Newton's Second Law was replaced with the partial derivative $\frac{\partial v}{\partial t}$. Because $v(x, t)$ is a function of space x as well as time t, the true acceleration of the air particles is given by [19]

$$\frac{dv}{dt} = \frac{\partial v}{\partial t} + v \frac{\partial v}{\partial x} \tag{4.2}$$

and therefore Equation (4.1) is correctly written as

$$-\frac{\partial p}{\partial x} = \rho \left(\frac{\partial v}{\partial t} + v \frac{\partial v}{\partial x} \right) \tag{4.3}$$

---

[5] Although the velocity of particles can change across the tube, we let $v(x, t)$ denote the average velocity in the cube, which becomes exact in the limit as $\Delta x \to 0$ [2].

which is a *nonlinear* equation in the variable v because the particle velocity v multiplies $\frac{\partial v}{\partial x}$. Consequently, it is difficult to determine a general solution. The approximation Equation (4.1) is accurate when the correction term $v\frac{\partial v}{\partial x}$ is small relative to $\rho\frac{\partial v}{\partial t}$. Typically, in speech production it is assumed that the particle velocity is very small and therefore the correction term introduces only second-order effects.[6] This approximation, in vector form, rules out rotational or jet flow [19], and thus the possibility of vortices along the oral cavity, as alluded to in Chapter 3 and further described in Chapter 11.

Equation (4.1) takes us halfway to the wave equation. Completing the derivation requires the use of the Gas Law and Conservation of Mass principle which can be shown to result in the relation [2]

$$-\frac{\partial p}{\partial t} = \rho c^2 \frac{\partial v}{\partial x} \tag{4.4}$$

where c is the velocity of sound and ρ is the air density assumed to be constant.[7] The pair of equations Equation (4.1) and Equation (4.4) represents one form of the wave equation. A second form is obtained by differentiating Equations (4.1) and (4.4) by x and t, respectively:

$$-\frac{\partial p^2}{\partial x^2} = \rho \frac{\partial^2 v}{\partial x \partial t}$$

$$-\frac{\partial p^2}{\partial t^2} = \rho c^2 \frac{\partial^2 v}{\partial x \partial t}$$

which, when combined to eliminate the mixed partials, can be written as a second-order partial differential equation in pressure only:

$$\frac{\partial p^2}{\partial x^2} = \frac{1}{c^2}\frac{\partial p^2}{\partial t^2}. \tag{4.5}$$

Likewise, the above equation pair can be alternatively combined to form the second-order partial differential equation in velocity only:

$$\frac{\partial v^2}{\partial x^2} = \frac{1}{c^2}\frac{\partial v^2}{\partial t^2}. \tag{4.6}$$

The alternate forms of the wave equation given by Equations (4.5) and (4.6) describe the pressure and velocity of air particles, respectively, as a function of position and time. In summary, the two different wave equation pairs are approximately valid under the following assumptions: (1) The medium is homogeneous (constant density), (2) The pressure change across a small distance can be linearized, (3) There is no viscosity of air particles, (4) The air particle velocity

---

[6] Portnoff [26] argues that when the DC component of the particle velocity v, i.e., the steady *net flow* component to v, is small relative to the speed of sound, then the correction term is negligible. The conditions under which the steady flow component of particle velocity through the vocal folds and within the vocal tract is small relative to the speed of sound are discussed in Chapter 11.

[7] As with Equation (4.1), the correct form of Equation (4.4) involves a nonlinear correction term and can be shown to be of the form $-\frac{\partial p}{\partial t} = \rho c^2 \frac{\partial v}{\partial x} - v\frac{\partial p}{\partial x}$ [19]. As before, the correction term is negligible if we assume the velocity multiplier is small.

is small (implying that the full derivative in Equation (4.3) is not necessary), and (5) Sound is adiabatic.

Our goal is to predict the speech sound pressure that we measure at the output of the lips. This must account for various sound sources, time and space variations of the vocal tract, nasal coupling, radiation at the lips, and energy losses. There are also other effects such as nonlinear coupling between airflow through the glottis and pressure in the vocal tract, and rotational flows that may arise, for example, with particle shearing pressure or when the particle velocity is not small enough to make nonlinear correction terms negligible. A formal theory of sound propagation for the complete configuration with all such considerations does not exist. Nevertheless, we can gain much insight into speech sound production with simplified analog models.

4.3   Uniform Tube Model

4.3.1   Lossless Case

In this section, following the development of Rabiner and Schafer [28], we use the wave equation to derive pressure and velocity relations in the uniform lossless tube of Figure 4.4. The uniform tube approximates an oral cavity with a roughly constant cross-section, the moving piston provides a model for glottal airflow velocity, and the open tube end represents the open lips.[8] This simple configuration approximates the vocal tract shape for the vowel /A/, as in the word "<u>up</u>." The tube has a time- and space-invariant cross-section $A(x, t) = A$ and planar sound waves are assumed to propagate longitudinally along the x axis. We assume, for the moment, that there is no friction along the walls of the tube. In reality, there is friction and the velocity is maximum at the center of the tube because the (minutely) corrugated surface of the tube will stop the wave and give zero velocity along the walls. The tube is open on the right side at $x = l$, l being the tube length. Recalling that $p(x, t)$ represents the incremental pressure about the average (atmospheric) pressure, at the open end of the tube, we assume there are no variations in air pressure, i.e., $p(x = l, t) = 0$, but there are variations in particle velocity. This open-tube configuration is analogous to a short-circuit in electrical

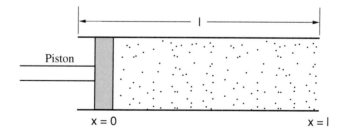

Figure 4.4   Uniform lossless tube. The piston provides an ideal airflow velocity source, and the pressure at the end of the tube is assumed zero.

---

[8] The bend in the vocal tract must also be taken into account. Sondhi [29] has found, however, that the curvature changes the location of resonances by only a few percent.

circuits. On the left end of the tube, the moving piston provides an ideal particle velocity source $v(x, t)$, i.e., the piston moves independently of pressure. This is analogous to an ideal current source in electrical circuits. We shortly further describe and refine these acoustic/electric analogies.

In the case of one-dimensional (planar) sound propagation in confined areas, it is convenient to use the velocity of a volume of air rather than particle velocity. We denote *volume velocity* by $u(x, t)$ and define it as the rate of flow of air particles perpendicularly through a specified area. Therefore, the relation between volume velocity and particle velocity for our uniform tube is given by $u(x, t) = Av(x, t)$. We can then rewrite the various forms of the wave equation in terms of volume velocity using the relation $v(x, t) = u(x, t)/A$. In particular, the second-order differential equation, Equation pair (4.5), (4.6), remains intact except for a replacement of $v(x, t)$ by $u(x, t)$, while the Equation pair (4.1), (4.4) becomes

$$-\frac{\partial p}{\partial x} = \frac{\rho}{A}\frac{\partial u}{\partial t}$$
$$-\frac{\partial u}{\partial x} = \frac{A}{\rho c^2}\frac{\partial p}{\partial t}.$$

$$(4.7)$$

We can show that two solutions of the following form (Exercise 4.1) satisfy Equation (4.7):

$$u(x, t) = u^+(t - x/c) - u^-(t + x/c)$$
$$p(x, t) = \frac{\rho c}{A}[u^+(t - x/c) + u^-(t + x/c)]$$

$$(4.8)$$

which represent velocity and pressure, each of which is comprised of a forward- and backward-traveling wave (Figure 4.5), denoted by the superscript $+$ and $-$, respectively. To see the forward-backward traveling wave property of Equation (4.8), consider a velocity wave at

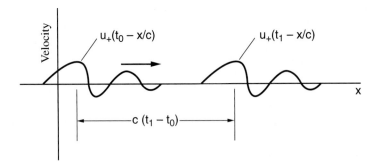

Figure 4.5 Forward-traveling velocity waves. The quantity $c(t_1 - t_0)$ is the distance moved in the time $t_1 - t_0$, where $t_1 > t_0$ and where c is the speed of sound. Backward-traveling waves are similar, but moving toward the left.

time t_0 and time t_1. To show that the wave travels a distance $c(t_1 - t_0)$, consider the forward going wave

$$u^+(t_1 - x/c) = u^+\left(t_0 + \frac{(t_1 - t_0)c}{c} - x/c\right)$$

$$= u^+\left(t_0 + \frac{(t_1 - t_0)c - x}{c}\right)$$

$$= u^+\left(t_0 - \frac{x - x_0}{c}\right)$$

where $x_0 = (t_1 - t_0)c$. Thus, the wave at time t_1 is the wave at time t_0 shifted in space by x_0 to the right. A similar argument can be made for the backward traveling wave. Therefore, the forward and backward waves propagate without change in shape.

It is instructive to observe that the wave equation for plane-wave sound propagation is similar to that for plane-wave propagation along an electrical transmission line, where voltage and current are analogous to pressure and volume velocity [2],[28]. The voltage $e(x, t)$ and current $i(x, t)$ for a lossless transmission line are related by the coupled equations

$$-\frac{\partial e}{\partial x} = L\frac{\partial i}{\partial t}$$

$$-\frac{\partial i}{\partial x} = C\frac{\partial e}{\partial t} \tag{4.9}$$

where L is inductance and C is capacitance per unit length of the transmission line. Figure 4.6 illustrates the acoustic and electrical analogies where $\frac{\rho}{A}$, defined as the *acoustic inductance*, is analogous to the electrical inductance L, and $\frac{A}{\rho c^2}$, defined as the *acoustic capacitance*, is analogous to the electrical capacitance C. These quantities reflect the "inertia" (mass or density) and the "springiness" (elasticity) of the air medium, respectively. In the context of electrical transmission lines, steady-state solutions are often written in the frequency domain using a complex sinewave representation. For example, a sinusoidal current wave of frequency Ω, traveling forward at speed c in the transmission line medium, is written as $i(x, t) = I(x, \Omega)e^{j\Omega(t-x/c)}$. Motivated by this electrical/acoustic analogy, we will use similar representations in describing pressure and volume velocity in acoustic propagation.

We have seen above the general form of the solution for a uniform-tube configuration. In order to determine a specific solution, we observe that at the boundary $x = 0$ the volume velocity is determined by the sinusoidally vibrating piston, i.e., $u(0, t) = U(\Omega)e^{j\Omega t}$, where by the above convention we use the complex form of the real driving velocity $\text{Re}[U(\Omega)e^{j\Omega t}]$. This is analogous to a sinusoidal current source for a transmission line. We write the source as a function of frequency because we are ultimately interested in the frequency response of the uniform tube. Because this volume velocity represents the glottal airflow velocity, we also write this boundary condition as $u(0, t) = u_g(t) = U_g(\Omega)e^{j\Omega t}$. The second boundary condition is that the pressure is zero at the tube termination (the lips), i.e., $p(l, t) = 0$, which is analogous to placing a short circuit at the end of a transmission line. We will see later in this section that neither of these boundary conditions is accurate in practice with a real oral tract, i.e., the

Acoustic	Electric
$-\dfrac{\partial p}{\partial x} = \dfrac{\rho}{A}\dfrac{\partial u}{\partial t}$	$-\dfrac{\partial v}{\partial x} = L\dfrac{\partial i}{\partial t}$
$-\dfrac{\partial u}{\partial x} = \dfrac{A}{\rho c^2}\dfrac{\partial p}{\partial t}$	$-\dfrac{\partial i}{\partial x} = C\dfrac{\partial v}{\partial t}$
p – Pressure	v – Voltage
u – Volume Velocity	i – Current
ρ/A – Acoustic Inductance	L – Electric Inductance
$A/(\rho c^2)$ – Acoustic Capacitance	C – Electric Capacitance

Figure 4.6   Acoustic tube/electric transmission line analogies. Inductance and capacitance represent quantities in per-unit length.

SOURCE: L.R. Rabiner and R.W. Schafer, *Digital Processing of Speech Signals* [28]. ©1978, Pearson Education, Inc. Used by permission.

glottal airflow velocity source is not ideal, and there is energy loss due to radiation from the lips, resulting in a nonzero pressure drop at the lips.

We have already seen the general solution to the wave equation in Equation (4.8). Because our velocity driving function is sinusoidal,[9] we assume the specific velocity and pressure wave solutions to be of the form

$$u(x,t) = k^+ e^{j\Omega(t-x/c)} - k^- e^{j\Omega(t+x/c)}$$

$$p(x,t) = \frac{\rho c}{A}[k^+ e^{j\Omega(t-x/c)} + k^- e^{j\Omega(t+x/c)}] \tag{4.10}$$

which are sinewaves traveling forward and backward in time and where k^+ and k^- represent the amplitudes of the forward- and backward-going waves, respectively.

---

[9] Observe that the vibrating piston can take on a more general form which, from the Fourier transform pair for continuous-time signals, can be written as a sum (integral) of complex exponentials of the form $U(\Omega)e^{j\Omega t}$, i.e., $u(0,t) = \frac{1}{2\pi}\int_{-\infty}^{\infty} U(\Omega)e^{j\Omega t}\,d\Omega$. When the uniform tube configuration is linear, the solution can be written as a sum of the individual solutions to the differential components $U(\Omega)e^{j\Omega t}\,d\Omega$. With the boundary condition at the tube termination of zero pressure $p(l,t) = 0$, the uniform tube can be shown to be linear (Exercise 4.2).

The next objective, given our two boundary conditions, is to solve for the unknown constants k^+ and k^-. The boundary condition at the source gives

$$u(0, t) = k^+ e^{j\Omega t} - k^- e^{j\Omega t} = U_g(\Omega)e^{j\Omega t}$$

resulting in

$$k^+ - k^- = U_g(\Omega). \tag{4.11}$$

Likewise, the boundary condition at the open end gives

$$p(l, t) = \frac{\rho c}{A}[k^+ e^{j\Omega(t-l/c)} + k^- e^{j\Omega(t+l/c)}] = 0$$

yielding

$$k^+ e^{-j\Omega l/c} + k^- e^{j\Omega l/c} = 0. \tag{4.12}$$

Solving simultaneously for the two expressions in k^+ and k^-, from Equation (4.11), $k^- = -U_g(\Omega) + k^+$, which we substitute into Equation (4.12) to obtain

$$k^+ e^{-j\Omega l/c} - [U_g(\Omega) - k^+]e^{j\Omega l/c} = 0,$$

which is rewritten as

$$k^+ = \frac{U_g(\Omega)e^{j\Omega l/c}}{e^{-j\Omega l/c} + e^{j\Omega l/c}}. \tag{4.13}$$

Likewise, from Equation (4.11), $k^+ = U_g(\Omega) + k^-$, which we substitute into Equation (4.12) to obtain

$$[U_g(\Omega) + k^-]e^{-j\Omega l/c} + k^- e^{j\Omega l/c} = 0,$$

which is rewritten as

$$k^- = \frac{-U_g(\Omega)e^{-j\Omega l/c}}{e^{j\Omega l/c} + e^{-j\Omega l/c}}. \tag{4.14}$$

Substituting Equations (4.13) and (4.14) into the Equation pair (4.10), we obtain

$$u(x, t) = \frac{\cos[\Omega(l-x)/c]}{\cos[\Omega l/c]}U_g(\Omega)e^{j\Omega t}$$

$$p(x, t) = j\frac{\rho c}{A}\frac{\sin[\Omega(l-x)/c]}{\cos[\Omega l/c]}U_g(\Omega)e^{j\Omega t} \tag{4.15}$$

which expresses the relationship between the sinusoidal volume velocity source and the pressure and volume velocity at any point in the tube. The particular solution is referred to as *standing waves* because the forward- and backward-traveling waves have added to make the wave shape stationary in time; the difference $t - \frac{x}{c}$ or the sum $t + \frac{x}{c}$ no longer appears as an argument in the complex exponential function. Since the volume velocity and pressure solutions are in complex form, we can think of the cosine and sine multiplier functions as the *envelope* of the volume velocity and pressure functions along the tube, given respectively by $\frac{\cos[\Omega(l-x)/c]}{\cos[\Omega l/c]}$ and

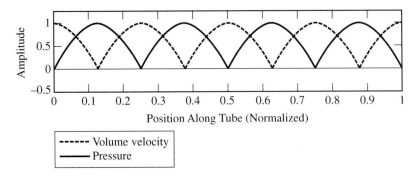

Figure 4.7 Envelopes of pressure and volume velocity standing waves resulting from sound propagation in a uniform tube with zero pressure termination.

$\frac{\rho c}{A} \frac{\sin[\Omega(l-x)/c]}{\cos[\Omega l/c]}$. We see that the two envelope functions are 90° out of phase (Figure 4.7), and so velocity and pressure are "orthogonal" in space. The two functions are also orthogonal in time because of the imaginary j multiplier of the pressure function. We can see this relation by writing the real counterpart of Equation (4.15):

$$\text{Re}[u(x, t)] = \frac{\cos[\Omega(l - x)/c]}{\cos[\Omega l/c]} U_g(\Omega) \cos(\Omega t)$$

$$\text{Re}[p(x, t)] = \frac{\rho c}{A} \frac{\cos[\Omega(l - x)/c + \pi/2]}{\cos[\Omega l/c]} U_g(\Omega) \cos(\Omega t + \pi/2), \qquad (4.16)$$

which reveals that at any time instant the velocity and pressure variations are 90° out of phase. An interesting consequence of this phase relation is that the acoustic kinetic and potential energies along the tube are also "orthogonal" (Exercise 4.3).

Consider now the specific relationship between the volume velocity at the open end of the tube (the lips) and the volume velocity at the source (the glottis). From the volume velocity component of Equation (4.15), we have at $x = l$

$$u(l, t) = \frac{\cos[\Omega(l - l)/c]}{\cos[\Omega l/c]} U_g(\Omega) e^{j\Omega t}$$

$$= \frac{1}{\cos[\Omega l/c]} U_g(\Omega) e^{j\Omega t}. \qquad (4.17)$$

Let $U(l, \Omega) = \frac{1}{\cos[\Omega l/c]} U_g(\Omega)$, i.e., the complex amplitude at the open tube end for complex input $U_g(\Omega) e^{j\Omega t}$. Then we can write

$$\frac{U(l, \Omega)}{U_g(\Omega)} = \frac{1}{\cos(\Omega l/c)}$$

$$= V_a(\Omega) \qquad (4.18)$$

where $V_a(\Omega)$ denotes the frequency response relating input to output volume velocities of the analog uniform-tube model of our simple speech production configuration. The roots of the denominator of $V_a(\Omega)$, i.e., the uniformly-spaced frequencies at which the peak (infinite) amplitude occurs, are the resonances of the analog system. These roots are given by solutions to $\frac{\Omega l}{c} = k\frac{\pi}{2}$, where k is an integer or

$$\Omega = k\frac{\pi}{2}\frac{c}{l}, \qquad k = 1, 3, 5 \ldots$$

EXAMPLE 4.2   Consider the frequency response in Equation (4.18) and suppose that a uniform tube has length $l = 35$ cm. With the speed of sound $c = 350$ m/s, the roots of our particular uniform tube are

$$f = \frac{\Omega}{2\pi} = \frac{k}{8}2000$$
$$= 250, 750, 1250, \ldots$$

The frequency response is shown in Figure 4.8. Observe that the resonances have zero bandwidth since there is no energy loss in the system (the reader should argue this using the definitions of bandwidth in Chapter 2). As we decrease the length l of the tube, the resonant frequencies increase. We have all had the experience of filling a tall bottle under a faucet. The bottle (tube) is open at the top, and we can think of the water hitting the bottom as providing a volume velocity source. As the water runs, the open volume decreases, reducing the tube length and increasing the aurally perceived resonances. An example is shown in Figure 4.8 in which the effective bottle length decreases from $l = 35$ cm to $l = 17.5$ cm. Resonances can also be thought of as "natural frequencies" which are excited by a disturbance. Why should a slug of air have natural frequencies? As we have seen, air has mass and springiness, and, as with lumped mechanical or electrical systems having inductance and capacitance, such a "mass-spring" system is more responsive to certain frequencies. An alternate analogy is the transmission line that we reviewed earlier as having inductance and capacitance per unit length; the acoustic "line" also has inductance and capacitance per unit length.   ▲

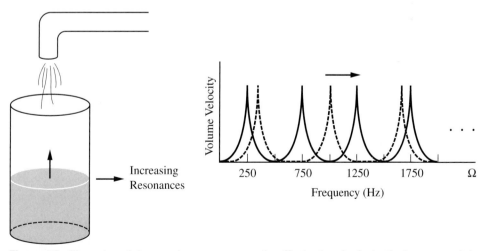

Figure 4.8 Illustration of time-varying resonances as the effective length of a bottle decreases as it is filled with running water (Example 4.2). We aurally perceive the increasing resonances as the tube length decreases.

An important quantity relating pressure and volume velocity is their ratio along the tube. We call this quantity the *acoustic impedance*, analogous to electric impedance given by voltage divided by current, and from Equation (4.15) we can express it as

$$Z_A(\Omega) = \frac{p(x,t)}{u(x,t)}$$

$$= j\frac{\rho c}{A}\tan[\Omega(l-x)/c].$$

When the tube is very short with length denoted by Δx, using a Taylor series expansion of the tangent function, we obtain the approximation (looking into the tube from $x = 0$)

$$Z_A(\Omega) \approx j\frac{\rho \Delta x}{A}\Omega$$

where $\frac{\rho \Delta x}{A}$ can be thought of as an acoustic mass, and the short plug of air acts as a lumped acoustic inductance[10] where $\frac{\rho}{A}$ is inductance per unit length. As with electrical and mechanical impedance, a large value of $Z_A(\Omega)$ implies that a large pressure is required to obtain a specified volume velocity, i.e., the medium "impedes" the flow of particles when acted upon. Impedance will become particularly important at glottal and lip boundaries because the impedance gives the pressure and velocity relations, coupled with the wave equation, that must be maintained in crossing these boundaries.

Returning to our solution Equation (4.18), we can change the frequency response to a transfer function by replacing Ω by $\frac{s}{j}$ to obtain[11]

$$V_a(s) = \frac{1}{\cos(\frac{sl}{jc})}$$

$$= \frac{1}{(e^{sl/c}+e^{-sl/c})/2}$$

$$= \frac{2e^{-sl/c}}{1+e^{-2sl/c}}. \tag{4.19}$$

From functional analysis, if a complex function $f(x)$ of a complex variable x meets certain restrictions, then, using a Taylor expansion, we can write $f(x)$ in terms of an infinite product of its zero factors, i.e., in terms of factors of the form $(x - x_k)$, where x_k is a zero. Flanagan has used this relation to show that Equation (4.19) can be written as [8]

$$V_a(s) = \frac{1}{\prod_{k=1}^{\infty}(s-s_k)(s-s^*_k)}$$

---

[10] When the short tube is closed on the right, rather than open, then its acoustic impedance is of the form $Z_A \approx -j\frac{\rho c^2}{\Omega A \Delta x}$ and acts rather as a lumped acoustic capacitance (Exercise 4.8).

[11] The Laplace transform of the continuous-time impulse response gives its transfer function. Recall that in the Laplace transform the complex variable $s = \sigma + j\Omega$. When s is evaluated on the imaginary axis $j\Omega$, i.e., $\sigma = 0$, then $s = j\Omega$. Therefore, $\Omega = \frac{s}{j}|_{\sigma=0}$.

where, because the corresponding response is real, the poles occur in complex conjugate pairs. There are then an infinite number of poles that occur on the $j\Omega$ axis and can be written as

$$s_n = \pm j \left[\frac{(2k + 1)\pi c}{2l} \right], \qquad k = 0, \pm 1, \pm 2, \ldots$$

corresponding to the resonant frequencies of the uniform tube. The poles fall on the $j\Omega$ axis, and thus the frequency response at resonant frequencies is infinite in amplitude, because there is no energy loss in the simple configuration. In particular, it was assumed that friction of air particles does not occur along the vocal tract walls, that heat is not conducted away through the vocal tract walls, and that the walls are rigid and so energy was not lost due to their movement. In addition, because we assumed boundaries consisting of an ideal volume velocity source and zero pressure at the lips, no energy loss was entailed at the input or output to the uniform tube. In practice, on the other hand, such losses do occur and will not allow a frequency response of infinite amplitude at resonance. Consider by analogy a lumped electrical circuit with a resistor, capacitor, and inductor in series. Without the resistor to introduce thermal loss into the network, the amplitude of the frequency response at resonance equals infinity. We investigate these forms of energy loss in the following two sections. Energy loss may also occur due to rotational flow, e.g., vortices and other nonlinear effects that were discarded in the derivation of the wave equation, such as from Equation (4.3). These more complex forms of energy loss are discussed in Chapter 11.

4.3.2 Effect of Energy Loss

Energy loss can be described by differential equations *coupled* to the wave equation that describes the pressure/volume velocity relations in the lossless uniform tube. These coupled equations are typically quite complicated and a closed-form solution is difficult to obtain. The solution therefore is often found by a numerical simulation, requiring a discretization in time and in space along the x variable.

Wall Vibration — Vocal tract walls are pliant and so can move under pressure induced by sound propagation in the vocal tract. To predict the effect of wall vibration on sound propagation, we generalize the partial differential equations of the previous section to a tube whose cross-section is nonuniform and time-varying. Under assumptions similar to those used in the derivation of Equation (4.7), as well as the additional assumption that the cross-section of a nonuniform tube does not change "too rapidly" in space (i.e., the x direction) and in time, Portnoff [26] has shown that sound propagation in a nonuniform tube with time- and space-varying cross-section $A(x, t)$ is given by

$$-\frac{\partial p}{\partial x} = \rho \frac{\partial (u/A)}{\partial t}$$
$$-\frac{\partial u}{\partial x} = \frac{1}{\rho c^2} \frac{\partial p A}{\partial t} + \frac{\partial A}{\partial t} \qquad (4.20)$$

which, for a uniform time-invariant cross-section, reduces to the previous Equation pair (4.1), (4.4).

Portnoff then assumed that small, differential pieces of the surface of the wall, denoted by $d\sum$, are independent, i.e., "locally reacting," and that the mechanics of each piece are modeled by a mass m_w, spring constant k_w, and damping constant b_w per unit surface area illustrated in Figure 4.9. Because the change in cross-section due to pressure change is very small relative to the average cross-section, Portnoff expressed the cross-section along the dimension x as $A(x, t) = A_o(x, t) + \Delta A(x, t)$, where $\Delta A(x, t)$ is a linear perturbation about the average $A_o(x, t)$. Using the lumped parameter model in Figure 4.9, the second-order differential equation for the perturbation term is given by

$$m_w \frac{d^2 \Delta A}{dt^2} + b_w \frac{d \Delta A}{dt} + k_w \Delta A = p(x, t) \tag{4.21}$$

where the mass, damping, and stiffness per unit surface area of the wall are assumed to be constant, i.e., not a function of the variable x along the tube, and where we have used the relation $\Delta A(x, t) = S_o(x, t)r$, $S_o(x, t)$ being the average vocal tract perimeter at equilibrium and r being the displacement of the wall perpendicular to the wall. (The reader should perform the appropriate replacement of variables in the model equation of Figure 4.9.) The pressure and perturbation relations in Equation (4.21) are coupled to the relations in the modified wave equation pair, Equation (4.20), and thus the three equations are solved simultaneously, under the two boundary conditions of the known volume velocity source $u(0, t)$ and the output pressure $p(l, t) = 0$. Discarding second-order terms in the expansion for $\frac{u}{A} = \frac{u}{A_o + \Delta A}$, the three coupled equations in variables p, u, and ΔA for the uniform tube reduce approximately to [26]

$$-\frac{\partial p}{\partial x} = \frac{\rho}{A_o} \frac{\partial u}{\partial t}$$

$$-\frac{\partial u}{\partial x} = \frac{A_o}{\rho c^2} \frac{\partial p}{\partial t} + \frac{\partial \Delta A}{\partial t}$$

$$p = m_w \frac{d^2 \Delta A}{dt^2} + b_w \frac{d \Delta A}{dt} + k_w \Delta A \tag{4.22}$$

where we assume that the average cross-section A_o is constant, i.e., the vocal tract is time-invariant being held in a particular configuration.

Under the steady-state assumption that sound propagation has occurred long enough so that transient responses have died out, and given that the above three coupled equations are linear and time-invariant, an input $u_g(t) = u(0, t) = U(\Omega)e^{j\Omega t}$ results in solutions of the form $p(x, t) = P(x, \Omega)e^{j\Omega t}$, $u(x, t) = U(x, \Omega)e^{j\Omega t}$, and $\Delta A(x, t) = \Delta \hat{A}(x, \Omega)e^{j\Omega t}$ along the tube. Therefore, the time-dependence of the three coupled equations is eliminated; the resulting three coupled equations are then functions of the variable x for each frequency Ω:

$$-\frac{\partial P(x, \Omega)}{\partial x} = \frac{\rho}{A_o} \Omega U(x, \Omega)$$

$$-\frac{\partial U(x, \Omega)}{\partial x} = \frac{A_o}{\rho c^2} P(x, \Omega) + \Omega \Delta \hat{A}(x, \Omega)$$

$$P(x, \Omega) = -\Omega^2 m_w \Delta \hat{A}(x, \Omega) + j\Omega b_w \Delta \hat{A}(x, \Omega) + k_w \Delta \hat{A}(x, \Omega). \tag{4.23}$$

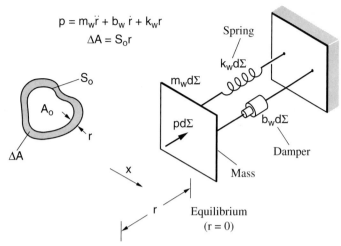

$$p = m_w \ddot{r} + b_w \dot{r} + k_w r$$
$$\Delta A = S_o r$$

Figure 4.9 Mechanical model of differential surface element $d\sum$ of vibrating wall. Adjacent wall elements are assumed uncoupled.

SOURCE: M.R. Portnoff, *A Quasi-One-Dimensional Digital Simulation for the Time-Varying Vocal Tract* [26]. ©1973, M.R. Portnoff and the Massachusetts Institute of Technology. Used by permission.

Portnoff has solved these coupled equations using standard numerical simulation techniques[12] invoking finite-difference approximations to the spatial partial differential operator $\frac{\partial}{\partial x}$. Suppose that the length of the tube is l. Then we seek the acoustic pressure and volume velocity at N equally spaced points over the interval $[0, l]$. In particular, $\frac{\partial}{\partial x}$ was approximated by a central difference with averaging; that is, for a partial differential with respect to x of a function $f(x)$, first compute the central difference over a spatial difference $\Delta x = \frac{l}{N-1}$:

$$g[n] = \frac{1}{2\Delta x}(f[(n-1)\Delta x] - f[(n+1)\Delta x])$$

and then perform a first backward average:

$$\frac{\partial}{\partial x} f(t) \approx \frac{1}{2}(g[n\Delta x] + g[(n-1)\Delta x]).$$

It can be shown that this transformation from the continuous- to the discrete-space variable is equivalent to a bilinear transformation.[13] This partial differential approximation leads to

---

[12] We can also obtain in this case an approximate closed-form, frequency-domain solution [26],[28], but we choose here to describe the numerical simulation because this solution approach is the basis for later-encountered nonlinear coupled equations.

[13] The bilinear transformation for going from continuous to discrete time is given by the s-plane to z-plane mapping of $s = \frac{2}{\Delta t}[\frac{z+1}{z-1}]$ and has the property that the imaginary axis in the s-plane maps to the unit circle in the z-plane [22]. The frequency Ω in continuous time is related to the discrete-time frequency by the tangential mapping $\frac{\Delta t \Omega}{2} = \tan(\frac{\omega}{2})$. Likewise, the bilinear mapping in going from continuous space to discrete space is given by $K = \frac{2}{\Delta x}[\frac{k+1}{k-1}]$ and has the same tangential mapping as the time variable.

$2N - 2$ equations in $2N$ unknown pressure and velocity variables. The two additional equations necessary for a unique solution are supplied by the boundary conditions at the lips and glottis.

The resulting frequency response $V_a(\Omega) = \frac{U(l,\Omega)}{U_g(\Omega)}$ of the numerical simulation is shown in Figure 4.10a (pressure and velocity computed at 96 samples in space along the x variable) for a uniform tube with yielding walls and no other losses, terminated in a zero pressure boundary condition [26]. The tube is 17.5 cm in length and 5 cm^2 in cross section. The yielding wall parameters are given by Flanagan [8] to be $m_w = 0.4$ gm/cm^2, $b_w = 6500$ dyne-sec/cm^3, and $k_w = 0$. The result is markedly different from the lossless case. Because of energy loss due to the wall vibration, the poles of the transfer function are no longer on the $j\Omega$ axis and so the bandwidth is nonzero. In addition, the resonant frequencies have slightly increased. Finally, these effects are most pronounced at low frequencies because the inertial mass of the wall results in less wall motion at high frequencies.

Viscosity and Thermal Loss — The effects of both viscous and thermal loss can be represented by modification of Equations (4.20) and (4.21) with the introduction of a resistive term, representing the energy loss due to friction of air particles along the wall, and a conductive term, representing heat loss through the vibrating walls, respectively [8],[26]. The resulting coupled equations are solved numerically for the steady-state condition, again using a central difference approximation to the spatial partial derivative [26]. With viscous and thermal loss only, effects are less noticeable than with wall vibration loss, being more pronounced at high frequencies where more friction and heat are generated. The addition of viscosity and thermal conduction to the presence of vibrating walls yields slight decreases in resonant frequencies and some broadening of the bandwidths. Figure 4.10b gives the result of Portnoff's numerical simulation with all three losses for the uniform tube with zero pressure termination [26].

4.3.3 Boundary Effects

We have up to now assumed that the pressure at the lips is zero and that the volume velocity source is ideal, and thus there is no energy loss at the output or input of our uniform tube. A more realistic picture for voiced speech is shown in Figure 4.11 which shows the addition of a glottal and radiation load.

A description of sound radiation at the lips and diffraction about the head is quite complicated. The effect of radiation, however, can be simplified by determining the acoustic impedance "seen" by the vocal tract toward the outside world. An approximate radiation impedance can be obtained by determining the impedance felt by a piston in a rigid sphere representing the head. The expression of this impedance as determined by Morse and Ingard [19], however, cannot be written in closed form. In a second approximation, the piston is assumed small compared to the sphere diameter, modeled by a piston set in an infinitely large wall (Figure 4.11). For this configuration, it has been shown [2],[8] that the acoustic impedance consists of a radiation resistance

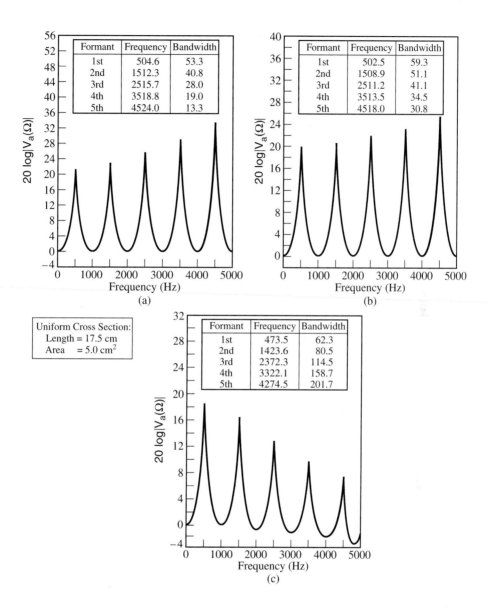

Figure 4.10 Frequency response of uniform tube with (a) vibrating walls with $p(l, 0) = 0$; (b) vibrating walls, and viscous and thermal loss with $p(l, 0) = 0$; (c) vibrating walls, viscous and thermal loss, and radiation loss [26],[28]. 3 dB bandwidths are given.

SOURCE: M.R. Portnoff, *A Quasi-One-Dimensional Digital Simulation for the Time-Varying Vocal Tract* [26]. ©1973, M.R. Portnoff and the Massachusetts Institute of Technology. Used by permission.

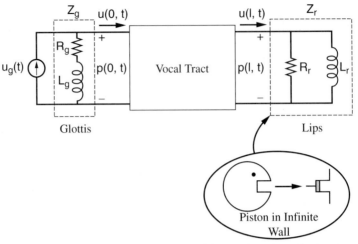

Figure 4.11 Glottal and lip boundary conditions as impedance loads, given by linearized serial and parallel electric circuit models, respectively. Piston in infinite wall model for radiation from lips is illustrated.

R_r (energy loss via sound propagation from the lips) in parallel with a radiation inductance L_r (inertial air mass pushed out at lips):

$$
Z_r(\Omega) = \frac{P(l, \Omega)}{U(l, \Omega)}
$$

$$
= \frac{1}{\frac{1}{R_r} + \frac{1}{j\Omega L_r}}
$$

$$
= \frac{j\Omega L_r R_r}{R_r + j\Omega L_r}. \tag{4.24}
$$

For the infinite baffle, Flanagan has given values of $R_r = 128/9\pi^2$ and $L_r = 8a/3\pi c$ where a is the radius of the opening and c the speed of sound.

Equation (4.24) can be represented in the time domain by a differential equation that is coupled to the wave equation solution of sound propagation within the vocal tract. As with other energy losses, Portnoff [26] has numerically simulated the effect of this coupling for the steady-state vocal tract condition. The frequency response $V_a(\Omega) = \frac{U(l,\Omega)}{U_g(\Omega)}$ with loss from the radiation load, as well as from vibrating walls, viscosity, and thermal conduction, is shown in Figure 4.10c. Consequences of the radiation load are broader bandwidths and a lowering of the resonances, the major effect being on the higher frequencies because radiation loss is greatest in this region. We can see this effect through Equation (4.24). For very small $\Omega \approx 0$,

$$
Z_r \approx 0
$$

so that the radiation load acts as a short circuit with $p(l, t) \approx 0$. For very large Ω with $\Omega L_r \gg R_r$

$$
Z_r \approx \frac{j\Omega L_r R_r}{j\Omega L_r} = R_r
$$

and so the radiation load takes on the resistive component at high frequencies. Because the energy dissipated in radiation is proportional to the real part of the complex impedance Z_r [2],[28] and because the real part of Z_r monotonically increases with the resistance R_r (the reader should confirm this monotonicity), we deduce that the greatest energy loss, and thus the greatest formant bandwidth increase, from radiation is incurred at high frequencies.

Our final boundary condition is the glottal source and impedance. This is the most difficult addendum to our simple uniform tube because the glottal volume velocity has been shown to be nonlinearly related to the pressure variations in the vocal tract. We return to the true complexity of the glottal impedance later in this chapter; for now it suffices to use a simplification to obtain a flavor for the glottal impedance effect. By simplifying a nonlinear, time-varying two-mass vocal fold model for predicting glottal airflow (Figure 3.5 in Chapter 3), Flanagan and Ishizaka [9] proposed a linearized time-invariant glottal impedance as a resistance R_g in series with an inductance L_g:

$$Z_g(\Omega) = R_g + j\Omega L_g. \tag{4.25}$$

As illustrated in Figure 4.11, this impedance is placed in parallel with an ideal volume velocity source $u_g(t)$ that, for the case of voiced speech, is a typical glottal airflow velocity function as given in Chapter 3. Applying Kirchoff's current law for electric circuits to the glottal source and impedance in Figure 4.11, we find the modified boundary condition is given by

$$U(0, \Omega) = U_g(\Omega) - \frac{P(0, \Omega)}{Z_g(\Omega)}$$

which in the time domain is a differential equation coupled to the partial differential wave equation and other differential loss equations described above. Portnoff [26] performed a numerical simulation of this more realistic glottal boundary condition and found for steady-state a broadening of bandwidths at low resonances; this is consistent with Equation (4.25) because $Z_g(\Omega)$ approaches an open circuit for high frequencies ($Z_g(\Omega) \approx j\Omega L_g$), and approaches a pure resistance for low frequencies ($Z_g(\Omega) \approx R_g$).

We have up to now found the frequency response relating volume velocity at the lips, $U(l, \Omega)$, to input volume velocity at the glottis, $U_g(\Omega)$. In practice, we measure the *pressure* at the lips with a pressure-sensitive microphone transducer. The pressure-to-volume velocity frequency response can be found as [28]

$$\begin{aligned} H(\Omega) &= \frac{P(l, \Omega)}{U_g(\Omega)} \\ &= \frac{P(l, \Omega)}{U(l, \Omega)} \frac{U(l, \Omega)}{U_g(\Omega)} \\ &= Z_r(\Omega) V_a(\Omega) \end{aligned} \tag{4.26}$$

where the radiation impedance $Z_r(\Omega) = \frac{P(l,\Omega)}{U(l,\Omega)}$, from Figure 4.11 and Equation (4.24), introduces a highpass filter effect onto the frequency response $V_a(\Omega)$.

4.3.4 A Complete Model

From the previous sections, the complete model of the uniform tube involves boundary condition effects, as well as energy loss due to vibrating walls, viscosity, and thermal conduction. The effects can be summarized as follows:

1. Resonances are due to the vocal tract; there is some shift of the resonant center frequencies due to the various sources of energy loss.
2. Bandwidths of the lower resonances are controlled by vibrating walls and glottal impedance loss. In practice, the glottal impedance is assumed to be infinite, i.e., $Z_g = \infty$, and the sources of broadening the bandwidths of lower resonances are lumped into the vibrating walls.
3. Bandwidths of the higher resonances are controlled by radiation, viscous, and thermal loss.

The systems described in the previous sections, from which these conclusions are drawn, were numerically simulated under certain simplifying assumptions such as a fixed uniform tube (except for the inclusion of yielding walls) and linearized and time-invariant networks representing glottal and radiation boundary conditions. In reality, these conditions are only approximate. For example, the vocal tract tube varies in time as well as in space, and the glottal airflow impedance model was significantly simplified, the nonlinear two-mass vocal fold model of Figure 3.5 being a more realistic basis. As such, the time-dependence cannot be taken out of the equations, as we did, for example, in taking Equation (4.22) to Equation (4.23), and therefore the resulting coupled equations, generally nonlinear,[14] must be solved in time as well as in space. Solving these nonlinear two-dimensional partial differential equations poses a considerable challenge.

Portnoff performed a numerical simulation of an even more complete model of sound propagation in the vocal tract using Flanagan and Ishizaka's nonlinear two-mass vocal fold model and with fixed nonuniform tubes representing various vowels. This model involves coupling of the vocal tract wave equation with differential equations of loss, as well as nonlinear differential equations describing the two-mass vocal fold model of Flanagan and Ishizaka and a nonlinear volume velocity/pressure relation at the glottis to be described in Section 4.5. Unlike in the numerical simulation of Equation (4.23), the time-dependence of these coupled equations cannot be eliminated because of the presence of nonlinearity. The numerical simulation, therefore, requires discretization in time Δt as well as in the spatial variable Δx. As illustrated in Figure 4.12, the discretization occurs over a semi-infinite strip in the time-space plane where the boundary condition in space on the line $x = -l$ is given by the glottal airflow source velocity and impedance model, and the boundary condition on the line $x = 0$ is given by the radiation load. The boundary condition on the time line $t = 0$ $(0 \leq x \leq l)$ is given by assuming zero initial conditions of pressure and volume velocity. Discretization in time is performed by replacing time differentials with a central difference operator followed by a first backward average, as was the discretization in space shown earlier. This discretization is performed over a net of uniformly spaced discrete points in time and space shown in Figure 4.12[15] and results in a

---

[14] For the moment, we are not addressing all nonlinear contributions that influence airflow and pressure, as described at the end of Section 4.2.

[15] For a lossless system, with time and spatial sampling constrained by the relation $\Delta x = c\Delta t$, the partial differential Equation (4.20) is simulated *exactly* [26].

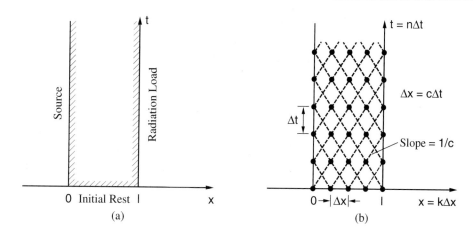

Figure 4.12   Numerical simulation specifications for a complete production model with glottal and radiation boundary conditions: (a) boundary conditions with semi-infinite strip in time-space (t, x) plane on which partial differential equations are solved; (b) rectangular net of time-space discretization.

SOURCE: M.R. Portnoff, *A Quasi-One-Dimensional Digital Simulation for the Time-Varying Vocal Tract* [26]. ©1973, M.R. Portnoff and the Massachusetts Institute of Technology. Used by permission.

set of difference equations whose solution gives values of acoustic volume velocity and pressure at points on the net.

EXAMPLE 4.3   The result of Portnoff's more complete numerical simulation for the vowel /o/ is shown in Figure 4.13 illustrating both the estimated frequency response $V_a(\Omega)$, as well as the estimated glottal volume velocity and lip pressure function [26]. The nonuniform area function was estimated from vocal tract measurements. Observe that the vocal tract impulse response is "truncated" roughly during the open phase of the vocal fold vibration, the envelope of the waveform abruptly falling at the onset of the glottal opening. In Section 4.5, we show that this envelope decay is typical and reflects a rapid decrease in the temporal envelope of primarily the first formant, and thus a corresponding increase in its bandwidth, due to a nonlinear coupling between the vocal tract pressure and glottal volume velocity.   ▲

The resulting simulations give a reasonable match[16] (with respect to formant center frequencies and bandwidths) of predicted frequency responses with those of natural vowels [26],[28]. Nevertheless, the need to invoke a time and spatial numerical simulation is cumbersome. A simpler alternative, but requiring additional approximations on pressure/volume velocity relations, is to estimate the desired transfer function $V_a(s)$ through a concatenated tube approximation of the cross-section function $A(x, t)$. This approach leads to a discrete-time model whose implementation can be made computationally efficient using digital signal processing techniques, and which is described in the following section.

---

[16] For turbulent noise sources, generated by high velocity flow through a constriction, the approximations leading to our coupled sets of differential equations are not necessarily valid [26].

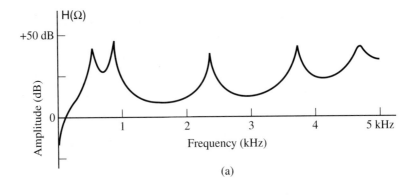

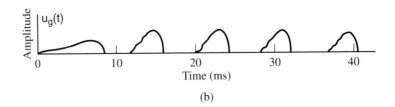

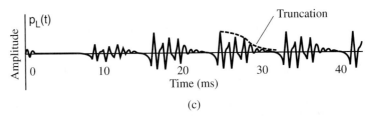

Figure 4.13 Complete numerical simulation of the vowel /o/: (a) frequency response of output lip pressure; (b) glottal airflow volume velocity; (c) lip output sound pressure [26].

4.4 A Discrete-Time Model Based on Tube Concatenation

A widely used model of sound propagation in the vocal tract during speech production is based on the assumption that the vocal tract can be represented by a concatenation of short *lossless uniform acoustic tubes*, as illustrated in Figure 4.14. We assume no energy loss within each tube and that loss is incurred only at the two boundaries, i.e., at the glottis and the lips. Furthermore, as alluded to earlier, it is typical to assume the glottal impedance to be infinite, so that loss is incurred only by way of radiation from the lips. Advantages of this model are that it is linear and easy to work with, providing a straightforward means to obtain a frequency response $V_a(\Omega) = \frac{U(l,\Omega)}{U_g(\Omega)}$, avoiding

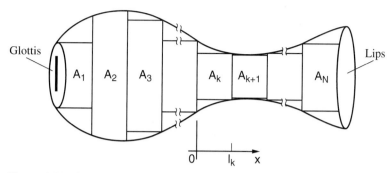

Figure 4.14   Concatenated tube model. The kth tube has cross-sectional area A_k and length l_k.

numerical simulation. Although assuming no loss in each tube segment, we will see that the radiation impedance can be modified to match observed formant bandwidths. The concatenated tube model also provides a convenient transition from a continuous-time to a discrete-time all-pole model, and the resulting all-pole model leads naturally into the linear prediction speech analysis of Chapter 5. On the other hand, although the frequency response predicted from the concatenated tube model can be made to approximately match spectral measurements, the concatenated tube model is less accurate in representing the physics of sound propagation than the coupled partial differential equation models of the previous sections. The contributions of energy loss from vibrating walls, viscosity, and thermal conduction, as well as nonlinear coupling between the glottal and vocal tract airflow, are not represented in the lossless concatenated tube model.

4.4.1 Sound Propagation in the Concatenated Tube Model

Consider the concatenated tube model in Figure 4.14 in which the kth tube has cross-sectional area A_k and length l_k. Because each tube is lossless and because we assume planar wave propagation, the pressure and volume velocity relations in each tube satisfy the wave equation, Equation (4.7), each quantity having a forward- and backward-traveling wave component as given in Equation (4.8). The volume velocity and presure solution for the kth tube is therefore written as

$$u_k(x, t) = u_k^+\left(t - \frac{x}{c}\right) - u_k^-\left(t + \frac{x}{c}\right)$$

$$p_k(x, t) = \frac{\rho c}{A_k}\left[u_k^+\left(t - \frac{x}{c}\right) + u_k^-\left(t + \frac{x}{c}\right)\right] \tag{4.27}$$

where the kth tube has length l_k, the spatial variable x falling in the range $0 \le x \le l_k$, as illustrated in Figure 4.14. To give a specific solution requires boundary conditions that are provided by the pressure and volume velocity at two adjacent tubes. In particular, we use the physical principle that pressure and volume velocity must be continuous both in time and in space everywhere in the system [28]. Therefore, at the kth/$(k + 1)$st junction we have

$$u_k(l_k, t) = u_{k+1}(0, t)$$

$$p_k(l_k, t) = p_{k+1}(0, t).$$

Applying the two boundary conditions with Equation (4.27), we then have

$$u_k^+ \left(t - \frac{l_k}{c} \right) - u_k^- \left(t + \frac{l_k}{c} \right) = u_{k+1}^+(t) - u_{k+1}^-(t)$$

$$\frac{A_{k+1}}{A_k} \left[u_k^+ \left(t - \frac{l_k}{c} \right) + u_k^- \left(t + \frac{l_k}{c} \right) \right] = u_{k+1}^+(t) + u_{k+1}^-(t). \qquad (4.28)$$

Now define $\tau_k = \frac{l_k}{c}$, which is the time of propagation down the length of the tube. Then the first equation in the Equation (4.28) pair can be written as

$$u_k^+(t - \tau_k) = u_k^-(t + \tau_k) + u_{k+1}^+(t) - u_{k+1}^-(t)$$

and substituting this expression for $u_k^+(t - \tau_k)$ into the second equation of our pair, we obtain, after some algebra,

$$u_{k+1}^+(t) = \left[\frac{2A_{k+1}}{A_{k+1} + A_k} \right] u_k^+(t - \tau_k) + \left[\frac{A_{k+1} - A_k}{A_{k+1} + A_k} \right] u_{k+1}^-(t). \qquad (4.29)$$

Then subtracting the top from the bottom component of the above modified equation pair, we obtain, after some rearranging,

$$u_k^-(t + \tau_k) = - \left[\frac{A_{k+1} - A_k}{A_{k+1} + A_k} \right] u_k^+(t - \tau_k) + \left[\frac{2A_k}{A_{k+1} + A_k} \right] u_{k+1}^-(t). \qquad (4.30)$$

Equations (4.29) and (4.30) illustrate the general rule that at a discontinuity along x in the area function $A(x, t)$ there occur *propagation and reflection of the traveling wave*, and so in each uniform tube, part of a traveling wave propagates to the next tube and part is reflected back (Figure 4.15). This is analogous to propagation and reflection due to a change in the impedance along an electrical transmission line.

From Equation (4.29), we can therefore interpret $u_{k+1}^+(t)$, which is the forward-traveling wave in the $(k + 1)$st tube at $x = 0$, as having two components:

1. A portion of the forward-traveling wave from the previous tube, $u_k^+(t - \tau_k)$, propagates across the boundary.
2. A portion of the backward-traveling wave within the $(k+1)$st tube, $u_{k+1}^-(t)$, gets reflected back.

Observe that $u_k^+(t - \tau_k)$ occurs at the $(k + 1)$st junction and equals the forward wave $u_k^+(t)$, which appeared at the kth junction τ_k seconds earlier. This interpretation is applied generally to the forward/backward waves at each tube junction. Likewise, from Equation (4.30), we can interpret $u_k^-(t + \tau_k)$, which is the backward-traveling wave in the kth tube at $x = l_k$, as having two components:

1. A portion of the forward-traveling wave in the kth tube, $u_k^+(t - \tau_k)$, gets reflected back.
2. A portion of the backward-traveling wave within the $(k + 1)$st tube, $u_{k+1}^-(t)$, propagates across the boundary.

Figure 4.15 further illustrates the forward- and backward-traveling wave components at the tube boundaries.

The above interpretations allow us to see how the forward- and backward-traveling waves propagate across a junction. As a consequence, we can think of $\frac{A_{k+1}-A_k}{A_{k+1}+A_k}$ as a *reflection coefficient* at the kth junction, i.e., the boundary between the kth and $(k+1)$st tube, that we denote by r_k:

$$r_k = \frac{A_{k+1} - A_k}{A_{k+1} + A_k} \tag{4.31}$$

which is the amount of $u_{k+1}^-(t)$ [or $u_k^+(t-\tau_k)$] that is reflected at the kth junction. It is straightforward to show that because $A_k > 0$, it follows that $-1 \leq r_k \leq 1$. From our definition Equation (4.31), we can write Equations (4.29) and (4.30) as

$$u_{k+1}^+(t) = (1 + r_k)u_k^+(t - \tau_k) + r_k u_{k+1}^-(t)$$
$$u_k^-(t + \tau_k) = -r_k u_k^+(t - \tau_k) + (1 - r_k)u_{k+1}^-(t) \tag{4.32}$$

which can be envisioned by way of a signal flow graph illustrated in Figure 4.16a. We will see shortly that these equations *at the junctions* lead to the volume velocity relations for multiple concatenated tubes. Before doing so, however, we consider the boundary conditions at the lips and at the glottis.

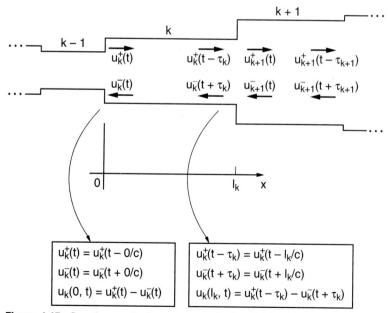

Figure 4.15 Sound waves in the concatenated tube model consist of forward- and backward-going traveling waves that arise from reflection and transmission at a tube junction.

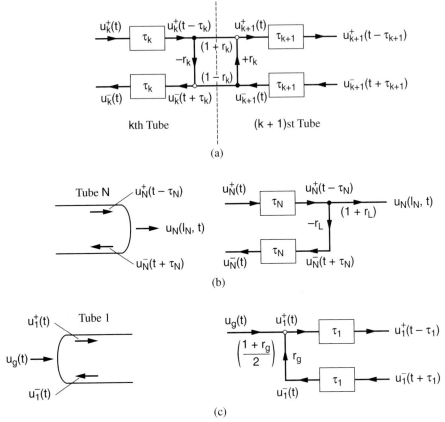

Figure 4.16 Signal flow graphs of (a) two concatenated tubes; (b) lip boundary condition; (c) glottal boundary condition. An open circle denotes addition.

SOURCE: L.R. Rabiner and R.W. Schafer, *Digital Processing of Speech Signals* [28]. ©1978, Pearson Education, Inc. Used by permission.

Suppose there are a total of N tubes. Then the boundary conditions at the lips relate pressure $p_N(l_N, t)$ and the volume velocity $u_N(l_N, t)$ at the output of the Nth tube [28]. From continuity $p_N(l_N, t) = p_L(t)$, which is the pressure at the lips output. Likewise, $u_N(l_N, t) = u_L(t)$, which is the volume velocity at the output of the lips. Suppose now that the radiation impedance is real, corresponding to a very large frequency Ω and thus, from Equation (4.24), a purely resistive load, i.e., $Z_r(\Omega) = R_r$. Then, from Figure 4.11, we can write the pressure/volume velocity relation at the lips as

$$p_N(l_N, t) = Z_r u_N(l_N, t), \tag{4.33}$$

where Z_r is not a function of Ω. From our earlier expressions for pressure and volume velocity within a tube, we then have

$$\frac{\rho c}{A_N}[u_N^+(t - \tau_N) + u_k^-(t + \tau_N)] = Z_r[u_N^+(t - \tau_N) - u_N^-(t + \tau_N)].$$

This expression can be written as

$$\left(\frac{\rho c}{A_N} + Z_r\right) u_N^-(t + \tau_N) = -\left(\frac{\rho c}{A_N} - Z_r\right) u_N^+(t - \tau_N)$$

and thus

$$u_N^-(t + \tau_N) = -\frac{\left(\frac{\rho c}{A_N} - Z_r\right)}{\left(\frac{\rho c}{A_N} + Z_r\right)} u_N^+(t - \tau_N)$$

$$= -r_L u_N^+(t - \tau_N) \qquad (4.34)$$

where the reflection coefficient at the lips is defined as

$$r_L = \frac{\frac{\rho c}{A_N} - Z_r}{\frac{\rho c}{A_N} + Z_r}.$$

Equation (4.34) states that the forward-going wave is reflected back at the lips boundary with reflection coefficient r_L. There is no backward-going wave contribution from free space to the right. The volume velocity at the lips output can then be expressed as

$$u_N(l_N, t) = u_N^+(t - \tau_N) - u_k^-(t + \tau_N)$$

$$= (1 + r_L) u_N^+(t - \tau_N)$$

which is shown schematically in Figure 4.16b. A complex radiation impedance $Z_r(\Omega)$ can be invoked by using a frequency-domain version of the previous derivation or by using a differential equation, the inductance resulting in a first-order ordinary differential equation.

We can also model the radiation from the lips with an additional tube of cross-section A_{N+1} and infinite length. The tube has no backward-propagating wave because of its infinite length, which is consistent with sound radiating out from the lips. In addition, if we select A_{N+1} such that $Z_r = \frac{\rho c}{A_{N+1}}$, then the expression for r_L becomes

$$r_L = \frac{\frac{\rho c}{A_N} - \frac{\rho c}{A_{N+1}}}{\frac{\rho c}{A_N} + \frac{\rho c}{A_{N+1}}}$$

$$= \frac{A_{N+1} - A_N}{A_{N+1} + A_N} \qquad (4.35)$$

which is our original definition of reflection coefficient r_N for two adjacent tubes of cross-sectional area A_{N+1} and A_N. Observe that when $Z_r(\Omega) = 0$ then $r_L = 1$, the pressure drop is zero, and the imaginary added tube becomes infinite in cross-section because A_N must be finite.

The boundary condition model for the glottis is shown in Figure 4.16c where the glottal impedance is assumed approximated as a series resistance and inductance. We can think of the glottal slit, anatomically, as sitting just below (schematically, to the left of) the first tube of the concatenated tube model, and so we denote the pressure at the output of the glottis by $p_1(0, t)$, i.e., the pressure at the left edge of the first tube, and the volume velocity output of the glottis by $u_1(0, t)$, i.e., the volume velocity at the left edge of the first tube. Suppose that the glottal impedance is real (purely resistive), i.e., $Z_g = R_g$, then using Kirchoff's current law, we can write in the time domain [28]

$$u_1(0, t) = u_g(t) - \frac{p_1(0, t)}{Z_g}.$$

Then from our earlier expressions for volume velocity and pressure within a uniform tube [Equation (4.27)], we have

$$u_1^+(t) - u_1^-(t) = u_g(t) - \frac{\rho c}{A_1} \left[\frac{u_1^+(t) + u_1^-(t)}{Z_g} \right].$$

Solving for the forward-going traveling wave

$$u_1^+(t) = \frac{(1 + r_g)}{2} u_g(t) + r_g u_1^-(t) \tag{4.36}$$

where

$$r_g = \frac{Z_g - \frac{\rho c}{A_1}}{Z_g + \frac{\rho c}{A_1}}.$$

If $Z_g(\Omega)$ is complex, then, as with a complex $Z_r(\Omega)$, a differential equation realization of the glottal boundary condition is required. A flow diagram of the flow relations in Equation (4.36) is illustrated in Figure 4.16c. As with the radiation impedance, we can also model the effect of glottal impedance with an additional tube of cross-section A_0 and infinite length. (We are looking down into the lungs.) If we make A_0 such that $Z_g = \frac{\rho c}{A_0}$, then our expression for r_g becomes

$$r_g = \frac{\frac{\rho c}{A_0} - \frac{\rho c}{A_1}}{\frac{\rho c}{A_1} + \frac{\rho c}{A_0}}$$

$$= \frac{A_1 - A_0}{A_1 + A_0}.$$

Consider now the special case of two concatenated lossless tubes of equal length, with radiation and glottal boundary conditions, depicted in the flow diagram of Figure 4.18a. In Exercise 4.9, using this flow diagram, you are asked to show that the transfer function relating the volume velocity at the lips to the glottis is given by

$$V_a(s) = \frac{b e^{-s2\tau}}{1 + a_1 e^{-s2\tau} + a_2 e^{-s4\tau}} \tag{4.37}$$

where $b = (1 + r_g)(1 + r_L)(1 + r_1)/2$, $a_1 = r_1 r_g + r_1 r_L$, and $a_2 = r_L r_g$. In this model, as well as more general models with an arbitrary number of lossless tubes, all loss in the system occurs at the two boundary conditions. Nevertheless, it is possible to select loss at the glottis and lips to give observed formant bandwidths. Furthermore, with this configuration Fant [7] and Flanagan [8] have shown that with appropriate choice of section lengths and cross-sectional areas, realistic formant frequencies can be obtained for vowels. As we noted earlier, however, the resulting model is not necessarily consistent with the underlying physics. In the following section, we describe a means of converting the analog concatenated tube model to discrete time. In this discrete-time realization, as in the analog case, all loss, and hence control of formant bandwidths, is introduced at the two boundaries.

4.4.2 A Discrete-Time Realization

The flow diagram in Figure 4.18a, because of the discrete delay elements, suggests that the concatenated tubes may be easily taken to a discrete-time realization. Consider a model consisting of N lossless concatenated tubes with total length l. As with the two-tube example, we make the time delays down each tube, and thus the length of each tube, equal. Each tube is of length $\Delta x = \frac{l}{N}$ and the time for the wave to propagate through one tube is $\tau = \frac{\Delta x}{c}$. Following Rabiner and Schafer [28], consider now the response of the N-tube model to a glottal input volume velocity $u_g(t)$ equal to a single impulse $\delta(t)$. If the impulse traveled down the tube without reflection, it would appear at the output of the final tube with a time delay of $N\tau$ seconds. The impulse, however, is partially reflected and partially propagated at each junction so that the impulse response is of the form

$$v_a(t) = b_0 \delta(t - N\tau) + \sum_{k=1}^{\infty} b_k \delta(t - N\tau - k2\tau) \qquad (4.38)$$

where 2τ is the round-trip delay within a tube. The earliest arrival is at time $N\tau$ and the following arrivals occur at multiples of 2τ after this earliest arrival due to multiple reflections and propagations. Because the Laplace transform (we are in continuous time) of a delayed impulse $\delta(t - t_o)$ is given by e^{-st_o}, we can write the Laplace transform of Equation (4.38) as

$$V_a(s) = b_0 e^{-sN\tau} + \sum_{k=1}^{\infty} b_k e^{-s(N+2k)\tau}$$

$$= e^{-sN\tau} \sum_{k=0}^{\infty} b_k e^{-sk2\tau}$$

where $e^{-sN\tau}$ corresponds to the delay required to propagate through N sections. Without this time delay, the impulse response is simply $v_a(t + N\tau)$; therefore, because we can always recover $v_a(t)$ with a simple time shift, we ignore the time delay of $N\tau$. Making the change of variables $s = j\Omega$, we obtain the frequency response

$$V_a(\Omega) = \sum_{k=0}^{\infty} b_k e^{-j\Omega k2\tau}$$

which can be shown to be periodic with period $\frac{2\pi}{2\tau}$ (Figure 4.17a):

$$V_a(\Omega + \frac{2\pi}{2\tau}) = \sum_{k=0}^{\infty} b_k e^{-j(\Omega + \frac{2\pi}{2\tau})k2\tau}$$

$$= \sum_{k=0}^{\infty} b_k e^{-j\Omega k2\tau} e^{-j(\frac{2\pi}{2\tau})k2\tau}$$

$$= V_a(\Omega).$$

The intuition for this periodicity is that we have "discretized" the continuous-space tube with space-interval $\Delta x = \frac{l}{N}$, and the corresponding time-interval $\tau = \frac{\Delta x}{c}$, so we expect periodicity to appear in the transfer function representation.

From Figure 4.17a, we see that $V_a(\Omega)$ has the form of a Fourier transform of a sampled continuous waveform with sampling time interval $T = 2\tau$. We use this observation to transform the analog filtering operation to discrete-time form with the following steps illustrated in Figure 4.17:

S1: Using the impulse-invariance method, i.e., replacing e^{sT} with the complex variable z where $T = 2\tau$, we transform the system function $V_a(\Omega)$ to discrete-time:

$$V_a(s) = \sum_{k=0}^{\infty} b_k (e^{s2\tau})^{-k}$$

which, with the replacement of e^{sT} by the complex variable z, becomes

$$V(z) = \sum_{k=0}^{\infty} b_k z^{-k}.$$

The frequency response $V(\omega) = V(z)|_{z=e^{j\omega}}$ will be designed to match desired formant resonances over the interval $[-\pi, \pi]$.

S2: Consider an excitation function $u_g(t)$ that is bandlimited with maximum frequency $\Omega_{max} = \frac{\pi}{2\tau}$ and sampled with a periodic impulse train with sampling interval $T = 2\tau$, thus meeting the Nyquist criterion to avoid aliasing. The Fourier transform of the resulting excitation is denoted by $U_g(\Omega)$. We then convert the impulse-sampled continuous-time input to discrete time. This operation is illustrated in the frequency-domain in Figure 4.17b.

S3: A consequence of using the impulse invariance method to perform filter conversion is a straightforward conversion of a continuous-time flow graph representation of the model to a discrete-time version. An example of this transformation is shown in Figure 4.18 for the two-tube case. Since the mapping of e^{sT} to z yields $V(z)$, the discrete-time signal flow graph can be obtained in a similar way. A delay of τ seconds corresponds to the continuous-time factor $e^{-s\tau} = e^{-s2\tau/2} = e^{-sT/2}$, which in discrete-time is a half-sample delay. Therefore, in a signal flow graph we can replace the delay τ by $z^{-1/2}$. Since a half-sample delay is difficult to implement (requiring interpolation), we move all lower-branch delays to

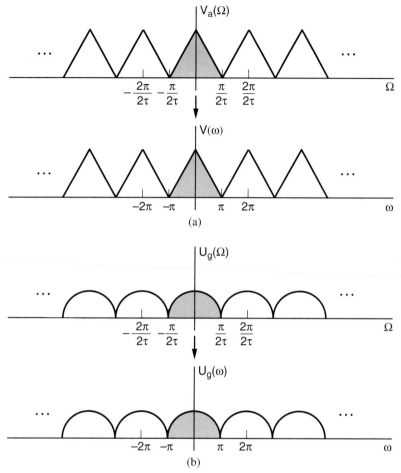

Figure 4.17 Frequency-domain view of discretizing analog filtering by concatenated tubes with equal length: (a) conversion of impulse-sampled continuous-time impulse response of concatenated tubes to discrete time; (b) conversion of impulse-sampled continuous-time glottal input to discrete time.

the upper branch, observing that delay is preserved in any closed branch with this change, as illustrated in Figure 4.18. The resulting delay offset can be compensated at the output [28].

S4: The final step in the conversion of continuous-time filtering to discrete-time is to multiply the discrete-time frequency responses of the excitation and the vocal tract impulse response to form the frequency response of the discrete-time speech output.

We now step back and view the discretization process from a different perspective. The original continuously spatial-varying vocal tract, assuming that it can be modeled as linear and time-invariant, has an impulse response that we denote by $\tilde{v}_a(t)$. By discretizing the vocal tract spatially, we have in effect sampled the impulse response temporally with sampling interval $T = 2\tau$. The constraint $\Delta x = c\tau$ indicates that spatial and time sampling are

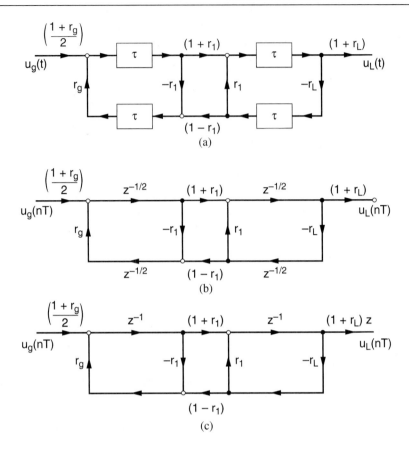

Figure 4.18 Signal flow graph conversion to discrete time: (a) lossless two-tube model; (b) discrete-time version of (a); (c) conversion of (b) with single-sample delays.

SOURCE: L.R. Rabiner and R.W. Schafer, *Digital Processing of Speech Signals* [28]. ©1978, Pearson Education, Inc. Used by permission.

intimately connected. The resulting frequency response is generally an *aliased* version of the (true) frequency response $\tilde{V}_a(\Omega)$ because the sampling may not meet the Nyquist criterion. The smaller the sub-tube length (increasing the number of tubes), then the less aliasing and the closer is the discrete-time frequency response to the (true) analog $\tilde{v}_a(t)$. We will see shortly that increasing the number of tubes implies that we increase the number of poles in the discrete model. The following example illustrates an interesting case:

EXAMPLE 4.4   Consider a uniform tube, itself modeled as a concatenation of uniform tubes, but with only a single tube element. The frequency response $\tilde{V}_a(\Omega)$ has an infinite number of resonances, as shown in Example 4.2. (Note that $V_a(\Omega)$ had previously denoted the true frequency response.) How can we possibly capture all of these resonances in discrete time? If we sample with $T = 2\tau = \frac{2l}{c}$, then we have only "half a resonance" over the discrete-time frequency range $[0, \pi]$; therefore, we need to divide the uniform tube into two equal tubes to allow a full resonance over $[0, \pi]$, i.e., we need to sample spatially twice as fast. This is an intriguing example which gives insight into the relation of the continuous- and discrete-time representations of resonant tubes. (The reader is asked to further elaborate on this example in Exercise 4.18.)   ▲

Our next objective is to derive a general expression for $V(z)$ in terms of the reflection coefficients. From the discrete-time flow graph in Figure 4.18, we know that

$$V(z) = \frac{U_L(z)}{U_g(z)} = f(r_g, r_1, r_2, \ldots r_L).$$

To obtain the function $f(r_g, r_1, r_2, \ldots r_L)$ from the flow graph directly is quite cumbersome. The flow graph, however, gives a modular structure by which the transfer function can be computed [28]. The resulting transfer function can be shown to be a stable all-pole function with bandwidths determined solely by the loss due to $Z_g(\Omega)$ and $Z_r(\Omega)$, i.e., $V(z)$ is of the form

$$V(z) = \frac{A z^{-N/2}}{D(z)}$$

where

$$D(z) = 1 - \sum_{k=1}^{N} a_k z^{-k}$$

and where the poles correspond to the formants of the vocal tract. Moreover, under the condition $r_g = 1$, i.e., infinite glottal impedance ($Z_g(\Omega) = \infty$) so that no loss occurs at the glottis, a recursion from the lips to the glottis can be derived whereby the transfer function associated with each tube junction is introduced recursively [28]. The recursion for the denominator polynomial $D(z)$ is given by

$$D_0(z) = 1$$
$$D_k(z) = D_{k-1}(z) + r_k z^{-k} D_{k-1}(z^{-1}), \qquad k = 1, 2, \ldots N$$
$$D(z) = D_N(z).$$

Because the vocal tract tube cross-sections (from which reflection coefficients are derived) $A_k > 0$, we can show that the poles are all inside the unit circle, i.e., the resulting system function is stable.[17] As before, we can imagine the $(N+1)$st tube is infinite in length with A_{N+1} selected so that $r_N = r_L$. When $Z_g(\Omega) = \infty$ and when $Z_r = 0$ so that $r_N = r_L = 1$, then there is a short circuit at the lips. Under this condition, there is no loss anywhere in the system and hence zero bandwidths arise; with $Z_g(\Omega) = \infty$, the radiation impedance $Z_r(\Omega)$ is the only source of loss in the system and controls the resonance bandwidths. The following example illustrates how to choose the number of tube elements to meet a desired bandwidth constraint.

---

[17] When $Z_g(\Omega) = \infty$ the recursion is associated with *Levinson's recursion*, which will be derived in the context of linear prediction analysis of Chapter 5. Using Levinson's recursion, we will prove that the poles of $D(z)$ lie inside the unit circle.

EXAMPLE 4.5   Let the vocal tract length $l = 17.5$ cm and the speed of sound $c = 350$ m/s. We want to find the number of tube sections N required to cover a bandwidth of 5000 Hz, i.e., the excitation bandwidth and the vocal tract bandwidth are 5000 Hz. Recall that $\tau = \frac{l}{cN}$ and that $\frac{2\pi}{4\tau}$ is the cutoff bandwidth. Therefore, we want $\frac{1}{4\tau} = 5000$ Hz. Solving for τ, the delay across a single tube, $\tau = \frac{1}{20000}$. Thus, from above we have $N = \frac{l}{c\tau} = 10$. Since N is also the order of the all-pole denominator, we can model up to $\frac{N}{2} = 5$ complex conjugate poles. We can also think of this as modeling one resonance per 1000 Hz.   ▲

We see that the all-pole transfer function is a function of only the reflection coefficients of the original concatenated tube model, and that the reflection coefficients are a function of the cross-sectional area functions of each tube, i.e., $r_k = \frac{A_{k+1} - A_k}{A_{k+1} + A_k}$. Therefore, if we could estimate the area functions, we could then obtain the all-pole discrete-time transfer function. An example of this transition from the cross-sectional areas A_k to $V(z)$ is given in the following example:

EXAMPLE 4.6   This example compares the concatenated tube method with the Portnoff numerical solution using coupled partial differential equations [26]. Because an infinite glottal impedance is assumed, the only loss in the system is at the lips via the radiation impedance. This can be introduced, as we saw above, with an infinitely-long $(N + 1)$th tube, depicted in Figure 4.19 with a terminating cross-sectional area selected to match the radiation impedance, according to Equation (4.35), so that $r_N = r_L$. By altering this last reflection coefficient, we can change the energy loss in the system and thus control the bandwidths. For example, we see in Figure 4.19 the two different cases of $r_N = 0.714$ (non-zero bandwidths) and $r_L = 1.0$ (zero bandwidths). This example summarizes in effect all we have seen up to now by comparing two discrete-time realizations of the vocal tract transfer function that have similar frequency responses: (1) A numerical simulation, derived with central difference approximations to partial derivatives in time and space, and (2) A (spatially) discretized concatenated tube model that maps to discretized time.   ▲

4.4.3 Complete Discrete-Time Model

We have in the previous sections developed a discrete-time model of speech production. The discrete-time model was derived from a set of concatenated lossless tubes approximating the spatially-varying vocal tract, with a glottal input given by an ideal volume velocity source; the radiation load at the concatenated tube output is characterized by a highpass frequency response. In this section, we revisit and extend the discrete-time model of speech production with the three speech sound sources we introduced in Chapter 3, i.e., speech sounds with periodic, noise, and impulsive inputs. Before proceeding, however, we need to account for a missing piece in our discrete-time model.

Looking back to Section 4.3.3, we see that in practice speech pressure is measured at the lips output, in contrast to volume velocity which is represented in the discrete-time transfer function of Section 4.3.1. Using Equation (4.26), showing the conversion from volume velocity to pressure at the lips, the discrete-time transfer function from the volume velocity input to the pressure output, denoted by $H(z)$, is given by

$$H(z) = V(z)R(z).$$

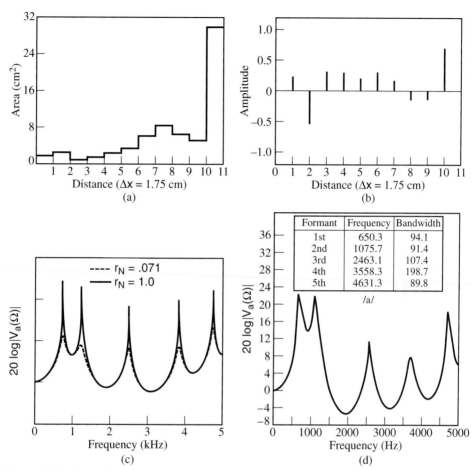

Figure 4.19 Comparison of the concatenated tube approximation with the "exact" solution for area function (estimated by Faut [7]) of the Russian vowel /a/ [26],[28]: (a) cross-section $A(x)$ for a vocal tract model with 10 lossless sections and terminated with a 30 cm^2 section that does not reflect; (b) reflection coefficients r_k for 10 sections; (c) frequency response of the concatenated tube model—the solid curve corresponds to the lossless termination (zero bandwidths) and the dashed curve corresponds to the condition with loss (finite bandwidths); (d) frequency response derived from numerical simulation of Portnoff.

SOURCE: M.R. Portnoff, *A Quasi-One-Dimensional Digital Simulation for the Time-Varying Vocal Tract* [26]. ©1973, M.R. Portnoff and the Massachusetts Institute of Technology. Used by permission.

$R(z)$ denotes the discrete-time radiation impedance and $V(z)$ is the discrete-time all-pole vocal tract transfer function from the volume velocity at the glottis to volume velocity at the lips. The radiation impedance $R(z) = Z_r(z)$ is a discrete-time counterpart to the analog radiation impedance $Z_r(s)$. You will show in Exercise 4.20 that $R(z) \approx 1 - z^{-1}$ and thus acts as approximately a differentiation of volume velocity to obtain pressure, introducing about a 6 dB/octave highpass effect. Although $R(z)$ is derived as a single zero on the unit circle, it is more realistically modeled as a zero slightly inside the unit circle, i.e.,

$$R(z) \approx 1 - \alpha z^{-1}$$

with $\alpha < 1$, because near-field measurements at the lips do not give quite the 6 dB/octave rolloff predicted by a zero on the unit circle [8]. The analog $Z_r(s)$ was derived by Flanagan under the assumption of pressure measurements in the far field, i.e., "sufficiently" far from the source [8]. Considering the pressure/volume velocity relation at the lips as a differentiator, the speech pressure waveform in continuous time, $x(t)$, measured in front of the lips can be expressed as

$$x(t) \approx A \frac{d}{dt}[u_g(t) * v(t)] = A\left[\frac{d}{dt}u_g(t)\right] * v(t),$$

where the gain A controls loudness. (The reader should prove this equality.) The effect of radiation is therefore typically included in the source function; the source to the vocal tract becomes the derivative of the glottal flow volume velocity often referred as the *glottal flow derivative*, i.e., the source is thought of as[18] $\frac{d}{dt}u_g(t)$ rather than $u_g(t)$.

The discrete-time speech production model for periodic, noise, and impulsive sound sources is illustrated in Figure 4.20. Consider first the periodic (voiced) speech case. For an input consisting of glottal airflow over a single glottal cycle, the z-transform of the speech output is expressed as

$$X(z) = A_v G(z) H(z)$$
$$= A_v G(z) V(z) R(z)$$

where A_v is again controlling the loudness of the sound and is determined by the subglottal pressure which increases as we speak louder, thus increasing the volume velocity at the glottis. $G(z)$ is the z-transform of the glottal flow input, $g[n]$, over one cycle and which may differ

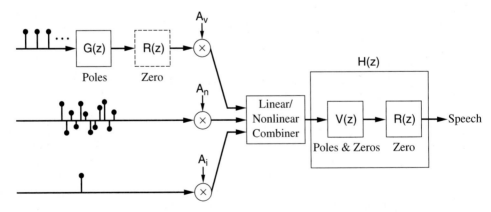

Figure 4.20 Overview of the complete discrete-time speech production model.

---

[18] Moving the differentiation to the source holds strictly only when the speech production components are linear. We saw in Chapter 2 that components of a nonlinear system are not necessarily commutative.

with the particular speech sound (i.e., phone), speaker, and speaking style. $R(z)$ is the radiation impedance that we model as a single zero $R(z) = 1 - \alpha z^{-1}$, and $V(z)$ is a stable all-pole vocal tract transfer function from the volume velocity at the glottis to the volume velocity at the lips, and which is also a function of the particular speech sound, speaker, and speaking style. We have seen in Chapter 3 that an approximation of a typical glottal flow waveform over one cycle is of the form

$$g[n] = (\beta^{-n}u[-n]) * (\beta^{-n}u[-n]),$$

i.e., two time-reversed exponentially decaying sequences, that has z-transform

$$G(z) = \frac{1}{(1 - \beta z)^2}$$

which for real $\beta < 1$ represents two identical poles outside the unit circle. This model assumes infinite glottal impedance, i.e., no loss at the glottis [no loss at the glottis allowed us in the previous section to obtain an all-pole model for $V(z)$]. All loss in the system is assumed to occur by radiation at the lips. For the voiced case, the z-transform at the output of the lips can then be written over one glottal cycle as a rational function with poles inside and outside the unit circle and a single zero inside the unit circle, i.e.,

$$\begin{aligned}
X(z) &= A_v G(z) V(z) R(z) \\
&= A_v \frac{(1 - \alpha z^{-1})}{(1 - \beta z)^2 \prod_{k=1}^{C_i}(1 - c_k z^{-1})(1 - c_k^* z^{-1})}
\end{aligned} \tag{4.39}$$

where we assume C_i pole pairs of $V(z)$ inside the unit circle. This rational function, along with a uniformly-spaced impulse train to impart periodicity to the input, is illustrated in the upper branch of Figure 4.20. Observe in this rational function that $V(z)$ and $R(z)$ are minimum-phase, while $G(z)$, as we have modeled it, is maximum-phase, having two poles outside the unit circle. Referring to our discussion in Chapter 2 of frequency-domain phase properties of a sequence, we can deduce that the glottal flow input is responsible for a gradual "attack" to the speech waveform within a glottal cycle during voicing. We return to this important characteristic in later chapters when we develop methods of speech analysis and synthesis.

If we apply the approximate differentiation of the radiation load to the glottal input during voicing, we obtain a "source" function illustrated in Figure 4.21 for a typical glottal airflow over one glottal cycle. In the glottal flow derivative, a rapid closing of the vocal folds results in a large negative impulse-like response, called the *glottal pulse*, which occurs at the end of the open phase and during the return phase of the glottal cycle, as shown in Figure 4.21. The glottal pulse is sometimes considered the primary excitation for voiced speech, and has a wide bandwidth due to its impulse-like nature [1]. (Note that we are using the term "glottal pulse" more strictly than in Chapter 3.) We have also illustrated this alternative source perspective in the upper branch of Figure 4.20 by applying $R(z)$ just after $G(z)$.

Two other inputs to the vocal tract are noise and impulsive sources. When the source is noise, as, for example, with fricative consonants, then the source is no longer a periodic glottal airflow sequence, but rather a random sequence with typically a flat spectrum, i.e., white noise,

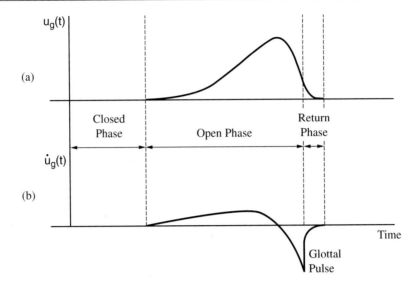

Figure 4.21 Schematic of relation between (a) the glottal airflow and (b) the glottal flow derivative over a glottal cycle.

although this noise may be colored by the particular constriction and shape of the oral tract. The output z-transform at the lips is then expressed by

$$X(z) = A_n U(z) H(z)$$
$$= A_n U(z) V(z) R(z)$$

where $U(z)$ denotes the z-transform of a noise sequence, $u[n]$. The third source is the "burst" which occurs during plosive consonants and which for simplicity we have modeled as an impulse. The output z-transform at the lips for the impulsive input is given by

$$X(z) = A_i H(z) = A_i V(z) R(z).$$

The noise and impulse source are shown in the lower two branches of Figure 4.20. Observe that all sources can occur simultaneously as with, for example, voiced fricatives, voiced plosives, and aspirated voicing. Furthermore, these sources may not be simply linearly combined, as we saw in the voiced fricative model of Example 3.4 of Chapter 3, a possibility that we have represented by the linear/nonlinear combiner element in Figure 4.20.

In the noise and impulse source state, oral tract constrictions may give zeros (absorption of energy by back-cavity anti-resonances) as well as poles. Zeros in the transfer function also occur for nasal consonants, as well as for nasalized vowels. Methods have been developed to compute the transfer function of these configurations [16],[17]. These techniques are based on concatenated tube models with continuity constraints at tube junctions and boundary conditions similar to those used in this chapter for obtaining the transfer function of the oral tract. (A number of simplifying cases are studied in Exercises 4.7 and 4.12.) In these cases, the vocal tract transfer function $V(z)$ has poles inside the unit circle, but may have zeros inside and outside the unit circle, and thus $X(z)$ is of the form

$$X(z) = A \frac{(1 - \alpha z^{-1}) \prod_{k=1}^{M_i} (1 - a_k z^{-1}) \prod_{k=1}^{M_o} (1 - b_k z)}{(1 - \beta z)^2 \prod_{k=1}^{C_i} (1 - c_k z^{-1})(1 - c_k^* z^{-1})} \qquad (4.40)$$

where $(1 - a_k z^{-1})$ and $(1 - b_k z)$ are zeros inside and outside the unit circle, respectively, and are due to the oral and/or nasal tract configurations. The vocal transfer function is therefore generally *mixed phase*, i.e., it is not necessarily an all-pole minimum-phase transfer function. The maximum-phase elements of the vocal tract can also contribute (i.e., in addition to the maximum-phase glottal flow alluded to earlier) to a more gradual attack of the speech waveform during voicing or at a plosive than is obtained by a minimum-phase sequence.

The discrete-time model given in this section, sometimes referred to as the "source/filter" model, can perform well in matching a measured spectrum or waveform or when only input/output relations are desired. The model output waveform and its spectrum are good matches to measurements with appropriate selection of the source, vocal tract filter parameters, and radiation. Nevertheless, much of the physics of our original model, as well as physics we have not yet attempted to model, such as effects of nonlinearities, are not represented by the simple source/filter model. Furthermore, different vocal tract shapes, sources, nonlinear effects, and subsystem coupling may yield a similar output, i.e., an output measurement may not be uniquely invertible. In Section 4.5, we look at one of the many fascinating possible model refinements: the nonlinear coupling between the glottal flow velocity and vocal tract pressure during voicing. A further glimpse beyond the simplicity of the source/filter model is saved for Chapter 11.

4.5 Vocal Fold/Vocal Tract Interaction

In the source/filter speech production model of Section 4.4.3, we assumed that the glottal impedance is infinite and that the glottal airflow source is not influenced by, or not "coupled" to, the vocal tract. This allowed us to model the glottal source as an ideal volume velocity that is convolved with a linear vocal tract filter impulse response to produce the speech waveform. In reality, there exists a *nonlinear* coupling between the glottal airflow velocity and the pressure within the vocal tract. In a qualitative sense, the pressure in the vocal tract cavity just above the glottis "backs up" against the glottal flow and interacts nonlinearly with the flow. In certain wind instruments, as in the trumpet, such a mechanism is even more pronounced and, indeed, is essential in determining the sound of the instrument; the vibration of the "lip reed" is strongly, and nonlinearly, coupled to the resonant frequencies of the cavity [3]. An accurate integrated vocal fold/vocal tract model would preserve the constraints between the glottal airflow volume velocity and pressure within the vocal tract chamber.

We presented in Section 4.3.4 one approach to obtaining a more accurate coupled source/ filter model that uses the partial differential wave equation description of pressure and volume velocity in the vocal tract, the differential equations of loss in the tract and from lip radiation, and Flanagan and Ishizaka's two-mass model of vocal fold vibration (requiring measurement of model parameters) that invokes a nonlinear relation between glottal volume velocity and pressure. The resulting coupled equations have time-varying parameters and are nonlinear, and thus require a numerical solution, as was performed by Portnoff [26]. We saw that one unexpected result from Portnoff's simulation is that the vocal tract impulse response can abruptly decay within a glottal cycle (Figure 4.13). In this section, we use a simplified model of vocal fold/vocal tract coupling that gives insight into and predicts this "truncation" phenomenon, as well as predicts the modulation of formant frequencies and bandwidths within a glottal cycle. These insights are gained without the need of a numerical simulation and without the requirement of measuring the vocal fold parameters of Flanagan and Ishizaka's two-mass model [1].

4.5.1 A Model for Source/Tract Interaction

Figure 4.22 shows an electrical analog of a model for airflow through the glottis. In this model, voltage is the analog of sound pressure, and current is the analog of volume velocity. P_{sg} at the left is the subglottal (below the glottis) pressure in the lungs that is the power source for speech; during voicing it is assumed fixed. $p(t)$ is the sound pressure corresponding to a single first formant[19] in front of the glottis. $Z_g(t)$ is the time-varying impedance of the glottis, defined as the ratio of *transglottal pressure* (across the glottis), $p_{tg}(t)$, to the *glottal volume velocity* (through the glottis), $u_g(t)$, i.e.,

$$Z_g(t) = \frac{p_{tg}(t)}{u_g(t)} \tag{4.41}$$

where all quantities are complex. The R, C, and L elements are resistance, capacitance, and inductance, respectively, of an electrical model for the first formant whose nominal frequency and (3 dB attenuation) bandwidth are given by

$$\Omega_o = \sqrt{\frac{1}{LC}}$$

$$B_o = \frac{1}{RC}. \tag{4.42}$$

From Figure 4.22, we see that if $Z_g(t)$ is comparable to the impedance of the first-formant model, there will be considerable "interaction" between the source and the vocal tract, in violation of the source/filter assumption of independence. This interaction will induce an effective change in nominal center frequency Ω_o and bandwidth B_o of the first formant.

Empirically, and guided by aerodynamic theories, it has been found that $Z_g(t)$ is not only time-varying, but also nonlinear. Specifically, it has been shown by van den Berg [31] that the transglottal pressure and glottal volume velocity are related by

$$p_{tg}(t) \approx \left[\frac{k\rho}{2A^2(t)} \right] u_g^2(t) \tag{4.43}$$

---

[19] Recall that the vocal tract transfer function has many complex pole pairs corresponding to speech formants. Only the first formant is included in Figure 4.22, because for higher formants the impedance of the vocal tract is negligible compared to that at the glottis. Numerical simulations, to be described, initially included multiple formants of the vocal tract and multiple subglottal resonances. These simulations verify that formants above the first formant and subglottal resonances negligibly influence glottal flow [1].

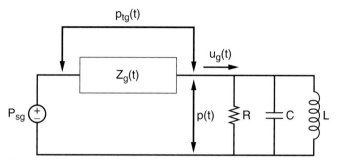

Figure 4.22 Diagram of a simple electrical model for vocal fold/vocal tract interaction. Only the first-formant model is included.

SOURCE: C.R. Jankowski, *Fine Structure Features for Speaker Identification* [10]. ©1996, C.R. Jankowski and the Massachusetts Institute of Technology. Used by permission.

where ρ is the density of air and $A(t)$ is the smallest time-varying area of the glottal slit. (Looking down the glottis, the cross-sectional glottal area changes with depth.) The term $k = 1.1$ and includes the effect of a nonuniform glottis with depth. Writing the current nodal equation of the circuit in Figure 4.22, and using Equation (4.43), we can show that (Exercise 4.19)

$$C\frac{dp(t)}{dt} + \frac{p(t)}{R} + \frac{1}{L}\int_0^t p(\tau)d\tau = A(t)\sqrt{\frac{2p_{tg}(t)}{k\rho}} \tag{4.44}$$

where

$$p(t) + p_{tg}(t) = P_{sg}.$$

Equations (4.43) and (4.44) represent a pair of coupled nonlinear differential equations with time-varying coefficients, the time variation entering through the changing glottal slit area and the nonlinearity entering through the van den Berg glottal pressure/volume velocity relation.

One approach to simultaneously solving the above Equation pair (4.43), (4.44) is through numerical integration [1]. In this simulation, it was determined that the time-domain *skewness* of the glottal flow over one cycle (we saw this skewness in Chapter 3) is due in part to the time-varying glottal area function and in part to the loading by the vocal tract first formant; pressure from the vocal tract against the glottis will slow down the flow and influence its skewness. In addition to an asymmetric glottal flow, the numerical simulation also revealed an intriguing sinusoidal-like "ripple" component to the flow. The timing and amount of ripple are dependent on the configuration of the glottis during both the open and closed phases [1],[24],[25]. For example, with folds that open in a zipper-like fashion, the ripple may begin at a low level early into the glottal cycle, and then grow as the vocal folds open more completely. We can think of the ripple as part of the *fine structure* of the glottal flow, superimposed on the more slowly-varying asymmetric *coarse structure* of the glottal flow, as illustrated in Figure 4.23, showing the two components in a schematic of a typical glottal flow derivative over one cycle. We describe the separation of coarse- and fine-structure glottal flow components in Chapter 5. Finally, the "truncation" effect over a glottal cycle, alluded to earlier, was also observed.

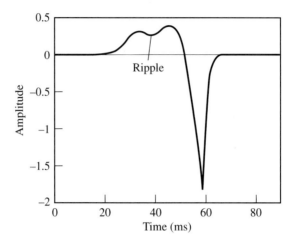

Figure 4.23 Glottal flow derivative waveform showing coarse and ripple component of fine structure due to source/vocal tract interaction.

SOURCE: M.D. Plumpe, T.F. Quatieri, and D.A. Reynolds, "Modeling of the Glottal Flow Derivative Waveform with Application to Speaker Identification" [25]. ©1999, IEEE. Used by permission.

There is an alternative way of looking at this problem that leads to an equivalent representation of the vocal fold/vocal tract interaction which gives similar observed effects, but more intuition about the interaction [1]. We again simplify the problem by assuming a one-formant vocal tract load. Then, from Equation (4.44) and with $p_{tg}(t) = P_{sg} - p(t)$ as the pressure across the glottis, we can write

$$C\frac{dp(t)}{dt} + \frac{p(t)}{R} + \frac{1}{L}\int_0^t p(\tau)d\tau = A(t)\sqrt{2p_{tg}(t)/k\rho}$$

$$= A(t)\sqrt{2(P_{sg} - p(t))/k\rho}$$

$$= A(t)\sqrt{2P_{sg}/k\rho}\sqrt{1 - \frac{p(t)}{P_{sg}}}.$$

Observing that from the Taylor series expansion of $\sqrt{1-x}$ we can linearize the square root function as $\sqrt{1-x} \approx 1 - \frac{1}{2}x$, and assuming $p(t) \ll P_{sg}$, we have

$$C\frac{dp(t)}{dt} + \frac{p(t)}{R} + \frac{1}{L}\int_0^t p(\tau)d\tau = A(t)\sqrt{2P_{sg}/k\rho}\left[1 - \frac{p(t)}{2P_{sg}}\right].$$

Then with some algebra, we have

$$C\frac{dp(t)}{dt} + \frac{p(t)}{R} + \frac{1}{L}\int_0^t p(\tau)d\tau + \frac{1}{2}p(t)A(t)\sqrt{2/k\rho\,P_{sg}} = A(t)\sqrt{2P_{sg}/k\rho}$$

that can be rewritten as

$$C\frac{dp(t)}{dt} + \frac{p(t)}{R} + \frac{1}{L}\int_0^t p(\tau)d\tau + \frac{1}{2}p(t)g_o(t) = u_{sc}(t) \qquad (4.45)$$

where

$$u_{sc}(t) = A(t)\sqrt{\frac{2P_{sg}}{k\rho}}$$

$$g_o(t) = \frac{u_{sc}(t)}{P_{sg}} = A(t)\sqrt{\frac{2}{k\rho\,P_{sg}}}.$$

Now differentiating Equation (4.45) with respect to time, we can show that (Exercise 4.19)

$$C\frac{d^2p(t)}{dt} + \left[\frac{1}{R} + \frac{1}{2}g_o(t)\right]\frac{dp(t)}{dt} + \left[\frac{1}{L} + \frac{1}{2}\dot{g}_o(t)\right]p(t) = \dot{u}_{sc}(t) \qquad (4.46)$$

which is represented by the Norton equivalent circuit of Figure 4.24 to the original circuit of Figure 4.22, where the equivalent time-varying resistance and inductance are given by

$$R_g(t) = \frac{2}{g_o(t)}$$

$$L_g(t) = \frac{2}{\dot{g}_o(t)} \qquad (4.47)$$

which are in parallel, respectively, with the first-formant resistance, capacitance, and inductance, R, C, and L. The new time-varying volume velocity source is given by $u_{sc}(t)$ of Equation (4.45).

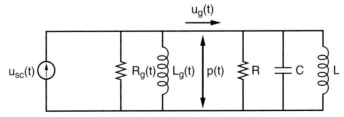

Figure 4.24 Transformed vocal fold/vocal tract first-formant interaction model that is Norton equivalent to circuit of Figure 4.22.

Source: C.R. Jankowski, *Fine Structure Features for Speaker Identification* [10]. ©1996, C.R. Jankowski and the Massachusetts Institute of Technology. Used by permission.

4.5.2 Formant Frequency and Bandwidth Modulation

Equation (4.46) can be interpreted as a linear differential equation with time-varying coefficients that cause a modulation of the formant frequency and bandwidth. To see how this modulation arises, we map Equation (4.46) to the frequency domain.

Standard Laplace transform methods cannot directly be applied because the coefficients of the equation are time-varying. Nevertheless, to obtain an approximate frequency-domain representation of the vocal cord/vocal tract interaction, we assume that at *each time instant* the glottal impedance is stationary, i.e., we pretend at each time instant that the linear differential equation represents a time-invariant system. Under this assumption, we can obtain a "pseudo-transfer function" from the volume-velocity source to the vocal tract pressure, $H(s, t) = \frac{P(s,t)}{U_{sc}(s)}$, with time-varying coefficients that are functions of the changing formant center frequency and bandwidth [1]. This transfer function is given by (Exercise 4.19)

$$H(s, t) = \frac{P(s, t)}{U_{sc}(s)} = \frac{s/C}{s^2 + B_1(t)s + \Omega_1^2(t)}, \tag{4.48}$$

with denominator polynomial coefficients given by the time-varying formant frequency $\Omega_1(t)$ and time-varying bandwidth $B_1(t)$:

$$\Omega_1^2(t) = \Omega_o^2 \left[1 + \frac{1}{2} L \dot{g}_o(t)\right]$$

$$B_1(t) = B_o \left[1 + \frac{1}{2} R g_o(t)\right] \tag{4.49}$$

where the nominal frequency and bandwidth, Ω_o and B_o, are defined in Equation (4.42), and $g_0(t)$ is proportional to the time-varying area of the glottis, as seen in Equation (4.45). Over a glottal cycle, the changing bandwidth $B_1(t)$ then follows that of the area function. The formant frequency $\Omega_1(t)$, being proportional to the derivative of the area function, rises at the onset of the glottal open phase and falls near the termination of this phase [1],[10],[24],[25].

We see in the expression for the bandwidth $B_1(t)$ that the nominal bandwidth, B_o, is multiplied by a time-dependent *bandwidth multiplier* factor, $[1 + \frac{1}{2} R g_o(t)]$. Because $A(t) \geq 0$ then $B_1(t) \geq B_o$, i.e., opening the glottis only increases the bandwidth of the formant. As with bandwidth modulation, the nominal formant frequency, Ω_o, is multiplied by a time-dependent factor. Due to the dependence of $\Omega_1(t)$ on the derivative of the glottal area, during a complete glottal cycle, $\Omega_1(t)$ will modulate both below and above its nominal value. The following example illustrates these properties:

EXAMPLE 4.7   Figure 4.25b gives the theoretically calculated $B_1(t)$ as a function of time over two glottal cycles for five Russian vowels /a/, /o/, /u/ , /i/, and /e/ for the glottal area function $A(t)$ in Figure 4.25a (from [1],[10]). The increase in bandwidth due to the opening glottis is higher for vowels with higher first formant. Even in the cases of minimum bandwidth modulation, i.e., for vowels with very low first formant (e.g., /i/,/u/), $B_1(t)$ is increased by a factor of about three to four.

Figure 4.25c gives the $\Omega_1(t)$ multiplier for the five vowels, again for the glottal area function $A(t)$ in Figure 4.25a. The $\Omega_1(t)$ multiplier ranges from about 0.8 to 1.2. It is not necessarily larger for vowels with high first formant, as with bandwidth modulation. An interesting property of the $\Omega_1(t)$

07/04/04
09:27 pm

ain Bernard Thierr Fourreau

JE DATE:
04-04-30 21:45:00

TLE:Discrete-time speech signal
ocessing : principles and practice /
omas F. Quatieri.

EM:6000525545

Please note that the date is in
the American format: year-month-day.

Final year students must return all
loans and pay any outstanding debts
before leaving the university.
Continuing students should remember
to return or renew their loans before

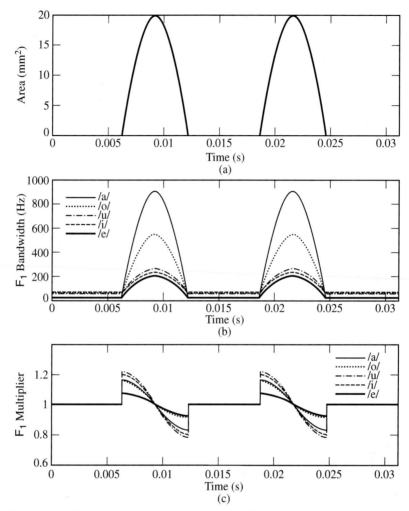

Figure 4.25   Illustration of time-varying first formant F_1 frequency and bandwidth for five Russian vowels [1]: (a) glottal area function; (b) bandwidth $B_1(t) = B_o \left[1 + \frac{1}{2} R g_o(t) \right]$; (c) formant frequency multiplier $\left[1 + \frac{1}{2} L \dot{g}_o(t) \right]$.

SOURCE: C.R. Jankowski, *Fine Structure Features for Speaker Identification* [10]. ©1996, C.R. Jankowski and the Massachusetts Institute of Technology. Used by permission.

multiplier is that at both glottal opening and closure, the change in $\Omega_1(t)$ is rather instantaneous, and can be from 10–20% of the formant frequency. Note that the glottal area function in Figure 4.25a is a simple approximation; we have seen that an actual glottal area does not increase as rapidly at glottal opening as at closing, due to physical constraints.   ▲

The increase of $B_1(t)$ within a glottal cycle is responsible for the effect we have been calling "truncation," and is due to a decrease in the impedance at the glottis as the glottis opens; the reduced glottal impedance diverts the glottal source velocity away from the vocal tract.

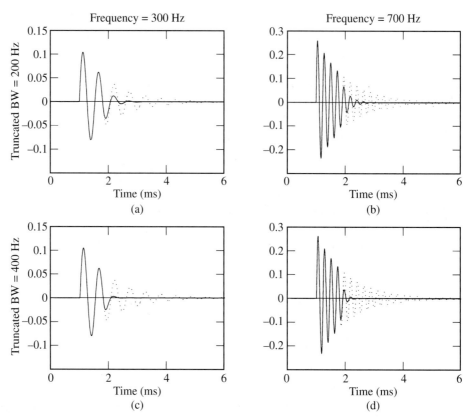

Figure 4.26 Illustration of the effect of truncation with synthetic waveforms. Truncated responses are denoted by solid lines and corresponding untruncated responses by dotted lines. The signal frequency is 300 Hz in panels (a) and (c) and 700 Hz in panels (b) and (d). The maximum bandwidth (referred to as truncated BW) during truncation is 200 Hz in panels (a) and (b) and 400 Hz in panels (c) and (d) for a nominal bandwidth of 60 Hz.

SOURCE: C.R. Jankowski, *Fine Structure Features for Speaker Identification* [10]. ©1996, C.R. Jankowski and the Massachusetts Institute of Technology. Used by permission.

Figure 4.26 shows the result of truncation for numerous synthetic waveforms [10]. The waveforms are generated with an implementation of the Klatt speech synthesizer which has the capability of continuously varying formant frequencies and bandwidths by updating resonator parameters every sample [12]. The bandwidth of the waveform shown is instantaneously changed from a nominal value of 60 Hz. All panels show the truncated response with a solid line, and the corresponding nontruncated response with a dotted line. Panels (a) and (c) are for the formant frequency equal to 300 Hz, while the formant frequency is 700 Hz in panels (b) and (d). The maximum bandwidth during truncation is 200 Hz in panels (a) and (b) and 400 Hz in panels (c) and (d). The truncation effect for an actual speech waveform is shown in Figure 4.27. The effect is shown from both a time- and frequency-domain perspective; a sudden drop in

Equation (2.14) from Chapter 2, i.e.,

$$B = \int_{-\infty}^{\infty} \left(\frac{da(t)}{dt} \right)^2 dt + \int_{-\infty}^{\infty} \left(\frac{d\phi(t)}{dt} - \bar{\omega} \right)^2 a^2(t) dt \qquad (4.54)$$

(with $\bar{\omega}$ the mean frequency) Cohen [5] proposed a definition of *instantaneous bandwidth* that is dependent *only* on the amplitude of $x(t)$ and not its frequency, i.e.,

$$BW(t) = \frac{1}{2\pi} \left(\frac{a'(t)}{a(t)} \right).$$

Show in the special case of a decaying exponential, i.e.,

$$x(t) = e^{-\alpha t},$$

that Cohen's definition of instantaneous bandwidth becomes

$$BW(t) = \frac{\alpha}{2\pi}. \qquad (4.55)$$

A more standard definition of bandwidth is the frequency distance between spectrum samples that are 3 dB lower than at the spectral peak. Show that the expression in Equation (4.55) is one-half of this standard definition. Likewise, again motivated in part by the bandwidth definition in Equation (4.54), as described in Chapter 2, Cohen defined $\frac{d\phi(t)}{dt}$ as the *instantaneous frequency* of $x(t)$. Describe qualitatively how the bandwidth and frequency multiplier factors in part (d) of this problem affect the bandwidth defined in Equation (4.54). In your description, consider how a rapid truncation corresponds to a larger instantaneous bandwidth $BW(t)$ and assume that $\frac{d\phi(t)}{dt}$ roughly follows the time-varying frequency $\Omega_1(t)$.

4.20 (MATLAB) The bilinear transformation used in mapping a Laplace transform to a z-transform is given by the s-plane to z-plane mapping of $s = \frac{2}{T}[\frac{z+1}{z-1}]$ and has the property that the imaginary axis in the s-plane maps to the unit circle in the z-plane [22]. The constant T is the sampling interval. Use the bilinear transformation to obtain the z-transform representation of the continuous-time radiation load of Equation (4.24). Specifically, with $T = 10^{-4}$, $c = 350$ m/s, $R_r = 1.4$, and $L_r = (31.5)10^{-6}$, argue that $Z_r(z) \approx 1 - z^{-1}$ where any fixed scaling is ignored. Use MATLAB to plot the magnitude of $Z_r(z)$ on the unit circle and show that $Z_r(z)$ introduces about a 6 dB/octave highpass effect. Argue from both a time- and frequency-domain viewpoint that the impulse response associated with $Z_r(z)$ acts as a differentiator.

4.21 (MATLAB)[20] In his work on nonlinear modeling of speech production, Teager [30] used the following "energy operator" on speech-related signals $x[n]$:

$$\psi(x[n]) = x^2[n] - x[n-1]x[n+1]. \qquad (4.56)$$

Kaiser [11] analyzed ψ and showed that it yields the energy of simple oscillators that generated the signal $x[n]$; this energy measure is a function of the frequency as well as the amplitude composition of a signal. Kaiser also showed that ψ can track the instantaneous frequency in single sinusoids and chirp signals, possibly exponentially damped. Teager applied ψ to signals resulting from bandpass filtering of speech vowels in the vicinity of their formants. The output from the energy operator fre-

---

[20] This problem can be considered a preview to Section 11.5.

quently consisted of several *pulses* per pitch period, with decaying peak amplitude. Teager suggested that these energy pulses indicate modulation of formants caused by nonlinear phenomena such as time-varying vortices in the vocal tract.

We can interpret these energy pulses by using a frequency modulation (FM) model for the time-varying formants. It is important to emphasize that we do not model speech production with FM, but rather the FM is a mathematical vehicle to model the acoustical consequences of nonlinear mechanisms of speech production. Specifically, consider the following exponentially-damped FM signal with sine modulation:

$$x[n] = Ae^{-an}\cos(\phi[n]) = Ae^{-an}\cos[\Omega_0 n + \beta\sin(\Omega_m n) + \theta] \tag{4.57}$$

where Ω_0 is the center (or carrier) frequency, $\beta = \frac{\Delta}{\Omega_m}$ is the modulation index, Δ is the frequency deviation, and Ω_m is the frequency of the modulating sinusoid with phase offset θ. The instantaneous frequency is $\Omega[n] = \frac{d\phi[n]}{dn} = \Omega_0 + \Delta\cos(\Omega_m n)$.

(a) Assume Ω_m is sufficiently small so that $\cos(\Omega_m) \approx 1$ and $\sin(\Omega_m) \approx \Omega_m$, and show that applying the energy operator ψ to $x[n]$ yields:

$$\psi(x[n]) \approx A^2 e^{-2an}\sin^2[\Omega_0 + \Delta\,\cos(\Omega_m n)] = A^2 e^{-2an}\sin^2(\Omega[n]). \tag{4.58}$$

Thus, ψ can track the instantaneous frequency $\Omega[n]$ of FM-sinewave signals. Using MATLAB, plot the (square root of the) energy operator output $\sqrt{\psi(x[n])}$ where $x[n]$ is the FM-sinewave signal in Equation (4.57) with $A = 10$, $a = 0.002$, $\Omega_0 = 0.2\pi$, $\Omega_m = 0.02\pi$, $\theta = 0$, and (a) $\Delta = \Omega_0$, (b) $\Delta = 0.2\Omega_0$, and (c) $\Delta = 0.01\Omega_0$.

(b) The outputs of Equation (4.58) due to input synthetic FM-sinewave signals correspond roughly to measurements made on actual speech using ψ. The instantaneous frequency $\Omega[n]$ plays the role of a time-varying formant. Thus, if we view Ω_0 as the center value of a formant, then the operator ψ followed by an envelope detector, followed by dividing with the envelope and square of the inverse sine, will yield the time-varying formant (within a single pitch period) as the instantaneous frequency $\Omega[n]$ of the FM signal. Figure 4.37 shows (a) a segment of a speech vowel /a/ sampled at $F_s = 10$ kHz and (b) the output from $\sqrt{\psi}$ when applied to a bandpass filtered version of (a) extracted around a formant at $F_0 = 1000$ Hz using a filter with impulse response $\exp(-b^2 n^2)\cos(\Omega_0 n)$, where $\Omega_0 = 2\pi F_0/F_s$ and $b = 800/F_s$. In Figure 4.37b there are two pulses per pitch period, and the exponentially-damped sine squared

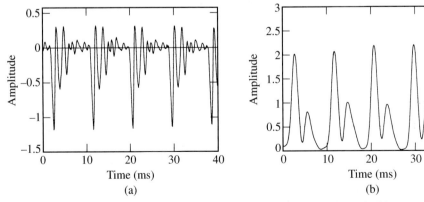

Figure 4.37 Result of energy operator on speech vowel /a/: (a) original speech; (b) output of energy operator.

model Equation (4.58) can be used to approximately represent the shape of these measured energy pulses. There are cases where we observe more than two pulses per glottal cycle, but also cases with only one major pulse per pitch period. Explain why these observations can be partially explained from the FM-sine model by comparing against your results in part (a) for different modulation indices.

BIBLIOGRAPHY

[1] T.V. Ananthapadmanabha and G. Fant, "Calculation of True Glottal Flow and Its Components," *Speech Communications*, vol. 1, pp. 167–184, 1982.

[2] L.L. Beranek, *Acoustics*, McGraw-Hill, New York, NY, 1954.

[3] M. Campbell and C. Greated, *The Musician's Guide to Acoustics*, Schirmer Books, New York, NY, 1987.

[4] D.G. Childers and C.F. Wong, "Measuring and Modeling Vocal Source-Tract Interaction," *IEEE Trans. Biomedical Engineering*, vol. 41, no. 7, pp. 663–671, July 1994.

[5] L. Cohen, *Time-Frequency Analysis*, Prentice Hall, Englewood Cliffs, NJ, 1995.

[6] P.B. Denes and E.N. Pinson, *The Speech Chain: The Physics and Biology of Spoken Language*, Anchor Press-Doubleday, Garden City, NY, 1973.

[7] G. Fant, *Acoustic Theory of Speech Production*, Mouton, The Hague, 1970.

[8] J.L. Flanagan, *Speech Analysis, Synthesis, and Perception*, Springer-Verlag, New York, NY, 1972.

[9] J.L. Flanagan and K. Ishizaka, "Computer Model to Characterize the Air Volume Displaced by the Vibrating Vocal Cords," *J. Acoustical Society of America*, vol. 63, pp. 1559–1565, Nov. 1978.

[10] C.R. Jankowski, *Fine Structure Features for Speaker Identification*, Ph.D. Thesis, Massachusetts Institute of Technology, Dept. of Electrical Engineering and Computer Science, June 1996.

[11] J.F. Kaiser, "On a Simple Algorithm to Calculate the 'Energy' of a Signal," *Proc. Int. Conf. Acoustics, Speech, and Signal Processing*, vol. 1, pp. 381–384, Albuquerque, NM, April 1990.

[12] D. Klatt, "Software for a Cascade/Parallel Formant Synthesizer," *J. Acoustical Society of America*, vol. 67, no. 3, pp. 971–995, 1980.

[13] D. H. Klatt and L.C. Klatt, "Analysis, Synthesis, and Perception of Voice Quality Variations among Female and Male Talkers," *J. Acoustical Society of America*, vol. 87, no. 2, pp. 820–857, 1990.

[14] P.K. Kundu, *Fluids Mechanics*, Academic Press, New York, NY, 1990.

[15] T. Kuhn, *The Structure of Scientific Revolution*, University of Chicago Press, 1970.

[16] I.T. Lim and B.G. Lee, "Lossless Pole-Zero Modeling of Speech Signals," *IEEE Trans. Speech and Audio Processing*, vol. 1, no. 3, pp. 269–276, July 1993.

[17] I.T. Lim and B.G. Lee, "Lossy Pole-Zero Modeling for Speech Signals," *IEEE Trans. Speech and Audio Processing*, vol. 4, no. 2, pp. 81–88, March 1996.

[18] M. Liu and A. Lacroix, "Pole-Zero Modeling of Vocal Tract for Fricative Sounds," *Proc. Int. Conf. Acoustics, Speech, and Signal Processing*, vol. 3, pp. 1659–1662, Munich, Germany, 1997.

[19] P.M. Morse and K.U. Ingard, *Theoretical Acoustics*, Princeton University Press, Princeton, NJ, 1986.

[20] L. Nord, T.V. Ananthapadmanabha, and G. Fant, "Signal Analysis and Perceptual Tests of Vowel Responses with an Interactive Source-Filter Model," *J. Phonetics*, vol. 14, pp. 401–404, 1986.

[21] C.L. Nikias and M.R. Raghuveer, "Bispectrum Estimation: A Digital Signal Processing Framework," *Proc. IEEE*, vol. 75, no. 1, pp. 869–891, July 1987.

[22] A.V. Oppenheim and R.W. Schafer, *Discrete-Time Signal Processing*, Prentice Hall, Englewood Cliffs, NJ, 1989.

[23] A. Pickles, *An Introduction to Auditory Physiology*, Second Edition, Academic Press, New York, NY, 1988.

[24] M.D. Plumpe, *Modeling of the Glottal Flow Derivative Waveform with Application to Speaker Identification*, Masters Thesis, Massachusetts Institute of Technology, Department of Electrical Engineering and Computer Science, Feb. 1997.

[25] M.D. Plumpe, T.F. Quatieri, and D.A. Reynolds, "Modeling of the Glottal Flow Derivative Waveform with Application to Speaker Identification," *IEEE Trans. Speech and Audio Processing*, vol. 7, no. 5, pp. 569–586, Sept. 1999.

[26] M.R. Portnoff, *A Quasi-One-Dimensional Digital Simulation for the Time-Varying Vocal Tract*, Masters Thesis, Massachusetts Institute of Technology, Dept. of Electrical Engineering and Computer Science, May 1973.

[27] A.D. Pierce, *Acoustics: An Introduction to Its Physical Principles and Applications*, McGraw-Hill, New York, NY, 1981.

[28] L.R. Rabiner and R.W. Schafer, *Digital Processing of Speech Signals*, Prentice Hall, Englewood Cliffs, NJ, 1978.

[29] M.M. Sondhi, "Resonances of a Bent Vocal Tract," *J. Acoustical Society of America*, vol. 79, pp. 1113–1116, April 1986.

[30] H.M. Teager and S.M. Teager, "Evidence for Nonlinear Production Mechanisms in the Vocal Tract," chapter in *Speech Production and Modeling*, W.J. Hardcastle and A. Marchal, eds., NATO Adv. Study Inst. Series D, vol. 55, Bonas, France; Kluwer Acad. Publ., Boston, MA, pp. 241–261, 1990.

[31] J. W. van den Berg, "On the Air Response and the Bernoulli Effect of the Human Larynx," *J. Acoustical Society of America*, vol. 29, pp. 626–631, 1957.

[32] K. Wang and S.A. Shamma, "Self-Normalization and Noise Robustness in Early Auditory Representations," *IEEE Trans. Speech and Audio Processing*, vol. 2, no. 3, pp. 421–435, July 1995.

CHAPTER **5**

Analysis and Synthesis of Pole-Zero Speech Models

5.1 Introduction

In Chapter 4, based on knowledge of speech production and the wave equation for sound propagation, we developed a continuous- and then a discrete-time transfer function for the relation between the acoustic pressure at the lips output and the volume velocity at the glottis. For the idealized voiced speech case, the transfer function contains poles that correspond to the resonances of the vocal tract cavity. Using a particular model of the glottal airflow, two additional poles were introduced outside the unit circle. The radiation load at the lips was represented by one zero inside the unit circle. More generally, the transfer function from the glottis to the lips also includes zeros that represent the energy-absorbing anti-resonances resulting from the back cavity during unvoiced plosives or fricatives, from the oral passage during nasal consonants, or from the nasal passage during nasalized vowels.

Speech sounds with periodic or impulsive sources are loosely categorized as "deterministic," while speech sounds with noise sources are loosely classified as "stochastic" sounds. In Sections 5.3 and 5.4 of this chapter we estimate parameters of an all-pole system function for both the deterministic and stochastic sound classes, respectively; we will see that the resulting solution equations for the two classes are similar in structure. The solution approach is referred to as *linear prediction analysis* from which we derive, in Section 5.6, a method of speech synthesis based on the all-pole model. We also see how this method of analysis is intimately associated with the concatenated lossless tube model of Chapter 4, and as such presents a mechanism for estimating the vocal tract shape from all-pole system parameters. In Section 5.5, we evaluate the "goodness" of the linear prediction solution both in the time domain with respect to the shape of the waveform obtained in synthesis, and in the frequency domain with respect to the magnitude and the phase of the resulting vocal tract impulse response. We then describe in Section 5.7 methods that generalize the all-pole system analysis to estimation of a model transfer function consisting of both poles and zeros. This leads to a "pitch synchronous" technique, based on the glottal closed phase, for separating the glottal flow waveform from the vocal tract impulse response. These methods do not constrain the glottal flow function to a maximum-phase two-pole model. Example estimates illustrate the typical glottal airflow, as well as speaker-dependent contributions such as aspiration and secondary glottal pulses, that occur in the breathy and diplophonic voices, described in Chapter 3. Because the methods require that we view speech through a sliding window, we begin in Section 5.2 with a brief return to short-time processing of speech.

5.2 Time-Dependent Processing

We have seen that an essential property of speech production is that the vocal tract and the nature of its source vary with time and that this variation can be rapid. Many analysis techniques, however, assume that these characteristics change relatively slowly, which means that, over a short-time interval, e.g., 20–40 ms, the vocal tract and its input are stationary. By "stationary," we mean that the vocal tract shape, and thus its transfer function, remain fixed (or nearly fixed) over this short-time interval. In addition, a periodic source is characterized by a steady pitch and glottal airflow function for each glottal cycle within the short-time interval; likewise, a noise source has fixed statistical properties over this time.

In analyzing the speech waveform, we apply a sliding window whose duration is selected to make the short-time stationarity assumption approximately valid. We can make the window arbitrarily short in time to meet this short-time criterion; this would provide a window short enough to resolve short-duration and rapidly-changing events without (most of the time) blending them with adjacent sounds, thus giving adequate *time resolution*. On the other hand, we want the window long enough to resolve fine structure in the speech spectrum such as individual harmonics and closely-spaced resonances, thus giving adequate *frequency resolution*. As we have seen in our discussion of the wideband and narrowband spectrograms in Chapter 3, these time-frequency objectives often cannot be met simultaneously, a reflection of the uncertainty principle that we reviewed in Chapter 2. We select a window duration, therefore, to trade off our objectives, typically of duration 20–40 ms. Our selected window slides at a *frame interval* sufficient to follow changing speech events, typically 5–10 ms, and thus adjacent sliding windows overlap in time. The specific shape of the window also contributes to the time and frequency resolution properties; typical windows are rectangular or tapered at their ends, and are characterized by different mainlobe and sidelobe structure. The rectangular window, for example, has a narrower mainlobe than the tapered Hamming window, but higher sidelobe structure.

In performing analysis over each window, we estimate the vocal tract transfer function parameters, e.g., vocal tract poles and zeros, as well as parameters that characterize the vocal tract input of our discrete-time model. The short-time stationarity condition requires that the parameters of the underlying system are nearly fixed under the analysis window and therefore that their estimation is meaningful.

5.3 All-Pole Modeling of Deterministic Signals

5.3.1 Formulation

We begin by considering a transfer function model from the glottis to the lips output for deterministic speech signals, i.e., speech signals with a periodic or impulsive source. From Chapter 4, during voicing the transfer function consists of glottal flow, vocal tract, and radiation load contributions given by the all-pole z-transform

$$H(z) = AG(z)V(z)R(z)$$

$$= \frac{A}{1 - \sum_{k=1}^{p} a_k z^{-k}}$$

where we have lumped the gain and glottal airflow (assumed modeled by two poles outside the unit circle) over a single glottal cycle into the transfer function $H(z)$. This corresponds to an idealized input $u_g[n]$ that is a periodic impulse train. (Observe that the notation $H(z)$ has taken on a different meaning from that in Chapter 4.) With a single-impulse input for a plosive, there is no glottal flow contribution, $G(z)$. Observe that we have made $R(z)$ all-pole, as well as $V(z)$ and $G(z)$. We saw in Chapter 4, however, that a simplified model for $R(z)$ is a single zero inside the unit circle. A zero inside the unit circle, however, can be expressed as an infinite product of poles inside the unit circle. To see this relation, first recall that the geometric series $\sum_{k=0}^{\infty} r^k$ converges to $\frac{1}{1-r}$ for $|r| < 1$ so that letting $r = az^{-1}$, we have

$$\sum_{k=0}^{\infty} (az^{-1})^k = \sum_{k=0}^{\infty} a^k z^{-k}$$

$$= \frac{1}{1 - az^{-1}}, \qquad |az^{-1}| < 1 \quad \text{or} \quad |z| > |a|.$$

Cross-multiplying, we then have

$$1 - az^{-1} = \frac{1}{\sum_{k=0}^{\infty} a^k z^{-k}}$$

$$= \frac{1}{\prod_{k=0}^{\infty}(1 - b_k z^{-1})}, \qquad |z| > |a|$$

which, in practice, is approximated by a finite set of poles as $a^k \to 0$ with $k \to \infty$. Consequently, $H(z)$ can be considered an all-pole representation, albeit representing a zero by a large number of poles is not efficient in light of estimating the zero directly. We study zero estimation methods later in this chapter. Our goal here is to estimate the filter coefficients a_k for a specific order p and the gain A. The poles can be inside or outside the unit circle, the outside poles corresponding to the maximum-phase two-pole model for $G(z)$.

The basic idea behind linear prediction analysis is that each speech sample is approximated as a *linear combination* of past speech samples. This notion leads to a set of analysis techniques for estimating parameters of the all-pole model. To motivate the idea, consider the

z-transform of the vocal tract input $u_g[n]$, $U_g(z)$, with a gain A [being embedded in $H(z)$], and let $S(z)$ denote the z-transform of its output. Then $S(z) = H(z)U_g(z)$ and we can write

$$H(z) = \frac{S(z)}{U_g(z)} = \frac{A}{1 - \sum_{k=1}^{p} a_k z^{-k}}$$

or

$$S(z)[1 - \sum_{k=1}^{p} a_k z^{-k}] = S(z) - \sum_{k=1}^{p} a_k S(z) z^{-k}$$
$$= AU_g(z)$$

which in the time domain is written as

$$s[n] = \sum_{k=1}^{p} a_k s[n - k] + A u_g[n]. \qquad (5.1)$$

Equation (5.1) is sometimes referred to as an *autoregressive* (AR) model because the output can be thought of as regressing on itself. The coefficients a_k are referred to as the linear prediction coefficients, and their estimation is termed *linear prediction analysis* [2]. Quantization of these coefficients, or of a transformed version of these coefficients, is called linear prediction coding (LPC) and will be described in Chapter 12 on speech coding.

Recall that in voicing we have lumped the glottal airflow into the system function so that $u_g[n]$ is a train of unit samples. Therefore, except for the times at which $u_g[n]$ is nonzero, i.e., every pitch period, from Equation (5.1) we can think of $s[n]$ as a linear combination of past values of $s[n]$, i.e.,

$$s[n] = \sum_{k=1}^{p} a_k s[n - k], \quad \text{when } u_g[n] = 0.$$

This observation motivates the analysis technique of linear prediction. Before describing this analysis method, we note a minor inconsistency in our formulation. We have assumed that the z-transform of the glottal airflow is approximated by two poles outside the unit circle, i.e., $G(z) = \frac{1}{(1-\beta z)^2}$. Because our all-pole transfer function is written in negative powers of z, and thus assumed causal, it is implied that any delay due to the glottal source has been removed. We will see shortly that removal of this delay is of no consequence in the analysis.

We first state some definitions [21]. A *linear predictor* of order p is defined by

$$\tilde{s}[n] = \sum_{k=1}^{p} \alpha_k s[n - k]. \qquad (5.2)$$

The sequence $\tilde{s}[n]$ is the prediction of $s[n]$ by the sum of p past weighted samples of $s[n]$. The system function associated with the pth order predictor is a finite-length impulse response (FIR) filter of length p given by

$$P(z) = \sum_{k=1}^{p} \alpha_k z^{-k}. \tag{5.3}$$

Prediction in the z-domain is therefore represented by $\tilde{S}(z) = P(z)S(z)$. The *prediction error sequence* is given by the difference of the sequence $s[n]$ and its prediction $\tilde{s}[n]$, i.e.,

$$e[n] = s[n] - \tilde{s}[n]$$

$$= s[n] - \sum_{k=1}^{p} \alpha_k s[n - k] \tag{5.4}$$

and the associated *prediction error filter* is defined as

$$A(z) = 1 - \sum_{k=1}^{p} \alpha_k z^{-k}$$

$$= 1 - P(z) \tag{5.5}$$

which is illustrated in Figure 5.1a.

Suppose now that a measurement $s[n]$ is the output of an all-pole system modeled as in Equation (5.1). Then if the predictor coefficients equal the model coefficients, i.e., $\alpha_k = a_k$,

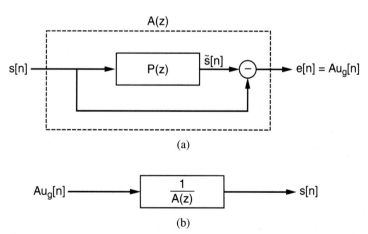

(a)

(b)

Figure 5.1 Filtering view of linear prediction: (a) prediction-error filter $A(z) = 1 - P(z)$; (b) recovery of $s[n]$ with $\frac{1}{A(z)}$ for $\alpha_k = a_k$. $A(z)$ is also considered the inverse filter because it can yield the input to an all-pole transfer function.

we can write the prediction error as

$$e[n] = s[n] - \sum_{k=1}^{p} \alpha_k s[n-k]$$

$$= \sum_{k=1}^{p} a_k s[n-k] + Au_g[n] - \sum_{k=1}^{p} \alpha_k s[n-k]$$

$$= Au_g[n]$$

and thus the input sequence $Au_g[n]$ can be recovered by passing $s[n]$ through $A(z)$. For this reason, under the condition $\alpha_k \approx a_k$, the prediction error filter $A(z)$ is sometimes called the *inverse filter*. Correspondingly, when we pass $Au_g[n]$ through the system $\frac{1}{A(z)}$, we obtain $s[n]$ as shown in Figure 5.1b. Because $A(z)$ is of finite order p, it consists of p zeros; likewise, its inverse $\frac{1}{A(z)}$ consists of p poles. Observe now that for voiced speech idealized with a periodic impulse train input, when $\alpha_k = a_k$, $e[n]$ is an impulse train and therefore equal to zero most of the time except at the periodic impulses. Likewise, when the speech waveform is a single impulse response, as for an idealized plosive sound, then $e[n]$ is zero except at the time of the impulsive input. In a stochastic context, when $u_g[n]$ is white noise, then the inverse filter "whitens" the input signal, as we will elaborate in Section 5.4.[1]

EXAMPLE 5.1   Consider an exponentially decaying impulse response of the form $h[n] = a^n u[n]$ where $u[n]$ is the unit step. Then the response to a scaled unit sample $A\delta[n]$ is simply

$$s[n] = A\delta[n] * h[n] = Aa^n u[n].$$

Consider now the prediction of $s[n]$ using a linear predictor of order $p = 1$. The prediction error sequence with $\alpha_1 = a$ is given by

$$e[n] = s[n] - as[n-1]$$

$$= A(a^n u[n] - aa^{n-1} u[n-1])$$

$$= Aa^n (u[n] - u[n-1])$$

$$= A\delta[n]$$

i.e., the prediction of the signal is exact except at the time origin. This is consistent with $A(z) = 1 - P(z) = 1 - az^{-1}$ being the inverse filter to $H(z)$.   ▲

These observations motivate linear prediction analysis that estimates the model parameters when they are unknown. The approach of linear prediction analysis is to find a set of prediction coefficients that minimizes the mean-squared prediction error over a short segment of the speech waveform. The minimization leads to a set of linear equations that can be solved efficiently for the linear predictor coefficients a_k and the gain A.

---

[1] Observe that our subscript notation "g" is not appropriate for plosive and noise sounds because the source of these sounds typically does not occur at the glottis.

5.3.2 Error Minimization

The prediction of a sample at a particular time n_o is illustrated in Figure 5.2a. Over all time, we can define a mean-squared prediction error as

$$E = \sum_{m=-\infty}^{\infty} (s[m] - \tilde{s}[m])^2$$

$$= \sum_{m=-\infty}^{\infty} e^2[m].$$

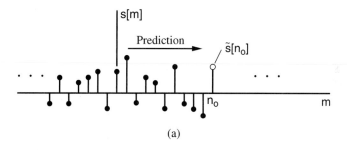

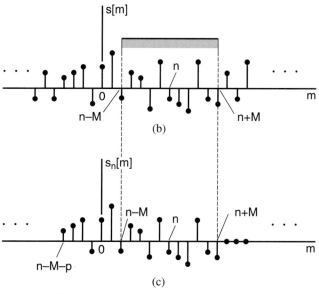

Figure 5.2 Short-time sequences used in linear prediction analysis: (a) prediction at time n_o; (b) samples in the vicinity of time n, i.e., samples over the interval $[n - M, n + M]$; (c) samples required for prediction of samples in the interval $[n - M, n + M]$. This set of samples, denoted by $s_n[m]$, includes samples inside and outside the interval $[n - M, n + M]$.

This definition, however, is not desirable because, as we have seen, we must assume short-time stationarity of the speech waveform. In practice, therefore, the prediction error is formed over a finite interval (Figure 5.2b), i.e.,

$$E_n = \sum_{m=n-M}^{n+M} e^2[m]$$

where the subscript n refers to "in the vicinity of time n." The time interval $[n - M, n + M]$ is called the *prediction error interval*. For notational simplicity, we have assumed an interval of odd length $2M + 1$ samples. Alternatively, we can write

$$E_n = \sum_{m=-\infty}^{\infty} e_n^2[m] \tag{5.6}$$

where

$$e_n[m] = s_n[m] - \sum_{k=1}^{p} \alpha_k s_n[m - k], \quad n - M \le m \le n + M \tag{5.7}$$

and zero elsewhere. The sequence $s_n[m]$ is defined in the vicinity of time n over the interval $[n - M - p, n + M]$ (and zero elsewhere), including p samples prior to time $n - M$ because these samples are needed in the prediction of the first p samples beginning at time $n - M$ (Figure 5.2c). The error E_n is a quadratic function in the unknowns α_k and as such can be visualized as a "quadratic bowl" in p dimensions. Our goal is to find the minimum of this function, i.e., the bottom of the quadratic bowl.

To minimize E_n over α_k, we set the derivative of E_n with respect to each variable α_k to zero [14],[21]:

$$\frac{\partial E_n}{\partial \alpha_i} = 0, \quad i = 1, 2, 3 \ldots p$$

resulting in

$$\frac{\partial E_n}{\partial \alpha_i} = \frac{\partial}{\partial \alpha_i} \sum_{m=-\infty}^{\infty} \left(s_n[m] - \sum_{k=1}^{p} \alpha_k s_n[m - k] \right)^2$$

$$= 2 \sum_{m=-\infty}^{\infty} \left(s_n[m] - \sum_{k=1}^{p} \alpha_k s_n[m - k] \right) \left(-\frac{\partial}{\partial \alpha_i} \sum_{k=1}^{p} \alpha_k s_n[m - k] \right).$$

Observing that $-s_n[m - i] = -\frac{\partial}{\partial \alpha_i} \sum_{k=1}^{p} \alpha_k s_n[m - k]$ because $\alpha_k s_n[m - k]$ is constant with respect to $\frac{\partial}{\partial \alpha_i}$ for $k \ne i$, we then have with $\frac{\partial E_n}{\partial \alpha_i}$ set to zero:

$$0 = 2 \sum_{m=-\infty}^{\infty} \left(s_n[m] - \sum_{k=1}^{p} \alpha_k s_n[m - k] \right) (-s_n[m - i]).$$

Multiplying through gives

$$\sum_{m=-\infty}^{\infty} s_n[m-i]s_n[m] = \sum_{k=1}^{p} \alpha_k \sum_{m=-\infty}^{\infty} s_n[m-i]s_n[m-k], \qquad 1 \le i \le p. \quad (5.8)$$

Define now the function

$$\Phi_n[i,k] = \sum_{m=-\infty}^{\infty} s_n[m-i]s_n[m-k], \qquad 1 \le i \le p, \qquad 1 \le k \le p.$$

Then we have

$$\sum_{k=1}^{p} \alpha_k \Phi_n[i,k] = \Phi_n[i,0], \qquad i = 1,2,3\ldots p. \quad (5.9)$$

The set of Equations (5.9) are sometimes referred to as the *normal equations* and can be put in matrix form as

$$\Phi \underline{\alpha} = \underline{b} \quad (5.10)$$

where the elements of the (i,k)th element of the $p \times p$ matrix Φ are given by $\Phi_n[i,k]$, the ith element of the $1 \times p$ vector $\underline{b}$ is given by $\Phi_n[i,0]$, and the ith element of the $1 \times p$ vector $\underline{\alpha}$ is given by α_i. The solution then requires obtaining $\Phi_n[i,k]$ from the speech samples, setting up the matrix equation, and inverting.

Finally, the minimum error for the optimal α_k can be derived as follows:

$$\begin{aligned}
E_n &= \sum_{m=-\infty}^{\infty} \left(s_n[m] - \sum_{k=1}^{p} \alpha_k s_n[m-k]\right)^2 \\
&= \sum_{m=-\infty}^{\infty} s_n^2[m] - 2\sum_{m=-\infty}^{\infty} s_n[m] \sum_{k=1}^{p} \alpha_k s_n[m-k] \\
&\quad + \sum_{m=-\infty}^{\infty} \sum_{k=1}^{p} \alpha_k s_n[m-k] \sum_{l=1}^{p} \alpha_l s_n[m-l].
\end{aligned} \quad (5.11)$$

Interchanging sums, the third term in Equation (5.11) can be rewritten as

$$\begin{aligned}
\sum_{m=-\infty}^{\infty} \sum_{k=1}^{p} \alpha_k s_n[m-k] \sum_{l=1}^{p} \alpha_l s_n[m-l] &= \sum_{l=1}^{p} \alpha_l \left\{ \sum_{k=1}^{p} \alpha_k \sum_{m=-\infty}^{\infty} s_n[m-l]s_n[m-k] \right\} \\
&= \sum_{l=1}^{p} \alpha_l \sum_{m=-\infty}^{\infty} s_n[m-l]s_n[m]
\end{aligned}$$

where the last step invoked the optimal solution [Equation (5.8)] for the bracketed term. Replacing the third term in Equation (5.11) by the above simplified form, we have

$$
\begin{aligned}
E_n &= \sum_{m=-\infty}^{\infty} s_n^2[m] - 2\sum_{k=1}^{p}\alpha_k \sum_{m=-\infty}^{\infty} s_n[m-k]s_n[m] \\
&\quad + \sum_{l=1}^{p}\alpha_l \sum_{m=-\infty}^{\infty} s_n[m-l]s_n[m] \\
&= \sum_{m=-\infty}^{\infty} s_n^2[m] - \sum_{k=1}^{p}\alpha_k \sum_{m=-\infty}^{\infty} s_n[m-k]s_n[m] \\
&= \Phi_n[0,0] - \sum_{k=1}^{p}\alpha_k \Phi_n[0,k].
\end{aligned}
\tag{5.12}
$$

When we don't know the order of an underlying all-pole (order p) transfer function and the data $s[n]$ follows an order p autoregression within the prediction error interval, the error E_n can be monitored to help establish p, because under this condition the error for a pth order predictor in theory equals that of a $(p+1)$st order predictor. The predictor coefficients can also be monitored in finding p because the value of predictor coefficients for $k > p$ equals zero. (The reader should argue both properties from Equations (5.9) and (5.12).)

In Equation (5.6), the prediction error $e_n[m]$ is non-zero only "in the vicinity" of the time n, i.e., $[n-M, n+M]$. We saw in Figure 5.2, however, in predicting values of the short-time sequence $s_n[m]$, p values outside (and to the left) of the prediction error interval $[n-M, n+M]$ are required. For a long stationary speech sequence, these values are available. This is in contrast to cases where data is not available outside of the prediction error interval as, for example, with rapidly-varying speech events. The solution that uses for prediction such samples outside the prediction error interval, but forms the mean-squared error from values only *within* the prediction interval, is referred to as the *covariance method* of linear prediction. When the signal measurement $s[n]$ corresponds to an underlying all-pole system function with a single or periodic impulse input, then the covariance method can give the exact model parameters from a short-time speech segment (Exercise 5.8). An alternative method assumes speech samples of zero value outside the prediction error interval, but forms the mean-squared error over all time. This method is referred to as the *autocorrelation method* of linear prediction. The zero-value approximation outside the short-time interval does not result in the exact all-pole model, as can the covariance method, under the above conditions. Nevertheless, the approximation leads to a very efficient computational procedure which, interestingly, we will see is tied to our concatenated tube model of Chapter 4.

Before leaving this section, we show that an alternative "geometric" approach, in contrast to the above "algebraic" approach, to solving the minimization problem leading to the normal equations is through the *Projection Theorem* [16],[25]. To do so, we interpret Equation (5.7) as an overdetermined set of equations in p unknowns:

$$
\begin{bmatrix}
s[n-M+0-1] & s[n-M+0-2] & \cdots & s[n-M+0-p] \\
s[n-M+1-1] & s[n-M+1-2] & \cdots & s[n-M+1-p] \\
s[n-M+2-1] & s[n-M+2-2] & \cdots & s[n-M+2-p] \\
\cdot & \cdot & \cdots & \cdot \\
\cdot & \cdot & \cdots & \cdot \\
\cdot & \cdot & \cdots & \cdot \\
s[n+M-1] & s[n+M-2] & \cdots & s[n+M-p]
\end{bmatrix}
\underbrace{}_{S_n}
\underbrace{
\begin{bmatrix}
\alpha_1 \\
\alpha_2 \\
\cdot \\
\cdot \\
\cdot \\
\alpha_p
\end{bmatrix}
}_{\alpha}
$$

$$
=
\underbrace{
\begin{bmatrix}
s_n[n-M] \\
s[n-M+1] \\
s[n-M+2] \\
\cdot \\
\cdot \\
\cdot \\
s[n+M]
\end{bmatrix}
}_{\underline{s}_n}
$$

The columns of the above $p \times (2M+1)$ matrix S_n are considered basis vectors for a vector space. The Projection Theorem states that the least-squared error occurs when the error vector $\underline{e}_n$ with components $e_n[m] = s_n[m] - \sum_{k=1}^{p} \alpha_k s_n[m-k]$, for m in the interval $[n-M, n+M]$, is orthogonal to each basis vector, i.e., $S_n^T \underline{e}_n = 0$, where "T" denotes transpose. This orthogonality condition can be shown to lead to the normal equations that can be equivalently written as [16],[26]

$$
(S_n^T S_n)\underline{\alpha} = -S_n^T \underline{s}_n.
$$

The beauty of this approach is its simplicity and its straightforward extension to other contexts such as stochastic signals.

5.3.3 Autocorrelation Method

In the previous section, we introduced the covariance method of linear prediction that uses samples outside the prediction error interval and can result in an exact solution when the signal follows an all-pole model. In this section, we describe an alternative approach to linear prediction analysis that is referred to as the *autocorrelation method* that is a suboptimal method, but a method that leads to an efficient and stable solution to the normal equations. The autocorrelation method assumes that the samples outside the time interval $[n-M, n+M]$ are all zero and extends the prediction error interval, i.e., the range over which we minimize the mean-squared error, to $\pm\infty$. For convenience, the short-time segment begins at time n and ends at time $n + N_w - 1$ ($N_w = 2M + 1$), along the time axis m, rather than being centered at time n as we had previously assumed. We then shift the segment to the left by n samples so that the first nonzero sample falls at $m = 0$. Alternately, we can view this operation as shifting the speech sequence $s[m]$ by n samples to the left and then windowing by an N_w–point rectangular window, $w[m] = 1$ for $m = 0, 1, 2 \ldots N_w - 1$, to form the sequence (Figure 5.3)

$$s_n[m] = s[m + n]w[m].$$

From this perspective, it can be seen with the help of the third-order predictor example in Figure 5.4 that the prediction error is nonzero only in the interval $[0, N_w + p - 1]$

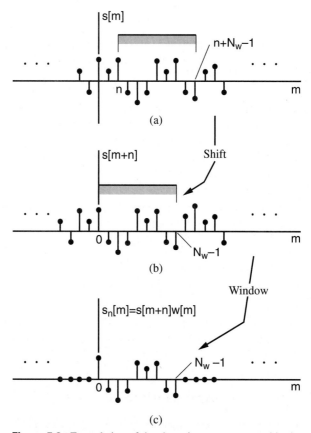

(a)

(b)

(c)

Figure 5.3 Formulation of the short-time sequence used in the autocorrelation method. In this interpretation, the waveform $s[m]$ [panel (a)] is shifted by n samples [panel (b)] and then windowed by an N_w–point rectangular sequence $w[m]$ [panel (c)]. The resulting sequence $s_n[m]$ is zero outside the interval $[0, N_w - 1]$.

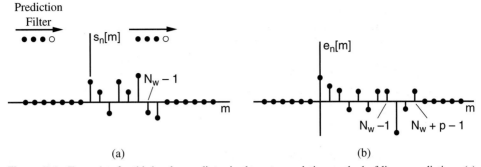

Figure 5.4   Example of a third-order predictor in the autocorrelation method of linear prediction: (a) sliding predictor filter; (b) prediction error. Prediction error is largest at the beginning and the end of the interval $[0, N_w + p - 1]$.

with N_w the window length and p the predictor order. The prediction error is largest at the left and right ends of the segment because we are predicting from and to zeros, respectively. Consequently, a tapered window is often used (e.g., Hamming). Without a tapered window, the mean-squared error Equation (5.6) may be dominated by end effects, tending to give the largest errors inside the window at the left edge, and outside the window at the right edge, as illustrated in Figure 5.4. On the other hand, with a tapered window, the data is distorted, hence biasing the estimates α_k. For example, an all-pole waveform may no longer follow an all-pole model when tapered. Tradeoffs exist, therefore, in the selection of a window for linear prediction analysis. Nevertheless, because of the finite-length window, the estimated α_k will never be exactly correct even when $s[n]$ follows an all-pole model.

With this formulation we now state the autocorrelation method of linear prediction [14],[21]. Let the mean-squared prediction error be given by

$$E_n = \sum_{m=0}^{N_w+p-1} e_n^2[m]$$

where the limits of summation refer to our new time origin and where the prediction error outside this interval is zero. We can then show that the normal equations take on the following form (Exercise 5.1)

$$\sum_{k=1}^{p} \alpha_k \Phi_n[i, k] = \Phi_n[i, 0], \qquad i = 1, 2, 3 \ldots p \qquad (5.13)$$

where

$$\Phi_n[i, k] = \sum_{m=0}^{N_w+p-1} s_n[m - i]s_n[m - k], \qquad 1 \leq i \leq p, \qquad 0 \leq k \leq p$$

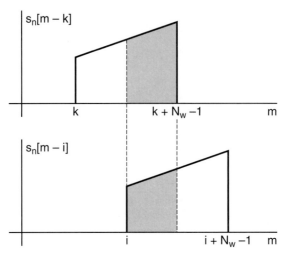

Figure 5.5 Construction of the autocorrelation function in the autocorrelation method. Overlapping regions of $s_n[m-k]$ and $s_n[m-i]$ are used in determining $\Phi_n[i,k]$.

with specific summation limits due to windowing the speech waveform. Now, from Figure 5.5, we can write $\Phi_n[i,k]$ as

$$\Phi_n[i,k] = \sum_{m=i}^{k+N_w-1} s_n[m-i]s_n[m-k]$$

recognizing that only the interval $[i, k + N_w - 1]$ contributes to the sum. With a change in variables $m \to m - i$, we can rewrite the function $\Phi_n[i,k]$ as

$$\Phi_n[i,k] = \sum_{m=0}^{N_w-1-(i-k)} s_n[m]s_n[m + (i - k)], \quad 1 \le i \le p, \quad 0 \le k \le p$$

which is a function of only the difference $i - k$ and so we denote it as

$$r_n[i - k] = \Phi_n[i,k].$$

Letting $\tau = i - k$, sometimes referred to as the correlation "lag," we obtain the *short-time autocorrelation function*

$$r_n[\tau] = \sum_{m=0}^{N_w-1-\tau} s_n[m]s_n[m + \tau]$$

$$= s_n[\tau] * s_n[-\tau] \tag{5.14}$$

which is the short-time sequence $s_n[m]$ *convolved* with itself flipped in time. The autocorrelation function is a measure of the "self-similarity" of the signal at different lags τ. When $r_n[\tau]$ is large, then signal samples spaced by τ are said to be highly correlated.

Some properties of $r_n[\tau]$ are as follows (Exercise 5.3):

P1: For an N–point sequence, $r_n[\tau]$ is zero outside the interval $[-(N-1), N-1]$.

P2: $r_n[\tau]$ is even in τ.

P3: $r_n[0] \geq r_n[\tau]$.

P4: $r_n[0]$ equals the energy in $s_n[m]$, i.e., $r_n[0] = \sum_{m=-\infty}^{\infty} |s_n[m]|^2$.

P5: If $s_n[m]$ is a segment of a periodic sequence, then $r_n[\tau]$ is periodic-like, reflecting $r_n[\tau]$ as a measure of periodic self-similarity. Because $s_n[m]$ is short-time, the overlapping data in the correlation decreases as τ increases, and thus the amplitude of $r_n[\tau]$ decreases as τ increases; e.g., with a rectangular window, the envelope of $r_n[\tau]$ decreases linearly.

P6: If $s_n[m]$ is a random white noise sequence, then $r_n[\tau]$ is impulse-like, reflecting self-similarity within a small neighborhood (see Appendix 5.A for a brief review of random process theory).

The following example illustrates these properties with speech waveforms:

EXAMPLE 5.2   Examples of autocorrelation functions for a vowel, unvoiced plosive, voiced plosive, and unvoiced fricative are shown in Figure 5.6. Speech segments were obtained by applying

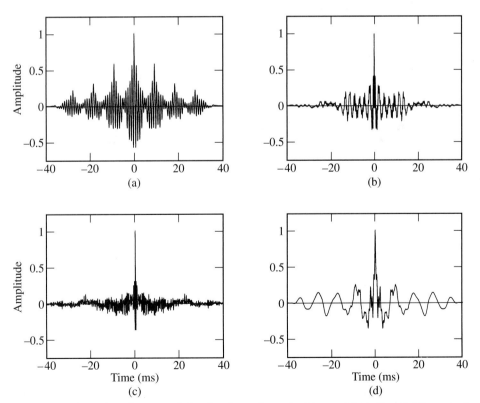

Figure 5.6 Illustration of autocorrelation functions of speech: (a) vowel /o/ in "p<u>o</u>p"; (b) unvoiced plosive /k/ in "ba<u>k</u>er"; (c) unvoiced fricative /f/ in "<u>f</u>ather"; (d) voiced plosive /g/ in "<u>g</u>o."

a short-time rectangular window of about 40 ms to various speech waveforms. The autocorrelation function for the fricative phone is noisy and non-periodic as well as impulse-like, while for the voiced phone it is periodic-like with decreasing amplitude. The autocorrelation function of the voiced plosive shows a high-frequency and a noise contribution superimposed on a periodic component, while its unvoiced counterpart is comprised mainly of a high-frequency component with overriding noise due to the turbulent source and aspiration following the burst.   ▲

With our new definitions, by letting $\Phi_n[i, k] = r_n[i - k]$, we can rewrite the normal equations as

$$\sum_{k=1}^{p} \alpha_k r_n[i - k] = r_n[i - 0], \quad 1 \le i \le p$$

or

$$\sum_{k=1}^{p} \alpha_k r_n[i - k] = r_n[i], \quad 1 \le i \le p$$

which represent p linear equations in p unknowns, α_k, for $1 \le k \le p$. Using the normal equation solution, it can then be shown that the corresponding minimum mean-squared prediction error is given by

$$E_n = r_n[0] - \sum_{k=1}^{p} \alpha_k r_n[k]. \tag{5.15}$$

As before, the normal equations can be put in matrix form:

$$R_n \underline{\alpha} = \underline{r}_n, \tag{5.16}$$

that is,

$$
\underbrace{\begin{bmatrix}
r_n[0] & r_n[1] & r_n[2] & \cdots & r_n[p-1] \\
r_n[1] & r_n[0] & r_n[1] & \cdots & r_n[p-2] \\
r_n[2] & r_n[1] & r_n[0] & \cdots & r_n[p-3] \\
\cdot & \cdot & \cdot & \cdots & \cdot \\
\cdot & \cdot & \cdot & \cdots & \cdot \\
\cdot & \cdot & \cdot & \cdots & \cdot \\
r_n[p-1] & r_n[p-2] & r_n[p-3] & \cdots & r_n[0]
\end{bmatrix}}_{R_n}
\underbrace{\begin{bmatrix}
\alpha_1 \\
\alpha_2 \\
\cdot \\
\cdot \\
\cdot \\
\alpha_p
\end{bmatrix}}_{\underline{\alpha}}
=
\underbrace{\begin{bmatrix}
r_n[1] \\
r_n[2] \\
\cdot \\
\cdot \\
\cdot \\
r_n[p]
\end{bmatrix}}_{\underline{r}_n}.
$$

The matrix R_n in Equation (5.16) has the property of being symmetric about the diagonal and all elements of the diagonal are equal. We will see that this structure, which is referred to as the "Toeplitz" property, implies an efficient algorithm for solution. Motivation for the name "autocorrelation method" now becomes clear; entries in the matrix R_n are the first p autocorrelation coefficients of $s_n[m]$. It can be shown that the columns of R_n are linearly independent and thus R_n is invertible [16],[25].

We now consider a number of properties of the autocorrelation method. As we stated earlier, for finite-length data the method can never give an exact solution, even when $s[n]$ follows an all-pole model. In certain cases, however, as the window length arbitrarily increases, the autocorrelation solution approaches the true solution. We illustrate this property in the following example adapted from McClellan [16]:

EXAMPLE 5.3   Consider an exponentially decaying impulse response of the form $h[n] = a^n u[n]$, with $u[n]$ the unit step. Then the response to a unit sample is simply

$$s[n] = \delta[n] * h[n] = a^n u[n]$$

and has a one-pole z-transform

$$S(z) = \frac{1}{1 - az^{-1}}, \quad |a| < 1.$$

We are interested in estimating the parameter a of a first-order all-pole model from the first N samples of $s[n]$ by using the autocorrelation method of linear prediction. This is equivalent to applying an N–point rectangular window in the interval $[0, N - 1]$. Using the closed-form solution for a finite geometric series, the first two autocorrelation coefficients can be computed from the first N samples of $s[n]$ as [16]

$$r_0[0] = \sum_{m=0}^{N-1} a^m a^m = \frac{1 - a^{2N}}{1 - a^2}$$

$$r_0[1] = \sum_{m=0}^{N-2} a^m a^{m+1} = a\frac{1 - a^{2N-2}}{1 - a^2}$$

where the subscript "0" refers to the data segment located at time $n = 0$. The solution for the first-order predictor coefficient α_1 should be a, but it is always biased[2] for finite N. From the autocorrelation normal equations, Equation (5.16), we can compute α_1 as

$$\alpha_1 = \frac{r_0[1]}{r_0[0]} = a\frac{1 - a^{2N-2}}{1 - a^{2N}}$$

which approaches the true solution a as $N \to \infty$. We can also show that the minimum squared error, Equation (5.15), for this solution is given by (Exercise 5.5)

$$E_0 = \frac{1 - a^{4N-2}}{1 - a^{2N}}$$

that as $N \to \infty$ gives the value of $s^2[0] = 1$, which is consistent with the first-order predictor's inability to predict the first nonzero sample of the sequence. We saw this property in Example 5.1 where the prediction error sequence for the true predictor is given by $e[n] = s[n] - as[n-1] = \delta[n]$, i.e., with $\alpha_1 = a$, the prediction of the signal is exact except at the time origin.

Source: J.H. McClellan, "Parametric Signal Modeling" [16]. ©1988, Pearson Education, Inc. Used by permission.   ▲

---

[2] It is important to note that the error incurred in predicting the beginning of a causal impulse response is not the cause of bias in either the autocorrelation or the covariance method because the "model's impulse response is assumed produced with zero initial conditions" [16].

The previous example illustrates that, with enough data, the autocorrelation method yields a solution close to the true single-pole model for an impulse input. When the underlying measured sequence is the impulse response of an arbitrary all-pole sequence, then this result can be shown to also hold. There are a number of scenarios in the speech context, however, in which the true solution is not obtained with the autocorrelation method even with an arbitrarily long data sequence. As an important example, consider a periodic sequence, simulating a steady voiced sound, formed by convolving a periodic impulse train $p[n]$ with an all-pole impulse response $h[n]$. The z-transform of $h[n]$ is given by

$$H(z) = \frac{1}{1 - \sum_{k=1}^{p} \alpha_k z^{-k}}$$

so that

$$h[n] = \sum_{k=1}^{p} \alpha_k h[n - k] + \delta[n]$$

where we have made the gain of the impulse input of value unity. By multiplying both sides of the above difference equation by $h[n - i]$ and summing over n and noting that $h[n]$ is causal, we can show that (Exercise 5.7)

$$\sum_{k=1}^{p} \alpha_k r_h[i - k] = r_h[i], \quad 1 \leq i \leq p \tag{5.17}$$

where the autocorrelation of $h[n]$ is denoted by $r_h[\tau] = h[\tau] * h[-\tau]$ and which represents an alternate approach to obtaining the normal equations when the signal of interest is known *a priori* to be an all-pole impulse response. Suppose now that a periodic waveform $s[n]$ is constructed by running a periodic impulse train through $h[n]$, i.e.,

$$s[n] = \sum_{k=-\infty}^{\infty} h[n - kP]$$

where P is the pitch period. The normal equations associated with $s[n]$, windowed over multiple pitch periods, for an order p predictor, are given by

$$\sum_{k=1}^{p} \alpha_k r_n[i - k] = r_n[i], \quad 1 \leq i \leq p \tag{5.18}$$

where $r_n[\tau]$ can be shown to equal periodically repeated replicas of $r_h[\tau]$, i.e.,

$$r_s[\tau] = \sum_{k=-\infty}^{\infty} r_h[\tau - kP]$$

but with decreasing amplitude due to the window (Exercise 5.7). The autocorrelation function of the windowed $s[n]$, $r_n[\tau]$, can thus be thought of as "aliased" versions of $r_h[\tau]$. When the aliasing is minor, the two solutions, Equations (5.17) and (5.18), are approximately equal. The accuracy of this approximation, however, decreases as the pitch period decreases because the autocorrelation function replicas, repeated every P samples, overlap and distort the underlying desired $r_h[\tau]$ and this effect is potentially more severe for higher-pitched speakers (Figure 5.7). To summarize, from Equation (5.17), if the autocorrelation function in the normal equations equals that of $h[n]$, then the solution to the normal equations must equal the α_k's of $H(z)$ and thus the true solution. This true solution is achieved approximately for a periodic waveform for large pitch periods.

In practice, in the autocorrelation method when a finite segment is extracted, and even with little autocorrelation aliasing present, distortion in the true autocorrelation can occur due to window truncation or tapering effects, as well as time-variation of the vocal tract (Exercise 5.10). We will see later in this chapter that the covariance method, on the other hand, can obtain the true solution even from a periodic waveform. The covariance method, however, has the disadvantage that, unlike the autocorrelation method, stability is not guaranteed when the underlying signal does not follow an all-pole model [16],[21].

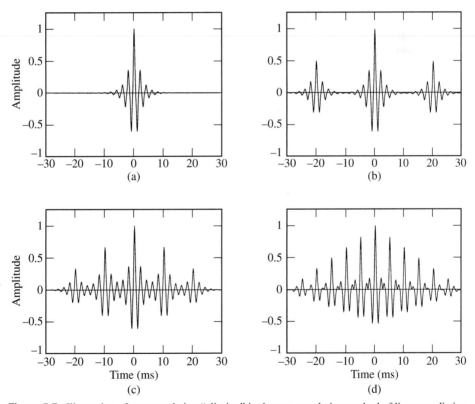

Figure 5.7 Illustration of autocorrelation "aliasing" in the autocorrelation method of linear prediction for a windowed periodic waveform: (a) $r_h[\tau]$; (b) $r_n[\tau]$ for 50-Hz pitch; (c) $r_n[\tau]$ for 100-Hz pitch; (d) $r_n[\tau]$ for 200-Hz pitch. Aliasing increases with rising pitch. For the 200-Hz pitch case, low-order autocorrelation coefficients are considerably different from those of $r_h[\tau]$, thus illustrating that accuracy of linear prediction analysis decreases with increasing pitch. In this example, $h[m]$ consists of two poles at 500 Hz and 1500 Hz and $s_n[m]$ is the convolution of $h[m]$ with a periodic impulse train, rectangularly windowed over a 40-ms duration.

Observe that we have not yet computed the unknown gain A of our model. We will return to this in a moment, but first we describe an efficient recursive method for solving the normal equations for the autocorrelation method. This recursive solution is the basis for a number of important properties of the autocorrelation method.

5.3.4 The Levinson Recursion and Its Associated Properties

Although we can solve for the predictor coefficients using a matrix inverse in Equations (5.10) and (5.16), the Toeplitz form of R_n in the normal equations (5.16) of the autocorrelation method leads to a more efficient recursive approach. A direct inverse, i.e., $\underline{\alpha} = R_n^{-1} \underline{r}_n$ requires on the order of p^3 multiplies and additions using a conventional inversion algorithm such as Gaussian elimination for matrix equations of the general form $A\underline{x} = \underline{b}$ [25]. On the other hand, the recursive solution requires on the order of p^2 multiplies and additions and, furthermore, also leads to a direct link to the concatenated lossless tube model of Chapter 4, and thus a mechanism for estimating the vocal tract area function from an all-pole-model estimation. The recursive method was first published in 1947 by Levinson [13] and is known as the *Levinson recursion*.

The Recursion — The basic idea of the Levinson recursion is to build an order $i + 1$ solution from an order i solution until we reach the desired order p. Each recursion solves the ith-pole (or ith-order prediction) problem, finding the solution that minimizes the mean-squared error for each order predictor, by updating the lower-order solution. Alternatively, using the Projection Theorem, we can interpret each solution as giving an error vector that is orthogonal to columns of a matrix similar in structure to S_n but of lower order; at each stage the error vector becomes orthogonal to a vector space of increasing dimensionality, becoming progressively decorrelated and eventually "whitened" when $i = p$ [26]. Specifically, the Levinson recursion is given in the following steps. For simplicity, we occasionally drop the subscript n notation on $s_n[m]$ and $r_n[\tau]$, unless otherwise stated.

S1: Initialize the prediction error and predictor coefficient of an order-zero predictor, i.e.,

$$\alpha_0^0 = 0$$
$$E^0 = r[0].$$

Because we assume that the initial predictor is zero, i.e., all predictor coefficients α_k are zero, the predicted sequence $\tilde{s}[n] = 0$. The mean-squared prediction error, therefore, is the energy in the sequence which, as we saw in a previous section, is the first value of the autocorrelation of the windowed $s[n]$, i.e., $r[0]$.

S2: Form a weighting factor for the ith pole model as

$$k_i = \left(r[i] - \sum_{j=1}^{i-1} \alpha_j^{i-1} r[i-j] \right) / E^{i-1}.$$

S3: Using the result of step **S2**, update coefficients for the ith pole model as

$$\alpha_i^i = k_i$$
$$\alpha_j^i = \alpha_j^{i-1} - k_i \alpha_{i-j}^{i-1}, \qquad 1 \le j \le i-1$$

where we have set the highest-order coefficient to k_i and where k_i has also been used as a weighting factor on the coefficients of order $i-1$ to obtain the coefficients of order i. We refer to the constants k_i as the *partial correlation coefficients* for which the motivation will shortly become clear. These coefficients are also called the *PARCOR* coefficients.

S4: Update the minimum mean-squared prediction error for the ith pole model as

$$E^i = (1 - k_i^2)E^{i-1}.$$

E^i, being a mean-squared prediction error, provides a way of monitoring the accuracy of prediction when the predictor order is unknown.

S5: Repeat steps **S2** to **S4** for $i = 1, 2, \ldots p$. At the pth step we have the desired pth order predictor coefficients

$$\alpha_j^* = \alpha_j^p, \quad 1 \le j \le p$$

where "*" denotes the (optimal) coefficients that minimize the mean-squared pth-order prediction error.

The derivation of the Levinson recursion can be found in [15],[16]. From step **S2**, we can show on each iteration that the predictor coefficients α_k can be written as solely functions of the autocorrelation coefficients (Exercise 5.11). Finally, the desired transfer function is given by

$$H(z) = \frac{A}{1 - \sum_{k=1}^{p} \alpha_k^* z^{-1}}$$

where the gain A has yet to be determined. Calculation of the gain depends on certain properties of the estimated prediction coefficients and will be described later in this section.

Properties — Properties of the autocorrelation method motivated by the Levinson recursion are given below.

P1: The magnitude of each of the partial correlation coefficients is *less than unity*, i.e., $|k_i| < 1$. To show this, we first recall that the mean-squared prediction error is always greater than zero because we cannot attain perfect prediction even when the sequence follows an all-pole model. Therefore, from the Levinson recursion, we have

$$E^i = (1 - k_i^2)E^{i-1} > 0$$

from which

$$(1 - k_i^2) = \frac{E^i}{E^{i-1}} > 0$$

and so it follows that

$$k_i^2 < 1 \quad \text{or} \quad |k_i| < 1.$$

P2: The roots of the predictor polynomial $A(z)$ lie *strictly inside the unit circle*, i.e., the transfer function $H(z)$ has poles inside the unit circle and thus is minimum-phase. This can be shown by using the previous property that $|k_i| < 1$ [16]. Therefore, the condition $|k_i| < 1$ is sufficient for stability. In contrast, with the covariance method, the poles are not restricted to lie inside the unit circle, and thus stability is not guaranteed.

P3: Suppose that a stable sequence $s[n]$ has a z-transform $S(z)$ which is all-pole, but with both a minimum-phase and a maximum-phase contribution:

$$S(z) = \frac{A}{\prod_{k=1}^{N_i}(1 - c_k z^{-1})(1 - c_k^* z^{-1}) \prod_{k=1}^{N_o}(1 - d_k z)(1 - d_k^* z)}, \quad |c_k|, |d_k| < 1$$

where we assume poles occur in complex conjugate pairs. If the prediction error interval is infinite in duration and two-sided, then the autocorrelation method yields the true minimum-phase poles. The resulting maximum-phase poles, however, fall in the conjugate reciprocal locations of their true counterparts, i.e., the maximum-phase poles are converted to minimum-phase poles, so that the resulting transfer function is of the form

$$\hat{S}(z) = \frac{A}{\prod_{k=1}^{N_i}(1 - c_k z^{-1})(1 - c_k^* z^{-1}) \prod_{k=1}^{N_o}(1 - d_k z^{-1})(1 - d_k^* z^{-1})}, \quad |c_k|, |d_k| < 1.$$

We can argue this property as follows. The sequence $s[n]$ follows an all-pole model, with some poles inside and some poles outside the unit circle. Observe that if we flip all maximum-phase poles inside the unit circle to their conjugate reciprocal locations, then the autocorrelation of $s[n]$, i.e., $r_s[\tau] = s[n] * s[-n]$ remains intact. This invariance can be seen by first observing that the spectral magnitude of $s[n]$, when squared and inverse Fourier transformed, gives the autocorrelation function. As we saw in Chapter 2, the squared spectral magnitude function does not change with conjugate reciprocal zero-flipping. Now, we know from our property (**P1**) above that the autocorrelation method of linear prediction must yield poles inside the unit circle. Using the fact that the autocorrelation method gives the true solution for a minimum-phase sequence with an infinitely-long prediction error interval (as in Example 5.3), the reader can then deduce the remainder of the argument.

It is interesting to observe the meaning of this property with respect to the spectral phase of a measurement and its all-pole estimate. Suppose that $s[n]$ follows an all-pole model, and suppose the prediction error function is defined over all time, i.e., we assume no window truncation effects. Let us write the Fourier transform of $s[n]$ as

$$S(\omega) = M_s(\omega)e^{j\theta_s(\omega)}$$
$$= M_s(\omega)e^{j[\theta_s^{min}(\omega)+\theta_s^{max}(\omega)]}$$

where $\theta_s^{min}(\omega)$ and $\theta_s^{max}(\omega)$ are the Fourier transform phase functions for the minimum- and maximum-phase contributions of $S(\omega)$, respectively. Then it is straightforward to argue (Exercise 5.14) that the autocorrelation method solution can be expressed as

$$\hat{S}(\omega) = M_s(\omega)e^{j[\theta_s^{min}(\omega)-\theta_s^{max}(\omega)]}$$
$$= M_s(\omega)e^{j[\theta_s(\omega)-2\theta_s^{max}(\omega)]}, \tag{5.19}$$

i.e., the maximum-phase component is made negative, thus distorting the desired phase function of $S(\omega)$. In the following example, we show an effect of this phase change in the linear prediction analysis of speech.

EXAMPLE 5.4   Consider the discrete-time model of the complete transfer function from the glottis to the lips derived in Chapter 4, but without zero contributions from the radiation and vocal tract:

$$H(z) = \frac{A}{(1 - \beta z)^2 \prod_{k=1}^{C_i}(1 - c_k z^{-1})(1 - c_k^* z^{-1})}.$$

In this model the glottal flow is represented by the double maximum-phase pole at $z = \beta$. Suppose we measure a single impulse response denoted by $h[n]$ which is equal to the inverse z-transform of $H(z)$. Then application of the autocorrelation method with a model order set to the number of poles of $H(z)$, i.e., $p = 2 + 2C_i$, with the prediction error defined over the entire extent of $h[n]$, yields a solution

$$\hat{H}(z) = \frac{A}{(1 - \beta z^{-1})^2 \prod_{k=1}^{C_i}(1 - c_k z^{-1})(1 - c_k^* z^{-1})}$$

i.e., the two maximum-phase poles due to the glottal flow are converted to their minimum-phase counterparts (Figure 5.8).

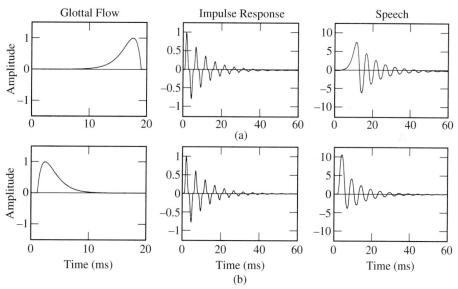

Figure 5.8   Transformation of glottal maximum-phase poles to minimum-phase poles by the autocorrelation method of linear prediction: (a) maximum-phase glottal flow (with displacement from the time origin), vocal tract impulse response, and resulting speech waveform for one excitation impulse; (b) minimum-phase glottal flow (with displacement from the time origin), vocal tract impulse response, and resulting speech waveform for one excitation impulse as obtained by the autocorrelation method of linear prediction. This example illustrates how the "attack" of an impulse response (and thus its perception) can be altered through linear prediction analysis. The simulated vocal tract consists of two resonances at 200 Hz and 600 Hz. The glottal flow waveform in (a) is the time-reversed decaying exponential $0.93^n u[n]$ (with $u[n]$ the unit step) convolved with itself.

Consider now the true frequency response written as

$$H(\omega) = M_h(\omega)e^{j\theta_h(\omega)}$$
$$= M_h(\omega)e^{j[\theta_v(\omega)+\theta_g(\omega)]}$$

where $\theta_v(\omega)$ is the minimum-phase contribution due to the vocal tract poles inside the unit circle, and $\theta_g(\omega)$ is the maximum-phase contribution due to the glottal poles outside the unit circle. From Equation (5.19) and the above expression for $\hat{H}(z)$, the resulting estimated frequency response can be expressed as

$$\hat{H}(\omega) = M_h(\omega)e^{j[\theta_v(\omega)-\theta_g(\omega)]}$$
$$= M_h(\omega)e^{j[\theta_h(\omega)-2\theta_g(\omega)]}$$

so that the original phase function has been modified by the subtraction of twice the glottal phase component. This phase distortion can have a perceptual consequence when constructing a synthetic speech waveform from $\hat{h}[n]$ [the inverse Fourier transform of $\hat{H}(\omega)$] because, as seen in Figure 5.8, the gradual onset of the glottal flow, and thus of the speech waveform during the open phase of the glottal cycle, is transformed to a "sharp attack," consistent with the energy concentration property of minimum-phase sequences that we reviewed in Chapter 2.   ▲

P4: There is a *one-to-one correspondence* between the two sets of parameters $[\alpha_1, \alpha_2, \dots , \alpha_p, A]$ and $[r_n[0], r_n[1], \dots , r_n[p]]$. We have already indicated that the predictor coefficients are uniquely determined by the first p autocorrelation coefficients (Exercise 5.11). It is also possible to show that we can go in the other direction [16], i.e., the autocorrelation can be obtained from the predictor coefficients. Likewise, it is possible to show that the set consisting of partial correlation coefficients along with the gain, i.e., $[k_1, k_2, \dots , k_p, A]$, have a one-to-one correspondence with either of the two sets. Although we will not formally prove these correspondences here, we can motivate them by reversing the Levinson recursion to obtain the partial correlation coefficients from the predictor coefficients [14]:

$$k_i = \alpha_i^i$$
$$\alpha_j^{i-1} = \left(\frac{1}{1-k_i^2}\right)(\alpha_j^i + k_i\alpha_{i-j}^i), \qquad j = 1, 2, \dots i - 1 \qquad (5.20)$$

thus obtaining lower-order models from higher-order models, rather than higher from lower order as in the (forward) Levinson recursion. You will prove this reverse Levinson recursion in Exercise 5.12 and use it in that problem to argue the one-to-one correspondence between different parameter sets.

P5: This next property is called *autocorrelation matching*. Consider a measurement $s_n[m] = s[n + m]w[m]$ with autocorrelation $r_n[\tau]$. Let $h[n]$ denote the impulse response associated with the estimated transfer function of order p, $H(z) = \frac{A}{1-\sum_{k=1}^p \alpha_k z^{-k}}$, obtained from the autocorrelation method of linear prediction using the measurement $s_n[m]$. We denote the autocorrelation of $h[n]$ by $r_h[\tau]$. Suppose now that the energy in the impulse response $h[n]$ equals the energy in the measurement $s_n[m]$, i.e., $r_h[0] = r_n[0]$. Then it follows from this constraint that the two autocorrelation functions are equal for $|\tau| < p$, i.e.,

$$r_h[\tau] = r_n[\tau], \qquad \text{for} \quad |\tau| \leq p. \qquad (5.21)$$

The proof of this property is quite involved and so we refer the interested reader to [16]. In the following subsection, we use the autocorrelation matching property to compute the gain A of the estimated system response; in particular, we show that when $r_h[0] = r_n[0]$ then the squared gain factor is given by the minimum average prediction error and thus can be expressed by

$$A^2 = E_n = r_n[0] - \sum_{k=1}^{p} \alpha_k r_n[k].$$

Computation of Gain in Autocorrelation Method — One approach[3] to select the gain A is to require that the energy in the all-pole impulse response $h[n]$ equals the energy in the measurement $s[n]$, i.e., $r_h[0] = r_n[0]$. As we have seen, this constraint results in the matching of the autocorrelation of the two sequences for $|\tau| < p$, a property that will be exploited in determining the value of A.

We begin with the all-pole transfer function

$$H(z) = \frac{A}{1 - \sum_{k=1}^{p} \alpha_k z^{-k}} \tag{5.22}$$

where α_k for $k = 1, 2 \ldots p$ are determined by the autocorrelation method using short-time autocorrelation coefficients $r_n[\tau]$, $\tau = 0, 1, 2 \ldots p$. Transforming Equation (5.22) to the time domain, we have

$$h[m] = \sum_{k=1}^{p} \alpha_k h[m - k] + A\delta[m] \tag{5.23}$$

where $\delta[n]$ is the unit sample function. Multiplying Equation (5.23) by $h[m - i]$ results in

$$h[m]h[m - i] = \sum_{k=1}^{p} \alpha_k h[m - k]h[m - i] + A\delta[m]h[m - i], \tag{5.24}$$

and letting $i = 0$ and summing over m gives

$$\sum_{m=-\infty}^{\infty} h[m]h[m] = \sum_{k=1}^{p} \alpha_k \sum_{m=-\infty}^{\infty} h[m - k]h[m] + A \sum_{m=-\infty}^{\infty} \delta[m]h[m]. \tag{5.25}$$

Because $h[0] = A$ from Equation (5.23), the last term in Equation (5.25) becomes A^2 and so Equation (5.25) can be written as

$$r_h[0] = \sum_{k=1}^{p} \alpha_k r_h[k] + A^2. \tag{5.26}$$

---

[3] We will see in a later section of this chapter that this particular selection of the gain A can be placed within a more rigorous framework in which our time-domain prediction error function E_n is mapped to the frequency domain. Nevertheless, in practice it may be desirable to require the energy in $h[n]$ to equal the energy over a pitch period, rather than over the complete short-time waveform, the energy in a pitch period perhaps more accurately reflecting the energy in the vocal tract impulse response.

Under the condition that

$$\text{Energy in } h[m] \; = \; \text{Energy in short time measurement } s_n[m]$$

or

$$r_h[0] \; = \; r_n[0] \tag{5.27}$$

then the autocorrelation matching property holds, i.e.,

$$r_h[\tau] \; = \; r_n[\tau] \qquad |\tau| \; \leq \; p. \tag{5.28}$$

Therefore, from Equations (5.26) and (5.28),

$$A^2 \; = \; r_h[0] \; - \; \sum_{k=1}^{p} \alpha_k r_h[k]$$

$$= \; r_n[0] \; - \; \sum_{k=1}^{p} \alpha_k r_n[k]. \tag{5.29}$$

Finally, from an earlier result for the prediction error (Equation (5.15)),

$$A^2 \; = \; E_n \; = \; r_n[0] \; - \; \sum_{k=1}^{p} \alpha_k r_n[k] \tag{5.30}$$

where E_n is the average minimum prediction error for the (optimal) pth-order predictor. Therefore, requiring that the energy in the all-pole impulse response $h[m]$ equals the energy in the measurement $s_n[m]$ leads to a squared gain equal to the minimum prediction error.

5.3.5 Lattice Filter Formulation of the Inverse Filter

Recall that we wrote the pth-order model as

$$H(z) \; = \; \frac{A}{A(z)}$$

where

$$A(z) \; = \; 1 \; - \; \sum_{k=1}^{p} \alpha_k z^{-k}$$

was referred to as the *inverse filter* or as the *prediction error filter*. In this section, we exploit the Levinson recursion, together with prediction error filters at different stages of the recursion, to obtain the *lattice filter* formulation of the autocorrelation method [15],[21].

Lattice Structure — Consider the ith-order prediction error filter

$$A^i(z) \; = \; 1 \; - \; \sum_{k=1}^{i} \alpha_k^i z^{-k}$$

which we can think of as the ith stage of the Levinson recursion. The output to $A^i(z)$ for the short-time input sequence $s_n[m]$ is given by (Henceforth, for simplicity in this section, we again drop the n subscript notation.)

$$e^i[m] = s[m] - \sum_{k=1}^{i} \alpha_k^i s[m-k].$$ (5.31)

We can think of $e^i[m]$ as the *forward* prediction error sequence for the ith-order prediction error (inverse) filter. In the z-domain, this forward prediction error is given by

$$E^i(z) = A^i(z)S(z).$$ (5.32)

Consider now the counterpart to the forward prediction error sequence, i.e., we define the *backward* prediction error sequence as

$$b^i[m] = s[m-i] - \sum_{k=1}^{i} \alpha_k^i s[m-i+k]$$ (5.33)

which represents the prediction of $s[m-i]$ from i samples of the future input, i.e., the samples $s[m-i+k]$ for $k = 1, 2, \ldots i$ *follow* the sample $s[m-i]$, as illustrated in Figure 5.9a. The z-transform of $b^i[m]$ can be found as follows. Taking the z-transform of Equation (5.33),

$$B^i(z) = z^{-i}S(z) - \sum_{k=1}^{i} \alpha_k^i z^{-i} z^k S(z)$$

$$= z^{-i}S(z)[1 - \sum_{k=1}^{i} \alpha_k^i z^k]$$

and because $A^i(z) = 1 - \sum_{k=1}^{i} \alpha_k^i z^{-k}$, then

$$B^i(z) = z^{-i} A^i(z^{-1}) S(z).$$ (5.34)

With the above time- and frequency-(z) domain expressions for the forward and backward prediction errors and using step **S3** of the Levinson recursion, i.e., $\alpha_j^i = \alpha_j^{i-1} - k_i \alpha_{i-j}^{i-1}$, $1 \le j \le i - 1$, with much algebra, we can arrive at the following relations (Appendix 5.B):

$$e^i[m] = e^{i-1}[m] - k_i b^{i-1}[m-1]$$
$$b^i[m] = b^{i-1}[m-1] - k_i e^{i-1}[m]$$ (5.35)

which can be interpreted as the forward and backward prediction error sequences, respectively, for the ith-order predictor in terms of prediction errors of the $(i-1)$th-order predictor. The recursion in Equation (5.35) begins with the 0th-order predictor giving

$$e^0[m] = s[m]$$
$$b^0[m] = s[m]$$

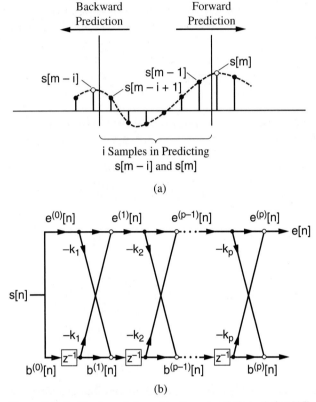

Figure 5.9 Lattice formulation of the linear prediction solution [15], [21]: (a) forward and backward prediction using an ith-order predictor; (b) flow graph of the lattice implementation.

SOURCE: L.R. Rabiner and R.W. Schafer, *Digital Processing of Speech Signals* [21]. ©1978, Pearson Education, Inc. Used by permission.

because the 0th-order predictor is equivalent to no prediction. These equations can be depicted by a flow graph which we refer to as a *lattice network* (Figure 5.9b). The output of the pth stage of the lattice equals the forward prediction error for the pth-order predictor, i.e., $e^P[m] = e[m]$ which equals the output of $A(z)$. When the underlying measurement follows an all-pole model, the output of the lattice is therefore an impulse for $\alpha_k = a_k$. The lattice represents an elegant implementation of the prediction error (inverse) filter $A(z)$.

The partial correlation coefficients k_i required in the lattice are obtained by way of the Levinson recursion. This requires computation of the predictor coefficients at each stage. On the other hand, it was shown by Itakura and Saito [11] that the partial correlation coefficients could (remarkably!) be found recursively through the prediction errors derived from the lattice. In particular, the partial correlation coefficients can be written as

$$k_i = \frac{\sum_{n=-\infty}^{\infty} e^{i-1}[n] b^{i-1}[n]}{\sum_{n=-\infty}^{\infty} (e^{i-1}[n])^2 \sum_{n=-\infty}^{\infty} (b^{i-1}[n])^2}$$

which is the correlation at lag zero between the forward and backward prediction errors at stage $i - 1$, normalized by the product of their energies, hence the terminology, "partial correlation (PARCOR) coefficients." This implies that we can compute the required k_i *without* computation of the predictor coefficients. We can then use these computed k_i's in the Levinson recursion without ever having to compute the autocorrelation coefficients $r_n[\tau]$. We return to this result in discussion of alternative lattice structures for linear prediction analysis and synthesis.

Relation of Linear Prediction to the Lossless Tube Model — The recursive solution to the linear prediction estimation bears a strong resemblance to the recursion given in Chapter 4 for the lossless tube model. Recall that for the lossless concatenated tube model, with glottal impedance $Z_g(z) = \infty$, i.e., an open circuit, we wrote a recursion for computing the denominator polynomial in the all-pole transfer function $V(z) = \frac{U_L(z)}{U_g(z)}$ from the reflection coefficients r_k, i.e., for

$$V(z) = \frac{A}{D(z)}, \qquad D(z) = 1 - \sum_{k=1}^{N} \alpha_k z^{-k}$$

the recursion for the denominator polynomial $D(z)$ is given by

$$D_0(z) = 1$$
$$D_k(z) = D_{k-1}(z) + r_k z^{-k} D_{k-1}(z^{-1}), \quad k = 1, 2, \ldots N$$
$$D(z) = D_N(z) \tag{5.36}$$

where N is the number of tubes and where the reflection coefficient r_k is a function of cross-sectional areas of successive tubes, i.e.,

$$r_k = \frac{A_{k+1} - A_k}{A_{k+1} + A_k}. \tag{5.37}$$

Likewise, the Levinson recursion for obtaining the denominator in the all-pole transfer function

$$H(z) = \frac{A}{A(z)}, \qquad A(z) = 1 - \sum_{k=1}^{p} \alpha_k z^{-k}$$

can be written in the z-domain by mapping the relation $\alpha_j^i = \alpha_j^{i-1} - k_i \alpha_{i-j}^{i-1}$ for $1 \le j \le i-1$ and its starting and ending conditions to obtain (Appendix 5.B)

$$A^0(z) = 1$$
$$A^i(z) = A^{i-1}(z) - k_i z^{-i} A^{i-1}(z^{-1}), \qquad i = 1, 2, \ldots p$$
$$A(z) = A_p(z)$$

where k_i are the partial correlation coefficients that are determined at each step in the Levinson recursion. In this recursion, we obtain the starting condition by mapping $a_0^0 = 0$ to $A^0(z) = 1 - \sum_{k=1}^{0} \alpha_k^0 z^{-k} = 1$. The two recursions are seen to be identical when the reflection coefficients in the former equal the negative of the partial correlation coefficients in the latter, i.e., when $r_i = -k_i$, then $D_i(z) = A_i(z)$, showing the relation of linear prediction analysis (the autocorrelation method) with the lossless concatenated tube model.

If we could measure the cross-sectional areas of the tubes in the concatenated tube model, then we would know the reflection coefficients because $r_i = \frac{(A_{i+1} - A_i)}{(A_{i+1} + A_i)}$. And suppose we set the partial correlation coefficients to the negative of the reflection coefficients, i.e., $k_i = -r_i$. Then, as we noted earlier, the two above recursions are identical, and so the Levinson recursion yields for its last (pth) step in the recursion the transfer function of the vocal tract, i.e.,

$$D(z) = A^p(z).$$

Alternatively, if we don't know the cross-sectional area ratios, we can first compute the autocorrelation coefficients from the speech measurement. We then solve the Levinson recursion from which we can obtain the relative cross-sectional areas given by $\frac{A_{i+1}}{A_i} = \frac{(1-k_i)}{(1+k_i)}$. The Levinson recursion thus provides a means for obtaining vocal tract shape from a speech measurement.

Observe that we have not included boundary conditions in the above transfer function, i.e., that the transfer function $V(z)$ represents the ratio between an ideal volume velocity at the glottis and at the lips:

$$V(z) = \frac{U_L(z)}{U_g(z)}.$$

Our speech pressure measurement at the lips output, however, has embedded within it the glottal shape $G(z)$, as well as the radiation load $R(z)$. Recall that in the voiced case (with no vocal tract zeros), the complete transfer function is given by

$$H(z) = AG(z)V(z)R(z)$$

$$= A\frac{(1 - \alpha z^{-1})}{(1 - \beta z)^2 \prod_{k=1}^{C_i}(1 - c_k z^{-1})(1 - c_k^* z^{-1})}$$

where the factors α and β represent a zero and double-pole due to the radiation and glottal shape, respectively. The presence of the glottal shape introduces poles that are not part of the vocal tract. As we saw earlier, the presence of the zero from $R(z)$ can also be thought of as a multiple-pole contribution. Because these poles are not part of the vocal tract, they distort the relative area function estimate. Wakita has observed that, because the net effect of the radiation and glottal shape is typically a 6 dB/octave fall-off contribution to the spectral tilt of $V(z) = \frac{U_L(z)}{U_g(z)}$, the influence of the glottal flow shape and radiation load can be approximately removed with a pre-emphasis of 6 dB/octave spectral rise [27]. Observe that because we have modeled the frequency response of the glottal flow shape with two maximum-phase poles, the pre-emphasis must account for this phase characteristic if a complete equalization is to be performed. However, because the autocorrelation method of linear prediction discards phase, the glottal phase component, in practice (for this application), need not be removed.

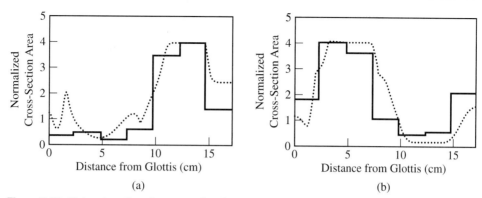

Figure 5.10   Estimation of vocal tract area functions using the autocorrelation method of linear prediction:
(a) the vowel /a/; (b) the vowel /i/. The dotted curve is the cross-section measurement made for a Russian
vowel (from Fant [27]), while the solid curve is the cross-section tube estimates based on linear prediction
using the American vowel counterpart.

SOURCE: H. Wakita, "Estimation of the Vocal Tract Shape by Optimal Inverse Filtering and Acoustic/Articulatory
Conversion Methods" [27]. ©1972, H. Wakita. Used by permission.

EXAMPLE 5.5   Two examples are given in Figure 5.10 that show good matches to measured
vocal tract area functions for the vowels /a/ and /i/ derived from estimates of the partial correlation
coefficients [27].   In these examples, a cross-sectional area function A_N for the last tube in the
concatenated tube model was obtained from an average cross-sectional area of the lip opening, thus
allowing for absolute area function measurements, rather than the above relative measures. With A_N,
we can obtain $A_{N-1} = A_N \frac{1+k_{N-1}}{1-k_{N-1}}$ from which we obtain A_{N-2} and so on. For the examples in
Figure 5.10, the speech bandwidth is 3500 Hz and a seven-tube model was assumed, allowing for three
resonances in the middle of the band and one at either band edge.[4] In this approach, Wakita found
that the area function estimates are sensitive to the spectral slope of the glottal/radiation equalization,
the area function of the lips used for normalization, and an accurate estimate of the parameters of
the all-pole model, so that improvements in these characteristics should give improved cross sections.
Furthermore, the lossless concatenated tube model does not allow distribution of energy loss along the
vocal tract, and this limitation, as well as other effects such as vibrating vocal tract walls, nonlinearities,
and distributed sources not represented in the model, may lead to refined area function estimates.   ▲

5.3.6 Frequency-Domain Interpretation

We have defined for linear prediction analysis a mean-squared prediction error (Equation (5.6))
in the time-domain.   In this section, we show that this time-domain error has a frequency-
domain counterpart, based on the difference in the log-magnitude of the measurement and
model frequency responses, that leads to our earlier solution for the all-pole model coefficients
and, in addition, to a rigorous way to justify our earlier (heuristic) selection of the gain A.

---

[4] Recall from Chapter 4 that two tubes are needed to represent one complex-conjugate pole pair associated
with a typical resonance. Thus, seven tubes can account for three complex-conjugate pole pairs and one solitary
pole at $\omega = 0$ or $\omega = \pi$.

Furthermore, in Section 5.5.2 this frequency-domain error leads to insights into the performance of the autocorrelation method of linear prediction, different from insights that can be obtained in the time domain. We here outline the proof of this time-frequency equivalence; the details are left as an exercise (Exercise 5.17).

Consider the all-pole model of speech production

$$H(\omega) = \frac{A}{A(\omega)}$$

with

$$A(\omega) = 1 - \sum_{k=1}^{P} \alpha_k e^{-j\omega k}.$$

Suppose we define a new frequency-domain error function [15]

$$I = \int_{-\pi}^{\pi} [e^{Q(\omega)} - Q(\omega) - 1] \frac{d\omega}{2\pi} \tag{5.38}$$

with $Q(\omega)$ defined by the difference of the log-magnitude of measured and modeled spectra, i.e.,

$$Q(\omega) = \log |S(\omega)|^2 - \log |H(\omega)|^2 \tag{5.39}$$

where $S(\omega)$ is the Fourier transform of the short-time segment of speech being modeled. Our goal is the minimization of the error criterion I, with respect to the unknown α_k's and unknown gain A.

Using Equation (5.39), we can write $Q(\omega)$ as

$$Q(\omega) = \log \left| \frac{E(\omega)}{A} \right|^2$$

where $E(\omega) = A(\omega)S(\omega)$ is the Fourier transform of the prediction error sequence. Suppose now that the zeros of $A(z)$ are constrained to lie inside the unit circle. Writing $|A(\omega)|^2 = A(\omega)A^*(\omega)$, it follows that

$$\int_{-\pi}^{\pi} \log |A(\omega)|^2 \frac{d\omega}{2\pi} = 0$$

from which we can show that

$$\int_{-\pi}^{\pi} \log \left| \frac{A}{A(\omega)} \right|^2 \frac{d\omega}{2\pi} = \log A^2.$$

Substituting $Q(\omega)$ into I, and using the previous result, we can then show that minimization of I, with respect to the unknown filter coefficients α_k, is equivalent to minimization of the average squared prediction-error filter output $\int_{-\pi}^{\pi} |E(\omega)|^2 d\omega$, and hence yields the same solution as the autocorrelation method of linear prediction! Finally, it is possible to show that minimization of I with respect to the unknown A^2 gives the optimal gain solution

$$A^2 = \frac{1}{2\pi} \int_{-\pi}^{\pi} |E(\omega)|^2 d\omega,$$

i.e., the mean-squared linear prediction error and, therefore, that the optimal A^2 is the same solution as obtained by the heuristic energy-matching solution of the previous section where we constrained the energy in our all-pole estimated impulse response $h[n]$ to equal the energy in the measurement $s[n]$. The frequency-domain error function I thus provides a rigorous way of obtaining both the unknown filter coefficients and the unknown gain.

Jumping ahead to Figure 5.14a, we see an illustration of the function $f(Q) = e^Q - Q - 1$ used in the error criterion I. We can argue from this view of the error criterion that linear prediction analysis tends to fit spectral peaks more accurately than spectral valleys. We return to this observation in our discussion of the performance of the autocorrelation method.

5.4 Linear Prediction Analysis of Stochastic Speech Sounds

We motivated linear prediction analysis with the observation that for a single impulse or periodic impulse train input to an all-pole vocal tract, the prediction error is zero "most of the time." Such analysis therefore appears not to be applicable to speech sounds with fricative or aspirated sources modeled as a stochastic (or random) process (Appendix 5.A), specifically, often as a white noise sequence. In this section we see that the autocorrelation method of linear prediction, however, can be formulated for the stochastic case where a white noise input takes on the role of the single impulse. The solution to a stochastic optimization problem, analogous to the minimization of the mean-squared error function E_n in Section 5.3, leads to normal equations which are the stochastic counterparts to our earlier solution [14].

5.4.1 Formulation

Consider a speech sound where turbulence is created at a constriction in the vocal tract or at a constricted glottis as, for example, with unvoiced fricatives and breathy sounds, respectively.[5] One model of the generation of this sound is the excitation of an all-pole vocal tract with a sample function of white noise $u_o[n]$ that represents the turbulence at the constriction and that is characterized by zero mean and unity variance, i.e., $E(u_o[n]) = 0$ and $E(u_o^2[n]) = 1$, where the operator $E()$ denotes expected value (Appendix 5.A) and where the subscript "o" denotes that the source turbulence can occur anywhere along the oral (or pharyngeal) tract and not necessarily at the glottis. The difference equation representation of the output waveform is given by

$$s[n] = \sum_{k=1}^{p} \alpha_k s[n - k] + Au_o[n] \tag{5.40}$$

where $s[n]$ and $u_o[n]$ are each sample functions of a random process and where A denotes the gain factor for the white noise source. The frequency-domain representation of the stochastic

---

[5] As we have seen in Example 3.4, for voiced fricatives, turbulence at a vocal tract constriction is *modulated* by the cyclic glottal flow. This nonlinear model and other nonlinear models of sound production, e.g., distributed stochastic sources due to vortical flow, for the moment, we do not consider.

output is given by the *power density spectrum* or *power spectrum* (for clarity, notationally different from in Appendix 5.A) of the output waveform $s[n]$ expressed as

$$P_s(\omega) = A^2|V(\omega)R(\omega)|^2 \qquad (5.41)$$

where there is no glottal pulse shape component as in the voiced speech case. Equation (5.41) is the stochastic counterpart to our frequency response model of voiced speech with an impulse input.

An estimate of the power spectrum is obtained by computing the squared magnitude of the Fourier transform of $s[n]$ over an analysis window of finite duration N_w and dividing by N_w, i.e.,

$$|S_{N_w}(\omega)|^2 = \frac{1}{N_w}\left|\sum_{n=0}^{N_w-1} s[n]e^{j\omega n}\right|^2$$

which is referred to as the *periodogram* and whose expected value is the desired power spectrum smoothed by the Fourier transform of the rectangular window $w[n] = 1$ for $0 \leq n < N_w$ (Appendix 5.A) [17]. The periodogram fluctuates about the power spectrum with a variance that is proportional to the square of the power spectrum for large N_w. Various smoothing techniques have been developed to reduce this variance, but at the expense of spectral resolution [17]. For large N_w, where end effects at the boundaries of the data segment are negligible [17], we can write

$$|S_{N_w}(\omega)|^2 \approx |H(\omega)|^2|U_{oN_w}(\omega)|^2$$

where $|U_{oN_w}(\omega)|^2$ is the N_w–point periodogram of the white noise input and is shaped by the square of the system frequency response magnitude, $|H(\omega)|^2 = |AV(\omega)R(\omega)|^2$, to form the N_w–point periodogram of the speech waveform.

As with the voiced case, we define the prediction error between $s[n]$ and its prediction $\tilde{s}[n]$

$$e[n] = s[n] - \tilde{s}[n] \qquad (5.42)$$

where the prediction $\tilde{s}[n]$ is given by

$$\tilde{s}[n] = \sum_{k=1}^{p} \alpha_k s[n - k].$$

And, as also with the voiced case, when the prediction coefficients equal the actual coefficients of the underlying all-pole model, i.e., $\alpha_k = a_k$, then the prediction error equals the source input $Au_o[n]$. Under this condition, the expected value of the squared prediction error can be shown to equal the variance of the input sequence. Specifically, because the unit variance white-noise input is weighted by the gain A, we have the prediction error variance $E(e^2[n]) = A^2$. As in the deterministic case, the resulting filter $A(z) = \sum_{k=1}^{p} \alpha_k z^{-1}$ is called the *inverse filter*. When the underlying system is all-pole of order p and $\alpha_k = a_k$, then $A(z)$ "whitens" the measurement, i.e., as we stated, the output of $A(z)$ with input $s[n]$ is the white-noise process $Au_o[n]$. This is analogous to the deterministic case where the given inverse filter results in an impulse or impulse train. Consequently, we can take a path to finding the unknown α_k's that is similar to that for the deterministic case. However, unlike the deterministic case, i.e., an impulse or impulse train as input, the prediction error is not zero over contiguous time intervals.

5.4.2 Error Minimization

In the stochastic case, we define an error function as the expected value of the squared prediction error

$$E(e^2[n]) = E[(s[n] - \tilde{s}[n])^2] \tag{5.43}$$

where $e[n]$ and $s[n]$ are viewed as sample functions from a random process and $\tilde{s}[n]$ is the prediction defined in Equation (5.42). Minimizing Equation (5.43) with respect to the unknown $\underline{\alpha}$, we obtain the normal equations for the stochastic case. We arrive at the optimizing solution by taking derivatives with respect to α_k, as in the deterministic case, or by using the Projection Theorem for linear least-squared-error optimization that states that the minimum error $e[n]$ must be orthogonal to the subspace spanned by the signal basis that makes up our prediction.[6] In the later case, the "basis vectors" are the observations $s[n - k]$, so we require the orthogonality condition

$$E(e[n]s[n - k]) = 0, \qquad k = 1, 2, \ldots p.$$

Because the white noise input is stationary in a stochastic sense (Appendix 5.A) and the vocal tract filter is linear and time-invariant, the output is also stationary. (The reader should argue the validity of this property.) Under this condition, the optimizing solution can be found as

$$R\underline{\alpha} = \underline{r} \tag{5.44}$$

where the elements of the $p \times p$ matrix R, denoted by $\Phi[i, k]$, are defined by

$$\Phi[i, k] = r[\tau]|_{\tau=i-k} = E(s[m]s[m + \tau]), \qquad 1 \le i \le p, \qquad 0 \le k \le p$$

and where $r[\tau]$ is the autocorrelation function of the random process (denoted by $\phi_{ss}[\tau]$ in Appendix 5.A). The expected value of the squared prediction error, i.e., its variance, for the minimization solution is given by

$$E(e^2[n]) = r[0] - \sum_{k=1}^{p} \alpha_k r[k].$$

Because the random process is stationary, all elements along diagonal elements are equal and the matrix is symmetric about the center diagonal. Thus the matrix is Toeplitz. Consequently, the normal equations for the stochastic case are identical in structure to those for the autocorrelation method solution in the deterministic case and can be solved similarly. When $s[n]$ follows a pth-order all-pole model, then Equation (5.44) for order p yields the true all-pole coefficients; this can be argued in a style similar to that for the deterministic case where Equation (5.17) was applied. Observe that $r[\tau] = E(s[m]s[m + \tau])$ represents an *ensemble average* or equivalently a time average (over all time) under an ergodicity constraint (Appendix 5.A). In practice, however, $r[\tau]$ must be *estimated* from a finite length sequence, and as with the deterministic case, the manner

---

[6] We could have used this same approach in optimization for the deterministic case by replacing the expected value operator by the sum operator.

in which the boundaries of the data segment are considered will lead to the autocorrelation or covariance method of solution.

5.4.3 Autocorrelation Method

In the autocorrelation method, the correlation coefficients are estimated from a truncated speech segment, i.e.,

$$r_n[\tau] = \frac{1}{N_w} \sum_{m=0}^{N_w-1-\tau} s_n[m]s_n[m+\tau]$$

$$= \frac{1}{N_w} s_n[m] * s_n[-m]$$

where the data is assumed zero outside of the truncation interval. The discrete-time Fourier transform of this autocorrelation estimate is the periodogram of $s_n[m]$, as defined in the previous section. Except for the scale factor $\frac{1}{N_w}$, this estimate is identical to that in the autocorrelation method for the deterministic case. As in the deterministic case, therefore, we form a matrix R_n which is identical (except for the scale factor $\frac{1}{N_w}$) to that given in Equation (5.16) of Section 5.3.3. As before, because the resulting matrix R_n is Toeplitz, the Levinson recursion can be used for the computation of α_k by constructing successively higher-order predictors. This leads again to a lattice filter implementation of the inverse filter $A(z)$ and lends itself to an interpretation of a "progressive whitening" by removing more and more of the correlation of the input signal at each stage. The formal stochastic development of this property is beyond the scope of our treatment and can be found in [26].

Finally, the gain A of the source remains to be determined. Given the stochastic nature of the signals, it is reasonable to select a gain such that the average power in the measurement equals the average power in the response to a white-noise input. In a stationary stochastic framework, the average power of a zero-mean sequence is its variance, i.e., the mean of the squared sequence (Appendix 5.A), so that we require that the variance of the measurement equal the variance of the model output. With a known autocorrelation function $r[\tau]$, this condition is equivalent to writing

$$r[0] = A^2 r_h[0]$$

where $r_h[\tau]$ is the autocorrelation function of the impulse response of the estimated all-pole transfer function. The squared gain is thus $A^2 = \frac{r[0]}{r_h[0]}$. When the autocorrelation $r[n]$ is not known then $r[0]$ is replaced by its estimate $r_n[0]$.

5.5 Criterion of "Goodness"

We can evaluate the accuracy of the linear prediction analysis in either the time domain or the frequency domain, and either through the system or source estimation.

5.5.1 Time Domain

Suppose that the underlying speech model is all-pole of order p and suppose that the autocorrelation method is used in the estimation of the coefficients of the predictor polynomial $P(z)$.

If we estimate the predictor coefficients exactly then, with a measurement $s[n]$ as input, the output of the inverse filter $A(z) = 1 - P(z)$, i.e., the prediction error, is a perfect impulse train for voiced speech, a single impulse for a plosive, and white noise for noisy (stochastic) speech. In practice, however, the autocorrelation method of linear prediction analysis does not yield such idealized outputs when the measurement $s[n]$ is inverse filtered by the estimated system function $A(z)$. We have seen that the true solution is not obtained exactly by the autocorrelation method, even when the vocal tract response follows an all-pole model. Furthermore, in a typical waveform segment, the actual vocal tract impulse response is not all-pole for a variety of reasons, including the presence of zeros due to the radiation load, nasalization, and a back vocal cavity during frication and plosives.[7] In addition, the glottal flow shape, even when adequately modeled by a two-pole function, is not minimum phase, introducing an undesirable phase component in the output of the inverse filter, as demonstrated in the following example:

EXAMPLE 5.6   Consider the vocal tract transfer function model $H(z)$ in Example 5.4. In that example, the speech waveform $s[n]$ consisted of simply a single impulse response $h[n]$, i.e., $s[n] = h[n]$, in which the glottal flow waveform was embedded. We saw that application of the autocorrelation method converts the two maximum-phase poles of the glottal flow contribution to their minimum-phase counterparts. Writing the true frequency response as

$$S(\omega) = H(\omega) = M_h(\omega)e^{j[\theta_v(\omega)+\theta_g(\omega)]}$$

where $\theta_v(\omega)$ is the minimum-phase contribution from vocal tract poles inside the unit circle and $\theta_g(\omega)$ is the maximum-phase contribution from glottal poles outside the unit circle, then the frequency response of the resulting inverse filter $A(z)$ from the autocorrelation method is given by

$$A(\omega) = AM_h(\omega)^{-1}e^{j[-\theta_v(\omega)+\theta_g(\omega)]}$$

where the maximum-phase glottal contribution becomes minimum-phase and where the source gain factor A is applied to cancel the gain in $H(\omega)$ because it is not part of $A(\omega)$. The frequency response of the output of the inverse filter $A(z)$ with input $s[n]$ is, therefore, given by

$$E(\omega) = S(\omega)A(\omega)$$
$$= M_h(\omega)e^{j[\theta_v(\omega)+\theta_g(\omega)]}A(\omega)$$
$$= Ae^{j2\theta_g(\omega)}.$$

The frequency response of the prediction error, therefore, can be viewed as having a flat Fourier transform magnitude distorted by an all-pass function $e^{j2\theta_g(\omega)}$. In the time domain, this is represented by a distorted impulse:

$$e[n] = A\delta[n] * a_g[n]$$

where $a_g[n]$ is the inverse Fourier transform of $e^{j2\theta_g(\omega)}$.   ▲

---

[7] We have seen that although a zero can be approximated by a large number of poles, this is not an efficient modeling strategy.

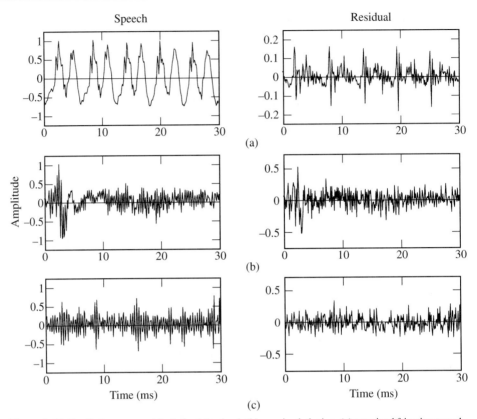

Figure 5.11 Prediction error residuals for (a) voiced; (b) unvoiced plosive; (c) unvoiced fricative sounds. A 14th-order predictor is used for 5000 Hz-bandwidth signals.

We conclude then that, either because our assumed minimum-phase all-pole model or our estimation method or both are not adequate, we cannot expect to see an idealized glottal source waveform at the output of the inverse filter. Figure 5.11 shows examples of inverse filtering segments of a vowel, unvoiced plosive, and unvoiced fricative with an $A(z)$ derived from the autocorrelation method of linear prediction (order 14). The figure shows pictorially that the estimated all-pole model is not an adequate representation, particularly for the vowel and plosive sounds. The estimation was performed on 20 ms (Hamming) windowed speech segments. In reconstructing a residual from an entire utterance (we discuss synthesis methods shortly), typically one hears in the prediction error not a noisy buzz, as expected from an idealized residual comprising impulse trains, solitary impulses, and white noise, but rather roughly the *speech itself*, implying that some of the vocal tract spectrum is passing through the inverse filter.

5.5.2 Frequency Domain

Alternatively, we can study the behavior of linear prediction analysis in the frequency domain with respect to how well the spectrum derived from linear prediction analysis matches the spectrum of a sequence that follows an all-pole model. We are also interested in evaluating the method when the measurement does not follow an all-pole model. In this section, we present properties of spectral estimation for periodic and stochastic sounds using the autocorrelation method of linear prediction analysis.

Consider a periodic impulse train source $u_g[n] = \sum_{k=-\infty}^{\infty} \delta[n - kP]$, with Fourier transform denoted by $U_g(\omega)$, and vocal tract impulse response with all-pole frequency response, $H(\omega)$. We have seen that the Fourier transform of the windowed output $s_n[m]$ is given by

$$S(\omega) = \frac{2\pi}{P} \sum_{k=-\infty}^{\infty} H(k\omega_o)W(\omega - k\omega_o)$$

where $W(\omega)$ is the window transform, $\omega_o = \frac{2\pi}{P}$ is the fundamental frequency, and where, for simplicity, we remove reference to the window location n. Linear prediction analysis attempts to estimate $|H(\omega)|$ (via the autocorrelation function) and hence the *spectral envelope* of the harmonic spectrum $S(\omega)$ (introduced in Chapter 3), as illustrated in Figure 5.12a.

For stochastic sounds, we saw that we can express the periodogram of our white-noise driven model as

$$|S_{N_w}(\omega)|^2 \approx |H(\omega)|^2 |U_{oN_w}(\omega)|^2$$

where $|U_{oN_w}(\omega)|^2$ is the periodogram of an N_w–point sample function of white noise. Typically the periodogram is "noisy" with high variance whose expectation is a smooth version of the desired power spectrum given by $|H(\omega)|^2$ (assuming a unity-variance input), i.e., the periodogram estimate is biased. For this case, we can interpret the goal of linear prediction analysis in the frequency domain as estimating the smoothed $|H(\omega)|^2$, which can be viewed as the *spectral envelope* of the periodogram, as illustrated in Figure 5.12b. This perspective of estimating the spectral envelope is thus analogous to that for periodic sounds. We now consider frequency-domain properties for these two signal classes.

Properties — P1: Recall the autocorrelation matching property: When the gain of the all-pole model is chosen so that energy in the model and the measurement are equal, then their respective autocorrelation functions match for the first p lags. This implies for large p that $|H(\omega)|$ matches the Fourier transform magnitude of the windowed signal, $|S(\omega)|$.

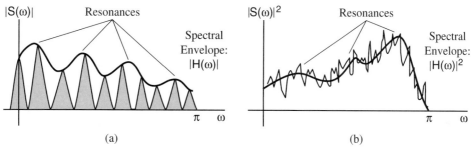

Figure 5.12 Schematics of spectra of periodic and stochastic speech sounds: (a) harmonic spectrum of a voiced sound. For simplicity, we assume $\frac{2\pi}{P}W(0) = 1$; (b) periodogram of a stochastic sound. For simplicity, we assume a unit variance white-noise source and that bias in the periodogram estimate is negligible.

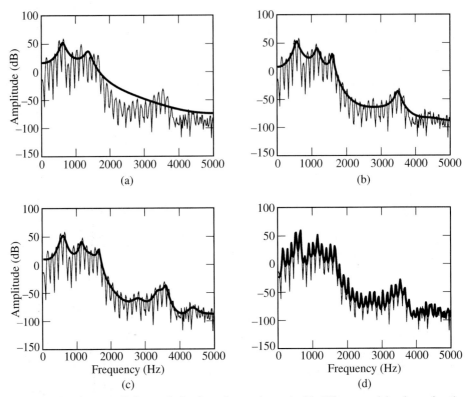

Figure 5.13 Linear prediction analysis of steady vowel sound with different model orders using the autocorrelation method: (a) order 6; (b) order 14; (c) order 24; (d) order 128. In each case, the all-pole spectral envelope (thick) is superimposed on the harmonic spectrum (thin), and the gain is computed according to Equation (5.30).

EXAMPLE 5.7   The spectral matching property for the autocorrelation method is shown in Figure 5.13 for a segment of a steady vowel sound. As the order p increases, the harmonic structure of $S(\omega)$ is revealed in the all-pole spectrum, indicating that we have overestimated the model order. Observe that this property can also be seen in the time domain; from Figure 5.7, as the order p increases, replicas of the autocorrelation function of $h[n]$ have an increasingly stronger influence on the solution. Figure 5.13 also illustrates the result of underestimating the model order; as p decreases, the fine resonant structure of the underlying all-pole system is lost, with only the largest resonances being represented. The order p therefore can be thought of as controlling the degree of smoothing of the measured spectrum. In sound synthesis, over-estimation of p results in a "tonal" quality, with spurious resonances coming and going in time, while under-estimation of the order results in a "muffled" quality because resonant bandwidths are broadened and closely-spaced resonances are smeared. ▲

From the autocorrelation matching property, we might expect $H(\omega)$ to converge to $S(\omega)$ as $p \to \infty$. However, this convergence does not occur. Suppose in the autocorrelation method, we let the model order p go to infinity. Then, from the autocorrelation matching property, we have

$$|S(\omega)| = |H(\omega)| = \left| \frac{A}{A(\omega)} \right|.$$

However, because the measurement will generally have both a minimum- and maximum-phase component,

$$\angle S(\omega) \neq \angle H(\omega) = \angle \frac{A}{A(\omega)}.$$

Therefore, the output of the inverse filter $A(z)$ with $p \rightarrow \infty$ and with input $s_n[m]$ is not the unit sample $A\delta[m]$.

P2: Recall in our frequency-domain interpretation of the mean-squared prediction error (Section 5.3.6) that we alluded to the fact that linear prediction spectral analysis tends to fit spectral peaks more accurately than spectral valleys. We now look at this property in more detail. Recall the definition of the function $Q(\omega)$ as

$$Q(\omega) = \log |S(\omega)|^2 - \log |H(\omega)|^2$$

$$= \log \left| \frac{S(\omega)}{H(\omega)} \right|^2$$

and recall the function

$$f(Q) = e^Q - Q - 1$$

which is the integrand of the frequency-domain error in Equation (5.38). We see in Figure 5.14a that $f(Q)$ is asymmetric with a steeper slope for positive Q. This implies that for $|S(\omega)| > |H(\omega|$ a larger error penalty is invoked than when $|S(\omega)| < |H(\omega|$. In other words, the linear prediction error favors a good spectral fit near spectral peaks, as illustrated schematically in Figure 5.14b. This interpretation has implications when the underlying signal does not follow the assumed all-pole model [14]. For example, it implies that resonant peaks are

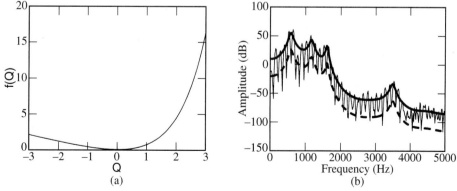

Figure 5.14 Illustration of favoring a match to spectral peaks in linear prediction analysis: (a) the asymmetric function $f(Q) = e^Q - Q - 1$; (b) schematic that shows favoring spectral peaks over spectral valleys. Upper (solid) envelope is favored over lower (dashed) envelope.

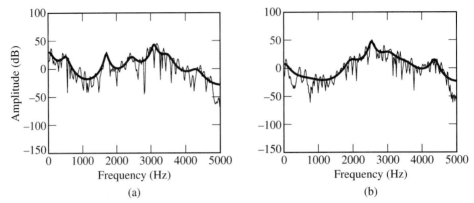

Figure 5.15 Spectral peak matching in linear prediction analysis for (a) an unvoiced plosive ("t" in "tea") and (b) an unvoiced fricative component of an affricate ("ch" in "which") of Figure 5.11. Order 14 was used in the autocorrelation method.

better matched than valleys when zeros are present in the vocal tract transfer function. It also implies that the highest resonances are best matched when the model order is under-estimated. When the order is over-estimated, harmonics are best matched for voiced speech (consistent with our earlier observation), while spurious spectral peaks are best matched for unvoiced speech. This spectral matching property for an unvoiced plosive and unvoiced fricative is illustrated in Figure 5.15.

5.6 Synthesis Based on All-Pole Modeling

We are now in a position to *synthesize* the waveform from model parameters estimated using linear prediction analysis. The synthesized signal is given by

$$s[n] = \sum_{k=1}^{p} \alpha_k s[n-k] + Au[n]$$

where $u[n]$ is either a periodic impulse train, an impulse, or white noise, and the parameters α_k and A are estimated by way of the normal equations and specific gain requirements. We use here the autocorrelation method because it gives a stable and efficient solution. Nevertheless, the covariance method does have its role in speech analysis/synthesis, an example of which we see later in the chapter. There remain a number of considerations in the design of the analyzer and synthesizer.

Window Duration: Typically, we select the window duration to be fixed and to fall in the range of 20–30 ms to give a satisfactory time-frequency tradeoff (Exercise 5.10). The duration can be refined to adapt to pitch, voicing state, and even phoneme class to account for different time-frequency resolution requirements.

Frame Interval: A typical rate at which to perform the analysis is 10 ms, the frame interval with which the analysis window slides, selected to approximately capture the time evolution of the speech signal. There may be more complicated strategies in which different parameters are updated and interpolated at different rates [21]. For example, the pitch and voicing state may be updated at a 10-ms rate while the system parameters are updated at a 20-ms rate, under the assumption that the vocal tract is changing more slowly than the source parameters.

Model Order: There are three components to be considered: the vocal tract, the source, and the radiation. For the vocal tract, from Chapters 3 and 4, we found it was typical to assume an average "resonant density" of one resonance per 1000 Hz. We assume for illustration a bandwidth of 5000 Hz; then 5 poles are required and hence a model order of $p = 10$ for the vocal tract representation. The radiation load is modeled in discrete time by one zero inside the unit circle, which we have shown is represented by a product of poles inside the unit circle. By convention, it has been found that about 4 poles are adequate in this representation [14]. Finally, we must account for the glottal flow; here we select our two-pole maximum-phase model. In summary, we require a total of 10 (vocal tract) + 4 (radiation) + 2 (glottal flow) = 16 poles. Keep in mind that, although the *magnitude* of the speech frequency response is approximately matched by this model, we lose the shape of the measured time-domain speech waveform since the frequency response *phase* of the actual glottal flow waveform is not preserved. This is because, as we saw in Example 5.4, the maximum-phase poles of the modeled glottal flow are transformed to their minimum-phase counterparts by the linear prediction analysis. In addition, any maximum-phase zeros that arise in the vocal tract transfer function are also transformed to their minimum-phase counterparts.

Voiced/Unvoiced State and Pitch Estimation: Estimation of these parameters is described in Chapter 10. Current analysis and synthesis strategies do not account for distinguishing between plosive and fricative unvoiced speech sound categories; they are typically lumped into the unvoiced state of a binary voiced/unvoiced state decision. The unvoiced state corresponds to the condition where the vocal folds are not vibrating. The speaker's pitch is estimated during voiced regions only. A *degree* of voicing may also be desired in a more complex analysis and synthesis, where voicing and turbulence occur simultaneously, as with voiced fricatives and breathy vowels; we save such refinements for later chapters. We assume then that we are provided a binary voiced/unvoiced decision on each frame and a pitch estimate on each voiced frame.

Synthesis Structures: In synthesis, the source excitation is created by concatenating an impulse train during voicing, for which spacing is determined by the time-varying pitch contour, and with white noise during unvoicing. In one strategy, the gain for the voicing state is determined by forcing the energy in the impulse response equal to the energy measured on each frame. In an unvoiced state, the gain is obtained by requiring that the variance of the measurement equal the variance of the model output. For each frame, either a series of periodic impulses or a random white noise sequence is convolved with an impulse response $h_k[n]$ that changes on each (kth) frame; the output of each convolution is then overlapped and added with adjacent frame outputs (Figure 5.16). Alternatively, the filtering with $h_k[n]$ can be performed with the direct-form implementation (Chapter 2) motivated by the difference equation representation of its all-pole z-transform (Figure 5.17a). Filter values are updated on each frame and a natural continuity exists in time across successive frames because old values are stored in delay elements. Other structures also exist, one derived directly from the concatenated-tube flow graph, and a second from an all-pole version of the all-zero lattice inverse filter in Figure 5.9 [16]. These

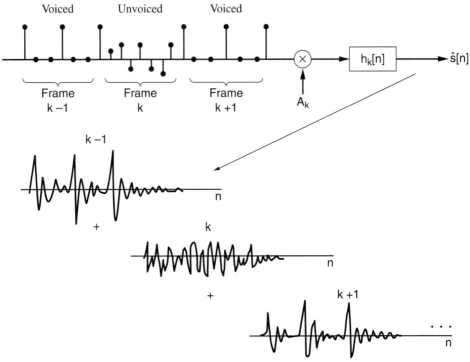

Figure 5.16 Overlap-add synthesis using an all-pole model. The waveform is generated frame by frame by convolutional synthesis and the filter output on each frame is overlapped and added with adjacent frame outputs.

two alternative implementations of $\frac{1}{A(z)}$ are illustrated in Figure 5.17b,c. An advantage of the lattice structures is that they are expressed in terms of the reflection coefficients r_i (or partial correlation coefficients k_i) that are more suitable to quantization (Chapter 12) because their magnitude, as we saw earlier, does not exceed unity, unlike the linear prediction coefficients α_k that have a large dynamic range. This property of not exceeding unity also makes the reflection coefficients amenable to interpolation, without the loss of the all-pole filter stability constraint (Exercise 5.15).

The synthesized waveform based on the autocorrelation method of linear prediction looks speech-like, but with loss of absolute phase structure because of its minimum-phase characteristic. As we see in the example of Figure 5.18, the reconstructed waveform is more "peaky" than the original; the idealized glottal flow, being made minimum-phase, is time-reversed with thus a sharper attack than the actual glottal flow. The peakiness at the origin is consistent with a minimum-phase sequence having the greatest energy concentration near the origin of all causal sequences with the same frequency response magnitude. (This property was expressed in Equation (2.23) of Chapter 2.). This peakiness is said to impart a "buzzy" quality to the linear prediction synthesis. A number of techniques have been applied to reduce this buzziness, such as post-filtering the synthesized speech by all-pass dispersion networks. In another method, pre-emphasis to reduce the glottal contribution prior to linear prediction analysis is followed by

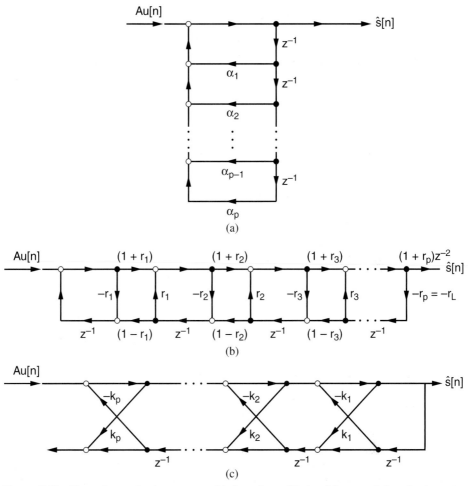

Figure 5.17 Alternative synthesis structures: (a) direct form; (b) signal flow graph from lossless tube model [21]; (c) all-pole lattice [16]. Open circles denote addition.

SOURCES: L.R. Rabiner and R.W. Schafer, *Digital Processing of Speech Signals* [21]. ©1978, Pearson Education, Inc. Used by permission; J.H. McClellan, "Parametric Signal Modeling" [16]. ©1988, Pearson Education, Inc. Used by permission.

postfiltering with a glottal flow-like waveform with a gradual attack [22]. In addition, in this method the glottal pulse duration was varied with the pitch period, the ratio of the open-to-closed glottal phase being maintained. Both the reduction in peakiness and the nonstationarity of the glottal pulse were observed to help reduce buzziness in the synthesis.

In doing speech synthesis, we can investigate the importance of the various linear prediction parameters and the limitations of the approach. Quality[8] degrades as, for example, the

---

[8] The term "quality" refers to speech attributes such as naturalness, intelligibility, and speaker recognizability. More discussion on quality is given in the introduction of Chapter 12 on speech coding.

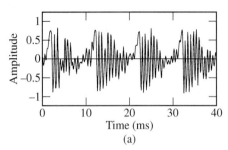 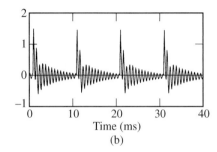

Figure 5.18 Speech reconstruction based on linear prediction analysis using the autocorrelation method: (a) original; (b) synthesized. The reconstruction was performed with overlap-add synthesis and using a 14th-order all-pole model.

predictor order decreases [2]. As the order decreases, the number of poles is underestimated, thus smearing the desired frequency response and removing the lower-energy formants (Figure 5.13), resulting in a muffled and lowpass character to the reconstructed sound. The synthesis quality is also found to be a function of the speaker's characteristics. For example, as the rate of articulation or the pitch of a speaker increases, the quality is found to degrade, the former illustrating the importance of the stationarity assumption in all-pole estimation over a 20-ms analysis window, and the later illustrating the problematic nature of high-pitched speakers due to increased aliasing in the autocorrelation function with increasing pitch.

5.7 Pole-Zero Estimation

In this section, we consider the more general problem of estimating parameters of a pole-zero rational z-transform model of a speech measurement $s[n]$. The rational transform is expressed as a quotient of a numerator and denominator polynomial, i.e.,

$$X(z) = \frac{B(z)}{A(z)}$$

$$= \frac{\sum_{k=0}^{q} b_k z^{-k}}{1 - \sum_{k=1}^{p} a_k z^{-k}} \tag{5.45}$$

where we assume that the signal model $x[n]$ is causal with initial value at the origin $n = 0$ and that the gain term A is embedded in the numerator polynomial. The zeros in the numerator polynomial can be introduced by coupling of the nasal and oral cavities, by a back cavity behind a constriction, or by the radiation load, and the poles can be introduced by the vocal tract and glottal flow. In this section, however, we take a different perspective and view the glottal flow as a finite-length sequence, thus contributing zeros in the numerator of Equation (5.45). This view allows more flexibility in the glottal flow model, motivated by the observation that the glottal flow differs with speaker and speaking condition and contains fine structure such as aspiration and other perturbations.

Our objective is to estimate the numerator and denominator coefficients b_k and a_k, respectively. One approach is to minimize the mean-squared error between the *model* and the *measurement*, i.e., minimize $\sum_{n=-\infty}^{\infty}(x[n] - s[n])^2$ with respect to the unknown coefficients. Using Parseval's relation, we can map this error function to the frequency domain in terms of $A(\omega)$ and $B(\omega)$. Minimizing this error with respect to the unknown polynominal coefficients leads to equations that are nonlinear in these parameters and thus difficult to solve (Exercise 5.21) [16].

5.7.1 Linearization

An alternative approach to solving for the above polynomial coefficients is to linearize the parameter estimation problem by first cross-multiplying the model in Equation (5.45) to obtain

$$X(z)A(z) = B(z) \tag{5.46}$$

which, in the time domain, is given by the difference equation

$$x[n] - \sum_{k=1}^{p} a_k x[n-k] = \sum_{k=0}^{q} b_k \delta[n-k]$$

from which we see that

$$x[n] - \sum_{k=1}^{p} a_k x[n-k] = 0, \quad n > q. \tag{5.47}$$

We can, therefore, set up an error minimization, involving the model $x[n]$ and the measurement $s[n]$, that solves for the predictor coefficients in Equation (5.47) and that is identical to the all-pole case. Specifically, let $\tilde{s}[n]$ be the prediction of the measurement $s[n]$, i.e., $\tilde{s}[n] = \sum_{k=1}^{p} \alpha_k s[n-k]$. We then minimize the error function $\sum_{n=q+1}^{\infty}(s[n] - \tilde{s}[n])^2$ with respect to the coefficients α_k. We have assumed here the correct model order p; in practice, we would set p equal to twice the expected number of resonances for a particular bandwidth. Let us now think of the estimated coefficients of $A(z)$ as a time sequence $-\alpha_n$ for $n = 0, 1, 2 \ldots p$ ($\alpha_o = -1$). With Equation (5.46) as the basis, we then "inverse filter" the measurement $s[n]$ by $-\alpha_n$ to obtain an estimate of the numerator "sequence," i.e.,

$$\beta_n = s[n] * (-\alpha_n) \tag{5.48}$$

where we denote the estimates of b_n, the coefficients of the numerator polynomial $B(z)$, by β_n. We have therefore separated the pole estimation problem from the zero estimation problem while keeping the solution linear in the unknown parameters.

In practice, however, the inverse filtering of $s[n]$ does not lead to reliable estimates of the numerator coefficients, and thus the true zeros of the underlying signals. This is because the method relies on an accurate estimate of $A(z)$ and because it assumes the relation $S(z)A(z) = B(z)$, both of which, in reality, suffer from inexactness. More reliable estimates rely on formulating a least-squared error estimation for the unknown β_n's. In one such approach, the impulse response corresponding to the estimated all-pole component of the system function, i.e., $\frac{1}{\hat{A}(z)} = \frac{1}{1-\sum_{k=1}^{p}\alpha_k z^{-1}}$, is first obtained. Denoting this impulse response by $h_{\underline{\alpha}}[n]$, we then define an error function

$$e_{\underline{\alpha}}[n] = s[n] - \beta_n * h_{\underline{\alpha}}[n] \tag{5.49}$$

where the sequence β_n to be estimated represents the unknown coefficients of $B(z)$ in the time domain, and where the term $\beta_n * h_{\underline{\alpha}}[n]$ represents an estimate of the rational z-transform model in the time domain. We then minimize the error function $\sum_{n=-\infty}^{\infty} e_{\underline{\alpha}}^2[n]$ with respect to the unknown β_n for $n = 0, 1, 2, \ldots q$. (As with the number of poles, the number of zeros is not known and must be surmised or estimated.) This leads to a set of normal equations that require the autocorrelation function of the impulse response $h_{\underline{\alpha}}[n]$ and the cross-correlation function of $h_{\underline{a}}[n]$ with $s[n]$ (Exercise 5.9). This approach was originally introduced by Shanks [12],[23]. Furthermore, it can be shown using Parseval's Theorem that the mean-squared error E in the frequency domain is given by

$$E = \int_{-\pi}^{\pi} \frac{1}{\hat{A}(\omega)} |\hat{A}(\omega)S(\omega) - B(\omega)|^2 d\omega. \tag{5.50}$$

Shanks' method thus can be interpreted as estimating the coefficients of the numerator $B(z)$ by fitting a polynomial to the Fourier transform of the linear prediction error $\hat{A}(\omega)S(\omega)$, i.e., of the estimated inverse filter output [12]. We also see in Equation (5.50) that the error function is given more weight in the important high energy resonant regions corresponding to the poles of $\frac{1}{\hat{A}(z)}$.

Another approach, motivated by Shanks' method, is to iteratively build up more accurate estimates of the poles and zeros *simultaneously*. Suppose on the ith iteration we have an estimate of $A(\omega)$ denoted by $A^i(\omega)$ and an estimate of $B(\omega)$ denoted by $B^i(\omega)$. Then we form the error function on the next iteration as [16]

$$E^{i+1}(\omega) = \frac{1}{A^i(\omega)} [A^{i+1}(\omega)S(\omega) - B^{i+1}(\omega)]$$

where, as before, we can view $\frac{1}{A^i(\omega)}$ as more heavily weighting the error $A^{i+1}(\omega)S(\omega) - B^{i+1}(\omega)$ at the higher-energy resonant regions of the signal. We then minimize $e^{i+1}[n]$ (the reader should make the mapping to the time domain) with respect to the unknown α_k's and β_k's to form estimates for the next iteration.

In this section, we have introduced a few of many methods of estimating zeros of a rational z-transform. Two other methods of zero estimation are, for example, "inverse linear prediction" [12] (Exercise 5.22) and a method proposed by Steiglitz [24] (Exercise 5.6).

5.7.2 Application to Speech

We now return to our transfer function for voiced speech from the glottis to the lips that we have expressed in the form

$$\begin{aligned} H(z) &= AG(z)V(z)R(z) \\ &= \frac{AG(z)R(z)}{A(z)} \end{aligned}$$

where $G(z)$ is the z-transform of the glottal flow over one glottal cycle, $V(z) = \frac{1}{A(z)}$ represents the minimum-phase all-pole (resonant) component (assuming here that the vocal tract does not introduce zeros), and $R(z)$ represents a zero inside the unit circle due to the radiation load. In the simplifying case when the glottal flow is an impulse, then $H(z) = AV(z)R(z)$ and the methods of the previous section can be applied to estimating the single zero of $R(z)$ and the poles of $V(z)$ (Exercise 5.4). These methods are not constrained to estimating zeros inside the unit circle; the zeros can be minimum- or maximum-phase.

With the presence of the glottal flow, however, we are faced with a new problem. Consider our idealized glottal flow model of two maximum-phase poles for a single glottal cycle. We have not yet discussed methods for estimating maximum-phase poles of a z-transform; we will see such methods in Chapter 6. For now, we take a different and less constraining approach to modeling the glottal flow as a finite-length sequence. Our idealized two-pole model is, after all, simply an approximate representation of a certain class of smooth glottal flow sequences. The glottal flow typically consists of additional components of fine structure as, for example, aspiration and ripple components. As such, we represent the glottal flow, together with the radiation load and gain, by a finite number of zeros, i.e.,

$$AG(z)R(z) = A\prod_{k=1}^{M_i}(1 - u_k z^{-1})\prod_{k=1}^{M_o}(1 - v_k z)$$

$$= \sum_{k=-M_o}^{M_i} b_k z^{-k}$$

with M_i zeros inside and M_o zeros outside the unit circle and where we embedded the gain A in the coefficients b_k. We now introduce the z-transform $P(z)$ of a periodic impulse train $p[n] = \sum_{k=-\infty}^{\infty}\delta[n - kP]$ with period P to represent the periodicity of a voiced sound. (Up to now in this section, we have assumed only one impulse response in the measurement $s[n]$.) Our discrete-time model of voiced speech then becomes in the z-domain

$$S(z) = H(z)P(z) = \frac{AG(z)R(z)P(z)}{A(z)}$$

$$= \frac{B(z)}{A(z)}P(z)$$

where $B(z) = AG(z)R(z)$. The z-transform $S(z)$ in the time-domain is then given by

$$s[n] = \sum_{k=1}^{p} a_k s[n - k] + b_n * \sum_{m=-\infty}^{\infty} \delta[n - mP].$$

The sequence b_n represents the convolution of the scaled all-zero glottal flow $g[n]$ [inverse z-transform of $AG(z)$] with the radiation operator $r_L[n] = \delta[n] - \gamma\delta[n - 1]$ [inverse z-transform of the single-zero radiation component $R(z)$], i.e., $b_n = Ag[n] * r_L[n]$. Because, as we have seen, the radiation sequence $r_L[n]$ acts as a differentiator, the sequence b_n represents the scaled glottal flow derivative over a single glottal cycle.

We denote the length of the sequence b_n by L that we assume is sufficiently less than a glottal cycle such that there exists a time region I within which a glottal cycle is "free of zeros." In other words, there exists a region I in which the above difference equation is not driven by b_n. This interval I is the closed-phase glottal region (minus one sample due to the radiation operator $r[n]$ that for simplicity we ignore in our discussion), as illustrated in Figure 5.19. Under this condition, we can write

$$s[n] = \sum_{k=1}^{p} a_k s[n-k], \quad n \in I.$$

Therefore, if we are to use the two-step methods of the previous section for pole-zero estimation, the initial pole estimation must be performed over the limited region I which is "free of zeros." This approach is referred to as *pitch synchronous analysis* [5],[19],[28],[30].

The duration of the interval I may be very short. For example, for a high-pitch female of 400 Hz pitch at a 10000 Hz sampling rate, the interval I consists of 25 samples minus the duration of b_n—very short, indeed! With the autocorrelation method of linear prediction, where $s[n]$ is set to zero outside the region I, the error function is therefore dominated by prediction error end effects. For this reason, i.e., under this condition of very limited data, the covariance method gives a more accurate solution to the a_k parameters. The glottal flow derivative estimate is then obtained by inverse filtering the speech waveform with this all-pole vocal tract filter. A consequence of the inverse filtering is an approximate separation of the vocal tract filter from the speech waveform. Indeed, the inverse-filtered speech is essentially unintelligible, perceived largely as a "buzz." Although we do not currently have a quantitative way of measuring the accuracy of this separation on real speech, a frequency-domain view confirms that negligible vocal tract formant energy is present [19]; the inverse-filtered spectrum typically consists of a smooth lowpass function with an occasional weak peak in the vicinity of the first formant, due

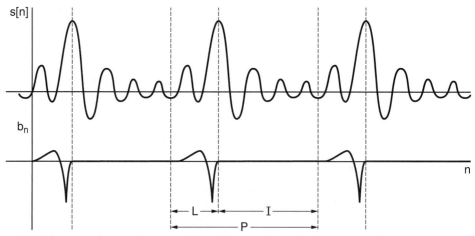

Figure 5.19 Schematic of the glottal closed-phase region I and its timing relation to the speech waveform, the glottal flow derivative duration L, and the pitch period P.

to the presence of a ripple component, consistent with the Ananthapadmanabha and Fant theory of nonlinear source/vocal tract interaction [1].

Of course, one must first estimate the glottal closed-phase region to obtain the interval I; this is a difficult task for which a number of methods have been developed. Detailed description of these methods is beyond our scope here, but we will give the reader a flavor for a few methods. Childers, for example, uses electroglottographic (EGG) analysis to identify the glottal closed phase over which the covariance method of linear prediction is applied to obtain an estimate of the vocal tract transfer function [4]. Wong, Markel, and Gray [28] and Cummings and Clements [5] perform, on the speech waveform itself, a sliding covariance analysis with a one sample shift, using a function of the linear prediction error to identify the glottal closed phase. This approach of relying on the prediction error was observed to have difficulty when the vocal folds do not close completely or when the folds open slowly.

Yet another method of glottal closed phase estimation relies also on a sliding covariance analysis but, rather than using the prediction error from this analysis, uses vocal tract formant modulation which, we have seen, is predicted by Ananthapadmanabha and Fant [1] to vary more slowly in the glottal closed phase than in its open phase and to respond quickly to a change in glottal area (Figure 4.25). A "stationary" region of formant modulation gives a closed-phase time interval, over which we estimate the vocal tract transfer function; a stationary region is present even when the vocal folds remain partly open[9] [19].

With an estimate of the inverse filter $A(z)$, we can apply the methods of the previous section to obtain an estimate of b_n and thus the glottal flow convolved with the radiation load, i.e., the glottal flow derivative. An approximate estimate of the glottal flow, unaffected by radiation, can be obtained by pre-emphasizing $s[n]$ at 12 dB/octave or by integrating the glottal flow derivative estimate. In the following example, the autocorrelation method applied over multiple pitch periods is compared against the above pitch synchronous covariance method for the purpose of inverse filtering the speech waveform.

EXAMPLE 5.8   An example of a glottal flow derivative estimate using the pitch-synchronous covariance method followed by inverse filtering is shown in Figure 5.20c. In this example, the vocal tract poles are first estimated over the glottal closed-phase region I using the covariance method of linear prediction with a 14th-order predictor. The closed-phase glottal region corresponds to a stationary formant region as described above [19]. The resulting $A(z)$ is then used to inverse filter $s[n]$ to obtain an estimate of $AG(z)R(z)$. Because of the differentiation by $R(z)$, a large negative glottal pulse excitation is observed in the glottal flow derivative to occur at glottal closure. We also see a ripple component, consistent with the Ananthapadmanabha and Fant theory of nonlinear source/vocal tract interaction [1]. In addition, we see, in Figure 5.20b, the result of the counterpart autocorrelation method using a 14th-order predictor and a 20-ms analysis window covering multiple pitch periods. The method gives a more pulse-like residual consisting of a series of sharp negative-going peaks at roughly the location of glottal pulses, very different from the actual glottal flow derivative, because it

---

[9] The formant frequencies and bandwidths are expected to remain constant during the closed phase, but shift during the open phase. For voices in which the vocal folds never completely close, such as breathy voices, a similar formant modulation occurs. For such voices, during the nominally closed phase, the glottis should remain approximately constant, resulting in a nearly fixed change in formant frequencies. When the vocal folds begin to open, the formants move from their relatively stationary values during the closed phase. The first formant is found to be more stationary than higher formants during the closed phase and exhibits a more observable change at the start of the open phase [19].

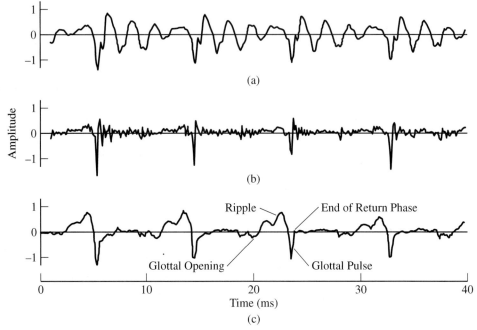

Figure 5.20 Example of estimation of glottal flow derivative: (a) original speech; (b) "whitened" speech using an inverse filter derived from the autocorrelation method over multiple pitch periods; (c) estimated glottal flow derivative using the pitch-synchronous covariance method over a closed glottal phase.

SOURCE: M.D. Plumpe, T.F. Quatieri, and D.A. Reynolds, "Modeling of the Glottal Flow Derivative Waveform with Application to Speaker Identification" [19]. ©1999, IEEE. Used by permission.

attempts to estimate the combined glottal flow/vocal tract/radiation frequency response. The result of the covariance method, on the other hand, shows a clear view of the closed, open, and return glottal phases as well as the glottal pulse and ripple component overriding the coarse glottal flow derivative.

▲

The next example illustrates a somewhat more atypical glottal flow contribution of prominent secondary glottal pulses that were described in Chapter 3.

EXAMPLE 5.9   Figure 5.21 shows glottal flow derivative estimates with the presence of secondary glottal pulses. As in the previous example, the vocal tract poles are first estimated over the glottal closed-phase region I using the pitch-synchronous covariance method of linear prediction with a 14th order predictor. As before, the closed-phase glottal region corresponds to a stationary formant region. Manifestation of the secondary pulses can also be seen in the speech waveform itself.   ▲

We note that the alternate zero estimation methods of Section 5.7.1 (e.g., Shanks' method) are also applicable to glottal flow estimation. In addition, the basic method and its extensions can be applied to estimating zeros in the vocal tract transfer function, as in nasalized vowels. The separation of glottal/radiation zeros and vocal tract zeros in this case, however, poses a new and challenging problem (Exercise 5.25). Observe that the methods of zero estimation of this chapter

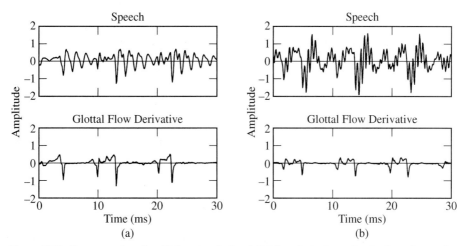

Figure 5.21 Two examples of multiple pulses in the glottal flow derivative estimate. In each case, the speech waveform is shown above the glottal flow derivative waveform.

SOURCE: M.D. Plumpe, T.F. Quatieri, and D.A. Reynolds, "Modeling of the Glottal Flow Derivative Waveform with Application to Speaker Identification" [19]. ©1999, IEEE. Used by permission.

cannot be extended to unvoiced speech modeled by stochastic processes because there does not exist a region "free of zeros" for this speech class. Finally, having a pole-zero representation of speech production, we can generalize the synthesis structures of Section 5.6. Lattice structures, for example, can be generalized to include zeros [18].

5.7.3 High-Pitched Speakers: Using Two Analysis Windows

For high-pitched speakers, it is possible that the above techniques invoke too small a glottal closed-phase interval I. For example, a speaker with a fundamental frequency of 200 Hz and an open phase 70% of a pitch period will have a closed phase of only 1.5 ms $= 0.30/200$ Hz. To address this problem, we use a covariance analysis which is based on two windows across two successive pitch periods [19]. Assuming that the rate of change of the vocal tract is dependent on time and not on the number of pitch periods, the vocal tract variation over two frames for a 200-Hz voice, for example, is approximately the same as one frame of a 100-Hz voice, since both last for 10 ms. By splitting the covariance method analysis window into two parts, we double the length of our interval I. Because this technique is dependent on stationarity of both the vocal tract and the source across multiple pitch periods, it is used only when the pitch period is small (chosen empirically at less than 6.5 ms).

As an extension to the covariance method of linear prediction, two windows of speech data can be used to calculate the matrix Φ i.e.,

$$\Phi[i, k] = \sum_{n=M_1}^{M_1+L_1-1} s[n-i]s[n-k] + \sum_{n=M_2}^{M_2+L_2-1} s[n-i]s[n-k], \quad 1 \le i, k \le p, \quad (5.51)$$

where M_1 is the start of the first region, L_1 is the length of the first region, M_2 is the start of the second region, and L_2 is the length of the second region. The only change required to convert the standard covariance linear prediction procedure into a two-window procedure is this change in the calculation of the matrix Φ. The properties of the matrix Φ with elements $\Phi[i, k]$ are the same as for the one-window case as long as the windows are non-overlapping, allowing efficient solution by Cholesky decomposition (Exercise 5.19).

5.8 Decomposition of the Glottal Flow Derivative

In previous sections of this chapter, we described a method of estimating the glottal flow derivative waveform from voiced speech that relies on inverse filtering the speech waveform with an estimate of the vocal tract transfer function. Estimation of the vocal tract transfer function is performed during the glottal closed phase within which the vocal folds are in a closed position and there is no dynamic source/vocal tract interaction. With the objective of modeling the temporal structure of the glottal flow and decomposing glottal flow components, in this section we develop a time-domain feature representation of the glottal flow derivative during voicing. The *coarse structure* of the flow derivative is represented by the piecewise-functional Liljencrants-Fant (LF) model [8] consisting of seven parameters, obtained by a nonlinear estimation method of the Newton-Gauss type. The coarse structure is then subtracted from the glottal flow derivative estimate to give its *fine-structure* component, reflecting characteristics not captured by the general flow shape such as aspiration and the ripple component we have previously described.

5.8.1 Model

The relation between the coarse structure of the glottal flow, denoted here by $u_{gc}(t)$, and its derivative, denoted by $v_{gc}(t) = \dot{u}_{gc}(t)$, was shown in Figure 4.21. The glottal flow derivative is characterized by a large negative impulse-like glottal pulse and the open- and closed-glottal phases. Fine structure of the glottal flow derivative, denoted by $v_{gf}(t)$, is the residual waveform $v_{gf}(t) = v_g(t) - v_{gc}(t)$, $v_g(t)$ being the complete glottal flow derivative. Two contributions of fine structure are considered, ripple and aspiration. As was illustrated in Figure 4.23, ripple is a sinusoidal-like perturbation that overlays the coarse glottal flow, and arises from the time-varying and nonlinear coupling of the glottal flow with the vocal tract cavity. Our second form of fine structure, aspiration at the glottis, arises when turbulence is created as air flows through constricted vocal folds, and is also dependent on the glottis for its timing and magnitude. For example, a long, narrow opening, which constricts the air flow along the entire glottal length, tends to produce more aspiration than, for example, a triangular-shaped opening with partial constriction. The creation of turbulence at the glottis is highly nonlinear and a satisfactory physical model has yet to be developed. A simplification is to model aspiration as a random noise process, which is the source to the vocal tract. We model the complete fine-structure source as the addition of the aspiration and ripple source components, although other fine-structure components may be present.

We now propose functional models of the coarse and fine structure useful for feature representation and, in particular, in speaker recognition described in Chapter 14. The features we want to capture through the coarse structure include the glottal open, closed, and return phases, the speeds of opening and closing, and the relationship between the glottal pulse and the peak glottal flow. To model the coarse component of the glottal flow derivative, $v_{gc}(t)$, we use the Liljencrants-Fant (LF) model [8], expressed over a single glottal cycle by (Figure 5.22)

$$
\begin{aligned}
v_{LF}(t) &= 0, & 0 &\leq t < T_o \\
&= E_o e^{\alpha(t-T_o)} \sin[\Omega_o(t - T_o)], & T_o &\leq t < T_e \\
&= -E_1[e^{-\beta(t-T_e)} - e^{-\beta(T_c-T_e)}], & T_e &\leq t < T_c
\end{aligned}
\tag{5.52}
$$

where $E_1 = \frac{E_e}{1-\exp[-\beta(T_c-T_e)]}$, and where the time origin, $t = 0$, is the start time of the closed phase (also the end of the return phase of the previous glottal cycle which we later also denote by T_{c-1}), T_o is the start time of the open phase (also the end of the closed phase), T_e is the start time of the return phase (also the end of the open phase and time of the glottal pulse), and T_c is the end time of the return phase (also the beginning of the closed phase of the next glottal cycle). Three of the parameters, E_o, Ω_o, and α, describe the shape of the glottal flow during the open phase. The two parameters E_e and β describe the shape of the return phase. Because at time $t = T_e$, E_o can be calculated from E_e using the relation $E_o = \frac{E_e}{e^{\alpha(T_e-T_o)} \sin[\Omega_o(T_e-T_o)]}$, we reduce the number of waveshape parameters to four, i.e., Ω_o, α, E_e, and β. For the LF model, we estimate E_e, not E_o or E_1; E_e is the absolute value of the negative peak for which an initial estimate is easily obtained. The resulting four waveshape parameters do not include the glottal timing parameters; therefore, the times T_o, T_e, and T_c must also be made variables. We thus have a 7-parameter model to describe the glottal flow, with the four LF model shape parameters, and three parameters indicating the timing of the flow components. A descriptive summary of the seven parameters of the LF model is given in Table 5.1.

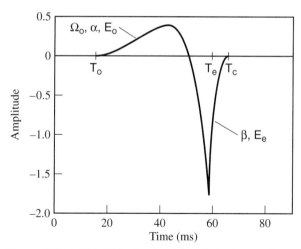

Figure 5.22   LF Model for the glottal flow derivative waveform.

Source: M.D. Plumpe, T.F. Quatieri, and D.A. Reynolds, "Modeling of the Glottal Flow Derivative Waveform with Application to Speaker Identification" [19]. ©1999, IEEE. Used by permission.

Table 5.1 Description of the seven parameters of the LF model for the glottal flow derivative waveform.

SOURCE: M.D. Plumpe, T.F. Quatieri, and D.A. Reynolds, "Modeling of the Glottal Flow Derivative Waveform with Application to Speaker Identification" [19]. ©1999, IEEE. Used by permission.

T_o	The time of glottal opening.
α	Factor that determines the ratio of E_e to the peak height of the positive portion of the glottal flow derivative.
Ω_o	Frequency that determines flow derivative curvature to the left of the glottal pulse; also determines how much time elapses between the zero crossing and T_e.
T_e	The time of the maximum negative value of the glottal pulse.
E_e	The value of the flow derivative at time T_e.
β	An exponential time constant which determines how quickly the flow derivative returns to zero after time T_e.
T_c	The time of glottal closure.

In characterizing the glottal flow derivative, we also define five time intervals within a glottal cycle. The first three intervals correspond to timing within the LF model of coarse structure, while the last two intervals come from timing measurements made on formant modulation described earlier. The time intervals are given as

1. $I_1 = [0, T_o]$ is the closed phase for the LF model.
2. $I_2 = [T_o, T_e]$ is the open phase for the LF model.
3. $I_3 = [T_e, T_c]$ is the return phase for the LF model.
4. $I_4 = [0, T_f]$ is the closed phase for the formant modulation.
5. $I_5 = [T_f, T_e]$ is the open phase for the formant modulation.

5.8.2 Estimation

With a glottal source waveform estimate, we can estimate parameters of the coarse and fine model components of the glottal flow derivative. The seven parameters of the LF model to be estimated for each glottal cycle were summarized in Table 5.1. A least-squared error minimization problem can be formulated to fit the LF model of Equation (5.52) to the glottal flow derivative waveform. For the glottal flow derivative estimate $v_g[n]$, the error criterion is given as

$$E(\underline{x}) = \sum_{n=0}^{N_o-1} v_g^2[n] + \sum_{n=N_o}^{N_e-1} \left(v_g[n] - E_o e^{\alpha(n-N_o)} \sin[\omega_o(n - N_o)] \right)^2$$

$$+ \sum_{n=N_e}^{N_c-1} \left(v_g[n] - E_1 \left[e^{-\beta(N_c-N_e)} - e^{-\beta(n-N_e)} \right] \right)^2$$

where N_o, N_e, and N_c are sample-time[10] counterparts to the continuous-time variables T_o, T_e, and T_c of Equation (5.52), ω_o is a discrete-time frequency, and $\underline{x}$ is a vector of the seven model parameters. The error $E(\underline{x})$ is a nonlinear function of the seven model parameters with no closed-form solution, and thus the problem is solved iteratively using a nonlinear least-squared error minimization with calculation of first- and second-order gradients.

Standard methods for solving nonlinear least-squares problems, such as the Gauss-Newton method, are not adequate when the minimum error $E(\underline{x})$ is large [6]. This is often the case in fitting the LF model to the glottal flow derivative waveform because ripple and aspiration, not represented by the LF model, manifest themselves in $E(\underline{x})$. An algorithm more amenable to large optimization error is an adaptive nonlinear least-squares regression technique, referred to as the NL2SOL algorithm.[11] For specifics on parameter estimation with the NL2SOL algorithm, the interested reader is referred to [19]. This algorithm also has the advantage of allowing bounds to enable parameters to be limited to physically reasonable values. For example, if the value ω_o is less than π, the model will have no negative flow derivative during the open phase, which is inconsistent with a negative going glottal pulse. Another example of an unrealistic condition is the parameter E_e's taking on a positive value or a value near zero. Therefore, π and 0 are the lower bounds for estimates of the model parameters ω_o and E_e, respectively. When a resulting model parameter estimate is too close to its bound, in particular, a constraint of 1% of its bound obtained experimentally, we consider data for that frame to be unreliable. We refer to such parameters as "singularities." Subtracting the estimated coarse structure from the glottal flow derivative waveform yields the fine structure with contributions of aspiration due to turbulence at the glottis and ripple due to source/vocal tract interaction, as illustrated in the following example:

EXAMPLE 5.10   Figure 5.23a shows an example of the coarse-structure estimate (dashed) superimposed on the glottal flow derivative estimate (solid), along with the timing estimates T_o, T_e, and T_c of the LF model. The closed phase interval $I_1 = [0, T_o)$, as determined by the LF model, comprises an interval of aspiration followed by ripple, the ripple continuing into the open phase $I_2 = [T_o, T_e)$, after which we see the occurrence of a sharp glottal pulse and a gradual return phase $I_3 = [T_e, T_c)$. The residual (Figure 5.23b), formed by subtracting the coarse structure from the glottal flow derivative estimate, forms the fine-structure estimate. The starting time of the open phase, T_f, according to formant modulation, is also shown.[12] In this case, the closed phase $I_4 = [0, T_f)$, as determined by formant modulation, consists of primarily aspiration, while the following open phase $I_5 = [T_f, T_e)$ is initially dominated by ripple, with no significant glottal flow, and then a slowly rising glottal flow with superimposed ripple.   ▲

---

[10] For notational convenience, the sampling time interval is normalized to unity.

[11] The NL2SOL algorithm is the Association for Computing Machinery algorithm 573 [6],[7]. The acronym derives from its being a NonLinear secant approximation To the Second Order part of the Least-squared Hessian.

[12] This example also illustrates the improved temporal resolution that can be gained by the timing parameter T_f in representing fine structure.

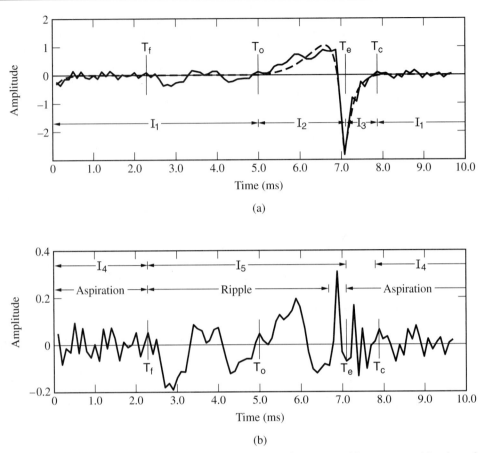

Figure 5.23 Example of a glottal flow derivative estimate and its coarse and fine structure: (a) estimated glottal flow derivative (solid) and overlaid LF model, i.e., the coarse structure (dashed); (b) the fine structure obtained by subtracting the coarse structure from the glottal flow derivative estimate. Aspiration and ripple are seen in different intervals of the glottal cycle.

SOURCE: M.D. Plumpe, T.F. Quatieri, and D.A. Reynolds, "Modeling of the Glottal Flow Derivative Waveform with Application to Speaker Identification" [19]. ©1999, IEEE. Used by permission.

Through experiments of this kind, it has been observed that coarse structure and fine structure of ripple and aspiration during voicing are speaker-dependent [19]. In addition, similar experiments have shown that such characteristics may have more general dependencies such as dialect [29]. We will exploit such dependencies in Chapter 14 in automatic speaker recognition.

5.9 Summary

In this chapter, we introduced the concept of linear prediction analysis of speech sounds whose underlying system function follows an all-pole model. We saw that linear prediction analysis can be formulated for both deterministic and stochastic sound classes using the autocorrelation or covariance method of solution. For the autocorrelation method, the solution is referred to as the normal equations. The autocorrelation method of linear prediction analysis assumes zero data outside of a short-time windowed speech segment and thus typically does not result in an exact solution, even when the data follows an all-pole model. Nevertheless, the autocorrelation method provides a stable and efficient solution to all-pole parameter estimation. Techniques of synthesis based on the linear prediction analysis were presented. Limitations of speech analysis/synthesis based on the autocorrelation method of linear prediction were also described. One problem that plagues this approach is the requirement of stationarity of the speech source and system under the analysis window. Exercise 5.13 gives an interesting alternate modeling scheme whereby predictor coefficients are themselves modeled as polynomials. This requirement of stationarity, however, is not exclusive to linear prediction-based analysis/synthesis; we will encounter it in all analysis/synthesis techniques throughout the text.

We also saw in this chapter that a more accurate approach to speech modeling is to allow zeros, as well as poles, in the model vocal tract transfer function, representing the "anti-resonances" of the vocal tract, as well as an arbitrary glottal flow function. We described a number of techniques for doing pole-zero estimation and decomposition. This modeling approach, together with the covariance method of linear prediction applied pitch-synchronously, provided a means for glottal flow estimation that led to a decomposition of the glottal flow derivative into its coarse and fine-structure (aspiration and ripple) components. These glottal components are used in Chapter 14 in an automatic speaker recognition application. Such methods of separating speech components represent just some of decomposition approaches we describe in the text. In the next chapter, for example, we describe a different approach to separating the pole and zero contributions of a vocal tract transfer function based on a method referred to as *homomorphic deconvolution*.

Appendix 5.A: Properties of Stochastic Processes

In this brief review, we assume familiarity with fundamental concepts of probability theory, including random variables, the probability density (distribution), and stochastic averages. Further review of this material can be found in [17],[26].

Random Processes

Consider tossing a coin. On each toss at time n, we assign value 1 to a sequence $x[n]$ (heads) and -1 (tails). Suppose the probability of heads is p and probability of tails is $1 - p$. An example is $p = \frac{1}{4}$ and $1 - p = \frac{3}{4}$. We interpret the sequence value $x[n]$ as a particular value of the *random variable* x_n. That is, the value $x[n]$ is the *outcome* of the experiment of tossing a coin. The random variable x_n has associated with it the probability $P_{x_n}(x[n])$. We now give numerous definitions associated with properties of random variables.

Definition: A *random process* is an indexed set of random variables

$$\{x_n\}, \qquad -\infty < n < \infty$$

with a probability of each random variable $P_{x_n}(x[n])$ which is a function of the index n.

Definition: A *sample sequence* (or "realization") of the random process is a *particular* sequence of values

$$\{x[n]\}, \qquad -\infty < n < \infty.$$

The number of possible sample sequences is infinite.

Definition: An *ensemble* (of sample sequences) is a collection of all possible sample sequences.

In applying random process theory to speech processing, we consider a particular sequence to be *one of an ensemble of sample sequences*. We don't always know the underlying probability description of the random process, and so it must be inferred or estimated.

In a generalization of the binary coin tossing experiment, each random variable x_n in the random process takes on a continuum of values, and has an associated *probability density function* (pdf) denoted by $p_{x_n}(x[n])$, where $x[n]$ is a particular value of the random variable. The pdf integrates to unity (with certainty that one of all possible events must occur):

$$\int_{-\infty}^{\infty} p_{x_n}(x[n])dx[n] = 1.$$

Inter-dependence between two random variables x_n and x_m of the process is given by the *joint pdf:*

$$p_{x_n,x_m}(x[n], x[m]).$$

A *complete characterization* of the random process would require all possible joint density functions.

Definition: *Statistical independence* of two random variables x_n and x_m must satisfy the condition

$$p_{x_n,x_m}(x[n], x[m]) = p_{x_n}(x[n])p_{x_m}(x[m]).$$

If this is true for all n, m $(n \neq m)$, then the process is "white."

Definition: *Stationarity* of a random process requires that all joint pdf's are independent of a shift in the time origin. For a second-order pdf, for example, we must have

$$p_{x_{n+k},x_{m+k}}(x[n + k], x[m + k]) = p_{x_n,x_m}(x[n], x[m]).$$

The pdf depends on the distance between variables, not absolute time (Figure 5.24).

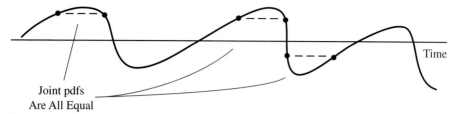

Figure 5.24 Illustration of stationarity of a random process. The distance between samples is the same for each depicted pair.

Ensemble Averages

Average properties of the random process often serve as a useful, but incomplete, characterization of the process. Below we consider some essential definitions.

Definition: The *mean* of a random variable x_n from a process is given by

$$m_{x_n} = E[x_n] = \int_{-\infty}^{\infty} x[n] p_{x_n}(x[n]) dx[n]$$

which satisfies the property of superposition, i.e.,

$$E[x_n + y_n] = E[x_n] + E[y_n]$$
$$E[ax_n] = aE[x_n].$$

m_{x_n} is sometimes called an *ensemble* average for a particular x_n, i.e., the average over all sample sequences at time n.

Definition: The *mean-squared value* of a random variable x_n from a process is given by

$$E[x_n^2] = \int_{-\infty}^{\infty} x^2 p_{x_n}(x[n]) dx[n]$$

which is sometimes called the "average power."

Definition: The *variance* of x_n is given by

$$\begin{aligned}
\sigma_{x_n}^2 &= E[(x_n - m_{x_n})^2] \\
&= E[x_n^2 + m_{x_n}^2 - 2x_n m_{x_n}] \\
&= E[x_n^2] + m_{x_n}^2 - 2m_{x_n} E[x_n] \\
&= E[x_n^2] - m_{x_n}^2.
\end{aligned}$$

Taking the square root, we obtain σ_{x_n} which is called the *standard deviation*.

Definition: The *autocorrelation* of two random variables x_n and x_m from a process is given by

$$\begin{aligned}
\phi_{xx}[n, m] &= E[x_n x_m] \\
&= \int_{-\infty}^{\infty} \int_{-\infty}^{\infty} x[n] x[m] p_{x_n, x_m}(x[n], x[m]) dx[n] dx[m],
\end{aligned}$$

which is the measure of the dependence between variables of the random process at different times.

Definition: The *autocovariance* of two random variables from a process is given by

$$v_{xx}[n, m] = E[(x_n - m_{x_n})(x_m - m_{x_m})]$$
$$= \phi_{xx}(n, m) - m_{x_n}m_{x_m},$$

which is the autocorrelation without the mean.

Definition: *Uncorrelated random variables* (or "linearly independent") must satisfy the condition

$$E[x_n y_n] = E[x_n]E[y_n].$$

A sufficient condition for linear independence is given by

$$p_{x_n, y_m}(x[n], y[m]) = p_{x_n}(x[n])p_{y_m}(y[m]),$$

i.e., the variables are statistically independent. In general, linear independence is not sufficient for statistical independence. If the mean of either random variable is zero, then uncorrelated implies

$$E[x_n y_n] = 0.$$

Stationary Random Process

Definition: For *wide sense stationarity*, both the mean and variance of the random process variables are constant:

$$m_x = E[x_n]$$
$$\sigma_x^2 = E[(x_n - m_x)^2]$$

and the autocorrelation is a function only of the time difference, τ, i.e.,

$$\phi_{xx}[n, n + \tau] = \phi_{xx}[\tau] = E[x_n x_{n+\tau}].$$

Definition: For stationarity in the *strict sense*, all joint pdf's are functions of time differences (corresponding to our previous definition of stationarity).

For a *white noise* sequence, we have

$$\phi_{xx}[\tau] = \sigma_x^2 \delta[\tau]$$

where $\delta[\tau]$ is the unit sample sequence. A white noise sequence is thus wide sense stationary and is uncorrelated for all time differences.

Time Averages

The notion of ensemble average (the above statistical averages) is conceptually important but not practical. Under a condition known as *ergodicity*, ensemble averages are equal to time averages.

Definition: An *ergodic random process* is a random process for which time averages equal ensemble averages. Here, wide-sense stationarity is implicit.

A consequence of ergodicity is that for any single sample sequence, $(< \cdot >$ denoting time average)

$$< x[n] > = \lim_{N \to \infty} \left(\frac{1}{2N+1} \right) \sum_{n=-N}^{N} x[n]$$

$$= E[x_n] = m_x$$

and

$$< x[n]x[n+\tau] > = \lim_{N \to \infty} \left(\frac{1}{2N+1} \right) \sum_{n=-N}^{N} x[n]x[n+\tau]$$

$$= E[x_n x_{n+\tau}] = \phi_{xx}[\tau].$$

With finite limits on the sums, we obtain only estimates of mean and autocorrelation.

Power Density Spectrum

Definition: The discrete-time Fourier transform of the autocorrelation function of a wide sense stationary random process is referred to as the *power density spectrum* or *power spectrum*. Denoting the power spectrum of a process $x[n]$ by $S_x(\omega)$, we have the Fourier transform pair

$$S_x(\omega) = \sum_{m=-\infty}^{\infty} \phi_{xx}[m] e^{-j\omega m}$$

$$\phi_{xx}[m] = \frac{1}{2\pi} \int_{-\pi}^{\pi} S_x(\omega) e^{j\omega m} d\omega.$$

The description "power density" is motivated by seeing from the above inverse Fourier transform that the average power in $x[n]$ is given by [17],[26]

$$E(x^2[n]) = \phi_{xx}[0] = \frac{1}{2\pi} \int_{-\pi}^{\pi} S_x(\omega) d\omega$$

and thus each frequency band of (infinitesimally small) width $d\omega$ makes a contribution $\frac{S_x(\omega)d\omega}{2\pi}$ to the total average power. The power spectrum is sometimes normalized by its total average power in order to have the properties of a probability density. The interpretation of $S_x(\omega)$ as a power density is analogous to our formulation of the energy density for deterministic signals through Parseval's Theorem in Chapter 2, i.e., $\sum_{n=-\infty}^{\infty} |x[n]|^2 = \frac{1}{2\pi} \int_{-\pi}^{\pi} |X(\omega)|^2 d\omega$.

Definition: An estimate of the power spectrum is obtained by computing the squared magnitude of the discrete-time Fourier transform of $x[n]$ over a finite interval N, i.e.,

$$|X_N(\omega)|^2 = \frac{1}{N} \left| \sum_{n=0}^{N-1} x[n] e^{j\omega n} \right|^2$$

which is referred to as the *periodogram*. The periodogram is related to the estimate of the autocorrelation $\phi_{xx}[\tau]$, $r_n[\tau] = \frac{1}{N} \sum_{m=0}^{N-1-\tau} x_n[m]x_n[m+\tau] = \frac{1}{N} x_n[\tau] * x_n[-\tau]$, through the Fourier transform. The periodogram fluctuates about the power spectrum with a variance that is proportional to the power spectrum for large N [17]. The expected value of the periodogram is given by

$$E[|X_N(\omega)|^2] = \frac{1}{2\pi} \int_{-\pi}^{\pi} S_x(\theta) W(\omega - \theta) d\theta,$$

i.e., it is the circular convolution of the power spectrum of $x[n]$ with the Fourier transform of the rectangular window $w[n] = 1$ for $0 \leq n < N$, and thus the periodogram is a biased estimate of the power spectrum in the sense that spectral resolution is lost. Various smoothing techniques have been developed to reduce the variance of the periodogram, such as windowing the autocorrelation estimate or applying non-rectangular windows to the time waveform, but at the expense of further loss in spectral resolution [17].

Often we are concerned with passing white noise through a linear time-invariant system. Because the power spectrum of a white noise input $x[n]$ with variance σ^2 is given by $S_x(\omega) = \sigma^2$, the power spectrum of the linear time-invariant system output $y[n]$ can be expressed as

$$S_y(\omega) = \sigma^2 |H(\omega)|^2$$

where $H(\omega)$ is the frequency response of the system. For an N-point measurement of the system output $y[n]$, with large N, where end effects at the boundaries of the data segment are negligible [17], we can write the periodogram of the output as

$$|Y_N(\omega)|^2 \approx |H(\omega)|^2 |X_N(\omega)|^2$$

where $|X_N(\omega)|^2$ is the N-point periodogram of the white noise input and is shaped by the system frequency response $|H(\omega)|^2$ to form the N-point periodogram of the waveform.

Appendix 5.B: Derivation of the Lattice Filter in Linear Prediction Analysis

Define the forward ith-order prediction error as

$$e^i[m] = s[m] - \sum_{k=1}^{i} \alpha_k^i s[m - k]$$

with z-transform

$$E^i(z) = A^i(z) S(z) \tag{5.53}$$

where

$$A^i(z) = 1 - \sum_{k=1}^{i} \alpha_k^i z^{-k} \tag{5.54}$$

and define the backward ith-order prediction error as

$$b^i[m] = s[m-i] - \sum_{k=1}^{i} \alpha_k^i s[m+k-i]$$

with z-transform

$$B^i(z) = z^{-i} A^i(z^{-1}) S(z). \tag{5.55}$$

From Levinson's recursion,

$$\alpha_j^i = \alpha_j^{i-1} - k_i \alpha_{i-j}^{i-1}, \qquad 1 \le j \le i-1$$

$$\alpha_i^i = k_i. \tag{5.56}$$

Substituting Equation (5.56) into Equation (5.54), we have

$$A^i(z) = 1 - \sum_{k=1}^{i-1} \alpha_k^i z^k - k_i z^{-i}$$

$$= 1 - \sum_{k=1}^{i-1} \left[\alpha_k^{i-1} z^{-k} - k_i \alpha_{i-k}^{i-1} z^{-k} \right] - k_i z^{-i}$$

$$= \left[1 - \sum_{k=1}^{i-1} \alpha_k^{i-1} z^{-k} \right] + k_i \sum_{k=1}^{i-1} \alpha_{i-k}^{i-1} z^{-k} - k_i z^{-i}. \tag{5.57}$$

In the second term of Equation (5.57) letting $k' = i - k$, Equation (5.57) becomes

$$A^i(z) = \left[1 - \sum_{k=1}^{i-1} \alpha_k^{i-1} z^{-k} \right] + k_i \sum_{k'=i-1}^{1} \alpha_{k'}^{i-1} z^{-i+k'} - k_i z^{-i}$$

$$= \underbrace{\left[1 - \sum_{k=1}^{i-1} \alpha_k^{i-1} z^{-k} \right]}_{A^{i-1}(z)} - k_i z^{-i} \underbrace{\left[1 - \sum_{k'=1}^{i-1} \alpha_{k'}^{i-1} z^{k'} \right]}_{A^{i-1}(z^{-1})}$$

or

$$A^i(z) = A^{i-1}(z) - k_i z^{-i} A^{i-1}(z^{-1}). \tag{5.58}$$

Substituting Equation (5.58) into Equation (5.53),

$$E^i(z) = \left[A^{i-1}(z) - k_i z^i A^{i-1}(z^{-1}) \right] S(z)$$

$$= A^{i-1}(z) S(z) - k_i z^{-i} A^{i-1}(z^{-1}) S(z). \tag{5.59}$$

From Equations (5.59), (5.53), and (5.55),

$$E^i(z) = \underbrace{A^{i-1}(z) S(z)}_{E^{i-1}(z)} - k_i z^{-i} \underbrace{z^{-(i-1)} A^{i-1}(z^{-1}) S(z)}_{B^{i-1}(z)}$$

so that

$$E^i(z) = E^{i-1}(z) - k_i z^{-i} B^{i-1}(z) \tag{5.60}$$

which, in the time domain, is given by

$$e^i[m] = e^{i-1}[m] - k_i b^{i-1}[m-1]. \tag{5.61}$$

Now, substituting Equation (5.58) into Equation (5.55),

$$
\begin{aligned}
B^i(z) &= z^{-i}\left[A^{i-1}(z) - k_i z^i A^{i-1}(z)\right] S(z) \\
&= z^{-i} A^{i-1}(z^{-1})S(z) - k_i A^{i-1}(z)S(z) \\
&= z^{-1}\underbrace{z^{-(i-1)}A^{i-1}(z^{-1})S(z)}_{B^{i-1}(z)} - k_i \underbrace{A^{i-1}(z)S(z)}_{E^{i-1}(z)}
\end{aligned}
$$

or

$$B^i(z) = z^{-1}B^{i-1}(z) - k_i E^{i-1}(z) \tag{5.62}$$

which, in the time domain, is given by

$$b^i[m] = b^{i-1}[m-1] - k_i e^{i-1}[m]. \tag{5.63}$$

EXERCISES

5.1 Show that for the autocorrelation method of linear prediction of Section 5.2.3, the normal equations are written as

$$\sum_{k=1}^{p} \alpha_k \Phi_n(i, k) = \Phi_n(i, 0), \qquad i = 1, 2, 3 \ldots p$$

where

$$\Phi_n(i, k) = \sum_{m=0}^{N+p-1} s_n[m - i]s_n[m - k], \qquad 1 \le i \le p, \qquad 0 \le k \le p$$

and where N is the data length.

5.2 In this problem, you show that if $s_n[m]$ is a segment of a periodic sequence, then its autocorrelation $r_n[\tau]$ is periodic-like. You also investigate a small deviation from periodicity. In particular, consider the periodic impulse train

$$x[n] = \sum_{k=-\infty}^{\infty} \delta[n - kP].$$

(a) Compute and sketch the autocorrelation $r_n[\tau]$ of the windowed sequence $x[n]$, i.e., $x[n]w[n]$, when the window applied to $x[n]$ is rectangular over the interval $0 \le n < N_w$ with length $N_w = 4P$, showing that $r_n[\tau]$ is periodic-like and falls off roughly linearly with increasing τ.

(b) How does your result from part (a) change if the pitch period increases by one sample on each period and the window length is long enough to cover the first three impulses of $x[n]$? Observe that the first three impulses occur at $n = 0$, $P + 1$, and $2P + 2$.

(c) Suppose now that $x[n]$ is convolved with a single decaying exponential

$$y[n] = x[n] * a^n u[n]$$

with $u[n]$ the unit step. How does your result in parts (a) and (b) change with $y[n]$? Do not compute; just sketch the results.

5.3 Prove the autocorrelation properties **(P1)**-**(P3)** of Section 5.3.3.

5.4 Occasionally the vocal cords vibrate with a "secondary flap" within a pitch period. Consider a model for the response of the vocal tract to one primary glottal impulse $\delta[n]$, and one secondary glottal impulse at time n_o, $\delta[n - n_o]$, within a pitch period (here we are assuming an idealized glottal pulse):

$$y[n] = h[n] * (\delta[n] + \alpha\delta[n - n_o])$$
$$= h[n] + \alpha h[n - n_o]$$

where the scale factor α on the secondary pulse is typically less than one.

(a) Suppose you are given $h[n]$ and the time delay of the secondary pulse n_o. Derive an estimate of the scaling factor α by minimization of the error criterion:

$$E(\alpha) = \sum_{-\infty}^{\infty}(s[n] - y[n])^2$$

with respect to the unknown α and where $s[n]$ is the speech measurement.

(b) Suppose now that the z-transform of the vocal tract impulse response is given by

$$H(z) = \frac{(1 - bz)}{A(z)}$$

consisting of a minimum-phase (all-pole) $A(z) = 1 - \sum_{k=1}^{p}\alpha_k z^{-1}$ and one maximum-phase zero $(1 - bz)$ with $|b| < 1$. Suppose you are given $A(z)$, the speech measurement $s[n]$, and the delay n_o. Estimate the maximum-phase zero.

(c) Suppose you do not know $A(z)$. Describe a two-step procedure for estimating the zero $(1 - bz)$ and the poles of $A(z)$ in part (b).

5.5 In this problem you complete the missing steps of Example 5.3 in which we build a first-order all-pole model from the first N samples of $x[n] = a^n u[n]$ by using the autocorrelation method of linear prediction.

(a) Show how the estimate α_1 is found in Example 5.3 from the autocorrelation normal equations.

(b) Give the steps in finding the minimum squared error E_0 in Example 5.3 and show that this error converges to unity as $N \to \infty$.

5.6 Suppose we want to derive a rational model for an unknown system S using the technique in Figure 5.25. A known input $x[n]$ is applied to the system and the output $y[n]$ is measured. Then the parameters of two FIR filters are chosen to minimize the energy in the error signal $e[n]$ in Figure 5.25.

(a) Write the normal equations that describe the optimal least-squared error solution for $A(z)$ and $B(z)$.

(b) The philosophy of this method is that if the error is small, then $\frac{B(z)}{A(z)}$ is a reasonable model for S. Suppose S is an LTI rational system; show that this method identifies the parameters of S exactly when the orders of the numerator and denominator polynomials of S are known.

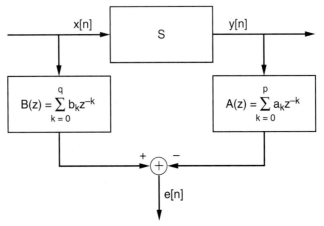

Figure 5.25 Pole-zero estimation for a rational model.

SOURCE: K. Steiglitz, "On the Simultaneous Estimation of Poles and Zeros in Speech Analysis" [24]. ©1977, IEEE. Used by permission.

5.7 This problem addresses the difficulty inherent in linear prediction analysis for high-pitched speakers.

(a) Suppose $h[n]$ is the impulse response of an all-pole system where

$$H(z) = \frac{1}{1 - \sum_{k=1}^{P} \alpha_k z^{-k}}$$

so that

$$h[n] = \sum_{k=1}^{p} \alpha_k h[n-k] + \delta[n].$$

Show that

$$\sum_{k=1}^{p} \alpha_k r_h[i-k] = r_h[i], \quad 1 \le i \le p.$$

Hint: Multiply both sides of the difference equation by $h[n-i]$ and sum over n. Note that $h[n]$ is causal.

(b) Assume $s[n]$ is a periodic waveform, given by

$$s[n] = \sum_{k=-\infty}^{\infty} h[n-kP]$$

where P is the pitch period. Show that the autocorrelation of $s[n]$, windowed over multiple pitch periods, consists of periodically repeated replicas of $r_h[\tau]$, i.e.,

$$r_s[\tau] = \sum_{k=-\infty}^{\infty} r_h[\tau - kP].$$

but with decreasing amplitude due to the window.

(c) Using your result in parts (a) and (b), explain the difference between your result in part (a) and the normal equations for the autocorrelation method using the windowed speech signal $s[n]$.

(d) Using your results from parts (b) and (c), explain why linear prediction analysis is more accurate for low-pitched speakers than high-pitched speakers.

5.8 Unlike the autocorrelation method, the covariance method of linear prediction can determine the exact all-pole model parameters from a finite data segment. Consider a signal of the form $x[n] = a^n u[n]$ with $u[n]$ the unit step.

(a) Suppose the rectangular window $w[n]$ has duration $N_w \geq 2$. Determine the normal equations for the covariance method for $w[n]x[n] = w[n]a^n$ (Figure 5.26). Solve for the one unknown prediction coefficient α_1, as a function of the true value a. Compute the mean-squared prediction error $E = \sum_{n=0}^{N_w-1} e^2[n]$.

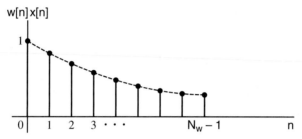

Figure 5.26 Sequence $w[n]x[n] = w[n]a^n$ with $w[n]$ a rectangular window.

(b) Suppose now a "quasi-periodic" sequence is formed by two replicas of $x[n]$:

$$\tilde{x}[n] = \sum_{k=0}^{1} x[n - kP]$$

as shown in Figure 5.27. Again use the covariance method to obtain an estimate of a, where the window is placed "pitch synchronously," beginning at time sample $n = P + 1$ with $N_w - 1 > P + 1$ as shown in Figure 5.27. What is the implication for your solution of not requiring "pitch synchrony?"

(c) If you were to repeat part (b) using the autocorrelation method, how would your estimate of a change? Do not solve. Explain your reasoning.

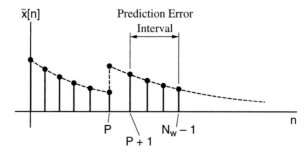

Figure 5.27 $\tilde{x}[n]$ formed from two replicas of $x[n]$.

5.9 Suppose the z-transform of a sequence $x[n]$ consists of both poles and zeros:

$$X(z) = \frac{B(z)}{A(z)}$$

where

$$B(z) = \sum_{k=0}^{q} b_k z^{-k}$$

$$A(z) = 1 - \sum_{k=1}^{p} a_k z^{-k}.$$

(a) Given a measurement $s[n]$, formulate a least mean-squared prediction error problem to solve for the coefficients of $A(z)$, free of the coefficients b_k. Assume you know the order p of $A(z)$. Select the prediction-error window to be right-sided of infinite length. Derive the normal equations for the covariance method with the appropriate limits of summation. Do your normal equations give the true coefficients of $A(z)$? Do not solve.

(b) Assume now an estimate $\hat{A}(z)$ of $A(z)$. Denote $\hat{h}_a[n]$ as the impulse response of $\frac{1}{\hat{A}(z)}$. We want to formulate a linear least mean-squared error problem to solve for the coefficients b_k of $B(z)$. To do so, we form the error

$$e[n] = s[n] - b[n] * \hat{h}_a[n]$$

where

$$b[n] = b_n, \quad 0 \le n \le q$$
$$= 0, \quad \text{otherwise.}$$

Find the normal equations by minimization of the error function

$$E = \sum_{n=-\infty}^{\infty} e^2[n]$$

with respect to the unknown $b[n]$ for $n = 0, 1 \ldots q$. Set up the normal equations in matrix form. Is the governing matrix Toeplitz? Using Parsevel's Theorem, show that the mean-squared error E has the frequency-domain interpretation Equation (5.50).

(c) Suppose now that $X(z) = \frac{B(z)}{A(z)}$ is the z-transform of the vocal tract output volume velocity where $\frac{1}{A(z)}$ represents the vocal tract poles and $B(z)$ represents the glottal flow of one glottal cycle during voicing. Suppose, however, that during the glottal open phase, i.e., the time over which the glottal flow occurs, the effective vocal tract shape changes as the glottal slit area increases and then decreases. Given that the glottal pulse width is q samples, and $A(z)$ is of order p, repeat part (a) estimating the poles (prediction coefficients) using only the waveform during the closed glottal phase, and also not allowing the time-varying vocal tract during the open glottal phase to influence the pole estimates. *Hint:* Consider a prediction error window in which $x[n]$ is not a function of the changing vocal tract parameters, but rather only the *stationary* vocal tract parameters.

5.10 Describe the effect on the short-time autocorrelation function $r_n[\tau]$ of a change in the vocal tract impulse response over successive periods when the analysis window falls over multiple pitch periods. What does this imply about the time resolution of the autocorrelation method of linear prediction analysis? Describe advantages and disadvantages in the autocorrelation method of using an analysis window one period in duration with respect to time and frequency resolution.

5.11 Consider a 2nd-order linear predictor where the autocorrelation method solution is used to obtain the predictor coefficients $[\alpha_1, \alpha_2]$ through:

$$\begin{bmatrix} r[0] & r[1] \\ r[1] & r[0] \end{bmatrix} \begin{bmatrix} \alpha_1 \\ \alpha_2 \end{bmatrix} = \begin{bmatrix} r[1] \\ r[2] \end{bmatrix}$$

where $r[n]$ represents the estimated autocorrelation coefficients. This matrix equation can be solved recursively for $\underline{\alpha} = [\alpha_1 \alpha_2]^T$ using the Levinson recursion. Begin with the 0th-order predictor coefficient and mean-squared error given, respectively, by

$$\alpha_0^0 = 0$$
$$E^0 = r[0]$$

and show that for the first two passes through the Levinson recursion ($i = 1, 2$) the partial correlation coefficients k_i, predictor coefficients $\underline{\alpha}^i$, and mean-squared prediction error E_i can each be written as functions of autocorrelation coefficients, the final pass ($i = 2$) being given by

$$k_2 = \frac{r[0]r[2] - r^2[1]}{r^2[0] - r^2[1]}$$

$$\alpha_2^2 = \frac{r[0]r[2] - r^2[1]}{r^2[0] - r^2[1]}$$

$$\alpha_2^1 = \frac{r[1]r[0] - r[1]r[2]}{r^2[0] - r^2[1]}$$

$$E^2 = \left[1 - \left(\frac{r[0]r[2] - r^2[1]}{r^2[0] - r^2[1]} \right)^2 \right] \left[r[0] - \frac{r^2[1]}{r[0]} \right]$$

where the solution for the optimal 2nd-order coefficients is given by

$$\alpha_1^* = a_1^2$$
$$\alpha_2^* = a_2^2$$

with * denoting the result of the autocorrelation method. It is not surprising that the prediction error, as well as the partial correlation coefficients, are functions of the autocorrelation coefficients, because the Levinson recursion solves the normal equations for the autocorrelation method.

5.12 We saw in Exercise 5.11 that the forward Levinson recursion can be viewed as a mapping of a set of p autocorrelation coefficients $r[\tau]$ into a set of predictor filter coefficients α_k for the autocorrelation method of linear prediction analysis. In this problem you are asked to look further into this correspondence and the correspondence among other all-pole representations, as well as the stability of the resulting all-pole filter $\frac{A}{1 - \sum_{k=1}^{p} \alpha_k z^{-1}}$.

(a) Derive the backward Levinson recursion in Equation (5.20) that maps a set of predictor filter coefficients into a set of partial correlation coefficients.

(b) Using the forward Levinson recursion and backward Levinson recursion, argue for a one-to-one correspondence between the autocorrelation coefficients $[r[0], r[1], \ldots r[p]]$ and predictor coefficients and gain $[\alpha_1, \alpha_2, \ldots \alpha_p, A]$. Then argue that a one-to-one mapping exists between the predictor coefficients along with the gain, $[\alpha_1, \alpha_2, \ldots \alpha_p, A]$, and the partial correlation coefficients along with the gain, $[k_1, k_2, \ldots k_p, A]$, i.e., one coefficient set can be obtained from the other.

(c) State a stability test for the all-pole filter $H(z) = \frac{A}{1 - \sum_{k=1}^{p} \alpha_k z^{-1}}$ in terms of the reflection coefficients (as derived from the prediction filter coefficients). Give your reasoning.

(d) Is the filter

$$H(z) = (1 - 2z^{-1} - 6z^{-2} + z^{-3} - 2z^{-4})^{-1}$$

stable? Use your stability test from part (c).

5.13 Suppose that a speech sound is characterized by time-varying pth order all-pole predictor coefficients that change linearly as

$$a_k[n] = b_k^0 + b_k^1 n.$$

Assume a speech segment is extracted with a rectangular window so that the segment is zero outside the interval $[0, N)$. Determine the normal equations for the autocorrelation method of linear prediction that give the unknown parameters b_k^0 and b_k^1. Do not solve.

5.14 Derive the minimum- and maximum-phase contributions in Equation (5.19) that arise when the autocorrelation method is applied to a sequence having a Fourier transform with a minimum- and maximum-phase component.

5.15 Suppose that we are given a linear prediction analysis/synthesis system whose synthesizer is shown in Figure 5.28. The control parameters available to the synthesizer on each frame are:

1. Pitch and voicing;

2. The first autocorrelation coefficient $r[0]$ that is the energy in the windowed speech signal;

3. $\underline{\alpha}$, a pth order set of predictor coefficients, that are guaranteed to represent a stable filter.

(a) Describe a method for determining the gain A from the received parameters.

(b) Suppose that the synthesizer filter is a direct-form realization of the transfer function $T(z)$ using the $\underline{\alpha}$. It is determined experimentally that interpolation of the filter parameters between frame updates is desirable. Given $\underline{\alpha}_1$ and $\underline{\alpha}_2$, the sets of predictor coefficients from two successive speech frames at a 20-ms frame update, as well as the first autocorrelation coefficients on the successive frames, denoted by $r_1[0]$ and $r_2[0]$, describe qualitatively a method that

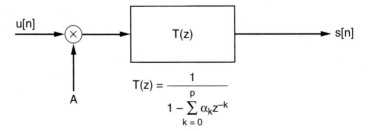

Figure 5.28 Linear prediction synthesizer.

produces a stable pth order set of interpolated filter coefficients at a 10-ms frame update. In developing your method, assume the partial correlation coefficients k_i can be generated from the reverse Levinson algorithm. Also assume there is a one-to-one correspondence between the predictor coefficients and partial correlation coefficients, i.e., we can derive one from the other. Do not prove this correspondence, but use it in the development of your method. *Hint:* If the partial correlation coefficients are such that $|k_i| < 1$, then the predictor coefficients represent a stable all-pole system. Note also that the interpolation of two sets of stable predictor coefficients does not necessarily result in a stable system.

(c) Consider the lattice realization of the inverse filter corresponding to your interpolated predictor coefficients from part (b). Is the lattice filter stable? Explain.

5.16 We have seen that the lattice formulation of linear prediction bears a strong resemblance to the lossless concatenated tube model. This relationship, based on the equivalence of the Levinson recursion and the vocal tract transfer function recursion, leads to a procedure for obtaining the vocal tract tube cross-sectional areas from the partial correlation coefficients. List reasons for possible inaccurate prediction of the cross-sectional areas. Consider, for example, the validity of the planar-wave equation in individual tubes, energy loss, glottal and radiation effects, and nonlinear phenomena.

5.17 Consider the all-pole model of speech production:

$$H(\omega) = \frac{A}{A(\omega)}$$

with

$$A(\omega) = 1 - \sum_{k=1}^{p} \alpha_k e^{-j\omega k}.$$

In this problem, you will show that the error function

$$I = \int_{-\pi}^{\pi} [e^{Q(\omega)} - Q(\omega) - 1] \frac{d\omega}{2\pi} \tag{5.64}$$

with

$$Q(\omega) = \log |S(\omega)|^2 - \log |H(\omega)|^2$$

where $S(\omega)$ is the Fourier transform of the short-time segment of speech being modeled, leads to the autocorrelation method of linear prediction.

(a) Show that $Q(\omega)$ can be written as

$$Q(\omega) = \log \left| \frac{E(\omega)}{A} \right|^2 \tag{5.65}$$

where $E(\omega) = A(\omega)S(\omega)$ is the Fourier transform of the prediction error sequence.

(b) Suppose that the zeros of $A(z)$ are constrained to lie inside the unit circle. Show[13] that

$$\int_{-\pi}^{\pi} \log |A(\omega)|^2 \frac{d\omega}{2\pi} = 0.$$

---

[13] One approach to this solution is through the complex cepstrum (described in Chapter 6) associated with $|A(\omega)|^2 = A(\omega)A^*(\omega)$.

(c) Using your result in part (b), show that

$$\int_{-\pi}^{\pi} \log \left| \frac{A}{A(\omega)} \right|^2 \frac{d\omega}{2\pi} = \log A^2.$$

(d) Substitute Equation (5.65) into Equation (5.64) and use your result from part (c) to show that minimization of I, with respect to the unknown filter coefficients α_k, is equivalent to minimization of the average squared filter output $\int_{-\pi}^{\pi} |E(\omega)|^2 d\omega$, and hence yields the same solution as the autocorrelation method of linear prediction.

(e) Show that minimization of I with respect to the unknown A^2 gives the optimal gain solution

$$A^2 = \frac{1}{2\pi} \int_{-\pi}^{\pi} |E(\omega)|^2 d\omega.$$

Argue, therefore, that the optimal A^2 is the same gain solution of Section 5.3.4 as obtained by requiring that the energy in the all-pole impulse response $h[n]$ equals the energy in the measurement $s[n]$. The error function I thus provides a rigorous way of obtaining both the unknown filter coefficients and the unknown gain.

(f) Using Figure 5.14a, illustrating the function $f(Q) = e^Q - Q - 1$, argue that linear prediction spectral analysis tends to fit spectral peaks more accurately than spectral valleys.

(g) Using a Taylor series expansion for $f(Q) = e^Q - Q - 1$, show that $f(Q)$ is symmetric for small Q. What does this imply for spectral matching?

5.18 It has been observed that the linear prediction residual (from the inverse filter $A(z)$) often contains "intelligible" speech. What may this imply about an all-pole model and the linear prediction estimate of the model parameters?

5.19 Derive Equation (5.51) for the use of two analysis windows in the covariance method of linear prediction analysis. Argue that the covariance properties of the matrix Φ, with elements $\Phi[i, k]$ of Equation (5.51), are the same as for the one-window analysis case provided that the two windows are non-overlapping.

5.20 Consider the two-pole model for the glottal flow over one cycle given by

$$G(z) = \frac{1}{(1 - \alpha z^{-1})(1 - \beta z^{-1})} \tag{5.66}$$

with α and β both real, positive, and less than one, and where the region of convergence includes the unit circle. In the time domain, $g[n]$ can be expressed as the convolution of two decaying exponentials:

$$g[n] = (\alpha^n u[n]) * (\beta^n u[n]). \tag{5.67}$$

(a) Sketch $g[n]$ and $|G(\omega)|$. Assume α and β are close to unity, say, about 0.95. Why are Equations (5.66) and (5.67) a reasonable model for the spectral magnitude, but not for the shape of the glottal flow?

(b) Explain why a better model for the glottal flow is given by

$$\tilde{g}[n] = g[-n]. \tag{5.68}$$

Derive the z-transform of $\tilde{g}[n]$. Where are the poles of $\tilde{g}[n]$ in relation to those of $g[n]$? Describe the difficulty in estimating the model parameters of $\tilde{g}[n]$ by using the autocorrelation method of linear prediction.

(c) Suppose now that the speech waveform for a voiced segment is given by

$$s[n] = \tilde{g}[n] * v[n] * u[n]$$

where $\tilde{g}[n]$ is given by Equations (5.67) and (5.68), where $u[n]$ consists of two pulses

$$u[n] = \delta[n] + b_1\delta[n - P]$$

with P the pitch period and b_1 real, positive, and less than unity, and where $v[n]$ is a minimum-phase vocal tract response with only one resonance:

$$V(z) = \frac{1}{(1 - a_1 z^{-1})(1 - a_1^* z^{-1})}.$$

Suppose that a fourth-order linear prediction analysis (using the autocorrelation method) is applied to $\tilde{g}[n] * v[n]$ (i.e., assume no influence by $u[n]$) resulting in a system function $\frac{1}{A(\omega)}$ such that

$$\left| \frac{1}{A(\omega)} \right| = |\tilde{G}(\omega)||V(\omega)|.$$

What are the prediction coefficients α_k, $k = 1, 2, 3, 4$ of the function $A(\omega)$? Now apply the inverse filter $A(\omega)$ to the speech signal $s[n]$. Determine an analytical expression for the Fourier transform of the resulting source estimate. Show that the Fourier transform magnitude of your result is only a function of the pulse train $u[n]$. Express the Fourier transform phase in terms of the phase of the glottal pulse $\tilde{g}[n]$ and the phase of $u[n]$.

5.21 Suppose we model the z-transform of a measured sequence with the rational function in Equation (5.45). Consider minimizing the mean-squared error $\sum_{n=-\infty}^{\infty}(x[n] - s[n])^2$ where $s[n]$ is the measurement and $x[n]$ is the inverse z-transform of our model. Map this error criterion to the frequency domain in terms of $A(\omega)$ and $B(\omega)$ using Parseval's Theorem. Then show that least mean-squared error minimization with respect to the unknown model numerator and denominator coefficients [b_k and a_k of Equation (5.45)] leads to equations that are nonlinear in the unknown coefficients.

5.22 Consider the problem of estimating zeros of the numerator polynomial of a rational z-transform model. Develop a method of "inverse linear prediction" by converting zero estimation to pole estimation.

5.23 (MATLAB) In this exercise, use the voiced speech signal *speech1_10k* (at 10000 samples/s) in the workspace *ex5M1.mat* located in companion website directory Chap_exercises/chapter5. This problem illustrates the autocorrelation method of linear prediction.

   (a) Window *speech1_10k* with a 25-ms Hamming window. Compute the autocorrelation of the resulting windowed signal and plot.

   (b) Assume that two resonances represent the signal and model the vocal tract with 4 poles. Set up the autocorrelation matrix R_n, using your result from part (a). The autocorrelation matrix is of dimension 4×4.

   (c) Solve for the linear predictor coefficients by matrix inversion.

   (d) Plot the log-magnitude of the resulting frequency response:

$$H(\omega) = \frac{A}{1 - \sum_{k=1}^{p} \alpha_k e^{-j\omega k}}$$

where the gain is given by Equation (5.30). Compare your result with the log-magnitude of the Fourier transform of the windowed signal. What similarities and differences do you observe?

(e) Using your estimates of the predictor coefficients from part (c), compute the prediction error sequence associated with *speech1_10k* and plot. From the prediction error sequence, what conclusions might one draw about the model (i.e., all-pole/impulse-train-driven) and estimation accuracy?

5.24 (MATLAB) In this problem you will use your results from Exercise 5.23 to perform speech synthesis of the speech waveform *speech1_10k* in the workspace *ex5M1.mat* located in companion website directory Chap_exercises/chapter5.

(a) Using your estimates of the predictor coefficients from Exercise 5.23, compute an estimate of the vocal tract impulse response.

(b) Using the prediction error sequence you computed in Exercise 5.23, estimate an average pitch period of the voiced sequence *speech1_10k*.

(c) Using your results from parts (a) and (b), synthesize an estimate of *speech1_10k*. How does your waveform estimate differ from the original? Consider the minimum-phase nature of the impulse response estimate.

(d) Using the MATLAB number generator *randm.m*, synthesize the "whispered" counterpart to your voiced synthesis of part (c). Using the MATLAB *sound.m* function, listen to your two estimates and compare to the original.

5.25 (MATLAB) Consider a nasalized vowel where the nasal tract introduces zeros into the vocal tract system function. The source to the vocal tract is a periodic glottal flow that we assume has an effective duration L over a single glottal cycle. Under this condition, the glottal flow over one period can be modeled as $L - 1$ zeros, and we will assume that all of the zeros lie outside the unit circle. Suppose the vocal tract consists of N poles inside the unit circle and M zeros (due to the nasal passage) also inside the unit circle. Then we can express the system function of the nasalized vowel as

$$H(z) = A \frac{\prod_{k=1}^{M}(1 - a_k z^{-1}) \prod_{k=1}^{L}(1 - b_k z)}{\prod_{k=1}^{N}(1 - c_k z^{-1})}.$$

Suppose also that the pitch period is P samples in duration. Our goal is to obtain the glottal waveform and poles and zeros of the vocal tract system function.

(a) Find a time interval I within a pitch period over which the speech waveform is "free" of the effect of zeros.

(b) Set up the covariance method normal equations over the interval I to solve for the vocal tract poles. (The corresponding polynomial coefficients is sufficient.) Comment on the drawback of this method when a speech formant has a very wide bandwidth.

(c) We have assumed that zeros of the glottal flow are maximum phase and the zeros due to the nasal branch are minimum phase. Using your result from part (b), describe a method for obtaining the unknown zeros by inverse filtering. (The corresponding polynomial coefficients are sufficient.) Then propose a method for separating the minimum and maximum phase zeros. *Hint:* Consider applying linear prediction analysis to the reciprocal of the zero polynomial.

(d) Implement parts (b) and (c) in MATLAB using the synthetic waveform *speech_10k* (at 10000 samples/s) in workspace *ex5M2.mat* located in the companion website directory Chap_exercises/chapter5. Assume 4 vocal tract poles, 2 vocal tract zeros, a glottal flow width of 20 samples, and a pitch period of 200 samples. You should compute the predictor coefficients associated with the vocal tract poles and then inverse filter. Then find the coefficients of the vocal tract numerator polynomial and the glottal flow. You will find in *ex5M2.readme* located in Chap_exercises/chapter5 a description of how the synthetic waveform was produced and some suggested MATLAB routines of possible use.

(e) Repeat the MATLAB exercise in part (d), but now use the basic covariance method, i.e., the prediction-error window lies over multiple pitch periods (not synchronized within a glottal cycle). Here your goal is to match the spectrum of the combined vocal tract response and glottal flow, and not to separate these components. Therefore, show only the result of your inverse filtering. Keep in mind that you need to use a large number of poles to model the vocal tract zeros. Plot the log-magnitude spectrum of one glottal cycle of the result of your inverse filter and compare (superimpose) it with the result of your inverse filter over one glottal cycle for the synchronized covariance method of part (d). Explain your results.

BIBLIOGRAPHY

[1] T.V. Ananthapadmanabha and G.Fant, "Calculation of True Glottal Flow and Its Components," *Speech Communications*, vol. 1, pp. 167–184, 1982.

[2] B.S. Atal and S.L. Hanauer, "Speech Analysis and Synthesis by Linear Prediction of the Speech Waveform," *J. Acoustical Society of America*, vol. 50, pp. 637–655, 1971.

[3] D.G. Childers and C. Ahn, "Modeling the Glottal Volume-Velocity Waveform for Three Voice Types," *J. Acoustical Society of America*, vol. 97, no. 1, pp. 505–519, Jan. 1995.

[4] D.G. Childers and C.K. Lee, "Vocal Quality Factors: Analysis, Synthesis, and Perception," *J. Acoustical Society of America*, vol. 90, no. 5, pp. 2394–2410, Nov. 1991.

[5] K.E. Cummings and M.A. Clements, "Analysis of Glottal Waveforms Across Stress Styles," *Proc. Int. Conf. Acoustics, Speech, and Signal Processing*, pp. 369–372, Albuquerque, NM, 1990.

[6] J.E. Dennis, D.M. Gay, and R.E. Welsch, "An Adaptive Nonlinear Least-Squares Algorithm," *ACM Trans. on Mathematical Software*, vol. 7, no. 3, pp. 348–368, Sept. 1981.

[7] J.E. Dennis, D.M. Gay, and R.E. Welsch, "Algorithm 573 NL2SOL—An Adaptive Nonlinear Least-Squares Algorithm," *ACM Transactions on Mathematical Software*, vol. 7, no. 3, pp. 369–383, Sept. 1981.

[8] G. Fant, "Glottal Flow: Models and Interaction," *J. Phonetics*, vol. 14, pp. 393–399, 1986.

[9] J.L. Flanagan, *Speech Analysis, Synthesis, and Perception*, Springer-Verlag, New York, NY, 1972.

[10] J.L. Flanagan and K. Ishizaka, "Computer Model to Characterize the Air Volume Displaced by the Vibrating Vocal Cords," *J. Acoustical Society of America*, vol. 63, pp. 1559–1565, Nov. 1978.

[11] F.I. Itakura and S. Saito, "A Statistical Method for Estimation of Speech Spectral Density and Formant Frequencies," *Elec. and Comm. in Japan*, vol. 53-A, no. 1, pp. 36–43, 1970.

[12] G.E. Kopec, A.V. Oppenheim, and J.M. Tribolet, "Speech Analysis by Homomorphic Prediction," *IEEE Trans. Acoustics, Speech, and Signal Processing*, vol. ASSP–25, no. 1, pp. 40–49, Feb. 1977.

[13] N. Levinson, "The Wiener RMS Error Criterion in Filter Design and Prediction," *J. Mathematical Physics*, vol. 25, pp. 261–278, Jan. 1947.

[14] J. Makhoul, "Linear Prediction: A Tutorial Review," *Proc. IEEE*, vol. 63, no. 4, pp. 561–580, April 1975.

[15] J.D. Markel and A.H. Gray, *Linear Prediction of Speech*, Springer-Verlag, New York, NY, 1976.

[16] J.H. McClellan, "Parametric Signal Modeling," chapter in *Advanced Topics in Signal Processing*, J.S. Lim and A.V. Oppenheim, eds., Prentice Hall, Englewood Cliffs, NJ, 1988.

[17] A.V. Oppenheim and R.W. Schafer, *Digital Signal Processing*, Prentice Hall, Englewood Cliffs, NJ, 1975.

[18] A.V. Oppenheim and R.W. Schafer, *Discrete-Time Signal Processing*, Prentice Hall, Englewood Cliffs, NJ, 1989.

[19] M.D. Plumpe, T.F. Quatieri, and D.A. Reynolds, "Modeling of the Glottal Flow Derivative Waveform with Application to Speaker Identification," *IEEE Trans. Speech and Audio Processing*, vol. 1, no. 5, pp. 569–586, Sept. 1999.

[20] M.R. Portnoff, *A Quasi-One-Dimensional Digital Simulation for the Time-Varying Vocal Tract*, SM Thesis, Massachusetts Institute of Technology, Dept. of Electrical Engineering and Computer Science, May 1973.

[21] L.R. Rabiner and R.W. Schafer, *Digital Processing of Speech Signals*, Prentice Hall, Englewood Cliffs, NJ, 1978.

[22] M.R. Sambur, A.E. Rosenberg, L.R. Rabiner, and C.A. McGonegal, "On Reducing the Buzz in LPC Synthesis," *J. Acoustical Society of America*, vol. 63, no. 3, pp. 918–924, March 1978.

[23] J.L. Shanks, "Recursion Filters for Digital Processing," *Geophysics*, vol. 32, pp. 33–51, 1967.

[24] K. Steiglitz, "On the Simultaneous Estimation of Poles and Zeros in Speech Analysis," *IEEE Trans. Acoustics, Speech, and Signal Processing,* vol. 25, pp. 229–234, June 1977.

[25] G. Strang, *Linear Algebra and Its Applications*, Academic Press, New York, NY, 1976.

[26] C.W. Therrien, *Discrete Random Signals and Statistical Signal Processing*, Prentice Hall, Englewood Cliffs, NJ, 1992.

[27] H. Wakita, "Estimation of the Vocal Tract Shape by Optimal Inverse Filtering and Acoustic/ Articulatory Conversion Methods," SCRL Monograph, no. 9, July 1972.

[28] D.Y. Wong, J.D. Markel, and A.H. Gray, Jr., "Least-Squares Glottal Inverse Filtering from the Acoustic Speech Waveform," *IEEE Trans. Acoustics, Speech, and Signal Processing*, vol. ASSP–27, no. 4, pp. 350–355, Aug. 1979.

[29] L.R. Yanguas, T.F. Quatieri, and F. Goodman, "Implications of Glottal Source for Speaker and Dialect Identification," *Proc. Int. Conf. Acoustics, Speech, and Signal Processing*, vol. 2, pp. 813–816, Phoenix, AZ, 1999.

[30] B.Yegnanarayana and R.N.J. Veldhuis, "Extraction of Vocal-Tract System Characteristics from Speech Signals," *IEEE Trans. Speech and Audio Processing*, vol. 6, no. 4, pp. 313–327, July 1998.

Homomorphic Signal Processing

6.1 Introduction

Signals that are added together and have disjoint spectral content can be separated by linear filtering. Often, however, signals are not additively combined. In particular, the source and system in the linear speech model are *convolutionally* combined and, consequently, these components cannot be separated by linear filtering. The speech signal itself may also be convolved with a system response such as when distorted by the impulse response of a transmission channel or by a flawed recording device. In addition, the speech signal may be *multiplied* by another signal as occurs, for example, with a time-varying fading channel or with an unwanted expansion of its dynamic range. In these cases, it is desired to separate the nonlinearly combined signals to extract the speech signal or its source and system components.

The linear prediction analysis methods of the previous chapter can be viewed as a process of *deconvolution* where the convolutionally combined source and system speech production components are separated. Linear prediction analysis first extracts the system component by inverse filtering then extracts the source component. This chapter describes an alternative means of deconvolution of the source and system components referred to as *homomorphic filtering*. In this approach, convolutionally combined signals are mapped to additively combined signals on which linear filtering is applied for signal separation. Unlike linear prediction analysis, which is a "parametric" (all-pole) approach to deconvolution, homomorphic filtering is "nonparametric" in that a specific model need not be imposed on the system transfer function in analysis.

We begin this chapter in Section 6.2 with the principles of *homomorphic systems* which form the basis for homomorphic filtering. Homomorphic systems for convolution are one of a number of homomorphic systems that map signals nonlinearly combined to signals combined by addition on which linear filtering can be applied for signal separation. As illustrated above, signals may also be combined by other nonlinear operations such as multiplication. Because our main focus in this chapter, however, is speech source and system deconvolution, Section 6.3 develops in detail homomorphic systems for convolution and, in particular, homomorphic systems that map convolution to addition through a logarithm operator applied to the Fourier transform of a sequence. Section 6.4 then analyzes the output of this homomorphic system for input sequences with rational z-transforms and for short-time impulse trains, the convolution of the two serving as a model for a voiced speech segment. Section 6.5 shows that homomorphic systems for convolution need not be based on the logarithm by introducing the spectral root homomorphic system, which relies on raising the Fourier transform of a sequence to a power. For some sequences, homomorphic root analysis can be of advantage over the use of the logarithm for signal separation. As a precursor to the analysis of real speech, in Section 6.6 we then look at the response of homomorphic systems to a windowed periodic waveform and its implications for homomorphic deconvolution. In particular, we address the difference between a windowed periodic waveform and an exact convolutional model. We will see that homomorphic analysis of windowed periodic waveforms benefits from numerous conditions on the analysis window and its location within a glottal cycle in the deconvolution of a mixed-phase system response (i.e., with both minimum- and maximum-phase components). Similar conditions on window duration and alignment for accurate system phase estimation appear in a variety of speech analysis/synthesis systems throughout the text, such as the phase vocoder and sinusoidal analysis/synthesis.

In the remainder of the chapter, we investigate the application of homomorphic systems to speech analysis and synthesis. The properties of homomorphic filtering for voiced and unvoiced speech are described first in Section 6.7. An important consideration in these systems is the phase one attaches to the speech transfer function estimate: zero, minimum, or mixed phase. We will see that the mixed-phase estimate requires a different and more complex method of analysis, alluded to above, from a minimum- or maximum-phase estimate when dealing with windowed periodic waveforms, and we explore the perceptual consequences of these different phase functions for synthesis in Section 6.8. Unlike linear prediction, homomorphic analysis allows for a mixed-phase estimate. This is one of a number of comparisons made of the two systems that leads in Section 6.9 to a homomorphic filtering scheme that serves as a preprocessor to linear prediction. This method, referred to as "homomorphic prediction," can remove the waveform periodicity that renders linear prediction problematic for high-pitched speakers. Finally, a number of the exercises at the end of the chapter explore some applications of homomorphic filtering that fall outside the chapter's main theme. This includes the restoration of old acoustic recordings and dynamic range compression for signal enhancement. The homomorphic filtering approach to these problems will be contrasted to alternative methods throughout the text.

6.2 Concept

An essential property of linear systems is that of *superposition* whereby the output of the system to an input of two additively combined sequences, $x[n] = x_1[n] + x_2[n]$, is the sum of the individual outputs; in addition, a scaled input results in a scaled output. The superposition property can be expressed explicitly as

$$L(x_1[n] + x_2[n]) = L(x_1[n]) + L(x_2[n])$$
$$L(\alpha x[n]) = \alpha L(x[n]) \tag{6.1}$$

where L represents a linear operator and α a scaling factor. A consequence of superposition is the capability of linear systems to separate, i.e., filter, signals that fall in disjoint frequency bands.

EXAMPLE 6.1   Figure 6.1 shows the Fourier transform magnitude of a sequence consisting of two additive components that fall in nonoverlapping frequency bands; i.e., $x[n] = x_1[n] + x_2[n]$ where $X_1(\omega)$ and $X_2(\omega)$ reside in the frequency bands $[0, \frac{\pi}{2}]$ and $[\frac{\pi}{2}, \pi]$, respectively. Application of the highpass filter $H(\omega)$ separates out $x_2[n]$ and can be expressed as

$$y[n] = h[n] * (x_1[n] + x_2[n])$$
$$= h[n] * x_2[n]$$
$$= x_2[n]$$

where $h[n]$ is the inverse Fourier transform of $H(\omega)$. ▲

To allow the separation of signals that are nonlinearly combined, Oppenheim [9] introduced the concept of *generalized superposition*, which leads to the notion of *generalized linear filtering*. In formulating the generalized principle of superposition, consider two signals $x_1[n]$ and $x_2[n]$ that are combined by some rule which we denote by ∘, i.e.,

$$x[n] = x_1[n] \circ x_2[n] \tag{6.2}$$

and consider a transformation on $x[n]$ denoted by ϕ. In addition, we define a generalized multiplicative operator ":". In generalizing the notion of superposition, we require ϕ to have the

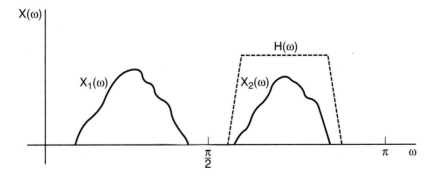

Figure 6.1 Signal with disjoint low- and high-frequency spectra $X_1(\omega)$ and $X_2(\omega)$.

Figure 6.2 Homomorphic system ϕ.

following two properties:

$$\phi(x_1[n] \circ x_2[n]) = \phi(x_1[n]) \circ \phi(x_2[n])$$

$$\phi(\alpha : x[n]) = \alpha : \phi(x[n]). \tag{6.3}$$

These can be viewed as an analogy to the linear superposition properties of Equation (6.1), where addition and scalar multiplication have been generalized. An even wider class of systems characterized by generalized superposition is given by the two properties

$$\phi(x_1[n] \circ x_2[n]) = \phi(x_1[n]) \triangle \phi(x_2[n])$$

$$\phi(\alpha : x[n]) = \alpha \natural \phi(x[n]) \tag{6.4}$$

and is illustrated in Figure 6.2. Systems that satisfy these two properties are referred to as *homomorphic systems* and are said to satisfy a *generalized principle of superposition* [9].[1]

Part of the practical importance of homomorphic systems for speech processing lies in their capability of transforming nonlinearly combined signals to additively combined signals so that linear filtering can be performed. This capability stems from the fact that homomorphic systems can be expressed as a cascade of three homomorphic sub-systems, which is referred to as the *canonic representation* of a homomorphic system [9]. The canonic representation of a homomorphic system ϕ is illustrated in Figure 6.3. The signals combined by the operation $\circ$ are transformed by the sub-system $D_\circ$ to signals that are additively combined. Linear filtering is performed by the linear system L (mapping addition to addition), and the desired signal is then obtained by the inverse operation $D_\triangle^{-1}$, which maps addition to the operation $\triangle$.

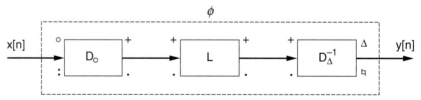

Figure 6.3 Canonical formulation of a homomorphic system.

SOURCE: A.V. Oppenheim and R.W. Schafer, *Discrete-Time Signal Processing* [13]. ©1989, Pearson Education, Inc. Used by permission.

---

[1] This notation and terminology stem from the study of vector spaces, which is a framework for abstract linear algebra [9]. In this framework, a sequence is considered a vector in a vector space. A vector space is characterized by vector addition, denoted by $\circ$, which is a rule for combining vectors in a vector space, and by scalar multiplication, denoted by :, which is a rule for combining vectors with scalars in a vector space. A linear transformation on a vector space, denoted by ϕ, maps the input vector space to an output vector space. The transformation is *homomorphic* if it satisfies the generalized principle of superposition of Equation 6.4 where $\triangle$ and $\natural$ define the output vector space.

EXAMPLE 6.2   Consider two sequences $x_1[n]$ and $x_2[n]$ of low-frequency and high-frequency content that are nonlinearly combined by the operation $\circ$. As in Example 6.1, the objective is to separate the high-frequency signal $x_2[n]$. Applying the homomorphic system ϕ and using the canonical representation, we have

$$\hat{x}[n] = D_\circ(x[n]) = D_\circ(x_1[n]) + D_\circ(x_2[n])$$
$$= \hat{x}_1[n] + \hat{x}_2[n]$$

that are linearly combined. If the operation $\circ$ is such that $\hat{x}_1[n]$ and $\hat{x}_2[n]$ have disjoint spectra, then the highpass component can be separated. If L denotes the highpass filter, then the output of L is given by

$$\hat{y}[n] = L(\hat{x}[n]) = \hat{x}_2[n]$$

and, therefore, when the operation $\Delta = \circ$

$$x_2[n] = D_\circ^{-1}(\hat{y}[n])$$

thus extracting the high-frequency component.   ▲

Example 6.2 illustrates the use of homomorphic systems in performing generalized linear filtering. The system of Example 6.2 is considered a *homomorphic filter* having the property that the desired component passes through the system unaltered while the undesired component is removed. Two typical nonlinear operators are convolution and multiplication. As illustrated in the introduction, many problems arise where signals are so combined, such as in speech deconvolution, waveform distortion, and dynamic range compression. Since our primary focus is speech deconvolution, the next section looks in detail at homomorphic systems for convolution [13].

6.3 Homomorphic Systems for Convolution

In homomorphic systems for convolution, the operation $\circ$ is convolution, i.e., $\circ = *$ and the resulting homomorphic system D_* maps convolution to addition and the inverse system D_*^{-1} maps addition to convolution. This class of homomorphic systems is useful in speech analysis [11],[12] as demonstrated in the following example:

EXAMPLE 6.3   Consider a sequence $x[n]$ consisting of a system impulse response $h[n]$ convolved with an impulse train $p[n] = \sum_{k=-\infty}^{\infty} a_k \delta[n - kP]$ (with P the impulse spacing), i.e., $x[n] = h[n] * p[n]$. The goal is to estimate the response $h[n]$. Applying the canonical representation for convolution, we have

$$\hat{x}[n] = D_*(x[n]) = D_*(h[n]) + D_*(p[n])$$
$$= \hat{h}[n] + \hat{p}[n]$$

that contains additively combined sequences. Suppose that D_* is such that $\hat{p}[n]$ remains a train of impulses with spacing P and suppose that $\hat{h}[n]$ falls between impulses. Then, if L denotes the operation of multiplying by a rectangular window (for extracting $\hat{h}[n]$), we have

$$\hat{y}[n] = L(\hat{x}[n]) = L(\hat{h}[n]) + L(\hat{p}[n])$$
$$= \hat{h}[n]$$

and, therefore,

$$h[n] = D_*^{-1}(\hat{y}[n])$$

thus separating the impulse response.   ▲

An approach for finding the components of the canonical representation and, in particular, the elements D_* and D_*^{-1}, is to note that if $x[n] = x_1[n] * x_2[n]$, then the z-transform of $x[n]$ is given by $X(z) = X_1(z)X_2(z)$ and because we want the property that convolution maps to addition, i.e., $D_*(x_1[n] * x_2[n]) = D_*(x_1[n]) + D_*(x_2[n])$, this motivates the use of the logarithm in the operators; i.e., $D_*[x] = \log(Z[x])$ and $D_*^{-1}[x] = Z^{-1}[\exp(x)]$ where Z denotes z-transform. However, if we want to represent sequences in the time domain, rather than in the z domain, then it's desirable to have the operations $D_* = Z^{-1}[\log(Z)]$ and $D_*^{-1} = Z^{-1}[\exp(Z)]$. The canonical system with the forward and inverse operators is summarized in Figure 6.4, showing that our selection of D_* and D_*^{-1} gives the desired properties of mapping convolution to addition and addition back to convolution, respectively. However, in this construction of D_* we have overlooked the definition of the logarithm of a complex z-transform which we refer to henceforth as the "complex logarithm." Because the complex logarithm is key to the canonical system, the existence of D_* relies on the validity of $\log[(X_1(z)X_2(z)] = \log[X_1(z)] + \log[X_2(z)]$ and this will depend on how we define

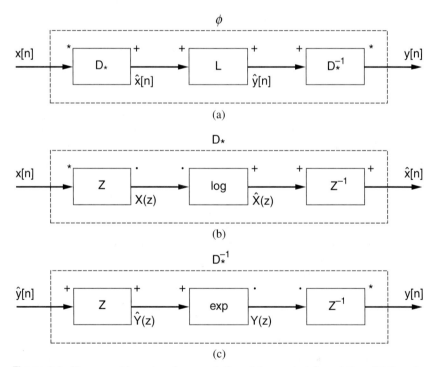

Figure 6.4 Homomorphic system for convolution: (a) canonical formulation; (b) the subsystem D_*; and (c) its inverse.

the complex logarithm.[2] For the trivial case of real and positive z-transforms, the logarithm, sometimes referred to as the "real logarithm," of the product is the product of the logarithms. Generally, however, this property is more difficult to obtain, as illustrated in the following example:

EXAMPLE 6.4   Consider two real and positive values a and b. Then $\log(ab) = \log(a) + \log(b)$. On the other hand, if $b < 0$, then $\log(ab) = \log(a|b|e^{jk\pi}) = \log(a) + \log(|b|) + jk\pi$, where k is an odd integer. Thus, the definition of $\log(ab)$ in this case is ambiguous.   ▲

Example 6.4 indicates that special consideration must be made in defining the logarithm operator for complex $X(z)$ in order to make the logarithm of a product the sum of the logarithms [13]. Suppose for simplicity that we evaluate $X(z)$ on the unit circle ($z = e^{j\omega}$), i.e., we evaluate the Fourier transform.[3] Then we consider the real and imaginary parts of the complex logarithm by writing the logarithm in polar form as

$$\log[X(\omega)] = \log(|X(\omega)|e^{j\angle X(\omega)})$$
$$= \log(|X(\omega)|) + j\angle X(\omega). \tag{6.5}$$

Then, if $X(\omega) = X_1(\omega)X_2(\omega)$, we want the logarithm of the real parts and the logarithm of the imaginary parts to equal the sum of the respective logarithms. The real part is the logarithm of the magnitude and, for the product $X_1(\omega)X_2(\omega)$, is given by

$$\log(|X(\omega)|) = \log(|X_1(\omega)X_2(\omega)|)$$
$$= \log(|X_1(\omega)|) + \log(|X_2(\omega)|) \tag{6.6}$$

provided that $|X_1(\omega)| > 0$ and $|X_2(\omega)| > 0$, which is satisfied when zeros and poles of $X(z)$ do not fall on the unit circle. In this case, there is no problem with the uniqueness and "additivity" of the logarithms. The imaginary part of the logarithm is the phase of the Fourier transform and requires more careful consideration. As with the real part, we want the imaginary parts to add

$$\angle X(\omega) = \angle[X_1(\omega)X_2(\omega)]$$
$$= \angle X_1(\omega) + \angle X_2(\omega). \tag{6.7}$$

The relation in Equation (6.7), however, generally does not hold due to the ambiguity in the definition of phase, i.e., $\angle X(\omega) = \text{PV}[\angle X(\omega)] + 2\pi k$, where k is any integer value, and where PV denotes the principle value of the phase which falls in the interval $[-\pi, \pi]$. Since an arbitrary multiple of 2π can be added to the principal phase values of $X_1(\omega)$ and $X_2(\omega)$, the additivity property generally does not hold. One approach to obtain uniqueness is to force continuity within the definition of phase, i.e., select the integer k such that the function $\angle X(\omega) = \text{PV}[\angle X(\omega)] + 2\pi k$ is continuous (Figure 6.5). Continuity ensures not only uniqueness, but

---

[2] There is no such problem with the inverse exponential operator since $e^{a+b} = e^a e^b$. Thus, along with the forward and inverse z-transforms, addition is unambiguously mapped back to convolution.

[3] We assume the sequence $x[n]$ is stable and thus that the region of convergence of $X(z)$ includes the unit circle.

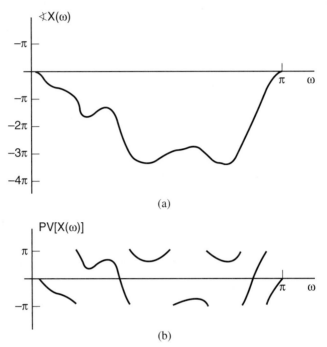

Figure 6.5 Fourier transform phase continuity: (a) typical continuous phase function; (b) its principal value.

SOURCE: A.V. Oppenheim and R.W. Schafer, *Discrete-Time Signal Processing* [13]. ©1989, Pearson Education, Inc. Used by permission.

also guarantees that the phase of the product $X_1(\omega)X_2(\omega)$ is the sum of the individual phase functions, i.e., Equation (6.7) is satisfied (Exercise 6.1).

An alternative approach to resolving ambiguity of the definition of phase is through the phase derivative[4] with which we define the phase as

$$\angle X(\omega) = \int_0^\omega \dot{\angle} X(\beta)d\beta$$

where the derivative of the phase with respect to ω, $\dot{\angle} X(\omega) = \frac{d}{d\omega}\angle X(\omega)$, is uniquely defined through the real and imaginary parts of $X(\omega)$, $X_r(\omega)$ and $X_i(\omega)$, respectively, and is shown in Exercise 6.2 to be given by

$$\dot{\angle} X(\omega) = \frac{d\angle X(\omega)}{d\omega} = \frac{X_r(\omega)\dot{X}_i(\omega) - X_i(\omega)\dot{X}_r(\omega)}{|X(\omega)|^2}. \qquad (6.8)$$

With this expression for $\dot{\angle} X(\omega)$, we can show (given that $|X(\omega)| \neq 0$) that $\angle X(\omega)$ is unique

---

[4] For a rational z-transform, the phase derivative is a measure of the rate of change of the continuous angle accumulated over all poles and zeros.

and that the additivity property of Equation (6.7) is satisfied (Exercise 6.1). The phase derivative eliminates any constant level in the phase, but this is simply π or $-\pi$ which is sign information.[5] The two means of removing phase ambiguity are also useful in motivating *phase
unwrapping* algorithms which are described in Section 6.4.4.

Observe that since $x[n]$ is real, then $|X(\omega)|$ and thus $\log(|X(\omega)|)$ are real and even.
Likewise, since $x[n]$ is real, then $j\angle X(\omega)$ is imaginary and odd. Hence, the inverse Fourier
transform of $\log[X(\omega)] = \log[|X(\omega)|] + j\angle X(\omega)$ is a real sequence and is expressed as

$$\hat{x}[n] = \frac{1}{2\pi} \int_{-\pi}^{\pi} \log[X(\omega)]e^{j\omega n}d\omega.$$

The sequence $\hat{x}[n]$ is referred to as the *complex cepstrum*. The even component of the complex
cepstrum, denoted as $c[n]$, is given by $c[n] = (\hat{x}[n] + \hat{x}[-n])/2$ and is referred to as the *real
cepstrum* because it is the inverse Fourier transform of the real logarithm, which is the real part of
the complex logarithm, i.e., $\log(|X(\omega)|) = \text{Re}\{\log[X(\omega)]\}$. The primary difference between
the complex and real cepstrum is that we have discarded the phase of the complex cepstrum.
Discarding the phase is useful, as we will see when dealing with minimum-phase sequences or
when the phase is difficult to compute. Observe that applying an inverse Fourier transform to
a log-spectrum makes the real and complex cepstra a function of the time index n. This time
index is sometimes referred to as "quefrency"; the motivation for this nomenclature will become
clear shortly.

6.4 Complex Cepstrum of Speech-Like Sequences

Following the development of Oppenheim and Schafer [13], we investigate the complex cepstrum
of two classes of sequences in preparation for deconvolving real speech signals: sequences with
rational z-transforms and periodic impulse trains. Homomorphic filtering is introduced for
separating sequences that convolutionally combine these two signal classes, and is given an
alternative interpretation as a spectral smoothing process.

6.4.1 Sequences with Rational z-Transforms

Consider the class of sequences with rational z-transforms of the form

$$X(z) = Az^{-r} \frac{\prod_{k=1}^{M_i}(1 - a_k z^{-1}) \prod_{k=1}^{M_o}(1 - b_k z)}{\prod_{k=1}^{N_i}(1 - c_k z^{-1}) \prod_{k=1}^{N_o}(1 - d_k z)} \tag{6.9}$$

where $(1 - a_k z^{-1})$ and $(1 - c_k z^{-1})$ are zeros and poles inside the unit circle and $(1 - b_k z)$ and
$(1 - d_k z)$ are zeros and poles outside the unit circle with $|a_k|, |b_k|, |c_k|, |d_k| < 1$ so that there
are no zeros or poles on the unit circle. The term z^{-r} represents a delay of the sequence with
respect to the time origin which we assume for the moment can be estimated and removed.[6] The
factor A is assumed positive; a negative A introduces a sign change which can be thought of as

---

[5] Sign information, however, is important in synthesis if the sign varies over successive analysis frames.

[6] In speech modeling, the delay often represents a shift of the vocal tract impulse response with respect to
the time origin.

an additive factor of π to the phase of $X(z)$ since $-1 = e^{j\pi}$. Taking the complex logarithm then gives

$$
\hat{X}(z) = \log(A) + \sum_{k=1}^{M_i} \log(1 - a_k z^{-1}) + \sum_{k=1}^{M_o} \log(1 - b_k z)
$$

$$
- \sum_{k=1}^{N_i} \log(1 - c_k z^{-1}) - \sum_{k=1}^{N_o} \log(1 - d_k z).
$$

Consider $\hat{X}(z)$ as a z-transform of a sequence $\hat{x}[n]$. We want the inverse z-transform to be a stable sequence, i.e., absolutely summable, so that the region of convergence (ROC) for $\hat{X}(z)$ must include the unit circle ($|z| = 1$). This is equivalent to the condition that the Laurent power series $\hat{X}(z) = \sum_{n=-\infty}^{\infty} \hat{x}[n] z^{-n}$ is analytic on the unit circle. This condition implies that all components of $\hat{X}(z)$, i.e., of the form $\log(1 - \alpha z^{-1})$ and $\log(1 - \beta z)$ with $|\alpha|, |\beta| < 1$, must represent z-transforms of sequences whose ROC includes the unit circle. With this property in mind, we write the following power series expansions for two generic terms:

$$
\log(1 - \alpha z^{-1}) = -\sum_{n=1}^{\infty} \frac{\alpha^n}{n} z^{-n}, \quad |\alpha z^{-1}| < 1
$$

$$
\log(1 - \beta z) = -\sum_{n=1}^{\infty} \frac{\beta^n}{n} z^n, \quad |\beta z| < 1. \tag{6.10}
$$

The ROC of the two series is illustrated in Figure 6.6a,b, the first converging for $|z| > |\alpha|$ and the second for $|z| < |\beta^{-1}|$. The ROC of $\hat{X}(z)$ is therefore given by an annulus which borders on radii corresponding to the poles and zeros of $X(z)$ closest to the unit circle in the z-plane and which includes the unit circle as shown in Figure 6.6c. From our z-transform properties reviewed in Chapter 2, the first z-transform corresponds to a right-sided sequence while the second corresponds to a left-sided sequence. The complex cepstrum associated with a rational $X(z)$ can therefore be expressed as

$$
\hat{x}[n] = \log(A)\delta[n] - \left[\sum_{k=1}^{M_i} \frac{a_k^n}{n} - \sum_{k=1}^{N_i} \frac{c_k^n}{n} \right] u[n-1]
$$

$$
+ \left[\sum_{k=1}^{M_o} \frac{b_k^{-n}}{n} - \sum_{k=1}^{N_o} \frac{d_k^{-n}}{n} \right] u[-n+1] \tag{6.11}
$$

where $u[n]$ is the unit step function. Therefore, the zeros and poles inside the unit circle contribute to the right side of the complex cepstrum, while the zeros and poles outside the unit circle contribute to the left side of the complex cepstrum; the value of $\hat{x}[n]$ at the origin is due to the gain term A (Figure 6.7). We see then that the complex cepstrum is generally two-sided and for positive or negative time is a sum of decaying exponentials that are scaled by $\frac{1}{n}$.

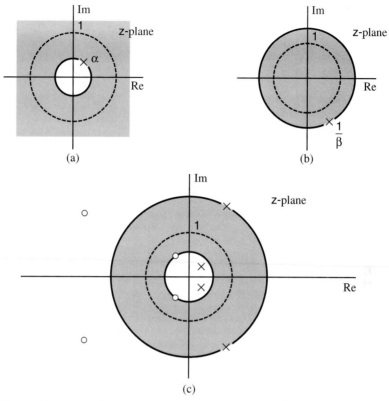

Figure 6.6 Region of convergence (ROC) for (a) $\log(1 - \alpha z^{-1})$ and (b) $\log(1 - \beta z)$, while (c) shows the annular ROC for a typical rational $X(z)$. For all cases, the ROC includes the unit circle; the unit circle is shown as a dashed line.

It was noted earlier that the linear phase term z^{-r} is removed prior to determining the complex cepstrum. The following example illustrates the importance of removing this term:

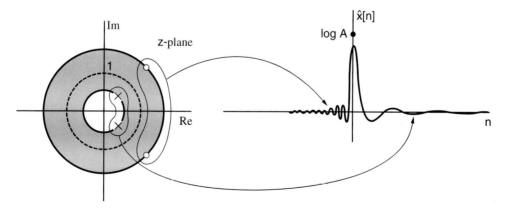

Figure 6.7 Schematized example illustrating right- and left-side contributions to the complex cepstrum.

EXAMPLE 6.5   Consider the z-transform

$$X(z) = z^{-r} \frac{(1 - az^{-1})(1 - bz)}{(1 - cz^{-1})} \tag{6.12}$$

where a, b, and c are real and less than unity. The ROC of $X(z)$ contains the unit circle so that $x[n]$ is stable. A delay term z^{-r} corresponds to a shift in the sequence, for example, a shift in the vocal tract impulse response relative to the time origin. The complex cepstrum is given by

$$\hat{x}[n] = -\left[\frac{a^n}{n} - \frac{c^n}{n}\right] u[n - 1] + \left[\frac{b^{-n}}{n}\right] u[-n + 1] + Z^{-1}[\log(z^{-r})]$$

where the inverse z-transform, denoted by Z^{-1}, of the shift term is given by [16]

$$Z^{-1}[\log(z^{-r})] = r \frac{\cos(\pi n)}{n}, \quad n \neq 0$$
$$= 0, \quad\quad\quad\quad n = 0.$$

The inverse z-transform of the linear phase term is a $\frac{\sin(x)}{x}$-type (sinc) function which can swamp the complex cepstrum. On the unit circle, $z^{-r} = e^{-j\omega r}$ contributes a linear ramp to the phase and thus, for a large shift r, dominates the phase representation and gives a large discontinuity at π and $-\pi$. By tracing the vector representation of the pole and zero components of $X(z)$, one sees (Exercise 6.3) that the phase has zero value at $\omega = 0$ and continuously evolves to a zero value at $\omega = \pi$; furthermore, each phase component must lie within the bounds $|\theta(\omega)| < \pi$ (Exercise 6.3). The phase of $X(z)$ is given by the sum of these three phase components with the linear term $-r\omega$, which, for $r \gg 1$, will dominate the sum of the three pole-zero phase components, i.e.,

$$\theta(\omega) = \theta_{pz}(\omega) - r\omega$$
$$\approx -r\omega$$

for $r \gg 1$, where $\theta_{pz}(\omega)$ denotes the sum of the nonzero pole-zero phase contributions. To illustrate these properties, the unwrapped phase of the pole component of $X(z)$ is shown in Figure 6.8, along with an unwrapped linear phase contribution, and the sum of the phase from the pole and the linear phase.   ▲

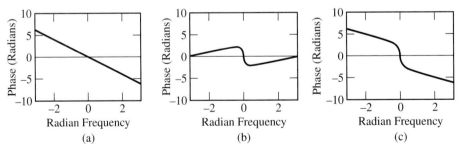

Figure 6.8   Illustration of phase contributions of Example 6.5: (a) unwrapped phase of the linear phase contribution; (b) unwrapped phase of the pole component; (c) sum of (a) and (b). The value of r in z^{-r} is negative (-2) and the pole occurs on the real z axis (0 Hz).

If $X(z)$ has no poles or zeros outside the unit circle, i.e., $b_k = d_k = 0$ so $x[n]$ is a *minimum-phase* sequence, then $\hat{x}[n]$ is right-sided ($\hat{x}[n] = 0$ for $n < 0$), and if $X(z)$ has no poles or zeros inside the unit circle, i.e., $a_k = c_k = 0$ so that $x[n]$ is a *maximum-phase* sequence, then $\hat{x}[n]$ is left-sided ($\hat{x}[n] = 0$ for $n > 0$). An implication is that if $x[n]$ has a rational z-transform and is minimum-phase, then the complex cepstrum can be derived from $\log(|X(\omega)|)$ (and thus from the real cepstrum) or from the phase $\angle X(\omega)$ (to within a scale factor). To show the former, recall that the even part of the complex cepstrum equals the real cepstrum and is given by

$$c[n] = \frac{\hat{x}[n] + \hat{x}[-n]}{2}$$

whose z-transform is given by $\log(|X(\omega)|)$. Therefore, because the complex cepstrum of a minimum-phase sequence with a rational z-transform is right-sided

$$\hat{x}[n] = l[n]c[n] \tag{6.13}$$

where

$$l[n] = 1, \quad n = 0$$
$$= 2, \quad n > 0$$
$$= 0, \quad n < 0.$$

A similar argument shows that $\hat{x}[n]$ can be determined from the magnitude of $X(\omega)$ for a maximum-phase sequence, and that the phase is sufficient to recover $\hat{x}[n]$ for minimum- or maximum-phase sequences within a scale factor (Exercise 6.4).

6.4.2 Impulse Trains Convolved with Rational z-Transform Sequences

The second class of sequences of interest in the speech context is a train of uniformly-spaced unit samples with varying weights

$$p[n] = \sum_{r=0}^{Q} \alpha_r \delta[n - rN]$$

whose z-transform can be expressed as a polynomial in z^N as

$$P(z) = \sum_{r=0}^{Q} \alpha_r z^{-rN}$$
$$= \sum_{r=0}^{Q} \alpha_r (z^N)^{-r}$$
$$= \prod_{r=0}^{Q-1} [1 - a_r (z^N)^{-1}].$$

$P(z)$ can thus be expressed as a product of factors of the familiar form $(1 - a_k\mu^{-1})$ where $\mu = z^N$. Therefore, if $p[n]$ is minimum-phase, assuming $|a_r\mu^{-1}| < 1$ and using Equation (6.10), we can express $\log[P(z)]$ as

$$\log[P(z)] = \sum_{r=0}^{Q-1} \log[1 - a_r(z^N)^{-1}]$$

$$= \sum_{r=0}^{Q-1} \left[-\sum_{k=1}^{\infty} \frac{a_r^k}{k}(z^N)^{-k} \right]$$

and so the resulting complex cepstrum $\hat{p}[n]$ is an infinite right-sided sequence of unit samples spaced N samples apart. More generally, for non-minimum-phase sequences of this kind, the complex cepstrum is two-sided with uniformly spaced impulses.

We next look at a specific example of a synthetic speech waveform derived by convolving a periodic impulse train with a sequence with a rational z-transform.

EXAMPLE 6.6   Consider a sequence $x[n] = h[n] * p[n]$ where the z-transform of $h[n]$ is given by

$$H(z) = \frac{(1 - bz)(1 - b^*z)}{(1 - cz^{-1})(1 - c^*z^{-1})}$$

where b, b^* and c, c^* are complex conjugate pairs, all with magnitude less than unity so that the zero pair is outside the unit circle and the pole pair is inside the unit circle. $p[n]$ is a periodic impulse train windowed by a decaying exponential:

$$p[n] = \beta^n \sum_{k=0}^{\infty} \delta[n - kP]$$

thus having z-transform

$$P(z) = \sum_{k=0}^{\infty} \beta^{kP} z^{-kP}$$

where β is selected so that $p[n]$ is minimum-phase. The complex cepstrum $\hat{y}[n] = \hat{h}[n] + \hat{p}[n]$ is illustrated in Figure 6.9 (for $b = 0.99e^{j0.12\pi}$ and $c = -1.01e^{j0.12\pi}$), showing that the two components $\hat{h}[n]$ and $\hat{p}[n]$ are approximately separated along the n axis. (The analytic expressions are left as an exercise.)   ▲

An important observation within this example is that the complex cepstrum allows for the possibility of separating or *deconvolving* the source and filter, which we investigate in the next section.

6.4.3 Homomorphic Filtering

We saw in the previous section that the complex cepstrum of speech-like sequences consists of the sum of a low-quefrency (the term "quefrency" was introduced at the end of Section 6.2) component due to the system response and a high-quefrency component due to the pulse train source. When the complex cepstrum of $h[n]$ resides in a quefrency interval less than a pitch period then the two components can be separated from each other [11],[12],[13].

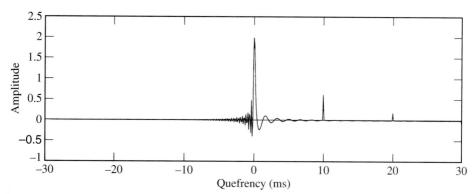

Figure 6.9   Complex cepstrum of $x[n] = p[n] * h[n]$ of Example 6.6. The sequence $p[n]$ is minimum-phase and $h[n]$ is mixed phase (zeros inside and poles outside the unit circle).

Further insight into this separation process can be gained with a spectral smoothing inter-pretation of homomorphic deconvolution. We begin by viewing $\log[X(\omega)]$ as a "time signal" and suppose it consists of low-frequency and high-frequency contributions. Then one might lowpass or highpass filter this signal to separate the two frequency components. One imple-mentation of the lowpass filtering is given in Figure 6.10a, which is simply the concatenation of our forward and inverse sub-systems D_* and D_*^{-1} with a "filter" $l[n]$ placed between the two operators. As illustrated in Figure 6.10b, the filtering operation on $\hat{X}(\omega) = \log[X(\omega)]$ can be implemented by (inverse) Fourier transforming the signal $\log[X(\omega)]$ to obtain the complex cepstrum, applying the filter $l[n]$ to the complex cepstrum, and then (forward) Fourier trans-forming back to a desired signal $\hat{Y}(\omega)$. In this operation, we have interchanged the time and frequency domains by viewing the frequency-domain signal $\log[X(\omega)]$ as a time signal to be filtered. This view originally led to the nomenclature [2] "cepstrum" since $\hat{x}[n]$ can be thought

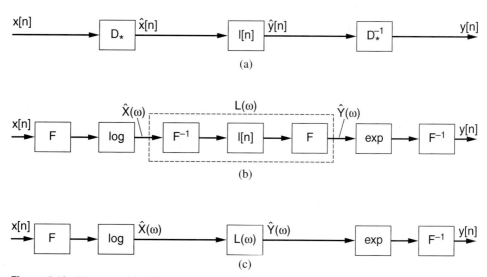

Figure 6.10   Homomorphic filtering interpreted as a linear smoothing of $\log[X(\omega)]$: (a) quefrency-domain implementation; (b) expansion of the operations in (a); (c) frequency-domain interpretation.

of as the "spectrum" of $\log[X(\omega)]$; correspondingly, the time-axis for $\hat{x}[n]$ is referred to as "quefrency," and filter $l[n]$ as the "lifter." Rather than transforming to the quefrency domain, we could have directly convolved $\log[X(\omega)]$ with the Fourier transform of the lifter $l[n]$, denoted as $L(\omega)$. The three elements in the dotted lines of Figure 6.10b can then be replaced by $L(\omega)$, which can be viewed as a smoothing function:

$$\hat{Y}(\omega) = L(\omega) \circledast \log[X(\omega)] \tag{6.14}$$

which is illustrated in Figure 6.10c and where $\circledast$ denotes circular convolution.

With this spectral smoothing perspective of homomorphic filtering, one is motivated to smooth $X(\omega)$ directly rather than through its logarithm. An advantage of smoothing the logarithm, however, is that the logarithm *compresses* the spectrum, thus reducing its dynamic range (i.e., its range of values) and giving a better estimate of low-energy spectrum regions after smoothing; without this "dynamic range compression," the low-energy regions, e.g., high-frequency regions in voiced speech, may be distorted by leakage from high-energy regions, e.g., low-frequency regions in voiced speech (Figure 6.11). In a speech processing context, the low-energy resonances and harmonics can be distorted by leakage from the high-energy regions. The logarithm is simply one compressive operator. In a later section we explore spectral root ho-

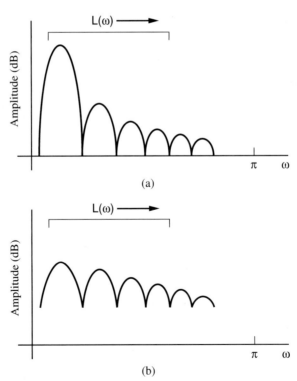

Figure 6.11 Schematic of smoothing (a) a harmonic spectrum in contrast to (b) the logarithm of a harmonic spectrum.

momorphic deconvolution that provides a generalization of the logarithm, and that is motivated by the spectral smoothing interpretation of homomorphic filtering.

6.4.4 Discrete Complex Cepstrum

In previous sections, we determined analytically the complex cepstrum of a variety of classes of discrete-time sequences $x[n]$ using the discrete-time Fourier transform $X(\omega)$ or z-transform $X(z)$. In practice, however, the discrete Fourier transform (DFT) is applied to a sequence $x[n]$ of finite-length N. An N-point DFT is then used to compute the complex cepstrum as

$$X(k) = \sum_{n=0}^{N-1} x[n]e^{-j\frac{2\pi}{N}kn}$$

$$\hat{X}(k) = \log[X(k)] = \log(|X(k)|) + j\angle X(k)$$

$$\hat{x}_N[n] = \frac{1}{N}\sum_{k=0}^{N-1} \hat{X}(k)e^{j\frac{2\pi}{N}kn}$$

where $\hat{x}_N[n]$ is referred to as the discrete complex cepstrum. Two computational issues arise: (1) Aliasing in $\hat{x}_N[n]$ since $\hat{x}[n]$ is infinitely long, i.e., $\hat{x}_N[n]$ is an aliased version of $\hat{x}[n]$, being of the form $\hat{x}_N[n] = \sum_{r=-\infty}^{\infty} \hat{x}[n-rN]$ and (2) *Phase unwrapping* from samples of the principal phase values to ensure a continuous phase function. To avoid significant distortion from aliasing, the DFT must be "large enough." Similar considerations hold for the real cepstrum. In Sections 6.7 and 6.8, we will see that in the context of speech analysis, a 512- to 1024-point DFT is adequate.

In addressing the second computational issue, we want the unwrapped phase samples to equal our analytic definition of continuous phase. We earlier saw in Section 6.3 two frameworks in which a phase unwrapping algorithm can be developed:[7] (1) Phase continuity by adding appropriate multiples of 2π to the principal phase value, and (2) Continuity by integration of the phase derivative. In the latter, we can analytically obtain the continuous phase function by integrating the phase derivative by way of the real and imaginary parts. In practice, however, we have only samples of the principal phase and the real and imaginary parts of the Fourier transform. The two frameworks motivate the following algorithms for phase unwrapping:

Modulo 2π Phase Unwrapper: This algorithm finds an integer multiple of 2π for each k, expressed as $2\pi r[k]$, to add to the principal phase function to yield a continuous phase [13]. That is, we find a phase function of the form

$$\angle X(k) = \text{PV}[X(k)] + 2\pi r[k] \tag{6.15}$$

such that $\angle X(k)$ is continuous. Let $r[0] = 0$; then to find $r[k]$ for $k > 1$, we perform the following steps:

---

[7] Phase unwrapping appears throughout the text in a number of other contexts, such as in the phase vocoder and sinewave analysis/synthesis.

S1: If $PV[X(k)] - PV[X(k-1)] > 2\pi - \epsilon$ (positive jump of 2π is detected) then subtract 2π, i.e.,

$$r[k] = r[k-1] - 1.$$

S2: If $PV[X(k)] - PV[X(k-1)] < -(2\pi - \epsilon)$ (negative jump of 2π is detected) then add 2π, i.e.,

$$r[k] = r[k-1] + 1.$$

S3: Otherwise

$$r[k] = r[k-1].$$

This approach to phase unwrapping yields the correct unwrapped phase whenever the frequency spacing $\frac{2\pi}{N}$ is small enough so that the difference between any two adjacent samples of the unwrapped phase is less than the threshold ϵ. There exist, however, numerous cases where this "small enough" condition is not satisfied, as demonstrated in the following example:

> **EXAMPLE 6.7**   Consider a sequence $x[n]$ that has a zero very close to the unit circle and is located midway between two DFT frequencies $\omega_z = \frac{\omega_k + \omega_{k-1}}{2}$. The phase will change by approximately $+\pi$ between ω_k and ω_{k-1} if the zero is inside the unit circle and by approximately $-\pi$ if the zero is outside the unit circle (Figure 6.12a).[8] With the DFT spacing shown in Figure 6.12, the above phase unwrapping algorithm cannot distinguish a natural discontinuity due to the closeness of the zero to the unit circle, and an artificial discontinuity due to the wrapping (modulo 2π) of a smooth phase function (Figure 6.12b). This situation can occur even when the zeros and poles are not close to the unit circle, but are clustered so that phase changes accumulate to create the above ambiguity.   ▲

As we saw in Example 6.7, zeros or poles close to the unit circle may cause natural phase discontinuities which are difficult to distinguish from modulo 2π jumps, requiring for accurate phase unwrapping a very large DFT that may be impractical. An alternative algorithm exploits the additional information one can obtain through the phase derivative, which gives the direction and rate at which the phase is changing.

Phase Derivative-Based Phase Unwrapper: An alternative phase unwrapping algorithm combines the information contained in both the phase derivative and principal value of the phase [19]. We saw earlier that the unwrapped phase can be obtained analytically as the integral of the phase derivative:

$$\angle X(\omega) = \int_0^\omega \dot{\angle} X(\beta) d\beta \tag{6.16}$$

where the phase derivative $\dot{\angle} X(\omega)$ is uniquely defined through the real and imaginary parts of $X(\omega)$, $X_r(\omega)$, and $X_i(\omega)$, respectively, and is given in Equation (6.8). Although the unwrapped phase can be precisely defined through Equation (6.16), in general it cannot be implemented in discrete time. Nevertheless, using Equation (6.16), one can attempt to compute the unwrapped phase by numerical integration. The accuracy of this approach depends on the accuracy of the

---

[8] Consider a vector argument using the example $1 - \alpha z^{-1} = z^{-1}(z - \alpha)$.

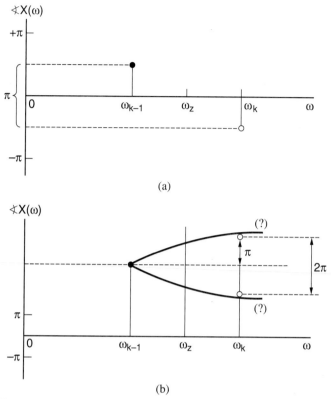

Figure 6.12 Phase unwrapping ambiguity: (a) natural unwrapped phase change of π across two DFT frequencies, with zero close to unit circle; (b) two possible phase values in performing phase unwrapping.

SOURCE: J.M. Tribolet, "A New Phase Unwrapping Algorithm" [19]. ©1977, IEEE. Used by permission.

derivative of the real and imaginary components, possibly approximated by first differences, and on the size of the integration step $\frac{2\pi}{N}$, but this approach can lead to significant errors since the numerical error can accumulate (Exercise (6.5)). In order to avoid the accumulation of error, the principal phase value can be used as a reference as we now show [19].

The phase unwrapping problem can be restated as determining the integer value $q(\omega_k)$ such that

$$\angle X(\omega_k) = \text{PV}[X(\omega_k)] + 2\pi q(\omega_k)$$

gives the integrated phase derivative at ω_k. Assume the phase has been correctly unwrapped up to ω_{k-1} with value $\theta(\omega_{k-1})$. Then the unwrapped phase at ω_k is given by

$$\theta(\omega_k) = \theta(\omega_{k-1}) + \int_{\omega_{k-1}}^{\omega_k} \dot\theta(\omega)d\omega$$

where $\dot{\theta}(\omega)$ denotes $\angle X(\omega)$. As indicated in Figure 6.12b, we can then think of numerical integration as "predicting" the next phase value from the previous one. We then compare that predicted value against the candidate phase values $PV[X(\omega_k)] + 2\pi q(\omega_k)$. For trapezoidal numerical integration, the unwrapped phase is estimated as

$$\hat{\theta}(\omega_k) = \theta(\omega_{k-1}) + \frac{(\omega_k - \omega_{k-1})}{2}[\dot{\theta}(\omega_k) + \dot{\theta}(\omega_{k-1})]$$

which improves as the DFT length increases (frequency spacing decreases). One then selects a value of $q(\omega_k)$ such that the difference between the predicted value and candiate value are minimized, i.e., we minimize

$$E[k] = |(PV[X(\omega_k)] + 2\pi q(\omega_k)) - \hat{\theta}(\omega_k)| \tag{6.17}$$

over $q(\omega_k)$. One can reduce this minimum error, and thus improve the accuracy of the phase unwrapper, by reducing the frequency spacing (by increasing the DFT length).

There have been numerous attempts to improve on these phase unwrapping algorithms, including a method of polynomial factoring [17] and another based on Chebyshev polynomials [6], both of which give closed-form solutions to the unwrapping problem rather than recursively use past values of unwrapped phase as in the above methods. These closed-form solutions, however, appear to lack computational robustness because they require impractical numerical precision.

6.5 Spectral Root Homomorphic Filtering

A different homomorphic system for convolution is motivated by mapping $x[n] = h[n] * p[n]$ to $\check{x}[n] = \check{h}[n] * \check{p}[n]$ such that $\check{p}[n]$ is a new pulse train with the same spacing as $p[n]$, and where $\check{h}[n]$ is more time-limited than $h[n]$. If $\check{h}[n]$ is sufficiently compressed in time, it can be extracted by time-liftering in the vicinity of $\check{h}[n]$. One such class of homomorphic systems replaces the logarithm with the γ power of the z-transform of $X(z)$, i.e., by the rooting operation $X(z)^\gamma$ [5]. As with time-liftering the complex cepstrum, this alternate means of homomorphic filtering, referred to as "spectral root homomorphic filtering" [5], can also be considered as a spectral smoother.

Spectral root homomorphic filtering, illustrated in Figure 6.13, is an analog to the log-based system; the difference is that the γ and $1/\gamma$ replace the logarithmic and exponential operations. If we consider real-valued γ, then we define

$$\check{X}(\omega) = X(\omega)^\gamma = |X(\omega)|^\gamma e^{j\gamma \angle(X(\omega))}. \tag{6.18}$$

As with the complex logarithm, in order to make our definition unique, the phase must be unambiguously defined, and this can be done through either of the approaches described in the previous section. Then, since $x[n]$ is a real and stable sequence, $\check{X}(z)$ is a valid z-transform with an ROC that includes the unit circle. Under this condition, we define a sequence, analogous to the complex cepstrum, as the inverse Fourier transform of $\check{X}(\omega)$ that we refer to as the "spectral root cepstrum":

$$\check{x}[n] = \frac{1}{2\pi} \int_{-\pi}^{\pi} \check{X}(\omega) e^{j\omega n} d\omega.$$

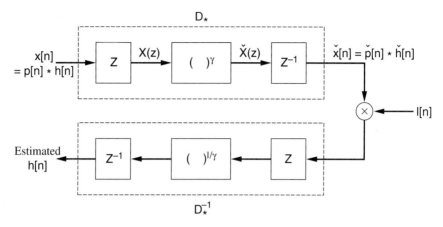

Figure 6.13 Spectral root homomorphic filtering.

SOURCE: J.S. Lim, "Spectral Root Homomorphic Deconvolution System" [5]. ©1979, IEEE. Used by permission.

Because $|\check{X}(\omega)|$ is even and since $\angle \check{X}(\omega)$ is odd, then $\check{x}[n]$ is a real and stable sequence. As with liftering the complex cepstrum, liftering the spectral root cepstrum can be used to separate the fast- and slow-varying components of $\check{X}(\omega)$.

As before, we consider a class of sequences with rational z-transforms of the form of Equation (6.9). Then $\check{X}(z)$ is expressed as

$$\check{X}(z) \ = \ A^{\gamma} \frac{\prod_{k=1}^{M_i}(1 - a_k z^{-1})^{\gamma} \prod_{k=1}^{M_o}(1 - b_k z)^{\gamma}}{\prod_{k=1}^{N_i}(1 - c_k z^{-1})^{\gamma} \prod_{k=1}^{N_o}(1 - d_k z)^{\gamma}} \tag{6.19}$$

where $(1 - a_k z^{-1})$ and $(1 - c_k z^{-1})$ are zeros and poles inside the unit circle and $(1 - b_k z)$ and $(1 - d_k z)$ are zeros and poles outside the unit circle with $|a_k|, |b_k|, |c_k|, |d_k| < 1$. The time-shift term z^{-r} has been removed, assuming it can be estimated, and the factor A is assumed positive. Each factor in Equation (6.19) can be rewritten using the following power series expansion:

$$(1 - a z^{-1})^{\gamma} \ = \ 1 + \sum_{l=1}^{\infty} \binom{\gamma}{l} (-a)^l z^{-l}$$

where $\binom{\gamma}{l} = \gamma \frac{(\gamma-1)}{2} \frac{(\gamma-2)}{3} \cdots \frac{(\gamma-l+1)}{l}$. Thus the spectral root cepstrum of, for example, the kth zero inside the unit circle is given by

$$\check{x}[n] \ = \ Z^{-1}[(1 - a z^{-1})^{\gamma}] \ = \ \delta[n] + \sum_{l=1}^{\infty} \binom{\gamma}{l} (-a)^l \delta[n - l]$$

and the spectral root cepstra for the three remaining factors $(1 - bz)^{\gamma}$, $(1 - cz^{-1})^{\gamma}$, and $(1 - dz)^{\gamma}$ are similarly derived (Exercise 6.7).

Many of the properties of the spectral root cepstrum are similar to those of the complex cepstrum because the former can be written in terms of the latter [5]. To see this, let

$\hat{X}(z) = \log[X(z)]$. Then $\check{X}(z) = X(z)^\gamma$ is related to $\hat{X}(z)$ by

$$\check{X}(z) = e^{\gamma \hat{X}(z)} = 1 + \gamma \hat{X}(z) + \frac{\gamma^2}{2} \hat{X}(z)^2 + \frac{\gamma^3}{3} \hat{X}(z)^3 \ldots.$$

Using the inverse z-transform, the relation between $\check{x}[n]$ and $\hat{x}[n]$ is then given by

$$\check{x}[n] = \delta[n] + \gamma \hat{x}[n] + \frac{\gamma^2}{2} \hat{x}[n] * \hat{x}[n] + \frac{\gamma^3}{3} \hat{x}[n] * \hat{x}[n] * \hat{x}[n] \ldots. \qquad (6.20)$$

From Equation (6.20), we see that if $x[n]$ is minimum-phase, then $\check{x}[n]$ is right-sided, i.e., $\check{x}[n] = 0$ for $n < 0$, since its complex cepstrum $\hat{x}[n]$ is right-sided. A similar observation can be made for maximum-phase, left-sided sequences. As with the complex cepstrum, for minimum- and maximum-phase sequences, $\check{x}[n]$ can be obtained from $|X(\omega)|$ (Exercise 6.7). Finally, to complete the analogy with the complex cepstrum, if $p[n]$ is a train of impulses with equal spacing of N then $\check{p}[n]$ remains an impulse train with the same spacing (Exercise 6.7).

When the unwrapped phase is defined unambiguously, e.g., through the phase derivative, then the spectral root cepstrum of a convolution of two sequences equals the convolution of their spectral root cepstra. That is, if $x[n] = x_1[n] * x_2[n]$, then $X(z) = X_1(z) X_2(z)$ so that $X(z)^\gamma = X_1(z)^\gamma X_2(z)^\gamma$, resulting in $\check{x}[n] = \check{x}_1[n] * \check{x}_2[n]$. This convolutional property is the basis for the spectral root deconvolution system which is analogous to the complex cepstrum deconvolution system that maps convolution to addition, i.e., $\hat{x}[n] = \hat{x}_1[n] + \hat{x}_2[n]$. To see how the spectral root cepstrum can be used for deconvolution, we look at the following example where the convolutional components are an impulse train $x_1[n] = p[n]$ and an all-pole response $x_2[n] = h[n]$:

EXAMPLE 6.8   Suppose $h[n]$ is a minimum-phase all-pole sequence of order q. Consider a waveform $x[n]$ constructed by convolving $h[n]$ with a sequence $p[n]$ where

$$p[n] = \delta[n] + \beta \delta[n - N], \text{ with } \beta < 1$$

so that

$$x[n] = p[n] * h[n]$$

where $q < N$ and where

$$P(z) = 1 + \beta z^{-N}.$$

Suppose we form the spectral root cepstrum of $x[n]$ with $\gamma = -1$. Then, using the Taylor series expansion for $\frac{1}{(1+\beta z^{-1})}$ and replacing z by z^N, it is seen that the inverse z-transform of $P^{-1}(z)$ is an impulse train with impulses spaced by N samples (Exercise 6.8). Also $H^{-1}(z)$ is all-zero, since $H(z)$ is all-pole, so that $\check{h}[n]$ is a q-point sequence. Because $q < N$, $h[n]$ can be deconvolved from $x[n]$ by inverting $X(z)$ to obtain $X^{-1}(z)$, and liftering $h^{-1}[n]$, the inverse z-transform of $H^{-1}(z)$, using a right-sided lifter of q samples (Exercise 6.8).   ▲

More generally, when $p[n]$ is of the form $\sum_{k=0}^{M} a_k \delta[n - kN]$ and when $\check{h}[n]$ is sufficiently low-time limited, then low-time liftering of $\check{x}[n]$ yields an estimate of $h[n]$ scaled by the value of the pulse train a_0 at the origin; as shown in Example 6.8, this estimate can be exact. In comparison with the complex cepstrum, low-time liftering the complex cepstrum does not recover the response $h[n]$ to within a scale factor since $\hat{h}[n]$ is always infinite in extent. In general, however, the relative advantages are not clear-cut since for a general pole-zero sequence, $\check{h}[n]$ is also infinitely long. In this situation, we select γ to maximally compress $\check{h}[n]$ such that it has the smallest energy "concentration" in the low-time region. One definition of energy concentration is the percentage of energy in the first n points of $\check{h}[n]$ relative to its total energy, i.e., $d[n] = \frac{\sum_{k=1}^{n} |\check{h}[k]|^2}{\sum_{k=1}^{\infty} |\check{h}[k]|^2}$ [5]. For an all-pole sequence, as we saw in the previous example, $\gamma = -1$ gives a tight concentration, while for an all-zero sequence, $\gamma = 1$ is preferred. For pole-zero sequences, the selection of γ depends on the pole-zero arrangement; for example, for voiced speech dominated by poles, a γ closer to $\gamma = -1$ is optimal. Empirically, it has been found that as the number of poles increases relative to the number of zeros, γ should be made closer to -1, and vice-versa when zeros dominate, as illustrated in the following example:

EXAMPLE 6.9   Consider a sequence of the form $x[n] = p[n] * x[n]$, as in Example 6.8. Figure 6.14 illustrates an example of extracting an $h[n]$ with ten poles and two zeros [5]. Figure 6.14a shows the logarithm of its spectral magnitude. The spectral log-magnitude of an estimate of $h[n]$ derived from low-time liftering the real cepstrum is shown in Figure 6.14b. The same estimate derived from

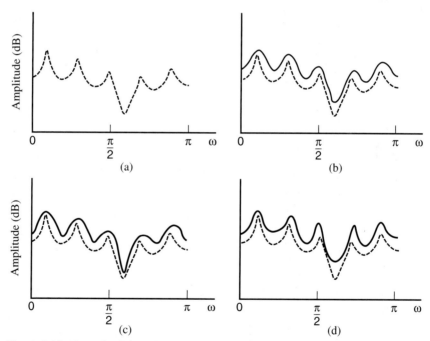

Figure 6.14   Example of spectral root homomorphic filtering on synthetic vocal tract impulse response: (a) log-magnitude spectrum of impulse response; (b) estimate of log-magnitude spectrum of $h[n]$ derived from low-time gating real cepstrum; (c) log-magnitude spectral estimate derived from low-time gating spectral root cepstrum with $\gamma = +0.5$; (d) same as (c) with $\gamma = -0.5$.

the real spectral root cepstrum is given in Figures 6.14c and 6.14d with $\gamma = +0.5$ and $\gamma = -0.5$, respectively. The negative value of γ results in better pole estimates, while the positive value results in better zero estimates. For this example, the poles dominate the spectrum and thus the negative γ value is preferred. $\blacktriangle$

As with a logarithmic cepstral-based analysis, a spectral smoothing interpretation of the spectral root cepstral-based system can be made. This spectral smoothing interpretation shows, for example, that $H(\omega)$ in Example 6.9 is estimated by lowpass filtering $X(\omega)^\gamma$. It is natural to ask, therefore, whether the speech spectrum may be smoothed based on some other transformation that is neither logarithmic nor a spectral root function. Finally, as with the complex cepstrum, computational considerations include the aliasing due to insufficient DFT length and inaccuracy in phase unwrapping. Unlike the complex cepstrum, phase unwrapping is required in both the forward $X(\omega)^\gamma$ and the inverse $X(\omega)^\gamma$ transformations (Exercise 6.9).

6.6 Short-Time Homomorphic Analysis of Periodic Sequences

Up to now we have assumed that a model of voiced speech is an *exact* convolution of an impulse train and an impulse response, i.e., $x[n] = p[n] * h[n]$, where the impulse train $p[n]$, with impulse spacing P, is of finite extent. In practice, however, a periodic waveform is windowed by a finite-length sequence $w[n]$ to obtain a short-time segment of voiced speech:

$$s[n] = w[n](p[n] * h[n]).$$

Ideally, we want $s[n]$ to be close to the convolutional model $\tilde{x}[n] = (w[n]p[n]) * h[n]$. We expect this to hold under the condition that $w[n]$ is smooth relative to $h[n]$, i.e., under this smoothness condition, we expect[9]

$$\tilde{x}[n] = (w[n]p[n]) * h[n]$$
$$\approx w[n](p[n] * h[n])$$

so that

$$\hat{s}[n] \approx \hat{p}[n] + \hat{h}[n] \tag{6.21}$$

where $\hat{p}[n]$ is the complex cepstrum of the windowed impulse train $w[n]p[n]$ and $\hat{h}[n]$ is the complex cepstrum of $h[n]$. It can be shown, however, that the complex cepstrum of the windowed sequence is given exactly by [21]

$$\hat{s}[n] = \hat{p}[n] + D[n] \sum_{k=-\infty}^{+\infty} \hat{h}[n - kP] \tag{6.22}$$

where $D[n]$ is a weighting function concentrated at $n = 0$ and is dependent on the window $w[n]$. In this section, we do not prove this general case, but rather look at a simplifying case to provide insight into the representation for performing deconvolution. This leads to a set of conditions

---

[9] Consider the special case where the impulse spacing P is very large so that there is no overlap among the sequences $h[n - kP]$. Suppose also that the window is piecewise flat and constant over each $h[n - kP]$. Then the approximation in Equation (6.21) becomes an equality.

on the analysis window $w[n]$ and the cepstral lifter $l[n]$ under which $\hat{s}[n] \approx \hat{p}[n] + \hat{h}[n]$ in the context of deconvolution of speech-like periodic signals.

6.6.1 Quefrency-Domain Perspective

Consider first a voiced speech signal modeled as a perfectly periodic waveform $x[n] = p[n] *$ $h[n]$ with source $p[n] = \sum_{k=-\infty}^{+\infty} \delta[n - kP]$ with P the pitch period, and with vocal tract impulse response $h[n]$; then only *samples* of the $\log[X(\omega)]$ are defined at multiples of the fundamental frequency $\omega_o = \frac{2\pi}{P}$, that is,

$$\log[X(\omega_k)] = \log[P(\omega_k)] + \log[H(\omega_k)]$$

where $\omega_k = k\omega_o$ and is undefined elsewhere because $X(\omega) = 0$ for $\omega_k \neq k\omega_o$. Suppose we define $\log[X(\omega)] = 0$ for $\omega_k \neq k\omega_o$. Then a system component of the form $\sum_{k=-\infty}^{+\infty} \hat{h}[n - kP]$ appears in the complex cepstrum, i.e., the system component consists of replicas of the desired complex cepstrum $\hat{h}[n]$. These replicas must occur at the pitch period rate because samples of the logarithm are available only at harmonics. Moreover, aliasing at the pitch period rate can occur as a consequence of the undersampling of the spectrum $H(\omega)$. From this perspective, we must account for the Nyquist sampling condition and the decay of $\hat{h}[n]$.

To further develop this concept analytically, let us now introduce a short-time window $w[n]$ and define

$$s[n] = w[n](p[n] * h[n])$$

where as before $p[n] = \sum_{k=-\infty}^{\infty} \delta[n - kP]$, and $h[n]$ is the system impulse response. We have seen in Chapter 2 that

$$S(\omega) = \frac{1}{2\pi}[P(\omega)H(\omega)] \circledast W(\omega)$$

$$= \frac{1}{P} \sum_{k=-\infty}^{\infty} H(k\omega_o)W(\omega - k\omega_o). \qquad (6.23)$$

We write $s[n]$ as

$$s[n] = (p[n]w[n]) * g[n] \qquad (6.24)$$

where $g[n]$ is a sequence assumed "close to" $h[n]$. Then, in this form $\hat{S}(\omega)$ can be expressed as

$$S(\omega) = \left[\frac{1}{P} \sum_{k=-\infty}^{\infty} W(\omega - k\omega_o) \right] G(\omega). \qquad (6.25)$$

Therefore, taking the logarithm of the both sides of Equations (6.23) and (6.25) and solving for $\log[G(\omega)]$, we have

$$\log[G(\omega)] = \log \left[\sum_{k=-\infty}^{\infty} H(k\omega_o)W(\omega - k\omega_o) \right] - \log \left[\sum_{k=-\infty}^{\infty} W(\omega - k\omega_o) \right]. \quad (6.26)$$

To simplify the expression in Equation (6.26), consider a rectangular $W(\omega)$ where $W(\omega) = 1$ for $|\omega| \leq \omega_o/2$ and otherwise is zero in the interval $[-\pi, \pi]$ (and periodic with period 2π). In the time domain, $W(\omega)$ corresponds to the sinc function $w[n] = \frac{\sin(\omega_o n/2)}{\pi n}$. The second term in Equation (6.26) then becomes zero and, with our choice of $W(\omega)$, the logarithm operator can be taken inside the summation of the first term, resulting in

$$\log[G(\omega)] = \sum_{k=-\infty}^{\infty} \log[H(k\omega_o)]W(\omega - k\omega_o)$$

$$= W(\omega) \circledast \left(\log[H(\omega)] \sum_{k=-\infty}^{\infty} \delta[\omega - k\omega_o] \right). \tag{6.27}$$

Therefore, from Equations (6.24) and (6.27), we can write the complex cepstrum of $s[n]$ as

$$\hat{s}[n] = \hat{p}[n] + \hat{g}[n] \tag{6.28}$$

where $\hat{p}[n]$ is the complex cepstrum of $p[n]w[n]$ and where the complex cepstrum of $g[n]$ is given by

$$\hat{g}[n] = w[n] \sum_{k=-\infty}^{\infty} \hat{h}[n - kP] \tag{6.29}$$

where $\hat{h}[n]$ is the complex cepstrum of $h[n]$ and $w[n]$ is the inverse Fourier transform of the rectangular function $W(\omega)$. The result is illustrated in Figure 6.15. We see that Equation (6.29) is a special case of Equation (6.22) with $D[n] = w[n]$.

As with the purely convolutional model, i.e., $\tilde{x}[n] = (w[n]p[n]) * h[n]$, the contributions of the windowed pulse train and impulse response are additively combined so that deconvolution is possible. Now, however, the impulse response contribution is repeated at the pitch period rate. This is a different sort of aliasing, dependent upon pitch, from the aliasing we saw earlier in

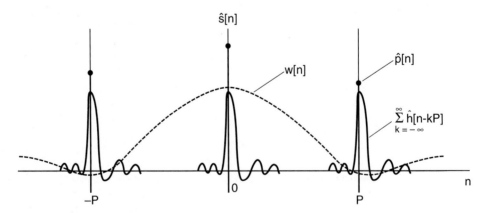

Figure 6.15 Schematic of complex cepstrum of windowed periodic sequence.

Section 6.4.4 due to an insufficient DFT length. We also see that both the impulse response contribution and its replicas are weighted by a "distortion" function $D[n] = w[n]$. In the particular case described, $D[n]$ is pitch-dependent because the window transform $W(\omega)$ is rectangular of width equal to the fundamental frequency. Furthermore, $D[n] = \frac{\sin(\omega_o n)}{\pi n}$ has zeros passing through integer pitch periods, thus reducing the effect of replicas of $\hat{h}[n]$. We have assumed in this derivation that the window $w[n]$ is centered at the origin so that the window contains no linear-phase component. Likewise, we have assumed that the impulse response $h[n]$ has no linear phase, thus avoiding the dominating effect of linear phase shown in Example 6.5. An implication of these two conditions is that the window and impulse response are "aligned." When the window is displaced from the origin, an expression similar to Equations (6.22) and (6.29) can be derived using the approach in [21]. The distortion term $D[n]$ becomes a function of the window shift r and thus is written $D[n, r]$; this distortion function, however, becomes increasingly severe as r increases [21].

We can now state conditions on our particular window $w[n] = \frac{\sin(\omega_o n/2)}{\pi n}$ under which $s[n] \approx (p[n]w[n]) * h[n]$ from the view of the the quefrency domain [Equations (6.28) and (6.29) and more generally Equation (6.22)]. First, we should select the time-domain window $w[n]$ to be long enough so that $D[n]$ is "smooth" in the vicinity of the quefrency origin and over the extent of $\hat{h}[n]$. Second, the window $w[n]$ should be short enough to reduce the effect of the replicas of $\hat{h}[n]$. Verhelst and Steenhaut [21] have shown that for typical windows (such as Hamming or Hanning), a compromise of these two conflicting constraints is a window $w[n]$ whose duration is about 2 to 3 pitch periods. Finally, the window $w[n]$ should be centered at the time origin and "aligned" with $h[n]$, i.e., there is no linear phase contribution from $h[n]$. In the context of homomorphic deconvolution, in selecting the low-time lifter $l[n]$, we also want to account for the general two-sidedness of $\hat{h}[n]$. Therefore, a low-time lifter width of half the pitch period should be used for deconvolution. Under these conditions, for $|n| < P/2$, the complex cepstrum is close to that derived from the conventional model $\tilde{x}[n] = (w[n]p[n]) * h[n]$. Finally, note that there are now two reasons for low-pitched waveforms to be more amenable to cepstral analysis than high-pitched waveforms: With high-pitched speakers, there is stronger presence of $\hat{p}[n]$ close to the origin, as noted earlier, but there is also more aliasing of $\sum_{k=-\infty}^{\infty} \hat{h}[n - kP]$ (Exercise 6.11).

In the next section, from the perspective of the frequency domain, we arrive at the above constraints on the analysis window $w[n]$ in a more heuristic way than in this section. This gives us insights different from those we have seen in the quefrency domain.

6.6.2 Frequency-Domain Perspective

Consider again a voiced speech signal modeled as a perfectly periodic waveform, i.e., as $x[n] = p[n] * h[n]$ where $p[n] = \sum_{k=-\infty}^{+\infty} \delta[n - kP]$ with P as the pitch period and where $h[n]$ is a vocal tract impulse response. This sequence corresponds in the frequency domain to impulses weighted by $H(\omega_k)$ at multiples of the fundamental frequency ($\omega_k = k\omega_o$). Windowing in the time domain can then be thought of as a form of interpolation across the harmonic samples $X(\omega_k) = P(\omega_k)H(\omega_k)$ by the Fourier transform of the window, $W(\omega)$. We want to determine conditions on the window under which this interpolation results in the desired convolution model, i.e., $\tilde{x}[n] = (w[n]p[n]) * h[n]$.

For a particular window, e.g., Hamming, one can use an empirical approach to selecting the window length and determining how well we achieve the above convolutional model. This entails looking at the difference between the desired convolutional model $\tilde{x}[n] = (w[n]p[n]) * h[n]$ and the actual windowed sequence $s[n] = w[n](p[n] * h[n])$. With $E(\omega) = S(\omega)/\tilde{X}(\omega)$, we define a measure of spectral degradation with respect to the spectral magnitude as

$$D = \frac{1}{2\pi} \int_{-\pi}^{\pi} \log |E(\omega)|^2 d\omega.$$

Over a representative set of pitch periods and speech-like sequences, for a Hamming window this spectral distance measure was found empirically to be minimized for window length in the range of roughly 2 to 3 pitch periods [14]. An implication of this result is that the length of the analysis window should be adapted to the pitch period to make the windowed waveform as close as possible (in the above sense) to the desired convolutional model.

An empirical approach can also be taken for determining window conditions for a phase measurement congruent with the convolutional model [14]. We intuitively see a problem with measuring phase in the frequency domain where $s[n] = w[n](p[n] * h[n])$ is mapped to

$$S(\omega) = \frac{1}{2\pi} W(\omega) \circledast [P(\omega)H(\omega)]$$

which is essentially the mainlobe of the window Fourier transform weighted and repeated at the harmonic frequencies. As we increase the window length beyond about 2 to 3 pitch periods, the phase of $S(\omega)$ between the main harmonic lobes becomes increasingly meaningless in light of the desired convolutional model. As we decrease the window length below 2 to 3 pitch periods, then the window transform mainlobes overlap one another and effectively provide an interpolation of the real and imaginary parts of $S(\omega)$ across the harmonics, resulting in a phase more consistent with the convolutional model [14],[15]. Using this viewpoint, a window length of roughly 2 to 3 pitch periods can be argued to be "optimal" for phase measurements to be consistent with the convolutional model (Exercise 6.12). A heuristic argument can also be made in the frequency domain for the requirement that the window be centered at the time origin under the condition that $h[n]$ has no linear phase, i.e., the window center and $h[n]$ are "aligned" (Exercise 6.12). These same conditions were established in the previous section using the more analytic quefrency-domain perspective.

EXAMPLE 6.10   Figure 6.16 illustrates examples of sensitivity to window length and alignment (position) for a synthetic speech waveform with a system function consisting of two poles inside the unit circle at 292 Hz and 3500 Hz and a zero outside the unit circle at 2000 Hz, and a periodic impulse source of 100-Hz pitch [14]. Figure 6.16a shows the unwrapped phase of the system function's pole-zero configuration. In panels (b) and (c), the analysis window is the square of the sinc function, i.e., $w[n] = [\frac{\sin((\pi/P)n)}{\pi n)}]^2$, so that its Fourier transform is a triangle with length equal to twice the fundamental frequency. Figure 6.16b shows an unwrapped phase estimate with the window time-aligned with $h[n]$, while Figure 6.16c shows the estimate from the window at a different position. Figure 6.16d,e shows the sensitivity to the length of a Hamming window. In all cases, the unwrapped phase is computed with the modulo 2π phase unwrapper of Section 6.4.4. This example illustrates fundamental problems in the phase representation, being independent of the reliability of the phase unwrapper; it is further studied in Exercise 6.12.   ▲

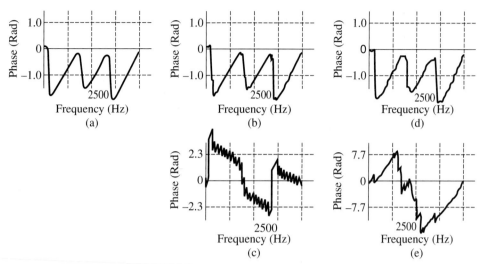

Figure 6.16 Sensitivity of system phase estimate to the analysis window in Example 6.10: (a) unwrapped phase of artificial vocal tract impulse response; (b) unwrapped phase of a periodic waveform with window $w[n] = [\frac{\sin(\pi n/P)}{\pi n)}]^2$ and time-aligned; (c) same as (b) with window displacement; (d) same as (b) with Hamming window 2 pitch periods in length; (e) same as (d) with Hamming window 3.9 pitch periods in length.

SOURCE: T.F. Quatieri, "Minimum- and Mixed-Phase Speech Analysis/Synthesis by Adaptive Homomorphic Deconvolution" [14]. ©1979, IEEE. Used by permission.

6.7 Short-Time Speech Analysis

6.7.1 Complex Cepstrum of Voiced Speech

Recall that the transfer function model from the glottal source to the lips output for voiced speech is given by

$$H(z) = AG(z)V(z)R(z)$$

where the glottal flow over a single cycle $G(z)$, the source gain A, and the radiation load $R(z)$ are all embedded within the system function. The corresponding voiced speech output in the time domain is given by

$$x[n] = p[n] * h[n]$$
$$= Ap[n] * g[n] * v[n] * r_L[n]$$

where $p[n]$ is the idealized periodic impulse train, $g[n]$ is the glottal volume velocity, $v[n]$ is the resonant (and anti-resonant) vocal tract impulse response, and $r_L[n]$ represents the lip radiation response. We assume a rational z-transform for the resonant vocal tract contribution which is stable (poles inside the unit circle) and which may have both minimum- and maximum-phase zeros, i.e.,

$$V(z) = \frac{\prod_{k=1}^{M_i}(1 - a_k z^{-1}) \prod_{k=1}^{M_o}(1 - b_k z)}{\prod_{k=1}^{N_i}(1 - c_k z^{-1})}. \qquad (6.30)$$

As we have seen, "pure" vowels contain only poles, while nasalized vowels and unvoiced sounds can have both poles and zeros. The radiation load is modeled by a single zero $R(z) \approx 1 - \beta z^{-1}$, so that with a typical $\beta \approx 1$, we obtain a high-frequency emphasis. The glottal flow waveform can be modeled by either a finite set of minimum- and maximum-phase zeros, i.e.,

$$G(z) = \prod_{k=1}^{L_i}(1 - \alpha_k z^{-1}) \prod_{k=1}^{L_o}(1 - \beta_k z)$$

or, more efficiently (but less generally), by two poles outside the unit circle:

$$G(z) = \frac{1}{(1 - cz)(1 - dz)}$$

which, in the time domain, is a waveform beginning with a slow attack and ending with a rapid decay, emulating a slow opening and an abrupt closure of the glottis. We have assumed that $H(z)$ contains no linear phase term so that $h[n]$ is "aligned" with the time origin.

One goal in homomorphic analysis of voiced speech is to separate $h[n]$ and the source $p[n]$ [11],[12]. The separation of $h[n]$ into its components is treated later in the chapter. In order to perform analysis, we must first extract a short-time segment of speech with an analysis window $w[n]$. Let

$$s[n] = w[n](p[n] * h[n])$$

which is assumed to be approximately equal to the convolutional model of the form

$$\tilde{x}[n] = (w[n]p[n]) * h[n]$$

i.e., $\tilde{x}[n] \approx s[n]$ under the conditions on $w[n]$ determined in the previous section: the window duration is 2 to 3 pitch periods and its center is aligned with $h[n]$; the latter condition is more important for good phase estimation of the transfer function than for good spectral magnitude estimation. The discrete complex cepstrum is then computed using an N-point DFT as

$$\hat{s}_N[n] \approx \hat{p}_N[n] + \hat{h}_N[n]$$

and a similar expression can be written for the real cepstrum. For a typical speaker, the duration of the short-time window lies in the range of 20 ms to 40 ms. We assume that the source and system components lie roughly in separate quefrency regions and, in particular, that negligible aliasing of the replicas of $\hat{h}[n]$ in Equation (6.22) occurs within $\frac{P}{2}$ of the origin. We also assume that the distortion function $D[n]$ is "smooth" in this same region $|n| < \frac{P}{2}$ so that $\hat{h}[n]$ is not significantly distorted, and that $D[n]$ attenuates replicas of $\hat{h}[n]$ for $|n| > \frac{P}{2}$ to make them negligible, as described in Section 6.6.1. We then apply a low-quefrency cepstral lifter, $l[n] = 1$ for $|n| < \frac{P}{2}$ and zero elsewhere, to separate $\hat{h}[n]$, and apply the complementary high-quefrency cepstral lifter to separate the input pulse train. The following example illustrates properties of this method:

EXAMPLE 6.11   Figure 6.17 illustrates homomorphic filtering for a speech waveform from a female speaker with an average pitch period of about 5 ms. The continuous waveform was sampled at 10000 samples/s and is windowed by a 15-ms Hamming window. A 1024-point FFT is used in computing the discrete complex cepstrum. The window center is aligned roughly with $h[n]$ and shifted to the time origin ($n = 0$) using the strategy to be described in Section 6.8.2. The figure shows the spectral log-magnitude and unwrapped phase estimates of $h[n]$ superimposed on the short-time Fourier transform measurements, as well as the time-domain source and system estimates. Observe that the unwrapped phase of the short-time Fourier transform is approximately piecewise-flat, resulting from the symmetry of the analysis window and the removal of linear phase through the alignment strategy. (The reader should argue this property.). We will see another example of this piecewise-flat phase characteristic again in Chapter 9 (Figure 9.32). A low-quefrency lifter of 6 ms duration

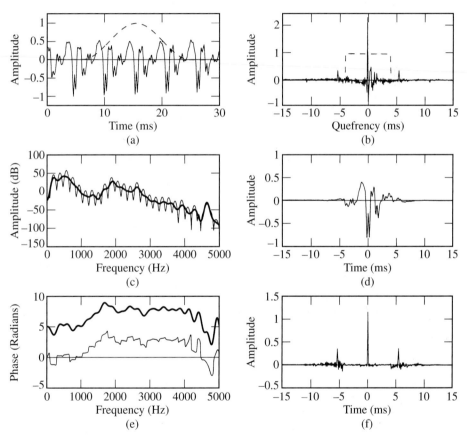

Figure 6.17 Homomorphic filtering of voiced waveform from female speaker: (a) waveform (solid) and aligned analysis window (dashed); (b) complex cepstrum of windowed speech signal $s[n]$ (solid) and low-quefrency lifter (dashed); (c) log-magnitude spectrum of $s[n]$ (thin solid) and of the impulse response estimate (thick solid); (d) impulse response estimate from low-quefrency liftering; (e) spectral unwrapped phase of $s[n]$ (thin solid) and of the impulse response estimate (thick solid) (The smooth estimate is displaced for clarity.); (f) estimate of windowed impulse train source from high-quefrency liftering.

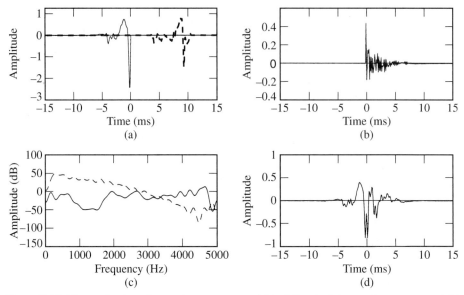

Figure 6.18 Deconvolved maximum-phase component (a) (solid) and minimum-phase component (b) in Example 6.11. The convolution of the two components (d) is identical to the sequence of Figure 6.17d. Panel (c) shows the log-magnitude spectra of the maximum-phase (dashed) and minimum-phase (solid) component. The maximum-phase component in panel (a) and its log-magnitude spectrum in panel (c) resemble those of a typical glottal flow derivative sequence. For reference, the dashed sequence in panel (a) is the glottal flow derivative derived from the pole/zero-estimation method of Section 5.7.2.

(unity in the interval $[-3, 3]$ ms and zero elsewhere) was applied to the cepstrum to obtain the $h[n]$ estimate and a complementary lifter (zero in the interval $[-3, 3]$ ms and unity elsewhere) was applied to obtain the windowed pulse train source estimate. The deconvolved maximum-phase and minimum-phase response components are shown in Figure 6.18 (using a left-sided and right-sided 3-ms lifter, respectively). The convolution of the two components is identical to the sequence of Figure 6.17d. The figure also shows the log-magnitude spectra of the maximum-phase and minimum-phase components. The maximum-phase component and its log-magnitude spectrum resemble those of a typical glottal flow derivative sequence, indicating that, for this particular example, the radiation (derivative) contribution may be maximum-phase, unlike our earlier minimum-phase zero model. For reference, the glottal flow derivative derived from the pole/zero-estimation method of Section 5.7.2 is also shown.   ▲

In this example, our spectral smoothing interpretation of homomorphic deconvolution is seen in the low-quefrency and high-quefrency liftering that separates the slowly-varying and rapidly-varying components of the spectral log-magnitude and phase. We see that the resulting smooth spectral estimate does not necessarily traverse the spectral peaks at harmonics, in contrast to the linear prediction analysis of Chapter 5 that more closely preserves the harmonic amplitudes. In addition, because the mixed-phase estimate is obtained by applying a symmetric lifter, of an extent roughly $-\frac{P}{2} \leq n \leq \frac{P}{2}$, to the complex cepstrum, the lifter length decreases with the pitch period. Homomorphic filtering by low-quefrency liftering will, therefore, impart more spectral smoothing with increasing pitch. The resulting smoothing widens the formant bandwidths which tends to result in a "muffled" quality in synthesis (Exercise 6.6). Therefore, the requirement that the duration of the low-quefrency lifter be half the pitch period leads to more artificial widening of the formants for females and children than for males.

As indicated earlier, depending on whether we use the real or complex cepstrum, the impulse response can be zero-, minimum-, or mixed-phase. To obtain the zero-phase estimate, the real cepstrum (or the even part of the complex cepstrum) is multiplied by a symmetric low-quefrency lifter $l[n]$ that extends over the region $-\frac{P}{2} \leq n \leq \frac{P}{2}$. The resulting impulse response estimate is symmetric about the origin (Exercise 6.14). On the other hand, the minimum-phase counterpart can be obtained by multiplying the real cepstrum by the right-sided lifter $l[n]$ of Equation (6.13). The resulting impulse response is right-sided and its energy is compressed toward the origin, a property of minimum-phase sequences [13] that was reviewed in Chapter 2. When the vocal tract is indeed minimum-phase, we saw earlier, at the end of Section 6.4.1, that the liftering operation yields the correct phase function. When the vocal tract is mixed-phase, however, which can occur with back cavity or nasal coupling, for example, then this phase function is only a rough approximation to the original; we have created a synthetic phase from the vocal tract spectral magnitude. In this case, for a rational z-transform, the effect is to flip all zeros that originally fall outside the unit circle to inside the unit circle. We now look at a "proof by example" of this property.

EXAMPLE 6.12   Consider the z-transform

$$X(z) = \frac{(1 - az^{-1})(1 - bz)}{\prod_{k=1}^{M_i}(1 - c_k z^{-1})}$$

where a, b, and c_k are less than unity. The ROC of $X(z)$ contains the unit circle so that $x[n]$ is stable. There is one zero outside the unit circle giving a maximum-phase component. The real cepstrum is written as

$$c[n] = c_{max}[n] + c_{min}[n]$$

where $c_{max}[n]$ and $c_{min}[n]$ are the real cepstra of the minimum- and maximum-phase components, respectively. Using the right-sided lifter $l[n]$ of Equation (6.13), we can then write the minimum-phase construction as

$$\bar{x}[n] = c_{max}[n]l[n] + c_{min}[n]l[n]$$
$$= c_{max}[n]l[n] + \hat{x}_{min}[n].$$

The component $\hat{x}_{min}[n]$ is obtained using Equation (6.13) that constructs the complex cepstrum of a minimum-phase sequence from its real cepstrum. We interpret the minimum-phase construction of the maximum-phase term, i.e., $c_{max}[n]l[n]$, as follows. From our earlier discussion, the complex cepstrum of the maximum-phase component can be written as

$$\hat{x}_{max}[n] = \frac{b^{-n}}{n}, \quad n < 0$$
$$= 0, \qquad n \geq 0.$$

Then, because

$$c_{max}[n]l[n] = \hat{x}_{max}[-n]$$

it follows that the minimum-phase construction of the maximum-phase term is given by

$$c_{max}[n]l[n] = \frac{-b^n}{n}, \quad n > 0$$
$$= 0, \quad n \leq 0$$

resulting in the zero $(1 - bz)$ flipped inside the unit circle.   ▲

This argument can be generalized for multiple poles and zeros outside the unit circle, i.e., all poles and zeros outside the unit circle are flipped inside. The accuracy of the resulting minimum-phase function depends, then, on the presence of zeros and poles outside the unit circle. A similar argument can be used for the counterpart maximum-phase construction (Exercise 6.15). We will see examples of the various phase options on real speech when we describe alternatives for speech synthesis in Section 6.8.

6.7.2 Complex Cepstrum of Unvoiced Speech

Recall the transfer function model from the source to the lips output for unvoiced speech:

$$H(z) = AV(z)R(z)$$

where, in contrast to the voiced case, there is no glottal volume velocity contribution. The resulting speech waveform in the time domain is given by

$$x[n] = u[n] * h[n]$$
$$= u[n] * v[n] * r[n]$$

where $u[n]$ is white noise representing turbulence at the glottal opening or at some other con-striction within the oral cavity. As before, we assume a rational z-transform for the vocal tract contribution which is stable, i.e., poles inside the unit circle, and which may have both minimum- and maximum-phase zeros introduced by coupling of a cavity behind the vocal tract constriction. As before, the radiation lip load is modeled by a single zero.

In short-time analysis, we begin with the windowing of a speech segment

$$s[n] = w[n](u[n] * h[n]).$$

The duration of the analysis window $w[n]$ is selected so that the formants of the unvoiced speech power spectral density are not significantly broadened.[10] In addition, in order to enforce

---

[10] The periodogram of a discrete-time random process is the magnitude squared of its short-time Fourier transform divided by the sequence length and fluctuates about the underlying power spectral density of the random process (Appendix 5.A). Since the window multiples the random process, the window's transform is convolved with, i.e., "smears," its power spectral density. This bias in the periodogram estimate of the power spectral density is reduced by making the window long, but this lengthening can violate waveform stationarity and increases the variance of the periodogram. This can be a particularly difficult choice also because unvoiced speech may consist of very long segments, e.g., voiceless fricatives, or very short segments, e.g., plosives. Furthermore, unvoiced plosives, as we saw in Chapter 3, have an impulse-like component (i.e., the initial burst) followed by an aspiration component.

a convolutional model, we make the assumption that $w[n]$ is "sufficiently smooth" so as to be seen as nearly constant over $h[n]$. Therefore, as with voiced speech, we assume the convolutional model:

$$s[n] = w[n](u[n] * h[n])$$
$$\approx (w[n]u[n]) * h[n].$$

Defining the windowed white noise as $q[n] = u[n]w[n]$, a discrete complex cepstrum is computed with an N-point DFT and expressed as

$$\hat{s}_N[n] \approx \hat{q}_N[n] + \hat{h}_N[n].$$

A similar expression can be found for the real cepstrum. Unlike the voiced case, there is overlap in these two components in the low-quefrency region and so we cannot separate $\hat{h}_N[n]$ by low-quefrency liftering; noise is splattered throughout the entire quefrency axis. Nevertheless, the spectral smoothing interpretation of homomorphic deconvolution gives insight as to why homomorphic deconvolution can still be applicable. As we have seen, we can think of $\log[X(\omega)]$ as a "time signal" which consists of low-frequency and high-frequency contributions. The spectral smoothing interpretation of low-quefrency liftering indicates that we may interpret it as smoothing the fluctuations in $\log[X(\omega)]$ that perturb the underlying system function $H(\omega)$, i.e., fluctuations due to the random source component that excites the vocal tract. Although this interpretation has approximate validity for the spectral log-magnitude, the smoothing viewpoint gives little insight for phase estimation. In practice, sensitivity of the unwrapped phase to small perturbations in spectral nulls prohibits a meaningful computation; for stochastic sequences, the unwrapped phase can jump randomly from sample-to-sample in discrete frequency and large trends can arise that do not reflect the underlying true phase function of $H(\omega)$. Moreover, the phase of the system function for unvoiced speech, excluding perhaps plosives, is not deemed to be perceptually important [3]. For these reasons, the real cepstrum has been typically used in practice for the unvoiced case,[11] resulting in a zero- or minimum-phase counterpart to the mixed-phase system response. Finally, the excitation noise component is typically not extracted by high-quefrency liftering because it is said to be emulated by a white-noise process. Nevertheless, the deconvolved excitation may contain interesting fine source structure, e.g., with voiced fricatives, diplophonic and breathy voices, and nonacoustic-based distributed sources (Chapter 11) that contribute to the quality and distinction of the sound and speaker.

6.8 Analysis/Synthesis Structures

In speech analysis, the underlying parameters of the speech model are estimated, and in synthesis, the waveform is reconstructed from the model parameter estimates. We have seen in short-time homomorphic deconvolution that, by liftering the low-quefrency region of the cepstrum, an

---

[11] A cepstrum has been defined based on high-order moments of a random process [8]. Under certain conditions on the process, the phase of the vocal tract can then be estimated. Alternatively, smoothing of the real and imaginary parts of the complex Fourier transform may provide a useful phase estimate (Exercise 13.4).

estimate of the system impulse response is obtained. If we lifter the complement high-quefrency region of the cepstrum, and invert with the homomorphic system D_*^{-1} to obtain an excitation function, then convolution of the two resulting component estimates yields the original short-time segment exactly. With an overlap-add reconstruction from short-time segments, the entire waveform is recovered; we have, therefore, an identity system with no "information reduction." (In linear prediction analysis/synthesis, this is analogous to reconstructing the waveform from the convolution of the all-pole filter and the output of its inverse filter.) In applications such as speech coding and speech modification, however, a more efficient representation is often desired. The complex or real cepstrum provides an approach to such a representation because pitch and voicing can be estimated from the peak (or lack of a peak) in the high-quefrency region of the cepstrum [7]. Other methods of pitch and voicing estimation will be described in Chapter 10 of the text. Assuming then that we have a succinct and accurate characterization of the speech production source, as with linear prediction-based analysis/synthesis, we are able to "efficiently" synthesize an estimate of the speech waveform. This synthesis can be performed based on any one of several possible phase functions: zero, minimum- (maximum-), or mixed-phase functions. The original homomorphic analysis/synthesis system, developed by Oppenheim [11], was based on zero- and minimum-phase impulse response estimates. In this section, we describe this analysis/synthesis system, its generalization with a mixed-phase representation, and then an extension using spectral root deconvolution.

6.8.1 Zero- and Minimum-Phase Synthesis

The general framework for homomorphic analysis/synthesis is shown in Figure 6.19. In zero- or minimum-phase reconstruction, the analyzer consists of Fourier transforming a short-time speech segment, computing the real logarithm $\log |X(\omega)|$, and inverse transforming to generate the real cepstrum. A 1024-point DFT is sufficient to avoid DFT aliasing and a typical frame interval is 10–20 ms. In the analysis stage, for voiced segments, a pitch-adaptive Hamming window of length of 2 to 3 pitch periods is first applied to the waveform. This choice of the window length is needed to make the windowed periodic waveform approximately follow a convolutional model, as described in Section 6.6. The pitch and voicing estimates are obtained from the cepstrum or by some other means, as indicated by the waveform input to the pitch and voicing estimation module of Figure 6.19. The analysis stage yields a real cepstrum and an estimate of pitch and voicing. The cepstral lifter $l[n]$ selects a zero- or minimum- (maximum-) phase cepstral representation. As discussed in Section 6.6, the cepstral lifter adapts to the pitch period and is of length $\frac{P}{2}$ to avoid aliasing of replicas of the cepstrum of the system impulse response, $\hat{h}[n]$.

In synthesis, the DFT of the liftered cepstral coefficients is followed by the complex exponential operator and inverse DFT. This yields either a zero- or minimum- (maximum-) phase estimate of the system impulse response. We denote this estimate by $h[n, pL]$ for the pth frame and a frame interval of L samples. The excitation to the impulse response is generated on each frame in a manner similar to that in linear prediction synthesis (Figure 5.16). During voicing, the excitation consists of a train of unit impulses with spacing equal to the pitch period. This train is generated over a frame interval beginning at the last pitch pulse from the previous frame. During unvoicing, a white noise sequence is generated over the frame interval. We denote the excitation for the pth frame by $u[n, pL]$. One can then construct for each frame a

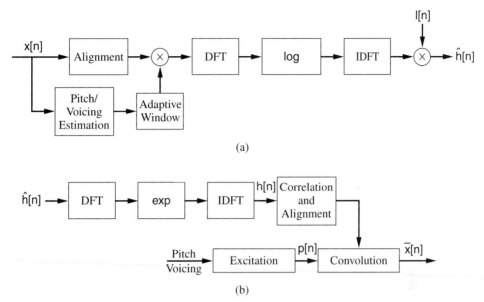

(a)

(b)

Figure 6.19 General framework for homomorphic analysis/synthesis: (a) analysis; (b) synthesis. The alignment and correlation operations are used for mixed-phase reconstruction.

SOURCE: T.F. Quatieri, "Minimum- and Mixed-Phase Speech Analysis/Synthesis by Adaptive Homomorphic Deconvolution" [14]. ©1979, IEEE. Used by permission.

short-time waveform estimate $\bar{x}[n, pL]$ by direct convolution

$$\bar{x}[n, pL] = u[n, pL] * h[n, pL] \tag{6.31}$$

and so the reconstructed waveform is given by

$$\bar{x}[n] = \sum_{p=-\infty}^{\infty} u[n, pL] * h[n, pL].$$

In order to avoid sudden discontinuities in pitch or spectral content, linear interpolation of the pitch period and impulse response is performed, with the convolution being updated at each pitch period. The pitch interpolation is performed by allowing each period to change according to its interpolated value at each pulse location. The impulse responses are constructed by linearly interpolating the response between $h[n, pL]$ and $h[n, (p - 1)L]$ at the interpolated pitch periods. This strategy leads to improved quality, in particular, less "rough," over use of the frame-based convolution of Equation (6.31) [11],[14]. In comparing phase selections in synthesis, the minimum-phase construction was found in informal listening tests to be of higher quality than its zero-phase counterpart, the maximum-phase version being least perceptually desirable; each has its own perceptual identity. The minimum-phase construction was considered most "natural"; the zero-phase rendition was considered most "muffled," while the maximum-phase was most "rough" [11],[14].

We noted above that a Hamming window of the length of 2 to 3 pitch periods was applied to satisfy our constraint of Section 6.6 for a convolutional model. This window length was also found empirically to give good time-frequency resolution. It was observed that with a window much longer than 2 to 3 pitch periods, time resolution was compromised, resulting in a "slurred" synthesis because nonstationarities are smeared, blurring and mixing speech events. On the other hand, with an analysis window much shorter than 2 to 3 pitch periods, frequency resolution is compromised, resulting in a more "muffled" synthesis because speech formants are smeared, abnormally widening formant bandwidths (Exercise 6.6).

6.8.2 Mixed-Phase Synthesis

The analysis/synthesis framework based on mixed-phase deconvolution, using the complex cepstrum, is also encompassed in Figure 6.19. The primary difference from the zero- and minimum-phase systems is the pre- and post-alignment. As shown in Section 6.6, to make the windowed segment meet the conditions for a convolutional model, the window should be aligned with the system impulse response. This alignment is defined as centering the window at the time origin and removing any linear phase term z^{-r} (in the z-domain) in the transfer function from the source to the lips output. The alignment removes any linear phase that manifests itself as a cepstral sinc-like function and can swamp the desired complex cepstrum (Example 6.5), corresponding to reducing distortion from $D[n]$ in the exact cepstral representation of Equation (6.22).

An approximate, heuristic alignment can be performed by finding the maximum value of the short-time segment and locating the first zero crossing prior to this maximum value. Alternative, more rigorous, methods of alignment will be described in Chapters 9 and 10, and are also given in [14],[15]. Ideally, the alignment removes the linear phase term z^{-r}; in practice, however, a small residual linear phase often remains after alignment. With low-quefrency liftering of the complex cepstrum, this residual linear phase appears as a time shift "jitter," n_p, in the impulse response estimate over successive frames. The presence of jitter results in a small but audible change in pitch which can give a "hoarseness" to the reconstruction, as discussed in Chapter 3. If we assume the estimated transfer function is represented by a finite number of poles and zeros, we have on the pth frame

$$H(z, pL) = z^{-n_p} \frac{\prod_{k=1}^{M_i}(1 - \hat{a}_k z^{-1}) \prod_{k=1}^{M_o}(1 - \hat{b}_k z)}{\prod_{k=1}^{N_i}(1 - \hat{c}_k z^{-1})} \tag{6.32}$$

where z^{-n_p} represents the time jitter after alignment. One approach to remove n_p uses the unwrapped phase at π; in practice, however, this method is unreliable (Exercise 6.16). A more reliable approach is motivated by observing that only *relative* delays between successive impulse response estimates need be removed. The relative delay can be eliminated with a method of post-alignment in synthesis that invokes the cross-correlation function of two successive impulse response estimates [14],[15]. Given that system functions are slowly varying over successive analysis frames except for the random time shifts n_p, we can express two consecutive impulse response estimates as

$$h_p[n] = h[n] * \delta[n - n_p]$$
$$h_{p+1}[n] = h[n] * \delta[n - n_{p+1}].$$

Their cross-correlation function, therefore, is given approximately by

$$
\begin{aligned}
r[n] &= h_p[n] * h_{p+1}[n] \\
&\approx (h[n] * \delta[n - n_p]) * (h[n] * \delta[n - n_{p+1}]) \\
&= r_h[n] * \delta[n - (n_p - n_{p+1})],
\end{aligned}
$$

where $r_h[n]$ is the autocorrelation function of $h[n]$. Therefore, the location of the peak in $r[n]$ is taken as an estimate of the relative delay $n_p - n_{p+1}$. We then perform post-alignment by shifting the $(p+1)$st impulse response estimate by $n_p - n_{p+1}$ points. When post-alignment is not performed, a "hoarse" quality is introduced in the reconstruction due to change in the pitch period by the difference $n_p - n_{p+1}$ over successive frames [15].

EXAMPLE 6.13   Figure 6.20 shows a comparison of zero-, minimum-, and mixed-phase syntheses for a voiced speech segment from a male speaker. (This is the same segment used in Figure 5.18 to illustrate the minimum-phase reconstruction from linear prediction analysis/synthesis.) The three systems are identical except for the manner of introducing phase and the elimination of post-alignment in the minimum- and zero-phase versions. Specifications of the analysis/synthesis parameters are given in the previous Section 6.8.1. Observe that the minimum-phase version is different from the minimum-phase reconstruction from linear prediction analysis because the homomorphic estimate does not impose an all-pole transfer function.   ▲

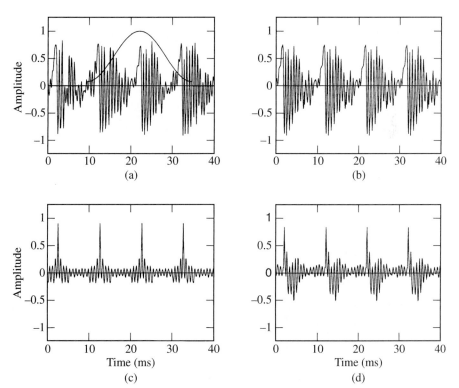

Figure 6.20 Homomorphic synthesis based on (b) mixed-phase, (c) zero-phase, and (d) minimum-phase analysis and synthesis. Panel (a) is the original.

For a database of five males and five females (3–4 seconds in duration), in informal listening (by ten experienced listeners), when compared with its minimum-phase counterpart, the mixed-phase system produces a small but audible improvement in quality [14],[15]. When preferred, the mixed-phase system was judged by the listeners to reduce "buzziness" of the minimum-phase reconstruction. Minimum-phase reconstructions are always more "peaky" or less dispersive than their mixed-phase counterparts because the minimum-phase sequences have energy that is maximally compressed near the time origin [13], as reviewed in Chapter 2 and discussed in the context of linear prediction analysis in Chapter 5. This peakiness may explain the apparent buzziness one hears in minimum-phase reconstructions when compared to their mixed-phase counterparts. One also hears a "muffled" quality in the minimum-phase system relative to the mixed-phase system, due perhaps to the removal of accurate timing and fine structure of speech events by replacing the original phase by its minimum-phase counterpart. These undesirable characteristics of the minimum-phase construction are further accentuated in the zero-phase construction. It is also interesting that differences in the perceived quality of zero-, minimum-, and mixed-phase reconstructions were found to be more pronounced in males than in females [15]. This finding is consistent with the human auditory system's having less phase sensitivity to high-pitched harmonic waveforms, which is explained by the *critical-band theory* of auditory perception (described in Chapters 8 and 13) and which is supported by psychoacoustic experiments. The importance of phase will be further described in the context of other speech analysis/synthesis systems and auditory models throughout the text.

6.8.3 Spectral Root Deconvolution

We saw earlier that a sequence estimated from spectral root deconvolution approaches the sequence estimated from the complex cepstrum as the spectral root γ approaches zero (Exercise 6.9). Thus, under this condition, the two deconvolution methods perform similarly. On the other hand, for a periodic waveform, by selecting γ to give the greatest energy concentration possible for $\check{h}[n]$, we might expect better performance from spectral root deconvolution in speech analysis/synthesis. In particular, we expect that, because voiced speech is often dominated by poles, we can do better with the spectral root cepstrum than the complex cepstrum by selecting a fractional power close to -1. In fact, we saw earlier for an all-pole $h[n]$ that perfect recovery can be obtained when the number of poles is less than the pitch period (Example 6.8); we also saw that, in general, spectral peaks (poles) are better preserved with smaller γ and spectral nulls (zeros) are better preserved with larger γ (Example 6.9). For real speech, similar results have been obtained [5]. As with cepstral deconvolution, in spectral root deconvolution, we choose the analysis window according to the constraints of Section 6.6 so that a windowed periodic waveform approximately follows the desired convolutional model.

A minimum-phase spectral root homomorphic analysis/synthesis system, with a $\gamma = -1/3$, was observed to preserve spectral formant peaks better than does the counterpart real-cepstral analysis/synthesis [5]. For a few American English utterances, this spectral root system was judged in informal listening to give higher quality, in the sense of a more natural sounding synthesis, than its real-cepstrum counterpart using the same window and pitch and voicing. For γ close to zero, the quality of the two systems is similar, as predicted.

6.9 Contrasting Linear Prediction and Homomorphic Filtering

6.9.1 Properties

In the introduction to this chapter we stated that homomorphic filtering is an alternative to linear prediction for deconvolving the speech system impulse response and excitation input. It is of interest, therefore, to compare the properties of these two deconvolution techniques. Linear prediction, being parametric, tends to give sharp smooth resonances corresponding to an all-pole model, while homomorphic filtering, being nonparametric, gives wider spurious resonances consistent with the spectral smoothing interpretation of cepstral liftering. Linear prediction gives an all-pole representation, although a zero can be represented by a large number of poles, while homomorphic filtering can represent both poles and zeros. Linear prediction is constrained to a minimum-phase response estimate, while homomorphic filtering can give a mixed-phase estimate by using the complex cepstrum. Speech analysis/synthesis by linear prediction is sometimes described as "crisper" but more "mechanical" than that by homomorphic filtering, which is sometimes perceived as giving a more "natural" but "muffled" waveform construction.

Nevertheless, many problems encountered by the two methods are similar. For example, both methods suffer more speech distortion with increasing pitch. Aliasing of the vocal tract impulse response at the pitch period repetition rate can occur in the cepstrum, as well as in the autocorrelation function, although aliasing can be better controlled in the complex cepstrum by appropriate time-domain windowing and quefrency liftering. In both cases, time-domain windowing can alter the assumed speech model. In the autocorrelation method of linear prediction, windowing results in the prediction of nonzero values of the waveform from zeros outside the window. In homomorphic filtering, windowing a periodic waveform distorts the convolutional model. Finally, there is the question of model order; in linear prediction, the number of poles is required, while in homomorphic filtering, the length of the low-quefrency lifter must be chosen. The best window and order selection in both methods is often a function of the pitch of the speaker. In the next section, we see that some of these problems can be alleviated by merging the two deconvolution methods.

6.9.2 Homomorphic Prediction

There are a number of speech analysis methods that rely on combining homomorphic filtering with linear prediction and are referred to collectively as *homomorphic prediction*. There are two primary advantages of merging these two analysis techniques: First, by reducing the effects of waveform periodicity, an all-pole estimate suffers less from the effect of high-pitch aliasing; Second, by removing ambiguity in waveform alignment, zero estimation can be performed without the requirement of pitch-synchronous analysis.

Waveform Periodicity — Consider the autocorrelation method of linear prediction analysis. Recall from Chapter 5 that the autocorrelation function of a waveform consisting of the convolution of a short-time impulse train and an impulse response, i.e., $x[n] = p[n] * h[n]$, equals the convolution of the autocorrelation function of the response and that of the impulse train

$$r_x[\tau] = r_h[\tau] * r_p[\tau].$$

Thus, as the spacing between impulses (the pitch period) decreases, the autocorrelation function of the impulse responses suffers from increasing distortion. If one can extract an estimate of the spectral magnitude of $h[n]$ then linear prediction analysis can be performed with an estimate of $r_h[\tau]$ free of the waveform periodicity. One approach is to first homomorphically deconvolve an estimate of $h[n]$ by lowpass liftering the real or complex cepstrum of $x[n]$. The autocorrelation function of the resulting impulse response estimate can then be used by linear prediction analysis. The following example demonstrates the concept:

EXAMPLE 6.14   Suppose $h[n]$ is a minimum-phase all-pole sequence of order p. Consider a waveform $x[n]$ constructed by convolving $h[n]$ with a sequence $p[n]$ where

$$p[n] = \delta[n] + \beta\delta[n - N], \quad \text{with } \beta < 1.$$

The complex cepstrum of $x[n]$ is given by

$$\hat{x}[n] = \hat{p}[n] + \hat{h}[n]$$

where $\hat{p}[n]$ and $\hat{h}[n]$ are the complex cepstra of $p[n]$ and $h[n]$, respectively, and which is of the form in Figure 6.9. The autocorrelation function, on the other hand, is given by

$$r_x[\tau] = (1 + \beta^2)r_h[\tau] + \beta r_h[\tau - N] + \beta r_h[\tau + N]$$

so that $r_h[\tau]$ is distorted by its neighboring terms centered at $\tau = +N$ and $\tau = -N$.   ▲

The example illustrates an important point: The first p coefficients of the real cepstrum of $x[n]$ are undistorted (with a long-enough DFT length used in the computation), whereas the first p coefficients of the autocorrelation function of the waveform $r_x[\tau]$ are distorted by aliasing of autocorrelation replicas (regardless of the DFT length used in the computation). Therefore, a cepstral lowpass lifter of duration less than p extracts a smoothed version of the spectrum, but not aliased. Moreover, the linear prediction coefficients can alternatively be obtained *exactly* through the recursive relation between the real cepstrum and the predictor coefficients of the all-pole model when $h[n]$ is all-pole (Exercise 6.13) [1].

Nevertheless, it is important to note that we have not considered a windowed periodic waveform. That is, as seen in Section 6.6, the cepstrum of a windowed periodic waveform does indeed experience aliasing distortion, as does the autocorrelation function; this distortion, however, is minimized by appropriate selection of the function $D[n]$.

Zero Estimation — Consider a transfer function of poles and zeros of the form

$$H(z) = \frac{N(z)}{D(z)}$$

and a sequence $x[n] = h[n] * p[n]$ where $p[n]$ is a periodic impulse train. Suppose that an estimate of $h[n]$ is obtained through homomorphic filtering of $x[n]$ and assume that the number of poles and zeros is known and that a linear-phase component z^{-r} has been removed. Then, following Kopec [4], we can estimate the poles of $h[n]$ by using the covariance method of linear prediction (Chapter 5) with a prediction-error interval that is free of zeros. The Shanks, or other methods described in Chapter 5, can then be applied to estimate the zeros. The following example illustrates the approach on a real speech segment:

EXAMPLE 6.15   In this example, a rational z-transform consisting of 10 poles and 6 zeros is used to model a segment of the nasalized vowel /u/ in the word "m<u>oo</u>n" [4]. Homomorphic prediction was performed by first applying homomorphic filtering on the complex cepstrum to obtain an impulse response estimate. The log-magnitude spectrum of this estimate is shown in Figure 6.21b. The covariance method of linear prediction analysis was then invoked to estimate the poles, and then the Shanks method was used to estimate the zeros. The method estimated a zero near 2700 Hz, which is typical for this class of nasalized vowels (Figure 6.21c).   ▲

Other zero estimation methods can also be combined with homomorphic filtering, such as the Steiglitz method (Exercise 5.6). In addition, forms of homomorphic prediction can be applied to deconvolve the vocal tract glottal source from the vocal tract transfer function. Moreover, this synergism of homomorphic filtering and linear prediction analysis allows, under certain conditions, this source/system separation when both zeros and poles are present in the vocal tract system function (Exercise 6.22).

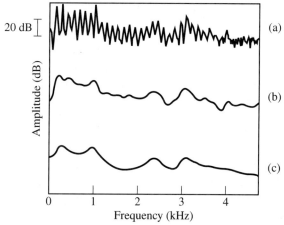

Figure 6.21 Homomorphic prediction applied to a nasalized vowel from /u/ in "m<u>oo</u>n": (a) log-magnitude spectrum of speech signal; (b) log-magnitude spectrum obtained by homomorphic filtering (low-time liftering the real cepstrum); (c) log-magnitude spectrum of 10-pole/6-zero model with zeros from Shanks method.

SOURCE: G.E. Kopec, A.V. Oppenheim, and J.M. Tribolet, "Speech Analysis by Homomorphic Prediction" [4]. ©1977, IEEE. Used by permission.

6.10 Summary

In this chapter, we introduced homomorphic filtering with application to deconvolution of the speech production source and system components. The first half of the chapter was devoted to the theory of homomorphic systems and the analysis of homomorphic filtering for convolution. Both logarithmic and spectral root homomorphic systems were studied. The complex cepstrum was derived for an idealized voiced speech waveform as a short-time convolutional model consisting of a finite-length impulse train convolved with an impulse response having a rational z-transform. The complex cepstrum of a windowed periodic waveform was also derived and a set of conditions was established for the accuracy of the short-time convolutional model. The second half of the chapter applied the theory of homomorphic systems to real speech and developed a number of speech analysis/synthesis schemes based on zero-, minimum-, and mixed-phase estimates of the speech system response. Finally, the properties of linear prediction analysis were compared with those of homomorphic filtering for deconvolving speech waveforms, and the two methods were merged in homomorphic prediction.

Although this chapter focuses on homomorphic systems for convolution for the purpose of source/system separation, the approach is more general. For example, homomorphic systems have been designed to deconvolve multiplicatively combined signals to control dynamic range (Exercise 6.19). In dynamic range compression, a signal, e.g., speech or audio, is modeled as having an "envelope" that multiplies the signal "fine structure"; the envelope represents the time-varying volume and the fine structure represents the underlying signal. The goal is to separate the envelope and reduce its wide fluctuations. A second application not covered in this chapter is recovery of speech from degraded recordings. For example, old acoustic recordings suffer from convolutional distortion imparted by an acoustic horn that can be approximated by a linear resonant filter. The goal is to separate the speech or singing from its convolution with the distorting linear system. This problem is similar to the source/system separation problem, except that here we are not seeking the components of the speech waveform, but rather the speech itself. This problem can be even more difficult since the cepstra of the horn and speech overlap in quefrency. A solution of this problem, developed in Exercise 6.20, represents a fascinating application of homomorphic theory.

Other applications of homomorphic filtering will be seen throughout the remainder of the text. For example, the style of analysis in representing the speech system phase will be useful in developing different phase representations in the phase vocoder and sinewave analysis/synthesis. Other extensions of homomorphic theory will be introduced as they are needed. For example, in the application of speech coding (Chapter 12) and speaker recognition (Chapter 14) where the cepstrum plays a major role, we will find that speech passed over a telephone channel effectively loses spectral content outside of the band 200 Hz to 3400 Hz, so that the cepstrum is not defined outside of this range. For this condition, we will, in effect, design a new homomorphic system that in discrete time requires a mapping of the high spectral energy frequency range to the interval $[0, \pi]$. This modified homomorphic system was first introduced in the context of seismic signal processing where a similar spectral distortion arises [20]. Also in the application of speaker identification, we will expand our definition of the cepstrum to represent the Fourier transform of a constant-Q filtered log-spectrum, referred to as the *mel-cepstrum*. The mel-cepstrum is interesting for its own sake because it is hypothesized to approximate signal processing in the early stages of human auditory perception [22]. We will see in Chapter 13 that homomorphic filtering, applied along the temporal trajectories of the mel-cepstral coefficients, can be used to remove convolutional channel distortions even when the cepstrum of these distortions overlaps the cepstrum of speech. Two such methods that we will study are referred to as *cepstral mean subtraction* and *RASTA* processing.

EXERCISES

6.1 Consider the phase of the product $X(z) = X_1(z)X_2(z)$. An ambiguity in the definition of phase arises since $\angle X(z) = \text{PV}[\angle X(z)] + 2\pi k$, where k is any integer value (PV = principle value).

    (a) Argue that when k is selected to ensure phase continuity, then the phase of the product $X_1(z)X_2(z)$ is the sum of the phases of the product components.

    (b) Argue that the phase additivity property in part (a) can also be ensured when the phase is defined as the integral of the phase derivative of $X(z)$ (see Exercise 6.2).

6.2 Show that the phase derivative $\dot{\angle}X(\omega)$ of the Fourier transform $X(\omega)$ of a sequence $x[n]$ can be obtained through the real and imaginary parts of $X(\omega)$, $X_r(\omega)$ and $X_i(\omega)$, respectively, as

$$\dot{\angle}X(\omega) = \frac{X_r(\omega)\dot{X}_i(\omega) - X_i(\omega)\dot{X}_r(\omega)}{|X(\omega)|^2}$$

where $|X(\omega)|$, the Fourier transform magnitude of $x[n]$, is assumed non-zero.

6.3 This problem addresses the nature of the phase function for poles and zeros of a rational z-transform in Equation (6.9).

    (a) Show that the phase on the unit circle of the terms $(1 - az^{-1})$, $(1 - bz)$, $(1 - cz^{-1})$, and $(1 - dz)$ all begin at zero (at $\omega = 0$) and end at zero (at $\omega = \pi$).

    (b) Show that the absolute value of the phase in (a) is less than π.

    Hint: Use a vector argument.

6.4 Show that the complex cepstrum of a minimum- or maximum-phase sequence can be recovered within a scale factor from the phase of its Fourier transform. *Hint:* Recall that the phase of $X(\omega)$ is the imaginary part of $\log[X(\omega)]$ that maps to the odd part of the complex cepstrum.

6.5 Suppose we are given samples of the phase derivative of the Fourier transform of a sequence. Consider a linear numerical integration scheme for computing the unwrapped phase in Equation (6.16) based on these samples. Argue why this scheme may be flawed when poles or zeros are located very close to the unit circle. How might this problem affect the phase derivative-based phase unwrapping algorithm of Section 6.4?

6.6 It has been observed that nasal formants have broader bandwidths than non-nasal voiced sounds. This is attributed to the greater viscous friction and thermal loss due to the large surface area of the nasal cavity. In linear prediction or homomorphic analysis of a short-time sequence $s[n]$, the resulting vocal tract spectral estimates often have wider bandwidths than the true underlying spectrum. Consequently, when the spectral estimates are used in synthesizing speech, the resulting synthesized speech is characterized by a "nasalized" or "muffled" quality.

(a) Using the frequency-domain linear prediction error expression given in Equation (5.38) of Chapter 5, give a qualitative explanation of bandwidth broadening by linear prediction analysis. *Hint:* Consider the error contribution in low-energy regions of $|S(\omega)|^2$.

(b) We have seen that low-quefrency liftering in homomorphic analysis is equivalent to filtering the complex logarithm. Suppose that $l[n]$ is the lifter with N-point discrete Fourier transform $L(k)$, and that $\log[S(k)]$ is the logarithm of the N-point discrete Fourier transform of the speech sequence $s[n]$. Write an expression for the modified logarithm (in the discrete frequency variable k) after application of the lifter $l[n]$. How might this expression explain formant broadening by homomorphic analysis?

(c) Referring to parts (a) and (b), explain qualitatively why formant broadening can be more severe for female (and children) than for male speakers. A distinct answer should be given for each technique (i.e., linear prediction and homomorphic analysis).

(d) Suppose the vocal tract impulse response $h[n]$ is all-pole and stable with system function

$$H(z) = \frac{A}{1 - \sum_{k=1}^{P} a_k z^{-k}}$$

that can be rewritten as

$$H(z) = \frac{A}{\prod_{k=1}^{P}(1 - b_k z^{-1})}$$

with poles $z = b_k$ that are located inside the unit circle. Formant bandwidth is related to the distance of a pole from the unit circle. As the pole approaches the unit circle, the bandwidth narrows. To compensate for formant bandwidth widening, in both linear prediction and homomorphic analysis, the poles are moved closer to the unit circle. Show that poles are moved closer to the unit circle by the transformation

$$\tilde{H}(z) = H(\frac{z}{\alpha}) = \frac{A}{\prod_{k=1}^{P}[1 - b_k(\frac{z}{\alpha})^{-1}]} \tag{6.33}$$

where α (real) is greater than unity and $|\alpha b_k| < 1$.

(e) Suppose $h[n]$ in part (d) has a z-transform that consists of a single pole of the form

$$H(z) = \frac{1}{1 - b_o z^{-1}}$$

and that b_o is real. Sketch the real cepstrum associated with $H(z)$ and $\tilde{H}(z)$ from part (d). For a general $H(z)$, how is the complex cepstrum modified by the transformation in Equation (6.33)? Give an expression in terms of the original complex cepstrum.

6.7 In Section 6.5, we introduced the spectral root homomorphic system. In this problem you are asked to develop some of the properties of this system for sequences with rational z-transforms.

(a) Derive the spectral root cepstrum for zeros outside the unit circle, and for poles inside and outside the unit circle.

(b) Show that if $x[n]$ is minimum phase, $\check{x}[n]$ can be obtained from $|X(\omega)|$.

(c) Let $p[n]$ be a train of equally-spaced impulses $p[n] = \sum_{k=-\infty}^{\infty} \delta[n - kN]$. Show that $\check{p}[n]$ remains an impulse train with the same spacing N. *Hint:* Use the relation between the spectral root cepstrum and the complex cepstrum derived in Section 6.5.

6.8 Suppose $h[n]$ is an all-pole sequence of order q. Consider a waveform $x[n]$ created by convolving $h[n]$ with a sequence $p[n]$ ($\beta < 1$):

$$p[n] = \delta[n] + \beta\delta[n - N]$$

so that

$$x[n] = p[n] * h[n]$$

where $q < N$, and where

$$P(z) = 1 + \beta z^{-N}.$$

(a) Argue that $P^{-1}(z)$ represents an impulse train with impulses spaced by N samples. *Hint:* Use the Taylor series expansion for $\frac{1}{(1+\beta z^{-1})}$ and replace z by z^N. Do not give an explicit expression.

(b) Show that $h[n]$ can be deconvolved from $x[n]$ by inverting $X(z)$ [i.e., forming $X^{-1}(z)$] and "liftering" $h^{-1}[n]$, the inverse z-transform of $H^{-1}(z)$. Sketch a block diagram of the deconvolution algorithm as a homomorphic filtering procedure. *Hint:* $H^{-1}(z)$ is all-zero.

(c) Suppose we add noise to $x[n]$ and suppose a resonance of $h[n]$ is close to the unit circle. Argue that the inversion process is very sensitive under this condition and thus the method may not be useful in practice.

6.9 This problem contrasts the spectral root and complex cepstrum.

(a) For $x[n] = p[n] * h[n]$, with $p[n]$ a periodic impulse train, discuss the implications of $p[n]$ not being aligned with the origin for both the complex and spectral root cepstrum.

(b) Using the relation between the spectral root cepstrum and the complex cepstrum, Equation (6.20), show that as γ approaches zero, the spectral root deconvolution system approaches the complex cepstral deconvolution.

(c) Show why it is necessary to perform two phase-unwrapping operations, one for the forward $X(\omega)^\gamma$ and one for the inverse $X(\omega)^{1/\gamma}$ transformations when γ is an integer. How does this change when γ is not an integer?

6.10 In this problem, a phase unwrapping algorithm is derived which does not depend on finding 2π jumps. Consider the problem of computing the unwrapped phase of the Fourier transform of a sequence $x[n]$:

$$X(\omega) = |X(\omega)|e^{j\theta(\omega)}.$$

The corresponding all-pass function is given by

$$X_a(\omega) = \frac{X(\omega)}{|X(\omega)|}$$

$$= \exp[j\theta(\omega)]$$

$$= \sum_{n=-\infty}^{\infty} x_a[n]\cos(\omega n) - j\sum_{n=-\infty}^{\infty} x_a[n]\sin(\omega n) \qquad (6.34)$$

where $x_a[n]$ is the inverse Fourier transform of $X_a(\omega)$.

(a) By determining the phase derivative associated with Equation (6.34), show that

$$\theta(\omega) = \sum_{n=-\infty}^{\infty} \sum_{m=-\infty}^{\infty} m x_a[m] x_a[n] \frac{\sin[\omega(n-m)]}{(n-m)}. \tag{6.35}$$

(b) Why might the "direct" unwrapped-phase algorithm in Equation (6.35) be flawed even when the sequence $x[n]$ is of finite extent? In answering this question, consider how you would compute the all-pass sequence $x_a[n]$.

6.11 This problem further investigates in the quefrency domain the complex cepstrum of windowed periodic waveforms which were described in Section 6.6.1. Assume $W(\omega)$ is rectangular with width equal to the fundamental frequency (pitch) as in Section 6.6.1.

(a) Under what conditions does $s[n] \approx (p[n]w[n]) * h[n]$? When these conditions do not hold, describe the kinds of distortions introduced into the estimate of $h[n]$ using homomorphic deconvolution. Why is the distortion more severe for high-pitched than for low-pitched speakers?

(b) In selecting a time-liftering window for deconvolving the vocal tract impulse response, how is its width influenced by minimum-, mixed-, and zero- phase sequences?

(c) Design an "optimal lifter" that compensates for the distortion function $D[n]$. What are some of the problems you encounter in an optimal design? How does the design change with pitch and with the sequence $h[n]$ being minimum-, mixed-, or zero-phase?

6.12 This problem further investigates in the frequency domain the complex cepstrum of windowed periodic waveforms which was described in Section 6.6.2. We have seen that to use homomorphic deconvolution to separate the components of the speech model, the speech signal, $x[n] = h[n] * p[n]$, is multiplied by a window sequence, $w[n]$, to obtain $s[n] = x[n]w[n]$. To simplify analysis, $s[n]$ is approximated by:

$$s[n] = (h[n] * p[n])w[n]$$
$$\approx h[n] * (p[n]w[n]).$$

Assume in the following that $h[n]$ has no linear phase:

(a) Suppose that the window Fourier transform $W(\omega)$ is a triangle symmetric about $\omega = 0$ so that $w[n]$ is the square of a sinc function and that $W(\omega)$ is zero-phase, i.e., $w[n]$ is centered at the time origin. Suppose that the width of $W(\omega)$ is much greater than the fundamental frequency. Discuss qualitatively the problem of extracting the unwrapped phase of $h[n]$. Repeat for the case where the width of $W(\omega)$ is much less than the fundamental frequency. Explain why a width of exactly twice the fundamental frequency provides an "optimal" interpolation across harmonic samples of $X(\omega)$ for extracting unwrapped phase.

(b) Consider $W(\omega)$ of part (a) but with a linear phase corresponding to a displacement of $w[n]$ from the time origin. Discuss qualitatively the problem of extracting the unwrapped phase of $h[n]$.

(c) Repeat parts (a) and (b) with a Hamming window whose width in the frequency domain is defined as the 3-dB width of its mainlobe.

6.13 This problem explores recursive relations between the complex cepstrum and a minimum-phase sequence.

(a) Derive a recursive relation for obtaining a minimum-phase sequence $h[n]$ directly from its complex cepstrum $\hat{h}[n]$. In determining the value $h[0]$, assume that the z-transform of $h[n]$ is rational with positive gain factor A. Derive a similar recursion for obtaining the complex cepstrum $\hat{h}[n]$ from $h[n]$. *Hint:* Derive a difference equation based on cross-multiplying in the relation $\frac{d}{dz} \log[X(z)] = \frac{1}{X(z)} \frac{d}{dz} X(z)$.

(b) Suppose $h[n]$ is all-pole (minimum-phase) of order p. Show that the all-pole (predictor) coefficients can be obtained recursively from the first p complex cepstral coefficients of $h[n]$ (not including $\hat{h}[0]$).

(c) Argue that, when an all-pole $h[n]$ is convolved with a pulse train $p[n]$, the all-pole (predictor) coefficients can be obtained from the complex cepstrum of $x[n] = h[n] * p[n]$, provided that $p < P$, where P is the pitch period in samples. Explain the "paradox" that truncating the complex cepstrum of $x[n]$ at p coefficients and inverting via the inverse Fourier transform and the complex exponential operator smears the desired spectrum $H(\omega)$, even though the first p coefficients contain all information necessary for recovering the minimum-phase sequence.

6.14 Suppose that the real cepstrum $c[n]$ is used to deconvolve a speech short-time segment $x[n]$ into source and system components. Argue that the resulting zero-phase impulse response is an even function of time. Suppose the odd part of the complex cepstrum $\hat{x}[n]$ is low-quefrency liftered. Argue that an all-pass transfer function is obtained for the system estimate and that the corresponding impulse response is an odd function of time.

6.15 Derive the maximum-phase counterpart to Example 6.12. That is, show by example that the maximum-phase construction of a mixed-phase sequence with a rational z-transform flips poles and zeros inside the unit circle to outside the unit circle. Show that the maximum- and minimum-phase constructions are time reversals of each other.

6.16 It was stated in Section 6.8.2 that one approach to remove the jitter n_p in a speech system impulse response, corresponding to the linear-phase residual in Equation (6.32), is to measure the unwrapped phase at π.

(a) For a rational transfer function of the form in Equation (6.32), propose a method to determine n_p from $H(z, pL)$ evaluated at $z = e^{j\pi}$.

(b) Argue why, in practice, this method is unreliable. *Hint:* Consider lowpass filtering prior to A/D conversion, as well as a mixed-source excitation, e.g., the presence of breathiness during voicing.

6.17 Consider a signal which consists of the convolution of a periodic impulse train $p[n]$, which is minimum-phase, and a mixed-phase signal $h[n] = h_{min}[n] * h_{max}[n]$ with poles configured in Figure 6.22.

(a) Using homomorphic processing on $y[n] = h[n] * p[n]$, describe a method to create a signal

$$y'[n] = h'[n] * p[n]$$

where

$$h'[n] = h_{min}[n] * \tilde{h}_{min}[n]$$

with $\tilde{h}_{min}[n]$ being the minimum-phase counterpart of $h_{max}[n]$ (i.e., poles outside the unit circle are flipped to their conjugate-reciprocal locations inside the unit circle).

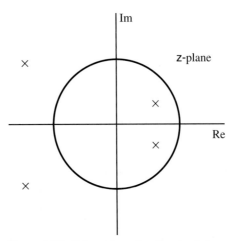

Figure 6.22 Pole configuration for mixed-phase signal of Exercise 6.17.

(b) Suppose that the period of $p[n]$ is long relative to the length of the real cepstrum of $h[n]$. Suppose also that homomorphic prediction (homomorphic filtering followed by linear prediction analysis) with the real cepstrum is performed on the signal $y[n] = h[n] * p[n]$, using a 4-pole model. Sketch (roughly) the pole locations of the (homomorphic prediction) estimate of the minimum-phase counterpart to $h[n]$.

6.18 In Exercise 5.20 of Chapter 5, you considered the analysis of a speech-like sequence consisting of the convolution of a glottal flow (modeled by two poles), a minimum-phase vocal tract impulse response (modeled by p poles), and a source of two impulses spaced by P samples. You are asked here to extend this problem to deconvolving the glottal flow and vocal tract impulse response using the complex and real cepstra. In working on this problem, please refer back to Exercise 5.20.

(a) Determine and sketch the complex cepstrum of $s[n]$ of Exercise 5.20 [part (c)]. Describe a method for estimating the glottal pulse $\tilde{g}[n]$ and the vocal tract response $v[n]$ from the complex cepstrum. Assume that the tails of the complex cepstra of $\tilde{g}[n]$ and $v[n]$ are negligibly small for $|n| > P$.

(b) Now we use the real cepstrum to estimate $\tilde{g}[n]$ and $v[n]$. First determine and sketch the real cepstrum of $s[n]$ of Equation (5.66). Next, suppose that you are given the spectral magnitude only of $\tilde{g}[n]$, i.e., you are given $|\tilde{G}(\omega)|$. Devise a procedure to recover both $\tilde{g}[n]$ and $v[n]$. Again assume that the tails of the real cepstra of $\tilde{g}[n]$ and $v[n]$ are negligibly small for $|n| > P$. Assume you know *a priori* that $\tilde{g}[n]$ is maximum-phase.

6.19 Consider a speech or audio signal $x[n] = e[n]f[n]$ with "envelope" $e[n]$ and "fine structure" $f[n]$. The envelope (assumed positive) represents a slowly time-varying volume fluctuation and the fine structure represents the underlying speech events (Figure 6.23).

(a) Design a homomorphic system for multiplicatively combined signals that maps the time-domain envelope and fine structure components of $x[n]$ to additively combined signals. In your design, consider the presence of zero crossings in $f[n]$, but assume that $f[n]$ never equals zero exactly. *Hint:* Use the magnitude of $x[n]$ and save the sign information.

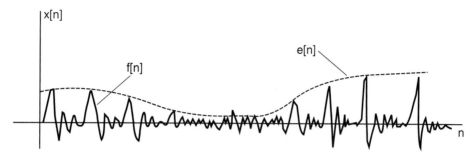

Figure 6.23 Acoustic signal with time-varying envelope.

(b) Suppose that $e[n]$ has a wide dynamic range, i.e., the values of $e[n]$ have wide fluctuations over time. Assume, however, that the fluctuations of $e[n]$ occur slowly in time, representing slowly-varying volume changes. Design a homomorphic filtering scheme, based on part (a), that separates the envelope from $x[n]$. Assume that the spectral content of $f[n]$ is high-frequency, while that of $e[n]$ is low-frequency and the two fall in roughly disjoint frequency bands. In addition, assume that the logarithm of these sequences (with non-overlapping spectra) yields sequences that are still spectrally non-overlapping. Design an inverse homomorphic system that reduces dynamic range in $x[n]$. Sketch a flow diagram of your system.

6.20 It has been of widespread interest to restore old acoustic recordings, e.g., Enrico Caruso recordings, that were made up to the 1920's with recording horns of the form shown in Figure 6.24. The recording horns distorted the source by introducing undesirable resonances. This problem walks you through the steps in using homomorphic filtering for restoration.

Engineers in the 1920's were excellent craftsmen and managed to avoid nonlinearities in the system of Figure 6.24. The signal $v(t)$, representing the grooves in the record, can therefore be approximately expressed as a convolution of the operatic singing voice $s(t)$ with a (resonant) linear

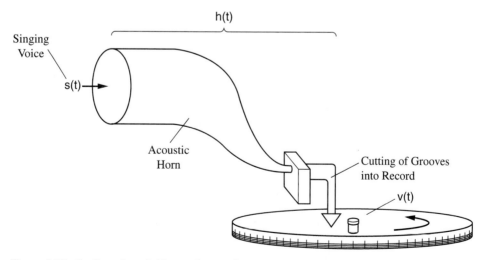

Figure 6.24 Configuration of old acoustic recordings.

acoustic horn $h(t)$:

$$v(t) = s(t) * h(t). \tag{6.36}$$

Our goal is to recover $s(t)$ without knowing $h(t)$, a problem that is sometimes referred to as "blind deconvolution." The approach in discrete time is to find an estimate of the horn $\hat{h}[n]$, invert its frequency response to form the inverse filter $\hat{H}^{-1}(\omega)$, and then apply inverse filtering to obtain an estimate of $s[n]$. (The notation "hat" denotes an estimate, and not a cepstrum as in this chapter.) We will accomplish this through a homomorphic filtering scheme proposed by Stockhom [18] and shown in Figure 6.25.

(a) Our first step is to window $v[n]$ with a sliding window $w[n]$ so that each windowed segment is given by

$$v_i[n] = v[n + iL]w[n]$$

where we slide $v[n]$ L samples at a time under the window $w[n]$. If $w[n]$ is "long and smooth relative to $s[n]$," argue for and against the following convolutional approximation:

$$
\begin{aligned}
v_i[n] &= w[n](s[n + iL] * h[n]) \\
&\approx (w[n]s[n + iL]) * h[n] \\
&= s_i[n] * h[n]
\end{aligned} \tag{6.37}
$$

where $s_i[n] = w[n]s[n + iL]$ and where we have ignored any shift in $h[n]$.

(b) Determine the complex cepstrum of $v_i[n]$ (from the approximation in Equation (6.37)) and argue why we cannot separate out $s[n]$ using the homomorphic deconvolution method that requires liftering the low-quefrency region of the complex cepstrum.

(c) Suppose that we average the complex logarithm (Fourier transform of complex cepstrum) over many segments, i.e.,

$$V_{avg} = \frac{1}{M} \sum_{i=1}^{M} \log[V_i(\omega)].$$

Give an expression for V_{avg} in terms of $S_{avg} = \frac{1}{M} \sum_{i=1}^{M} \log[S_i(\omega)]$ and $\log[H(\omega)]$. Suppose that $S_{avg} = $ constant. Describe a procedure for extracting $H(\omega)$ to within a gain factor.

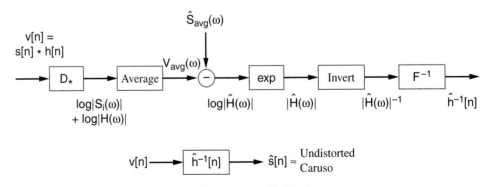

Figure 6.25 Restoration of Caruso based on homomorphic filtering.

(d) In reality, the operatic singing voice does not average to a constant, but to a spectrum with some resonant structure. Suppose we had formed an average $\hat{S}_{avg}$ from a contemporary opera singer with "similar" voice characteristics as Enrico Caruso. Describe a method for extracting $H(\omega)$ to within a gain factor (assuming no horn distortion on the contemporary opera singer). Devise a homomorphic filtering scheme for recovering Enrico Caruso's voice.

(e) Write out your expression V_{avg} in part (c) in terms of the spectral log-magnitude and phase. Explain why in practice it is very difficult to estimate the phase component of V_{avg}. Consider spectral harmonic nulls, as well as the possibility that the frequency response of the singer and the acoustic horn may have a large dynamic range with deep spectral nulls, and that the spectrum of the horn may have little high-frequency energy. Derive the zero-phase and minimum-phase counterparts to your solutions in parts (c) and (d). Both of these methods avoid the phase estimation problem. After inverse filtering (via homomorphic filtering), you will be left with an estimate of the singer's frequency response in terms of the actual singer's frequency response and the phase of the acoustic horn filter (that you did not estimate). If the horn's frequency response is given by

$$H(\omega) = |H(\omega)|e^{j\theta_H(\omega)},$$

and assuming you can estimate $|H(\omega)|$ exactly, then write the singer's frequency response estimate $\hat{S}(\omega)$ in terms of the actual singer's spectrum $S(\omega)$ and the phase of the horn $\theta_H(\omega)$. Do this for both your zero-phase and minimum-phase solutions. Assume that the horn's phase $\theta_H(\omega)$ has both a minimum- and maximum-phase component:

$$\theta_H(\omega) = \theta_H^{min}(\omega) + \theta_H^{max}(\omega)$$

What, if any, may be the perceptual significance of any phase distortion in your solution?

(f) Denote an estimate of the horn's inverse filter by $g[n]$. Using the result from Exercise 2.18, propose an iterative scheme for implementing $g[n]$ and thus for restoring Enrico Caruso's voice. Why might this be useful when the horn filter has little high-frequency energy?

6.21 (MATLAB) In this problem, you use the speech waveform *speech1_10k* (at 10000 samples/s) in the workspace *ex6M1.mat* located in the companion website directory Chap_exercises/chapter6. The exercise works through a problem in homomorphic deconvolution, leading to the method of homomorphic prediction.

(a) Window *speech1_10k* with a 25-ms Hamming window. Using a 1024-point FFT, compute the real cepstrum of the windowed signal and plot. For a clear view of the real cepstrum, set the first cepstral value to zero (which is the DC component of the log-spectral magnitude) and plot only the first 256 cepstral values.

(b) Estimate the pitch period (in samples and in milliseconds) from the real cepstrum by locating a distinct peak in the quefrency region.

(c) Extract the first 50 low-quefrency real cepstral values using a lifter of the form

$$\begin{aligned} l[n] &= 1, \quad n = 0 \\ &= 2, \quad 1 \leq n \leq 49 \\ &= 0, \quad \text{otherwise.} \end{aligned}$$

Then Fourier transform (using a 1024-point FFT) and plot the first 512 samples of the resulting log-magnitude and phase.

(d) Compute and plot the minimum-phase impulse response using your result from part (c). Plot just the first 200 samples to obtain a clear view. Does the impulse response resemble one period of the original waveform? If not, then why not?

(e) Use your estimate of the pitch period (in samples) from part (b) to form a periodic unit sample train, thus simulating an ideal glottal pulse train. Make the length of the pulse train 4 pitch periods. Convolve this pulse train with your (200-sample) impulse response estimate from part (d) and plot. You have now synthesized a minimum-phase counterpart to the (possibly mixed-phase) vowel *speech1_10k*. What are the differences between your construction and the original waveform?

(f) Compute and plot the autocorrelation function of the impulse response estimate from part (d). Then, using this autocorrelation function, repeat parts (b) through (d) of Exercise 5.23 of Chapter 5 to obtain an all-pole representation of the impulse response. Does the log-magnitude of the all-pole frequency response differ much from the one obtained in Exercise 5.23? You have now performed "homomorphic prediction" (i.e., homomorphic deconvolution followed by linear prediction), which attempts to reduce the effect of waveform periodicity in linear prediction analysis.

6.22 (MATLAB) In Exercise 5.25 you considered separating glottal source and vocal tract components of a nasalized vowel where the nasal tract introduces zeros into the vocal tract system function. In this problem, you are asked to extend this problem to deconvolving the glottal flow and vocal tract impulse response using the complex cepstrum. In working this problem, please refer back to Exercise 5.25.

(a) We assume that the zeros of the glottal pulse are maximum phase and the zeros due to the vocal tract nasal branch are minimum phase. Using your result from parts (a) and (b) of Exercise 5.25, propose a method based on homomorphic deconvolution for separating the minimum-phase vocal tract zeros, the maximum-phase glottal pulse zeros, and the minimum-phase vocal tract poles.

(b) Implement part (a) in MATLAB using the synthetic speech waveform *speech1_10k* (at 10000 samples/s) in workspace *ex6M2.mat* located in the companion website directory Chap_exercise/ chapter6. Assume 4 vocal tract poles, 2 vocal tract zeros, a glottal pulse width of 20 samples, and a pitch period of 200 samples. You should compute the predictor coefficients associated with the vocal tract poles using results of part (d) of Exercise 5.25 and inverse filter the speech waveform. Then find the coefficients of the vocal tract numerator polynomial and the glottal pulse using homomorphic filtering. You will find in *ex6M2.readme*, located in Chap_exercises/chapter6, a description of how the synthetic waveform was produced and some suggested MATLAB functions of possible use.

BIBLIOGRAPHY

[1] B.S. Atal, "Effectiveness of Linear Prediction Characteristics of the Speech Wave for Automatic Speaker Identification and Verification," *J. Acoustical Society of America*, vol. 55, no. 6, pp. 1304–1312, June 1974.

[2] B. Bogert, M. Healy, and J. Tukey, "The Quefrency Analysis of Time Series for Echoes," chapter in *Proc. Symposium on Time Series Analysis*, M. Rosenblatt, ed., pp. 209–243, John Wiley and Sons, New York, 1963.

[3] G. Kubin, B.S. Atal, and W.B. Kleijn, "Performance of Noise Excitation for Unvoiced Speech," *Proc. IEEE Workshop on Speech Coding for Telecommunications*, Sainte-Adele, Quebec, pp. 35–36, 1993.

[4] G.E. Kopec, A.V. Oppenheim, and J.M. Tribolet, "Speech Analysis by Homomorphic Prediction," *IEEE Trans. Acoustics, Speech, and Signal Processing*, vol. ASSP–25, no. 1, pp. 40–49, Feb. 1977.

[5] J.S. Lim, "Spectral Root Homomorphic Deconvolution System," *IEEE Trans. Acoustics, Speech, and Signal Processing*, vol. ASSP–27, no. 3, pp. 223–232, June 1979.

[6] R. McGowan and R. Kuc, "A Direct Relation Between a Signal Time Series and its Unwrapped Phase," *IEEE Trans. Acoustics, Speech, and Signal Processing*, vol. ASSP–30, no. 5, pp. 719–724, Oct. 1982.

[7] P. Noll, "Cepstrum Pitch Determination," *J. Acoustical Society of America*, vol. 41, pp. 293–309, Feb. 1967.

[8] C.L. Nikias and M.R. Raghuveer, "Bispectrum Estimation: A Digital Signal Processing Framework," *Proc. IEEE*, vol. 75, no. 1, pp. 869–891, July 1987.

[9] A.V. Oppenheim, "Superposition in a Class of Nonlinear Systems," Ph.D. Thesis, Massachusetts Institute of Technology, Dept. of Electrical Engineering and Computer Science, Cambridge, MA, 1965.

[10] A.V. Oppenheim, "Generalized Superposition," *Information Control*, vol. 11, nos. 5–6, pp. 528–536, Nov.–Dec. 1967.

[11] A.V. Oppenheim, "Speech Analysis/Synthesis Based on Homomorphic Filtering," *J. Acoustical Society of America*, vol. 45, pp. 458–465, Feb. 1969.

[12] A.V. Oppenheim and R.W. Schafer, "Homomorphic Analysis of Speech," *IEEE Trans. Audio and Electroacoustics*, vol. AU–16, pp. 221–228, June 1968.

[13] A.V. Oppenheim and R.W. Schafer, *Discrete-Time Signal Processing*, Prentice Hall, Englewood Cliffs, NJ, 1989.

[14] T.F. Quatieri, "Minimum- and Mixed-Phase Speech Analysis/Synthesis by Adaptive Homomorphic Deconvolution," *IEEE Trans. Acoustics, Speech, and Signal Processing*, vol. ASSP–27, no. 4, pp. 328–335, Aug. 1979.

[15] T.F. Quatieri, "Phase Estimation with Application to Speech Analysis/Synthesis," Sc.D. Thesis, Massachusetts Institute of Technology, Dept. of Electrical Engineering and Computer Science, Cambridge, MA, Nov. 1979.

[16] R.W. Schafer, "Echo Removal by Discrete Generalized Linear Filtering," Ph.D. Thesis, Massachusetts Institute of Technology, Dept. of Electrical Engineering, Cambridge, MA, Feb. 1968.

[17] K. Steiglitz and B. Dickinson, "Computation of the Complex Cepstrum by Factorization of the z-Transform," *Proc. IEEE Int. Conf. Acoustics, Speech, and Signal Processing*, pp. 723–726, May 1977.

[18] T.G. Stockholm, T.M. Cannon, and R.B. Ingebretsen, "Blind Deconvolution through Digital Signal Processing," *Proc. IEEE*, vol. 63, no. 4, pp. 678–692, April 1975.

[19] J.M. Tribolet, "A New Phase Unwrapping Algorithm," *IEEE Trans. Acoustics, Speech, and Signal Processing*, vol. ASSP–25, no. 2, pp. 170–177, April 1977.

[20] J.M. Tribolet, *Seismic Applications of Homomorphic Signal Processing*, Prentice Hall, Englewood Cliffs, NJ, 1979.

[21] W. Verhelst and O. Steenhaut, "A New Model for the Short-Time Complex Cepstrum of Voiced Speech," *IEEE Trans. Acoustics, Speech, and Signal Processing*, vol. ASSP–34, no. 1, pp. 43–51, Feb. 1986.

[22] K. Wang and S.A. Shamma, "Self-Normalization and Noise Robustness in Early Auditory Representations," *IEEE Trans. Speech and Audio Processing*, vol. 2, no. 3, pp. 421–435, July 1995.

Short-Time Fourier Transform Analysis and Synthesis

7.1 Introduction

In analyzing speech signal variations with the discrete-time Fourier transform, we encounter the problem that a single Fourier transform cannot characterize changes in spectral content over time such as time-varying formants and harmonics. In contrast, the discrete-time *short-time Fourier transform* (STFT), introduced in our discussion of discrete-time spectrographic analysis in Chapter 3, consists of a separate Fourier transform for each instant in time. In particular, we associate with each instant the Fourier transform of the signal in the neighborhood of that instant, so that spectral evolution of the signal can be traced in time. For practical implementation, each Fourier transform in the STFT is replaced by the discrete Fourier transform (DFT). The resulting STFT is discrete in both time and frequency. We call this the *discrete STFT* to distinguish it from the discrete-time STFT, which is continuous in frequency. These two transforms, their properties and inter-relationships, and a glimpse into their use in speech processing analysis/synthesis applications are the primary focus of this chapter.

The discrete-time STFT and discrete STFT were implicitly used in the previous two chapters in two analysis/synthesis techniques, based on linear prediction and homomorphic filtering, in applying a short-time window to the waveform $x[n]$ to account for time variations in the underlying source and system. These two analysis/synthesis methods are *model-based* in the sense that they rely on a source/filter speech model and often on a further parameterization with rational (pole-zero) z-transforms to represent the vocal tract system function and glottal source. In this chapter, we develop a new set of analysis/synthesis techniques that largely do not require such strict source/filter speech models.

In Section 7.2 of this chapter, we give a formal introduction to the discrete-time and the discrete STFTs for analysis, as well as a brief look into their time-frequency resolution properties. We then consider, in Section 7.3, the problem of synthesizing a sequence from its STFT. While this is straightforward for the discrete-time STFT, a number of important STFT concepts are introduced for addressing the more challenging task of synthesis from the discrete STFT. The basic theory part of the chapter is essentially concluded in Section 7.4 in treating the magnitude of the STFT as a transform in its own right. Next, in Section 7.5, we consider the important practical problem of estimating a signal from a modified STFT or STFT magnitude that does not satisfy the definitional constraints of the STFT, leading to many practical applications of the STFT and STFT magnitude in speech processing. The particular applications of time-scale modification and noise reduction are introduced in Section 7.6.

7.2 Short-Time Analysis

In this section, following the development in [13],[20], we explore two different views of the STFT: (1) the Fourier transform view and (2) the filter bank view. We begin with the first perspective, which draws on the STFT representation of a sequence being analogous to that of the Fourier transform.

7.2.1 Fourier Transform View

The expression for the discrete-time STFT at time n was given in Chapter 3 as[1]

$$X(n, \omega) = \sum_{m=-\infty}^{\infty} x[m]w[n - m]e^{-j\omega n}, \tag{7.1}$$

where $w[n]$ is assumed to be non-zero only in the interval $[0, N_w - 1]$ and is referred to as the *analysis window* or sometimes as the *analysis filter* for reasons that will become clear later in this chapter. The sequence $f_n[m] = x[m]w[n - m]$ is called a short-time section of $x[m]$ at time n. This sequence is obtained by time-reversing the analysis window, $w[m]$, shifting the result by n points, and multiplying it with $x[m]$. With the short-time section for time n, we can take its Fourier transform to obtain the frequency function $X(n, \omega)$. This series of operations is illustrated in Figure 7.1. To obtain $X(n + 1, \omega)$, we slide the time-reversed analysis window one point from its previous position, multiply it with $x[m]$, and take the Fourier transform of the resulting short-time section. Continuing this way, we generate a set of discrete-time Fourier transforms that together constitute the discrete-time STFT. Typically, the analysis window is selected to have a much shorter duration than the signal $x[n]$ for which the STFT is computed; as we have seen for a speech waveform, the window duration is typically set at about 20–30 ms or a few pitch periods.

By analogy with the discrete Fourier transform (DFT), the discrete STFT is obtained from the discrete-time STFT through the following relation:

$$X(n, k) = X(n, \omega)|_{\omega = \frac{2\pi}{N}k}, \tag{7.2}$$

where we have sampled the discrete-time STFT with a *frequency sampling interval*[2] of $\frac{2\pi}{N}$ in order to obtain the discrete STFT. We refer to N as the *frequency sampling factor*. Substituting

---

[1] We have changed our notation slightly from that in Chapter 3, where we denoted the STFT by $X(\omega, \tau)$.

[2] More strictly, the sampling occurs over one period so that $X(n, k) = X(n, \omega)|_{\omega = \frac{2\pi}{N}k}$ for $k = 0, 1, \ldots$ $N - 1$, and zero elsewhere, but here we think of the DFT as periodic with period N [14].

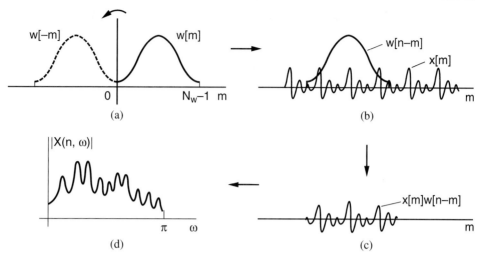

Figure 7.1   Series of operations required to compute a short-time section and STFT: (a) flip window; (b) slide the window sample-by-sample over the sequence (*Note:* $w[-(m-n)] = w[n-m]$); (c) multiply the sequence by the displaced window; (d) take the Fourier transform of the windowed segment.

Equation (7.1) into Equation (7.2), we obtain the following relation between the discrete STFT and its corresponding sequence $x[n]$:

$$X(n, k) = \sum_{m=-\infty}^{\infty} x[m]w[n-m]e^{-j\frac{2\pi}{N}km}. \qquad (7.3)$$

The following example illustrates the distinction between the discrete STFT and the discrete-time STFT.

EXAMPLE 7.1   Consider the periodic unit sample sequence $x[n] = \sum_{l=-\infty}^{\infty} \delta[n-lP]$ and analysis window $w[n]$ a triangle of length P. The discrete-time STFT is given by

$$X(n, \omega) = \sum_{m=-\infty}^{\infty} x[m]w[n-m]e^{-j\omega m}$$

$$= \sum_{m=-\infty}^{\infty} \left(\sum_{l=-\infty}^{\infty} \delta[m-lP] \right) w[n-m]e^{-j\omega m}$$

$$= \sum_{l=-\infty}^{\infty} w[n-lP]e^{-j\omega l P}$$

where, in the last step, the delta function picks off the values of $m = lP$. The discrete-time STFT is a series of windows translated in time by lP samples and with linear phase $-\omega lP$. For each time n, because the translated windows are nonoverlapping, $X(n, \omega)$ has a constant magnitude and linear phase along the frequency dimension. Also, because the $w[n-lP]$ are non-overlapping, for each frequency ω, $X(n, \omega)$ has a magnitude that follows the replicated triangular window and a phase that is fixed in frequency, i.e., $\angle X(n, \omega) = -\omega lP$, along the time dimension.

Consider now the discrete STFT and suppose that the frequency sampling factor $N = P$. Then the discrete STFT is given by

$$X(n, k) = \sum_{m=-\infty}^{\infty} x[m]w[n - m]e^{-j\frac{2\pi}{N}km}$$

$$= \sum_{m=-\infty}^{\infty} \left(\sum_{l=-\infty}^{\infty} \delta[m - lP] \right) w[n - m]e^{-j\frac{2\pi}{P}km}$$

$$= \sum_{l=-\infty}^{\infty} w[n - lP]e^{-j\frac{2\pi}{P}klP}$$

$$= \sum_{l=-\infty}^{\infty} w[n - lP]$$

and so consists of translated non-overlapping windows. For each discrete frequency k, the magnitude of $X(n, k)$ follows the translated windows and the phase is zero; because the frequency sampling interval equals the fundamental frequency of the periodic sequence $x[n]$, we see no phase change in time.   ▲

Recall that the two-dimensional function $|X(n, \omega)|^2$ was called the *spectrogram* in Chapter 2. For a "short" window $w[n]$, as in Example 7.1 where the window duration is one pitch period, $|X(n, \omega)|^2$ is referred to as the *wideband spectrogram*, exhibiting periodic temporal structure in $x[n]$ as "vertical striations." In Example 7.1, these vertical striations correspond to the repeated, non-overlapping windows $w[n - lP]$. When the window $w[n]$ is "long" in duration, e.g., a few pitch periods in duration, we refer to $|X(n, \omega)|^2$ as the *narrowband spectrogram*, exhibiting the harmonic structure in $x[n]$ as "horizontal striations" (Exercise 7.2).

In many applications, the time variation (the n dimension) of $X(n, k)$ is decimated by a temporal decimation factor, L, to yield the function $X(nL, k)$. Just as the discrete-time STFT can be viewed as a set of Fourier transforms of the short-time sections $f_n[m] = x[m]w[n - m]$, the discrete STFT in Equation (7.3) is easily seen to be a set of DFTs of the short-time sections $f_n[m]$. When the time dimension of the discrete STFT is decimated, the corresponding short-time sections $f_{nL}[m]$ are a subset of $f_n[m]$ obtained by incrementing time by multiples of L. This notion is illustrated in Figure 7.2. How we chose sampling rates in time and frequency for a unique representation of $x[n]$ is a question that we will return to later.

In viewing $X(n, \omega)$ as a Fourier transform for each fixed n, we see that the frequency function $X(n, \omega)$ for each n has all the general properties of a Fourier transform [13]. For example, with respect to the frequency variable ω, $X(n, \omega)$ is periodic with period 2π and Hermetian symmetric for real sequences. Another property, analogous to that of the Fourier transform, is that a time shift in a sequence leads to a linear phase factor in the frequency domain. Suppose we shift $x[n]$ by n_o samples. Then with a change in variables $q = m - n_o$, we have

$$\tilde{X}(n, \omega) = \sum_{m=-\infty}^{\infty} x[m - n_o]w[n - m]e^{-j\omega m} = \sum_{q=-\infty}^{\infty} x[q]w[n - n_o - q]e^{-j\omega(q+n_o)}$$

$$= e^{-j\omega n_o} \sum_{q=-\infty}^{\infty} x[q]w[n - n_o - q]e^{-j\omega q} = e^{-j\omega n_o} X(n - n_o, \omega).$$

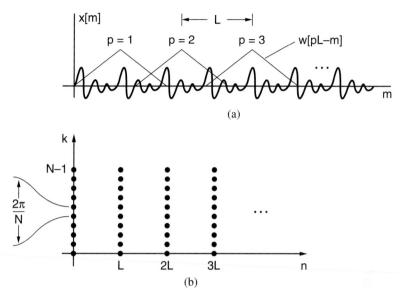

Figure 7.2 Time and frequency decimation used in computing the discrete STFT $X(nL, k)$: (a) analysis window positions; (b) time-frequency sampling.

SOURCE: S.H. Nawab and T.F. Quatieri, "Short-Time Fourier Transform" [13]. ©1987, Pearson Education, Inc. Used by permission.

Thus, a shift by n_o in the original time sequence introduces a linear phase, but also a shift in time, corresponding to a shift in each short-time section by n_o. Likewise, most of the properties of the discrete-time STFT have a straightforward extension to the discrete STFT [13]. For the above shifting property, however, it should be noted that if the discrete STFT is decimated in time by a factor L, when the shift is not an integer multiple of L, there is no general relationship between the discrete STFTs of $x[n]$ and $x[n-n_o]$. This happens because the short-time sections corresponding to $X(nL, k)$ cannot be expressed as shifted versions of the short-time sections corresponding to the discrete STFT of $x[n - n_o]$ [13].

In the implementation of the discrete STFT from the Fourier transform view, the FFT can be used to efficiently compute $X(n, k)$ by computing the time-aliased version of each short-time section and then applying the N-point FFT to each of those sections [13]. If N is greater than or equal to the analysis window length, the computation of the time-aliased version is eliminated.

7.2.2 Filtering View

The STFT can also be viewed as the output of a filtering operation where the analysis window $w[n]$ plays the role of the filter impulse response, and hence the alternative name *analysis filter* for $w[n]$. For the filtering view of the STFT, we fix the value of ω at ω_o (in the Fourier transform view, we had fixed the value of n), and rewrite Equation (7.1) as

$$X(n, \omega_o) = \sum_{m=-\infty}^{\infty} (x[m]e^{-j\omega_o m})w[n - m]. \tag{7.4}$$

We then recognize from the form of Equation (7.4) that the STFT represents the convolution of the sequence $x[n]e^{-j\omega_o n}$ with the sequence $w[n]$. Using convolution notation in Equation (7.4), we obtain

$$X(n, \omega_o) = (x[n]e^{-j\omega_o n}) * w[n]. \tag{7.5}$$

The product $x[n]e^{-j\omega_o n}$ can be interpreted as the modulation of $x[n]$ up to frequency ω_o. Thus, $X(n, \omega_o)$ for each ω_o is a sequence in n which is the output of the process illustrated in Figure 7.3a. The signal $x[n]$ is modulated with $e^{-j\omega_o n}$ and the result passed through a filter whose impulse response is the analysis window, $w[n]$. We can view this as a modulation of a band of frequencies in $x[n]$ around ω_o down to baseband, and then filtered by $w[n]$ (Figure 7.3b).

A slight variation on the filtering and modulation view of the STFT is obtained by manipulating Equation (7.5) into the following form (Exercise 7.1):

$$X(n, \omega_o) = e^{-j\omega_o n}(x[n] * w[n]e^{j\omega_o n}). \tag{7.6}$$

In this case, the sequence $x[n]$ is first passed through the same filter as in the previous case except for a linear phase factor. The filter output is then modulated by $e^{-j\omega_o n}$. This view of the time variation of the STFT for a fixed frequency is illustrated in Figure 7.4a. We can view this as filtering out a band of frequencies that is then demodulated down to baseband (Figure 7.4b).

The discrete STFT of Equation (7.3) can also be interpreted from the filtering viewpoint. In particular, having a finite number of frequencies allows us to view the discrete STFT as the output of the filter bank shown in Figure 7.5a, i.e.,

$$X(n, k) = e^{-j\frac{2\pi}{N}kn}(x[n] * w[n]e^{j\frac{2\pi}{N}kn}).$$

Observe that each filter is acting as a bandpass filter centered around its selected frequency. Thus, the discrete STFT can be viewed as a collection of sequences, each corresponding to the frequency components of $x[n]$ falling within a particular frequency band, as illustrated in the following example. This implementation is similar to that of the filter banks used in the early analog spectrograms [8].

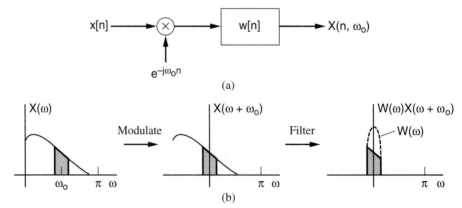

Figure 7.3 Filtering view of STFT analysis at frequency ω_o: (a) block diagram of complex exponential modulation followed by a lowpass filter; (b) operations in the frequency domain.

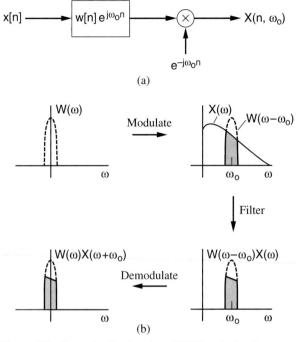

Figure 7.4 Alternative filtering view of STFT analysis at frequency ω_o: (a) block diagram of bandpass filtering followed by complex exponential modulation; (b) operations in the frequency domain.

EXAMPLE 7.2 Consider a window that is Gaussian in shape, i.e., of the form $w[n] = e^{a(n-n_o)^2}$, shown in Figure 7.6a (for $n_o = 0$). (For the STFT, the window is made causal and of finite length.) The discrete STFT with DFT length N, therefore, can be considered as a bank of filters with impulse responses

$$h_k[n] = e^{a(n-n_o)^2} e^{j\frac{2\pi}{N}kn}.$$

If the input sequence $x[n] = \delta[n]$, then the output of the kth bandpass filter is simply $h_k[n]$. Figure 7.6b,c,d shows the real component of the impulse response for three bandpass filters ($k = 5, 10, 15$) for a discrete STFT where $N = 50$ corresponds to bandpass filters spaced by 200 Hz for a sampling rate of 10000 samples/s. Each impulse response has an *envelope*, i.e., the Gaussian window, that multiplies the modulation function $e^{j\frac{2\pi kn}{N}}$. Observe that the output of filter $k = 0$ equals the Gaussian window of Figure 7.6a. ▲

As with the Fourier transform interpretation, the filtering view also allows us to easily deduce a number of STFT properties. In particular, we view $X(n, \omega)$ as a filter output for each fixed frequency. Therefore, the time variation of $X(n, \omega)$ for each ω has all the general properties of a filtered sequence. We list a few of these properties below [13].

P1: If $x[n]$ has length N and $w[n]$ has length M, then $X(n, \omega)$ has length $N + M - 1$ along n.

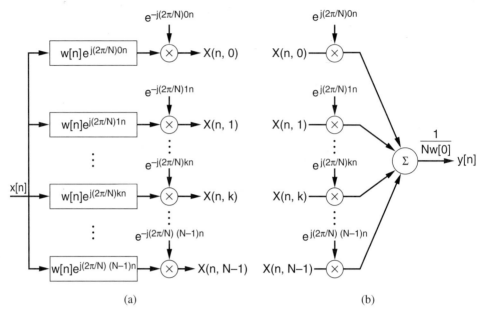

Figure 7.5   The filtering view of analysis and synthesis with the discrete STFT: (a) the discrete STFT (analysis) as the output of a filter bank consisting of bandpass filters; (b) filter bank summation procedure for signal synthesis from the discrete STFT.

SOURCE: S.H. Nawab and T.F. Quatieri, "Short-Time Fourier Transform" [13]. ©1987, Pearson Education, Inc. Used by permission.

P2: The bandwidth of the sequence (along n) $X(n, \omega_o)$ is less than or equal to the bandwidth of $w[n]$.

P3: The sequence $X(n, \omega_o)$ has the spectrum centered at the origin.

   In the first property, we make use of a standard result for the length of a sequence obtained through the convolution of any two sequences of lengths N and M. For the second property, we note that $X(n, \omega)$ as a function of n is the output of a filter whose bandwidth is the bandwidth of the analysis window. The third property follows from the modulation steps used in obtaining $X(n, \omega_o)$ (Figures 7.3 and 7.4). The STFT properties from the filtering viewpoint remain the same for the discrete STFT since, for a fixed frequency, the time variation of the discrete STFT is the same as the time variation of the discrete-time STFT at that frequency. The next example illustrates properties **P2** and **P3** for the filter bank of Example 7.2.

EXAMPLE 7.3   Consider the filter bank of Example 7.2 that was designed with a Gaussian window of the form $w[n] = e^{a(n-n_o)^2}$, shown in Figure 7.6a. Figure 7.7 shows the Fourier transform magnitudes of the output of the four complex bandpass filters $h_k[n]$ for $k = 0, 5, 10,$ and 15 of Figure 7.6. Each spectral bandwidth is identical to the bandwidth of the Gaussian window. Indeed, in each case the spectral magnitude equals that of the window transform, except for a frequency shift, and must itself be Gaussian in shape. (The reader should verify this property). After demodulation by $e^{-j\frac{2\pi}{N}kn}$ (as in Equation (7.6) and its discrete version), the resulting bandpass outputs have the same spectral shapes as in the figure, but centered at the origin.   ▲

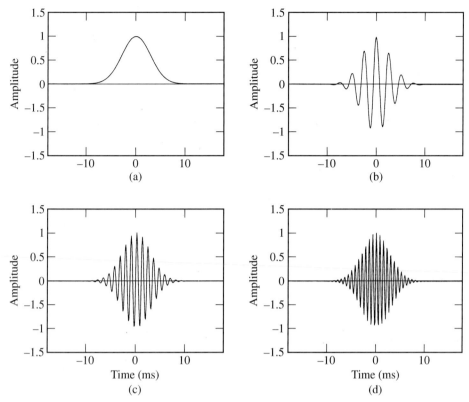

Figure 7.6   Real part of the bandpass filter outputs with unit sample input $\delta[n]$ for the discrete filter bank of Example 7.2 prior to demodulation: (a) Gaussian window $w[n]$ (also output of filter $k = 0$); (b) discrete frequency $k = 5$; (c) discrete frequency $k = 10$; (d) discrete frequency $k = 20$. Frequency of the output increases with increasing k, while the Gaussian envelope remains intact.

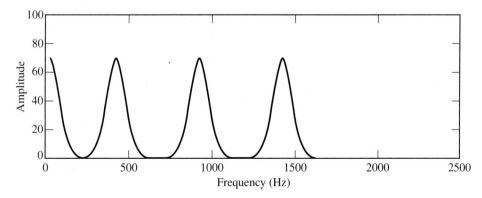

Figure 7.7   Superimposed spectra of bandpass filter outputs of Figure 7.6.

Finally, in the implementation of the discrete STFT from the filtering view, the signal $x[n]$ is passed through a bank of filters, shown in Figure 7.5a, where the output of each filter is the time variation of the STFT at frequency ω_k. If the output of each filter is decimated in time by a factor L, then we obtain the discrete STFT $X(nL, k)$.

7.2.3 Time-Frequency Resolution Tradeoffs

We have seen in Chapter 3 that a basic issue in analysis window selection is the compromise required between a long window for showing signal detail in frequency and a short window for representing fine temporal structure. We first recall that the STFT, $X(n, \omega)$, is the Fourier transform of the short-time section $f_n[m] = x[m]w[n - m]$. From Fourier transform theory, we know that the Fourier transform of the product of two sequences is given by the convolution of their respective Fourier transforms. With $X(\omega)$ as the Fourier transform of $x[m]$ and $W(-\omega)e^{j\omega n}$ as the Fourier transform of $w[n - m]$ with respect to the variable m, we can write the STFT as [13],[17]

$$X(n, \omega) = \frac{1}{2\pi} \int_{-\pi}^{\pi} W(\theta)e^{j\theta n} X(\omega + \theta)d\theta. \tag{7.7}$$

Thus, the frequency variation of the STFT for any fixed time may be interpreted as a smoothed version of the Fourier transform of the underlying signal. Thus, for faithful reproduction of the properties of $X(\omega)$ in $X(n, \omega)$, the function $W(\omega)$ should appear as an impulse with respect to $X(\omega)$. The closer $W(\omega)$ is to an impulse (i.e., with narrow bandwidth), $X(n, \omega)$ is said to have better *frequency resolution*. However, for a given window, frequency resolution varies inversely with the effective length of the window. Thus, good frequency resolution requires long analysis windows, whereas the desire for short-time analysis, and thus good *time resolution*, requires short analysis windows, as illustrated in the following example.

> **EXAMPLE 7.4**   An example of the time-frequency resolution tradeoff is given in Figure 7.8, which shows the Fourier transform magnitude of a section of a chirp signal whose frequency is a linear function of time [13]. The aim is to measure the instantaneous frequency at the center of the chirp. This is performed by using rectangular analysis windows of various lengths. The very short window of duration 5 ms gives good temporal resolution, but low frequency resolution because of spectrum smoothing by a wide window in frequency. The very long window of duration 20 ms yields a wide spectrum, reflecting the time-varying frequency of the chirp, giving neither good temporal nor spectral resolution. An intermediate window length of 10 ms provides a good tradeoff in time and frequency resolution.   ▲

We see implied in the previous example that Equation (7.7) is not a valid smoothing interpretation when the sequence $x[n]$ is not stationary. One approach, however, to apply this interpretation in the speech context is to *pretend* that a short-time segment comes from a longer stationary signal. For example, we can pretend that a quasi-periodic speech segment comes from an infinitely long periodic sequence. With this viewpoint, we can use the smoothing interpretation of Equation (7.7) and deduce the time-frequency tradeoff that we described in Chapter 3 in relation to the narrowband and wideband spectrograms for voiced speech. We illustrate the particular case of harmonic spectra in Figure 7.9 for a long and a short window. The long window better represents the spectral harmonicity and results in harmonic amplitudes that better reflect the underlying vocal tract spectral envelope. The short window blurs the spectral harmonicity,

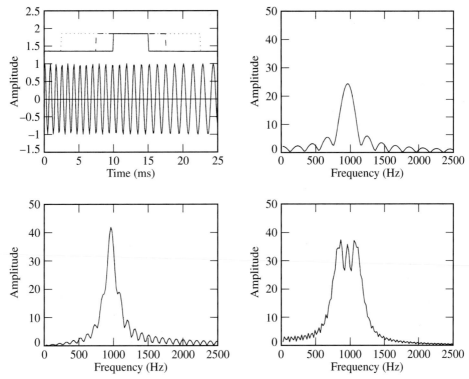

Figure 7.8 Effect of the length of the analysis window on the discrete Fourier transform of a linearly frequency-modulated sinusoid of length 25 ms whose frequency decreases from 1250 Hz to 625 Hz. The discrete Fourier transform uses a rectangular window centered at 12.5 ms, as illustrated in panel (a). Transforms are shown for three different window lengths: (b) 5 ms [solid in (a)]; (c) 10 ms [dashed in (a)]; (d) 20 ms [dotted in (a)].

SOURCE: S.H. Nawab and T.F. Quatieri, "Short-Time Fourier Transform" [13]. ©1987, Pearson Education, Inc. Used by permission.

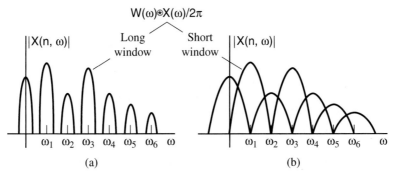

Figure 7.9 Schematic of convolutional view of time-frequency resolution tradeoff with long and short analysis windows for harmonic spectra: (a) long window; (b) short window.

and degrades the harmonic amplitudes, but better captures changes in the harmonicity and the spectral envelope. More generally, a similar view holds for unvoiced sounds (e.g., plosives and fricatives). Nevertheless, we are still faced with the problem that some sounds are stationary over only very short time spans, or are never stationary, as we saw in Chapter 3 with diphthongs and with rapid conversational speech. A similar problem arises when sounds of very different spectral and temporal character are closely spaced, as with a voiced plosive preceding a vowel. In these cases, poor frequency resolution with the STFT may be our only option when good time resolution is desired. Chapter 11 will address these tradeoffs in more detail, as well as alternative time-frequency distributions that provide resolution tradeoffs different from that with the spectrogram.

7.3 Short-Time Synthesis

In this section, we first consider the problem of obtaining a sequence back from its discrete-time STFT. The inversion is represented mathematically by a *synthesis equation* which expresses a sequence in terms of its discrete-time STFT. Whereas the discrete-time STFT is always invertible in this sense, the discrete STFT requires certain constraints on its sampling rate for invertibility. A common approach to developing synthesis methods for the discrete STFT has been to start from one of the many synthesis equations [13],[20]. A discretized version of such an equation is then considered as the basis for a candidate synthesis method and conditions are derived under which such a method can synthesize a sequence from its discrete STFT. We describe two classical methods, the Filter Bank Summation (FBS) method and the Overlap-Add (OLA) method.

7.3.1 Formulation

The invertibility of the discrete-time STFT is easily seen through the Fourier transform interpretation of the STFT, where the discrete-time STFT is viewed for each value of n as a function of frequency obtained by taking the Fourier transform of the short-time section $f_n[m] = x[m]w[n-m]$. It follows that if, for each n, we take the inverse Fourier transform of the corresponding function of frequency, then we obtain the sequence $f_n[m]$. If we evaluate this short-time section at $m = n$, we obtain the value $x[n]w[0]$. Assuming that $w[0]$ is non-zero, we can divide by $w[0]$ to recover $x[n]$. The process of taking the inverse Fourier transform of $X(n, \omega)$ for a specific n and then dividing by $w[0]$ is represented by the following relation:

$$x[n] = \frac{1}{2\pi w[0]} \int_{-\pi}^{\pi} X(n, \omega)e^{j\omega n}d\omega. \tag{7.8}$$

This equation represents a *synthesis equation* for the discrete-time STFT. In fact, there are numerous synthesis equations that map $X(n, \omega)$ uniquely back to $x[n]$. Observe that there is much redundancy in the STFT where the analysis window slides one sample at a time; therefore, the use of one inverse Fourier transform per time sample in Equation (7.8) is inefficient.

In contrast to the discrete-time STFT, the discrete STFT $X(n, k)$ is not always invertible. For example, consider the case when $w[n]$ is bandlimited with bandwidth of B. Figure 7.10 shows the filter regions used to obtain $X(n, k)$ for the case when the sampling interval $\frac{2\pi}{N}$ is greater than B. Note that in this case there are frequency components of $x[n]$ which do not pass through any of the filter regions of the discrete STFT. Those frequency components can have arbitrary values and yet we would have the same discrete STFT. Thus, in these cases, the

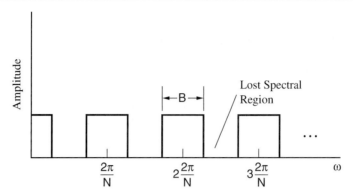

Figure 7.10   Undersampled STFT when the frequency sampling interval $\frac{2\pi}{N}$ is greater than the analysis-filter bandwidth B.

SOURCE: S.H. Nawab and T.F. Quatieri, "Short-Time Fourier Transform" [13]. ©1987, Pearson Education, Inc. Used by permission.

discrete STFT is not a unique representation of $x[n]$ and therefore cannot be invertible. The invertibility problem is also of interest when the discrete STFT has been decimated in time. For example, consider the case when the analysis window $w[n]$ is non-zero over its length N_w. When temporal decimation factor L is greater than N_w, there are samples of $x[n]$ which are not included in any short-time section of the discrete STFT. These samples can have arbitrary values and yet we would have the same time-decimated discrete STFT. Thus, in such cases the discrete STFT is again not a unique representation of $x[n]$ and therefore cannot be invertible.

By selecting appropriate constraints on the frequency-sampling and time-decimation rates, the discrete STFT becomes invertible. As an example, we consider the case of a finite-length analysis window. We have already seen that in such cases the discrete STFT is not invertible if the temporal decimation factor L is greater than the analysis window length N_w. We now see that if the temporal decimation factor is less than or equal to the analysis window length, then the discrete STFT is invertible, provided we impose constraints on the frequency sampling interval. Suppose that the temporal decimation factor is equal to the analysis window length. The discrete STFT in this case consists of the DFTs of adjacent but non-overlapping short-time sections. Thus, to reconstruct $x[n]$ from its discrete STFT, we must require that each N_w-point short-time section be recoverable from its DFT. However, the DFT of an N_w point sequence is invertible, provided its frequency sampling interval is less than or equal to $\frac{2\pi}{N_w}$ [14]. It follows that the discrete STFT is invertible when the analysis window is non-zero over its finite-length N_w, the temporal decimation factor $L \leq N_w$, and the frequency sampling interval $\frac{2\pi}{N} \leq \frac{2\pi}{N_w}$. We will see in the following sections that more relaxed bounds can be derived in the sense that the above conditions are sufficient but not necessary for invertibility.

7.3.2 Filter Bank Summation (FBS) Method

In this section we present a traditional short-time synthesis method that is commonly referred to as the Filter Bank Summation (FBS) method [1],[13],[20]. This method is best described in terms of the filtering interpretation of the discrete STFT. In this interpretation, the discrete STFT is considered to be the set of outputs of a bank of filters. In the FBS method, the output of each filter is modulated with a complex exponential, and these modulated filter outputs are summed at each instant of time to obtain the corresponding time sample of the original sequence, as shown in Figure 7.5b. For dealing with temporal decimation, the traditional strategy is to perform a temporal interpolation filtering on the discrete STFT in order to restore the temporal decimation factor to unity. The FBS method is then performed on the interpolated output.

The FBS method is motivated by the relation between a sequence and its discrete-time STFT given in Equation (7.8), derived in the previous section for showing the invertibility of the discrete-time STFT. The FBS method carries out a discretized version of the operations suggested by this relation. That is, given a discrete STFT, $X(n, k)$, the FBS method synthesizes a sequence $y[n]$ satisfying the following equation:

$$y[n] = \frac{1}{Nw[0]} \sum_{k=0}^{N-1} X(n, k)e^{j\frac{2\pi}{N}nk}. \tag{7.9}$$

We are interested in deriving conditions for the FBS method such that the sequence $y[n]$ in Equation (7.9) is the same as the sequence $x[n]$. We can conjecture conditions by viewing the discretized equation from the filter bank viewpoint of Figure 7.5. We first see that the modulation resulting from the complex exponentials in analysis is undone by the demodulation of the complex exponentials at the filter bank outputs in synthesis. Therefore, for $y[n] = x[n]$ the sum of the filter bank outputs must add to unity. We now show this condition more formally.

Substituting $X(n, k)$ in Equation (7.9) for the FBS method, we obtain

$$y[n] = \frac{1}{Nw[0]} \sum_{k=0}^{N-1} \underbrace{\left[\sum_{m=-\infty}^{\infty} x[m]w[n-m]e^{-j2\pi km/N} \right]}_{X(n,k)} e^{j\frac{2\pi}{N}nk}. \tag{7.10}$$

Using the linear filtering interpretation of the STFT [the discrete version of Equation (7.6)], this equation reduces to

$$y[n] = \frac{1}{Nw[0]}x[n] * \sum_{k=0}^{N-1} w[n]e^{j\frac{2\pi}{N}nk}.$$

Taking $w[n]$ out of the summation and noting that the finite sum over the complex exponentials reduces to an impulse train with period N, we obtain

$$y[n] = \frac{1}{w[0]}x[n] * w[n] \sum_{r=-\infty}^{\infty} \delta[n-rN].$$

In the above expression, we note that $y[n]$ is obtained by convolving $x[n]$ with a sequence that is the product of the analysis window with a periodic impulse train. It follows that if we desire $y[n] = x[n]$, then the product of $w[n]$ and the periodic impulse train must reduce to $w[0]\delta[n]$. That is

$$w[n] \sum_{r=-\infty}^{\infty} \delta[n-rN] = w[0]\delta[n]. \tag{7.11}$$

This is satisfied for any causal analysis window whose length N_w is less than or equal to the number of analysis filters N, i.e., *any* finite-length analysis window can be used in the FBS

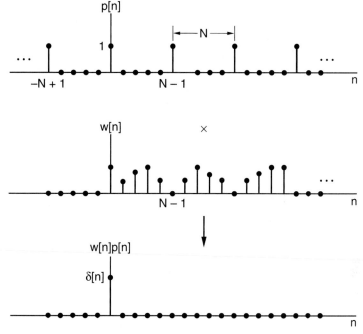

Figure 7.11   Example of an analysis window and how it satisfies the FBS constraint. The analysis-window length is longer than the frequency sampling factor. The sequence $p[n] = \sum_{r=-\infty}^{\infty} \delta[n, n-r]$.

SOURCE: S.H. Nawab and T.F. Quatieri, "Short-Time Fourier Transform" [13]. ©1987, Pearson Education, Inc. Used by permission.

method provided the length of the window is less than the frequency sampling factor N. We can even have $N_w > N$, provided $w[n]$ is chosen such that every Nth sample is zero, i.e.,

$$w[rN] = 0 \quad \text{for } r = -1, 1, -2, 2, -3, 3, \cdots \tag{7.12}$$

as illustrated in Figure 7.11.

Equation (7.12) is known as the *FBS constraint* because this is the requirement on the analysis window that ensures exact signal synthesis with the FBS method. This constraint is more commonly expressed in the frequency domain. Taking the Fourier transform of both sides of Equation (7.11), we obtain (Exercise 7.3)

$$\sum_{k=0}^{N-1} W\left(\omega - \frac{2\pi}{N}k\right) = Nw[0]. \tag{7.13}$$

This constraint essentially states that the frequency responses of the analysis filters should sum to a constant across the entire bandwidth. We have already seen that any finite-length analysis window whose length is less than or equal to the frequency sampling factor N satisfies this constraint. We conclude that a filter bank with N filters, based on an analysis filter of a length less than or equal to N, is *always* an all-pass system. This is not surprising because, from the synthesis equation $x[n] = \frac{1}{2\pi w[0]} \int_{-\pi}^{\pi} X(n, \omega)e^{j\omega n} d\omega$, at time n the inverse DFT of $X(n, k)/Nw[0]$ must give $x[n]$ when the DFT length is longer than the sequence.

A generalization of the FBS method is motivated by the following relation between a sequence and its discrete-time STFT:

$$x[n] = \frac{1}{2\pi} \int_{-\pi}^{\pi} \left[\sum_{r=-\infty}^{\infty} f[n, n - r]X(r, \omega) \right] e^{j\omega n} d\omega, \qquad (7.14)$$

where the "smoothing" function $f[n, m]$ is referred to as the time-varying *synthesis filter*. It can be shown that any $f[n, m]$ that satisfies

$$\sum_{m=-\infty}^{\infty} f[n, -m]w[m] = 1 \qquad (7.15)$$

makes Equation (7.14) a valid synthesis equation (Exercise 7.6). It should be noted that the motivating equation for the basic FBS method can be obtained by setting the synthesis filter to be a non-smoothing filter, i.e., $f[n, m] = \delta[m]$ in Equation (7.14) (Exercise 7.6).

The *generalized FBS* method carries out a discretized version of the operations suggested by Equation (7.14). That is, given a discrete STFT which is decimated in time by a factor L, the generalized FBS method synthesizes a sequence $y[n]$ satisfying the following equation:

$$y[n] = \frac{L}{N} \sum_{r=-\infty}^{\infty} \sum_{k=0}^{N-1} f[n, n - rL]X(rL, k)e^{-j\frac{2\pi}{N}nk}. \qquad (7.16)$$

Although Equation (7.16) contains the time-varying synthesis filter $f[n, n - rL]$, we consider here the time-invariant case, $f[n, n - rL] = f[n - rL]$, because of its practical importance in interpolating a time-decimated STFT. For a time-invariant synthesis filter, Equation (7.16) reduces to

$$y[n] = \frac{L}{N} \sum_{r=-\infty}^{\infty} \sum_{k=0}^{N-1} f[n - rL]X(rL, k)e^{-j\frac{2\pi}{N}nk}. \qquad (7.17)$$

This equation holds when the following constraint is satisfied by the analysis and synthesis filters as well as the temporal decimation and frequency sampling factors [13]:

$$L \sum_{r=-\infty}^{\infty} f[n - rL]w[rL - n + pN] = \delta[p], \quad \text{for all } n. \qquad (7.18)$$

The constraint in Equation (7.18) reduces to the constraint Equation (7.13) for the basic FBS method when the synthesis filter for the generalized FBS method is $f[n, m] = \delta[m]$ and $L = 1$ [13]. Finally, it should also be noted that if $L \neq 1$, and if we let $f[n]$ be the interpolating filter preceding the FBS synthesis, then the synthesis of Equation (7.17) with the synthesis filter $f[n]$ is equivalent to the entire process of interpolation and filter bank summation. Moreover, there exist fast computational methods based on this generalized FBS formulation. Some of these methods such as *helical interpolation* introduced by Portnoff and the *weighted overlap-add method* introduced by Crochiere are described in [4],[13],[18],[20].

7.3.3 Overlap-Add (OLA) Method

Just as the FBS method was motivated from the filtering view of the STFT, the OLA method is motivated from the Fourier transform view of the STFT [1],[13],[20]. The simplest method obtainable from the Fourier transform view is, in fact, not the OLA method. It is instead a method known as the Inverse Discrete Fourier Transform (IDFT) method. In this method, for each fixed time, we take the inverse DFT of the corresponding frequency function and divide the result by the analysis window. This method is generally not favored in practical applications because a slight perturbation in the STFT can result in a synthesized signal very different from the original.[3] For example, consider the case where the STFT is multiplied by a linear phase factor of the form $e^{j\omega n_o}$ with n_o unknown. Then the IDFT for each fixed time results in a shifted version of the corresponding short-time section. Since the shift n_o is unknown, dividing by the analysis window without taking the shift into account, introduces a distortion in the resulting synthesized signal. In contrast, the OLA method, which we describe in this section, results in a shifted version of the original signal without distortion.

In the OLA method, we take the inverse DFT for each fixed time in the discrete STFT. However, instead of dividing out the analysis window from each of the resulting short-time sections, we perform an overlap and add operation between the short-time sections. This method works provided the analysis window is designed such that the overlap and add operation effectively eliminates the analysis window from the synthesized sequence. The intuition here is that the redundancy within overlapping segments and the averaging of the redundant samples remove the effect of windowing.

The OLA method is motivated by the following relation between a sequence and its discrete-time STFT:

$$x[n] = \frac{1}{2\pi W(0)} \int_{-\pi}^{\pi} \sum_{p=-\infty}^{\infty} X(p, \omega) e^{j\omega p} d\omega, \qquad (7.19)$$

where

$$W(0) = \sum_{n=-\infty}^{\infty} w[n].$$

This synthesis equation can be thought of as the synthesis equation in Equation (7.8), averaged over many short-time segments and normalized by $W(0)$. For the derivation of this synthesis equation, the reader is referred to [13]. The OLA method carries out a discretized version of the operations suggested on the right of Equation (7.19). That is, given a discrete STFT $X(n, k)$, the OLA method synthesizes a sequence $y[n]$ given by

$$y[n] = \frac{1}{W(0)} \sum_{p=-\infty}^{\infty} \left[\frac{1}{N} \sum_{k=0}^{N-1} X(p, k) e^{j\frac{2\pi}{N}kn} \right].$$

The term inside the rectangular brackets is an inverse DFT which for each p gives us

$$f_p[n] = x[n]w[p - n],$$

---

[3] One might consider averaging out this effect by summing many inverted DFTs, with the analysis window divided out, at each time n. This is in fact the strategy of the OLA method, but without the need to divide out each IDFT by the window shape.

provided that the DFT length N is longer than the window length N_w, i.e., there is no aliasing from the DFT inversion [14]. [Observe also that the inverse DFT is implicitly zero outside the interval $[0, N)$.] The expression for $y[n]$ therefore becomes

$$y[n] = \frac{1}{W(0)} \sum_{p=-\infty}^{\infty} x[n]w[p - n],$$

which can be rewritten as

$$y[n] = x[n] \left(\frac{1}{W(0)} \right) \sum_{p=-\infty}^{\infty} w[p - n].$$

In the above expression, we note that $y[n]$ is equal to $x[n]$ provided

$$\sum_{p=-\infty}^{\infty} w[p - n] = W(0),$$

which we observe is *always true because the sum of values of a sequence must always equal the first value of its Fourier transform.* Furthermore, if the discrete STFT had been decimated in time by a factor L, it can be similarly shown that if the analysis window satisfies (Exercise 7.4)

$$\sum_{p=-\infty}^{\infty} w[pL - n] = \frac{W(0)}{L}, \tag{7.20}$$

then $x[n]$ can be synthesized using the following relation:

$$x[n] = \frac{L}{W(0)} \sum_{p=-\infty}^{\infty} \left[\frac{1}{N} \sum_{k=0}^{N-1} X(pL, k)e^{j\frac{2\pi}{N}kn} \right]. \tag{7.21}$$

Equation (7.20) is the general constraint imposed by the OLA method on the analysis window. Unlike the case for $L = 1$, this constraint does not hold for any arbitrary window. It requires the sum of all the analysis windows (obtained by sliding $w[n]$ with L-point increments) to add up to a constant, as shown in Figure 7.12. It is interesting to note the duality between this constraint and the FBS constraint in Equation (7.13), where the shifted versions of the Fourier transform of the analysis window were required to add up to a constant. For the FBS method, we also saw that all finite-length analysis windows whose length N_w is less than the number of analysis filters N satisfy the FBS constraint. Analogously, we can show that the OLA constraint in Equation (7.20) is satisfied by all finite-bandwidth analysis windows whose maximum frequency is less than $\frac{2\pi}{L}$, where L is the temporal decimation factor. In addition, this finite-bandwidth constraint can be relaxed by allowing the shifted window transform replicas to take on value zero at the frequency origin $\omega = 0$. In particular, if $W(\omega - \frac{2\pi}{L}k)$ has zeros at $\omega = \frac{2\pi}{L}k$, then the OLA constraint holds. This is remarkably analogous to the relaxation of the FBS constraint $N_w < N$ by allowing the window $w[n]$ to take on value zero at $n = \pm N, \pm 2N, \ldots$. A proof of this frequency-domain view of the OLA constraint is given in [13].

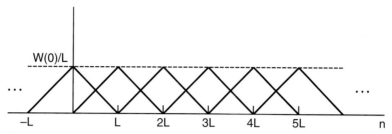

Figure 7.12 The OLA constraint visualized in the time domain.

SOURCE: S.H. Nawab and T.F. Quatieri, "Short-Time Fourier Transform" [13]. ©1987, Pearson Education, Inc. Used by permission.

The duality[4] between the FBS and OLA constraints is summarized in Figure 7.13. The FBS method depends on a sampling relation in frequency while the OLA method depends on a sampling relation in time. With the FBS method, each time sample is obtained by summing filter outputs, while with the OLA method each time sample is obtained by summing different short-time sections. The OLA constraint with no time decimation always holds, provided, of course, that the DFT length is longer than the window length. This is true because the sum of the window's samples is equal to its first Fourier transform value. However, with time decimation, the OLA constraint does not always hold. We saw one situation where it does hold:

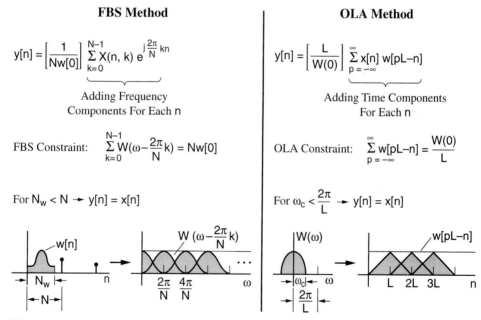

Figure 7.13 Duality between the FBS and OLA constraints. Relaxation of constraints to allow zero crossings in time and in frequency is not shown.

---

[4] Observe that the duality is not exact because a finite bandwidth window implies an infinite window duration. Nevertheless, we assume the effective duration is finite.

The window bandwidth is less than $\frac{2\pi}{L}$. We can relax this constraint if the shifted window transforms pass through zero at the frequency origin; in particular, if the window transform equals zero at $\omega = \frac{2\pi}{L}k$, the constraint holds. This is analogous to relaxing the FBS constraint that the window length N_w is less than the frequency sampling factor N by letting the analysis window pass through zero at $n = kN$.

7.3.4 Time-Frequency Sampling

We now give a different qualitative discussion of the above time-frequency sampling concepts for the OLA and FBS constraints from the perspective of classical time- and frequency-domain aliasing [20]. This discussion also serves to further summarize the sampling issues for these methods, and gives motivation for our earlier statement that sufficient but not necessary conditions for invertibility of the discrete STFT are that the analysis window is non-zero over its finite length N_w, the temporal decimation factor $L \leq N_w$, and the frequency sampling interval $\frac{2\pi}{N} \leq \frac{2\pi}{N_w}$.

Consider a short-time segment $f_n[m] = w[m]x[n - m]$ and its Fourier transform $X(n, \omega)$ with the analysis window of duration N_w. From the Fourier transform view, recovering time sequence $f_n[m]$ (with respect to the variable m) by an inverse DFT of the discrete STFT $X(n, k)$ requires a frequency sampling interval of $\frac{2\pi}{N_w}$ or finer to avoid time-domain aliasing of the time segment $f_n[m]$. Consider now a time decimation factor L. From the filtering view of the STFT, recovering the time sequence $X(n, k)$ (with respect to the time variable n) requires that this time sampling interval L meets the Nyquist criterion based on the bandwidth of the "filter" $w[n]$. This implies that we sample $X(n, k)$ at a time interval $L \leq \frac{2\pi}{\omega_c}$, where ω_c is the filter bandwidth (i.e., $W(\omega)$ is zero outside $[-\omega_c, \omega_c]$ within the interval $[-\pi, \pi]$), to avoid frequency-domain aliasing of the time sequence $X(n, \omega)$. This time-frequency sampling is illustrated in Figure 7.14. Selecting a Hamming window, typically used in speech analysis, and defining the bandwidth with respect to the 3 dB attenuation points, we find that the above sampling requirements, over all N filters in our filter bank, imply four times the number of samples in the original representation of the sequence $x[n]$ [20].

The above time-frequency sampling constraints, derived from simple aliasing considerations, are consistent with the OLA constraint (filter bandwidth $\omega_c \leq \frac{2\pi}{L}$, L being the time decimation factor) and the FBS constraint (window duration $N_w \leq N$, $\frac{2\pi}{N}$ being the frequency sampling interval). These window length and bandwidth constraints can, however, be relaxed, as we have seen in the OLA and FBA constraints, by allowing zeros in the window or its transform at the appropriate time or frequency points, respectively. This implies that we can avoid the four-fold increase in sampling requirements in the above example with a Hamming window analysis. We return to this issue later in Chapter 12 in our discussion of time-frequency analysis with application to speech coding.

Integrating our discussion of aliasing with the OLA and FBS methods, we summarize the following time-frequency constraint considerations from the perspective of each method.

OLA Method

1. For a window length N_w and with a DFT length chosen to give sufficient frequency sampling, i.e., a frequency sampling factor less than $\frac{2\pi}{N_w}$, then each short-time segment can be recovered because there is no time-domain aliasing.

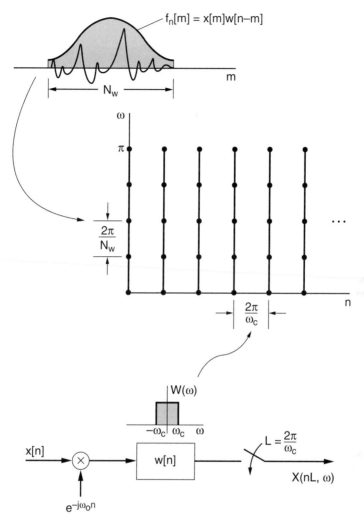

Figure 7.14 Time-frequency sampling constraints from the perspective of classical time- and frequency-domain aliasing. The time sampling must satisfy the Nyquist criterion to avoid aliasing in frequency (but the OLA constraint allows relaxing the finite filter bandwidth constraint), while the frequency sampling must be fine enough to avoid aliasing in time (but the FBS constraint allows relaxing the finite window duration condition).

2. The time decimation factor L is fine enough if the filter bandwidth $\omega_c \leq \frac{2\pi}{L}$. This results in no frequency-domain distortion from the window transform and is the strict form of the OLA constraint.

3. We can relax (2) if we allow zeros in the window transform. In this case, we can undersample in time and still recover $x[n]$.

FBS Method

1. With time decimation of filter outputs by the decimation factor L, we assume that the time sampling meets the Nyquist criterion to recover each filter output, i.e., $L \leq \frac{2\pi}{\omega_c}$, so that frequency-domain aliasing does not occur.

2. The frequency sampling is sufficient if $\frac{2\pi}{N_w} \geq \frac{2\pi}{N}$, i.e., $N_w \leq N$, giving no time-domain distortion from the window (as indicated in Equation (7.11)). This is the strict form of the FBS constraint.

3. We can relax (2) if we allow zeros in the window. In this case, we can undersample in frequency and still recover $x[n]$.

7.4 Short-Time Fourier Transform Magnitude

In speech applications, the spectrogram, which is the squared STFT magnitude (STFTM), has played a major role. For example, visual cues in the spectrogram have been related to parameters important for speech perception. In fact, it has been suggested [2] that the human ear extracts perceptual information strictly from a spectrogram-like representation of speech. Alternatively, experienced speech researchers have trained themselves to "read" the spectrogram itself [23], which indicates that, at least on the phonetic level, the speech signal is largely retained in the spectrogram. One might question, then, how much "information" actually has been lost in the spectrogram, and whether a signal can be recovered from this time-frequency magnitude representation.

By removing a (possibly) unnecessary phase function (the STFT is a complex-valued function), a magnitude-only representation may have uses in a variety of applications, such as time-scale modification and enhancement of speech, where estimation of the phase of the STFT is difficult [16]. For example, as we will see in Chapter 13, the estimation of the phase of a signal's frequency response in the presence of noise is more difficult than the estimation of its magnitude. Indeed, a number of techniques have been developed that obtain a STFT phase estimate from a STFT magnitude estimate, thus circumventing the more difficult phase estimation problem. In this section we introduce the magnitude of the STFT as an alternative time-frequency signal representation. We shall see that many signals can be uniquely represented by the real-valued and non-negative STFTM. Furthermore, we develop analysis and synthesis techniques for the STFTM just as we did for the STFT and show that, while STFTM analysis is similar to STFT analysis, short-time synthesis is very different for the two transforms.

The STFTM is related to another function, the short-time autocorrelation, $r[n, m]$, through the following Fourier transform relationship:[5]

$$r[n, m] = \frac{1}{2\pi} \int_{-\pi}^{\pi} |X(n, \omega)|^2 e^{j\omega m} d\omega$$

$$|X(n, \omega)|^2 = \sum_{m=-\infty}^{\infty} r[n, m] e^{-j\omega m}$$

---

[5] We use here a slightly different notation than that in Chapter 5, where the short-time autocorrelation was written as $r_n[\tau]$.

where m is the autocorrelation "lag" introduced in Chapter 5. The autocorrelation $r[n, m]$ is given by the convolution of the short-time section $f_n[m] = x[m]w[m - n]$ with its time-reversed version, i.e.,

$$r[n, m] = f_n[m] * f_n[-m]$$

with "∗" denoting convolution. Generally, the short-time section $f_n[m] = x[m]w[m - n]$ cannot be obtained from its short-time autocorrelation function [14]. Because $f_n[m]$ is of finite length, its z-transform consists of zeros that can lie inside or outside the unit circle. A conjugate reciprocal flip of a zero preserves the sequence's Fourier transform magnitude and thus its autocorrelation. That is, we saw in our review in Chapter 2 that applying the all-pass function

$$H_{ap}(z) = \frac{z^{-1} - a^*}{1 - az^{-1}}$$

flips the zero at $z = \frac{1}{a^*}$ to its conjugate reciprocal location $z = a$ without a change in the magnitude of the frequency response because $|H_{ap}(z)| = 1$. However, as we shall see shortly, the autocorrelations of short-time sections that have partial overlap in time can be used jointly to solve for the underlying short-time sections, thus removing this inherent ambiguity. This will enable us, under certain conditions, to use the STFTM as a unique representation of the underlying signal.

7.4.1 Signal Representation

We now consider the problem of determining when the discrete-time STFTM can be used to represent a sequence uniquely. After demonstrating uniqueness, we then proceed to determine an algorithm for sequence recovery from the STFTM. That the STFTM is not a unique representation in all cases is easily seen from the simple observation that $x[n]$ and its negative, $-x[n]$, have the same STFTM. We will also demonstrate that there are other kinds of situations where the STFTM is not a unique signal representation. We will then proceed to show that, by imposing certain mild restrictions on the analysis window and the signal, unique signal representation is indeed possible with the discrete-time STFTM [13].

To develop insight into the kinds of situations where a sequence cannot be represented uniquely by its discrete-time STFTM, let us consider the case of a sequence $x[n]$ with a gap of zero samples between two non-zero portions. That is, suppose $x[n]$ is the sum of two signals, $x_1[n]$ and $x_2[n]$, occupying different regions of the n-axis, as depicted in Figure 7.15 (upper). Suppose further that the gap of zeros between $x_1[n]$ and $x_2[n]$ is large enough so that there is no analysis window position for which the corresponding short-time section includes non-zero contribution from $x_1[n]$ as well as $x_2[n]$. Clearly, in such a situation the STFTM of $x[n]$ is the sum of the STFT magnitudes of $x_1[n]$ and $x_2[n]$. However, we have previously observed that a signal and its negative have the same STFTM. It follows that $x[n]$ has the same STFTM as the signals obtained from the differences $x_1[n] - x_2[n]$ and $x_2[n] - x_1[n]$, shown in Figure 7.15 (middle and bottom). We conclude that if there is a large enough gap of zero samples, there will be sign ambiguities on either side of the gap. Consequently, any uniqueness conditions must include a restriction on the length of the zero gaps between non-zero portions of the signal $x[n]$. In particular, the sufficient uniqueness conditions we show are the following:

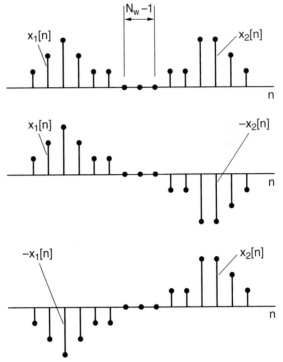

Figure 7.15 Three sequences with the same STFTM.

Source: S.H. Nawab and T.F. Quatieri, "Short-Time Fourier Transform" [13]. ©1987, Pearson Education, Inc. Used by permission.

1. The analysis window $w[n]$ is a known sequence of finite length N_w, with no zeros over its duration.
2. The sequence $x[n]$ is one-sided with at most $N_w - 2$ consecutive zero samples, and the sign of its first non-zero value is known.

The key to showing the uniqueness of $x[n]$ under the above conditions is the observation that $|X(n, \omega)|$ has additional information about the short-time sections of $x[n]$ besides their spectral magnitudes. This information is contained in the overlap of the analysis window positions, i.e., there is much redundancy between adjacent segments. If the short-time section at time n is known, then the signal corresponding to the spectral magnitude of the adjacent section at time $n + 1$ must be *consistent* in the region of overlap with the known short-time section. By consistent, we mean that if the analysis window were non-zero and of length N_w, then after dividing out the analysis window, the first $N_w - 1$ samples of the segment at time $n + 1$ must equal the last $N_w - 1$ samples of the segment at time n, as illustrated in Figure 7.16. Therefore, if we could *extrapolate* the last sample of a segment from its first $N_w - 1$ values, we could repeat this process to obtain the entire signal $x[n]$.

To develop the procedure for extrapolating the next sample of a sequence using its STFTM, assume that the sequence $x[n]$ has been obtained up to some time $n - 1$. Thus, the first $N_w - 1$ samples under the analysis window positioned at time n are known. The goal is to compute the

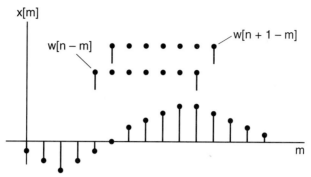

Figure 7.16 Illustration of the consistency required among adjacent short-time sections. Note the samples that are common to the adjacent sections. A rectangular analysis window is assumed.

<small>SOURCE: S.H. Nawab and T.F. Quatieri, "Short-Time Fourier Transform" [13]. ©1987, Pearson Education, Inc. Used by permission.</small>

sample $x[n]$ from these initial samples and the STFT magnitude $|X(n, \omega)|$ or, equivalently, $r[n, m]$. Note that the last value of the short-time autocorrelation function, $r[n, N_w - 1]$, is given by the product of the first and last value of the segment:

$$r[n, N_w - 1] = (w[0]x[n])(w[N_w - 1]x[n - (N_w - 1)])$$

as illustrated in Figure 7.17. Therefore, $x[n]$ is given by

$$x[n] = \frac{r[n, N_w - 1]}{w[0]w[N_w - 1]x[n - (N_w - 1)]}. \tag{7.22}$$

If the first value of the short-time section, i.e., $x[n - (N_w - 1)]$, happens to equal zero, we must then find the first non-zero value within the section and again use the product relation given by Equation (7.22). We can always find such a sample because we have assumed at most

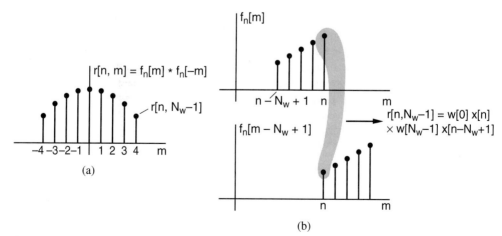

Figure 7.17 Computation of the last non-zero autocorrelation sample of a five-point sequence: (a) auto-correlation function; (b) product of first and last sequence values.

$N_w - 2$ consecutive zero samples between any two non-zero samples of $x[n]$. This completes our argument for the sufficiency of our two conditions for uniquely representing $x[n]$ with $|X(n, \omega)|$. We will see in the next section that the uniqueness conditions still hold with at least N_w uniformly spaced samples of $|X(n, \omega)|$ in frequency over $[0, \pi]$, i.e., if a $(2N_w - 1)$-point (or greater) DFT is computed. The above argument for uniqueness, therefore, leads to the reconstruction algorithm:

S1: Initialize with $x[0]$.

S2: Update time n.

S3: Compute $r[n, N_w - 1]$ from the inverse DFT of $|X(n, k)|^2$.

S4: Compute $x[n] = \frac{r[n, N_w - 1]}{w[0]w[N_w - 1]x[n - (N_w - 1)]}$.

S5: Return to Step (**S2**) and repeat.

We refer to this as a *sequential extrapolation* algorithm.

7.4.2 Reconstruction from Time-Frequency Samples

In order to carry out STFTM analysis on a digital computer, we need to introduce the *discrete* STFTM. By sampling the frequency dimension of the STFTM, $|X(n, \omega)|$, we obtain the discrete STFTM, which is defined as $|X(n, k)|$, the magnitude of the discrete STFT. In the last section, we saw that, under certain conditions, the discrete-time STFTM is a unique signal representation. We noted that the theory can be easily extended to the discrete STFTM. In particular, the uniqueness conditions of the previous section relied on using the short-time autocorrelation functions of adjacent short-time sections which are overlapping in time. These autocorrelation functions can be obtained even if the STFTM is sampled in frequency. That is, if the analysis window is N_w points long, then each short-time autocorrelation function is, at most, $2N_w - 1$ points long and thus can be obtained (without aliasing) from $2N_w - 1$ frequency samples of the STFTM (Exercise 7.7). Therefore, the uniqueness conditions of the discrete-time STFTM extend without change to the discrete STFTM with adequate frequency sampling.

To consider the effects of temporal decimation with factor L, we note that adjacent short-time sections now have an overlap of $N_w - L$ instead of $N_w - 1$. The successive extrapolation procedure discussed in the previous section can be extended to this case by requiring the extrapolation of the L last samples of a short-time section, using the first $N_w - L$ samples and the short-time autocorrelation function of that section (Exercise 7.7). This can be accomplished provided the overlap between adjacent short-time sections is greater than $\frac{N_w}{2}$ and there are no zero-gaps of length greater than $N_w - 2L$. In addition, to initialize the extrapolation procedure, L initial samples of the underlying sequence must be known. We summarize the sufficient uniqueness conditions for the partial overlap case as follows:

1. The analysis window $w[n]$ is a known sequence of finite length N_w, with no zeros over its duration.

2. The sequence $x[n]$ is one-sided with, at most, $N_w - 2L$ consecutive zero samples. L consecutive samples of $x[n]$ (from the first non-zero sample) are known. This is a sufficient but not a necessary condition.

As with the $L = 1$ case, a sequential extrapolation algorithm can be implemented for the more general case. For synthesizing a sequence $x[n]$ from $|X(nL, k)|$ under the above conditions, we assume that the first non-zero sample of $x[n]$ falls at $n = 0$ and that the L samples of $x[n]$ for $0 \leq n < L$ are known. The L known samples of $x[n]$ completely determine the short-time section corresponding to $|X(nL, k)|$ for $n = 1$. The short-time section corresponding to $|X(nL, k)|$ for $n = 2$ can then be extrapolated from its DFT magnitude and its known samples in the region of overlap with the previously determined short-time section. This process continues as the complete extrapolation of each new short-time section makes possible the extrapolation of the next overlapping short-time section. Although valid in theory, this algorithm has computational difficulty as L increases. For L greater than about four samples, accuracy of the reconstruction degrades rapidly in time because of a computational round-off error as with, for example, quantization error in the FFT; the sequential nature of the algorithm causes the effect of this error to accumulate over time [12]. Thus, more robust reconstruction algorithms are needed [6],[12]. We discuss one such algorithm in Section 7.5.2.

There are also a variety of other uniqueness conditions that express the tradeoff between time decimation and frequency sampling [7],[12],[19]. For example, for a time-decimation factor of $L = 1$, it can be shown that, under certain mild conditions, at each time instant two samples of $|X(n, \omega)|$ in frequency, but not necessarily at the same samples, ω_1 and ω_2, are sufficient to recover $x[n]$ [19]. More generally, $2L$ frequency samples are required for an L-sample time decimation [7]. A number of these conditions are derived in Exercises 7.8 and 7.10.

7.5 Signal Estimation from the Modified STFT or STFTM

In many applications, it is desired to synthesize a signal from a time-frequency function consisting of a modified STFT or STFTM. Such modifications may arise due to quantization errors (e.g., from speech coding) or due to desired time-varying filtering in signal processing applications such as speech enhancement (e.g., noise reduction) and speech modification (e.g., change in articulation rate). In noise reduction, each frequency slice of the STFT is reduced in certain regions, while in changing articulation rate, we modify the STFT by discarding or adding spectral slices. We study such STFT modifications in later sections of this chapter and throughout the text.

An arbitrary function of time and frequency, however, does not necessarily represent the STFT or STFTM of a signal. This is because the definitions of these transforms impose a structure on their time and frequency variations. In particular, because of the overlap between short-time sections, adjacent short-time segments cannot have arbitrary variations. A necessary but not sufficient condition on these variations is that the short-time section corresponding to each time instant must lie within the duration of the corresponding analysis window. For example, the short-time section corresponding to $X(0, \omega)$ is given by $f_0[n] = x[n]w[-n]$ and, therefore, it must lie within the duration of $w[-n]$, as illustrated in the following example.

EXAMPLE 7.5   Consider modifying a valid $X(n, \omega)$ by inserting a zero gap where there is known to lie an unwanted interfering sinewave component. This modification is illustrated in Figure 7.18 with multiplication by the filter $H(n, \omega)$. Then the resulting $Y(n, \omega) = H(n, \omega) \times X(n, \omega)$, when inversed transformed at a specific time n to obtain the modified short-time sequence, denoted by $g_n[m]$, is non-zero beyond the extent of the original short-time segment $f_n[m] = x[m]w[n - m]$.   ▲

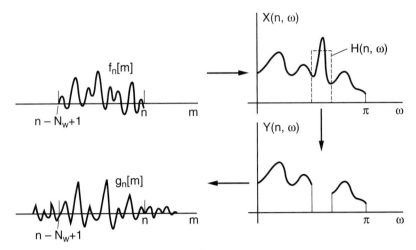

Figure 7.18 Schematic of violation of STFT duration constraint after modification. The original STFT $X(n, \omega)$ is modified by a filter $H(n, \omega)$ that removes an interfering sinewave component.

Another condition that the STFT or STFTM must satisfy is that adjacent short-time sections must be consistent in their region of overlap. When the STFT or STFTM of a signal is modified, the resulting time-frequency function does not generally satisfy such constraints. That is, if the short-time segment $f_n[m] = x[m]w[n - m]$ corresponds to $X(n, \omega)$, then, after modification, the inverse Fourier transform of the resulting STFT $Y(n, \omega)$, denoted by $g_n[m]$, is not always consistent with its adjacent short-segments, e.g., $g_{n+1}[m]$, because after dividing out the respective analysis window functions, the resulting sequences are not necessarily equal in the region of window overlap. Consider the following example.

EXAMPLE 7.6   Suppose a time-decimated STFT, $X(nL, \omega)$, is multiplied by a linear phase factor $e^{j\omega n_o}$ at time n to obtain $Y(nL, \omega) = X(nL, \omega)e^{jn_o\omega}$. Likewise, the following spectral slice $X((n + 1)L, \omega)$ is multiplied by the negative of this linear phase factor, i.e., $e^{-j\omega n_o}$, to obtain $Y((n + 1)L, \omega) = X((n + 1)L, \omega)e^{-jn_o\omega}$. Then, as illustrated in Figure 7.19, the inverse Fourier transforms, denoted by $g_{nL}[m]$ and $g_{(n+1)L}[m]$, respectively, are not consistent in their region of overlap.   ▲

The synthesis methods we discussed in Sections 7.3 and 7.4 were derived with the assumption that the time-frequency functions to which they are applied satisfy the constraints in the definitions of the STFT or STFTM, i.e., the definitions of the STFT and STFTM, as we have just seen, impose structure on the time-frequency functions. Given a function which does not satisfy those constraints, the synthesis methods have no theoretical validity for their application. However, under certain conditions, those methods can be shown to yield "reasonable" results in the presence of modifications. For example, in Section 7.5.1 we illustrate conditions under which the FBS and OLA methods yield intuitively satisfying results when the STFT has been modified with a multiplicative factor. We can think of these methods as "heuristic" in that they blindly use the FBS, OLA, and signal extrapolation synthesis methods. In Section 7.5.2, on the other hand, we discuss a "rigorous" theoretically-based approach to signal synthesis from the modified STFT that invokes least-squared-error approximation methods. A similar approach is then discussed in Section 7.5.3 for signal estimation from the modified STFTM.

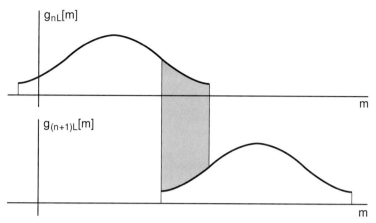

Figure 7.19   Consistency must be satisfied in adjacent short-time segments after modification for a valid STFT. This figure illustrates the violation of the consistency constraint with linear phase modification. After dividing out the window, the resulting sequences are not equal in their region of overlap.

7.5.1   Heuristic Application of STFT Synthesis Methods

Historically, signal estimation from the modified STFT has been performed by applying the FBS and OLA synthesis methods of Section 7.3 on time-frequency functions which are not valid STFT functions. Even though a modified STFT, $Y(n, \omega)$, is not a valid STFT, it is desirable that applying a synthesis method to $Y(n, \omega)$ should yield a "reasonable" result. Such heuristic use of the synthesis methods has been of practical importance in many signal processing applications. Since the synthesis methods have no theoretical basis for their application in such situations, it is common to analyze the effects that the methods have on the synthesized signal [1]. In this section, we contrast the FBS and OLA synthesis methods when they are applied to the STFT which has been modified through multiplication with another time-frequency function. For both the methods, the resulting synthesized signal can be shown to be a time-varying convolution between $x[n]$ and a function $\hat{h}[n, m]$.

Let us assume that the STFT $X(n, \omega)$ has been modified by a function $H(n, \omega)$ to give $Y(n, \omega)$, i.e.,

$$Y(n, \omega) = X(n, \omega)H(n, \omega).$$

This corresponds to a new short-time segment $g_n[m] = f_n[m] * h[n, m]$, where $h[n, m]$ can be thought of as a time-varying system impulse response (introduced in Chapter 2). We first investigate the use of the FBS synthesis method. In the FBS method, we discretize frequency to obtain

$$Y(n, k) = Y(n, \omega)|_{\omega = \frac{2\pi}{N}k} = X(n, k)H(n, k)$$

where $\frac{2\pi}{N}$ is the frequency sampling interval. Letting $\tilde{h}[n, m]$, for each n, represent the N-point inverse DFT of $H(n, k)$, we have $\tilde{h}[n, m] = \sum_{l=-\infty}^{\infty} h[n, m - lN]$. Keep in mind that although we have written $\tilde{h}[n, m]$ as periodic, we take only inverse DFT samples in the interval $[0, N - 1]$ [14]. For the FBS method, it can be shown that the resulting sequence

$y[n]$ can be written as

$$y[n] = \sum_{m=-\infty}^{\infty} x[n - m]\hat{h}[n, m] \tag{7.23}$$

where

$$\hat{h}[n, m] = w[m] \sum_{l=-\infty}^{\infty} h[n, m - lN]$$

which is equivalent to filtering $x[n]$ with the time-varying impulse response $\hat{h}[n, m]$. We see that the time-varying impulse response $\hat{h}[n, m]$ is obtained by multiplying $\tilde{h}[n, m]$, for each n, by the window $w[m]$ (Appendix 7.A and Exercise 7.11). Note that periodic replicas of the sequence $h[n, m]$ are used because of the discretized frequencies. We thus window $\sum_{l=-\infty}^{\infty} h[n, m - lN]$ and then convolve the resulting sequence $\hat{h}[n, m]$ at each time instant, n, with respect to the variable m.

Using the OLA synthesis method, it can be similarly shown that $y[n]$ can be obtained by convolving $x[n]$ with a time-varying impulse response as in Equation (7.23). For the OLA method, however, the time-varying impulse response $\hat{h}[n, m]$ is given by (Exercise 7.11)

$$\hat{h}[n, m] = w[n] * \sum_{l=-\infty}^{\infty} h[n, m - lN] \tag{7.24}$$

which, for the mth value of impulse response $\hat{h}[n, m]$, is the convolution of $h[n, m]$ with $w[n]$ with respect to the variable n. Therefore, each coefficient of the time-varying response is *smoothed* (along the time variable n) by the window $w[n]$ that typically has a lowpass frequency response [20]. For the FBS method, on the other hand, each time-varying response $h[n, m]$ is multiplied by the window $w[m]$ (along the time variable m). An important implication is that the FBS method allows instantaneous modification, while the OLA method restricts the modification to the bandwidth of the window [20]. It is interesting to note that if $h[n, m]$ is independent of n, i.e., $h[n, m] = h[m]$, and so represents a time-invariant impulse response, then the FBS method results in a synthesized signal which is the convolution of $x[n]$ with $h[n]w[n]$. On the other hand, the time-invariant case for the OLA method results in a synthesized signal which is the convolution of $x[n]$ with $h[n]$. The following example illustrates the OLA approach for this case.

EXAMPLE 7.7   Suppose we want to deliberately introduce reverberation into a signal $x[n]$ by convolution with the filter

$$h[n] = \delta[n] + \alpha\delta[n - n_o]$$

which has as its Fourier transform

$$H(\omega) = 1 + \alpha e^{-j\omega n_o}.$$

(Reverberation is often used as a special effect in speech and music modification.) But we will introduce reverberation through time-invariant modification of the STFT of $x[n]$:

$$Y(n, \omega) = X(n, \omega)H(\omega)$$

where

$$X(n, \omega) = \sum_{m=-\infty}^{\infty} x[m]w[n - m]e^{-j\omega m}.$$

Suppose we use the OLA method, i.e.,

$$y[n] = \frac{1}{W(0)} \sum_{p=-\infty}^{\infty} \left[\frac{1}{N} \sum_{k=0}^{N-1} Y(p, k)e^{j\frac{2\pi}{N}kn} \right].$$

It is then possible to derive an expression for $y[n]$ in terms of the original sequence $x[n]$ and the filter $h[n]$ by writing

$$y[n] = \frac{1}{W(0)} \sum_{m=-\infty}^{\infty} x[m] \underbrace{\left[\frac{1}{N} \sum_{k=0}^{N-1} H(k)e^{j\frac{2\pi}{N}k(n-m)} \right]}_{\text{IDFT}\rightarrow\sum_{r=-\infty}^{\infty} h[n-m+rN]} \underbrace{\left[\sum_{p=-\infty}^{\infty} w[p - m] \right]}_{W(0)}$$

$$= \sum_{m=-\infty}^{\infty} x[m]\hat{h}[n - m]$$

where IDFT denotes the inverse DFT, and where for $h[n]$, we have the modified echo-generating function

$$\hat{h}[n] = \sum_{r=-\infty}^{\infty} h[n + rN]$$

$$= \sum_{r=-\infty}^{\infty} (\delta[n + rN] + \alpha\delta[n - n_0 + rN])$$

from which we take values only in the interval $[0, N - 1]$ because of the inverse DFT operation (represented in Figure 7.20 by the sequence $R[n]$ which is a rectangular function, non-zero in the interval $[0, N - 1]$). Therefore, STFT-based filtering with OLA synthesis does indeed yield the desired reverberation when $n_o < N$, as seen in Figure 7.20.   ▲

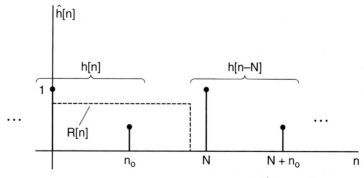

Figure 7.20 Illustration of the echo-generating function $\hat{h}[n]$ that results from the OLA method in Example 7.7. $R[n]$ is the rectangular function, non-zero in the interval $[0, N - 1]$, invoked by the inverse DFT.

In this section we have seen how the effects of applying the FBS and OLA methods to the modified STFT may be analyzed for the case of multiplicative modifications. A similar analysis can also be carried out for situations where a time-frequency function has been added to a valid STFT [1].

7.5.2 Least-Squared-Error Signal Estimation from the Modified STFT

Rather than applying the FBS and OLA methods in a heuristic manner, we now consider a different approach that is specifically designed for signal estimation from the modified STFT. In this approach, we estimate a signal whose STFT is closest in some sense to the modified STFT (Figure 7.21)[6] [6]. More specifically, we want to minimize the mean-squared-error between the discrete-time STFT, $X_e(n, \omega)$, of the signal estimate, $x_e[n]$:

$$X_e(n, \omega) = \sum_{m=-\infty}^{\infty} w[n - m] x_e[m] e^{-j\omega m}$$

and the modified discrete-time STFT, which we denote by $Y(n, \omega)$. This optimization results in the following solution for the estimated signal $x_e[n]$ [6]:

$$x_e[n] = \frac{\sum_{m=-\infty}^{\infty} w[m - n] f_m[n]}{\sum_{m=-\infty}^{\infty} w^2[m - n]} \tag{7.25}$$

where $f_m[n]$ is the inverse Fourier transform of the short-time segment at time m, corresponding to the modified STFT, $Y(m, \omega)$. The specific distance measure used in the minimization is the squared error between $X_e(n, \omega)$ and $Y(n, \omega)$ integrated over all ω and summed over all n:

$$D[X_e(n, \omega), Y(m, \omega)] = \sum_{m=-\infty}^{\infty} \frac{1}{2\pi} \int_{-\pi}^{\pi} |X_e(m, \omega) - Y(m, \omega)|^2 d\omega. \tag{7.26}$$

Proof of Equation (7.25) can be made by using Parseval's theorem, making the error criterion a function of the desired signal, $x_e[n]$, and minimizing with respect to the desired signal for each time n (Exercise 7.15). Note that although the distance measure is defined over continuous frequency, the implementation of the solution for $x_e[n]$ that minimizes the distance measure

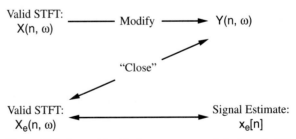

Figure 7.21 Approach of finding a sequence $x_e[n]$ whose (valid) STFT is "close" to a desired $Y(n, \omega)$ in some sense.

---

[6] This approach has also been used to recover a sequence from other time-frequency distributions such as the wavelet transform and Wigner distribution that we describe in Chapters 8 and 11.

involves frequency samples of $Y(n, \omega)$; therefore, it is required that the frequency sampling factor $\frac{2\pi}{N}$ be small enough so that unaliased versions of the short-time sections $f_m[n]$ are obtained.

The solution in Equation (7.25) extends in a simple manner to the case involving temporal decimation. Specifically, if L is the temporal decimation factor, then the solution in Equation (7.25) becomes:

$$x_e[n] = \frac{\sum_{m=-\infty}^{\infty} w[mL - n] f_{mL}[n]}{\sum_{m=-\infty}^{\infty} w^2[mL - n]} \qquad (7.27)$$

which is illustrated graphically in Figure 7.22, where we see that for each time n a set of weighted short-time segments contributes to the solution. Finally, it is interesting to observe that when no modification is made, then the optimal solution is $x[n]$ as expected, provided that $\sum_{p=-\infty}^{\infty} w^2[pL - n] \neq 0$ (Exercise 7.15). We refer to Equation (7.27) as the least-squared-error (LSE) solution.

In general, the sum in the denominator of the right side of Equation (7.27) is a function of n. However, there exist analysis windows $w[n]$ such that the sum in the denominator is independent of n. It should be noted that the sum in the denominator has the same form as the sum in the constraint Equation (7.20) for the OLA method except that the analysis window is replaced by its square. That is, any window whose square satisfies the OLA constraint will make the

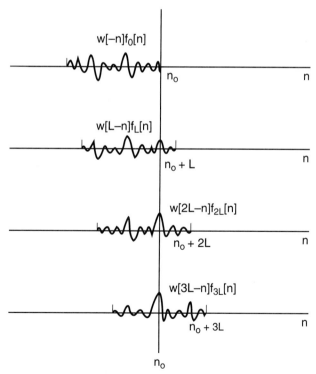

Figure 7.22   Least-squared-error (LSE) solution from the modified STFT with time decimation: $x_e[n] = \frac{\sum_{m=-\infty}^{\infty} w[mL-n] f_{mL}[n]}{\sum_{m=-\infty}^{\infty} w^2[mL-n]}$. Weighted short-time segment contributions are shown at time n_o.

Table 7.1 Comparison of the OLA and LSE solutions.

	OLA	LSE
Synthesis	$y[n] = \frac{L}{W(0)} \sum_{m=-\infty}^{\infty} \underbrace{x[n]w[mL-n]}_{f_{mL}[n]}$	$x_e[n] = \frac{\sum_{m=-\infty}^{\infty} w[mL-n] f_{mL}[n]}{\sum_{m=-\infty}^{\infty} w^2[mL-n]}$
Constraint	$\sum_{m=-\infty}^{\infty} w[mL-n] = \frac{W(0)}{L}$	$\sum_{m=-\infty}^{\infty} w^2[mL-n] \neq 0$

denominator sum in Equation (7.27) independent of n. If this happens, then Equation (7.27) can be simply interpreted as an overlap-add operation among the short-time sections corresponding to $Y(n, \omega)$, but with an additional weighting of each short-time section by the window. For the particular case of a rectangular analysis window that satisfies the OLA constraint, the LSE method reduces to the OLA method. A comparison of these relations is given in Table 7.1.

7.5.3 LSE Signal Estimation from Modified STFTM

The LSE approach can also be used for signal estimation from the modified STFTM which may or may not have come from a valid STFT. The approach is to estimate a signal $x_e[n]$ whose STFTM is "closest" in a least-squared-error sense to the modified STFTM [6]. More specifically, the resulting method estimates a sequence $x_e[n]$ from a desired time-frequency function $|Y(n, \omega)|$, which is a modified version of an original STFTM, $|X(n, \omega)|$, similar to that illustrated in Figure 7.21 but with STFT magnitudes replacing the STFTs. The method iteratively reduces the following distance measure between the STFTM, $|X_e(n, \omega)|$, of the signal estimate and the modified STFTM, $|Y(n, \omega)|$:

$$D[|X_e(n, \omega)|, |Y(n, \omega)|] = \sum_{m=-\infty}^{\infty} \frac{1}{2\pi} \int_{-\pi}^{\pi} \left[|X_e(m, \omega)| - |Y(m, \omega)|\right]^2 d\omega \quad (7.28)$$

where the minimization occurs with respect to the unknown signal $x_e[n]$ embedded within $X_e(n, \omega)$. The solution is found iteratively because as yet no closed-form solution has been discovered for $x_e[n]$ using the distance criterion in Equation (7.28). The iteration takes place as follows [6]. An arbitrary sequence (usually white noise) is selected as the first estimate $x_e^1[n]$ of $x_e[n]$. We then compute the STFT of $x_e^1[n]$ and modify it by replacing its magnitude by the desired magnitude $|Y(n, \omega)|$. From the resulting modified STFT, we can obtain a signal estimate (closest to this modified STFT) using the method based on Equation (7.25) in the previous section. This process then continues iteratively. In particular, the $(i+1)$st estimate $x_e^{i+1}[n]$ is first obtained by computing the STFT, $X_e^i(n, \omega)$, of $x_e^i[n]$ and replacing its magnitude by $|Y(n, \omega)|$ to obtain $Y^i(n, \omega)$. The signal with the STFT closest to $Y^i(n, \omega)$ is found by using Equation (7.25). All steps in the iteration can be summarized in the following update equation:

$$x_e^{i+1}[n] = \frac{\sum_{m=-\infty}^{\infty} w[m-n] \frac{1}{2\pi} \int_{-\pi}^{\pi} Y^i(m, \omega) e^{j\omega n} d\omega}{\sum_{m=-\infty}^{\infty} w^2[m-n]}$$

where

$$Y^i(m, \omega) = |Y(m, \omega)| \frac{X_e^i(m, \omega)}{|X_e^i(m, \omega)|}.$$

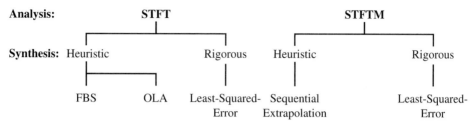

Figure 7.23 Overview of "heuristic" and "rigorous" synthesis methods corresponding to a modified STFT or SFTM analysis approach.

This algorithm is particularly important when phase is not available or difficult to measure or estimate, as demonstrated in the applications of the following section.

It has been shown [6] that this iterative procedure reduces the distance measure of Equation (7.28) on every iteration. Furthermore, the process converges to a local minimum, not necessarily the global minimum, of that distance measure. Although we restricted our above discussion to the discrete-time STFT, these results are easily extendable to the case where the STFT has been decimated in time. Furthermore, with discrete frequency the method iteratively reduces the distance measure in Equation (7.28), provided the frequency sampling factor is sufficiently large to avoid aliasing when determining the short-time sections corresponding to $Y^i(n, \omega)$.

Figure 7.23 summarizes our signal estimation methods from the modified STFT or STFTM. In each case, we have looked at both "heuristic" and "rigorous" methods of signal construction. For the STFT, the heuristic methods involved a brute-force application of the FBS and OLA synthesis methods, while the rigorous methods involved a closed-form solution to the LSE approach. For the STFTM, the heuristic methods involved a brute-force application of the sequential extrapolation method, while the rigorous method again involved the LSE approach, but using an iterative algorithm for solution.

7.6 Time-Scale Modification and Enhancement of Speech

The signal construction methods of this chapter can be applied in a variety of speech applications. In this section, we consider both the heuristic and rigorous synthesis methods from a modified STFT and STFTM in two applications: time-scale modification (i.e., changing the articulation rate of a speaker) and noise reduction. Time-scale modification is studied in a variety of other contexts throughout the text, while noise reduction is introduced more formally in a stochastic filtering framework in Chapter 13.

7.6.1 Time-Scale Modification

Model of Articulation Rate Change — Figure 7.24 shows a simple linear source/filter model for a change in the articulation rate of a speaker. The source input consists of impulses during voicing or plosives, and white noise during noisy speech. The glottal flow during voicing is embedded within the vocal tract response $h[n]$. In this model, the pitch changes at a faster or slower rate than the "nominal" (or normal) pitch so that the basic pitch *trajectory* is preserved; while the vocal tract moves faster or slower, the spectral *evolution* being preserved. In other

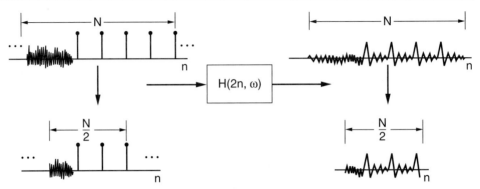

Figure 7.24 Model for uniform change in articulation rate. With a change in articulation rate, the source and system stay in (or move through) a certain state for longer or shorter time durations. In this illustration, the articulation rate is increased by a factor of two, so the source and speech waveforms are reduced to half their original length. In the source, the pitch trajectory is preserved during voicing, but the number of glottal cycles is reduced by a factor of two. The duration of the noise excitation is also reduced by a factor of two. The vocal tract moves twice as fast, but its frequency response is preserved, as indicated by the modified time-varying system function $H(2n, \omega)$.

words, the source and system pass through the same states as in nominal articulation, but they pass through these states for longer or shorter time durations. If the articulation rate is modified uniformly by a factor ρ, then, according to our model, the speech waveform is compressed or expanded in time by this factor ρ. The compression and expansion, however, do not occur as in speeding up or slowing down a tape recorder, because this would change pitch and spectral structure, but rather the waveform maintains its shape, simply decreasing or increasing the number of pitch periods during voicing and making unvoiced sounds last shorter or longer.

From our idealized convolutional model of speech production, the STFT of the speech waveform $X(n, \omega)$ can be written as a product of source and system components. In particular, for a sufficiently long window $w[n]$ (but intended to not be long enough to violate stationarity), we can think of each spectral slice of $X(n, \omega) \approx U(n, \omega) H(n, \omega)$, where $U(n, \omega)$ and $H(n, \omega)$ are the STFT of the source and the time-varying vocal tract system function, respectively. The duration of the window $w[n]$ is typically selected as two or three times the average pitch period, so that $U(n, \omega)$ is characterized by harmonic structure during voicing and by random fluctuations about the power spectrum during unvoiced regions (excluding an idealized impulsive source for a plosive sound category). In the frequency domain, the goal of time-scale modification is to modify the rate at which $U(n, \omega)$ and $H(n, \omega)$ vary with time and, therefore, how fast $X(n, \omega)$ varies with time. This coincides with how a change in articulation rate influences the speech spectrogram $|X(nL, \omega)|$. Suppose that a person speaks twice as slowly as his/her normal articulation rate. Then the spectral evolution of the speech would stretch over roughly twice the time of a normally spoken passage. This implies that the spectrogram is stretched out like an accordion. A similar argument can be made in speeding up articulation rate.

The model of articulation rate change is approximate; it is not known how articulation rate change takes place precisely, but it is clear that it does not occur uniformly, as assumed by our simple model. This nonuniform rate change was illustrated in Figure 3.30 of Chapter 3, which showed that voiced sounds tend to be altered more than fricative sounds or plosives.

For example, one does not expect a plosive to expand without limit because such expansion would alter the essence of the sound. We consider this more complex model in Chapter 9. Furthermore, the mechanism is probably more complicated than involving a simple temporal scaling of pitch and spectral trajectories. Time-scale modification may also entail changes in glottal volume velocity at the source and vocal tract shape which are not predictable by our idealized model. For example, in our model we have embedded the glottal flow during voicing in the vocal tract impulse response. With change in articulation rate, however, the flow may alter the spectral structure of this composite system response. For example, when speaking faster it may be less likely that we completely close the vocal folds (than in a nominal speaking state) because all articulators are moving faster; this incomplete closure affects the trend of the speech spectrum.

STFT Synthesis — One of the first time-scale modification systems uses the "cut and paste" method of Fairbanks [5].[7] To do time-compression by, for example, a factor of two, we extract successive short-time segments of the speech waveform and then discard every other time slice. Likewise, to expand the waveform, we repeat successive time slices. The modified temporal slices are then overlapped and added (Figure 7.25). In the frequency domain with time-decimation factor L, this corresponds to first forming the STFT, $X(nL, \omega)$, and then discarding or repeating spectral slices to form a new STFT $Y(nL, \omega)$. For example, with time-scale compression by a factor of two, $Y(nL, \omega) = X(2nL, \omega)$ (Figure 7.26). As we argued earlier, for a sufficiently long window $w[n]$, we can think of each spectral slice $X(nL, \omega) \approx U(nL, \omega)H(nL, \omega)$, where $U(nL, \omega)$ and $H(nL, \omega)$ are the STFTs of the source and the time-varying vocal tract spectrum, respectively. Thus, the source (pitch and voicing state) and system (vocal tract spectrum) characterization are roughly preserved in the modified STFT $Y(nL, \omega)$.

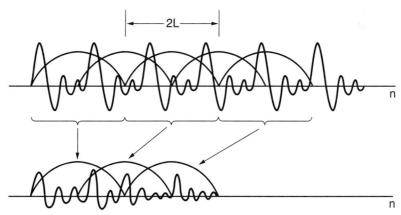

Figure 7.25   Time-scale compression by the Fairbanks technique. Alternating short-time segments are discarded and the remaining ones are overlapped and added.

---

[7] Observe that the methods in this section, including the Fairbanks and STFT- or STFTM-based approaches, are largely "non-model based," in contrast to "model-based" approaches. Examples of model-based approaches to time-scale modification use linear prediction and homomorphic analysis/synthesis (Exercise 7.16).

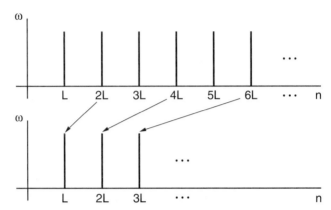

Figure 7.26   A STFT view of the Fairbanks technique, discarding every other frame for time-scale compression by a factor of two.

One can inverse Fourier transform the modified STFT, $Y(nL, \omega)$, and then, according to the (heuristic use of) OLA method, synthesize a modified sequence by overlapping and adding the resulting time slices. Although this transformation to the frequency domain and back is clearly superfluous, it gives us a different perspective on Fairbank's technique from the view of the OLA method. Observe that this synthesis is similar to the more rigorous LSE approach of Equation (7.27), except that, in the LSE approach, the inverted short-time segments are weighted by the analysis window, and the sum of weighted short-time segments is normalized by the sum of the squared overlapping windows. There is a problem, however, with both the heuristic and rigorous approaches: voiced segments are not "pitch synchronized," reflecting that the desired $Y(nL, \omega)$ is not a valid STFT in the sense that corresponding consecutive short-time segments are not consistent. For example, simply discarding every other frame gives a set of time slices that are not pitch synchronized when overlapped and added, as illustrated in Figure 7.27, producing a waveform construction with an erratic pitch perceived as "hoarseness" (Figure 7.25). A similar problem arises in time expansion where time slices are repeated to give a modified STFT. Alternatively, one can align the window on each frame at the same relative location within a pitch period, such as at a glottal pulse time, the time of glottal opening or closing, or a time instant at which there is no linear phase in the system impulse response (as we did in Section 6 of Chapter 6 in the context of homomorphic synthesis). You are asked to design such a system in Exercise 7.22.

This pitch-synchronized approach was introduced by Scott and Gerber [22], who used a glottal pulse time estimate to align pitch periods prior to overlapping and adding short-time segments, thus improving the quality of the time-scaled waveform. Successive frames are shifted so that they align at glottal pulse times. Rather than explicitly measuring glottal pulse times (or other times characteristic of the glottal flow), measures of "waveform similarity" are also used to select splicing points. In one such method, *synchronized overlap add* (SOLA), successive frames to be overlapped are cross-correlated. The peak of the cross-correlation function gives the time shift to make the two overlapping frames synchronize and thus add coherently[8] [21]. Other extensions of this approach have also been developed [9],[11].

---

[8] Conceptually, the measurement and application of this time difference is similar to the post-alignment process used in the mixed-phase homomorphic synthesis in Section 6 of Chapter 6.

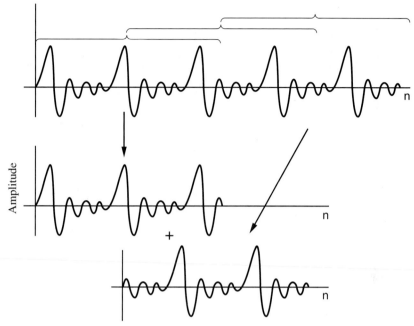

Figure 7.27 The problem of "pitch synchrony" in the OLA and LSE synthesis approaches. Pitch periods in the modified short-time segments are not synchronized.

The problems we have been describing here with pitch synchrony relate directly to the STFT phase function and, in particular, to the linear phase component of the STFT, because a time shift maps to a linear phase change. We now take an alternative approach to synthesis for time-scale modification, invoking the magnitude only of the STFT.

STFTM Synthesis — One way to avoid the need of pitch synchrony is to use only the magnitude of the STFT (STFTM) in time-scale modification. This approach is also motivated, in part, by the observation that when a person talks more rapidly or more slowly, the spectrogram is compressed or expanded in time like an accordion. As we argued above, with a sufficiently long window, $X(nL, \omega)$ captures essential source and system properties for articulation rate change. As with the Fairbanks technique, we discard or add frames, but now perform waveform synthesis from a modified STFTM, $|Y(nL, \omega)|$, using the iterative LSE method described in the previous section [6]. For time-compression by a factor of two, for example, we have the desired $|Y(nL, \omega)| = |X(2nL, \omega)|$, which was illustrated in Figure 7.26. Alternatively, to avoid the discarding of time slices, and thus the loss of pieces of the spectrogram, we propose the following algorithm for time-scale modification by a factor $\rho = \frac{M}{L}$ illustrated with $\rho = \frac{1}{2}$:

S1: Compute $|X(nL, \omega)|$ at an appropriate frame interval (e.g., $L = 128$ at 10000 samples/s).

S2: Pretend $|X(nL, \omega)|$ was determined with time decimation $M = \frac{L}{2}$ and form the desired STFTM (Figure 7.28):

$$|Y(nM, \omega)| = |X(nL, \omega)|.$$

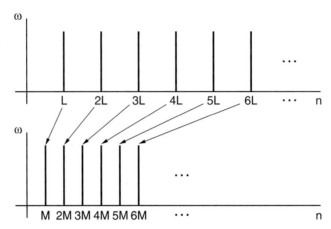

Figure 7.28 Alternative modified STFT for time-scale modification where no frames are discarded. In the example, $M = \frac{L}{2}$.

S3: Apply the LSE iterative algorithm with the desired $|Y(nM, \omega)|$.

EXAMPLE 7.8   Figure 7.29 shows an example with an original spectrogram [panel (a)] computed with $L = 128$ and the desired spectrogram [panel (b)] formed by compressing the original so that $M = \frac{L}{2} = 64$ [6]. The Fourier transforms were implemented with 512-point FFTs and the LSE algorithm ran for 100 iterations. Panel (c) in the figure gives the spectrogram of the resulting compressed waveform. The initial waveform estimate is Gaussian white noise. The distance measure $D[|X_e(n, \omega)|, |Y(n, \omega)|]$ of Equation (7.28) for this example, as a function of iteration number, does not approach zero, but does decrease monotonically. ▲

In time-scale modification of voiced speech with this method, although judged to be generally of good quality, a reverberant characteristic is occasionally perceived due to lack of STFT phase control. In other words, the STFT phase of the modified speech can differ from that of the original speech because we have constrained only the STFT magnitude. This phase change corresponds to a dispersed waveform that, visually, may bear little resemblance to the original speech signal (Exercise 7.17). In addition, in slowing down noise-like unvoiced speech (e.g., fricatives), a metallic quality is perceived. This effect can be understood by recalling that the periodogram of a random process fluctuates about an underlying power spectral density. With time-scale expansion using the LSE method, random peaks in the periodogram are stretched and thus held over time, resulting in unnatural "tones" in the synthesis, not characteristic of the original random process. Another important property of the LSE method is that the synthesis is highly dependent on the length of the analysis window $w[n]$. A long window, e.g., 2–3 pitch periods in duration, and thus a narrowband spectrogram are necessary to capture harmonic structure; this ensures a time-scale modified waveform for which harmonics are stretched or compressed in time. (Recall that we want $X(n, \omega) \approx U(n, \omega)H(n, \omega)$, where for voiced speech $U(n, \omega)$ is harmonic.) In contrast, with a short window, e.g., about a pitch period in duration, the method modifies the pitch of the speaker and not the time scale (Exercise 7.18).

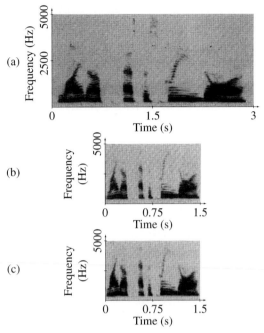

Figure 7.29   Time-scale modification with iterative LSE estimation: (a) original STFTM with $L = 128$;   (b) modified STFTM with $L = 64$;   (c) STFTM of LSE estimate. Speech utterance is, "Line up at the screen door."

Source: D. Griffin and J.S. Lim, "Signal Estimation from Modified Short-Time Fourier Transform" [6]. ©1984, IEEE. Used by permission.

7.6.2 Noise Reduction

A number of STFT processing techniques have been developed for the reduction of additive noise in speech signals. The noise corrupted signal $y[n]$ is given by

$$y[n] = x[n] + b[n]$$

where $x[n]$ is the speech and $b[n]$ is the noise sequence. The approach is to modify each spectral slice of the STFT of $y[n]$ to remove noise and, from the resulting modified STFT, construct an enhanced waveform. As with time-scale modification, both STFT- and STFTM-based synthesis methods can be applied.

STFT Synthesis — In one of the first approaches to noise reduction, the magnitude of the STFT is modified by subtracting off an approximate noise power spectrum $\hat{S}_b(\omega)$ from $|Y(nL, \omega)|^2$ and retaining the noisy phase [3], i.e., the desired modified STFT is formed as

$$\hat{X}(nL, \omega) = [|Y(nL, \omega)|^2 - \alpha \hat{S}_b(\omega)]^{\frac{1}{2}} e^{j \angle Y(nL, \omega)}$$

where if $|Y(nL, \omega)|^2 - \alpha \hat{S}_b(\omega) < 0$, then the difference is set to zero, and where the phase is retained because of difficulty in its estimation. In this *spectral subtraction* method, the parameter

α serves as a control for the degree of noise reduction. Alternatively, in another early approach to noise reduction, the magnitude of the STFT is modified by a time-varying "optimal filter" $H(nL, \omega)$. This filter reduces the measured spectral slice when the noise spectrum is high relative to the time-varying speech spectrum, and preserves the spectral measurement when the noise spectrum is low relative to the speech spectrum, but again keeps the measured phase, i.e.,

$$\hat{X}(nL, \omega) = [|Y(nL, \omega)|H(nL, \omega)]e^{j\angle Y(nL,\omega)}.$$

In either case, the OLA or LSE method can then be applied to obtain an enhanced waveform construction. Such processing has been performed with the above modifications with a resulting drop in noise level. The primary processing artifacts with these basic approaches are the presence of short tone-bursts of varying frequency in the noise, sometimes referred to as "musicality," and some distortion of the speech spectrum [10]. These noise reduction methods are described in greater detail in Chapter 13.

STFTM Synthesis — Alternatively, we can avoid retaining the noisy measured phase and obtain a signal estimate from the modified STFT magnitude, using either the sequential extrapolation algorithm or the LSE estimation method. In one example, spectral subtraction was applied to the STFTM of a noisy speech waveform using a temporal decimation factor $L = 64$ and Hamming window of length $N_w = 128$. An enhanced speech waveform was then obtained using a variation of the sequential extrapolation algorithm described in Section 7.4.2. With a variety of signal-to-noise ratios (SNR) between 0 to 20 dB, for SNR above about 10 dB the signal estimates from the modified STFTM had a reduced noise level and retained natural speech quality and speaker identifiability [12]. As with the use of the STFT-based synthesis above, the primary artifacts are "musicality" and some distortion of the speech spectrum, typical of short-time spectral subtraction. A potential advantage of the STFTM-based approach over the STFT-based synthesis methods is that a STFT phase estimate is obtained (indirectly) rather than retaining the noisy phase function. Formal listening tests comparing the STFT- and STFTM-based synthesis, however, have yet to be performed using spectral subtraction or other, more refined methods that modify the spectral magnitude.

7.7 Summary

In this chapter, we presented both a Fourier transform and a filtering perspective of the discrete-time and the discrete STFT in signal analysis. Corresponding to these two analysis viewpoints are the overlap-add (OLA) and filter bank summation (FBS) methods of signal synthesis from the discrete STFT. Likewise, we considered signal analysis from the STFT magnitude (STFTM) for which we developed a sequential extrapolation approach to synthesis. Analysis and synthesis with the STFTM is useful when the STFT phase is not appropriate or difficult to measure. We also considered the important practical problem of estimating a signal from a processed STFT or STFTM which does not satisfy the definitional constraints of the STFT. Under this condition, blind use of the OLA, FBS, and sequential extrapolation methods can be applied. A more "rigorous" approach, however, was also developed whereby a signal estimate is obtained that has an STFT or STFTM close to the desired modified time-frequency function in a least-squared-error sense. Finally, we touched upon two particular applications of STFT and STFTM analysis and synthesis: time-scale modification and noise reduction. These applications illustrate how the STFT and STFTM analysis/synthesis techniques can provide a framework for speech signal processing. We will encounter this framework in many other speech processing areas throughout the text.

Appendix 7.A: FBS Method with Multiplicative Modification

Let the discrete STFT be modified by a function $H(n, k)$, i.e.,

$$y[n] = \frac{1}{Nw[0]} \sum_{k=0}^{N-1} X(n, k) H(n, k) e^{j\frac{2\pi}{N}nk}$$

and consider the time-invariant case

$$H(n, k) = H(k).$$

Letting $w[0] = 1$, and substituting the discrete STFT, $X(n, k)$:

$$y[n] = \frac{1}{N} \sum_{k=0}^{N-1} \left[\sum_{m=-\infty}^{\infty} w[n-m]x[m]e^{-j\frac{2\pi}{N}km} \right] H(k)e^{j\frac{2\pi}{N}nk}$$

$$= \sum_{m=-\infty}^{\infty} w[n-m]x[m] \left[\frac{1}{N} \sum_{k=0}^{N-1} H(k)e^{j\frac{2\pi}{N}k(n-m)} \right]$$

where the bracketed term is the inverse DFT of $H(k)$ evaluated at $n - m$. Therefore,

$$y[n] = \sum_{m=-\infty}^{\infty} w[n-m]x[m] \sum_{l=-\infty}^{\infty} h[n-m-lN]$$

$$= \sum_{m=-\infty}^{\infty} x[m] \left(w[n-m] \sum_{l=-\infty}^{\infty} h[n-m-lN] \right).$$

Because the bracketed term is a function of $n - m$, $y[n]$ can be written:

$$y[n] = x[n] * \hat{h}[n]$$

where

$$\hat{h}[n] = w[n] \sum_{l=-\infty}^{\infty} h[n-lN]$$

and where we take only values in the interval $[0, N-1]$ due to the inverse DFT operation. But if we assume the FBS constraint $N_w < N$ is satisfied, we see in Figure 7.30 that there is no aliasing of the $h[n]$ replicas and $\hat{h}[n]$ reduces to $\hat{h}[n] = w[n]h[n]$. We then need to select a window $w[n]$ to achieve some desired fixed filter impulse response. In particular, we see that a rectangular window does not result in a change in the original impulse response $h[n]$.

More generally, it can be shown for a time-varying multiplicative modification $H(n, k)$:

$$y[n] = \sum_{m=-\infty}^{\infty} x[n-m]\hat{h}[n, m] \tag{7.29}$$

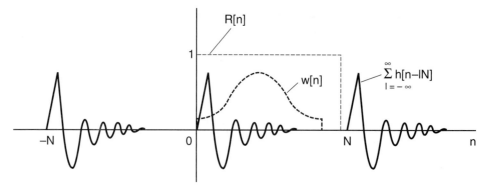

Figure 7.30 Schematic of relation of periodic sequence $\sum_{l=-\infty}^{\infty} h[n-lN]$ and the window $w[n]$ for the FBS method with a multiplicative modification. $R[n]$ is the rectangular function, non-zero in the interval $[0, N-1]$, invoked by the inverse DFT.

where

$$\hat{h}[n,m] = w[m] \sum_{l=-\infty}^{\infty} h[n, m - lN]$$

which is equivalent to filtering $x[n]$ with a time-varying impulse response $\hat{h}[n, m]$. We window $\sum_{l=-\infty}^{\infty} h[n, m - lN]$ and then convolve at each time instant, n, to obtain a single value $y[n]$. When aliasing of the response replicas does not occur, then $\hat{h}[n, m]$ reduces to the windowed time-varying filter response.

In summary, we see that for a time-varying $h[n, m]$, at each time n we invoke a different filter in the variable m. In essence, in Equation (7.29), we are filtering $x[n]$ with a time-varying impulse response. This operation was first introduced in Chapter 2 in Equations (2.26) and (2.27), where we described the time-varying unit sample response $h[n, m]$ and its related Green's function $g[n, m]$. This connection leads to other interesting relations between time-varying filtering and STFT analysis/synthesis derived by Portnoff [17], as well as its application to time-scale modification of speech [16].

EXERCISES

7.1 Show that the STFT expression in Equation (7.4) can be rewritten as

$$X(n, \omega_o) = e^{-j\omega_o n}(x[n] * w[n]e^{j\omega_o n}).$$

7.2 Repeat Example 7.1 using a window of length two pitch periods, corresponding to a narrowband spectrogram case. Compare your result with the wideband spectrogram case of Example 7.1.

7.3 Derive the FBS constraint Equation (7.13) in the frequency domain from Equation (7.11).

7.4 Show that for the OLA synthesis method, under OLA constraint Equation (7.20), the reconstruction Equation (7.21) follows.

7.5 Explain qualitatively the *duality principle* in both the time and frequency domains, which is associated with the FBS constraint Equation (7.13) and the OLA constraint Equation (7.20).

7.6 We stated in Section 7.3 that a generalized synthesis equation for recovering a sequence $x[n]$ from its STFT $X(n, \omega)$ is given by

$$x[n] = \frac{1}{2\pi} \int_{-\pi}^{\pi} \left[\sum_{r=-\infty}^{\infty} f[n, n-r] X(r, \omega) \right] e^{j\omega n} d\omega \qquad (7.30)$$

where

$$X(r, \omega) = \sum_{m=-\infty}^{\infty} x[m] w[r-m] e^{-j\omega m}$$

and where the function $f[n, m]$ is referred to as the synthesis filter.

(a) Show that a sequence $x[n]$ can be recovered from Equation (7.30) under the constraint:

$$\sum_{m=-\infty}^{\infty} f[n, -m] w[m] = 1 \qquad (7.31)$$

thus ensuring that Equation (7.30) is a valid synthesis equation.

(b) Show that with $f[n, m] = \delta[n]$, Equation (7.30) yields the synthesis equation for the basic FBS method.

(c) Show that with $f[n, m] = \frac{1}{W(0)}$, Equation (7.30) yields the synthesis equation for the OLA method with time decimation factor $L = 1$.

7.7 Consider a sequence $x[n]$ and an analysis window $w[n]$, which is non-zero over its length, N_w (assumed even). Suppose the first L samples of $x[n]$ are known. In this problem, you consider the representation of $x[n]$ by time and frequency samples of its STFT magnitude.

(a) Show that each short-time autocorrelation function $r[n, m]$ of $x[m] w[m-n]$ is at most $2N_w - 1$ points long and thus can be obtained (without aliasing) from $2N_w - 1$ frequency samples of the STFT magnitude.

(b) Consider the effects of temporal decimation with factor L (i.e., $r[nL, m]$) for which adjacent short-time sections have an overlap of $N_w - L$ samples. Prove that the sequence $x[n]$ can be uniquely recovered from the STFT magnitude, provided that:

1. Overlap between short-time sections is greater than $N_w/2$.
2. There exist no zero gaps in $x[n]$ of length greater than $N_w - 2L$.

7.8 The purpose of this problem is to show that, under certain conditions, only one frequency sample of the STFT magnitude is needed for unique representation of a non-negative sequence $x[n]$ for a time-decimation factor of unity. The STFT for a time-decimation factor of unity is defined by

$$X(n, \omega) = \sum_{m=-\infty}^{\infty} x[m] w[n-m] e^{-j\omega m}$$

where $w[n]$ is the analysis window of length N_w. Suppose that the following conditions on $x[n]$ and $w[n]$ hold (Figure 7.31):

1. $x[n]$ is a non-negative (i.e., $x[n] \geq 0$) right-sided sequence whose first non-zero value falls at $n = n_0$.

2. The analysis window $w[n]$ is N_w points long and positive (i.e., $w[n] > 0$) over the interval $0 \leq n \leq N_w - 1$.

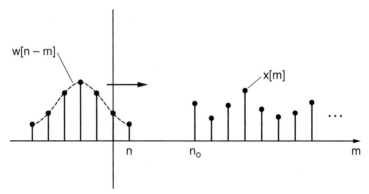

Figure 7.31 Relation between the analysis window and the sequence in generating the STFT for Example 7.8.

You are now asked to show that the sequence $x[n]$ is specified by one appropriately chosen frequency sample of $|X(n, \omega)|$.

(a) Consider the smallest value of n, namely n_o, such that $x[n]$ is non-zero. Show that $x[n_o]$ can be determined by the expression (Figure 7.31):

$$x[n_o] = \frac{|X(n_o, 0)|}{w[0]}.$$

(b) Suppose now that $x[n]$ is known up to time $n - 1$ and we want to compute the sample $x[n]$ from the previous values and $|X(n, \omega)|$. Show that

$$X(n, \omega) = Y(n, \omega) + x[n]w[0]e^{-j\omega n}$$

where

$$Y(n, \omega) = \sum_{m=n-N+1}^{n-1} x[m]w[n - m]e^{-j\omega m}.$$

$Y(n, \omega)$ is known, since it is a function of samples prior to time n.

(c) Under the above two conditions, show that $x[n]$ can be evaluated as

$$x[n] = \frac{1}{w[0]}(|X(n, 0)| - \sum_{m=n-N+1}^{n-1} x[m]w[n - m]).$$

Argue then that $x[n]$ can be recovered recursively for $n > n_o$ from only the DC value, i.e., $|X(n, 0)|$, of the STFT magnitude. It is interesting to note that this result can be generalized to require two frequency samples when the non-negative restriction is taken off of $x[n]$.

7.9 We saw in Section 7.6.1 a time-scale modification method based on a modified STFT magnitude, using the least-squared-error estimation method. This problem asks you to consider time-scale modification using the STFT.

(a) Suppose the STFT of $x[n]$ is computed with a time-decimation factor $L = 128$. Propose and describe a method based on least-squared-error estimation, from a modified STFT, to time-scale compress $x[n]$ by a factor of two. Briefly describe the steps of your method.

(b) Why might the least-squared-error estimation method be flawed? Carefully explain your reasoning. *Hint:* Consider the STFT phase.

7.10 Show that for a frame interval of L samples (i.e., time decimation of L samples), $2L$ samples of $|X(nL, k)|$ over $[0, \pi]$ in frequency are required to uniquely recover a sequence $x[n]$. Assume the window length $N_w \geq L/2$. For simplicity, assume N_w and L are even.

7.11 We derived in Section 7.5.1 (Appendix 7.A) the signal $y[n]$ resulting when the discrete STFT $X(n, k)$ of a signal $x[n]$ is modified by a time-invariant multiplicative modifier $H(n, k) = H(k)$. This was done for the FBS method. Derive the case of a time-varying multiplicative modifier $H(n, k)$ for the FBS method. Then derive the signal synthesis from a multiplicatively modified STFT using the OLA method with first the time-invariant and then the time-varying case. We will then have derived two cases for each method: (1) A time-invariant modification $H(k)$ and (2) A time-varying modification $H(n, k)$, for both FBS and OLA synthesis.

7.12 Consider modifying the STFT to obtain

$$Y(n, \omega) = X(n, k\omega)H(\omega)$$

where the modifying function is given by

$$H(\omega) = e^{jn_0\omega}$$

i.e., a linear-phase modification. Suppose a sequence $y[n]$ is computed by the filter bank summation (FBS) synthesis method:

$$y[n] = \left[\frac{1}{NW(0)}\right] \sum_{k=0}^{N-1} Y(n, k)e^{j\frac{2\pi}{N}kn}.$$

Derive an expression for $y[n]$ in terms of the original sequence $x[n]$ and the window $w[n]$. Consider two different cases: (1) The length of $w[n]$ less than N, and (2) The length of $w[n]$ greater than or equal to N. Give $y[n]$ for each case.

7.13 Figure 7.32 illustrates an idealization of the spectrogram, i.e., the short-time Fourier transform magnitude (STFTM), of the word "ate," consisting of the voiced phoneme /e/, the unvoiced phoneme /t/,

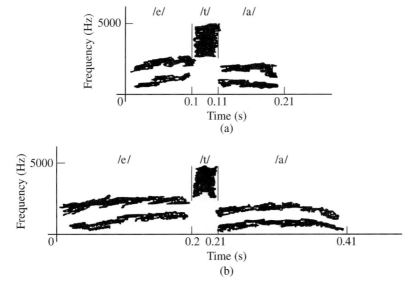

Figure 7.32 Spectrograms of the word "ate": (a) case I; (b) case II.

followed by the voiced phoneme /a/. The two different spectrograms correspond to two different articulation rates.

(a) Describe the differences in the articulation rate of the speaker with respect to the two spectrograms.

(b) Suppose you are given only the spectrogram of Figure 7.32(a) and the time-decimation factor $L = 10$ samples with a speech sampling rate of 10000 samples/s. In a few steps, describe a method to construct a waveform whose spectrogram approximately equals that of Figure 7.32(b). Assume a 200-sample Hamming analysis window. Are there any special considerations at the voiced/unvoiced and unvoiced/voiced transitions?

7.14 Consider a discrete-time signal $x[n]$ passed through a bank of filters $h_k[n]$ where each filter is given by a modulated version of a baseband prototype filter $h[n]$, i.e.,

$$h_k[n] \;=\; h[n] \exp[j(2\pi/N)kn]$$

where $h[n]$, a Hamming window, is assumed causal and lies over a duration $0 \le n < N_w$, and $2\pi/N$ is the frequency sampling factor. In this problem, you are asked to time-scale expand some simple input signals by time-scale expanding the filter bank outputs.

(a) State the constraint (with respect to the values N_w and N) such that the input $x[n]$ is recovered when the filter bank outputs are summed.

(b) If the input to the filter bank is the unit sample $\delta[n]$, then the output of each filter is a complex exponential with "envelope" $a_k[n] = h[n]$ and phase $\theta_k[n] = (2\pi/N)kn$. Suppose each complex exponential output is time-expanded by two by interpolation of its envelope and phase (as in Chapter 2, Exercise 2.19). Derive a new constraint (with respect to values N_w and N), so that the summed filter bank outputs equal $\delta[n]$.

(c) Suppose now that the filter bank input equals

$$x[n] \;=\; \delta[n] + \delta[n - n_o],$$

and that the filter bank outputs are time-expanded as in part (b). Derive a sufficient condition on N_w, N, and n_o so that the summed filter bank output is given by

$$y[n] \;=\; \delta[n] + \delta[n - 2n_o],$$

i.e., the unit samples are separated by $2n_o$ samples rather than n_o samples.

7.15 In Section 7.5.2, we considered minimizing the mean-squared error between a modified STFT $Y(mL, \omega)$ and a valid STFT $X_e(mL, \omega)$. In this problem, you investigate the solution to this optimization problem and some related properties.

(a) Derive Equation (7.27), i.e.,

$$x_e[n] \;=\; \frac{\sum_{m=-\infty}^{\infty} w[mL - n] f_{mL}[n]}{\sum_{m=-\infty}^{\infty} w^2[mL - n]}$$

by minimization of Equation (7.26), generalized with the time decimation factor L. $f_{mL}[n]$ is the inverse Fourier transform of the modified STFT, $Y(mL, \omega)$, and $w[n]$ is the analysis window. *Hint:* Use Parseval's Theorem, make the error criterion a function of the desired signal $x_e[n]$, and minimize with respect to the desired signal for each time n.

(b) Suppose that $f_{mL}[n]$ is obtained from the unmodified STFT of a signal $x[n]$, i.e., $Y(mL, \omega) = X(mL, \omega)$. Show that the solution of part (a) recovers $x[n]$, i.e., $x_e[n] = x[n]$, provided

$$\sum_{m=-\infty}^{\infty} w^2[mL - n] \neq 0.$$

(c) Now suppose that $f_{mL}[n]$ is obtained from the following modified STFT

$$Y(mL, \omega) = X(mL, \omega)e^{-j\omega n_o}.$$

Under what condition on $w[n]$ and L does $x_e[n] = x[n]$ within a shift n_o and constant scale factor, i.e., $x_e[n] = cx[n - n_o]$?

(d) Suppose in part (c) that $w[n]$ is unity over a duration $N_w = L$. Derive a simplified expression for $x_e[n]$, starting with your solution in part (c).

(e) Discuss the similarities and differences between the OLA synthesis and the least-squared-error synthesis (part (a)). Give one set of conditions under which they give the same solution. Consider window constraints for each method. Discuss the relative benefits of the least-squared error method against the OLA method.

7.16 We have seen the application of short-time Fourier transform analysis/synthesis of the speech waveform to time-scale modification. This problem asks you to consider the application of homomorphic analysis/synthesis to time-scale modification.

(a) Consider the voiced speech case in the context of homomorphic analysis/synthesis. Suppose that pitch, voicing state, and a minimum-phase impulse response, $h[n]$ have been estimated every 10 ms. How would one modify the synthesis structure so that a voiced 10-ms segment is mapped to a 20-ms duration without change in pitch or vocal tract spectral characteristics?

(b) Now consider the unvoiced speech case, again with a 10-ms analysis interval. How would one modify the synthesis structure so that a 10-ms unvoiced segment is mapped to a 20-ms unvoiced segment without a change in the noise-like excitation or vocal tract spectral characteristics?

7.17 This problem addresses the least-squared-error approach to time-scale modification from a modified STFT.

(a) Suppose that the discrete STFT magnitude $|X(nL, k)|$ (time-decimated) of a speech signal is computed with a time-decimation factor $L = 32$ samples. Give a modified STFT magnitude from which time-scale expansion by a factor of two can be performed. Propose an algorithm for estimating the time-scale expanded signal whose STFT magnitude is a least-squared-error estimate of the modified STFT magnitude.

(b) What can you conclude about the phase of the STFT of the resulting time-scale expanded signal from part (a)? What are the implications of your answer for the shape of the modified (expanded) waveform?

7.18 Suppose you are given a wideband spectrogram created with a short analysis window of duration less than the average pitch period for the speaker being analyzed. In this problem, you study the use of the least-squared-error (LSE) signal estimation (from spectral magnitude-only) method of Section 7.5.3 to perform time-scale modification.

(a) Explain why the LSE signal estimation method, using the above wideband spectrogram, results in pitch modification and not the intended time-scale modification.

(b) How small a time decimation (i.e., the value of L) of the wideband spectrogram does one need to sufficiently track energy variations within a pitch period and thus, using the above LSE method, avoid a resulting irregular pitch change and thus a hoarse quality to the speech synthesis.

7.19 This problem considers changes in the spectrogram of an utterance with time-scale and pitch modification. The utterance is spoken by a female.

    (a) In Figure 7.33, select the spectrograms that are narrowband spectrograms (generated with a wide time-domain analysis window) and the spectrograms that are wideband spectrogram (generated with a narrow time-domain analysis window). Briefly explain your reasoning.

    (b) In Figure 7.33, select the spectrogram that is the time-scaled compressed version of the spectrogram in panel (a). Which spectrogram is the time-scaled expanded version of the spectrogram in panel (a)? What speech transformations were invoked to obtain the spectrograms you have not selected? Briefly explain your reasoning.

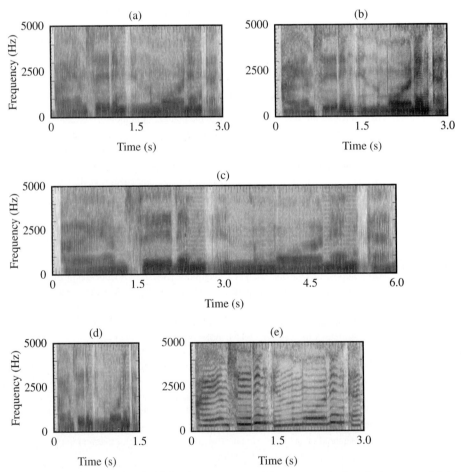

Figure 7.33 Narrowband and wideband spectrograms: (a) spectrogram of original passage; (b)–(e) spectrograms of modified passage.

(c) Suppose the time-scaled modified spectrograms in Figure 7.33 represent the results using STFT magnitude (STFTM) analysis/synthesis. In doing time-scale expansion by a factor of two with STFTM analysis/synthesis, suppose the synthesis interval in the synthesis stage for modification is 7 ms. What was the analysis frame duration in the analysis stage? In answering this question, use your result from Exercise 7.17.

7.20 We have seen that in idealized time-scale modification of a steady-state vowel, the pitch is maintained and the number of pitch periods changes. Suppose that the combined glottal flow (over one glottal cycle) and the vocal tract impulse response is denoted by $h[n] = g[n] * v[n]$. Then the steady-state vowel of duration N samples is expressed by

$$x[n] = r[n](p[n] * h[n])$$

$$= r[n] \sum_{k=-\infty}^{\infty} h[n - kP]$$

where $r[n]$ is an N-sample rectangular window, i.e., $r[n] = 1$, for $0 \leq n \leq N - 1$, and zero otherwise, and $p[n]$ is a periodic impulse train, $p[n] = \sum_{k=-\infty}^{\infty} \delta[n - kP]$.

(a) Write an expression for the steady-state vowel, time-scaled expanded by a factor of two. Sketch your result in contrast to the original waveform. Assume that the lumped response does not change in the modification, implying that neither the vocal tract impulse response nor the glottal flow shape changes.

(b) In reality, the glottal flow shape may change with a change in articulation rate. Suppose that the glottal flow during normal articulation is maximum-phase and given by

$$g[n] = \alpha^{-n}u[-n] * \alpha^{-n}u[-n]$$

i.e., a time-reversed decaying exponential convolved with itself. And suppose that when the articulation rate is slowed by a factor of two, then the glottal flow function widens as

$$g'[n] = g[n/2] = \alpha^{-n/2}u[-n] * \alpha^{-n/2}u[-n].$$

Assume now you are given the discrete STFT magnitude of $x[n]$, i.e., $|X(nL, k)|$, computed at a 10-sample frame interval (i.e., for nL where $L = 10$ is the time decimation factor) and with a 1024-point DFT. Assume also that

$$|X(nL, k)| \approx |P(nL, k)||G(k)||V(k)|$$

where $|P(nL, k)|$ is the STFT magnitude of $p[n]$, and assume you are given $|G(\omega)|$ (you can obtain this approximately from the complex cepstrum).

Propose an iterative algorithm from a modified STFT magnitude that yields a steady-state vowel, time-scale expanded by a factor of two, but having approximately a glottal flow with spectral magnitude $|G'(\omega)|$. You must first define the desired STFT magnitude and determine a way to find it. Can the resulting waveform be expressed as in your result of part (a)? (*Hint:* Consider the STFT phase of the resulting waveform, and whether your algorithm can discern a difference between $g'[n]$ and $g'[-n]$, or between $v[n]$ and $v[-n]$.)

7.21 (MATLAB) In this problem you use the speech waveform *speech2_10k* in the workspace *ex7M1.mat* located in companion website directory Chap_exercises/chapter7. This problem helps you to develop an understanding of the limitations of the STFT in achieving good time-frequency resolution. The speech was sampled at 10000 samples/s.

(a) Plot the speech signal *speech2_10k*, an unvoiced/voiced speech transition, and triangular windows of durations 30 ms and 5 ms, respectively, created using MATLAB function *triang.m*.

(b) Plot the first eight STFT log-magnitudes of *speech2_10k* by sliding the 30-ms triangular window in 15-ms intervals. Use a 1024-point FFT and display only the first 512 points of the STFT log-magnitude. Use the command *subplot (221)* so you can display two sets of four functions. Also, make a matrix of your eight STFT log-magnitudes (512 points each), and then use the *mesh.m* MATLAB function to plot the 2-D time-frequency function.

(c) Repeat part (b) with the 5-ms triangular analysis window.

(d) Repeat parts (b) and (c), but display spectrograms using the MATLAB function *spectrogram.m* or *specgram_ex7p21.m* in Chap_exercises/chapter7, rather than a mesh plot.

(e) Comment on the time-frequency resolution tradeoffs in using the long and short triangular windows.

7.22 (MATLAB) In this problem you use the voiced speech waveform *speech2_10k* in the workspace *ex7M1.mat* located in companion website directory Chap_exercises/chapter7. You are asked to design time-scale modification systems based on the OLA and LSE synthesis methods. The speech was sampled at 10000 samples/s.

(a) Write a MATLAB function to compute the STFT of the sequence *speech2_10k* using a 30-ms triangular analysis window, created using MATLAB function *triang.m*, at a 15-ms frame interval. Then reconstruct the original waveform from the STFT using the OLA approach to synthesis. Ignore tapering end effects of the first and last frames.

(b) Design in MATLAB an OLA-based synthesis method to time-scale expand the speech signal by a factor of two by repeating every frame.

(c) Repeat parts (a) and (b) using the LSE-based synthesis approach. Compare the time-scaled signals from each approach and comment on the differences.

(d) Repeat parts (b) and (c) using "pitch-synchronized" OLA- and LSE-based synthesis approaches. This will require that you estimate a consistent time instant (e.g., the waveform peak or the glottal pulse time) within a glottal cycle in order to synchronize consecutive frames. How does this approach improve your synthesis from parts (b) and (c)?

(e) Repeat parts (a)–(d) with a speech utterance from your own voice recording. For part (d), consider manually marking voiced and unvoiced regions and invoke pitch synchrony during the voiced regions.

7.23 (MATLAB) In this MATLAB exercise, use the workspace *ex7M2.mat*, as well as the function *uniform_bank.m*, both located in companion website directory Chap_exercises/chapter7. This exercise explores the conditions for recovery of a sequence using the filter bank summation (FBS) method.

(a) There are four different analysis windows (or "analysis filters") in the workspace *ex7M2.mat*: *filter1*, *filter2*, *filter3*, and *filter4*. Using the MATLAB command *subplot*, plot all four filters. Note that *filter4* is simply a shifted version of *filter3*.

(b) The function *uniform_bank.m* creates a filter bank $h_k[n]$. The output of the function is a 2-D array of modulated filter impulse responses, but without the demodulation term $e^{-j\frac{2\pi}{N}kn}$ shown in Figure 7.5. Also, the impulse responses are real because the complex conjugate response pairs have been combined. Note that the first and last filters do not correspond to complex conjugate pairs. Explain why.

(c) Do a *help* command on *uniform_bank.m* and run the function with a 250-Hz spacing between filters, the analysis filter *filter1*, and a plot factor of 100 (scaling the output for a good display). The function plots the frequency response magnitude of the filters using a 1024-point DFT. Assume the time sampling rate is 10000 samples/s so that the 512th frequency bin corresponds to 5000 Hz. Having the impulse responses $h_k[n]$ of your filter bank, write a MATLAB function to filter the unit sample *impulse* in *ex7M2.mat* with your filter bank, and then combine all impulse responses to create the composite (summed) filter bank impulse response. Explain the observed composite impulse response using the FBS constraint. Finally, plot the impulse response to the second and fifteenth bandpass filters and explain the difference in time structure of the two signals [recall that the filter bank is missing the STFT demodulation term $e^{-j\frac{2\pi}{N}kn}$ (Figure 7.5)].

(d) Repeat part (c) with analysis filter *filter2*. Using the FBS constraint, explain the reason for the deviation in the composite response from an impulse.

(e) Repeat part (c) with *filter3* (using a plot factor of 2). Superimpose the composite impulse response on *filter3* (plotting the first 400 samples) and again explain your observation using the FBS constraint.

(f) Repeat part (c) with *filter4*, which is *filter3* shifted by fifty samples to the right. Superimpose the composite impulse response on *filter4* (plotting the first 450 samples) and again explain your observation using the FBS constraint. Also comment on the difficulty in creating an impulsive composite response by simply shifting the analysis filter.

BIBLIOGRAPHY

[1] J.B. Allen and L.R. Rabiner, "A Unified Theory of Short-Time Spectrum Analysis and Synthesis," *Proc. IEEE*, vol. 65, no. 11, pp. 1558–1564, Nov. 1977.

[2] J.C. Anderson, *Speech Analysis/Synthesis Based on Perception*, Ph.D. Thesis, Massachusetts Institute of Technology, Dept. of Electrical Engineering and Computer Science, Cambridge, MA, Sept. 1984.

[3] S.F. Boll, "Suppression of Acoustic Noise in Speech Using Spectral Subtraction," *IEEE Trans. Acoustics, Speech, and Signal Processing*, vol. ASSP–27, no. 2, pp. 113–120, April 1979.

[4] R.E. Crochiere, "A Weighted Overlap-Add Method of Short-Time Fourier Analysis/Synthesis," *IEEE Trans. Acoustics, Speech and Signal Processing*, vol. ASSP–28, no. 1, pp. 99–102, Feb. 1980.

[5] G. Fairbanks, W.L. Everitt, and R.P. Jaeger, "Method for Time or Frequency Compression-Expansion of Speech," *IEEE Trans. Audio and Electroacoustics*, vol. AU–2, pp. 7–12, Jan. 1954.

[6] D. Griffin and J.S. Lim, "Signal Estimation from Modified Short-Time Fourier Transform," *IEEE Trans. Acoustics, Speech, and Signal Processing*, vol. ASSP–32, no. 2, pp. 236–243, April 1984.

[7] D. Israelevitz, "Some Results on the Time-Frequency Sampling of the Short-Time Fourier Transform Magnitude," *IEEE Trans. Acoustics, Speech, and Signal Processing*, vol. 33, no. 6, pp. 1611–1613, Dec. 1985.

[8] W. Koening, H.K. Dunn, and L.Y. Lacey, "The Sound Spectrogram," *J. Acoustical Society of America*, vol. 18, pp. 19–49, Feb. 1946.

[9] J. Laroche, "Time and Pitch Scale Modification of Audio Signals," chapter in *Applications of Digital Signal Processing to Audio and Acoustics*, M. Kahrs and K. Brandenburg, eds., Kluwer Academic Publishers, Boston, MA, 1998.

[10] J.S. Lim and A.V. Oppenheim, "Enhancement and Bandwidth Compression of Noisy Speech," *Proc. IEEE*, vol. 67, no. 12, pp. 1586–1604, Dec. 1979.

[11] E. Moulines and F. Charpentier, "Pitch-Synchronous Waveform Processing Techniques for Text-to-Speech Synthesis Using Diphones," *Speech Communication*, vol. 9, no. 5–6, pp. 453–467, 1990.

[12] S.H. Nawab, T.F. Quatieri, and J.S. Lim, "Signal Reconstruction from Short-Time Fourier Transform Magnitude," *IEEE Trans. Acoustics, Speech, and Signal Processing*, vol. ASSP–31, no. 4, pp. 986–998, Aug. 1983.

[13] S.H. Nawab and T.F. Quatieri, "Short-Time Fourier Transform," Chapter in *Advanced Topics in Signal Processing*, J.S. Lim and A.V. Oppenheim, eds., Prentice Hall, Englewood Cliffs, NJ, Oct. 1987.

[14] A.V. Oppenheim and R.W. Schafer, *Discrete–Time Signal Processing*, Prentice Hall, Englewood Cliffs, NJ, 1989.

[15] R.K. Potter, G.A. Kopp, and H.G. Kopp, *Visible Speech*, Dover Publications, Inc., New York, 1966.

[16] M.R. Portnoff, "Time-Scale Modification of Speech Based on Short-Time Fourier Analysis," *IEEE Trans. Acoustics, Speech,and Signal Processing*, vol. ASSP–30, no. 3, pp. 374–390, June 1981.

[17] M.R. Portnoff, "Time-Frequency Representation of Digital Signals and Systems Based on Short-Time Fourier Analysis," *IEEE Trans. Acoustics, Speech, and Signal Processing*, vol. ASSP–28, no. 1, pp. 55–69, Feb. 1980.

[18] M.R. Portnoff, "Implementation of the Digital Phase Vocoder Using the Fast Fourier Transform," *IEEE Trans. Acoustics, Speech, and Signal Processing*, vol. ASSP–24, no. 3, pp. 243–248, Feb. 1976.

[19] T.F. Quatieri, S.H. Nawab, and J.S. Lim, "Frequency Sampling of the Short-Time Fourier Transform Magnitude for Signal Reconstruction," *J. Optical Society of America*, Special Issue on Signal Recovery, vol. 73, pp. 1523–1526, Nov. 1983.

[20] L.R. Rabiner and R.W. Schafer, *Digital Processing of Speech Signals*, Prentice Hall, Englewood Cliffs, NJ, 1978.

[21] S. Roucos and A.M. Wilgus, "High-Quality Time-Scale Modification of Speech," *Proc. IEEE Int. Conf. Acoustics, Speech, and Signal Processing*, Tampa, FL, pp. 493–496, April 1985.

[22] R. Scott and S. Gerber, "Pitch-Synchronous Time-Compression of Speech," *Proc. Conf. Speech Communications Processing*, pp. 63–65, April 1972.

[23] V.W. Zue, "Acoustic-Phonetic Knowledge Representation: Implications from Spectrogram Reading Experiments," *Proc. 1981 NATO Advanced Summer Institute on Automatic Speech Analysis and Recognition*, Bonas, France, 1981.

Filter-Bank Analysis/Synthesis

8.1 Introduction

In the previous chapter, we introduced the filter bank summation (FBS) and overlap-add (OLA) methods of speech analysis and synthesis. In this chapter, we focus on extensions of the FBS method, in particular, beginning in Section 8.2 with its additional properties and practical design considerations. In Section 8.3 we interpret the filter-bank outputs in the FBS method for speech signals. Specifically, a sinewave model of the filter-bank outputs is developed for quasi-periodic speech signals, a perspective that leads to the *phase vocoder* for speech analysis and synthesis. Although we provide interpretation of the filter-bank outputs with respect to speech signals, the approach remains largely non-model-based, in contrast to the model-based approaches to speech analysis/synthesis studied earlier, such as those using linear prediction and homomorphic filtering. The phase vocoder is shown to be applicable in a number of areas, including speech coding and time-scale modification. We also describe limitations of this approach in the context of these applications, including the problem of achieving *phase coherence*, i.e., preserving the phase relation across sinewave outputs in synthesis. Loss of phase coherence is known to give a reverberant quality in the phase vocoder synthesis. Such limitations lead to a need for a more explicit formulation of sinewave components of speech; this sinewave representation will be described in Chapter 9.

In Section 8.4, we continue to address the problem of phase coherence and describe an approach to controlling individual sinewave phases in the phase vocoder so as to reduce loss of coherence in synthesis. As an example, we show how the shape of transient sounds can be approximately preserved in time-scale modification with appropriate phase control. In Section 8.5, we then take the FBS and phase vocoder methods to a generalization of filter-bank analysis/synthesis that involves constant-Q filters. We describe this generalization in the framework of the *wavelet transform*. The wavelet transform can be thought of as an extension of the STFT that provides good frequency resolution but poor time resolution for low frequency regions, and good time resolution but poor frequency resolution for high frequency regions. This leads us into the final Section 8.6 that gives a brief look at the relation of the wavelet transform to a front-end auditory filter-bank model. An AM-FM sinewave interpretation of the filter-bank outputs is used to speculate on why the human auditory system is phase-sensitive, especially for low-pitched speakers. We also describe how auditory processing elements may enhance joint time and frequency resolution, as well as sensitivity to temporal and spectral change in a signal.

These, as well as other principles of auditory signal processing in this section, provide a basis for auditory-motivated speech processing techniques later in the text.

8.2 Revisiting the FBS Method

Recall that an analysis/synthesis system based on a filter-bank representation of a signal $x[n]$ can be derived from the time-dependent short-time Fourier transform (STFT)

$$X(n, \omega) = \sum_{m=-\infty}^{\infty} w[n - m]x[m]e^{-j\omega m}. \tag{8.1}$$

Specifically, we saw in Chapter 7 that by replacing the expression $n - m$ by m, Equation (8.1) becomes

$$X(n, \omega) = e^{-j\omega n} \sum_{m=-\infty}^{\infty} x[n - m]w[m]e^{j\omega m}$$

$$= e^{-j\omega n}(x[n] * [w[n]e^{j\omega n}]) \tag{8.2}$$

where * denotes convolution. Equation (8.2) can be viewed as first a modulation of the window $w[n]$ to frequency ω, thus producing a bandpass filter $w[n]e^{j\omega n}$, followed by a filtering of $x[n]$ through this bandpass filter. The output is then demodulated back down to baseband.

When the frequency is sampled uniformly to form a bank of filters, i.e., $\omega_k = \frac{2\pi}{N}k$, we can express each bandpass filter as

$$h_k[n] = w[n]e^{j\frac{2\pi}{N}kn}$$

where the analysis window (filter) $w[n]$ is zero outside the interval $0 \leq n < N_w$ and $\frac{2\pi}{N}$ is the frequency spacing between bandpass filters, N being the number of filters. The output of each filter $h_k[n]$ can be written as

$$y_k[n] = x[n] * h_k[n]$$

$$= x[n] * (w[n]e^{j\omega_k n})$$

$$= e^{j\omega_k n}X(n, \omega_k) \tag{8.3}$$

which is Equation (8.2) without the final demodulation and which was illustrated in Figure 7.5 of Chapter 7. The discrete frequency samples $\omega_k = \frac{2\pi}{N}k$ can be thought of as center frequencies for each of the N "channels" of the filter bank. As was shown in Figure 7.5, the FBS synthesis is then given by

$$y[n] = \frac{1}{Nw[0]} \sum_{k=0}^{N-1} X(n, k)e^{j\omega_k n}$$

$$= \frac{1}{Nw[0]} \sum_{k=0}^{N-1} y_k[n]$$

$$= x[n] * \sum_{k=0}^{N-1} h_k[n]\left(\frac{1}{Nw[0]}\right) \tag{8.4}$$

where we have written $X(n, \omega_k)$ as $X(n, k)$, the discrete STFT. It is desirable to have the term to the right of the convolution sign equal to the unit sample $\delta[n]$ so that $y[n] = x[n]$. From Chapter 7, the resulting FBS constraint requires in the frequency domain that the composite frequency response be flat [Equation (7.13)]; the corresponding constraint in the time domain is that the duration of $w[n]$, N_w, be less than N or, less strictly, that $w[rN]$ be zero at $r = -1, +1, -2, +2 \dots$ [Equation (7.12)].

Consider now the design of such a filter bank where we specify a desired frequency response of the analysis window $w[n]$, which is sometimes called the *prototype filter* because all other filters in the filter bank are derived from it by modulation. For example, we may want $w[n]$, and thus each filter $h_k[n]$, to take on a certain bandwidth to achieve a certain frequency resolution. For voiced speech, in particular, it is often desired that $H_k(\omega) = W(\omega - \frac{2\pi}{N}k)$ pass only one harmonic. (We will see the importance of this constraint in our description of the phase vocoder.) Therefore, we now have two constraints on the window: (1) $w[n]$ must satisfy the FBS constraint of being short in the time domain, e.g., $N_w \leq N$, and (2) $w[n]$ must be narrow in the frequency domain. These are conflicting constraints and are difficult to meet, as seen in the uncertainty principle described in Chapter 2.

In order to help relieve these constraints on $w[n]$, consider a slightly modified filter bank in which each channel output is multiplied by the complex constant $p_k = e^{j\phi_k}$, as shown in Figure 8.1 [42]. The modified composite output becomes (dropping the constant scale factor for simplicity)

$$\tilde{y}[n] = \sum_{k=0}^{N-1} p_k y_k[n]$$

$$= \sum_{k=0}^{N-1} p_k (h_k[n] * x[n]). \tag{8.5}$$

The factor p_k provides a gain and phase adjustment for each channel in the bank. The effect of p_k on the composite response can be seen by writing $h_k[n] = w[n]e^{j\frac{2\pi}{N}kn}$ and expressing Equation (8.5) as

$$\tilde{y}[n] = \tilde{h}[n] * x[n] \tag{8.6}$$

where

$$\tilde{h}[n] = w[n]p[n]$$

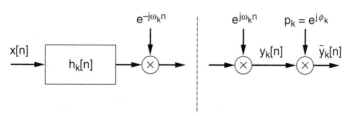

Figure 8.1   Phase adjustment factor of kth channel in FBS synthesis.

SOURCE: L.R. Rabiner and R.W. Schafer, *Digital Processing of Speech Signals* [42]. ©1978, Pearson Education, Inc. Used by permission.

with

$$p[n] = \sum_{k=0}^{N-1} p_k e^{j\frac{2\pi}{N}kn}.$$

We have seen this result earlier in Chapter 7 (Section 7.3.2) in the context of multiplicative modification of the STFT, i.e., we can interpret the introduction of the adjustment factor p_k as a multiplication of $X(n, \omega)$ by a function $P(\omega)$ at the uniformly spaced frequencies $\omega_k = \frac{2\pi}{N}k$. (The reader should carry through this interpretation.) The case $p_k = 1$ takes us back to our FBS constraint that for $\tilde{y}[n] = x[n]$, i.e., $\tilde{h}[n] = \delta[n]$, then $w[rN] = 0$ for $r = \pm 1, \pm 2, \ldots$, or the more useful constraint that the duration of $w[n]$, N_w, is less than or equal to N (Exercise 8.1).

As we mentioned earlier, the Fourier transform of $w[n]$, $W(\omega)$, is often selected to be lowpass and narrow and so it is difficult to obtain a window $w[n]$ that has short length $N_w < N$. When the FBS constraint is not satisfied, FBS synthesis results in multiple copies of the input (Chapter 7 and Exercise 8.1), i.e.,

$$y[n] = \sum_{r=-\infty}^{\infty} w[rN]x[n - rN]$$

which results in a reverberant quality to the reconstruction. In order to remove the reverberation, we might shorten the analysis window $w[n]$, but this would broaden its Fourier transform and sacrifice frequency resolution. On the other hand, we might increase the DFT length, i.e., the number of filters in our filter bank; this, however, increases computation significantly. An alternative is to select the factor p_k to allow control of the time shift in the impulse-train sequence $p[n]$, as illustrated in the following example adapted from Rabiner and Schafer [42]:

EXAMPLE 8.1   Consider a discrete STFT generated with a rectangular window $w[n] = 1$ for $0 \le n < N_w$, whose length $N_w = 60$, and computed at discrete uniform frequencies $\omega_k = \frac{2\pi}{N}k$ with the DFT length $N = 50$. The resulting composite impulse response associated with FBS synthesis is $\tilde{h}[n] = \delta[n] + \delta[n - 50]$ and so a second copy of the input signal $x[n]$ is introduced in the output at a 50-sample delay. The resultant waveform is distinctly reverberant. On the other hand, by applying the adjustment factor with a linear phase factor $p_k = e^{-j\frac{2\pi}{N}kn_o}$ with $n_o = 20$, for example, we remove the reverberation. The composite filter can be shown to be (Exercise 8.1)

$$\tilde{h}[n] = w[n] \sum_{r=-\infty}^{\infty} \delta[n - 20 - rN]$$

$$= w[20]\delta[n - 20] = \delta[n - 20] \tag{8.7}$$

which simply delays the input.   ▲

We have introduced the notion of phase adjustment in FBS synthesis, first, because it helps illustrate the time-frequency resolution tradeoffs in FBS analysis and synthesis and, secondly, because this concept will be used more generally in a number of contexts in reducing forms of reverberation in other filter bank-based analysis/synthesis. One such system is the next focus of this chapter, the phase vocoder.

8.3 Phase Vocoder

A particular formulation of FBS analysis and synthesis capitalizes on the underlying harmonic speech spectrum during voicing. This filter-bank analysis approach, which exploits the harmonic nature of a signal, originated in the context of music sound processing as a Fourier series. The Fourier series was computed over a sliding window of a single pitch period duration and provided a measure of amplitude and frequency trajectories of the musical tones [26],[39]. This technique evolved into a filter-bank-based processor and, ultimately, to signal analysis/synthesis referred to as the *phase vocoder* for both speech and music processing [11],[26]. In this section, we describe the fundamentals of the phase vocoder analysis and synthesis and its use in a number of applications, including speech coding, to reduce information for transmission, and speech transformations such as time-scale modification.

8.3.1 Analysis/Synthesis of Quasi-Periodic Signals

We now show that, under certain conditions on the analysis window (filter) $w[n]$, the output of each filter in the filter bank of the FBS method can be interpreted as discrete-time sinewaves that are both amplitude- and phase-modulated by the time-dependent Fourier transform [39],[42].

Consider a sequence $x[n]$ passed through the discrete bank of filters $h_k[n]$ of the FBS method. We have seen that each filter is given by the modulated version of the baseband prototype filter $w[n]$, i.e.,

$$h_k[n] = w[n]e^{j\frac{2\pi}{N}kn} \tag{8.8}$$

where $w[n]$ is assumed to be zero outside the interval $0 \leq n < N_w$, and where the frequency sampling interval $\frac{2\pi}{N}$, i.e., the frequency spacing between bandpass filters, is determined by the number of filters N. The output of each filter can be written as in Equation (8.3), which is Equation (8.2) without the final demodulation, evaluated at discrete frequency samples $\omega_k = \frac{2\pi}{N}k$ that can be thought of as center frequencies for each of the N "channels."

Because each filter impulse response $h_k[n]$ in Equation (8.3) is complex, each filter output $y_k[n]$ in Equation (8.3) is complex, so we can write the temporal envelope $a_k[n]$ and phase $\phi_k[n]$ of the output of the kth channel as

$$a_k[n] = |y_k[n]|$$

$$\phi_k[n] = \tan^{-1}\left[\frac{\text{Im}(y_k[n])}{\text{Re}(y_k[n])}\right]. \tag{8.9}$$

Thus, the output of each filter can be viewed as an amplitude- and phase-modulated complex sinewave (exponential)[1]

$$y_k[n] = a_k[n]e^{j\phi_k[n]} \tag{8.10}$$

and reconstruction of the signal (via the FBS method without the modulation factors in Figure 7.5) can be viewed as a sum of complex sinewaves

$$y[n] = \frac{1}{Nw[0]}\sum_{k=0}^{N-1}a_k[n]e^{j\phi_k[n]} \tag{8.11}$$

---

[1] The amplitude and phase functions are not necessarily equal to those derived from an analytic signal formulation of the corresponding real filter output because the complex filter may contain negative frequencies.

with amplitude and phase components given by Equation (8.9). The resulting analysis/synthesis structure is referred to as the *phase vocoder* [11]. When the FBS constraint is satisfied by the analysis filter $w[n]$, we have $y[n] = x[n]$.

The amplitudes and phases in Equations (8.10) and (8.11) can correspond to physically meaningful parameters for quasi-periodic signals typical of voiced speech. In order to see this property, the STFT is first written as

$$X(n, \omega_k) = |X(n, \omega_k)|e^{j\theta(n, \omega_k)} \tag{8.12}$$

where $\omega_k = \frac{2\pi}{N}k$ is the center frequency of the kth channel. Then, from Equations (8.3) and (8.12), the output of the kth channel filter is expressed as

$$y_k[n] = |X(n, \omega_k)|e^{j[\omega_k n + \theta(n, \omega_k)]} \tag{8.13}$$

and, therefore, from Equations (8.10) and (8.13), the temporal envelope $a_k[n] = |X(n, \omega_k)|$ and phase $\phi_k[n] = \omega_k n + \theta(n, \omega_k)$.

Consider now filters that are symmetric about π so that $\omega_{N-k} = 2\pi - \omega_k$, where as before $\omega_k = \frac{2\pi}{N}k$, and assume for simplicity that N is even (Figure 8.2). Then it is straightforward to show that (Exercise 8.2)

$$X(n, \omega_k) = X^*(n, \omega_{N-k}). \tag{8.14}$$

From Equations (8.13) and (8.14), the sum of two symmetric channels k and $N - k$ can be written as (Exercise 8.2)

$$\hat{y}_k[n] = 2|X(n, \omega_k)| \cos[\omega_k n + \theta(n, \omega_k)] \tag{8.15}$$

which can be interpreted as a real sinewave that is *amplitude- and phase-modulated* by the STFT, the "carrier" of the latter being the kth filter's center frequency. The changing, amplitude-modulated envelope (more strictly, amplitude modulation (AM) refers to a changing deviation about a steady amplitude component) is given by $2|X(n, \omega_k)|$ and the phase modulation (PM) about the carrier is given by $\theta(n, \omega_k)$.

Before investigating the response of the filter bank to a periodic input, we look at a useful interpretation of the output $\hat{y}_k[n]$ that comes from the concept of instantaneous frequency. To do so, we return for the moment to the analog realm where we write the STFT of a continuous time signal as

$$X(t, \Omega) = \int_{-\infty}^{\infty} x(\tau)w(t - \tau)e^{-j\Omega\tau}d\tau \tag{8.16}$$

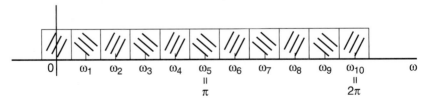

Figure 8.2 Filters whose center frequencies are symmetric about π, i.e., $\omega_{N-k} = 2\pi - \omega_k$ with $\omega_k = \frac{2\pi}{N}k$. In this example $N = 10$.

which can be expressed as

$$X(t, \Omega) = |X(t, \Omega)|e^{j\theta(t,\Omega)}$$
$$= a(t, \Omega) + jb(t, \Omega)$$

where

$$|X(t, \Omega)| = [a^2(t, \Omega) + b^2(t, \Omega)]^{\frac{1}{2}}$$
$$\theta(t, \Omega) = \tan^{-1}\left[\frac{b(t, \Omega)}{a(t, \Omega)}\right].$$

As in discrete time, we can derive an amplitude/phase modulated sinewave representation for each channel output (again using two symmetrically-placed filters):

$$\hat{y}_k(t) = 2|X(t, \Omega_k)|\cos[\Omega_k t + \theta(t, \Omega_k)]$$

and we then define the phase function of each filter output as

$$\Phi(t, \Omega_k) = \Omega_k t + \theta(t, \Omega_k).$$

Consider next the phase derivative

$$\frac{d}{dt}\Phi(t, \Omega_k) = \dot{\Phi}(t, \Omega_k)$$
$$= \Omega_k + \dot{\theta}(t, \Omega)$$

where

$$\dot{\theta}(t, \Omega_k) = \frac{d}{dt}\theta(t, \Omega_k).$$

We refer to $\dot{\Phi}(t, \Omega_k)$ as the *instantaneous frequency* at the output of each (kth) bandpass filter with center frequency Ω_k. Observe that the phase can be recovered as

$$\Phi(t, \Omega_k) = \int_0^t \dot{\Phi}(\tau, \Omega_k)d\tau + \Phi(0, \Omega_k)$$

where $\Phi(0, \Omega_k)$ is an initial condition. The signal $2|X(t, \Omega_k)|$ is likewise referred to as the *instantaneous amplitude* for each channel. The resulting filter-bank output is a sinewave with generally a time-varying amplitude and frequency modulation, as illustrated in Figure 8.3.

We can think of $\dot{\theta}(t, \Omega_k)$ as the deviation of the instantaneous frequency from the center frequency Ω_k of the kth filter, provided that $\dot{\Phi}(t, \Omega_k)$ is "slowly varying," as under certain conditions that we will see shortly. This deviation is also called *frequency-modulation* (FM) and is illustrated graphically in Figure 8.3a. We will see in a moment why we might expect this condition to hold for voiced speech. An alternative expression for the instantaneous frequency deviation, not requiring the explicit calculation of the phase derivative, is given by (Exercise 8.3)

$$\dot{\theta}(t, \Omega_k) = \frac{b(t, \Omega_k)\dot{a}(t, \Omega_k) - a(t, \Omega_k)\dot{b}(t, \Omega_k)}{a^2(t, \Omega_k) + b^2(t, \Omega_k)} \tag{8.17}$$

which is the time-domain counterpart to the frequency-domain phase derivative that we have seen in the context of homomorphic processing [Equation (6.8)].

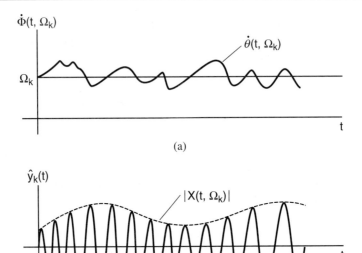

Figure 8.3 Interpretation of instantaneous frequency and amplitude in continuous time: (a) $\dot{\theta}(t, \Omega_k)$ is the deviation of the instantaneous frequency from the center frequency Ω_k of the kth filter (the frequency modulation) and $\dot{\Phi}(t, \Omega_k) = \Omega_k + \dot{\theta}(t, \Omega)$ is the instantaneous frequency; (b) the instantaneous amplitude $2|X(t, \Omega_k)|$ and instantaneous frequency $\dot{\Phi}(t, \Omega_k)$ characterize each filter bank sinewave output.

Based on this formulation, we seek the instantaneous frequency in discrete time. To obtain this function, suppose that $X(t, \Omega_k)$ as a function of time is bandlimited. Then we can sample the continuous-time STFT, with sampling interval T, to obtain the discrete-time STFT, i.e., [42]

$$X(n, \omega_k) = \frac{1}{T}X(t, \Omega_k)|_{t=nT} = X\left(nT, \frac{\omega_k}{T}\right)$$

where we have used the continuous-/discrete-time sampling relation reviewed in Chapter 2. Likewise, the phase derivative associated with $X(n, \omega_k)$ can be defined as a sampled version of $\dot{\theta}(t, \Omega_k)$, i.e.,

$$\dot{\theta}(n, \omega_k) = \frac{1}{T}\dot{\theta}(t, \Omega_k)|_{t=nT} = \dot{\theta}\left(nT, \frac{\omega_k}{T}\right)$$

where we assume $\dot{\theta}(t, \Omega_k)$ is also bandlimited. The continuous-time function $\dot{\theta}(t, \Omega_k)$, however, is not available because the STFT is computed in discrete time. One possibility is to discretize the continuous-time phase derivative expression given in Equation (8.17), i.e.,

$$\dot{\theta}(n, \omega_k) = \frac{b(n, \omega_k)\dot{a}(n, \omega_k) - a(n, \omega_k)\dot{b}(n, \omega_k)}{a^2(n, \omega_k) + b^2(n, \omega_k)}$$

where the corresponding $a(t, \Omega_k)$, $b(t, \Omega_k)$, and their derivatives are assumed to be bandlimited. Although $a(n, \omega_k)$ and $b(n, \omega_k)$ are available through the discrete-time STFT, this is not the case for their derivatives. Nevertheless, estimates of $\dot{a}(n, \omega_k)$ and $\dot{b}(n, \omega_k)$ can be obtained by discrete-time filtering of $a(n, \omega_k)$ and $b(n, \omega_k)$, such as by first forward or backward differencing. (Note that we cannot obtain the phase derivative by explicit phase differencing methods without first performing the difficult task of unwrapping the phase in time from principle phase measurements.)

We now return to the goal of gaining insight into the nature of the bandpass filter outputs for a speech input. In particular, we look at the response of our filter bank to exactly periodic and quasi-periodic inputs. We consider first a perfectly periodic input with stationary vocal tract and then a quasi-periodic input generated with a glottal excitation of slowly varying pitch and with a slowly changing vocal tract. The prototype filter of the filter bank is assumed to be narrowband and flat in the region in which each input sinewave component lies, as illustrated in Figure 8.4. Our goal is to compute the instantaneous amplitude and frequency of each filter output. We begin the analysis in continuous time because of the need to differentiate or integrate functions of phase.

Periodic Case — Consider the pth sinewave component of frequency Ω_p, denoted by $x(t) = A_p \cos(\Omega_p t + \phi_p)$, that passes through the kth channel filter without distortion, as illustrated in Figure 8.4. For this periodic case, $\Omega_p = p\Omega_o$ where Ω_o is the fundamental frequency. The pth sinewave component can be written as

$$x(t) = A_p \cos(\Omega_p t + \phi_p)$$
$$= \frac{A_p}{2}[e^{j(\Omega_p t + \phi_p)} + e^{-j(\Omega_p t + \phi_p)}].$$

The modulated output of the kth filter, i.e., the STFT, can be expressed as

$$X(t, \Omega_k) = \frac{A_p}{2} e^{j(\Omega_p t + \phi_p)} e^{-j\Omega_k t}$$
$$= \frac{A_p}{2} e^{j[(\Omega_p - \Omega_k)t + \phi_p]}.$$

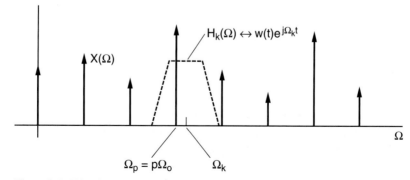

Figure 8.4 Filter-bank response to perfectly periodic sequence. One harmonic component is passed through $H_k(\Omega)$.

Then it can be seen that for the kth channel

$$|X(t, \Omega_k)| = \frac{A_p}{2}$$
$$\theta(t, \Omega_k) = (\Omega_p - \Omega_k)t + \theta(0, \Omega_k) \tag{8.18}$$

and thus the instantaneous frequency deviation is constant and given by

$$\dot{\theta}(t, \Omega_k) = \Omega_p - \Omega_k$$

which is the deviation of the harmonic component from the center frequency. Observe that by integrating $\dot{\theta}(t, \Omega_k)$ with the appropriate initial condition, we can recover $\theta(t, \Omega_k)$, i.e.,

$$\theta(t, \Omega_k) = \int_0^t \dot{\theta}(\tau, \Omega_k)d\tau + \phi_p$$

with $\phi_p = \theta(0, \Omega_k)$. Thus, we can (with $\frac{A_p}{2}$) recover $X(t, \Omega_k)$. Finally, the output of the kth real channel (the sum of two symmetric channels), after demodulation by $e^{j\Omega_k t}$, is given by

$$\hat{y}_k(t) = A_p \cos(\Omega_p t + \theta_p) \tag{8.19}$$

i.e., each signal component passes intact.

A similar analysis can be made for quasi-periodic signals which consist of a sum of sinewaves with slowly varying instantaneous amplitude and frequency, each of which is assumed to pass through a single filter.

Quasi-Periodic Case — We now look at the bandpass filter response to a sinewave with varying instantaneous frequency $\Omega_p(t)$ and amplitude $A_p(t)$, i.e,

$$x_p(t) = A_p(t) \cos[\psi_p(t)]$$

with

$$\psi_p(t) = \int_0^t \Omega_p(\tau)d\tau + \phi_p$$

which corresponds to a glottal excitation function with slowly varying pitch and amplitude and to a slowly varying vocal tract.[2] Suppose $x_p(t)$ is the input to a bandpass filter $h_k(t) = w(t)e^{j\Omega_k t}$ where $w(t)$ represents the analysis window. We impose the following constraints on the amplitude and frequency functions, illustrated in Figure 8.5:

1. $\Omega_p(t) \approx \Omega_p(t')$ over the duration of $w(t)$, i.e., $\Omega_p(t)$ remains at nearly its initial value over the interval $[t', t'']$.

2. $A_p(t) \approx A_p(t')$ over the duration of $w(t)$, i.e., $A_p(t)$ remains at nearly its initial value over the interval $[t', t'']$.

---

[2] We show in Chapter 9 how a changing source and vocal tract contribute to these time-varying amplitude and frequency functions.

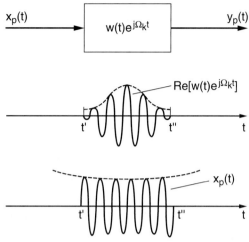

Figure 8.5 A sinewave component corresponding to slowly varying pitch and vocal tract. The input to a bandpass filter is a single sinewave $x_p(t)$ with a slowly varying amplitude (envelope) and instantaneous frequency. The instantaneous amplitude $A_p(t)$ and instantaneous frequency $\Omega_p(t)$ are assumed to appear as constant under the analysis (filter) window $w(t)$.

Then $x_p(t)$ appears to the filter over the time interval $[t', t'']$ as a steady sinewave, i.e.,

$$x_p(t) \approx A_p(t') \cos[\Omega_p(t')t + \phi_p], \qquad \text{over } w[n]$$

and thus the input appears as an eigenfunction of the linear time-invariant (LTI) filter $h_k(t)$ (Chapter 2). It follows that the magnitude and phase derivative of the STFT, $X(t, \Omega)$, are given by

$$|X(t, \Omega_k)| \approx \frac{A_p(t)}{2}$$

$$\dot{\theta}(t, \Omega_k) \approx \Omega_p(t) - \Omega_k. \tag{8.20}$$

Therefore, we can show that the output of the bandpass filter is of the form

$$y_p(t) \approx \frac{A_p(t)}{2} e^{j[\int_0^t \Omega_p(\tau)d\tau + \phi_p]}. \tag{8.21}$$

The reader is stepped through an informal proof of this result in Exercise 8.4.

Our general solution in the quasi-periodic case can be written in discrete time for a pth-harmonic input $x_p[n] = A_p[n]\cos(\psi_p[n])$ with $\psi_p[n] = \int_o^n \omega_p[\tau]d\tau + \phi_p$. With time sampling the continuous-time expression for the STFT in Equation (8.20), we have (Exercise 8.4)

$$|X(n, \omega_k)| \approx \frac{A_p[n]}{2}$$

$$\dot{\theta}(n, \omega_k) \approx \omega_p[n] - \omega_k. \tag{8.22}$$

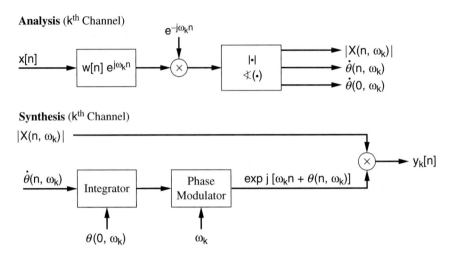

Figure 8.6 Analysis/synthesis structure in the phase vocoder.

We can then synthesize $x[n]$ given $|X(n, \omega_k)|$, $\dot{\theta}(n, \omega_k)$, and $\theta(0, \omega_k)$ with appropriate numerical integration of the phase derivative, e.g., through a cumulative sum (Exercise 8.18). The complete analysis/synthesis scheme is shown in Figure 8.6. Observe that error incurred in this cumulative sum will cause phase drift across the sinewave outputs of adjacent channels and thus a loss in the original channel phase relations. We refer to the preservation of this phase relation as *phase coherence*. Loss of phase coherence results in a change in the shape of the original signal. (We alluded to a different, but related phase distortion in our discussion of speech synthesis from the STFTM in Chapter 7.) The importance of phase drift warrants a more detailed discussion; in Section 8.4 we elaborate on the problem and present a scheme for maintaining phase coherence.

As an aside, it is interesting to note that the approximations used in developing the phase vocoder analysis and synthesis for a sinewave with time-varying amplitude and frequency, as an input to a filter with frequency response flat over the instantaneous frequency, are a special case of a more general solution output for an arbitrary filter. For an input of the form $x[n] = A[n]\cos(\psi[n])$, the output of an LTI filter $H(\omega)$ can be approximated as [2],[40]

$$y[n] \approx A[n]e^{j\psi[n]}H(e^{j\dot{\psi}[n]}) = x[n]H(e^{j\dot{\psi}[n]}) \tag{8.23}$$

where it is assumed that $H(\omega) = 0$ for $\omega < 0$, making the resulting signal analytic. If $x[n]$ has neither AM nor FM, then the approximation Equation (8.23) is exact, $x[n]$ being an eigenfunction of $H(\omega)$. In the time-domain, error bounds for the approximation have been derived by Bovik, Havlicek, and Desai [2]. Let $z[n]$ be the exact output of an LTI filter with frequency response $H(\omega)$ and impulse response $h[n]$ for input $x[n]$, i.e., $z[n] = x[n] * h[n]$, and let $y[n]$ be given by the approximation Equation (8.23). Then the error, defined as $\varepsilon[n] = |z[n] - y[n]|$ is bounded by

$$\varepsilon[n] \leq \sum_{p \neq 0} |h[p]| \int_{n-p}^{n} \left(|\dot{A}(v)| + A_{max}|p||\ddot{\psi}(v)| \right) dv, \tag{8.24}$$

where A_{max} is the maximum value of $A[n]$ and $A(v)$ and $\psi(v)$ are the continuous time signals corresponding to $A[n]$ and $\psi[n]$. From the above upper bound, we want the energy in the impulse response to be concentrated around $n = 0$. In addition, observe that AM, as well as FM, causes error in the approximation, and that this error increases with increasing modulation. An alternative frequency-domain approach to determining error bounds in the approximation has also been developed and shows different considerations [50]. These error bounds can give us a quantitative way to determine the accuracy of our approximations for specific filters $h_k[n]$ and amplitude and frequency modulating functions. Even when the approximation is accurate, however, Equation (8.23) reveals that the filter $H(\omega)$ can change the channel amplitude and phase when it deviates from a flat, zero-phase response in the region of the input frequency (Exercise 8.6).

Before proceeding to our discussion of phase coherence, we digress from theory for a moment to describe a number of fascinating applications of the phase vocoder.

8.3.2 Applications

In applying the phase vocoder, it is advantageous to express each analysis output in terms of the channel phase derivative $\dot{\theta}(n, \omega_k)$ and initial phase offset $\theta(0, \omega_k)$. For a single sinewave $x_p[n] = A_p[n]\cos(\psi_p[n])$ of a harmonic set (assumed to enter one filter of the filter bank), we have seen that these two quantities are given approximately by $\dot{\theta}(n, \omega_k) = \omega_p[n] - \omega_k$ and $\theta(0, \omega_k) = \phi_p$, respectively. A phase function can then be obtained by integration of the phase derivative which is added to the carrier phase $\omega_k n$. This approach makes the filter outputs amenable to speech coding , i.e., to bit-rate reduction through quantization,[3] and also amenable to speech modification, e.g., time-scale modification.

Speech Coding — An overview of a speech coder based on a filter bank approach is given in Figure 8.7. In this scheme, the demodulated output of each filter is time-decimated at the transmitter and quantized in the encoder module. The quantized values are encoded into a bit stream and transmitted over a channel. At the receiver, the bitstream is decoded and the quantized values are interpolated. Finally, the filter-bank outputs are modulated and summed to form the synthesized received signal. To understand the limitations of this approach, and thus to motivate the use of the phase vocoder in speech coding, we revisit our discussion in Chapter 7 of time-frequency sampling requirements on the STFT that follow from its Fourier transform view, corresponding to OLA synthesis, and from its filtering view, corresponding to FBS synthesis. Consider the Fourier transform viewpoint. If the window is of duration N_w, then we require a frequency sampling interval no greater than $\frac{2\pi}{N_w}$. From the filtering point of view, we require a sampling interval L that meets the Nyquist criterion based on the bandwidth of each bandpass filter. This implies that we sample at an interval $L \leq \frac{2\pi}{\omega_c}$, where ω_c is the filter bandwidth, to avoid frequency-domain aliasing of the time sequence $X(n, \omega_k)$. Thus, there results a sampling rate larger than the original sampling rate, e.g., four times the original sampling for a Hamming window [42]. These window length and bandwidth constraints can, however, be relaxed according to the FBS and OLA constraints of Chapter 7 by allowing zeros

---

[3] Formal definitions of bit rate and quantization are given in Chapter 12. For the moment, we can think of a decrease in bit rate as allowing a decrease in channel transmission bandwidth, while quantization is the dividing of a signal value into "quanta." Bit rate is reduced when decreasing the number of quanta to represent signal values (one bit corresponding to two quanta) and when increasing time decimation.

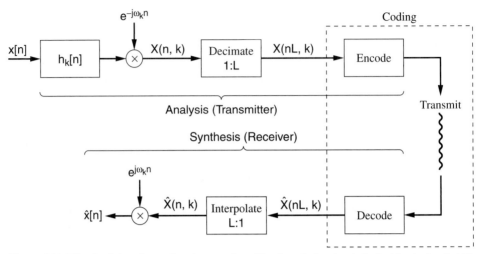

Figure 8.7 Filter-bank-based speech coder overview. The time decimation is limited by the bandwidth
of each analysis filter, according to the Nyquist criterion, to avoid aliasing in frequency. In addition, the
frequency decimation is limited by the duration of the analysis filter, i.e., the number of filters must be
large enough to avoid aliasing in time. The number of samples/s over all filter bank channels (in time and
frequency) is consequently larger than the input waveform sampling rate.

SOURCE: L.R. Rabiner and R.W. Schafer, *Digital Processing of Speech Signals* [42]. ©1978, Pearson Education,
Inc. Used by permission.

in the window or its transform. Although this relaxation of the constraints on window duration
and bandwidth may conceivably avoid this increase in the sampling requirement, it is not an
effective way to reduce bits in speech coding systems. Consequently, the filter-bank scheme in
Figure 8.7 is not an effective coder. Alternatively, when the signal of interest is speech, or any
signal with harmonic structure such as certain music and biological signals, we can exploit the
phase vocoder filter-bank output structure of the previous section.

The idea is based on the observation that magnitude and phase derivative functions at the
filter outputs can vary more slowly and therefore are characterized by a smaller bandwidth than
the filter waveform output, i.e., $|X(n, \omega_k)|$ and $\dot{\theta}(n, \omega_k)$ vary more slowly than $X(n, \omega_k)$ for
each channel k. We assume that we can design the analysis filter $w[n]$ to pass a single (pth)
harmonic of voiced speech within the passband of the kth channel.[4] Then, as we have seen, the
filter output magnitude and phase derivative are given approximately as

$$|X(n, \omega_k)| \approx \frac{A_p[n]}{2}$$

$$\dot{\theta}(n, \omega_k) \approx \omega_p[n] - \omega_k, \tag{8.25}$$

---

[4] This implies that for low pitch, the filter bandwidth is very small and therefore the window duration is
very large, thus perhaps violating the FBS constraint. Nevertheless, if indeed one sinewave (harmonic) is passed
by each filter, then signal synthesis is achieved. This apparent paradox is left for the reader to ponder.

both of which vary slowly if the pitch and vocal tract of the speaker vary slowly in time. Consequently, we can significantly time-decimate $|X(n, \omega_k)|$ and $\dot{\theta}(t, \omega_k)$. Observe that we could have also sampled the corresponding phase obtained by integrating the phase derivative. This unwrapped phase, however, can grow without bound, and its principal phase value contains sharp discontinuities at 2π wrapping points, and thus both functions are difficult to efficiently quantize. At the extreme, when the pitch and vocal tract are time-invariant, so that the harmonic amplitudes and frequencies are fixed, then *only one sample* of the amplitude and phase derivative is required to represent the two functions. Therefore, along with the phase offset, these parameters only are required in the representation of each filter output. This is not surprising because we are assuming that each filter output is a single sinewave.

To obtain a flavor for bit-rate reduction by quantizing the slowly varying magnitude and phase derivative functions, consider an early 28-channel phase vocoder with a channel spacing of 100 Hz [4],[42]. The log-magnitude and phase derivative signals were quantized non-uniformly with fewer bits being allocated to the higher channels because, as we will discuss in later chapters, the human auditory system is less sensitive to noise due to quantization in the high-frequency region. In addition, more bits were allocated to the channel phase derivative than to the magnitude. Specifically, 60 samples/s were used to represent the channel magnitude and phase derivative signals. For magnitude, two bits were used for the lower channels and one bit for the higher channels. For the phase derivative, three bits were used for the lower channels and two bits for the higher channels, resulting in speech coded at 7200 bits/s judged to be "good" quality, but not quality transparent from the original. On the other hand, if the output waveform from each filter channel is quantized directly as in Figure 8.7, rather than through the magnitude and phase derivative, then about 16000 bits/s are required for "good" quality. These rates are in contrast to about 64000 bits/s required to represent the speech waveform itself with quality transparent from the original. In spite of the reduction in bit rate, these rates are relatively high when compared to the lower rates that can be achieved with approaches that use the speech production model more explicitly.

In this section, we have hopefully stirred the reader's curiosity for ways in which dramatic reductions in bit rate can be achieved. Chapter 12 will provide a much more thorough look into this fascinating application area.

Speech Modification — The phase vocoder has been widely used in modification of speech and speech-like signals. In time-scale modification, for example, as we have seen in Chapter 7, the goal is to maintain the perceptual characteristics of the original signal and speaker (e.g., pitch and vocal tract spectrum) while changing the articulation rate of the speaker. Two approaches to performing time-scale modification with the phase vocoder are described. The original approach [11],[42] combines with time scaling a method of compressing (or expanding) the speech spectrum along the frequency axis. With the filter-bank output magnitude and phase derivative functions as given in Equation (8.25), the modification steps, illustrated in Figure 8.8, are as follows:

S1: Frequency compress (or expand) each channel by dividing the phase derivative $\dot{\theta}(n, \omega_k)$ of each channel by the rate change factor ρ, i.e.,

$$\hat{\dot{\theta}}(n, \omega_k) = \rho \dot{\theta}(n, \omega_k).$$

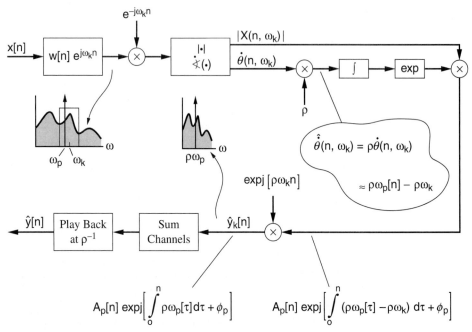

Figure 8.8 Time-scale modification with the phase vocoder using frequency compression/expansion and fast/slow playback.

This operation has the effect of compressing the spectrum for $\rho < 1$ and expanding the spectrum for $\rho > 1$ along the frequency axis because when one harmonic (the pth harmonic) enters each filter, we have

$$\hat{\dot{\theta}}(n, \omega_k) = \rho \dot{\theta}(n, \omega_k)$$
$$\approx \rho \omega_p[n] - \rho \omega_k$$

where the modified channel center frequency is at $\rho \omega_k$.

S2: Integrate (using a running sum) and exponentiate $\hat{\dot{\theta}}(n, \omega_k)$.

S3: Apply amplitude modulation to form the complex sequence for each channel given approximately by

$$\tilde{y}_k[n] = A_p[n] e^{j[\int_0^n (\rho \omega_p[\tau] - \rho \omega_k) d\tau + \phi_p]}$$

where we represent the running sum of the changing frequency over time by the integration operation (implemented with numerical integration).

S4: Modulate the channel signal by the new channel center frequency to form

$$\hat{y}_k[n] = \tilde{y}_k[n] e^{j \int_0^n \rho \omega_k \, d\tau}$$
$$= A_p[n] e^{j[\int_0^n \rho \omega_p[\tau] d\tau + \phi_p]}.$$

S5: Sum all N channels to form

$$\hat{y}[n] = \sum_{p=0}^{N-1} A_p[n]e^{j\left[\int_0^n \rho\omega_p[\tau]\,d\tau+\phi_p\right]}$$

where we assume one harmonic per channel. We then play back the signal $\hat{y}[n]$ at ρ times the original rate. This final operation has the effect of restoring the correct frequency composition and modifying the time scale and can be implemented by either changing the sampling rate of the output D/A converter or by playing back the analog rendition of $\hat{y}[n]$ at a speed different from the original recording.

Examples of time-scale expansion and compression using the above phase vocoder-based technique are given in [11],[42], illustrating an accordion-like modification of the original speech spectrograms. Observe in Step 5 that if we do not modify the time scale, then the synthesized waveform is characterized by frequency compression or expansion which can be useful in its own right.

In a second approach to time-scale modification with the phase vocoder, the phase and amplitude of each channel are interpolated or decimated directly to a new time scale, in contrast to relying on a change in time scale during play back. A rate change of an arbitrary rational value can be performed by combined interpolation and decimation. In one form of this technique, illustrated in Figure 8.9, demodulation by $e^{j\omega_k n}$ does not occur and the phase of each filter output in Equations (8.10) and (8.11) is obtained by integrating the phase derivative, i.e., the instantaneous frequency of each channel. The channel amplitude and phase functions are then time-scaled by time-decimation and/or interpolation. With time-scale modification by a factor ρ, the modified filter output for each channel is given by

$$\tilde{y}_k[n] = \tilde{a}_k[n]\cos(\rho\tilde{\phi}_k[n]) \tag{8.26}$$

where $\tilde{a}_k[n]$ and $\tilde{\phi}_k[n]$ are the decimated/interpolated amplitude and phase functions, respectively. The modified phase is scaled by ρ to maintain the original instantaneous frequency of each filter output. We can see the need for this scaling by writing

$$\rho\frac{d\phi(\frac{t}{\rho})}{dt} = \rho\left(\frac{1}{\rho}\right)\dot{\phi}(\frac{t}{\rho}). \tag{8.27}$$

which is the time-scaled instantaneous frequency.

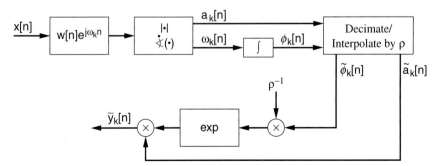

Figure 8.9 A direct approach to time-scale modification with the phase vocoder. The instantaneous frequency is integrated to form a phase function, decimated and/or interpolated to the new time scale, and finally scaled by the rate change factor ρ to restore the correct frequency.

In a variation of the technique, we unwrap the principle phase value along the time axis without computation of the phase derivative. We can do the unwrapping, for example, by detecting 2π jumps in the principle phase values over time, as in the algorithm described in Chapter 6 for phase unwrapping over frequency.

8.3.3 Motivation for a Sinewave Analysis/Synthesis

In spite of the many successes of the phase vocoder, it suffers from a number of limitations [31],[32],[39],[45]. Consider first the analyzer. In the applications of speech coding and time-scale modification, for example, it is assumed that only one sinewave enters each bandpass filter within the filter bank. When more than one sinewave enters a bandpass filter, our interpretation of a filter output as a sinewave with slowly varying amplitude- and frequency-modulation breaks down. In addition, a particular sinewave may not be adequately estimated when the filter response shape is non-flat, so that $H(\dot{\psi}[n])$ in Equation (8.23) contributes additional AM, or when the filter bandwidth and the sinewave modulation result in a large upper bound in Equation (8.24). Likewise, the estimation may fail when a sinewave frequency falls between two adjacent filters of the filter bank. In addition, sinewaves with rapidly varying frequency due to large vibrato or fast pitch change are difficult to track as their frequencies move across multiple filters. Although these measurement problems may be resolved by an appropriate combining of adjacent filter-bank outputs, such solutions for a speech input are likely to be cumbersome [45]. An example of a filter-bank structure with a non-flat prototype filter for a single FM-sinewave input is given in Exercise 8.17.

Consider now the synthesizer stage of the phase vocoder. In speech coding, for example, if we are to quantize samples of the phase derivative, then we need to recover the phase from these samples. In doing so, we encounter the problem of obtaining the phase from the phase derivative samples. As we saw, we can do this by numerical integration, but the resulting estimate deviates from the absolute phase because the initial phase offset may not be available and because numerical integration introduces error, thus causing a drift from the original phase even if the initial phase offset was known. This phase drift results in an incorrect phase relation across bandpass filters, i.e., loss of phase coherence, and thus a change in the waveform shape, sometimes referred to as "dispersion." Furthermore, the phase relation across channels changes with time because the phase error is continuously being introduced. A consequence of this dispersion is an annoying reverberant or what is also referred to as a "choral" effect. A similar waveform dispersion problem occurs in time-scale modification when using samples of the phase derivative. Integration of the phase derivative, as well as scaling of the resulting phase function, results in a loss of the original phase relation among sinewaves, thus giving an objectionable similar reverberant characteristic to the synthesis. This problem also occurs with the alternative direct implementation of time scaling where the unwrapped phase is obtained from the principal phase value (Exercise 8.5). We will describe in the following section a method for reducing dispersion in the phase vocoder. In Chapter 9, a method in a similar spirit is described in the different context of sinewave analysis/synthesis. Finally, we note that the phase vocoder was formulated for discrete sinewaves and hence was not designed for the representation of noise components of a sound. Under this input class, the filter output approximation of Equation (8.25) is not meaningful.

A number of refinements of the phase vocoder have addressed these problems [10],[20], [21],[31],[39]. For example, the assumption that only one sinewave passes through each filter motivates a filter bank with filter spacing equal to the fundamental frequency, thus allowing one harmonic to pass through each filter [20]. An alternative is to oversample in frequency, i.e., increase the number of filters, with filters of very narrow bandwidth with the hope that only one harmonic passes through each filter. One approach to prevent waveform dispersion is to use an overlap-add rendition of the synthesis with windows of length such that the overlap is always in phase [21]. Other ways to prevent dispersion are given in the following section. Another refinement of the phase vocoder was developed by Portnoff, who represented each sinewave component by a source and vocal tract filter contribution, thus introducing some control on the phase in synthesis [31], although the phase functions are still computed via a cumulative sum on phase derivatives and more than one sinewave can enter a filter. The sinewave frequencies in this model are constrained to be harmonically related. Portnoff also provided a rigorous analysis of the stochastic properties of the phase vocoder to a noise-like input.

The analysis stage of the original phase vocoder and its refinements views sinewave components as outputs of a bank of uniformly-spaced bandpass filters. Rather than relying on a filter bank to extract the underlying sinewave parameters, an alternate approach is to explicitly model and estimate time-varying parameters of sinewave components by way of spectral peaks in the short-time Fourier transform [25],[39]. It will be shown in Chapter 9 that this approach lends itself to sinewave tracking through frequency matching, phase coherence through a source and vocal tract filter phase model, and estimation of a stochastic component by use of an additive model of deterministic and stochastic signal components. As a consequence, the resulting sinewave analysis/synthesis scheme resolves many of the problems encountered by the phase vocoder, and provides a useful framework for a large range of speech and audio signal processing applications.

8.4 Phase Coherence in the Phase Vocoder

In this section, we describe approaches to achieve phase coherence in the phase vocoder. We previously defined phase coherence as the preservation of the original sinewave phase relations in the synthesized speech and saw that loss of phase coherence resulted in a reverberant or "choral" effect. One approach to achieving phase coherence is to apply a phase offset correction to each sinewave phase that attempts to make phase relations in the modified signal at time t correspond to those in the original signal at that time or, with time-scale modification, at a time t' that maps back to the original time scale. Phase coherence can be achieved at specific signal event times or at regular intervals over time. This concept was first introduced in the context of sinewave analysis/synthesis [38] and a specific method of achieving phase coherence at uniformly spaced frame boundaries will be described in this context in Chapter 9. For the phase vocoder, on the other hand, we first illustrate the principle of achieving phase coherence, by example, with time-scale modification of signals consisting of successive short-duration decaying sinewaves. By preserving phase coherence at specific event times, we can approximately maintain in the time-scaled signal the shape of the original temporal envelope [33],[34],[37] which may play an important role in auditory discrimination of such sounds. We then briefly describe a related approach by Puckette [32] and Laroche and Dolson [18] for reducing the reverberant quality of quasi-periodic waveforms in the phase vocoder.

8.4.1 Preservation of Temporal Envelope

We saw in Chapter 2 that the temporal envelope of a signal is sometimes defined, typically in the context of bandpass signals, as the magnitude of the analytic signal representation [29]. We also saw in the previous section that the temporal envelope was defined as the magnitude of the complex filter bank output. Other definitions of temporal envelope have been proposed based on estimates of attack and release dynamics [1]. The quality of a sound is sometimes associated with its temporal envelope. We saw, for example, in previous chapters that the conversion of a mixed-phase vocal cord/vocal tract impulse response to its minimum-phase counterpart can decrease its attack time and increase its peakiness, thus significantly altering the signal's temporal envelope and giving the synthesized speech a "buzzy" quality during voicing. More generally, assigning different Fourier transform phase functions to a given Fourier transform magnitude results in a large variety of temporal envelopes.

To further illustrate the relation between the frequency-domain phase and temporal structure of a signal, consider the following thought experiment. Suppose we are given the temporal envelope and the Fourier transform magnitude of a signal. We want to generate a time-scaled signal that has the given spectral magnitude and, with an appropriate selection of the Fourier transform phase, has a time-scaled version of the original temporal envelope. Although iterative methods can be applied to attempt to meet these time-frequency constraints [34],[53] a close match to both the spectral magnitude and modified temporal envelope, however, may not be consistent with the relationship between a sequence and its Fourier transform (Exercise 8.7). Nevertheless, we have proposed this thought experiment because it exemplifies the general approach of altering phase to preserve the temporal structure of a signal.

Consider now an analogous strategy in the context of the phase vocoder and, specifically, consider time-scale modification as given by Equation (8.26). Here we are given the desired time-scaled sinewave instantaneous amplitudes $\tilde{a}_k[n]$. We also have the desired time-scaled sinewave instantaneous frequencies by way of the phase derivatives $\rho \frac{d}{dn} \tilde{\phi}_k[n]$. The time-scaled phase functions $\rho \tilde{\phi}_k[n]$, however, typically result in a modified speech waveform whose temporal envelope is very different from that of the original waveform. Although Equation (8.26) does maintain amplitude and phase derivative (frequency) relations, it does not maintain absolute phase relations, i.e., it loses phase coherence. Our approach is to attempt to preserve the envelope by applying a phase offset correction to each channel. Clearly, however, a single phase correction to each function $\rho \tilde{\phi}_k[n]$ cannot preserve the phase relations over all time (Exercise 8.5). Within the framework of the phase vocoder rather than attempting to maintain the temporal envelope over all time, therefore, a different approach is to maintain the channel phase relations at time instants that are associated with distinctive features of the envelope [33],[34],[37],[39]. As a stepping stone to the approach, the notion of *instantaneous invariance* is introduced.

Instantaneous Invariance — It is assumed that the temporal envelope of a waveform near a particular time instant $n = n_o$ is determined by the amplitude and phase of its channel components at that time (i.e., $a_k[n_o]$ and $\phi_k[n_o]$), and by the time rate of change of these amplitude and phase functions. Suppose that we want to time-scale modify the sequence by rate-change factor ρ. To preserve the temporal envelope in the new time scale near $n = \rho n_o$, these amplitude and phase relations are maintained at that time. The phase relations can be maintained by adding an offset to each channel's phase, guaranteeing that the resulting phase trajectory takes on the desired phase at the specified time $n = \rho n_o$. In other words, a phase correction is introduced in each channel that sets the phase of the modified filter output $\tilde{y}_k[n]$ at $n = \rho n_o$ to the phase at $n = n_o$ in the original time scale. Denoting the phase correction by ϕ_k, the modified channel signal becomes

$$\tilde{y}_k[n] = \tilde{a}_k[n]\cos(\rho \tilde{\phi}_k[n] + \phi_k) \tag{8.28}$$

where $\phi_k = \phi_k[n_o] - \rho\tilde{\phi}_k[\rho n_o]$ and where $\tilde{a}_k[n]$ and $\tilde{\phi}_k[n]$ are the interpolated versions of the original amplitude and phase functions. An inconsistency arises, however, when preservation of the temporal envelope is desired at more than one time instant. One approach to resolving this inconsistency is to allow specific groups of channel components to contribute to different instants of time at which invariance is desired [33],[34],[37],[39].

The approach to invariance can be described by time-expanding the signal in Figure 8.10a that has a high- and low-frequency component, each with a different starting time. If all channels are "phase-synchronized" (also referred to as "phase-locked"), as above, near the low-frequency event, the phase relations at the high-frequency event are changed and vice versa. For this signal, with two events of different frequency content, it is preferable to distribute the phase synchronization over the two events; the high-frequency channels being phase-synchronized at the first event and the low-frequency channels being phase-synchronized at the second event. Equation (8.28) can then be applied to each channel group using the time instant for the respective event, thus phase-locking phases of channels that most contribute to each event.

One approach to assign channels to time instants uses the individual envelopes of the filter bank outputs [33]. Accordingly, the filter bank is designed with a short prototype filter such that each filter output reflects distinctive events that characterize the temporal envelope of the input signal. Channels are then clustered according to their similarity in envelope across frequency. The *onset time* of an event is defined within each channel as the location of the maximum of the channel envelope $a_k[n]$ and is denoted by $n_o(k)$. It is assumed that the signal is of short duration with no more than two events and that only one onset time is assigned to each channel; more generally, multiple onset times would be required. A histogram of onset times is formed, and the average values within each of the two highest bins are selected as the *event times*. These times are denoted by n_o^1 and n_o^2, and each of the k channels is assigned to n_o^1 or n_o^2 based on the minimum distance between $n_o(k)$ and the two possible event times. The distance is given by $D(p; k) = |n_o(k) - n_o^p|$ where $p = 1, 2$. The resulting two clusters of channels are denoted by $\{y_{k_p}^p[n]\}$ with $p = 1, 2$ and where for each p, k_p runs over a subset of the total number of bands. (For simplicity, the subscript p on k will henceforth be dropped.)

Finally, based on the channel assignment, a phase correction is introduced in each channel, making the phase of the modified filter output $\tilde{y}_k^p[n]$ at time $n = \rho n_o^p$ equal to the phase at the

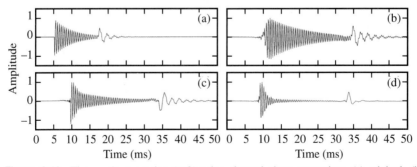

Figure 8.10   Time-scale expansion ($\times 2$) using channel phase correction: (a) original; (b) expansion with phase correction at 5 ms; (c) expansion with phase correction in clustered channels; (d) expansion without phase correction.

SOURCE: T.F. Quatieri, R.B. Dunn, and T.E. Hanna, "A Subband Approach to Time-Scale Modification of Complex Acoustic Signals" [33]. ©1995, IEEE. Used by permission.

event time $n = n_o^p$ in the original time scale. Denoting the phase correction for each cluster by ϕ_k^p, the modified channel signal becomes

$$\tilde{y}_k^p[n] = \tilde{a}_k^p[n]\cos\left(\rho\tilde{\phi}_k^p[n] + \phi_k^p\right), \qquad (8.29)$$

where $\phi_k^p = \phi_k^p[n_o^p] - \rho\tilde{\theta}_k^p[\rho n_o^p]$ and where p refers to the first or second cluster.

Short-Time Processing — To process a waveform over successive frames, we extract a signal segment every L samples, perform time-scale modification and phase correction on each segment using the phase vocoder, and then overlap and add the modified segments to give the final synthesis [33],[34],[37],[39]. We can thus think of this sequence of operations as combining the OLA and FBS synthesis on the modified STFT. Specifically, the filter-bank (satisfying the FBS constraint) modification is first applied to each windowed segment $f_{mL}[n] = w[mL-n]x[n]$. The frame length L is set to half the window length, i.e., $N_w = 2L$, and the window $w[n]$ is chosen such that $\sum_{m=-\infty}^{\infty} w[mL - n] = 1$, i.e., the overlapping windows form an identity (thus satisfying the OLA constraint). The two event times for the mth frame are selected as above, and saved. The procedure is repeated for frame $m + 1$. However, if the most recent event from frame m falls at least $L/4$ samples inside the current frame $m + 1$, then this event is designated the first event of frame[5] $m + 1$. With this condition, the second event time is found via the maximum of the histogram of the channel event onset times on frame $m + 1$ (excluding the previously chosen event time). Each channel is then assigned to a time instant based on the minimum distance between the two event times and the measured onset time $n_o(k)$. In addition, a frame is allowed to have no events by setting a histogram bin threshold below which a no-event condition is declared. In this case, channel phase offsets are selected to make the channel phases continuous across frame boundaries, i.e., the phase is allowed to "coast" from the previous frame.

EXAMPLE 8.2   Time-scale expansion can result in improved audibility of closely-spaced components for a variety of complex acoustic signals consisting of sums of rapidly damped sinewaves such as the sounds from mechanical impacts. An example of time-scale expansion of a sequence of transients from a closing stapler is shown in Figure 8.11, demonstrating the temporal and spectral fidelity in the time-scaled reconstruction [33],[39]. The above short-time processing approach was applied with a triangular window $w[n]$ of duration 10 ms and a 5-ms frame update, satisfying the OLA constraint. Each short-time segment was passed through a filter bank with 21 uniformly spaced filters $h_k[n]$, designed using a 2-ms prototype filter with Gaussian shape, satisfying the FBS constraint. Two event times were estimated on each frame using the histogram analysis described above. A goal in this example is to preserve the spectral envelope while time-expanding the temporal envelope of the signal. Although for the signal illustrated, the original spectrum was approximately preserved in the time-scaled signal, an observed difference is the narrowing of resonant bandwidth, a change which is consistent with stretching the temporal envelope.   ▲

---

[5] This approach is similar in style to the pitch-synchronous overlap-add method of time-scale modification introduced in Chapter 7. However, as described in Chapter 7, methods to achieve this synchrony rely on cross-correlating adjacent frames or determining consistent time instants within a glottal cycle, and not on synchronizing the phases of a filter-bank decomposition.

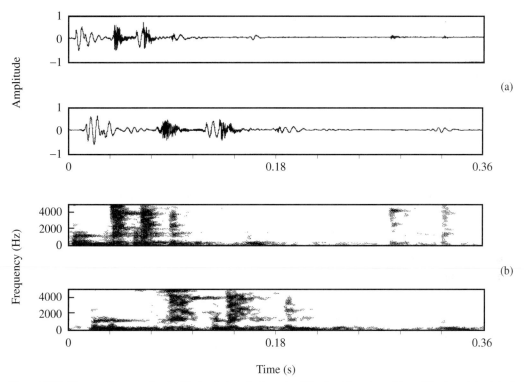

Figure 8.11 Time-scale expansion (×2) of a closing stapler using filter-bank/overlap-add modification: (a) original and time-expanded waveform; (b) spectrograms of part (a).

Source: T.F. Quatieri, R.B. Dunn, and T.E. Hanna, "A Subband Approach to Time-Scale Modification of Complex Acoustic Signals" [33]. ©1995, IEEE. Used by permission.

8.4.2 Phase Coherence of Quasi-Periodic Signals

We saw in Example 8.2 that our time-scale modification technique increases the distance between short-duration events in a signal, event time instants being estimated from a very short (2 ms) duration analysis filter in the filter bank. Consequently, applying this technique to speech will modify the pitch of the speaker as well as the articulation rate during quasi-periodic voiced regions (Exercise 8.8). Nevertheless, the technique can be extended to time-scale quasi-periodic signals by lengthening the analysis filter and by phase locking within every frame.

Puckette [32] and Laroche and Dolson [18] observed that for quasi-periodic signals, reduced reverberance in synthesis can be obtained by achieving phase coherence across filter bank channels that correspond to channel clusters. Each cluster is defined by dominant spectral regions for successive short-time segments. Specifically, for each short-time segment, a dominant spectral region is given by a "peak channel," which is a channel whose amplitude, $|X(nL, k)|$, is larger than its four nearest neighbors with respect to the frequency variable k. Channel clusters are then formed around each peak channel according to the (smallest) distance of a channel center frequency from each peak. For each cluster, the channel phases are then locked to the phase of the peak. Phase locking in this case means that the original phase relations, i.e., the difference between the phase of the peak channel and channels within its cluster in the original time scale, are preserved in the new time scale. For an analysis filter length of about a few pitch periods, e.g., about 20 ms, a channel peak often occurs near high-amplitude harmonics and channel clusters are formed around these dominant spectral values. In the context of the above short-time processing approach, consider the short-time segment $f_{mL}[n] = w[mL - n]x[n]$. Let the phase of the kth filter-bank output of this segment be denoted by $\phi_k[n, mL]$ and the corresponding phases of the pth cluster be denoted by $\phi_k^p[n, mL]$. The phases of the pth cluster of the time-scaled version of this segment are then given by $\rho\tilde{\phi}_k^p[n, mL]$. Denote the peak channel of the pth cluster by the index k_o^p. Then phase coherence of the modified signal can be achieved in the pth cluster by applying a phase offset correction ϕ_k^p to $\rho\tilde{\phi}_k^p[n, mL]$ such that for the kth channel within the pth cluster

$$(\rho\tilde{\phi}_k^p[n, mL] + \phi_k^p) - \rho\tilde{\phi}_{k_o}^p[n, mL] = \phi_k^p[n, mL] - \phi_{k_o}^p[n, mL], \quad \text{for } k \neq k_o$$

where n is taken at the center of the analysis and synthesis frames. The phase differences are thus preserved across channels within the pth cluster (but only at the center of the short-time segment). Because the phase of the dominant channel k_o^p of each cluster is preserved, time-scaled short-time segments add essentially coherently across consecutive frames.

Another approach to accomplish phase coherence in synthesis of quasi-periodic sequences by the phase vocoder was proposed by Sylvestre and Kabal [17],[49]. In time-scale modification, with short-time processing, the phase of a channel is reset at the onset of each frame to its value in the original time scale; this results, however, in a discontinuity in the unwrapped phase at a frame boundary, not necessarily a multiple of 2π. To remove the discontinuity, a fixed phase offset is first added to the resulting channel phase followed by a perturbation to the instantaneous frequency to enforce phase continuity (Exercise 8.5). The desired sequence is then obtained by concatenating successive modified short-time segments rather than by an overlap-add synthesis.

8.5 Constant-Q Analysis/Synthesis

We have seen repeatedly throughout the text the time-frequency resolution tradeoffs associated with the short-time Fourier transform (STFT). That is, according to the uncertainty principle reviewed in Chapter 2, we cannot obtain unlimited good time and frequency resolution simultaneously. A short analysis window $w[n]$ giving excellent temporal resolution implies poor frequency resolution and vice versa. Therefore, we select one "reasonable" time-frequency resolution tradeoff that is used in analyzing an entire speech waveform regardless of the speech event. Stationary sounds, e.g, steady vowels, and nonstationary sounds, e.g., plosive to vowel transitions, are processed with the same analysis window, typically about 20 ms in duration. This limitation can result in excessive smearing of transitions and transient sounds.

Ideally, we desire a time-frequency representation whose resolution can be adapted to the time and frequency characteristics of the sound, for example, giving good temporal resolution with rapidly changing and short-lived events and good frequency resolution in spectrally sharp regions. In this section, we do not attempt this general representation, but rather present an alternative time-frequency distribution called the *wavelet transform* which is one step toward more flexible time-frequency resolution. The wavelet transform achieves *constant-Q* resolution whereby time resolution increases and frequency resolution decreases with increasing frequency. In the context of speech processing, the importance of this transform lies in its providing a model of front-end auditory filter analysis. The goal of this section is twofold: first to describe the essential theory of the wavelet transform and compare it to the STFT from a filter bank perspective, and then to look briefly at a few of its applications to speech processing.

We begin this section with revisiting a problem in time-frequency analysis that motivates the wavelet transform. This leads to the wavelet transform approach and its theory for continuous time. We then describe a discrete-time filter-bank implementation of the wavelet transform that ties us back to the theme of this chapter. Finally we take a brief look at its application to speech processing, including time-scale modification, pitch estimation, and coding of speech and through these applications we show the link to auditory filter-bank front-end models. This leads us to an enticing glimpse of auditory modeling in Section 8.6, the final topic of this chapter.

8.5.1 Motivation

The problem that motivates the wavelet transform is the measurement of local frequency content in nonstationary signals. To illustrate the problem, consider the signal of Figure 8.12a consisting of two high-frequency short-duration tones superimposed on the sum of two low-frequency tones. The limitation of standard Fourier analysis on signals of the type in Figure 8.12a is that it measures the frequency content of the *entire* signal and so does not characterize the change in frequency content with time. This observation motivated the short-time Fourier transform (STFT) of Chapter 7 given in continuous time by

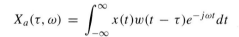

$$X_a(\tau, \omega) = \int_{-\infty}^{\infty} x(t)w(t - \tau)e^{-j\omega t} dt$$

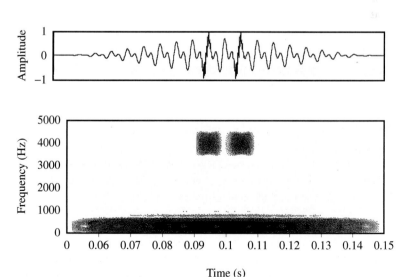

Figure 8.12   Low-frequency signal with superimposed high-frequency bursts: (a) waveform; (b) wideband spectrogram of (a).

whose squared magnitude $|X_a(\tau, \omega)|^2$ we have referred to as the spectrogram. [For later convenience, we have made a slight notational change from Equation (8.16).] This involves, as we have seen, looking at the signal through a sliding window and computing the Fourier transform under the window for each window shift τ. A limitation of the STFT is the fixed duration of the window $w(t)$; with $w(t)$, we cannot always *simultaneously* resolve short-lived events and closely-spaced long-duration tones, as illustrated in the spectrogram of Figure 8.12b. We have seen this tradeoff more quantitatively stated in the uncertainty principle that limits time-frequency resolution such that the product of duration $D(x)$ and bandwidth $B(x)$ of a signal $x(t)$ must exceed a constant (Chapter 2), i.e.,

$$D(x)B(x) \geq \frac{1}{4}. \tag{8.30}$$

In speech processing, such a limitation is of importance in identifying, for example, closely-spaced harmonic frequencies simultaneously with occurrence times of glottal pulses and short-duration plosive events (Exercise 8.12).

In defining an "ideal" time-frequency transform, acknowledging that we cannot defeat the uncertainty principle in the context of the Fourier transform,[6] we seek to minimize the limitations of the uncertainty principle for time-frequency localization. We also require that the transform be invertible (as with the invertibility of the STFT) and be a useful representation for numerous applications.

8.5.2 Wavelet Transform

An approach to dealing with the uncertainty principle, if not beating it, is to compute many spectrograms with different analysis window durations. As illustrated in Figure 8.13, the wavelet transform can be thought of as a collage of pieces of spectrograms based on different analysis windows and thus different time-frequency resolutions. Specifically, short windows are used at high frequency for good time resolution, and long windows are used at low frequency for good frequency resolution. Figure 8.14 shows how the wavelet transform time-frequency *resolution cell* changes with frequency, in contrast to the fixed time-frequency resolution of the STFT. We see from this perspective that we don't defeat the uncertainty principle, i.e., the area of the cell is fixed, but rather its relative dimensions change as we move around the time-frequency plane. This concept of analysis with different resolutions was developed independently in many fields including speech, image, and seismic processing, as well as quantum mechanics. The mathematics of this *multi-resolution analysis* was developed in the early 1980s and gave a strong foundation for the wavelet transform, as well as provided a unifying framework for the ideas formulated in the context of different applications [8],[22].

The *continuous wavelet transform* (CWT) formalizes the notion of adapting time resolution to frequency. In beginning the mathematical development, we define a set of functions as the time-scaled and shifted versions of a prototype $h(t)$, i.e.,

$$h_{\tau,a}(t) = \frac{1}{\sqrt{a}} h\left(\frac{t - \tau}{a}\right)$$

---

[6] We will see in Chapter 11 that uncertainty is a function of framework (e.g., time or scale [inverse of frequency]) and definition (e.g., local or global quantities). Indeed, with a different framework and quantities of interest, the uncertainty principle need not exist or may be characterized by a different resolution constraint.

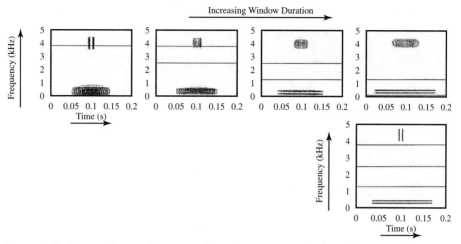

Figure 8.13 The wavelet transform as a collage of spectrograms for the multi-component signal of Figure 8.12a. A short window at high frequency gives good time resolution, while a long window at low frequency gives good frequency resolution. The lower right panel is obtained by piecing together regions of the upper four panels.

where $h(t)$ is the *basic wavelet*, $h_{\tau,a}(t)$ are the associated *wavelets*, τ is the time shift, and a is the scaling factor, as illustrated in Figure 8.15. We then define the continuous wavelet transform (CWT) as (where $*$ here denotes complex conjugation)

$$X_w(\tau, a) = \frac{1}{\sqrt{a}} \int_{-\infty}^{\infty} x(t) h^* \left(\frac{t - \tau}{a} \right) dt \tag{8.31}$$

which we can think of as a measure of "similarity" of $x(t)$ with $h(t)$ at different scales and time shifts. The factor $\frac{1}{\sqrt{a}}$ is present to normalize the energy in the wavelets. With a scale factor $a < 1$, the basic wavelet is contracted and the resulting $h(\frac{t-\tau}{a})$ is shifted past the signal $x(t)$, multiplied, and integrated, i.e., convolved with $x(t)$. We can think of the wavelet transform $X_w(\tau, a)$ from a filtering viewpoint (as we did with the STFT), i.e.,

$$X_w(\tau, a) = \frac{1}{\sqrt{a}} \int_{-\infty}^{\infty} x(t) h^* \left(\frac{t - \tau}{a} \right) dt$$

$$= \frac{1}{\sqrt{a}} x(\tau) * h^* \left(-\frac{\tau}{a} \right)$$

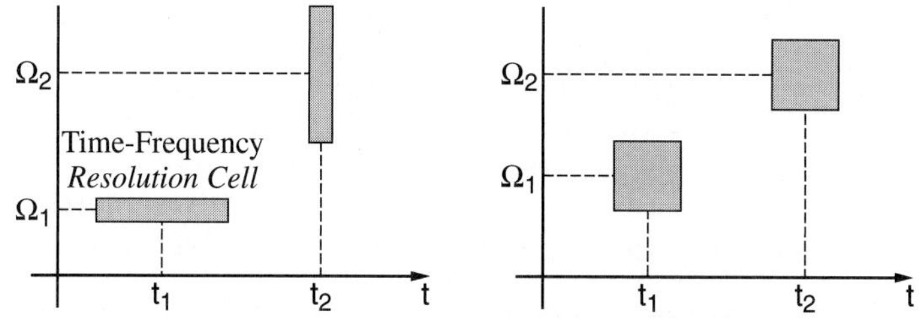

Figure 8.14 Adaptation of window size to frequency in the wavelet transform (left panel) in contrast to a fixed window in the STFT (right panel).

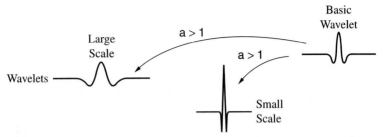

Figure 8.15 Schematic of a basic wavelet and its associated wavelets at different scales.

where * denotes convolution and where a smaller scale corresponds to wider-bandwidth filters. We refer to $|X_w(\tau, a)|^2$ as the *scalogram*,[7] in contrast to the spectrogram $|X(\tau, \omega)|^2$. An alternative interpretation of the wavelet transform is as a "zoom lens" at different time scales, i.e., we can rewrite Equation (8.31) as (Exercise 8.15)

$$X_w(\tau, a) = \frac{1}{\sqrt{a}} \int_{-\infty}^{\infty} x(at)h^*\left(t - \frac{\tau}{a}\right) dt \tag{8.32}$$

where now we keep the filter $h(t)$ unscaled and scale the signal $x(t)$. The concept of scale in general relates to time scale, but is analogous to frequency, as illustrated in the following example:

EXAMPLE 8.3   Consider a basic wavelet in the form of a modulated window, i.e.,

$$h(t) = w(t)e^{j\omega_o t}.$$

Then the CWT becomes

$$X_w(\tau, a) = \frac{1}{\sqrt{a}} \int_{-\infty}^{\infty} x(t)h^*\left(\frac{t - \tau}{a}\right) dt$$

$$= \frac{1}{\sqrt{a}} e^{j\frac{\omega_o}{a}\tau} \int_{-\infty}^{\infty} x(t)w\left(\frac{t - \tau}{a}\right) e^{-j\frac{\omega_o}{a}t} dt.$$

For each scale a, this expression can be thought of as an STFT at the frequency $\omega = \frac{\omega_o}{a}$ with a sliding window $w'(t - \tau) = w(\frac{t-\tau}{a})$. The CWT adapts the window size to frequency. At a small scale a, corresponding to a high frequency $\frac{\omega_o}{a}$, the wavelet is narrow in time and so has a wide bandwidth. At a large scale a, corresponding to a low frequency $\frac{\omega_o}{a}$, the wavelet is wide in time and so has a narrow bandwidth. As we increase scale (decrease frequency) the wavelet transform analyzes the signal with a window of decreasing bandwidth, thus giving good frequency resolution for low frequency and good time resolution for high frequency. Therefore, loosely speaking, as we have discussed, we can think of the scalogram as a collage of pieces of different spectrograms. We also see from this example that scale varies inversely with frequency. A comparison of the spectrogram and the scalogram, using a 10-ms Hamming window $w[n]$, is illustrated in Figure 8.16 for the signal of Figure 8.12 consisting of two low-frequency tones and two high-frequency clicks (bursts). We see that the scalogram reveals both the clicks and the tones, while the spectrogram, in resolving the clicks, is forced to merge the two tones. ▲

---

[7] The motivation for the squaring operation will become clear in Chapter 11, where we interpret both the spectrogram and scalogram as *energy densities* that are part of a larger class of time-frequency distributions.

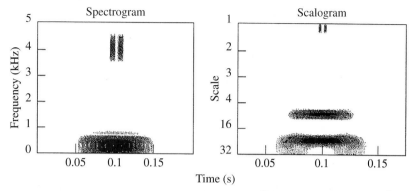

Figure 8.16 Comparison of the spectrogram $|X(\tau, \omega)|^2$ and scalogram $|X_w(\tau, a)|^2$ for the multi-component signal of Figure 8.12.

As with the STFT, we want to consider the problem of the invertibility of the wavelet transform, i.e., the conditions under which we can recover a signal $x(t)$ from its wavelet transform. It can be shown that for a large class of basic wavelets, $h(t)$, $x(t)$ can be recovered from a superposition of wavelets $h_{\tau,a}(t)$, i.e., the inverse continuous wavelet transform (ICWT)

$$x(t) = \frac{1}{C_h} \int_{-\infty}^{\infty} \int_{0}^{\infty} X_w(\tau, a) h_{\tau,a}^*(t) \frac{da d\tau}{a^2} \tag{8.33}$$

under the condition

$$C_h = \int_{-\infty}^{\infty} \frac{|H(\omega)|^2}{|\omega|} d\omega < \infty$$

which is called the *admissibility condition* (Exercise 8.13). This condition implies that $h(t)$ has zero mean, i.e., $\int_{-\infty}^{\infty} h(t) dt = H(\omega = 0) = 0$, because at $\omega = 0$ the denominator in the above condition is zero, i.e., having zero mean $h(t)$ must "wiggle" [8]. In addition, the admissibility condition requires that $H(\omega)$ not decay "too slowly" in frequency and thus has a bandpass characteristic; furthermore, it can be shown (from Parseval's theorem) that its time response $h(t)$ also cannot decay "too slowly" [8]. Therefore, we see the motivation for the nomenclature *wavelet*. Equation (8.33) implies that $x(t)$ can be written as a superposition of shifted and dilated wavelets.

The relation in Equation (8.33) also lends itself to a basis function interpretation whereby the continuous wavelet transform $X_w(\tau, a)$ measures the projection of $x(t)$ onto the basis $h_{\tau,a}(t)$, i.e., $X_w(\tau, a)$ is the inner product of $x(t)$ and $h_{\tau,a}(t)$. (In this chapter, we use the term "basis" loosely. Strictly, a function set is a basis for a vector space of signals if it provides a unique representation of each signal in the space [3].) The above invertibility result holds even though the elements of the wavelet basis $h_{\tau,a}(t)$ are generally not orthogonal;[8] to be *orthogonal*, the inner product of any two different wavelets must be zero.

---

[8] A similar synthesis formula can be found for the STFT which is different from our inversion formula Equation (7.8) of Chapter 7 [8]. The basis here is the shifted and modulated window $w(t)$, fixed in its time/frequency resolution. As with the continuous wavelet transform, the STFT basis is not orthogonal.

8.5.3 Discrete Wavelet Transform

Implementation of the CWT and ICWT requires discretizing the scale a, shift τ, and time t. As with the discrete STFT, we also face the issue of synthesis from the discretized CWT. We begin by discretizing shift and scale. We can think of sampling the CWT in scale a and shift τ to form a set of wavelet coefficients $c_{n,m}$ called the *discrete wavelet transform* of $x(t)$:

$$c_{n,m} = \int_{-\infty}^{\infty} x(t) h_{n,m}^*(t) dt$$

where

$$h_{n,m}(t) = \frac{1}{\sqrt{a_m}} h\left(\frac{t - \tau_n}{a_m}\right).$$

The wavelet coefficients $c_{n,m}$ represent the inner product of $x(t)$ with the discretized wavelet basis $h_{n,m}(t)$, which are the original wavelets sampled in scale and in shift. The wavelet coefficients are analogous to the coefficients of the discrete STFT, $X(n, k)$. Under certain conditions the discretized basis is orthogonal and complete[9] and so reconstruction of $x(t)$, i.e., the inverse discrete wavelet transform, is given by (Exercise 8.13)

$$x(t) = \sum_{n=-\infty}^{\infty} \sum_{m=-\infty}^{\infty} c_{n,m} h_{n,m}(t) \tag{8.34}$$

where the orthogonality condition is expressed as

$$\int_{-\infty}^{\infty} h_{n,m}(t) h_{p,q}^*(t) dt = \delta[n - p]\delta[m - q]$$

i.e., the inner product of $h_{n,m}(t)$ and $h_{p,q}(t)$ is zero when the wavelets are different. The sampling requirements on shift and scale for the wavelet basis $h_{n,m}(t)$ to satisfy orthogonality, and thus invertibility through Equation (8.34), are very different than for invertibility with the STFT because the (standard) STFT requires a uniform sampling in frequency and a uniform sampling in time (time decimation) (Figure 8.18). In general, it is not easy to find a basis (derived from the basic wavelet), together with a sampling strategy, to meet the orthogonality condition. Nevertheless, under certain conditions, it is possible to invert the discrete wavelet transform even when the basis is not orthogonal [8]. Orthogonality, however, makes the inversion process straightforward [via Equation (8.34)] and leads to an efficient filtering implementation of the discrete wavelet transform and its inverse [22].

If the wavelet basis $h_{n,m}(t)$ constitutes what is referred to as a *frame*, the reconstruction of a signal $x(t)$ is always possible [8],[22]. The basis $h_{n,m}(t)$ is a frame if there exists some $A > 0$ and $B > 0$ such that for all $x(t)$

$$A||x||^2 \leq \sum_{n,m} |\langle h_{n,m}, x\rangle|^2 \leq B||x||^2 \tag{8.35}$$

---

[9] A set of wavelets defines a *complete* basis if any signal can be reconstructed from linear combinations of the wavelet basis.

where the inner product

$$\langle f, g \rangle = \int_{-\infty}^{\infty} f(u)g(u)du$$

and the norm

$$||f||^2 = \int_{-\infty}^{\infty} f^2(u)du.$$

The values of A and B reflect the degree of *redundancy* of the basis. When $A = B$ we have a *tight frame* (and thus not redundant) and the wavelet basis can be shown to be orthogonal and thus to have a reconstruction formula given by Equation (8.34). When the frame condition Equation (8.35) holds, but orthogonality does not, the basis is redundant and a more involved reconstruction formula is required [8],[22][22].

One particular natural sampling in scale and shift is that of *dyadic* (or *octave*) sampling where for each scale $a_m = 2^m$ for $m = 1, 2, 3, \ldots$ we shift at $\tau_n = na_m$ for $n = 1, 2, 3 \ldots$. This is considered a "natural" sampling because as the scale increases by a factor of two (i.e., the bandwidth of the wavelet decreases by a factor of two), the sampling rate of shift decreases by a factor of two (half the bandwidth requires half the sampling rate). In other words, the wavelets are partitioned in octave bands and the time (shift) sampling is commensurate with bandwidth. A view of this dyadic sampling is shown in Figure 8.17. The dyadic wavelet basis is then given by

$$h_{n,m}(t) = 2^{-m/2}h(2^{-m}(t - n2^m))$$
$$= 2^{-m/2}h(2^{-m}t - n).$$

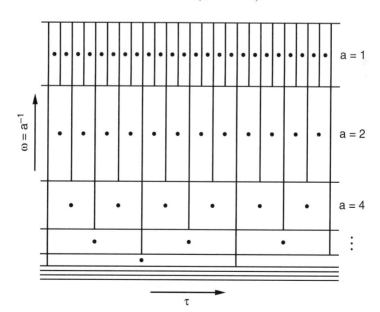

Figure 8.17   Sampling of scale and shift in a dyadic wavelet basis. The wavelets are partitioned in octave bands and the time shift is commensurate with bandwidth: $a_m = 1, 2, 4, \ldots 2^m \ldots$ and $\tau_n = na_m$.

From a signal processing perspective, the dyadic discrete wavelet transform can be considered as the output of a filter bank with constant-Q, octave-band, bandpass filters with impulse responses $2^{-m/2}h(2^{-m}t-n)$. A comparison of the increasing filter bandwidths of dyadic wavelets and the uniform filter bandwidths of the discrete STFT (discrete in frequency and continuous in time), as well as the corresponding filter impulse responses, is shown in Figure 8.18. In part because of the efficient implementation and auditory- and visual cortex-like time-frequency properties of dyadic wavelets, a large part of wavelet theory has involved finding dyadic wavelet bases that are orthogonal and that are useful in a variety of applications [22].

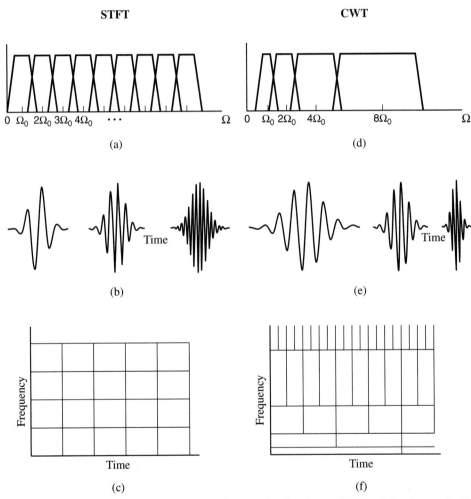

Figure 8.18 Comparison of the sampling requirements for the discrete STFT and the discrete dyadic wavelet transform from a filtering perspective. Panels (a) and (d) show the required filters in frequency, while (b) and (e) show their counterparts in time. The discrete STFT filters have constant bandwidth while the discrete dyadic wavelets have constant-Q bandwidth. Panels (c) and (f) give the respective time-frequency "tiles" that represent the essential concentration of the basis in the time-frequency plane.

SOURCE: O. Rioul and M. Vetterli, "Wavelets and Signal Processing" [43]. ©1991, IEEE. Used by permission.

EXAMPLE 8.4   One basic wavelet that has found widespread use is a function that approximates
a differentiator, illustrated in Figure 8.19a. A fascinating observation by Mallat [22] is that this wavelet
(among others) [10] satisfies certain properties that allow a signal to be reconstructed from local maxima
(with respect to shift τ) of $|X_w(\tau, a)|$ over a sampled scale, i.e., from the maxima of the discrete

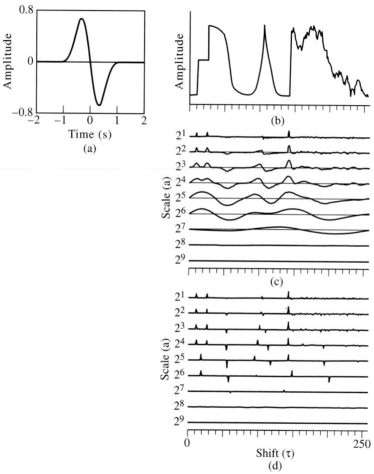

Figure 8.19 Signal representation by maxima of the discrete wavelet transform with
respect to shift: (a) a wavelet chosen to approximate a differentiator; (b) a signal and
its superimposed (essentially indistinguishable) reconstruction from wavelet maxima;
(c) wavelet transform outputs $X_w(\tau, a)$ for sampled at a dyadic scale; (d) points of
wavelet maxima of (c), i.e., $\max |X_w(\tau, a)|$ with respect to τ.

SOURCE: S. Mallat and W.L. Hwang, "Singularity Detection and Processing with Wavelets"
[23]. ©1992, IEEE. Used by permission.

---

[10] It is sufficient that the basic wavelet be the derivative of a function with energy concentrated near the
frequency origin [22].

set of filter-bank outputs. The maxima of the wavelet channel outputs indicate fast variations in the signal. An example of $X_w(\tau, a)$ sampled at a dyadic scale and the resulting maxima is shown in Figure 8.19. The original signal and its superimposed reconstruction are also given. The reconstruction uses an iterative algorithm similar in style to the STFT magnitude-only iterative reconstruction algorithm of Chapter 7. As stated, the wavelet filter maxima correspond to sharp changes and transients in a signal (as can be seen in the figure) and as such are useful in a variety of contexts, such as pitch and glottal closure estimation described in the following section, and in modeling aural data reduction via sampling the auditory front-end filter-bank outputs that are described later in this chapter.   ▲

Our last sampling consideration is that of continuous- to discrete-time conversion of the wavelet analysis and synthesis. For a discrete-time sequence $x[n]$, the wavelet transform, discrete in time as well as shift and scale, is denoted here by $X_w(n, m)$ (analogous to the discrete STFT):

$$X_w(n, m) = c_{n,m} = \sum_{p=-\infty}^{\infty} x[p]h_{n,m}^*[p].$$

We give without proof a filtering implementation for the particular discrete-time dyadic wavelet transform. Consider a discrete-time signal $x[n]$ derived by sampling $x(t)$. The filtering implementation of the forward transform (Figure 8.20) is given by an iterative cascade of identical stages, each stage consisting of a lowpass [by $P(\omega)$] and highpass [by $Q(\omega)$] decomposition of the signal followed by two-to-one downsampling. The sequences $p[n]$ and $q[n]$ are derived from the basic discrete-time wavelet $h[n]$; specifically, $p[n] = (-1)^n h[-n + 1]$ and $q[n] = h[n]$ [8]. It is remarkable that downsampling the highpass filter outputs at each stage (each stage representing different scales) gives the wavelet coefficients at different scales of the original continuous-time function $x(t)$ [8]. This operation can be thought of as half-band splitting of the signal at each stage and thus is equivalent to a constant-Q, octave band analysis, as illustrated in Figure 8.18d. A similar iterative structure can be used for inverting the wavelet transform from the wavelet coefficients. The condition for invertibility on the basic discrete-time wavelet $h[n]$, when dyadically sampled in shift and scale, is intimately related to the "perfect reconstruction" constraint on quadrature mirror and conjugate mirror digital filters

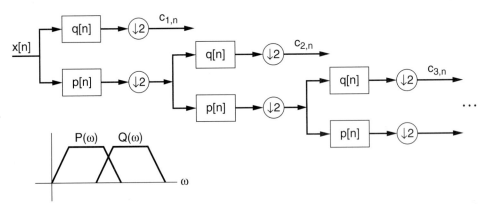

Figure 8.20 Iterative filtering implementation of discrete wavelet transform with dyadic orthogonal wavelets. Each filter in 8.18d represents the output of the highpass filter at each stage prior to downsampling. A similar iterative structure exists for the discrete inverse wavelet transform.

(i.e., the output of the filter bank configuration equals the input) [7],[48]. The equivalence of the constraint for invertibility of a dyadic wavelet basis and for perfect reconstruction of quadrature and conjugate mirror digital filter banks gives an important relation of wavelet theory to more traditional concepts in digital signal processing [3],[8],[22].

8.5.4 Applications

In this section, we briefly describe a few applications of the wavelet transform, beginning with a counterpart to the magnitude-only STFT speech modification algorithm developed in Chapter 7.

Time-Scale Modification — We will see in the following section that a simple model of front-end auditory processing is that of a wavelet transform along the basilar membrane which is located within the cochlear duct of the inner ear. The cochlar duct is a coiled tube, consisting of hard bone, and is filled with fluid. The basilar membrane is coiled within the cochlear duct and vibrates with an input stimulus. Each place along the membrane is characterized by a different resonant-like frequency response with the peak frequency and bandwidth decreasing as we move away from the opening of the cochlear duct, the auditory filters being approximately constant-Q. The reader may want to look ahead to Figures 8.24 and 8.25 for illustrations of the cochlear anatomy and associated filter frequency responses.

We have seen that the wavelet transform requires wavelet filters derived by scaling a basic wavelet. Irino and Kawahara [14],[15] used the above cochlear model for selecting a basic "auditory wavelet." The basic auditory wavelet is selected as the impulse response along the basilar membrane of a cochlear filter with peak frequency about 873 Hz, which is almost the center of the audible range on a logarithmic frequency scale. This function satisfies the admissibility condition and thus provides for a continuous wavelet basis for signal representation. This basic wavelet is then scaled[11] to form a discrete wavelet basis consisting of 128 channels from 55 Hz to 15 kHz. In the Irino-Kawahara auditory wavelet basis, the shift occurs every 0.5 ms. The resulting wavelet basis is not orthogonal, being highly redundant. The transform based on this wavelet filter set is referred to as the *auditory wavelet transform* (AWT), and the wavelet coefficients obtained from the AWT analysis of the signal are referred to as the AWT coefficients [14],[15]. Figure 8.21 shows frequency responses of the AWT on a logarithmic frequency scale over eight octaves compared against that of a cochlear filter model based on physical principles [15]. The amplitudes of the AWT were selected to match those of the cochlear filters at peak frequency. It is seen in Figure 8.21 that the essential characteristics of the cochlear filters are captured by the AWT. The analytic signal of each filter output is computed via the Hilbert transform (Chapter 2) and the resulting magnitude and phase are used for signal analysis. A possible advantage of the AWT over the STFT is that it may better represent information in auditory channels used in human perception.

Irino and Kawahara developed algorithms for reconstructing a signal from the modified wavelet transform for the purpose of time-scale modification. Suppose we are given a discrete auditory wavelet transform $X_w(nL, m)$, where nL refers to the uniform shift with L being the number of samples in 0.5 ms. One approach to obtain a modified transform for time-scale

---

[11] Because the scaling values are not necessarily rational, simple sampling rate conversion is not used to obtained the scaled wavelets. A cubic spline interpolation is invoked instead. Details of this scaling technique are given in [14],[15].

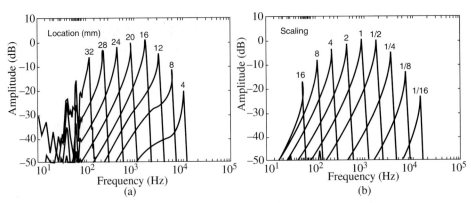

Figure 8.21 Frequency response at each octave of (b) the AWT on a log-linear scale compared against that of (a) a cochlear filter model based on physical principles [14],[15]. The frequency responses of the cochlear model are specified by locations along the basilar membrane relative to the cochlear duct opening. The amplitudes of the AWT were selected to match those of the cochlear filter model at peak frequency.

SOURCES: T. Irino and H. Kawahara, "Signal Reconstruction from Modified Auditory Wavelet Transformation" [14]. ©1993, IEEE; "Signal Reconstruction from Modified Wavelet Transformation—An Application to Auditory Signal Modeling" [15]. ©1992, IEEE. Used by permission.

modification is to pretend that $X_w(nL, m)$ was determined with time shift $M = \frac{L}{2}$ and to form the desired modified auditory wavelet transform

$$Y_w(nM, m) = X_w(nL, m).$$

This formulation is analogous to the approach we took in Chapter 7 for time-scale modification from the STFT. Also similar to the approach in Chapter 7, we then form a mean-squared-error distance metric

$$D[X_w^e, Y] = \sum_{n=-\infty}^{\infty} \sum_{m=-\infty}^{\infty} |X_w^e(nM, m) - Y_w(nM, m)|^2 \qquad (8.36)$$

that is minimized with respect to the unknown signal $x_e[n]$ embedded within its AWT $X_w^e(n, m)$. Unlike the counterpart signal estimation from the STFT, the resulting inverse problem is essentially impossible to solve due to the changing window (filter) associated with each wavelet channel (Exercise 8.16). In addition, as with the STFT, the phase of the modified AWT is such that it results in inconsistent discrete-wavelet transform slices at successive time shifts. Thus, a magnitude-only AWT reconstruction algorithm was proposed [14],[15], similar to that for magnitude-only STFT iterative reconstruction in Chapter 7, for a least-squared-error solution (Exercise 8.16). The following example illustrates the technique:

EXAMPLE 8.5   The AWT magnitude $|X_w(nL, m)|$ with a 0.5 ms time shift was computed for the speech utterance "kanojowa." For a time-scale compression by a factor of two, the modified magnitude function $|Y_w(nM, m)|$ was formed with $M = \frac{L}{2}$ so that the time shift is assumed to be half the original. The initial starting sequence in the magnitude-only (least-squared-error) iteration was white random noise. Figure 8.22 shows an example of the original and time-compressed AWT magnitude along time, as well as the AWT magnitude of the reconstruction after 20 iterations. The modified waveform was judged to be smooth and free of artifacts [15].   ▲

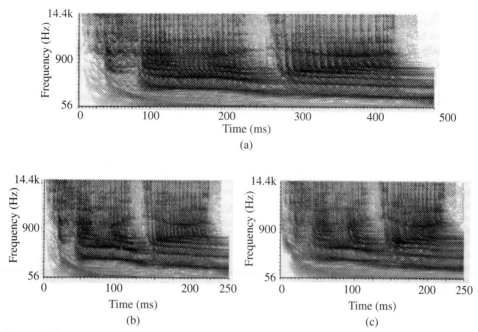

Figure 8.22 Time-scale modification with iterative least-squared-error estimation from a modified AWT magnitude: (a) original AWT magnitude for 12 channels with a 0.5 ms time shift; (b) modified AWT magnitude with an assumed 0.25 time shift; (c) AWT magnitude of the least-squared-error estimate of (b) for 20 iterations.

SOURCES: T. Irino and H. Kawahara, "Signal Reconstruction from Modified Auditory Wavelett Transformation" [14]. ©1993, IEEE. "Signal Reconstruction from Modified Wavelett Transformation—An Application to Auditory Signal Modeling" [15]. ©1992, IEEE. Used by permission.

Pitch Estimation — Kadambe and Boudreaux-Bartels [16] have applied a dyadic wavelet transform to the problem of estimating pitch and glottal closure time. We saw earlier in Example 8.4 that, with an appropriate selection of a basic wavelet, i.e., the first derivative of a low-frequency energy signal, the local maxima indicate abrupt changes in a signal and these local maxima are manifested across several scales in the transform.[12] We have seen in Chapters 3 and 5 that vocal cords can close abruptly, thus introducing a sharp negative pulse in the glottal flow derivative at the time of glottal closure. Kadambe and Boudreaux-Bartels found that this abrupt change manifests itself as local maxima in a dyadic wavelet basis across several dyadic scales, and that these local maxima can be exploited for estimation of glottal closure time and pitch. This concept is consistent with certain models of auditory processing [5],[9],[46],[47],[51], as we will see later in this chapter, and also consistent with one of the earliest successful pitch estimators by Gold and Rabiner [12].

---

[12] This same property has been found empirically for certain uniformly-spaced filter banks and was used (implicitly) earlier in Section 8.4.1 in determining event times for phase locking in the phase vocoder.

Using a cubic spline function [22] as the basic wavelet (centered at 8000 Hz), Kadambe and Boudreaux-Bartels determined maxima at five different scales, $a = 2^1, 2^2, \ldots 2^5$, with wavelet channel outputs sampled at 0.1 ms time shifts. If the locations of thresholded maxima agree to within a small difference across two scales, then an average of these locations is said to be a time of glottal closure. The pitch period is then estimated as the time difference between two glottal closure instants. Kadambe and Boudreaux-Bartels compared pitch estimation using the cubic spline wavelet with two standard pitch estimators: (1) The cepstral-based pitch estimator described in Chapter 6, and (2) The autocorrelation-based pitch estimator that we will describe in Chapter 10. Both of these classic pitch estimators require stationarity of pitch and vocal tract over an analysis window of about 2–3 pitch periods in duration, in contrast to the wavelet-based approach that uses *local* maxima of the wavelet transform magnitude as its feature. As such, the wavelet-based approach was able to more accurately track a time-varying pitch, while performing comparably with stationary pitch [16]. This property is illustrated in the following example:

> **EXAMPLE 8.6**   Using the basic wavelet of Figure 8.19a (i.e., an approximate differentiator), wavelet channel outputs are computed at five different scales, $a = 2^1, 2^2, \ldots 2^5$, with 0.01-ms time shifts. The input is a nasal speech sound with rapid pitch modulation. Figure 8.23 shows the maxima of the wavelet transform magnitude for the five channels. We see a correlation of maxima across these scales that correspond to the rapidly varying pitch period.   ▲

Subband Speech Coding — We saw in Section 8.5.3 that the discrete wavelet transform, using a dyadic wavelet basis, performs constant-Q analysis to obtain the wavelet coefficients, and can be implemented with an iterative lowpass- and highpass-band splitting of the waveform. Almost two decades prior to the wide use of the wavelet transform, a similar analysis and implementation was performed with *subband filtering* for the application of speech coding using the quadrature mirror and conjugate mirror filters and conditions for perfect reconstruction alluded to earlier [7],[22],[48]. We return to subband coding in Chapter 12.

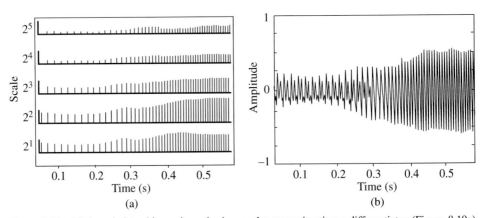

Figure 8.23 Pitch period tracking using a basic wavelet approximating a differentiator (Figure 8.19a) for a nasal with rapid pitch modulation: (a) maxima of wavelet transform magnitude for five different scales $a = 2^1, 2^2, \ldots 2^5$ with 0.01-ms time shifts; (b) voiced nasal speech waveform corresponding to (a). Maxima are given by sample values greater than two adjacent neighboring sample values.

8.6 Auditory Modeling

The major divisions of the peripheral auditory system—the outer ear, middle ear, and inner ear—are shown in Figure 8.24 [13],[27],[30]. Sound first enters the outer ear through the pinna, situated external to the head, and helps to localize sounds. Sound then travels down the auditory canal and results in vibration of the eardrum, which is the component of the outer ear that connects to the middle ear. The middle ear consists of three bones—the malleus, incus, and stapes—that act as a transformer to efficiently transport the vibrations of the eardrum to the inner ear. The middle ear is connected to the inner ear by way of the oval window. From the perspective of aural perception, the major component of the inner ear is the cochlea, which is a coiled tube having the appearance of a snail and is filled with fluid. A schematic of the uncoiled cochlea is shown in Figure 8.25a. Running about midway along the length of the cochlea and within the cochlear fluid is the basilar membrane that is held to the cochlea by bone.

Vibrations of the eardrum result in movement of the oval window that generates a compression sound wave in the cochlear fluid. This compression wave, in turn, causes a vertical vibration of the basilar membrane. Along the basilar membrane are located nearly 10000 inner hair cells in a regular geometric pattern. Embedded within the membrane are the cell bodies, above which protrude short stiff hairs (Figure 8.25b) that deflect when the basilar membrane vibrates. This deflection causes a chemical reaction within the cell bodies which finally leads to a "firing" of short-duration electrical (voltage) pulses in the nerve fibers that connect to the bottom of each inner hair cell. The nerve fibers from all inner hair cells are bunched together to form the auditory nerve (Figure 8.24). Electrical pulses run along the auditory nerve and ultimately reach the higher levels of auditory processing in the brain, where they are perceived as *sound*.

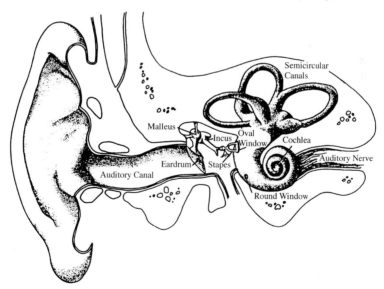

Figure 8.24 Primary anatomical components of the peripheral auditory system are the outer, middle, and inner ear.

Source: P.H. Lindsay and D.A. Norman, *Human Information Processing: An Introduction to Psychology*, Academic Press, New York, NY, 1972. ©1972, Harcourt, Inc. Reproduced by permission of the publisher.

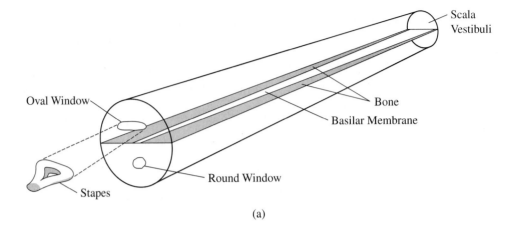

(a)

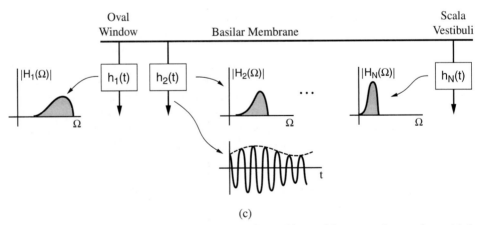

(b)

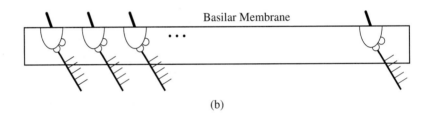

(c)

Figure 8.25 Schematic of front-end auditory processing and its model as a wavelet transform: (a) the uncoiled cochlea; (b) the transduction to neural firings of the deflection of hairs that protrude from the inner hair cells along the basilar membrane; (c) a signal processing abstraction of the cochlear filters along the basilar membrane. The filter tuning curves, i.e., frequency responses, are roughly constant-Q with bandwidth decreasing logarithmically from the oval window to the scala vestibuli.

Source for panel (a): D.M. Green, *An Introduction to Hearing*, [13]. ©1976, D.M. Green. Used by permission.

When the ear is excited by an input stimulus, different regions of the basilar membrane respond maximally to different frequencies, i.e., a frequency "tuning" occurs along the membrane. We can therefore think of the response patterns as due to a bank of cochlear filters along the basilar membrane (Figure 8.25c). Measurements show a roughly logarithmic increase in bandwidth of these filters, i.e., the filters are approximately[13] constant-Q in their frequency response, with bandwidth decreasing as we move away from the cochlear opening at the oval window and to the scala vestibuli which reside at the end of the basilar membrane. We saw earlier a set of modeled cochlear filters (Figure 8.21a) over an eight-octave range from 55 Hz to 15 kHz. The filters are characterized by an asymmetric frequency response with a steeper falloff to the right of their peak frequency than to the left (even on a linear frequency scale). The peak frequency at which maximum response occurs is referred to as the *characteristic frequency* of the cochlear filter. As we saw in our discussion of time-scale modification of Section 8.5.4, a simple model of the inner ear front-end auditory processing, therefore, is that of a wavelet transform along the vertically oscillating basilar membrane. This wavelet representation of the cochlear filters was introduced by Yang, Wang, and Shamma [51]. The constant-Q cochlear filter bank thus provides a range of analysis window (filter) durations and bandwidths with which to analyze the signal at different frequencies (Figure 8.25c). Rapidly varying signal components (e.g., the attack of the glottal pulse and plosives) are better analyzed with shorter windows than those of slower components and events (e.g., low-frequency harmonics of vowels). As with our characterization of the filter-bank outputs in the phase vocoder, the output of each cochlea filter can be thought of as an amplitude- and frequency-modulated sinewave. The envelope of each output and its instantaneous frequency and phase are important features that are responsible for exciting particular neural firing patterns used by the higher levels of auditory processing in the perception of speech.

Our goal in this section is to give three different perspectives of signal processing in the auditory system that relate to the filter-bank analysis theme of this chapter. We look first at the output of the cochlear filter bank as amplitude- (AM) and frequency-(FM) modulated sinewaves and describe a simple model as to how this AM and FM generally influences firing rate of the inner hair cells. We next add to this model auditory processing elements that give improved joint time and frequency resolution, as well as enhanced sensitivity to temporal and spectral change in a signal. Finally, we investigate how different rates of change in signals within the AM-FM sinewave representation may be used and processed by different auditory components within the auditory pathway. These different perspectives are useful in motivating certain speech signal processing techniques to follow, and also lay the groundwork for auditory principles, such as critical band theory and perceptual masking, that will be described and used as needed in other parts of the text.

8.6.1 AM-FM Model of Auditory Processing

Consider a quasi-periodic waveform, in continuous time, consisting of a sum of sinewaves with slowly varying amplitude $A_p(t)$ and frequency $\Omega_p(t)$, each sinewave of the form $x_p(t) = A_p(t) \cos[\psi_p(t)]$ with $\psi_p(t) = \int_0^t \Omega_p(\tau)d\tau \phi_p$. If a single sinewave were to enter each cochlear filter, then the amplitude and phase of each filter output are governed by the single sinewave input. Let us suppose that one objective of auditory processing is to measure (and

---

[13] Below about 800 Hz, the cochlear filters have almost equal bandwidth.

perceive) the amplitude and frequency of the sinewave. The nerve fibers attached to the bottom of each inner hair cell fire when the envelope of the output is large enough, i.e., if $A_p(t) > T$ (T being a threshold of firing); furthermore, an increase in the envelope gives an increase in the firing rate. The average firing rate associated with a particular inner hair cell is obtained by integrating the firing rates over many nerve fibers that are attached to each hair cell. This property implies that the spectral content of the input sound is traced out by measuring the average firing rate of the fiber bundles along the basilar membrane because the amplitude of a sinewave $A_p(t)$ entering each cochlear filter is determined by samples of the vocal tract spectrum (Figure 8.26). This model is called the *place theory* of hearing because the input spectrum is reflected in the average firing rate associated with each cochlear filter tuned to a particular frequency along the basilar membrane [9],[13],[27],[30]. This simple model is appropriate for high-pitched speakers because the harmonic spacing is large enough so that one sinewave may indeed effectively enter only one cochlear filter. If there is no interaction at higher auditory levels across cochlear channels, then sinewaves in the input are viewed independently of one another.

For lower-pitched speakers, on the other hand, more than one sinewave will enter a cochlear filter. (This is more likely to occur at high frequency than at low frequency where the filters are more narrow.) In this case, the envelope is determined by a sum of sinewaves over the cochlear filter bandwidth. Nevertheless, the envelope still reflects the average energy in the input spectrum over each cochlear band and thus the vocal tract spectrum. However, with more than one sinewave entering a cochlear filter, the output envelope becomes pitch-dependent, especially for high-frequency cochlear filters of wide bandwidth (Figure 8.27). Specifically, the envelope will peak more at the glottal pulse instant, particularly for high-frequency cochlear filters, which give better *time resolution* than low-frequency filters because of their short impulse responses. This abrupt change in the envelope corresponds to a (temporally) local increase in nerve firings and thus the possible encoding of pitch in these nerve firings. Observe that the relative phase of the input sinewaves will influence the shape of the envelope in this case and this is more likely to occur with decreasing pitch and with cochlear filters of increasing characteristic frequency. The effect of phase change on firing rate has been demonstrated using a discrete-time cochlear model [19]. In speech signal processing, therefore, changing of the phase may influence auditory

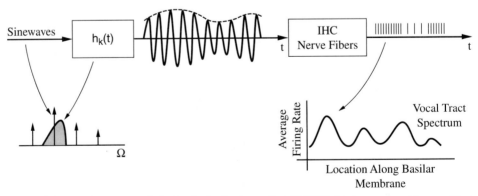

Figure 8.26   Auditory processing of a single slowly varying AM-FM sinewave as input to a cochlear filter, according to the place theory of hearing. The average firing rate is obtained by integrating the firing rate over many nerve fibers for a particular inner hair cell (IHC) and is roughly proportional to the input sinewave amplitude $A_p(t)$.

perception, especially of low-pitched sounds. For example, we saw in the phase vocoder that original harmonic phase relations can be lost in synthesis and result in a reverberant character to the speech. This loss of phase coherence can change the firing rate pattern (Figure 8.27). For example, the coherence loss can result in a less peaky envelope at the output of a cochlear filter and thus a "blurred" firing rate as seen by higher auditory levels. Preservation of phase may also be important for the second auditory processing mechanism we now describe.

This second mechanism is called *phase synchrony*. Until now we have ignored the absolute phase of the filter output and used only its envelope. The absolute phase, however, is also used in auditory processing because for low-frequency filters (about < 1000 Hz), nerve fibers fire *in synchrony with sinewave peaks*, the distance between firings therefore providing an estimate of the frequency of an input dominated by a single sinewave. If a single harmonic enters the cochlear filter, then the synchrony can result in a measure of the harmonic frequency. This is called the *temporal theory* of hearing [9],[13],[27],[30]. When multiple sinewaves enter the cochlear filter, there is evidence that, in the speech context, the synchrony occurs on the formant frequency dominating the cochlear band, and not necessarily on harmonic frequencies, or occurs on harmonic frequencies closest to the dominant formant frequency. Phase synchrony, therefore, can give measurements of either harmonic or formant frequency, depending on the number of sinewaves entering a cochlear filter and on the cochlear filter bandwidth. Observe, however, that although neural firings occur on peaks of low-frequency sinewaves, this firing *will not occur on every* sinewave peak. Thus one needs to look *across* multiple nerve fibers to make the appropriate measurement; the integration of neural firing patterns across adjacent fibers can provide the desired frequency measurement. Furthermore, phase synchrony across cochlear filter channels (in contrast to across nerve fibers within a cochlear channel) may also be exploited. This phenomenon gives further motivation to preserve sinewave phase relations in speech processing.

In summary, it appears then that sinewave phase relations are important both at high frequencies and at low frequencies, but for different reasons. At high frequencies, the envelope of cochlear filter outputs, and thus corresponding nerve firing patterns, are more prone to distortion

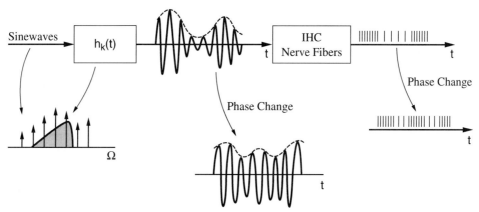

Figure 8.27 Auditory processing of a multiple of slowly varying AM-FM sinewaves as input to a cochlear filter, likely to occur with low pitch and at cochlear filters of high characteristic frequency. In this case, change in the phase relations of the input can alter the envelope shape and firing patterns of inner hair cell (IHC) nerve fibers, and thus, perhaps, perception of the sound.

with a change in sinewave phase relations. At low frequencies, a change in sinewave phase relations can degrade frequency measurements that rely on phase synchrony and the resulting correlation of firing patterns within and across cochlear channels.

8.6.2 Auditory Spectral Model

The auditory system has a unique sensitivity to rapid signal variation and can maintain this sensitivity under harsh conditions [30]. We look now at a mathematical model of the auditory processing functions that may be responsible for this capability. From above, the front-end stage of the auditory system can be modeled by a filter bank of the form

$$y(t, s) = h(t, s) * x(t)$$

where $h(t, s)$ represents the impulse responses of the (roughly) constant-Q cochlear filters at location s along the basilar membrane and where $*$ denotes convolution. Due to the near constant-Q nature of the cochlear filters, as we have seen, the filters being related by a simple dilation along the basilar membrane, the output $y(t, s)$ can be approximated by a wavelet transform of the input signal $x(t)$.

In this auditory model of Yang, Wang, and Shamma [51], the differentials of $y(t, s)$, both in time and frequency, are represented in higher levels of auditory processing. The differential in time at the output of each filter is introduced by the inner hair cell transduction, and is followed by a compressive nonlinearity and lowpass filtering (also introduced by the inner hair cell transduction):

$$z(t, s) = g[\partial_t y(t, s)] * w(t)$$

where $w(t)$ denotes the lowpass filter, $g(t)$ is the nonlinearity, and for simplicity, ∂_t denotes the partial derivative with respect to time t. The nonlinearity and lowpass filtering operations can be approximated by half-wave rectification and smoothing to form an envelope of the temporal derivative of each filter output.[14] The next operation is a differential in frequency, across cochlear filters, introduced by *lateral inhibition* [13],[30].

Lateral inhibition is a mechanism implemented by neural networks found in many biological systems. In vision, such a network exists in the retina and highlights fast transitions in an image as with edges. Likewise, in the auditory pathway (specifically, in the anteroventral cochlear nucleus), it has been shown that lateral inhibition enhances discontinuities of a signal's spectral content along the frequency (place) axis. Numerous lateral inhibitory networks exist, the simplest of which consists of mutual inhibition (subtraction) of the activity of two nearest-neighbor neural fibers. Mutual inhibition is implemented by neural shunting circuits that provide ratio processing of their inputs, which is a typical neural process throughout the body. We will look briefly at such networks in Chapter 11 in a different perspective on AM-FM estimation by the auditory system.

We approximate the lateral inhibition (subtraction) operation by differentiation with respect to location along the basilar membrane, which can be thought of as enhancing spatial

---

[14] Observe that when $y(t, s)$ is a slowly varying modulated sinewave, then the envelope of $\partial_t y(t, s)$ is approximately the product of AM and FM. For $y(t, s) = A \cos(\omega t)$, $\partial_t y(t, s) = A\omega \sin(\omega t)$ so that $z(t, s) = g[\partial_t y(t, s)] * w(t) \approx A\omega$.

changes across the basilar membrane channels. In the model of Yang, Wang, and Shamma, this differentation is followed by smoothing in frequency by a filter $v(s)$, so that the next stage becomes

$$z(t, s) = \partial_s g[\partial_t y(t, s)] * w(t) * v(s)$$
$$= \dot{g}[\partial_t y(t, s)]\partial_s \partial_t y(t, s) * w(t, s) \qquad (8.37)$$

where $\dot{g}(t)$ is the derivative of $g(t)$ with respect to its argument, and the two-dimensional smoothing function $w(t, s) = w(t) * v(s)$. This pattern is referred to as the *auditory spectrum* which may enhance perceptually useful features of the input waveform [51]. Under certain conditions, the function $\dot{g}(t) \approx \delta(t)$, and results in the sampling of $\partial_t y(t, s)$ at its extrema (i.e., maxima and minima) scaled by the mixed partial derivative $\partial_s \partial_t y(t, s)$ and smoothed by $w(t, s)$. Yang, Wang, and Shamma [51] and Slaney [47] have shown that, under certain conditions, a signal can be recovered from these samples (i.e., at extrema) or related features using an iterative approach similar to that alluded to in Example 8.4 of Section 8.5.3. (Nevertheless, there is not evidence that the auditory pathway *requires* this reconstruction.)

The auditory spectrum was analyzed with and without the compressive nonlinearity $g(t)$. From Equation (8.37), removing the nonlinearity, results in the time differential input $\partial_t x(t)$ passed through the *differential filter* $\partial_s h(t, s)$ that represents the derivative of $h(t, s)$ with respect to frequency (place) along the basilar membrane. The procedure is equivalent to a filter bank much more narrowly tuned than the wideband cochlear filters, with filters centered around the characteristic frequencies. An example, taken from Wang and Shamma [52] is given in Figure 8.28. An interesting hypothesis put forth by Wang and Shamma is that the differential

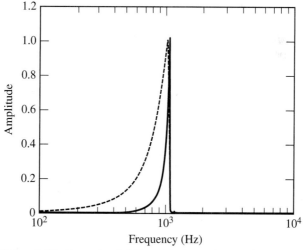

Figure 8.28 Example of differential filter. The solid line is the wideband cochlear filter, while the dotted line is the corresponding narrowband differential filter.

Source: K. Wang and S.A. Shamma, "Self-Normalization and Noise-Robustness in Early Auditory Representations" [52]. ©1994, IEEE. Used by permission.

cochlear filter bank operates in parallel with the original cochlear filter bank, thus providing simultaneously a narrowband and wideband time-frequency representation to higher levels of the auditory pathway at each place along the basilar membrane.

When the hair cell nonlinearity is taken into account, the resulting auditory spectrum is a function of both the original cochlear filters and differential cochlear filters, and, in particular, the output reflects the ratio of the energy of its narrowband differential filter to that of its corresponding wideband cochlear filter and is implemented by the neural shunting circuits discussed above. This *self-normalization* has the effect of enhancing a spectral line above a background noise level, and may in part be responsible for the superior robustness of the auditory system in noise [52]. The enhancement is greater when the cochlear bandwidth increases relative to the bandwidth of its corresponding differential filter. We will explore the robustness of the auditory system to noise from different perspectives in Chapter 13.

8.6.3 Phasic/Tonic View of Auditory Neural Processing

In this section, basic principles of neural processing are described with respect to "fast" and "slow" sound components. In particular, we describe Chistovich's [5] auditory model which, in the various low- and high-level stages of auditory processing, consists of *phasic* and *tonic* systems that operate in parallel and respond, respectively, to fast and slow sound types within an auditory channel. Phasic systems respond primarily to dynamic features such as changing temporal envelope characteristics and they detect, for example, sound onsets and offsets in frequency bands. Tonic systems respond primarily to the slowly varying spectral content of a signal within a band such as spectral patterns and formant transitions. Our formulation here is based in part on an exposition by Delgutte [9] on phasic and tonic neural mechanisms for the perception of speech. For each phasic and tonic mechanism, we look at both low-level auditory nerve and high-level cochlear nucleus neural processing. There is a great deal more interconnection and parallism in the cochlear nucleus than in the auditory nerve fibers which, by contrast, are simply a two-dimensional array of fibers organized according to the characteristic frequency (CF).

Phasic Systems — Auditory Nerve: There are often rapid changes in the amplitude envelopes of filtered sound events, and their *temporal relations* provide important information. A plosive consonant, for example, has an abrupt increase in high frequency spectral energy in the release of the burst, followed by an abrupt increase in low-frequency energy during the onset of the vowel; the time difference between the events was referred to in Chapter 3 as the voice onset time. Different voice onset times cue different plosive consonants. The degree of change in the amplitude envelope in a single frequency band is also important; for example, affricate consonants (e.g., "ch" in "<u>ch</u>op") have an abrupt temporal envelope, while fricative consonants (e.g., "sh" in "<u>sh</u>op") have a gradual envelope in high-frequency bands. In Chistovich's model, these different dynamic envelope characteristics lead to distinct patterns in neural firings of the phasic system component of the auditory nerve. The burst onset and vowel onset of a plosive, for example, have a large and abrupt discharge of nerve firings, at high-CF and low-CF phasic fibers, respectively. In the affricate and fricative consonants, an abrupt amplitude envelope yields a larger neural discharge peak than a gradual one in the same high-CF phasic fiber.

Following a large increase of nerve firings is a gradual decay in the firing rate called the *adaptation* [13],[30]. During adaptation, the response of the auditory nerve to future stimuli is much suppressed, especially to steady spectral components following the initial firing increase. Adaptation can occur on different time scales, ranging from a few milliseconds to several seconds to several minutes. An important property of adaptation is that it *enhances change* in a spectrum over time because a steady spectrum maintains the adaptation and contributes little to the discharge pattern, giving less responsiveness over time, while spectral *contrast* will revitalize nerve firings. One simple model of this contrast enhancement is that the past input to the nerve fiber is subtracted from the present so that only the change remains [27]. During adaptation, auditory nerves need change to respond. Steady or slowly changing spectral components are thus largely unseen by the phasic neural components of the auditory nerve, their measurement made by the tonic system components. A simplified view of the phasic response to a schematized plosive/vowel transition is shown in Figure 8.29. Observe that neural discharge also occurs at the modulation rate of an amplitude-modulated envelope, e.g., the pitch period of a speech waveform. Although we show in Figure 8.29 the discharge near peaks in the AM, the firing may occur only sporadically at these times. The reader should argue these neural response patterns based on the bandwidth of low-CF and high-CF cochlear filters.

Cochlear Nucleus: As firing patterns from acoustic transients travel up the auditory pathway, they are further enhanced by certain neural cells in the cochlear nucleus which reside in the lower part of the brain beyond the auditory nerve; these cells respond primarily to the onset

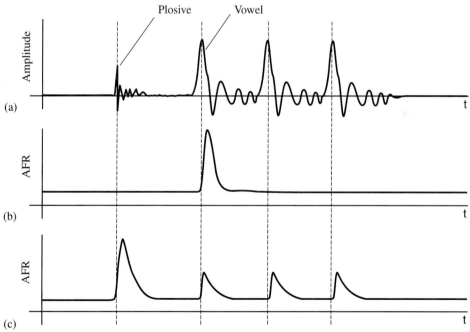

Figure 8.29   Schematized view of the phasic response of neural fibers in the auditory nerve to a plosive/vowel transition. Average firing rates (AFR) are illustrated for (b) a low-CF and (c) a high-CF channel of the auditory nerve for the speech waveform in panel (a). There is an average background discharge rate due to spontaneous emission of firings when no stimulus is present. In this schematic, the low-CF fiber responds to the vowel onset, while the high-CF fiber responds to the plosive and glottal pulse onsets. Observe that steady spectral components are suppressed in a phasic response.

envelope of a stimulus, giving little or no sustained response thereafter. Such cells are called *onset cells* and are found at almost all stages in the auditory pathway beginning at the cochlear nucleus. There is evidence that onset cells also tend to discharge at the modulation rate of the envelope of an amplitude-modulated tone (e.g., the pitch period of a speech waveform), and, in particular, synchronize to every peak in the AM envelope cycle, integrating information from closely-spaced neural channels.[15] Onset cells are also capable of a "double onset" mechanism, as with firing at both the burst release *and* the onset of a following vowel, thus perhaps directly encoding the voice onset time of a plosive consonant in their response pattern. These cells thus can respond to information from very different frequency bands, as well as neighboring channels.

We have seen that at both the auditory nerve and cochlear nucleus levels, the shape of an amplitude envelope of an auditory channel determines the phasic response of the neural fibers and cells. In speech signal processing, therefore, we must preserve the amplitude envelope shape within frequency bands by maintaining phase relations at the waveform level, as we saw earlier in Section 8.6.1; for example, *a blurring of the attack or modifying the AM of an envelope alters firing patterns along an auditory channel*. The envelope relation between channels (a different kind of phase relation) must also be preserved because onset cells, which respond to the envelope, gather information from both adjacent and distant frequency bands.

Tonic Systems — Auditory Nerve: There are a number of ways in which the auditory pathway may represent the slowly varying spectral content of a signal over time. We introduced earlier one approach in the *place theory* of hearing in which the input spectrum is reflected in the average firing rate associated with each cochlear filter tuned to a particular frequency along the basilar membrane, the average firing rate tracing out the formant energy across CF. The place theory is applicable to voiced as well as unvoiced sounds. There are many fascinating subtleties to this model [9]. The dynamic range and level of the input, for example, play a major role in the representation. Nerve fibers with a low spontaneous firing rate (i.e., the firing rate with no stimulus present) are activated at a low threshold and over a small dynamic range (20–40 dB), while nerve fibers with a high spontaneous firing rate are activated at a higher threshold and over a larger dynamic range (40–80 dB). High-threshold fibers provide a place representation of the formant pattern for moderate and high stimulus levels, while low threshold fibers provide the formant pattern for low stimulus levels. Another important feature of the place theory is that through feedback from the brainstem, which provides stimuli to the outer hair cells (the second large group of hair cells along the basilar membrane), the dynamic range of the auditory nerve fibers of some inner hair cells is shifted by about 15–30 dB in favor of higher-intensity stimuli. In addition, when this feedback is activated, frequency resolution of an auditory channel is improved. Such activation occurs particularly in noise and for specific channels associated with spectral bands that have low signal-to-noise ratio.

The above representation gives an *average* firing rate at each CF which is proportional to the stimulus level within spectral bands, and thus the spectrum is traced out across CF. Alternatively, in the temporal theory of hearing that we introduced earlier, in contrast to the place theory, phase synchrony may be used to measure the frequency of the stimulus (< 1 kHz).

---

[15] An onset cell does not necessarily discharge on the peak of a single channel envelope so that its synchronizing to peaks in the envelope requires integration of neighboring neural channels. Observe also that the onset cell, synchronizing on the amplitude envelope, does not phase lock to peaks in the underlying sinewave from a cochlear-filter output, as we described takes place at the auditory nerve level.

In this case, the tonic system takes on two forms at the auditory nerve level. Interspike intervals can be used to make very precise frequency measurements. We saw earlier that, with phase synchrony, either formant or harmonic (close to a dominant formant) frequency of the input signal can be measured, depending on the complexity (e.g., number of harmonics) of the input and the bandwidth of the channel. Remarkably, this information is coded in processors that are almost instantaneous in the sense that they have *temporal resolution that equals the inverse of the formant or harmonic frequency*, at least up to about 3000 Hz, or on the order of less than 1 millisecond of temporal resolution [9]. Because at the auditory nerve level the phase synchrony occurs at sinewave peaks, fine changes in the formant or harmonic frequencies can be tracked (Figure 8.30).

This high resolution of the auditory system presents an important implication for speech signal processing. Typical short-time Fourier transform analysis, with analysis windows of duration 10–20 ms or greater, cannot achieve such temporal resolution. *It follows that analysis tools of greater time resolution than the short-time Fourier transform are needed to exploit temporal patterns that are essential in human perception.* We return to this problem in Chapter 11. We have already encountered a second important implication: Preservation of channel phase relations is required to maintain the ability of the auditory nerve to provide phase synchrony across channels for formant and harmonic frequency estimation.

Cochlear Nucleus: The tonic component of the cochlear nucleus is characterized by two important cell classes: the *primary unit cells* and the *chopper unit cells*. Place and temporal functions are similar in style to those of the tonic component of the auditory nerve fibers. The primary unit cells provide the better time resolution, with very precise phase locking to harmonic frequencies or band-dominant formant frequencies (or harmonics closest to dominant formant frequencies), but can integrate across many auditory nerve fibers. Primary cells have dynamic range problems similar to low-threshold auditory nerve fiber cells. Chopper unit cells, on the other hand,

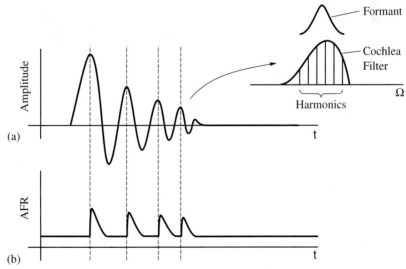

Figure 8.30   Schematic of average firing rate (AFR) [panel (b)] in estimation of changing formant frequency [panel (a)] by tonic auditory nerve component. Phase synchrony with respect to formant peaks allows fine time resolution.

have better dynamic range properties than primary unit cells (40 dB range of stimulus levels), in spite of the limited dynamic range of their input. This property is probably due to the fact that they integrate information from many auditory nerve fibers, while perhaps also invoking mutual inhibition among inputs from auditory nerve fibers. On the other hand, chopper cells are characterized by poorer temporal resolution, phase locking at only < 1 kHz. Thus, primary unit and chopper cells complement each other: Primary unit cells give precise representation of frequencies by their temporal discharge pattern, while chopper cells provide a place representation of spectra over a wide dynamic range. The implications for speech signal processing are similar to those for the tonic response patterns of auditory nerve fibers. Because at this high auditory level the cells integrate information from different auditory nerve fibers, timing and spectral content representations at the auditory nerve fiber level should be controlled with the higher levels in mind.

8.7 Summary

In this chapter, we developed a sinewave model of the filter-bank outputs for the FBS method for quasi-periodic speech signals. This channel model led to the phase vocoder for speech analysis and synthesis. In this context, we introduced the concept of phase coherence across sinewave outputs in synthesis. Loss of phase coherence is known to give a reverberant quality in the phase vocoder synthesis and a number of techniques were described for preserving this property, particularly in the application of time-scale modification. Various limitations of the phase vocoder lead to a speech analysis/synthesis method based on explicit modeling and estimation of sinewave components to be described in Chapter 9.

In this chapter we also introduced the wavelet transform, which is a generalization of filter-bank analysis/synthesis that involves constant-Q filters. We saw that the wavelet transform can serve as an approximate model of the front-end auditory cochlear filter bank. In addition, we used this model to speculate on the cause of aural phase sensitivity. Moreover, this model provided a stepping stone to more elaborate models of auditory neural processing in both the low-level auditory nerve fiber and the high-level cochlear nucleus. For example, we introduced Chistovich's hypothesis of phasic and tonic neural response patterns to fast and slow sound components, respectively, of the speech waveform. These auditory models provide a framework for various auditory-motivated speech signal processing throughout the remainder of the text.

EXERCISES

8.1 In this exercise you are asked to explore the effect of the channel phase adjustment of Equation (8.5) in FBS synthesis.

    (a) Show that making $p_k = 1$ results in the FBS constraint that for $\tilde{y}[n] = x[n]$, i.e., $\tilde{h}[n] = \delta[n]$, then $w[rN] = 0$ for $r = \pm 1, \pm 2, \ldots$, or the more useful constraint that the duration of the analysis window $w[n]$, N_w, is less than N.

    (b) Show that with $p_k = 1$, multiple copies of the input signal arise when the FBS constraint is not satisfied, specifically,

$$y[n] = \sum_{r=-\infty}^{\infty} w[rN]x[n - rN].$$

(c) Derive the expression in Equation (8.7) of Example 8.1 that results by applying the linear-phase adjustment factor $p_k = e^{-j\frac{2\pi}{N}kn_o}$ with $n_o = 20$, removing multiple copies of $x[n]$ and thus reverberation in synthesis.

8.2 Consider filters of a filter bank (Section 8.3, Figure 8.2) that are symmetric about π so that $\omega_{N-k} = 2\pi - \omega_k$ where $\omega_k = \frac{2\pi}{N}k$ and assume for simplicity that N is even. Show Equation (8.14), i.e.,

$$X(n, \omega_k) = X^*(n, \omega_{N-k}).$$

From Equations (8.13) and (8.14), show that the sum of two symmetric channels k and $N - k$ can be written as [Equation (8.15)]

$$\hat{y}_k[n] = 2|X(n, \omega_k)| \cos[\omega_k n + \theta_k(n, \omega_k)].$$

8.3 Show that the derivative of the continuous-time STFT phase $\theta(t, \Omega_k)$ with respect to time can be expressed as

$$\dot{\theta}(t, \Omega_k) = \frac{b(t, \Omega_k)\dot{a}(t, \Omega_k) - a(t, \Omega_k)\dot{b}(t, \Omega_k)}{a^2(t, \Omega_k) + b^2(t, \Omega_k)}$$

where $a(t, \Omega_k)$ and $b(t, \Omega_k)$ are the real and imaginary parts of the STFT $X(t, \Omega_k)$, respectively. *Hint:* Recall the analogous frequency-domain expression of Equation (6.8) from Chapter 6.

8.4 In this exercise we derive the output of the bandpass filter, $h_k(t) = w(t)e^{j\Omega_k}$ (of Section 8.3.1) to a sinewave with frequency $\Omega(t)$ and amplitude $A(t)$, i.e.,

$$x(t) = A(t)\cos[\psi(t)]$$

with

$$\psi(t) = \int_0^t \Omega(\tau)d\tau + \phi.$$

We think of $x(t)$ as one sinewave component of a quasi-periodic signal. The kth filter output is given by

$$y_k(t) = \int_{t'}^t \frac{A(\tau)}{2} \exp[j\psi(\tau)]h_k(t - \tau)d\tau$$

where t' is the starting non-zero point of $w(t - \tau)$. We assume that the amplitude and frequency functions satisfy the slowly varying conditions of Section 8.3.1 over the time interval $[t', t'']$ given for the quasi-periodic case and illustrated in Figure 8.5.

(a) Writing $\psi(t)$ as

$$\psi(t) = \int_0^t \Omega(\beta)d\beta$$

$$= \int_0^{t'} \Omega(\beta)d\beta + \int_{t'}^t \Omega(\beta)d\beta + \phi$$

so that

$$\psi(t) = \int_{t'}^t \Omega(\beta)d\beta + \phi',$$

given the slowly varying conditions of Section 8.3.1 for quasi-periodicity, show that

$$y_k(t) \approx \frac{A(t')}{2} \int_{t'}^t \exp j[\Omega(t')(\tau - t') + \phi']h_k(t - \tau)d\tau, \qquad t \in [t', t''].$$

(b) Assume that the frequency response of $h_k(t)$, $H_k(\Omega) = W(\Omega - \Omega_k)$, is flat with unity response near $\Omega = \Omega(t')$. Given your result in part (a), show that $y_k(t)$ can be written as

$$y_k(t) \approx \frac{A(t')}{2} \exp j[\Omega(t')(t - t') + \phi']$$

or

$$y_k(t) \approx \frac{A(t')}{2} \exp j[\Omega(t')t + \phi''], \qquad t \in [t', t'']$$

where $\phi'' = -\Omega(t')t' + \phi'$.

(c) Given your result in part (b), show that the STFT of $x(t)$ at $\Omega = \Omega_k$ is given by

$$X(t, \Omega_k) \approx \frac{A(t')}{2} \exp j[(\Omega(t') - \Omega_k)t + \phi''].$$

The phase of $X(t, \Omega_k)$ is then given by

$$\angle X(t, \Omega_k) \approx (\Omega(t') - \Omega_k)t + \phi''$$

and the phase derivative by

$$\dot{\theta}(t, \Omega_k) \approx \Omega(t') - \Omega_k.$$

Noting that t' can be replaced by t under the assumption that $\Omega(t)$ changes negligibly over the duration of the analysis window $w(t)$, argue therefore that

$$|X(t, \Omega_k)| \approx \frac{A(t)}{2}$$

$$\dot{\theta}(t, \Omega_k) \approx \Omega(t) - \Omega_k \tag{8.38}$$

and thus the kth channel output is given by Equation (8.21) when the pth slowly varying harmonic component of a quasi-periodic input falls in the bandwidth of the kth bandpass filter.

(d) Argue for the discrete-time counterpart of Equation (8.38) given in Equation (8.22).

8.5 We described in Section 8.3 the loss in phase coherence that occurs in time-scale modification when using samples of the phase derivative in the phase vocoder implementation of Figure 8.8 in which the frequency compressed (or expanded) signal is time-scaled during playback. Consider now the alternative "direct" implementation of time-scale modification with the phase vocoder given in Figure 8.9 in which the time-scaling occurs by decimation/interpolation of the sinewave amplitude and frequency functions.

(a) Explain why there results in the system of Figure 8.9 a loss of the original phase relations among sinewaves, thus giving an objectionable reverberant quality to the synthesis. Argue that a loss of phase coherence occurs even when the unwrapped phase is obtained from the principal phase value rather than the phase derivative as in Figure 8.9. *Hint:* Consider Equation (8.27).

(b) Argue that a single phase offset correction to each function $\rho\tilde{\phi}_k[n]$ in Equation (8.26) cannot preserve the original phase relations in the channel outputs over all time.

(c) Given the original principal phase function, $PV[\phi_k[n]]$, for each channel, propose an approach to correct the time-scaled phase function $\rho\tilde{\phi}_k[n]$ in Equation (8.26) so that loss in phase coherence does not occur, i.e., so that the original phase relations are maintained. *Hint:* Consider adding a phase offset and frequency perturbation to each channel. Also, consider that there may be no solution to this exercise if you attempt to maintain the original phase relations at all time samples.

8.6 Suppose that the amplitude of the filter frequency response $H(\omega)$ in Equation (8.23) follows a Gaussian function and that the filter phase is zero. Give an expression for the amplitude and phase of the sequence $y[n]$ in Equation (8.23) for a sinewave input with constant amplitude A and linearly changing frequency $\omega[n] = \alpha n$. How does your result effect analysis and synthesis in the phase vocoder? What are the implications for the phase vocoder when the phase of $H(\omega)$ is nonzero?

8.7 Suppose we are given the temporal envelope and the Fourier transform magnitude of a signal. We want to generate a time-scaled signal that has the given spectral magnitude and, with an appropriate selection of the Fourier transform phase, has a time-scaled version of the original temporal envelope.

   (a) Consider a single decaying sinewave of the form

$$x[n] = a^n \cos(\omega n) u[n]$$

   with $u[n]$ the unit step, and suppose we stretch the temporal envelope $a^n u[n]$ so that it becomes $a^{n/2} u[n]$. Explain why this altered temporal envelope is not "consistent" with the Fourier transform magnitude of the original signal.

   (b) Generalize your argument in part (a) for signals consisting of sums of decaying sinewaves, thus further explaining why methods that attempt to achieve a close match to both a spectral magnitude and a modified temporal envelope by Fourier transform phase selection may not be consistent with the relationship between a sequence and its Fourier transform.

8.8 Explain why the time-scale modification technique used in Example 8.2, when applied on speech, will modify the pitch of the speaker as well as the articulation rate. Consider the speech events that are detected by the 2-ms analysis filter. (See also Exercise 7.18.)

8.9 Argue qualitatively that excitation-driven techniques such as those based on linear prediction or homomorphic deconvolution, provided that the source and vocal tract impulse response are estimated accurately, do not suffer from loss of phase coherence in time-scale or pitch modification.

8.10 The oldest form of speech coding device is the channel vocoder which was invented by Homer Dudley [6]. Figures 8.31 and 8.32 show the analysis/synthesis structure for the channel vocoder.

   (a) Suppose the input $x(t)$ is a slowly varying steady-state voiced sound and that the bandpass filters are complex and ideal each with bandwidth of 200 Hz, and that taken together, they cover the input speech bandwidth of 5000 Hz. Assuming that only one harmonic of the input passes through each bandpass filter, roughly sketch the real part of the output of a bandpass filter which is located near 2500 Hz. (The specific number, 2500 Hz, is not important.)

   (b) The cascade of the magnitude operator and lowpass filter yields an estimate of the amplitude envelope of the output of each bandpass filter. Suppose the lowpass filter is ideal and has a bandwidth of about 100 Hz. Roughly sketch the output of the lowpass filter with input being the magnitude of the complex bandpass output from part (a). What is the required decimation rate, i.e., samples per second, at the bandpass output? What is the total number of channel filter parameters b_k required per second from all of the filters?

   (c) Denote each bandpass filter by $h_k(t)$ and the source in the synthesis by $e(t)$. Derive an expression for the output of the synthesizer, assuming the magnitude signals b_k are held constant. (In practice, interpolation of b_k will be required.) Suppose each bandpass filter is zero-phase. Sketch the real part of the output of the synthesizer, assuming a voiced input.

   (d) Describe conceptually the difference between the channel vocoder and the phase vocoder.

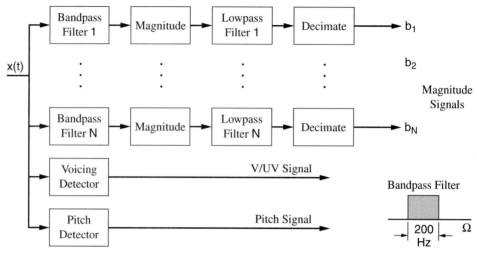

Figure 8.31 Block diagram of channel vocoder analyzer.

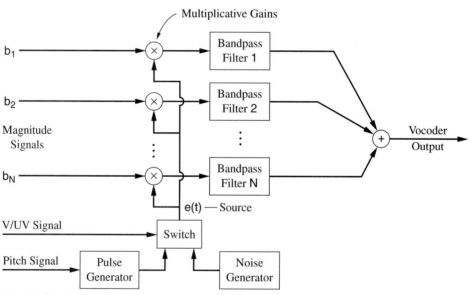

Figure 8.32 Block diagram of channel vocoder synthesizer.

8.11 Consider a discrete-time signal $x[n]$ passed through a bank of filters $h_k[n]$ where each filter is given by a modulated version of a baseband prototype filter $h[n]$, i.e.,

$$h_k[n] = h[n]e^{j\frac{2\pi}{N}kn}$$

where $h[n]$, a Hamming window, lies over a duration $0 \leq n < N_w$, and $\frac{2\pi}{N}$ is the frequency sampling interval. In this exercise, you are asked to time-scale expand some simple input signals by time-scale expanding the filter-bank outputs.

(a) State the perfect reconstruction constraint, i.e., the output equals the input, with respect to the values N_w and N, when the filter-bank outputs are summed.

(b) If the input to the filter bank is the unit sample $\delta[n]$, then the output of each filter is a complex exponential with envelope $a_k[n] = h[n]$ and phase $\theta_k[n] = \frac{2\pi}{N}kn$. Suppose that each complex exponential output is time-expanded by two by interpolation of its envelope and phase (i.e., place a zero between every other time sample and lowpass filter to create a signal of twice the original length). Derive a new perfect reconstruction constraint, with respect to values N_w and N, so that the summed filter-bank outputs equal $\delta[n]$.

(c) Suppose now that the filter-bank input equals

$$x[n] = \delta[n] + \delta[n - n_o],$$

and that the filter-bank outputs are time-expanded as in part (b). Derive a sufficient condition on N_w, N, and n_o so that the summed filter-bank output is given by

$$y[n] = \delta[n] + \delta[n - 2n_o],$$

i.e., the unit samples are separated by $2n_o$ samples rather than n_o samples.

8.12 In this exercise you investigate some of the properties of the continuous wavelet transform and its discrete counterpart on speech signals.

(a) Suppose you are given a signal consisting of two sinewaves at 3500 Hz and 3600 Hz, added to two impulses spaced apart by 5 milliseconds. Design a discrete STFT whose magnitude (spectrogram) reveals all four signal components. Then design a discrete wavelet transform that reveals all four signal components. Describe your STFT and wavelet transform designs qualitatively, considering, for example, the duration and shape of the analysis window and basic wavelet, respectively, for the required time-frequency resolution. In each case, sketch the approximate two-dimensional function (in time-frequency for the STFT and time-scale for the wavelet transform). Discuss the relative advantages for each two-dimensional representation.

(b) Repeat part (a) for the speech signal of Figure 8.33 consisting of a high-frequency voiced plosive having a low-frequency voice bar, followed by a low-pitched (100 Hz) vowel. Assume the voice bar consists of two tones at 100 Hz and 200 Hz. Design your STFT and wavelet transform to resolve the harmonics of the vowel and of the voice bar, while also resolving the onset of the plosive and onset of voicing (i.e., the onset of the vowel).

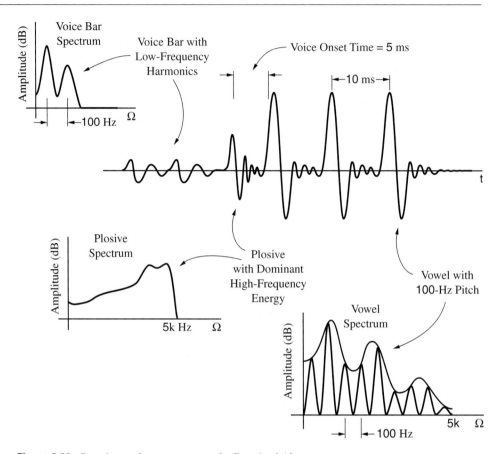

Figure 8.33 Speech waveform components for Exercise 8.12.

8.13 In this exercise you show the invertibility of the continuous wavelet transform and its discrete counterpart.

    (a) Prove the invertibility of the continuous wavelet transform under the admissibility condition on the basic wavelet given in Section 8.5.2. *Hint:* First determine the Fourier transform of $h_{\tau,a}(t)$, $H_{\tau,a}(\omega)$. Then use the generalized Parseval's theorem:

$$X_w(\tau, a) = \int_{-\infty}^{\infty} h_{\tau,a}^*(t) x(t)\, dt$$

$$= \int_{-\infty}^{\infty} H_{\tau,a}^*(\omega) X(\omega)\, d\omega.$$

    Finally, substitute the expression for the wavelet transform $X_w(\tau, a)$ into the continuous inverse wavelet transform Equation (8.33), assume that interchanging the resulting integrals is valid, and then work out the algebra.

    (b) Prove the invertibility of the discrete wavelet transform under the orthogonality condition on the discretized wavelet basis given in Section 8.5.3.

8.14 Argue whether it is possible to design a wavelet-like transform that has good time resolution at low frequencies and good frequency resolution at high frequencies. If so, then propose such a wavelet-like basis. *Hint:* Consider scaling a wideband low-frequency "basic wavelet" up to high-frequency "wavelets."

8.15 Show that the wavelet transform can be expressed as

$$X_w(\tau, a) = \frac{1}{\sqrt{a}} \int_{-\infty}^{\infty} x(at) h^* \left(\frac{t - \tau}{a} \right) dt$$

so that the transform acts as a "zoom lens," expanding or contracting the signal relative to the basic wavelet.

8.16 This exercise addresses the construction of a time-scaled sequence from a modified wavelet transform as described in Section 8.5.

(a) Set up equations for a closed form solution that minimizes the distance metric Equation (8.36) between the modified wavelet transform and the wavelet transform of the time-scaled signal estimate. Explain why these equations are difficult, if not impossible, to solve in terms of the changing wavelet window (filter) for each channel.

(b) Propose a magnitude-only iterative solution to estimate a time-scaled sequence from a modified wavelet transform. The method should attempt to minimize the distance metric of the form of Equation (8.36), but between a modified wavelet transform magnitude and the wavelet transform magnitude of the signal estimate. *Hint:* Consider the iterative solution for time-scale modification obtained from a modified STFT magnitude in Chapter 7.

8.17 In this exercise, we explore an approach to time-scale modification that relies on a filter-bank signal representation in the phase vocoder method. The output of each filter is viewed as an amplitude- and phase-modulated sinewave, the amplitude and unwrapped phases of which are interpolated to perform time-scale modification. Unlike in our phase vocoder of Section 8.3, however, the bandpass filters $H_k(\omega)$ do not have a flat response in the frequency vicinity of a sinewave input. As before, with time-scale modification by a factor ρ, the time-scale modified filter output is given by

$$\tilde{y}_k[n] = \tilde{a}_k[n] e^{j(\rho \tilde{\phi}_k[n])} \tag{8.39}$$

where $\tilde{a}_k[n]$ is the channel envelope and $\tilde{\phi}_k[n]$ is the unwrapped phase, both interpolated/decimated by the factor ρ. In this exercise you show that for a single AM-FM sinewave input, this filter-bank transformation results in a new AM-FM sinewave for which the AM and FM have been stretched. An example of using this method to time-expand a sinewave with flat amplitude and linear FM is illustrated in Figure 8.34, where the bandpass filters are Gaussian in shape and satisfy the FBS constraint. The FM begins at 1000 Hz and has a 15000 Hz/s sweep rate.

(a) Show that scaling the interpolated/decimated phase function in Equation (8.39) by ρ maintains the original instantaneous frequency, i.e., the phase derivative, of each filter output, but stretched or compressed in time.

(b) Consider now an AM-FM sinewave input of the form $x[n] = a[n] e^{j\theta[n]}$, where $\theta[n] = \int_0^n \omega[\tau] d\tau$. Argue that under certain "slowly varying" conditions on $a[n]$ and $\omega[n]$, $x[n]$ is approximately an eigenfunction of a linear time-invariant system, and thus of each filter in the filter bank. Argue that this approximation results in

$$y_k[n] \approx a[n] |H_k(\omega[n])| e^{j\angle H_k(\omega[n])} e^{j\theta[n]}$$

where the FM ($\omega[n]$) of the signal has been *transduced* to an AM ($|H_k(\omega[n])|$) within each channel.

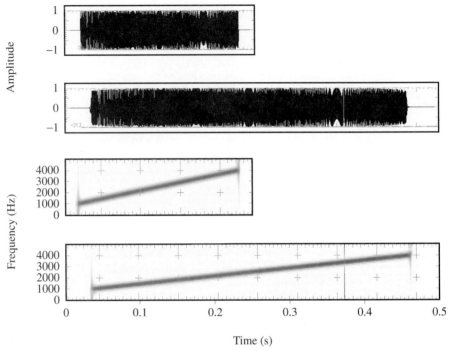

Figure 8.34 Time-scale modification of FM sinewave using a Gaussian-based filterbank: original waveform and time-scaled version (upper pair); spectrograms of original waveform and time-scaled version (lower pair).

SOURCE: T.F. Quatieri and T.E. Hanna, "Perfect Reconstruction Time-Scaling Filterbanks" [41]. ©1999, IEEE. Used by permission.

(c) Assume now that each filter in our filter bank is zero phase, i.e., $\angle H_k(\omega) = 0$. Also assume that the filter bank satisfies the FBS constraint, i.e., $\sum_{k=0}^{N-1} H_k(\omega) = 1$. With time-scaling the filter amplitude and phase output as in Equation (8.39), show that the sum of modified filter outputs is given by

$$\tilde{y}[n] = \sum_{k=0}^{N-1} y_k[n]$$
$$\approx \tilde{a}[n] e^{j\rho\tilde{\theta}[n]}$$

that is, the amplitude and frequency of the FM signal are approximately time-scaled without distortion. The duration of the sinewave has thus been increased while slowing the rate of change of its AM and FM.

(d) Consider a complex filter bank designed with a prototype filter and with some specified filter spacing over a 5000-Hz bandwidth. The prototype filter is selected to meet the FBS constraint (i.e., the output equals the input) when no modification is applied. If the number of filters over the full 5000 Hz band is $N = 25$, what is the maximum filter length for your result in part (c) to hold. (The example in Figure 8.34 uses this filter bank with a Gaussian prototype filter.)

8.18 (MATLAB) In this MATLAB exercise, use the workspace *ex8M1.mat*, as well as the function *uniform_bankx.m* located in the companion website directory *Chap_exercises/chapter8*. This exercise steps you through a design of the phase vocoder and explores the issue of phase coherence, as well as time-scale modification, for a speech waveform.

  (a) The first step is to design a filter bank that satisfies the FBS constraint, specifically with filter spacing of 50 Hz. We assume a sampling rate of 10000 samples/s and design a prototype 20 ms Hamming analysis filter. The function *uniform_bankx.m* can be used to compute the desired filter bank using the code:

   1. prototype = hamming(200)

   2. fbank = uniform_bankx(50, prototype,' 100, 10000, 2).

  Observe that the 100 bandpass filters are complex and that the last argument in *uniform_bankx.m* selects for plotting either the sum of the resulting filter-bank impulse responses (2) or the frequency response magnitude of this sum (1). Now repeat the above design with a filter spacing of 500 Hz and plot the filter-bank response sum. Why is this larger filter spacing not appropriate for the phase vocoder? With the filter bank in the 2-D array *fbank*, write a MATLAB function to perform analysis and synthesis. Use the speech waveforms *mtea_10k* and *ftea_10k* in the workspace *ex8M1.mat* as input to the filter bank. Perform this operation with both a 50 Hz and 500 Hz filter spacing and describe differences after listening to the reconstructions (using the MATLAB *sound.m* function).

  (b) In this part use the complex filter bank with 50 Hz spacing that resides in the array *fbank* from part (a). Write a MATLAB function to compute the amplitude envelopes and phase derivatives of the filter bank outputs, numerically integrate the phase derivative, and perform synthesis. One approach to computing the phase derivative is to first unwrap the phase of each channel (measured modulo 2π) and then compute the first difference. Define the first phase difference as the initial phase value so that a running sum of the phase differences will reproduce the original (unwrapped) phase exactly. You have now designed a phase vocoder. Use the speech waveforms *mtea_10k* and *ftea_10k* in the workspace *ex8M1.mat* as input to the filter bank. How do the reconstructions differ from that of part (a) both visually and aurally? Has phase coherence been maintained? Now modify the initial phase offset of each channel using the MATLAB function *rand.m* or *randn.m*, with normalization to the interval $[-\pi, \pi]$ and comment on the reconstruction, again visually and aurally. Has phase coherence been maintained? Do you find any perceptual differences in the reconstructions from male and female speech?

  (c) With your phase vocoder from part (b) design a time-scale modification system for an integer rate-change factor using the "direct" approach described in Section 8.3.2. Synthesize both time-compressed and time-expanded versions of the waveforms *mtea_10k* and *ftea_10k* for a variety of integer rate-change factors, and comment on the reconstruction quality. Has phase coherence been preserved?

8.19 (MATLAB) In this MATLAB exercise, you use the function *aud_transform.m*, the workspace *ex8M2.mat*, and the auditory filter responses located in the companion website directory *Chap_exercises/chapter8*. This exercise investigates the channel outputs for a filter bank consisting of measured cochlear filter responses.

  (a) The subdirectory *Auditory_filters* contains frequency responses measured along an actual basilar membrane of a cat (similar to that of a human). The suffix on each filename gives the characteristic frequency in Hertz, spanning an 8000-Hz range. The function *aud_transform.m* plots the frequency response from a desired file and converts the frequency response into a minimum- or zero-phase impulse response. Plot frequency responses and create minimum-phase impulse response for a variety of cochear filters of low- and high-characteristic frequency. (Auditory filters are said to be minimum phase.)

(b) Now use the speech waveforms *mtea_10k* and *ftea_10k* (at 10000 samples/s) in *ex8M2.mat* as input to your selected filters and plot the amplitude envelope for the filter outputs. Comment on the envelope characteristics for the filters with different speech sound classes (e.g., plosives vs. voiced sounds) and with the male and female speech. Describe how these envelopes might affect neural firing patterns in the auditory nerve and in the cochlear nucleus.

(c) Generate the minimum-phase impulse response for the lowest-characteristic frequency cochlear filter. Consider the fact that the cochlear filters are approximately constant-Q down to about 800 Hz, after which the bandwidths remain about constant. Describe the implication for time-frequency resolution of an auditory filter bank whose filter bandwidths were to continue to decrease logarithmically below 800 Hz.

BIBLIOGRAPHY

[1] B.A. Blesser, "Audio Dynamic Range Compression for Minimum Perceived Distortion," *IEEE Trans. Audio and Electro. Acoustics*, vol. AU–17, no. 1, pp. 22–32, March 1969.

[2] A.C. Bovik, J.P. Havlicek, and M.D. Desai, "Theorems for Discrete Filtered Modulated Signals," *Proc. IEEE Int. Conf. Acoustics, Speech, and Signal Processing*, vol. 3, pp. 153–156, Minneapolis, MN, April 1993.

[3] C.S. Burrus, R.A. Gopinath, and H. Guo, *Introduction to Wavelets and Wavelet Transforms*, Prentice Hall, Englewood Cliffs, NJ, 1998.

[4] J.P. Carlson, "Digitized Phase Vocoder," *Proc. Conf. on Speech Communications and Processing*, Boston, MA, Nov. 1967.

[5] L.A. Chistovich, V.V. Lubinskaya, T.G. Malinnikova, E.A. Ogorodnikova, E.I. Stoljarova, and S.J.S. Zhukov, "Temporal Processing of Peripheral Auditory Patterns of Speech," chapter in *The Representation of Speech in the Peripheral Auditory System*, R. Carlson and B. Granstrom, eds., pp. 165–180, Elsevier, Amsterdam, 1982.

[6] H. Dudley, R. Riesz, and S. Watkins, "A Synthetic Speaker," *J. Franklin Inst.*, vol. 227, no. 739, 1939.

[7] A. Croisier, D. Esteban, and C. Garland, "Perfect Channel Splitting by Use of Interpolation/Decimation-Tree Decomposition Techniques," *Int. Conf. Information Sciences and Systems*, pp. 443–446, Patras, Greece, Aug. 1976.

[8] I. Daubechies, *Ten Lectures on Wavelets*, SIAM, 1992.

[9] B. Delgutte, *Auditory Neural Processing of Speech*, chapter in *The Handbook of Phonetic Sciences*, W.J. Hardcastle and J. Laver, eds., Blackwell (Oxford), 1997.

[10] M. Dolson, "The Phase Vocoder: A Tutorial," *Computer Music Journal*, vol. 10, no. 4, pp. 14–27, Winter 1986.

[11] J.L. Flanagan and R.M. Golden, "Phase Vocoder," *Bell System Technical Journal*, vol. 45, no. 9, pp. 1493–1509, Nov. 1966.

[12] B. Gold and L.R. Rabiner, "Parallel Processing Techniques for Estimating Pitch Periods of Speech in the Time Domain," *J. Acoustical Society of America*, vol. 34, no. 7, pp. 916–921, 1962.

[13] D.M. Green, *An Introduction to Hearing*, John Wiley and Sons, New York, NY, 1976.

[14] T. Irino and H. Kawahara, "Signal Reconstruction from Modified Auditory Wavelet Transform," *IEEE Trans. Signal Processing*, vol. 41, no. 12, pp. 3549–3554, Dec. 1993.

[15] T. Irino and H. Kawahara, "Signal Reconstruction from Modified Wavelet Transform—An Application to Auditory Signal Modeling," *Proc. IEEE Int. Conf. Acoustics, Speech, and Signal Processing*, vol. 1, pp. 85–88, Minneapolis, MN, March 1992.

[16] S. Kadambe and G.F. Boudreaux-Bartels, "Application of the Wavelet Transform for Pitch Detection of Speech Signals," *IEEE Trans. on Information Theory*, vol. 38, no. 2, pt. 2, pp. 917–924, March 1992.

[17] J. Laroche, "Time and Pitch Scale Modification of Audio Signals," chapter in *Applications of Digital Signal Processing to Audio and Acoustics*, M. Kahrs and K. Brandenburg, eds., Kluwer Academic Publishers, Boston, MA, 1998.

[18] J. Laroche and M. Dolson, "Improved Phase Vocoder Time-Scale Modification of Audio," *IEEE Trans. Speech and Audio Processing*, vol. 7. no. 3, pp. 323–332, May 1999.

[19] E. Lindemann and J.M. Kates, "Phase Relationships and Amplitude Envelopes in Auditory Perception," *Proc. IEEE Workshop on Applications of Signal Processing to Audio and Acoustics*, Mohonk Mountain House, New Paltz, NY, Oct. 1999.

[20] D. Malah, "Time-Domain Algorithms for Harmonic Bandwidth Reduction and Time Scaling of Speech Signals," *IEEE Trans. Acoustics, Speech, and Signal Processing*, vol. 27, no. 2, pp. 121–133, April 1979.

[21] D. Malah and J.L. Flanagan, "Frequency Scaling of Speech Signals by Transform Techniques," *Bell System Technical Journal*, vol. 60, no. 9, pp. 2107–2156, Nov. 1981.

[22] S. Mallat, *A Wavelet Tour of Signal Processing*, Academic Press, New York, NY, 1998.

[23] S. Mallat and W.L. Hwang, "Singularity Detection and Processing with Wavelets," *IEEE Transactions on Information Theory*, vol. 38, no. 2, pp. 617–643, March 1992.

[24] P. Maragos, J.F. Kaiser, and T.F. Quatieri, "Energy Separation in Signal Modulations with Application to Speech Analysis," *IEEE Trans. Signal Processing*, vol. 41, no. 10, pp. 3024–3051, Oct. 1993.

[25] R.J. McAulay and T.F. Quatieri, "Speech Analysis-Synthesis Based on a Sinusoidal Representation," *IEEE Trans. Acoustics, Speech, Signal Processing*, vol. ASSP–34, no. 4, pp. 744–754, Aug. 1986.

[26] J.A. Moorer, "Signal Processing Aspects of Computer Music," *Proc. IEEE*, vol. 65, no. 8, pp. 1108–1137, Aug. 1977.

[27] B.C.J. Moore, *An Introduction to the Psychology of Hearing*, 2nd Edition, Academic Press, Boston, MA, 1988.

[28] H. Nawab and T.F. Quatieri, "Short-Time Fourier Transform," chapter in *Advanced Topics in Signal Processing*, J.S. Lim and A.V. Oppenheim, eds., Prentice Hall, Englewood Cliffs, NJ, 1988.

[29] A.V. Oppenheim and R.W. Schafer, *Digital Signal Processing*, Prentice Hall, Englewood Cliffs, NJ, 1975.

[30] A. Pickles, *An Introduction to Auditory Physiology*, Academic Press, 2nd Edition, New York, NY, 1988.

[31] M.R. Portnoff, "Time-Scale Modification of Speech Based on Short-Time Fourier Analysis," *IEEE Trans. Acoustics, Speech, and Signal Processing*, vol. 29, no. 3, pp. 374–390, June 1981.

[32] M. Puckette, "Phase-Locked Vocoder," *Proc. IEEE 1995 ASSP Workshop on Applications of Signal Processing to Audio and Acoustics*, Mohonk Mountain House, New Paltz, NY, Oct. 1995.

[33] T.F. Quatieri, R.B. Dunn, and T.E. Hanna, "A Subband Approach to Time-Scale Modification of Complex Acoustic Signals," *IEEE Trans. Speech and Audio Processing*, vol. 3, no. 6, pp. 515–519, Nov. 1995.

[34] T.F. Quatieri, R.B. Dunn, and T.E. Hanna, "Time-Scale Modification with Temporal Envelope Invariance," *Proc. IEEE 1991 Workshop on Applications of Signal Processing to Audio and Acoustics*, Mohonk Mountain House, New Paltz, NY, Oct. 1993.

[35] T.F. Quatieri, R.B. Dunn, and R.J. McAulay, "Signal Enhancement in AM–FM Interference," Technical Report TR–993, Lincoln Laboratory, Massachusetts Institute of Technology, May 1994.

[36] T.F. Quatieri, R.B. Dunn, R.J. McAulay, and T.E. Hanna, "Underwater Signal Enhancement Using a Sinewave Representation," *Proc. IEEE Oceans92*, pp. 449–452, Newport, RI, Oct. 1992.

[37] T.F. Quatieri, R.B. Dunn, R.J. McAulay, and T.E. Hanna, "Time-Scale Modification of Complex Acoustic Signals in Noise," Technical Report TR–990, Lincoln Laboratory, Massachusetts Institute of Technology, Jan. 1994.

[38] T.F. Quatieri and R.J. McAulay, "Phase Coherence in Speech Reconstruction for Enhancement and Coding Applications," *Proc. IEEE Int. Conf. Acoustics, Speech, and Signal Processing*, pp. 207–210, Glasgow, Scotland, May 1989.

[39] T.F. Quatieri and R.J. McAulay, "Audio Signal Processing Based on Sinusoidal Analysis/Synthesis," chapter in *Applications of Digital Signal Processing to Audio and Acoustics*, M. Kahrs and K. Brandenburg, eds., Kluwer Academic Publishers, Boston, MA, 1998.

[40] T.F. Quatieri, T.E. Hanna, and G.C. O'Leary, "AM–FM Separation Using Auditory-Motivated Filters," *IEEE Trans. Speech and Audio Processing*, vol. 5, no. 5, pp. 465–480, Sept. 1997. Also in *Proc. IEEE Int. Conf. Acoustics, Speech, and Signal Processing*, Atlanta, GA, vol. 2, pp. 977–980, May 1996.

[41] T.F. Quatieri and T.E. Hanna, "Perfect Reconstruction Time-Scaling Filterbanks," *IEEE Int. Conf. Acoustics, Speech, and Signal Processing*, vol. 2, pp. 495–498, Phoenix, AZ, March 1999.

[42] L.R. Rabiner and R.W. Schafer, *Digital Processing of Speech Signals*, Prentice Hall, Englewood Cliffs, NJ, 1978.

[43] O. Rioul and M. Vetterli, "Wavelets and Signal Processing," *IEEE Signal Processing Magazine*, vol. 8, no. 4, pp. 14–38, Oct. 1991.

[44] S. Roucos and A.M. Wilgus, "High-Quality Time-Scale Modification of Speech," *Proc. IEEE Int. Conf. Acoustics, Speech, and Signal Processing*, pp. 490–493, Tampa, FL, March 1985.

[45] X. Serra, "A System for Sound Analysis/Transformation/Synthesis Based on a Deterministic Plus Stochastic Decomposition," Ph.D. Thesis, CCRMA, Department of Music, Stanford University, 1989.

[46] S. Seneff, "Pitch and Spectral Estimation of Speech Based on Auditory Synchrony Model," *Proc. IEEE Int. Conf. Acoustics, Speech, and Signal Processing*, pp. 36.2.1–36.2.4, San Diego, CA, March 1994.

[47] M. Slaney, D. Naar, and R.F. Lyon, "Auditory Model Inversion for Signal Separation," *Proc. IEEE Int. Conf. Acoustics, Speech, and Signal Processing*, vol. 2, pp. 77–80, Adelaide, Australia, April 1994.

[48] M.J. Smith and T.P. Barnwell, "Exact Reconstruction for Tree-Structured Subband Coders," *IEEE Trans. Acoustics, Speech, and Signal Processing*, vol. ASSP–34, no. 3, pp. 431–441, June 1986.

[49] B. Sylvestre and P. Kabal, "Time-Scale Modification of Speech Using an Incremental Time-Frequency Approach with Waveform Structure Compensation," *Proc. IEEE Int. Conf. Acoustics, Speech, and Signal Processing*, vol. 1, pp. 81–84, San Francisco, CA, 1992.

[50] W. Torres and T.F. Quatieri, "Estimation of Modulation Based on FM-to-AM Transduction: Two-Sinusoidal Case," *IEEE Trans. Signal Processing*, vol. 47, no. 11, pp. 3084–3097, Nov. 1999.

[51] X. Yang, K. Wang, and S.A. Shamma, "Auditory Representations of Acoustic Signals," *IEEE Trans. Information Theory*, vol. 38, no. 2, pp. 824–839, March 1992.

[52] K. Wang and S.A. Shamma, "Self-Normalization and Noise-Robustness in Early Auditory Representations," *IEEE Trans. Speech and Audio*, vol. 2, no. 3, pp. 421–435, July 1994.

[53] S.G. Zauner, "Deterministic Signal Reconstruction from the Temporal and Spectral Correlation Functions," Thesis, Technischen Universität Wien, Sept. 1997.

CHAPTER 9

Sinusoidal Analysis/Synthesis

9.1 Introduction

We have seen in Chapters 5 and 6 that one approach to the analysis and synthesis of speech signals is to use the speech production model in which speech is viewed as the result of passing a source excitation waveform through a time-varying linear filter that models the resonant cavities of the vocal tract. In certain applications, it suffices to assume that the source function can be in one of two possible states corresponding to voiced or unvoiced speech. We have referred to these approaches of analysis/synthesis as "modeled-based." In Chapters 7 and 8, on the other hand, we introduced methods of analysis and synthesis that depend on Fourier transform and filter-bank representations and are less modeled-based in the sense of having less dependence on the state of the source and system. In the filter-bank-based phase vocoder, in particular, filter-bank outputs were assumed to be AM-FM sinewaves regardless of the source, although more meaningful mathematical models of filter outputs were derived under the assumption that one input sinewave falls within a filter. The analysis in the phase vocoder does not explicitly model and estimate the sinewave components, but rather views them as outputs of a bank of uniformly-spaced bandpass filters. It was observed that conditions under which this sinewave assumption holds were often violated for voiced speech, and a mathematical characterization of unvoiced speech in this framework was complex, thus motivating a more explicit representation and estimation of sinewaves in speech signals.

The goal of this chapter is the description of such a sinewave representation that is valid irrespective of the source state. This model, originally introduced in [30], is composed of sinusoidal components of arbitrary amplitudes, frequencies, and phases, and it results in an analysis/synthesis system based on explicit sinewave estimation. Figure 9.1 shows a schematic of the sinewave model *frequency tracks* that are the frequency trajectories of sinewaves from their onset to their termination. Each sinewave is represented by a time-varying envelope and

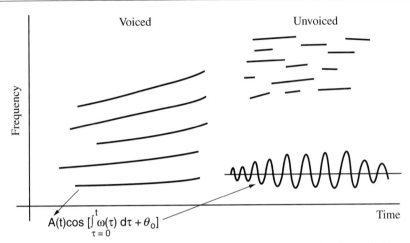

Figure 9.1   The sinewave model encompasses both voiced and unvoiced speech. Frequency tracks of voiced sinewaves are approximately harmonically related, while unvoiced tracks typically have no such relation and come and go randomly over short durations.

a phase equal to the integral of a time-varying frequency track. In voiced speech, such as vowels, the sinewave frequency tracks are roughly harmonically related and linger over long durations. Noise-like and transient sounds such as fricatives and plosives, although aharmonic, can nevertheless also be represented approximately by a sum of sinewaves, but generally without coherent phase structure; the model sinewave frequencies have arbitary value, their tracks coming and going randomly in time over shorter durations.

A number of other approaches to analysis/synthesis that are based on explicit sinewave models, in addition to the approach of this chapter, have been discussed in the literature. Hedelin [16] proposed a sinewave model for use in quantizing the baseband signal for speech coding. The amplitudes and phases of the underlying sinewaves are explicitly estimated using Kalman filtering techniques, and each sinewave phase is defined to be the integral of the associated instantaneous frequency. As in the original phase vocoder, absolute phase information is lost. Another sinewave speech coding system has been developed by Almeida and Silva [1]. In contrast to Hedelin's approach, their system uses a pitch estimate during voiced speech to establish a harmonic set of sinewaves. The sinewave phases are computed at harmonic frequencies from the short-time Fourier transform. In another implementation of this system, Marques and Almeida model unvoiced speech using a set of narrowband basis functions [25]. Yet another approach to modeling unvoiced speech in the context of the sinewave model is to explicitly generate noise via the linear filtering of white noise whenever unvoiced speech components are detected in different frequency bands. This approach, developed by Griffin and Lim [14], is referred to as the "multi-band excitation vocoder" and uses overlap-add reconstruction for synthesizing speech in unvoiced bands. For voiced bands the system uses a sinewave analysis/synthesis similar to that described in this chapter.

In this chapter, we describe an approach to sinewave modeling and analysis/synthesis that lends itself to determining sinewave frequency tracks through frequency matching [30], phase coherence during voicing through a source and vocal tract filter phase model [39],[40], and estimation of noise-like components by the use of an additive model of deterministic and stochastic signal contributions [50],[51]. In Section 9.2, the sinewave model is developed. In

Section 9.3, in the analysis stage, the amplitudes, frequencies, and phases of the sinewave model are estimated, while in Section 9.4, in the synthesis stage, these parameters are processed to produce a synthetic waveform that is essentially perceptually indistinguishable from the original. This sinewave analysis/synthesis *baseline* system forms the foundation for the remainder of the chapter. In Section 9.4, some applications of this baseline system, including signal modification, splicing, and estimation of vibrato, are also presented. This section ends with an overview of time-frequency resolution considerations for sinewave analysis that arise in these applications, including a wavelet-based front end analysis. Although the baseline sinewave analysis/synthesis is applicable to arbitrary signals, tailoring the system to a specific class can improve performance. In Section 9.5, a source/filter phase model for quasi-periodic signals is introduced within the sinewave representation. This model is important in numerous applications, including signal modification, such as time-scale modification, and reducing the peakiness (i.e., "peak-to-rms" valve) in a signal. The source/filter phase model provides *phase coherence*, i.e., preserving phase relations among sinewaves, which is essential for high quality in these applications. Finally, in Section 9.7, an *additive* model of deterministic and stochastic components is introduced within the sinewave representation. This two-component model is particularly important for representing speech sounds with simultaneous harmonic and aharmonic contributions.

9.2 Sinusoidal Speech Model

In the linear speech production model, the continuous-time[1] speech waveform $s(t)$ is assumed to be the output of passing a source excitation waveform $u(t)$ through a linear time-varying filter with impulse response $h(t, \tau)$ that models the characteristics of the vocal tract (Figure 9.2). Mathematically, the speech signal is expressed as

$$s(t) = \int_0^t h(t, t - \tau) u(\tau) d\tau$$

where the excitation is convolved with a different impulse response at each time t. It is proposed that the excitation $u(t)$ be represented by a sum of sinusoids of various amplitudes, frequencies and phases:[2]

$$u(t) = \text{Re} \sum_{k=0}^{K(t)} a_k(t) \exp[j\phi_k(t)] \tag{9.1}$$

where the phase function

$$\phi_k(t) = \int_0^t \Omega_k(\sigma) d\sigma + \phi_k$$

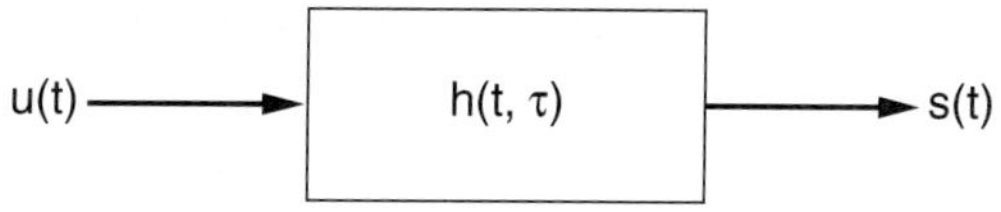

Figure 9.2 Speech waveform $s(t)$ modeled as the output of a linear time-varying vocal tract with source excitation $u(t)$.

---

[1] As in Chapter 8, a continuous-time model helps in the development of phase representations because of the need to differentiate and integrate.

[2] For clarity, in this chapter, we often replace the complex exponential notation $e^{j\phi}$ by $\exp[j\phi]$.

and $K(t)$ is the number of sinewave components at time t. Our sinewave model represents an arbitary source and is not constrained to a periodic, impulsive, or white noise form. For the kth sinewave component, $a_k(t)$ and $\Omega_k(t)$ represent the time-varying amplitude and frequency, and ϕ_k is the fixed phase offset to account for the fact that at time $t = 0$, the sinewaves are generally not in phase.[3] The representation of sinewaves as the real part ("Re") of complex exponentials is used because it simplifies the analysis that will be required for derivation of the sinewave phase model.

The vocal tract transfer function in terms of its time-varying magnitude $M(t, \Omega)$ and phase $\Phi(t, \omega)$ components is written as

$$H(t, \Omega) = M(t, \Omega) \exp[j\Phi(t, \Omega)]. \tag{9.2}$$

Then if the parameters of the excitation, $a_k(t)$ and $\Omega_k(t)$, are constant over the duration of the impulse response of the vocal tract filter, the speech waveform can be written as

$$s(t) = \text{Re} \sum_{k=1}^{K(t)} a_k(t) M[t, \Omega_k(t)] \exp\left[j \left(\int_0^t \Omega_k(\sigma)d\sigma + \Phi[t, \Omega_k(t)] + \phi_k \right) \right] \tag{9.3}$$

where each sinewave appears at time t as an eigenfunction of the system $h(t, \tau)$ (Appendix 9.A). By combining the excitation and vocal tract amplitudes and phases, the representation[4] can be written more concisely as

$$s(t) = \sum_{k=1}^{K(t)} A_k(t) \exp[j\theta_k(t)] \tag{9.4}$$

where

$$A_k(t) = a_k(t) M[t, \Omega_k(t)]$$
$$\theta_k(t) = \phi_k(t) + \Phi[t, \Omega_k(t)]$$
$$= \int_0^t \Omega_k(\sigma)d\sigma + \Phi[t, \Omega_k(t)] + \phi_k \tag{9.5}$$

represent the amplitude and phase of the kth sinewave along the frequency trajectory $\Omega_k(t)$ (referred to as an instantaneous frequency in Chapter 8). To simplify notation, the system amplitude and phase along each frequency trajectory $\Omega_k(t)$ are occasionally written as

$$M_k(t) = M[t, \Omega_k(t)]$$
$$\Phi_k(t) = \Phi[t, \Omega_k(t)]. \tag{9.6}$$

---

[3] Sinewaves are *in phase* at a time $t = t_o$ when the sinewave peaks all occur at this time, i.e., the phase of all sinewaves is a multiple of 2π at time $t = t_o$.

[4] In the following discussion, we often remove the "Re" notation and work with the complex version of $s(t)$. This quadrature representation is approximately equal, as we saw in Chapter 2, to its analytic signal representation that can be obtained through the Hilbert transform.

Equation (9.4) is the *basic* sinewave model that can be thought of as speech-independent, i.e., the model can be applied to *any* signal. The decomposition in Equation (9.5), on the other hand, is speech-dependent and will be useful for a number of applications where we require the contributions separately. The next step is to develop a robust procedure for extracting the amplitudes, frequencies, and phases of the component sinewaves from the speech waveform. Before doing so, however, we look at a few examples.

EXAMPLE 9.1   Consider an idealized voiced speech signal where the excitation frequencies are harmonically related with fundamental frequency Ω_o, in phase (with $\phi_k = 0$), and have unity amplitude (with $a_k(t) = 1$); in addition, the vocal tract is time-invariant. Then we can write

$$\Omega_k(t) = k\Omega_o$$

$$a_k(t) = 1; \quad \phi_k = 0$$

$$H(\omega, t) = H(\omega)$$

and the excitation phase as

$$\phi_k(t) = \int_0^t \Omega_k(\sigma)d\sigma + \phi_k$$

$$= \int_0^t k\Omega_o d\sigma + 0$$

$$= k\Omega_o t.$$

The vocal tract magnitude and phase along the (fixed) excitation frequency trajectories are given by

$$M[t, \Omega_k(t)] = M(k\Omega_o)$$

$$\Phi[t, \Omega_k(t)] = \Phi(k\Omega_o)$$

so that

$$s(t) = \sum_{k=1}^{K(t)} a_k(t)M[t, \Omega_k(t)] \exp\left[j\left(\int_0^t \Omega_k(\sigma)d\sigma + \Phi[t, \Omega_k(t)] + \phi_k \right)\right]$$

$$= \sum_{k=1}^{K} M(k\Omega_o) \exp\left[j\left(k\Omega_o t + \Phi[k\Omega_o]\right)\right].$$

We see then that each excitation sinewave has been shaped by the gain and phase of the system function, i.e., each input sinewave is shifted by the phase of $H(\Omega)$ and multiplied by the magnitude of $H(\Omega)$. ▲

EXAMPLE 9.2   Consider the idealized voiced speech signal of Example 9.1 where the excitation frequencies are harmonically related and the vocal tract is time-invariant, but now suppose that the pitch is linearly changing in time, i.e.,

$$\Omega_k(t) = k(\Omega_o ct)$$

where c is a constant that determines the rate of change of pitch. Then the excitation phase is expressed as

$$
\begin{aligned}
\phi_k(t) &= \int_0^t \Omega_k(\sigma)d\sigma + \phi_k \\
&= \int_0^t k\Omega_o c\sigma \, d\sigma + 0 \\
&= kc\Omega_o \frac{t^2}{2}
\end{aligned}
$$

which is a quadratically changing phase function. Therefore,

$$
s(t) = \sum_{k=1}^{K(t)} M(kc\Omega_o t) \exp\left[j\left(kc\Omega_o \frac{t^2}{2} + \Phi[kc\Omega_o t] \right) \right]
$$

where the system function is sampled along the time-varying frequencies of the excitation sinewaves. This model is valid only if $\Omega_k(t)$ is "slowly varying" over the duration of $h(t, \tau)$ (Appendix 9.A).   ▲

9.3 Estimation of Sinewave Parameters

To obtain a flavor for the analysis and synthesis problem, we begin with a simple example of analyzing and synthesizing a single sinewave.

EXAMPLE 9.3   Consider the discrete-time counterpart to our continuous-time sinewave model. In particular, consider a single discrete-time sinewave of the form

$$
x[n] = A\cos(\omega_o n + \phi)
$$

derived from a 500-Hz continuous-time signal at 10000 samples/s. We analyze $x[n]$ with the discrete STFT at a frame interval of $L = 200$ samples (20 ms), and the analysis window N_w has the same length. From Figure 9.3, which shows one slice of the STFT magnitude of the signal, we are motivated to form an estimate of the amplitude and frequency of the sinewave as the amplitude and frequency of the spectral maximum. Denote these estimates for the lth frame by $\hat{A}^l$ and $\hat{\omega}_o^l$, respectively. We can then synthesize the sinewave over the lth frame as

$$
\hat{x}^l[n] = \hat{A}^l \cos(\hat{\omega}_o^l n), \qquad lL \le n < (l+1)L
$$

which represents an estimate of $x[n]$ to within a phase offset. Therefore, if we perform this analysis and synthesis over successive frames, we obtain a waveform discontinuity at frame boundaries. This discontinuity would yield an annoying 100 Hz "buzz" corresponding to the frame rate. Alternatively, we can retain an estimate of the original (unwrapped) phase function, $\theta[n] = \omega_o n + \phi$, by a cumulative sum of the frequency estimate over successive frames, i.e.,

$$
\begin{aligned}
\hat{\theta}^l[n] &= \sum_{m=lL}^{n} \hat{\omega}_o^l + \hat{\theta}^{l-1} \\
&= (n - lL)\hat{\omega}_o^l + \hat{\theta}^{l-1}, \quad lL \le n < (l+1)L
\end{aligned} \tag{9.7}
$$

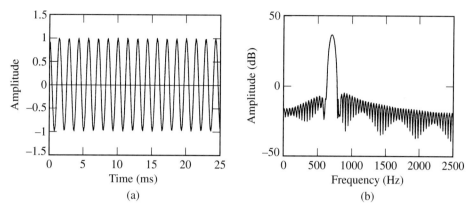

Figure 9.3 Spectral analysis of a single sinewave: (a) waveform; (b) spectral magnitude of the short-time segment in (a). A 25-ms Hamming window was applied.

where l denotes the frame number and where $\hat{\theta}^{l-1}$ is the phase of the right boundary of the previous frame. Our new estimate for the lth frame is then given by

$$\hat{x}^l[n] = \hat{A}^l \cos(\hat{\theta}^l[n]).$$

This reconstruction is inaccurate by an absolute phase offset ϕ when we initialize the process with zero phase. An alternative is to estimate the phase offset on each frame as the phase of the spectral peak. This method, however, does not ensure waveform continuity because of limited frequency accuracy as obtained with the DFT. Moreover, when sinewave parameters are time-varying, even more acute waveform discontinuity arises (Exercise 9.2).

Observe that we can extend the approach in Equation (9.7) to stationary multiple sinewaves. With voiced speech, for example, the magnitude peaks of the STFT, with a long enough analysis window, provide estimates of all the harmonically related sinewave amplitudes and phases. In the reconstruction of each sinewave, however, we lose not only absolute phase, but also the phase relations across sinewaves when the initial sinewave phases are set to zero (Exercise 9.20). Recall that we encountered a similar loss of *phase coherence* in the phase vocoder synthesis and thus a change in the waveform shape that we earlier in the text referred to as waveform "dispersion." Even when synthesis is performed with the correct phase offsets, inaccuracies in the measurements $\hat{\omega}_o^l$ made with the DFT result in dispersion. ▲

The general analysis/synthesis problem is to take a speech waveform, extract the parameters that represent a quasi-stationary portion of that waveform, and use those parameters to reconstruct an approximation that is "as close as possible" to the original speech. The general estimation problem in which the speech signal is to be represented by a sum of sinewaves is a difficult one to solve analytically; hence the approach taken here is pragmatic in the sense that an estimator is derived based on a set of idealized assumptions; then, once the structure of the ideal estimator is understood, modifications are made as the assumptions are relaxed to better match the real speech waveform. Looking ahead to Figure 9.7 we see that the speech waveform is short-time Fourier analyzed to obtain amplitudes, frequencies, and phases at multiple spectral peaks. The waveform is then reconstructed by interpolating these parameters across successive frames and modulating the sinewaves with the resulting functions.

As a first step, and as indicated in Example 9.3, the continuous-time axis is subdivided into a contiguous sequence of frames analyzed with a time window of length T. The center of the analysis window for the lth frame occurs at time t_l. We assume that the source excitation and vocal tract parameters are constant over the window duration and so are constant over the interval $I = [t_l - \frac{T}{2}, t_l + \frac{T}{2}]$.[5] Then we can write the kth amplitude and frequency function of our sinewave model in Equations (9.4) and (9.5) as constant for the lth frame, i.e.,

$$A_k^l(t) = A_k^l$$
$$\Omega_k^l(t) = \Omega_k^l.$$

Then the sinewave phase in Equation (9.5) can be written as

$$\theta_k^l(t) = \Omega_k^l(t - t_l) + \theta_k^l \tag{9.8}$$

where θ_k^l denotes the phase of the kth sinewave at the center of the lth frame at time $t = t_l$ (Figure 9.4a)[6] and is a function of the source phase offset ϕ_k and the vocal tract phase $\Phi(t, \Omega)$ at time t_l, i.e., $\theta_k^l = \phi_k^l + \Phi(t_l, \Omega_k^l)$. For frame l, over the time interval I, we then write the waveform as

$$s(t) = \sum_{k=1}^{K^l} A_k^l(t) \exp[j\theta_k^l(t)],$$

$$= \sum_{k=1}^{K^l} A_k^l \exp[j(\Omega_k^l(t - t_l) + \theta_k^l)]$$

$$= \sum_{k=1}^{K^l} A_k^l \exp(j\theta_k^l) \exp[j\Omega_k^l(t - t_l)], \quad t_l - \frac{T}{2} \le t \le t_l + \frac{T}{2}$$

where $K^l = K(t_l)$ and where, because we assume the frequency is constant over the analysis interval, the phase is linear with slope Ω_k^l.

To arrive at a discrete-time formulation, we perform the following two operations:

1. Shift the time interval to the origin, i.e., with $t' = t - t_l$, we have

$$s'(t') = \sum_{k=1}^{K^l} A_k^l \exp(j\theta_k^l) \exp(j\Omega_k^l t'), \quad -\frac{T}{2} \le t' \le \frac{T}{2}.$$

---

[5] In the speech context, this implies that the source and vocal tract parameters are constant over a time that includes the duration of the analysis window and the duration of the vocal tract impulse response.

[6] This point may cause confusion. Observe that the phase of the short-time Fourier transform at spectral peaks is a measurement of the phase offset of each sinewave relative to the frame center t_l.

2. Convert to discrete time (eliminating prime notation)

$$s[n] = \sum_{k=1}^{K^l} A_k^l \exp(j\theta_k^l) \exp(j\omega_k^l n), \quad -\frac{N_w - 1}{2} \leq n \leq \frac{N_w - 1}{2} \quad (9.9)$$

where N_w is the discrete-time window duration (assumed odd for simplicity of presentation). Equation (9.9) is a discrete-time stationary model for analysis with which the goal is to estimate the sinewave parameters A_k^l, θ_k^l, and ω_k^l on each frame. Finally, Equation (9.9) leads to an expression for the synthetic speech waveform over frame l as

$$s[n] = \sum_{k=1}^{K^l} \gamma_k^l \exp(j\omega_k^l n), \quad -\frac{N_w - 1}{2} \leq n \leq \frac{N_w - 1}{2} \quad (9.10)$$

where $\gamma_k^l = A_k^l \exp(j\theta_k^l)$ represents the kth complex amplitude for the kth component of the $K^l = K(t_l)$ sinewaves, and can be thought of as the measurements at continuous time t_l which is at the center of the analysis window. Figure 9.4 illustrates the entire process of going from continuous time to discrete time.

The problem now is to fit the synthetic speech waveform in Equation (9.10) to the original measured waveform, denoted here by $y[n]$. In particular, and to summarize the above idealization, our goal in analysis is to estimate the sinewave parameters that we have assumed constant over each frame. Our stationary assumption corresponds to amplitude, frequency, and phase functions in discrete time given by

$$A_k^l[n] = A_k^l$$
$$\omega_k^l[n] = \omega_k^l$$
$$\theta_k^l[n] = \omega_k^l n + \theta_k^l$$

where θ_k^l is the phase offset measured at the center of the lth frame. The parameters A_k^l, ω_k^l, and θ_k^l are the unknown parameters in the discrete-time model Equation (9.10). In order to specify a criterion for the goodness of fit of the sinewave model $s[n]$ to the speech measurement $y[n]$, it is necessary to address whether the speech source is either quasi-periodic, noisy, or impulsive, which are the three basic source classes introduced in Chapter 3.

9.3.1 Voiced Speech

A useful criterion for judging the goodness of fit is the mean-squared error (MSE) defined as[7]

$$\epsilon^l = \sum_{n=-(N_w-1)/2}^{(N_w-1)/2} | y[n] - s[n] |^2$$

$$= \sum_{n=-(N_w-1)/2}^{(N_w-1)/2} | y[n] |^2 - 2\text{Re} \sum_{n=-(N_w-1)/2}^{(N_w-1)/2} y[n]s^*[n] + \sum_{n=-(N_w-1)/2}^{(N_w-1)/2} | s[n] |^2 \quad (9.11)$$

---

[7] In this section, because the sinewaves in the model $s[n]$ are considered in complex quadrature form (approximately analytic signals), $y[n]$ is also considered in this complex form. As we saw in Chapter 2, an analytic signal representation is obtained by removing negative frequencies from a signal.

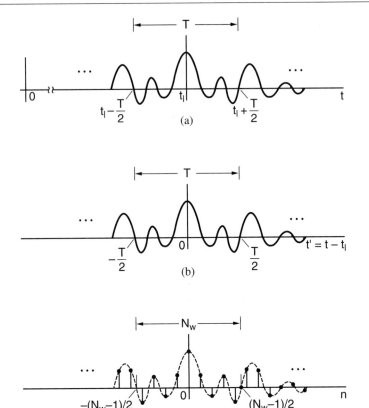

Figure 9.4 Discrete-time conversion of speech segment used in sinewave analysis: (a) original waveform over the analysis window duration. The segment center occurs at time t_l; (b) short-time segment shifted by t_l; (c) discrete-time version of (b) with window duration N_w assumed odd.

where $s[n]$ is the sinewave model, $y[n]$ is the measurement, and the superscript in ϵ^l refers to the lth frame. Substituting the speech model of Equation (9.10) into Equation (9.11) leads to the following expression for the MSE:

$$\epsilon^l = \sum_{n=-(N_w-1)/2}^{(N_w-1)/2} |y[n]|^2 - 2\mathrm{Re}\sum_{k=1}^{K^l}(\gamma_k^l)^* \sum_{n=-(N_w-1)/2}^{(N_w-1)/2} y[n]\exp(-jn\omega_k^l)$$

$$+ N_w \sum_{k=1}^{K^l}\sum_{m=1}^{K^l}\gamma_k^l(\gamma_m^l)^*\mathrm{sinc}(\omega_k^l - \omega_m^l) \tag{9.12}$$

where

$$\mathrm{sinc}(x) = \frac{\sin(N_w\frac{x}{2})}{N_w\sin(\frac{x}{2})}.$$

The problem now is to try to identify a set of sinewaves that minimizes Equation (9.12); this identification problem, in general, is difficult to solve because the desired sinewave parameters are nonlinear in ϵ^l. Insights into the development of a suitable estimator can be obtained by restricting the class of input speech signals to perfectly voiced speech, in which case $s[n]$ is not only deterministic but also harmonic. In this ideal case, we derive an estimator and then use this estimator in non-ideal cases, for example, nearly harmonic voiced speech and aharmonic unvoiced speech. We finally modify the estimator (if needed) to handle real speech; here we rely on our intuition and experience.

In the ideal voiced case, Equation (9.10) can be written as

$$s[n] = \sum_{k=1}^{K^l} \gamma_k^l \exp(jnk\omega_o^l),$$

where $\omega_o^l = \frac{2\pi}{P^l}$ and where P^l is the pitch period, which is assumed to be constant over the analysis window duration of the lth frame. For the purpose of establishing the structure of the ideal estimator, it is further assumed that the pitch period is known and that the width of the analysis window is a multiple of P^l. Under these idealized conditions, $\omega_k^l = k\omega_o^l$, and the sinc($\cdot$) function in the last term in Equation (9.12) reduces to

$$\text{sinc}(\omega_k^l - \omega_m^l) = \text{sinc}[(k - m)\omega_o^l] = \delta[k - m]$$

where $\delta[k - m] = 1$ if $l = m$ and $\delta[k - m] = 0$ if $k \neq m$. Then the expression for the MSE becomes

$$\epsilon^l = \sum_{n=-(N_w-1)/2}^{(N_w-1)/2} \mid y[n] \mid^2 -2N_w \text{Re} \sum_{k=1}^{K^l} (\gamma_k^l)^* Y(\omega_k^l) + N_w \sum_{k=1}^{K^l} \mid \gamma_k^l \mid^2 \qquad (9.13)$$

where

$$Y(\omega) = \frac{1}{N_w} \sum_{n=-(N_w-1)/2}^{(N_w-1)/2} y[n] \exp(-jn\omega) \qquad (9.14)$$

that can be thought of as one time slice of the short time Fourier transform (STFT) of the speech measurement for a rectangular analysis window centered at the time origin. By completing the square in Equation (9.13), the MSE can be written as

$$\epsilon^l = \sum_{n=-(N_w-1)/2}^{(N_w-1)/2} \mid y[n] \mid^2 +N_w \sum_{k=1}^{K^l} \left(\mid Y(\omega_k^l) - \gamma_k^l \mid^2 - \mid Y(\omega_k^l) \mid^2 \right),$$

which can be minimized by choosing as the estimate for the complex amplitudes,

$$\hat{\gamma}_k^l = Y(k\omega_0^l) \qquad (9.15)$$

which reduces the MSE to

$$\epsilon^l = \sum_{n=-(N_w-1)/2}^{(N_w-1)/2} \mid y[n] \mid^2 -N_w \sum_{k=1}^{K^l} \mid Y(k\omega_0^l) \mid^2.$$

From this it follows that the error is minimized by selecting all of the available harmonic frequencies in the speech bandwidth (where $K^l = $ bandwidth$/\omega_o^l$).

Equations (9.14) and (9.15) completely specify the structure of the ideal estimator and show that the optimum estimator is obtained through the discrete-time Fourier transform. Although these results are equivalent to a Fourier series representation of a periodic waveform, the above equations lend themselves to an intuitive generalization for the practical case. To see this, consider the STFT, that we are denoting by $|Y(\omega)|^2$. For the perfectly voiced speech case, this function is pulse-like in nature, with peaks occurring at all of the harmonics $k\omega_o^l$. Therefore, the frequencies of the underlying sinewaves correspond to the location of the peaks of $|Y(\omega)|^2$, and the estimates of the amplitudes and phases are obtained by evaluating the real and imaginary parts of the STFT at the frequencies of the peaks. The advantage of this latter interpretation of the estimator structure is that it can be applied when the ideal voiced speech assumption is no longer valid, namely when the sinewaves are aharmonic but with constant frequencies. To support this assumption, consider the STFT for the general sinusoidal speech model in Equation (9.10). In this case, the STFT (for the one time slice) is simply

$$Y(\omega) = \sum_{k=1}^{K^l} \gamma_k^l \text{sinc}(\omega_k^l - \omega).$$

Provided the analysis window is "wide enough" so that

$$| \omega_m^l - \omega_k^l | \ge \frac{4\pi}{N_w}, \quad m \neq k \tag{9.16}$$

then the Fourier transform magnitude squared can be written as

$$| Y(\omega) |^2 \approx \sum_{k=1}^{K^l} | \gamma_k^l |^2 \text{sinc}^2(\omega_k^l - \omega).$$

As before, the location of the peaks of the spectral magnitude corresponds to the sinewave frequencies, and the samples of the STFT at these frequencies correspond to the complex amplitudes. Therefore, the structure of the ideal estimator applies to a more general class of speech waveforms as long as Equation (9.16) holds. Since, during steady voicing, neighboring frequencies are separated by the fundamental frequency, Equation (9.16) suggests that the desired resolution can be achieved "most of the time" by requiring that the analysis window be at least two pitch periods wide.

EXAMPLE 9.4   Figure 9.5a shows an example of the spectral magnitude of a voiced speech waveform over a 25-ms analysis window. In this example, the spectral peaks, estimated as DFT values larger than two adjacent neighbors, occur primarily near the harmonics. However, beyond about 3000 Hz, peak locations become aharmonic due to the presence of aspiration noise. Peaks can also occur between harmonics when a noise component is present simultaneously with voicing. The sinewave representation of noise is addressed in the following section.   ▲

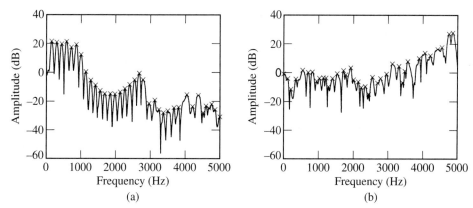

Figure 9.5 Typical STFT magnitude of voiced and unvoiced (fricative) speech: (a) voiced with an aspiration component; (b) unvoiced. Spectral peaks, whose locations are denoted by the crosses, determine which frequencies are selected to represent the speech waveform.

9.3.2 Unvoiced Speech

Speech generated with stochastic (noisy) and impulsive sources make up the unvoiced speech category. We have seen in Chapter 3 that stochastic inputs arise, for example, with aspiration at the glottis and frication at a vocal tract constriction. In these cases, the waveform is treated as a sample function of a random process. We saw in Chapter 5 that the STFT magnitude squared, i.e., the periodogram, fluctuates about the power spectrum underlying a random process (Appendix 5.A). If we are to apply the same analysis procedure used in developing the sinewave representation for voiced speech, then we must show that sinewaves corresponding to the sample peaks of our periodogram sum to a random process whose power spectrum adequately approximates the underlying power spectrum of the unvoiced sound.[8] An argument that indeed shows conditions for this property exploits the Karhunen-Loève expansion that allows constructing a random process over a finite interval from a series expansion of harmonic sinusoids with uncorrelated complex amplitudes (thus characterized by magnitude and phase) [55].

A mathematical analysis based on the Karhunen-Loève expansion shows that a sinusoidal representation is valid, and that we can use the same analysis as in the voiced speech case above, provided the frequencies are "close enough" so that the power spectrum changes slowly over consecutive frequencies. If the window width is constrained to be at least 20 ms, there will be "on the average" a set of periodogram peaks that are approximately 100 Hz apart. This condition has been shown to provide a sufficiently dense sampling to satisfy the necessary constraints while also providing samples that are roughly uncorrelated, i.e., the Karhunen-Loève expansion constraint can be ensured by not allowing samples to fall more than 100 Hz apart during unvoiced speech segments.

---

[8] An alternate approach, developed in a speech coding context and described in Chapter 12, uses a harmonically-dependent set of sinewaves with a random phase modulation [33],[34]. Yet another technique represents the signal by a sum of adjacent narrowband sinewaves of uniformly spaced center frequency with random amplitude and frequency modulation [27].

EXAMPLE 9.5   Figure 9.5b shows an example of a typical periodogram for a frame of unvoiced speech along with the amplitudes and frequencies that are estimated using the above peak-picking procedure. The analysis window is 20 ms in duration and the periodogram peaks are "on the average" no more than about 100 Hz apart in frequency, thus roughly satisfying the Karhunen-Loève expansion constraint for representing the underlying power spectrum. Synthesis experiments using the Karhunen-Loève expansion constraints will later be described that show empirically that unvoiced speech can be modeled as a sum of sinewaves derived from periodogram peaks.   ▲

Finally, we must address the representation and analysis of transient sounds, such as plosives, generated with impulsive sources which are neither quasi-periodic nor random. Here the justification of a sinewave model and the use of peaks in the STFT magnitude as an estimator for the model parameters are more empirical. The validity of the approach for this signal class is based on the observation that peak picking the STFT magnitude captures most of the spectral energy so that, together with the corresponding STFT phase, the short-time waveform character is approximately preserved. This interpretation will become more clear when one sees in the following sections that sinewave synthesis is roughly equivalent to an overlap-add of triangular-weighted short-time segments derived from the STFT peaks. A problem arises, however, in selecting an analysis window that provides adequate time-frequency resolution for both periodic and impulsive sounds. Later in this chapter, we revisit this resolution issue and also describe alternative approaches to a sinewave representation and analysis of transient sounds, such as replacing the STFT with a constant-Q wavelet transform.

9.3.3 Analysis System

This section details the specific methods used to implement the sinusoidal analyzer which are common to all of the applications discussed later in the chapter. The analysis in the preceding section implicitly assumed that the STFT was computed using a rectangular window. Returning to the standard STFT notation, we write the STFT of $y[n]$ as

$$Y(lL, \omega) = \sum_{m=-\infty}^{\infty} w[lL - m]y[m]\exp(-jm\omega)$$

where $w[n]$ represents the analysis window. Since the poor sidelobe performance of the rectangular window compromises the performance of the estimator, in all of the experiments described in this chapter the Hamming window is used to weight the measured data. While this results in a satisfactory sidelobe structure, it does so at the expense of broadening the main lobes of the spectral estimator. Therefore, in order to maintain the resolution properties that are needed to justify the optimality properties of the spectral processor, the constraint implied by Equation (9.16) is revised to require that the window width, N_w, be two and one-half times the average pitch period or 20 ms, whichever is the larger. Then the pitch-adaptive Hamming window is computed using

$$w[n] = 0.54 + 0.46\cos\left(2\pi\frac{n}{N_w}\right), \qquad 0 \le n < N_w$$

and normalized according to

$$\sum_{n=-\infty}^{\infty} w[n] = 1$$

so that the spectral peak estimate $\hat{A}_k^l$ yields the amplitude of an underlying sinewave (Exercise 9.1). The Hamming window is then applied to the speech segment and zero padded to the specified length of the DFT. For a high-quality analysis/synthesis, a 1024-point DFT is found adequate.

It should be noted that the placement of the analysis window $w[n]$ relative to the time origin is important for computing the phases. Typically, in frame-sequential processing, the window, $w[n]$, lies in the interval $0 \leq n < N_w$ and is symmetric about $(N_w - 1)/2$, a placement and shape that gives it a linear Fourier transform phase equal to $-\omega(N_w - 1)/2$. Since N_w is on the order of 200–500 discrete time samples, any error in the estimated frequencies results in a large random phase error and consequent distortion, perceived as hoarseness, in the reconstruction. An error of one DFT sample, for example, results in a $\frac{2\pi}{N}\frac{N_w-1}{2}$ phase error (where N is the DFT length) which could be on the order of π. Experiments have shown that the ear is very sensitive to short-term phase jitter whenever the phase error is greater than $\sim \pi/16$ radians. To eliminate the linear-phase term, *the windowed speech must be circularly shifted* before the DFT is taken so that its center is at $n = 0$. To do this, the Hamming window is placed symmetric relative to an origin defined as the center of the current analysis frame; hence the window takes on values over the interval $-(N_w - 1)/2 \leq n \leq (N_w - 1)/2$, rather than the more common interval $0 \leq n < N_w$. As shown in Figure 9.6, the speech values at negative values of n are wrapped around to the end of the FFT buffer.

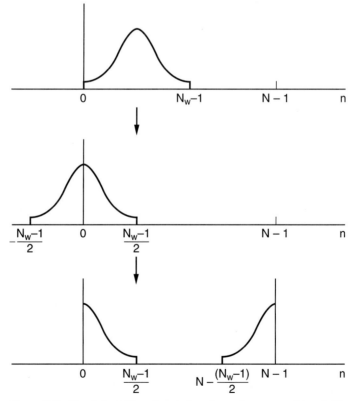

Figure 9.6 Circularly shifting the windowed speech segment of length N_w to reduce linear phase error due to DFT frequency sampling. The DFT length is denoted by N.

Analysis

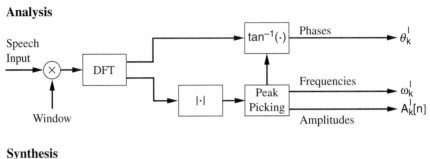

Synthesis

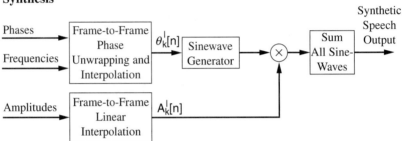

Figure 9.7   Block diagram of the baseline sinusoidal analysis/synthesis system.

SOURCE: R.J. McAulay and T.F. Quatieri, "Speech Analysis-Synthesis Based on a Sinusoidal Representation" [30]. ©1986, IEEE. Used by permission.

After the circular shift, the practical version of the idealized estimator obtains the frequencies of the underlying sinewaves as the locations of the peaks of $\mid Y\left(lL, \frac{2\pi}{N}k\right)\mid$, i.e., the DFT samples at which the slope changes from positive to negative. Denoting these frequency estimates by $\hat{\omega}_k^l$, the corresponding amplitudes and phases are given by

$$\hat{A}_k^l =\mid Y(lL, \hat{\omega}_k^l)\mid, \quad \hat{\theta}_k^l = \angle Y(lL, \hat{\omega}_k^l).$$

Henceforth, we drop the "hat" notation, for simplicity, unless needed. A block diagram of the sinewave analysis system is shown in Figure 9.7.

9.3.4 Frame-to-Frame Peak Matching

The above analysis provides a heuristic justification for the representation of the speech waveform in terms of the amplitudes, frequencies, and phases of a set of sinewaves that applies to one analysis frame. As speech evolves from frame to frame, different sets of these parameters will be obtained. The next problem to address is the association of amplitudes, frequencies, and phases measured on one frame with those that are obtained on a successive frame in order to define a set of sinewaves that will be continuously evolving in time. If the number of peaks were constant from frame to frame, the problem of matching the parameters estimated on one frame with those on a successive frame would simply require a frequency-ordered assignment of peaks. In practice, however, the locations of the peaks changes as the pitch changes, and there are rapid changes in both the location and the number of peaks during rapidly varying speech, such as at voiced/unvoiced transitions.

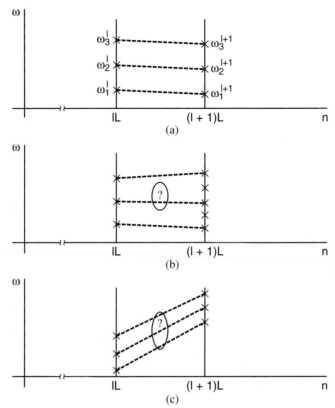

Figure 9.8 Problem of frequency matching: (a) slowly varying pitch;
(b) rapidly varying pitch; (c) rapid voiced/unvoiced transition.

To illustrate the problem of peak matching, suppose that three spectral peaks are found on two consecutive frames l and $l+1$ in a slowly varying voiced region with a small change in pitch, as shown in Figure 9.8a. As indicated, the matching requires a simple ordered alignment of the frequency estimates $[\omega_1^l, \omega_2^l, \omega_3^l]$ with $[\omega_1^{l+1}, \omega_2^{l+1}, \omega_3^{l+1}]$. The matching of the amplitude and phase values implicitly follows. In practice, however, peak locations change as the pitch and spectrum change. For example, suppose that the pitch quickly decreases over two consecutive frames. Then, as shown in Figure 9.8b, the number of peaks abruptly increases. A rapid change in not only the number of peaks but also their location can occur in rapidly varying regions such as voiced to unvoiced transitions, as shown in Figure 9.8c.

In order to account for such rapid movements in the spectral peaks and unequal number of peaks from frame-to-frame, the concepts of *birth* and *death* of sinusoidal components are introduced. The problem of matching spectral peaks in some optimal sense, while allowing for this birth/death process, is generally a difficult problem. One method, which has proven to be successful for speech reconstruction, is to define sinewave tracks for frequencies[9] that are successively nearest neighbors, conditioned on a frequency on the current frame to fall

---

[9] A more sophisticated matcher may use amplitude and phase, as well as frequency, information.

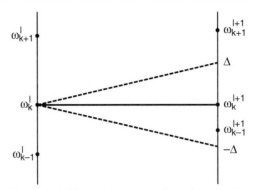

Figure 9.9 The matching interval condition used in nearest-neighbor sinewave frequency matching.

within a *matching interval*, $[-\Delta, \Delta]$, relative to its matched frequency on the previous frame (Figure 9.9). The matching procedure is made dynamic by allowing for tracks to begin at any frame (a birth) and to terminate at any frame (a death)—events which are determined when successive frequencies do not fall within the matching interval. The algorithm, although straightforward, is a rather tedious exercise in rule-based programming, and the reader is referred to [30] for a detailed understanding of the algorithm for actual software development. An illustration of the matching algorithm showing how the birth/death procedure accounts for rapidly varying peak locations is given in Figure 9.10.

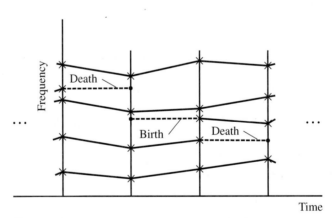

Figure 9.10 Different modes used in the birth/death frequency-matching process for determining frequency tracks. Note the death of two tracks during frames one and three and the birth of a track during the second frame.

SOURCE: R.J. McAulay and T.F. Quatieri, "Low Rate Speech Coding Based on the Sinusoidal Speech Model," chapter in *Advances in Speech Signal Processing* [34]. ©1992, Marcel Dekker, Inc. Courtesy of Marcel Dekker, Inc.

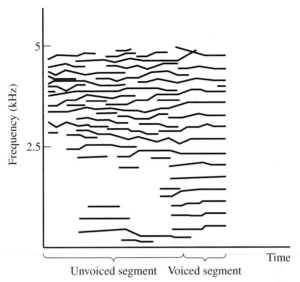

Figure 9.11 Typical frequency tracks for real speech.

SOURCE: R.J. McAulay and T.F. Quatieri, "Speech Analysis-Synthesis Based on a Sinusoidal Representation" [30]. ©1986, IEEE. Used by permission.

EXAMPLE 9.6   The result of applying the frequency matcher to a segment of real speech is shown in Figure 9.11, which illustrates the ability of the matcher to adapt quickly through transitory speech behavior such as voiced/unvoiced transitions and mixed voiced/unvoiced regions. The frame interval here is 10 ms and the frequency-matching interval width 2Δ is set at 100 Hz.   ▲

EXAMPLE 9.7   Suppose there are P frequencies (peaks) in frame l and Q frequencies (peaks) in frame $l + 1$. Suppose also that no frequency pairs meet the frequency-matching interval constraint. Then as many as $P + Q$ frequency tracks can exist across any given frame. For example, if $P = Q = 70$, then a maximum of 140 sinewaves can exist in any one frame.   ▲

9.4 Synthesis

Since a set of amplitudes, frequencies, and phases is estimated for each frame, it might seem reasonable to estimate the original speech waveform on the lth frame by generating synthetic speech using the equation

$$\hat{s}[n] = \sum_{k=1}^{K^l} A_k^l \cos(n\omega_k^l + \theta_k^l), \quad n = 0, 1, 2, \cdots, L - 1$$

where L is the length of the synthesis frame. As we indicated in Example 9.3, due to the time-varying nature of the parameters, however, this straightforward approach leads to discontinuities at the frame boundaries which seriously degrade the quality of the synthetic speech. Therefore, some provision must be made for smoothly interpolating the parameters measured from one frame to those that are obtained on the next. Several methods have been developed to accomplish the necessary interpolation.

9.4.1 Cubic Phase Interpolation

As a result of the frequency-matching algorithm described in the previous section, all of the parameters measured for an arbitrary frame l have been associated with a corresponding set of parameters for frame $l+1$. Letting $(A_k^l, \omega_k^l, \theta_k^l)$ and $(A_k^{l+1}, \omega_k^{l+1}, \theta_k^{l+1})$ denote the successive sets of parameters for the kth frequency track, an obvious solution to the amplitude interpolation problem is to take

$$A_k^l[n] = A_k^l + (A_k^{l+1} - A_k^l)\left(\frac{n}{L}\right), \quad n = 0, 1, \cdots, L - 1 \tag{9.17}$$

where the analysis time origin is the center of the analysis frame so that in Equation (9.17) we are reconstructing half of each of the lth and $(l + 1)$st frames.

Unfortunately, such a simple approach cannot be used to interpolate the phase because the phases θ^l and θ^{l+1} are measured modulo 2π. We illustrate this problem with the following example:

EXAMPLE 9.8   Consider a time-invariant vocal tract driven by a source consisting of sinewaves with fixed frequencies and amplitudes so that (in continuous time) $\Omega_k(t) = \Omega_k$. Then the phase of the speech sinewaves is expressed as

$$\theta_k(t) = \int_0^t \Omega_k(\sigma)d\sigma + \phi_k + \Phi[t, \Omega_k(t)]$$

$$= \int_0^t \Omega_k d\sigma + \phi_k + \Phi[t, \Omega_k]$$

$$= \Omega_k t + \phi_k + \Phi_k$$

where $\Phi_k = \Phi[t, \Omega_k]$ is the time-invariant vocal tract phase component of the phase offset. Figure 9.12 shows that when Ω_k is large, the sinewave phase $\theta_k(t)$ moves rapidly; therefore, the phase over successive frames moves through multiples of 2π, which results in principal phase measurements with 2π discontinuities. Interpolation cannot be performed with samples of such a wrapped phase function.   ▲

Hence, phase unwrapping must be performed jointly with interpolation to ensure that the frequency trajectories are meaningful across frame boundaries.

The first step in solving this problem is to postulate a phase interpolation function that is a cubic polynomial.[10] For simplicity, we omit the sinewave index k and for convenience, we write the phase as a continuous function of the time variable t, with $t = 0$ corresponding to the center of frame l and $t = T$ corresponding to the center of frame $l + 1$:

$$\theta(t) = \zeta + \gamma t + \alpha t^2 + \beta t^3. \tag{9.18}$$

---

[10] The idea of applying a cubic polynomial to interpolate the phase between frame boundaries was independently proposed in [30] for synthesis of sinewaves of arbitrary frequency and in [1] for harmonic sinewave synthesis.

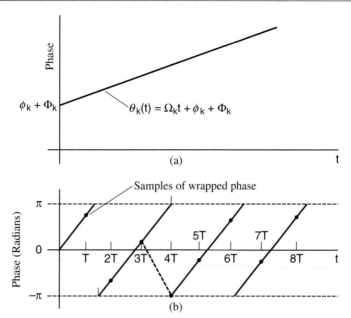

Figure 9.12 Phase functions for Example 9.8: (a) phase moves quickly when Ω_k is large; (b) samples of wrapped phase computed modulo 2π. The thick dashed line in panel (b) represents interpolation from wrapped phase samples that would result in an incorrect phase derivative and thus an incorrect frequency trajectory.

The choice of a cubic polynomial is motivated by four constraints that must be satisfied on each frame.[11] Before describing these constraints, we observe that when the vocal tract is slowly varying, the derivative of the sinewave phase is approximately the excitation frequency that we have measured from the location of spectral peaks. We can see this by expressing the sinewave phase function as

$$\theta(t) = \int_0^t \Omega(\sigma)d\sigma + \phi + \Phi[t, \Omega(t)]$$

so that its derivative under the assumption of a slowly varying $\Phi[t, \Omega(t)]$ becomes

$$\dot{\theta}(t) = \Omega(t) + 0 + \dot{\Phi}[t, \Omega(t)]$$
$$\approx \Omega(t).$$

It follows that, at the center of the lth and $(l+1)$st frames, the phase derivatives are given by the frequency measures on these consecutive frames, i.e.,

$$\dot{\theta}^l \approx \Omega^l$$
$$\dot{\theta}^{l+1} \approx \Omega^{l+1}.$$

We can now formulate the four constraints on the phase polynomial.

---

[11] Other interpolation schemes satisfying these constraints also exist. Exercises 9.4 and 9.5 work through, respectively, piecewise quadratic and piecewise linear interpolating functions over a frame.

Because the derivative of the phase is the frequency, it is necessary that the cubic phase function and its derivative equal the phase and frequency measured at the frame l. Therefore, it follows that at $t = 0$

$$\theta(0) = \zeta = \theta^l$$
$$\dot{\theta}(0) = \gamma = \Omega^l,$$

and, as a result, Equation (9.18) can be written as

$$\theta(t) = \theta^l + \Omega^l t + \alpha t^2 + \beta t^3.$$

Similarly, the cubic phase function and its derivative must equal the phase and frequency measured at the frame $l + 1$. Therefore, it follows that at $t = T$

$$\theta(T) = \theta^l + \Omega^l T + \alpha T^2 + \beta T^3 = \theta^{l+1} + 2\pi M$$
$$\dot{\theta}(T) = \Omega^l + 2\alpha T + 3\beta T^2 = \Omega^{k+1}. \tag{9.19}$$

Because the terminal phase θ^{l+1} is measured modulo 2π, it is necessary to augment it by the term $2\pi M$ where M is an unknown integer. This results in two equations in three unknowns: α, β, and M. Nevertheless, suppose we had the correct value of M. Then we can solve for α and β. In other words, at this point M is unknown, but for each value of M, Equation (9.19) can be solved for $\alpha(M)$ and $\beta(M)$, the dependence on M being made explicit. The solution is shown to satisfy the matrix equation (Appendix 9.B)

$$\begin{bmatrix} \alpha(M) \\ \beta(M) \end{bmatrix} = \begin{bmatrix} \frac{3}{T^2} & \frac{-1}{T} \\ \frac{-2}{T^3} & \frac{1}{T^2} \end{bmatrix} \begin{bmatrix} \theta^{l+1} - \theta^l - \Omega^l T + 2\pi M \\ \Omega^{l+1} - \Omega^l \end{bmatrix}. \tag{9.20}$$

In order to determine M and, ultimately, the solution to the phase unwrapping problem, an additional constraint needs to be imposed. Our final constraint is to make the resulting frequency function "maximally smooth," a concept that is now quantified. Figure 9.13 illustrates a typical set of cubic phase interpolation functions for a number of values of M. It seems clear on intuitive grounds that the best phase function to pick is the one that would have the least variation. This is what is meant by a maximally smooth frequency trajectory. In fact, if the frequencies were constant and the vocal tract were stationary, the true phase would be linear, its phase derivative constant, i.e., $\dot{\theta}(t) = \Omega^l$, and so its second derivative would be zero, i.e., $\ddot{\theta}(t) = 0$. Therefore, a reasonable criterion for smoothness is to choose M such that

$$f(M) = \int_0^T [\ddot{\theta}(t; M)]^2 dt$$

is a minimum where $\ddot{\theta}(t; M)$ denotes the second derivative of $\theta(t; M)$ with respect to the time variable t. Although M is integer valued, since $f(M)$ is quadratic in M, the problem is most easily solved by minimizing $f(x)$ with respect to the continuous variable x and then choosing M to be the integer closest to x. After straightforward but tedious algebra, it can be shown that the minimizing value of x is (Appendix 9.B)

$$x^* = \frac{1}{2\pi} \left[(\theta^l + \Omega^l T - \theta^{l+1}) + (\Omega^{l+1} - \Omega^l)\frac{T}{2} \right]$$

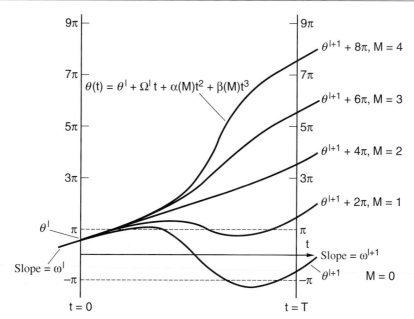

Figure 9.13 Typical set of cubic phase interpolation functions. The phase function for $M = 2$ is "maximally smooth."

Source: R.J. McAulay and T.F. Quatieri, "Speech Analysis-Synthesis Based on a Sinusoidal Representation" [30]. ©1986, IEEE. Used by permission.

from which M^* is determined as the nearest integer to x^*. M^* is then used in Equation (9.20) to compute $\alpha(M^*)$ and $\beta(M^*)$ and, in turn, the unwrapped phase interpolation function is denoted by

$$\theta(t) \;=\; \theta^l \,+\, \Omega^l t \,+\, \alpha(M^*)t^2 \,+\, \beta(M^*)t^3. \tag{9.21}$$

This phase function not only satisfies all of the measured phase and frequency endpoint constraints, but also unwraps the phase in such a way that $\theta(t)$ is maximally smooth.

Because the above analysis began with the assumption of an initial unwrapped phase θ^l corresponding to frequency Ω^l at the start of frame l, it is necessary to specify the initialization of the frame interpolation procedure. This is done by noting that, at some point in time, the track under study was born. When this event occurred, an amplitude, frequency, and phase were measured at frame $l + 1$, and the parameters at frame l to which these measurements correspond are defined by setting the amplitude to zero (i.e., $A^l = 0$) while maintaining the same frequency (i.e., $\Omega^l = \Omega^{l+1}$). In order to ensure that the phase interpolation constraints are satisfied initially, the unwrapped phase is defined to be the measured phase θ^{l+1} and the startup phase is defined to be

$$\theta^l \;=\; \theta^{l+1} \,-\, \Omega^{l+1} T$$

where T (in discrete time) is the number of samples traversed in going from frame $l + 1$ back to frame l. This startup procedure is consistent with assuming the frequency is constant over the initial frame interval and so the cubic polynomial reduces to a linear-phase trajectory over this interval. Observe that in this case, $M^* = 0$ because the measured phase becomes the desired unwrapped phase. We summarize the conditions at the birth of a track as [12]

$$A^l = 0$$
$$\Omega^l = \Omega^{l+1}$$
$$\theta^l = \theta^{l+1} - \Omega^{l+1}T.$$

A similar procedure is used for terminating a frequency track.

As a result of the above phase unwrapping procedure, each frequency track has associated with it an instantaneous unwrapped phase, which accounts for both the rapid phase changes due to the source and the slowly varying phase changes due to the shape of the glottal flow and the vocal tract transfer function. Returning to discrete time, the final synthetic waveform for the lth frame is given by

$$\hat{s}[n] = \sum_{k=1}^{K_l} A_k^l[n] \cos(\theta_k^l[n])$$

where, for the kth sinewave, $A_k^l[n]$ is given by Equation (9.17), $\theta_k^l[n]$ is the sampled data version of Equation (9.21), and K_l is the number of sinewaves. A block diagram description of the sinewave synthesis system is shown in Figure 9.7. We henceforth refer to this as the *baseline* sinusoidal analysis/synthesis system.

9.4.2 Overlap-Add Interpolation

While the cubic phase interpolation system is the recommended synthesis technique in that it produces very high-quality synthetic speech, it is computationally expensive to implement due to the fact that every component sinewave must be synthesized on a per sample basis. An alternate procedure which has proven to be satisfactory in some speech applications is to overlap and add weighted, i.e., windowed, segments of the reconstructed waveform from one frame to the next [35].

The sinewave parameters estimated on frame l can be used to generate the waveform

$$\hat{s}^l[n] = \sum_{k=1}^{K^l} A_k^l \cos(n\omega_k^l + \theta_k^l) \tag{9.22}$$

over the interval $0 \leq n < L$. Another representation of the waveform on this interval can be obtained using the sinewave parameters measured at frame $l + 1$:

$$\hat{s}^{l+1}[n] = \sum_{k=1}^{K^{l+1}} A_k^{l+1} \cos[(n - L)\omega_k^{l+1} + \theta_k^{l+1}]. \tag{9.23}$$

If the frame interval L is "short enough," $s^l[n]$ and $s^{l+1}[n]$ will be similar, and smoothly interpolated synthetic speech can be generated by weighting the above waveforms by a triangular window. As shown in Figure 9.14, the leading portion (downward slope of the triangular window)

---

[12] It is interesting to observe that one can obtain a faster attack and decay than a frame length by beginning or ending the amplitude interpolation within a frame duration.

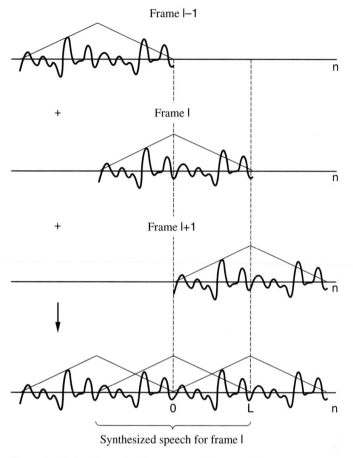

Figure 9.14 Overlap and adding segments in sinusoidal synthesis.

of the sinewave generated during the frame l is overlapped and added to the trailing portion (upward slope of the triangular window) of the sinewave in frame $l + 1$. This operation can be expressed as

$$\hat{s}[n] = \left(1 - \frac{n}{L}\right)\hat{s}^l[n] + \left(\frac{n}{L}\right)\hat{s}^{l+1}[n], \qquad n = 0, 1, \dots L - 1.$$

Since the component sinewaves are now of constant frequency, FFT techniques can be used to synthesize the component waveforms in Equation (9.22) and Equation (9.23) which results in reduced computational complexity [35]. Provided that the FFT frequency quantization is small enough ($\sim$ 30 Hz) and the frame rate is high enough ($\sim$ 100 Hz) this procedure can produce very high-quality output speech. It should be noted the method can be implemented without recourse to the frame-to-frame peak matching algorithm, which further reduces the computational complexity.

It is interesting to observe that the above overlap-add operation is also implemented by setting the frequency-matching window to zero in the matching algorithm. This is because, under this condition, sinewaves are born and die on every two consecutive frames, and over these two frames the birth/death process holds the frequency constant. Since the amplitude interpolation is linear, beginning and ending at zero for a birth and death, respectively, a triangular window of length twice the frame interval is effectively applied to each steady sinewave. This is an interesting alternative to direct overlap-add; although it is clearly not computationally efficient, it may serve to help explain the ability of the sinewave analysis/synthesis to represent short-lived and rapidly varying speech events such as plosives and unvoiced-to-voiced transitions (Exercise 9.8).

Finally, observe also that the overlap-add method can be interpreted as synthesizing a signal from a modified STFT. As a result, we might consider the least-squared error approach of Chapter 7, which resulted in a weighting of the short-time frame by the window and a normalization by the summed squared, shifted windows (Exercise 9.15).

Unfortunately, there are situations in practice when the constraint on the frame rate cannot be met and straightforward application of the overlap-add procedure leads to synthetic speech that is quite "rough." This scenario occurs in the design of low-rate sinewave speech coders that quantize the measured phases. Since many bits must be allocated to the phase-coding procedure, the frame rate must be $\sim$ 50 Hz, which is too slow to ensure effective interpolation of the sinewave phases. The reason for this is that the sinewave frequencies are implicitly assumed constant over the extent of the triangular window. At the 100-Hz frame rate, the triangular window is 20 ms wide, which is an interval over which the sinewave parameters can reasonably be assumed to be stationary. If the frame rate is less than 100 Hz, however, the triangular window exceeds 20 ms and the stationarity assumption begins to break down. This is not to say that sinewave synthesis cannot be performed at frame rates less than 100 Hz; quality can be regained by using the frame-to-frame peak-matching algorithm and the cubic phase interpolation technique. This suggests that there exists an interpolated set of sinewave parameters that can be used with the STFT overlap-add synthesizer and operate at an effective 100-Hz frame rate using the desired 20-ms triangular windows (Exercise 9.6) [35].

9.4.3 Examples

Non-real-time floating-point simulations have determined the effectiveness of the sinusoidal approach in modeling real speech. In one simulation, the speech is lowpass-filtered at 5 kHz, digitized at 10 kHz, and analyzed at 10-ms frame intervals [30]. The speech segments are weighted using a fixed or pitch-adaptive Hamming window, circularly rotated (Figure 9.6) and the STFT is computed using a 1024-point FFT. The sinewave amplitude and frequency are estimated by locating the peaks of the magnitude of the STFT. The phases are then computed using the corresponding real and imaginary parts. The frequency-matching and linear amplitude and cubic phase interpolation techniques are used for synthesis. A large speech data base was processed with this system, and it was found that the synthetic speech is perceived to be essentially indistinguishable from the original. Visual examination of many of the reconstructed passages shows that the waveform structure is essentially preserved, as seen in the following examples:

> **EXAMPLE 9.9**   An example of the reconstruction fidelity of sinewave analysis/synthesis is shown in Figure 9.15, which compares the waveforms for the original speech and the reconstructed speech during a vowel and two plosives ("go to"). In this case of a female speaker, the window duration is 15 ms and the frame interval is 10 ms. Some slight temporal smearing is seen at the onset of the two plosives; the first is the voiced plosive "g" in the word "go" and the second is the unvoiced plosive "t" in the word "to."   ▲

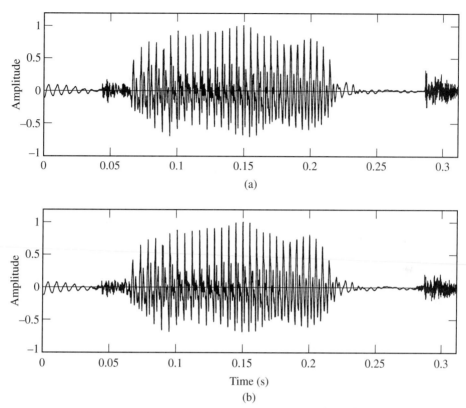

Figure 9.15 Reconstruction (b) of speech waveform (a) from female speaker using sinusoidal analysis/synthesis with 15-ms window and 10-ms frame.

EXAMPLE 9.10 A second example of the reconstruction fidelity is shown in Figure 9.16, which compares the waveforms for the original speech and the reconstructed speech for the "z" in the end of the word "jaz*z*" and the word "hour" (from the phrase "jazz hour"). In this case of a very low-pitched male speaker, the window duration is 60 ms and the frame interval is 10 ms. Some slight temporal smearing is seen at the onset of the vowel "o" in the word "h*ou*r." The waveform is well-reconstructed both visually and aurally in spite of the diplophonic behavior. The diplophonia exhibits itself in the measured phase of the harmonics, as well as in amplitude modulation of the harmonics (Exercise 3.3 in Chapter 3) that is captured by the peak-picking in sinewave analysis. ▲

The fidelity of the reconstructions suggests that the quasi-stationarity conditions seem to be satisfactorily met and that the use of the parametric model based on the amplitudes, frequencies, and phases of a set of sinewave components appears to be justifiable for both voiced and unvoiced speech. Although the sinusoidal model was originally designed in the speech context for a single speaker, it can represent almost any waveform. Successful reconstruction is obtained for multi-speaker waveforms, complex musical pieces, and biologic signals such as bird and whale sounds. Other signals tested include complex acoustic signals from mechanical impacts such as a bouncing can, a slamming book, and a closing stapler [41]. These signals were selected to

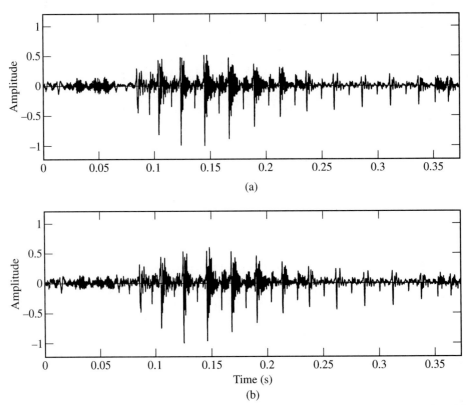

Figure 9.16 Reconstruction (b) of speech waveform (a) from a low-pitched male speaker using sinusoidal analysis/synthesis with 60-ms window and 10-ms frame.

have a variety of time envelopes, spectral resonances, and attack and decay dynamics. In each case, the window length and frame interval are tailored to the signal, and the reconstruction is both visually and aurally nearly imperceptible from the original; small discrepancies are found primarily at transitions and nonstationary regions where temporal resolution is limited due to the analysis window extent, as we saw with plosives and vowel onsets in the previous two examples. In addition, numerous synthetic and real background signals, including random signals (e.g., synthetic colored noise or an ocean squall) and AM-FM tonal interference (e.g., a blaring siren) were tested. The synthesized waveforms are essentially perceptually indistinguishable from the originals with little modification of the background [31],[44].

Although high-quality analysis/synthesis of speech has been demonstrated using amplitudes, frequencies, and phases at the spectral peaks of the STFT, it is often argued that the ear is insensitive to phase. The folklore about phase insensitivity dates back to as early as von Helmoltz [17]. This proposition, however, is not consistent with the experiments in Chapter 8 that show aural sensitivity to phase modification by the phase vocoder. The proposition is also inconsistent with auditory models of perception described in that chapter. The importance of phase measurements has also been shown to be essential to a high-quality sinewave synthesis. This property can be demonstrated by performing "magnitude-only" reconstruction by replacing

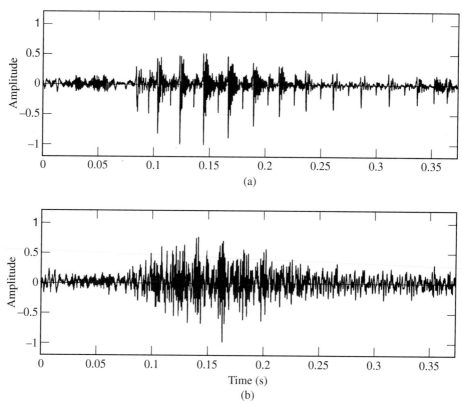

Figure 9.17 Magnitude-only reconstruction of speech (b) is compared against original (a) from a low-pitched male speaker. The synthesized waveform is dispersed compared to that of the synthesis with measured phases shown in Figure 9.16.

the cubic phase function in Equation (9.21) by a phase that is simply the integral of the instantaneous frequency, analogous to certain experiments performed with the phase vocoder. One way to do this is to make the instantaneous frequency be the linear interpolation of the frequencies measured at the frame boundaries and then perform the integration. Alternately, one can simply use the quadratic frequency derived from the cubic phase via initiating the cubic phase offset at zero upon the birth of a frequency track. A waveform synthesized by the magnitude-only system for the low-pitched speaker of Example 9.10 is given in Figure 9.17. While the speech from the magnitude-only synthesis is very intelligible and free of artifacts, it is quite different not only visually, i.e., dispersed, but also aurally because of the failure to maintain the true sinewave phases; the synthetic speech is reverberant during voicing and tonal during unvoiced speech, corresponding to the loss of *phase coherence*. We defined phase coherence in Chapter 8 as the preservation of the original component phase relations. For voiced speech, the aural sensitivity to loss in phase coherence is greater for low-pitched than high-pitched speech, consistent with the front-end auditory filter-bank model described in Chapter 8 in which filter output envelopes are modified more with decreasing pitch and increasing cochlear filter characteristic frequency.

9.4.4 Applications

We now briefly describe a number of applications of the baseline sinewave analysis/synthesis that illustrate the generality of the approach.

Time-Scale Modification — In time-scale modification, the magnitudes, frequencies, and phases of the sinewave components are modified to expand the time scale of a signal without changing its spectral or frequency (pitch) characteristics. Consider a time-scale modification by a rate-change factor ρ. By time-warping the sinewave frequency tracks, i.e., forming $\Omega_k(t\rho^{-1}) = \dot{\theta}_k(t\rho^{-1})$, the instantaneous frequency locations are preserved while modifying their rate of change in time [38]. Since $\frac{d}{dt}[\rho\theta_k(t\rho^{-1})] = \Omega_k(t\rho^{-1})$, this modification can be represented by a phase change in Equation (9.4) and thus the time-scaled signal can be expressed as

$$\tilde{s}(t) = \sum_{k=1}^{K(t)} A_k(t\rho^{-1}) \exp[j\rho\theta_k(t\rho^{-1})], \tag{9.24}$$

where the amplitude functions are also time-warped. (Recall similar operations on filter-bank outputs in the phase vocoder.) Suppose that in the discrete-time baseline analysis/synthesis system, the analysis and synthesis frame intervals are L samples. In the discrete-time analysis/synthesis system based on the model in Equation (9.24), the synthesis interval is mapped to $L' = \rho L$ samples. L' is constrained to be an integer value since the synthesis frame requires an integer number of discrete samples. The modified cubic phase and linear amplitude functions, which are derived for each sinewave component, are then sampled over this longer frame interval. This modification technique has been successful in time-scaling of speech, as well as a larger class of signals such as music, biological, and mechanical impact signals [38]. Nevertheless, a problem arises due to the inability of the system to maintain the original sinewave phase relations through $\rho\theta_k(t\rho^{-1})$; some signals can suffer from the reverberance typical of other modification systems, such as the phase vocoder and the "magnitude-only" reconstruction described in Section 7.6.1 of Chapter 7. An approach to preserve *phase coherence*, and thus improving quality, imparts a source/filter phase model on the sinewave components and is described in Section 9.5. In spite of its limitations, however, the technique of Equation (9.24) and its variations remain the most general for sinewave-based time-scale modification [38]. Similar approaches, using the baseline sinewave analysis/synthesis, have also been used for frequency transformations, including frequency compression and pitch modification [38].

Other Applications — Other applications of the baseline system are sound *splicing* and *morphing* [41],[50],[51] using the linear amplitude and cubic phase interpolators of Section 9.4.1. Splicing the plosive of one speech segment with the sustained portion of a vowel from a different speech segment, for example, can test the relative importance of the temporal components in characterizing the sound. This is performed by matching frequencies and then interpolating the amplitudes and phases of the two synthesized sounds at a splice point. In sound morphing, in contrast to splicing temporal segments of sounds, entire frequency tracks are blended together to transform one sound into another; the functional form for amplitude, frequency, and phase gives a natural means for moving from one sound or voice into another. A new frequency track is created as the interpolation of tracks $\omega_1[n]$ and $\omega_2[n]$ from the two sounds, represented by $\omega[n] = (\frac{N-n}{N})\omega_1[n] + (\frac{n}{N})\omega_2[n]$, over a time interval $[0, N]$. A similar operation is performed on the corresponding amplitude functions [53]. Such a time-varying blend of different signals can also be performed in the framework of the phase vocoder [36].

Another interesting application of sinewave analysis is the analysis of *vibrato* in the singing voice. Vibrato is the quasi-sinusoidal fluctuation of pitch and is crucial for the richness and naturalness of the sound [26],[41]. With vibrato, the resonant character of the vocal tract remains approximately fixed while the excitation from the vocal folds changes frequency in a quasi-sinusoidal manner. The output spectrum of the singing voice shows frequency modulation due to the activity of the vocal folds and amplitude modulation, i.e., *tremolo*, due to the source harmonics being swept back and forth through the vocal tract resonances[13] [26]. For quasi-periodic waveforms with time-varying pitch, each harmonic frequency varies synchronously. Because each harmonic is a multiple of the time-varying fundamental, with vibrato, higher harmonics have a larger bandwidth than lower harmonics. With rapid pitch vibrato, *the temporal resolution required for frequency estimation increases with harmonic number*. The sinusoidal analysis is useful (although, as we will see shortly, limited by a fixed analysis window) in tracking such harmonic frequency and amplitude modulation and can have an advantage over the phase vocoder that requires the modulated frequency to reside in a single channel [26]. The presence of vibrato in the analyzed tone may cause unwanted "cross-talk" between the bandpass filters of the phase vocoder; in other words, a sinewave may appear in the passband of two or more analysis filters during one vibrato cycle. Sinewave analysis, on the other hand, was found by Maher and Beauchamp [26] to provide an improvement over fixed filter-bank methods for the analysis of vibrato since it is possible to track changing frequencies and thereby avoid the cross-talk problem.

9.4.5 Time-Frequency Resolution

For some signal processing applications, it is important that the sinewave analysis parameters represent the actual signal components. Although a wide variety of sounds have been successfully analyzed and synthesized based on the sinusoidal representation, constraints on the analysis window and assumptions of signal stationarity do not allow accurate estimation of the underlying components for some signal classes. For example, with sinewave analysis of signals with closely spaced frequencies such as a low-pitched male voice, it is difficult to achieve adequate temporal resolution with a window selected for adequate frequency resolution. On the other hand, for signals with rapid modulation or signals with short duration and/or a sharp attack, such as with rapid pitch or formant transitions and plosives, it is difficult to attain adequate frequency resolution with a window selected for adequate temporal resolution. A long window can result in temporal smearing of transient or rapidly varying signal components and may be perceived as a mild dulling or distortion of the sound. A short window, on the other hand, may prevent accurate representation of low frequencies and closely spaced sinewaves. These problems reflect the tradeoff of time-frequency resolution that arises in the uncertainty principle (Chapter 2).

With the presence of pitch modulation, for example, to obtain equivalent time resolution in following each frequency trajectory by spectral peak locations, we would need to decrease the window duration with increasing frequency. By "equivalent" time resolution, we mean (loosely) the same frequency change under each analysis window. (Also keep in mind that the time span of one sinewave cycle decreases with increasing frequency.) With plosives, likewise, we would need to decrease the window length, to give better time resolution of the primarily

---

[13] McAdams [29] hypothesizes that the tracing of resonant amplitude by frequency modulation contributes to the distinctness of the sound in the presence of competing sound sources.

high-frequency energy. Sinewave analysis may, therefore, benefit from multi-resolution windowing. Such a multi-resolution sinewave analysis can be provided by the wavelet transform that we described in Chapter 8. The basic approach, introduced in [8],[9], is to first decompose the signal by a wavelet transform. Each wavelet filter output is then represented by a sum of sinewaves. A number of formal ways of representing the subband outputs in terms of sinewaves were developed by Goodwin [13], who modeled each output using progressively shorter windows with increasing frequency, as illustrated in Figure 9.18. Specifically, each window duration decreases by a factor of two for each frequency octave increase. Goodwin also introduced a more general "atomic decomposition" in which subband bandwidths and analysis window durations adapt to the signal's time-frequency characteristics [13]. Approaches in this style have been developed for a number of specific applications. Anderson [2], Goodwin [13], and Levine, Verma, and Smith [22] have developed multi-resolution approaches to sinewave analysis, particularly useful in preserving signal transients and transitions in signal modification and wideband coding applications. Hamdy, Ali, and Tewfik [15] have shown that the wavelet transform can be combined with a "residual" sinewave model that we describe later in this chapter, also for the wideband coding application.

Similar approaches exploit auditory perception principles. Although a short window for temporal tracking reduces frequency resolution, the human ear may not require as high a frequency resolution in perceiving high frequencies as for low frequencies. Ellis used this property of auditory perception in developing a constant-Q wavelet analysis within the sinewave

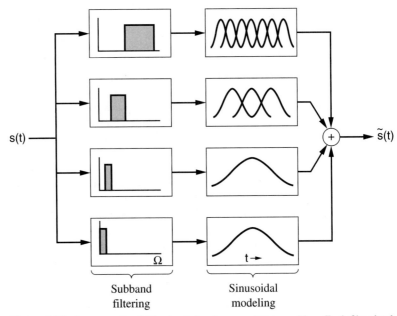

Figure 9.18 An octave-band filter bank for sinewave decomposition. Each filter-bank output is sinewave-analyzed with a window whose duration decreases by a factor of two for each frequency octave increase [13].

SOURCE: M.M. Goodwin, *Adaptive Signal Models: Theory, Algorithms, and Audio Applications* [13] (Figure 13.9). ©1992, Kluwer Academic Publishers. Used by permission.

framework for improving the perception of fine structure at signal onsets [8],[9]. Ghitza [12] and Anderson [2] have exploited auditory spectral masking (to be described in Chapters 12 and 13) with constant-Q filters to reduce the number of sinewaves required in sinewave synthesis for speech coding.

A number of approaches have also been proposed for modeling the nonstationarity of sinewave parameters over an analysis window.[14] One technique relies on a time-varying frequency model with a linear evolution of frequency. For a Gaussian analysis window, Marques and Almeida [28] have shown that the windowed signal can be written in complex form as

$$s(t) = \sum_{k=1}^{K} A_k o_k(t - t_o)$$

with

$$o_k(t) = e^{\mu_k t^2} e^{\lambda_k t} e^{j(\Delta_k t^2 + \Omega_k t)} \tag{9.25}$$

where the center frequency for each component is Ω_k, the frequency slope is $2\Delta_k$, and the Gaussian envelope is characterized by the parameters μ_k and λ_k. The Fourier transform of $s(t)$ in Equation (9.25) is given by

$$S(\Omega) = \sum_{k=1}^{K} A_k O_k(\Omega)$$

where, for a Gaussian window, $O_k(\Omega)$ can be evaluated analytically and also takes on a Gaussian form. This convenient form allows for estimation of the unknown parameters by iterative least-squared-error minimization using a log spectral error. To relieve the multiple sinewave estimation problem, estimation is performed one sinusoid at a time, successively subtracting each estimate from the signal's spectrum.[15] Improvement in an average reconstruction error over multiple frames was observed in speech signals whose pitch varies rapidly.

This problem of representing frequency variation is particularly severe for high-frequency sinewaves. We have seen that, because for periodic waveforms each harmonic frequency is an integer multiple of the fundamental frequency, higher frequencies experience greater variation than low frequencies. We mentioned earlier that with rapid pitch vibrato in the singing voice, for example, the temporal resolution required for frequency estimation increases with harmonic number. This high-frequency variation can be so great that the short-time *signal spectrum can appear noise-like in high-frequency regions*, thus reducing the efficacy of spectral peak peaking. (Look ahead to Figure 11.2.) This observation motivated an alternative approach by Ramalho to addressing sinewave frequency variations [46]. In this approach, the waveform is temporally warped according to an evolving pitch estimate, resulting in a nearly monotone-pitch synthesis. A fixed analysis window is then selected for a desired frequency resolution; dewarping the frequency estimate yields an estimate of the desired sinewave frequency trajectory.

---

[14] Observe that there is an inconsistency in assumptions made in baseline analysis and synthesis. In analysis, we assume the sinewave amplitudes and frequencies are fixed (stationary), while in synthesis we assume they are time-varying (nonstationary). However, we have faced this inconsistency in all parametric analysis/synthesis systems of the text thus far (see Exercise 5.13 for an alternative model).

[15] An iterative approach was also developed by George [11] for improving the estimation of low-level sinewaves in wideband spectra of large dynamic range. A least-squared-error minimization of sinewave parameters was formulated as an analysis-by-synthesis procedure in which sinewave estimates are successively subtracting from the signal; at each iteration, the component with the largest magnitude is subtracted and the estimation is carried out on the remainder.

9.5 Source/Filter Phase Model

The baseline sinewave model and analysis/synthesis system are applicable to arbitrary signals. For signals represented approximately by the output of a linear system driven by periodic, impulsive, or noise excitations, however, we saw in Section 9.2 that the baseline sinewave model can be refined by imposing a source/filter representation on the sinewave components. Within this framework, we now introduce the concept of sinewave *phase coherence*, which becomes the basis for a number of applications, including time-scale modification and peak-to-rms reduction.

9.5.1 Signal Model

We begin by reviewing the sinewave representation derived earlier for a source/filter speech model. We saw in Section 9.2 that we can write the speech excitation model as a sum of sinewaves of the form in Equation (9.1), i.e.,

$$u(t) = \sum_{k=1}^{K(t)} a_k(t) \exp[j\phi_k(t)] \tag{9.26}$$

where $K(t)$ represents the number of sinewaves at time t and $a_k(t)$ is the time-varying amplitude associated with each sinewave. For the kth sinewave, the excitation phase $\phi_k(t)$ is the integral of the time-varying frequency $\Omega_k(t)$ or

$$\phi_k(t) = \int_0^t \Omega_k(\sigma)d\sigma + \phi_k$$

where ϕ_k is a fixed phase offset which accounts for the fact that the sinewaves are generally not in phase. Since the system impulse response is also time-varying, the system transfer function is written in terms of its time-varying amplitude and phase as

$$H(t, \Omega) = M(t, \Omega) \exp[j\Phi(t, \Omega)]. \tag{9.27}$$

To simplify notation, the system amplitude and phase along each frequency trajectory $\Omega_k(t)$ are written as

$$M_k(t) = M[t, \Omega_k(t)]$$
$$\Phi_k(t) = \Phi[t, \Omega_k(t)]. \tag{9.28}$$

Passing the excitation described in Equation (9.26) through the linear time-varying system of Equation (9.27) results in the sinusoidal representation for the waveform:

$$s(t) = \sum_{k=1}^{K(t)} A_k(t) \cos[\theta_k(t)],$$

where

$$A_k(t) = a_k(t)M_k(t)$$

and

$$\theta_k(t) = \phi_k(t) + \Phi_k(t) \tag{9.29}$$

represent the amplitude and phase of each sinewave component along the frequency trajectory $\Omega_k(t)$. We saw that the accuracy of this representation is subject to the caveat that the parameters are slowly varying relative to the duration of the system impulse response.

In developing a source/filter phase model, we initially assume voiced speech and that the glottal flow contribution is embedded within the vocal tract impulse response. For a periodic voiced segment, the excitation function therefore reduces to a periodic impulse train. The excitation phase representation in Equation (9.26) can then be simplified by introducing a parameter representing the time at which an impulse occurs; we refer to this as an *onset time* of the periodic excitation. In the context of the sinewave model, an onset time corresponds to the time at which sinewaves are "in phase," by which we mean that they all reach their peak amplitude value, each sinewave having a multiple of 2π phase value. The excitation waveform is modeled over the analysis window duration as

$$u(t) = \sum_{k=1}^{K(t)} a_k(t) \exp[j(t - t_o)\Omega_k], \quad t \in [0, T], \tag{9.30}$$

where t_o is an onset time of the source and where the excitation frequency Ω_k is assumed constant over the duration of the analysis window. Comparison of Equation (9.26) with Equation (9.30) shows that here the excitation phase $\phi_k(t)$ is linear with respect to frequency. With this representation of the excitation, the excitation phase can be written in terms of the onset time t_o as

$$\phi_k(t) = (t - t_o)\Omega_k.$$

According to Equation (9.29), the system phase for each sinewave frequency is given by the phase function obtained when the linear excitation phase $(t - t_o)\Omega_k$ is subtracted from the composite phase $\theta_k(t)$, which consists of both excitation and system components, i.e.,

$$\Phi_k(t) = \theta_k(t) - (t - t_o)\Omega_k.$$

Similarly, from Equation (9.29), the system amplitude is obtained by dividing excitation amplitude, $a_k(t)$ into the composite amplitude $A_k(t)$. Alternatively, excitation components are obtained by removing system components from composite values.

9.5.2 Applications

We saw in Chapter 8 an auditory model that offers an explanation for perceptual phase sensitivity. Motivated by this model and empirical observations of phase sensitivity, in this section, we describe a method for preserving phase coherence (i.e., the original sinewave phase relations) or a desired phase relation (other than the original) in the context of sinewave analysis/synthesis for a number of applications.

Time-Scale Modification — The simplified linear model of the generation of speech predicts that a time-scaled modified waveform takes on the appearance of the original except for a change in time scale, e.g., we simply increase or decrease the number of pitch periods during voicing, as we saw in the time-scale modification model of Chapter 7. This section develops a time-scale modification system that preserves this *shape invariance* property for quasi-periodic signals by preserving the phase coherence across sinewaves in modification [39]. A similar approach can be applied to pitch modification [39] (Exercise 9.11).

Source/Filter Model: For a uniform change in the time scale, the time t_o corresponding to the original articulation rate is mapped to the transformed time t_o' through the mapping

$$t_o' = \rho t_o. \tag{9.31}$$

The case $\rho > 1$ corresponds to slowing down the rate of articulation by means of a time-scale expansion, while the case $\rho < 1$ corresponds to speeding up the rate of articulation by means of a time-scale compression. Events which take place at a time t_o' according to the new time scale will have occurred at $\rho^{-1} t_o'$ in the original time scale.

In an idealized sinewave model for time-scale modification, the "events" which are modified are the amplitudes and phases of the system and excitation components of each underlying sinewave. The rate of change of these events is a function of how fast the system moves and how fast the excitation characteristics change. In this simplified model, a change in the rate at which the system moves corresponds to a time scaling of the amplitude $M(t, \Omega)$ and the phase $\Phi(t, \Omega)$. The excitation parameters must be modified so that frequency trajectories are stretched and compressed while maintaining pitch. While the excitation amplitudes $a_k(t)$ can be time-scaled, a simple time scaling of the excitation phase $\phi_k(t)$ will alter pitch. Alternatively, the transformation given by $\rho\phi(t\rho^{-1})$ maintains the pitch but results in waveform dispersion. As in the baseline sinewave modification system of Equation (9.24), the phase relation among sinewaves is continuously being altered. A different approach to modeling the modification of the excitation phase function relies on the representation of the excitation in terms of its impulse locations, i.e., *onset times*, which were introduced in the previous section. In time-scale modification, the excitation onset times extend over longer or shorter time durations relative to the original time scale. This representation of the time-scaled modified excitation function is a primary difference from time-scale modification using the baseline system of Equation (9.24). The model for time-scale modification is illustrated in Figure 9.19.

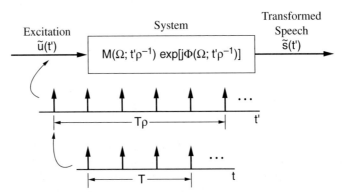

Figure 9.19 Onset-time model for time-scale modification where $t' = t\rho$ with ρ the rate-change factor. For a periodic speech waveform with pitch period P, the excitation is given by a train of periodically spaced impulses, i.e., $u(t) = \sum_{r=-\infty}^{\infty} \delta(t - t_o - rP)$, where t_o is a displacement of the pitch impulse train from the origin. This can also be written as a sum of complex sinewaves of the form $u(t) = \sum_{k=1}^{K(t)} a_k(t) \exp[j(t - t_o)\Omega_k]$.

SOURCE: T.F. Quatieri and R.J. McAulay, "Shape-Invariant Time-Scale and Pitch Modification of Speech" [39]. ©1992, IEEE. Used by permission.

Equations (9.26)-(9.31) form the basis for a mathematical model for time-scale modification [38],[39],[40]. To develop the model for the modified excitation function, suppose that the time-varying pitch period $P(t)$ is time-scaled according to the parameter ρ. Then the time-scaled pitch period is given by

$$\tilde{P}(t') = P(t'\rho^{-1})$$

from which a set of new onset times can be determined. The model of the modified excitation function is then given by

$$\tilde{u}(t') = \sum_{k=1}^{K(t)} \tilde{a}_k(t') \exp[j\tilde{\phi}_k(t')], \tag{9.32}$$

where

$$\tilde{\phi}_k(t') = (t'\rho^{-1} - t'_o)\Omega_k \tag{9.33}$$

and where t'_o is the modified onset time. The excitation amplitude in the new time scale is the time-scaled version of the original excitation amplitude function $a_k(t)$ and is given by

$$\tilde{a}_k(t') = a_k(t'\rho^{-1}).$$

The system function in the new time scale is a time-scaled version of the original system function so that the magnitude and phase are given by

$$\tilde{M}_k(t') = M_k(t'\rho^{-1})$$
$$\tilde{\Phi}_k(t') = \Phi_k(t'\rho^{-1}),$$

where $M_k(t)$ and $\Phi_k(t)$ are given in Equation (9.28). The model of the time-scaled waveform is completed as

$$\tilde{s}(t') = \sum_{k=1}^{K(t)} \tilde{A}_k(t') \exp[j\tilde{\theta}_k(t')]$$

where

$$\tilde{A}_k(t') = \tilde{a}_k(t')\tilde{M}_k(t')$$
$$\tilde{\theta}_k(t') = \tilde{\phi}_k(t') + \tilde{\Phi}_k(t') \tag{9.34}$$

which represent the amplitude and phase of each sinewave component.

The above time-scale modification model was developed for a periodic source and is applicable to quasi-periodic waveforms with slowly varying pitch and for which sinewave frequencies are approximately harmonic. There also exist, of course, many sounds that are aharmonic. Consider plosive consonants generated with an impulsive source. In these cases, a change in the articulation rate of our model is less desired and so the model should include a rate change that adapts to speech events [39]. This is because, as we saw in Chapter 3 (Section 3.5), a natural change in articulation rate incurs less modification for this sound class; compression or expansion of a plosive is likely to alter its basic character. A smaller degree of change was also observed in Chapter 3 to take place naturally with fricative consonants.

A change in the articulation rate of noisy aharmonic sounds may be useful, however, in accentuating the sound. In this case, the spectral and phase characteristics of the original waveform, and, therefore, the noise character of the sound are roughly preserved in an analysis/synthesis system based on the rate-change model in Equation (9.34), as long as the synthesis interval is 10 ms or less to guarantee sufficient decorrelation of sinewaves in time from frame to frame, and as long as the analysis window is 20 ms or more to guarantee approximate decorrelation in frequency of adjacent sinewaves [30],[39]. For time-scale expansion, this noise characteristic, however, is only approximate. Some slight "tonality" is sometimes perceived due to the temporal stretching of the sinewave amplitude and phase functions. In this case, and when the 10-ms synthesis frame condition is not feasible (computationally) due to a very expanded time scale (e.g., ρ greater than 2 with an analysis frame no less than 5 ms), then frequency and phase dithering models can be used to satisfy the decorrelation requirements [23],[39]. One extended model adds a random phase to the system phase in Equation (9.34) in only those spectral regions considered aharmonic.[16] For the kth frequency, the phase model is expressed as

$$\tilde{\theta}_k(t') = \tilde{\phi}_k(t') + \tilde{\Phi}_k(t') + b_k(\Omega_c)\varphi_k(t'), \tag{9.35}$$

where $b_k(\Omega_c)$ is a binary weighting function which takes on a value of unity for a frequency declared "aharmonic" and a value of zero for a "harmonic" frequency:

$$b_k(\Omega_c) = 1, \quad \text{if } \Omega_k > \Omega_c$$
$$= 0, \quad \text{if } \Omega_k \le \Omega_c, \tag{9.36}$$

where Ω_k are sinewave frequencies estimated on each frame and $\varphi_k(t')$ is a phase trajectory derived from interpolating random phase values over each frame, and differently for each sinewave; the phase values are selected from a uniformly distributed random variable on $[-\pi, \pi]$. The cutoff frequency Ω_c is the harmonic/aharmonic cutoff for each frame and varies with a "degree of harmonicity" measure (or "probability of voicing") P_v:

$$\Omega_c = P_v B \tag{9.37}$$

over a bandwidth B and where the harmonicity measure P_v, falling in the interval $[0, 1]$ (the value 1 meaning most harmonic), must be obtained. One approach to obtain P_v uses a sinewave-based pitch estimator [32] that will be described in Chapter 10. Figure 9.20 illustrates an example of frequency track designations in a speech voiced/unvoiced transition. This phase-dithering model can provide not only a basis for certain modification schemes, but also a basis for speech coding that we study in Chapter 12.

Analysis/Synthesis: With estimates of excitation and system sinewave amplitudes and phases at the center of the new time-scaled synthesis frame, the synthesis procedure becomes identical to that of the baseline system of Section 9.4. The goal then is to obtain estimates of the amplitudes $\tilde{A}_k(t)$ and phases $\tilde{\theta}_k(t)$ in Equation (9.34) on the synthesis frame of length $L' = \rho L$ where L is the analysis frame length. In the time-scale modification model, because the vocal tract system and its excitation amplitudes are simply time-scaled, we see from Equation (9.34) that the composite amplitude need not be separated and therefore the required time-scaled amplitude

---

[16] In adding synthetic harmonic and aharmonic components, it is important that the two components "fuse" perceptually [18], i.e., that the two components are perceived as emanating from the same sound source. Informal listening tests suggest that sinewave phase randomization yields a noise component that fuses with the harmonic component of the signal.

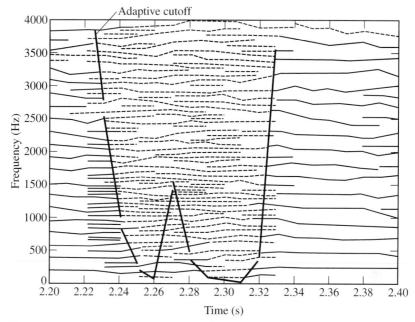

Figure 9.20   Transitional properties of frequency tracks with adaptive cutoff. Solid lines denote voiced frequencies, while dashed lines denote unvoiced frequencies.

Source: T.F. Quatieri and R.J. McAulay, "Peak-to-rms Reduction of Speech Based on a Sinusoidal Model" [42]. ©1991, IEEE. Used by permission.

can be obtained from the sinewave amplitudes measured on each frame l. The amplitudes for each frame in the new time scale are thus given by (re-introducing frame index notation that we earlier implicitly removed)

$$\tilde{A}_k^l = a_k^l M_k^l.$$

In discrete time, the time-scaled sinewave amplitudes are then obtained by linearly interpolating the values $\tilde{A}_k^l$ and $\tilde{A}_k^{l+1}$ over the synthesis frame duration L', identical to the scheme in Equation (9.17). The system and excitation phases, however, must be separated from the measured phases because the components of the composite phase $\tilde{\theta}_k(t)$ in Equation (9.34) are manipulated in different ways.

We first estimate the required system phase in the original time scale, i.e., relative to the original analysis frame, by subtracting the measured excitation from the measured composite phase, as according to Equation (9.29). The initial step in estimating the excitation phase is to obtain the onset time with respect to the center of the lth frame, denoted in discrete time by $n_o(l)$. Determining this *absolute* onset time[17] is not an easy task and a number of methods

---

[17] Recall that for voiced speech we have lumped the glottal flow function with the vocal tract impulse response. The resulting sequence without a linear phase term is concentrated at the time origin. One can think of the absolute onset time as the time displacement of this sequence relative to the center of a frame. Thus, in estimating the absolute onset time, we are in effect estimating the linear phase term of the system function. Other definitions of absolute onset time are with respect to glottal flow characteristics such as the time of glottal opening or the sharp, negative-going glottal pulse.

have been developed [33],[52]. One method of estimating the absolute onset times is based on a least-squared-error approach to finding the unknown $n_o(l)$ [33] and will be described in Chapter 10. Although this method can yield onset times to within a few samples, this slight inaccuracy is enough to degrade quality of the synthesis. Specifically, the measured absolute onset time can introduce pitch jitter, rendering the synthetic modified speech hoarse. (Recall this association in Chapter 3.)

An alternative perspective of onset time, however, leads to a way to avoid this hoarseness. Because the function of the onset time is to bring the sinewaves into phase at times corresponding to the sequence of excitation impulses, it is possible to achieve the same effect simply by keeping record of successive onset times generated by a succession of pitch periods. If $n_o(l - 1)$ is the onset time for frame $l - 1$ and if P^l is the pitch period estimated for frame l, then a succession of onset times can be specified by

$$q_0(l; j) = n_o(l - 1) + jP^l, \qquad j = 1, 2, 3 \ldots$$

If $q_0(l; J)$ is the onset time closest to the center of the frame l, then the onset time for frame l, is defined by

$$n_o(l) = q_0(l; J) = n_o(l - 1) + JP^l. \tag{9.38}$$

We call this time sequence the *relative* onset times. An example of a typical sequence of onset times is shown in Figure 9.21a. It is implied in the figure that, in general, there can be more

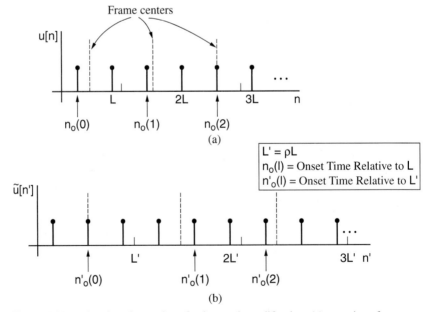

Figure 9.21 Estimation of onset times for time-scale modification: (a) onset times for system phase; (b) onset times for excitation phase.

Source: T.F. Quatieri and R.J. McAulay, "Shape-Invariant Time-Scale and Pitch Modification of Speech" [39]. ©1992, IEEE. Used by permission.

than one onset time per analysis frame. Although any one of the onset times can be used, in the face of computational errors due to discrete Fourier transform (DFT) quantization effects, it is best to choose the onset time which is nearest the center of the frame, since then the resulting phase errors will be minimized. This procedure determines a relative onset time, which is in contrast to finding the absolute onset time when excitation impulses actually occur. Because the relative onset time is obtained from the pitch period, pitch estimation is required. Given the importance of avoiding pitch jitter, fractional pitch periods are desired. One approach, consistent with a sinewave representation and described in Chapter 10, is based on fitting a set of harmonic sinewaves to the speech waveform that allows a fractional pitch estimate and thus yields accurate relative onset times [32].

Having an onset time estimate, from Section 9.5.1, the excitation phase for the lth frame is then given in discrete time by

$$\phi_k[n] = -[n - n_o(l)]\omega_k^l$$

where we assume a constant frequency ω_k^l and where $n_o(l)$ takes on the meaning of an onset time closest to the frame center time. To avoid growing frame and onset times, we shift each frame to time $n = 0$. This can be done by subtracting and adding lL (disregarding a half-frame length) in the above expression for $\phi_k[n]$ to obtain

$$\begin{aligned}
\phi_k[n] &= -[n - n_o(l)]\omega_k^l \\
&= -[(n - lL) - (n_o(l) - lL)]\omega_k^l \\
&= -[(n - lL) - \hat{n}_o(l)]\omega_k^l
\end{aligned}$$

where $\hat{n}_o(l) = n_o(l) - lL$ is the onset time relative to the lth frame center shifted to time $n = 0$. The excitation phase at the frame center is therefore given by

$$\phi_k^l = -[0 - \hat{n}_o(l)]\omega_k^l.$$

Finally, an estimate of the system phase at the measured frequencies is computed by subtracting the estimate of the excitation phase from the measured phase at the sinewave frequencies, i.e.,

$$\tilde{\Phi}_k^l = \theta_k^l - \phi_k^l.$$

When the excitation phase, derived from the relative onset time, is subtracted, some residual linear phase will be present in the system phase estimate because the relative linear phase is not equal to the absolute linear phase. This linear phase residual, however, is consistent at harmonics over successive frames and therefore does not pose a problem to the reconstruction since the ear is not sensitive to a linear phase shift. To see this property, consider a periodic voiced waveform with period P and suppose that the absolute onset time on the lth frame is m_l. And suppose the relative phase onset is given by $J_l P$ (relative to time $n = 0$) where P is the pitch period and J_l an integer referring to the lth frame. (For convenience, we have slightly changed our notation from above.) The linear phase difference at harmonics, denoted by Δ_k^l, is given by

$$\begin{aligned}
\Delta_k^l &= m_l \frac{2\pi k}{P} - J_l P \frac{2\pi k}{P} \\
&= m_l \frac{2\pi k}{P} - J_l 2\pi k.
\end{aligned}$$

And for the following frame, the linear phase residual is given by

$$\Delta_k^{l+1} = \left(m_l \frac{2\pi k}{P} + MP \frac{2\pi k}{P} \right) - J_{l+1} P \frac{2\pi k}{P}$$

$$= m_l \frac{2\pi k}{P} - (M - J_{l+1}) 2\pi k$$

where M is the number of pitch periods traversed in going from frame l to frame $l+1$. We see, therefore, that the phase residual remains constant (modulo 2π) over the periodic waveform. Furthermore, with small changes in pitch and with aharmonicity, the change in linear phase is much reduced because $|M - J_{l+1}| \ll J_{l+1}$.

The remaining step is to compute the excitation phase relative to the new synthesis interval of L' samples. As illustrated in Figure 9.21b, the pitch periods are accumulated until a pulse closest to the center of the lth synthesis frame is achieved. The location of this pulse is the onset time with respect to the new synthesis frame and can be written as

$$n_o'(l) = n_o'(l-1) + J' P^l,$$

where J' corresponds to the first pulse closest to the center of the synthesis frame of duration L'. The phase of the modified excitation $\tilde{u}(n')$, at the center of the lth synthesis frame, is then given by

$$\tilde{\phi}_k^l = -[lL' - n_o'(l)]\omega_k^l.$$

Finally, in the synthesizer the sinewave amplitudes over two consecutive frames, $\tilde{A}_k^{l-1}$ and $\tilde{A}_k^l$, are linearly interpolated over the frame interval L', as described above. The excitation and system phase components are summed and the resulting sinewave phases, $\tilde{\theta}_k^{l-1}$ and $\tilde{\theta}_k^l$, are interpolated across the duration L' using the cubic polynomial interpolator. A block diagram of the complete analysis/synthesis system is given in Figure 9.22. Finally, we note that similar techniques can be applied to frequency compression and pitch modification, as well as to such operations jointly (Exercises 9.10 and 9.11).

An important feature of the sinewave-based modification system is its straightforward extension to time-varying rate change, details of which are beyond the scope of this chapter [39]. Consider, for example, application to listening to books from audio recordings. Here we want time-varying control of the articulation rate, i.e., a "knob" that slows down the articulation rate in important or difficult-to-understand passages and speeds up speech in unimportant or uninteresting regions. In addition, we mentioned earlier that unvoiced sounds naturally are modified less than voiced sounds. As a consequence, the corresponding analysis/synthesis system can be made to adapt to the events in the waveform, i.e., the degree of voicing, which better emulates speech modification mechanisms, as discussed in the previous section. We saw that one way to achieve a measure of voicing, and thus the desired adaptivity, is through a measure of "harmonicity" P_v [39]. This event-adaptive rate change can be superimposed on a contextual-dependent rate change as given, for example, in the above application. The reconstructions are generally of high quality, maintaining the naturalness of the original, and are free of artifacts. Interfering backgrounds are also reproduced at faster and slower speeds. Although the phase model is pitch-driven, this robustness property is probably due to using the original sinewave amplitudes, frequencies, and phases in the synthesis rather than forcing a harmonic structure onto the waveform.

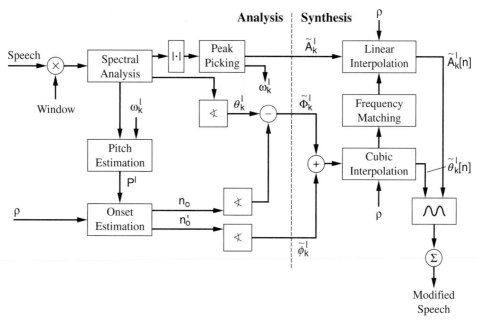

Figure 9.22   Sinusoidal analysis/synthesis for time-scale modification.

SOURCE: T.F. Quatieri and R.J. McAulay, "Shape-Invariant Time-Scale and Pitch Modification of Speech" [39]. ©1992, IEEE. Used by permission.

EXAMPLE 9.11   An example of time-scale expansion of a speech waveform is shown in Figure 9.23, where the time scale has been expanded by a "target" (i.e., desired maximum) factor of 1.7 and has been made to adapt according to the harmonicity measure P_v referred to in Equation (9.37); specifically, the rate change is given by $\rho = [1 - P_v] + 1.7P_v$. The analysis frame interval is 5 ms so that the synthesis frame interval is no greater than 8.5 ms. Observe that the waveform shape is preserved during voicing, as a result of maintaining phase coherence, and that the unvoiced segment is effectively untouched because $\rho \approx 1$ in this region, although some temporal smearing occurs at the unvoiced/voiced transition. Spectrogram displays of time-scaled signals show preservation of formant structure and pitch variations [39].   ▲

EXAMPLE 9.12   An example of time-scale modification of speech using sinewave analysis/synthesis is shown in Figure 9.24, where the rate change is controlled to oscillate between a compression with $\rho = 0.5$ and an expansion with $\rho = 1.5$. The figure shows that details of the temporal structure of the original waveform are maintained in the reconstruction despite the time-varying rate change; waveform dispersion, characteristic of the time-scale modification with the baseline sinewave analysis/synthesis, does not occur.   ▲

Peak-to-rms Reduction — In reducing the ratio of the peak value of a signal to its average power, the concern of sinewave phase coherence again arises. Here, however, the goal is not to preserve the original sinewave phase relations, but rather to intentionally modify them to yield a response with minimum "peakiness," relying on a phase design technique derived in a radar signal context [40],[42],[43].

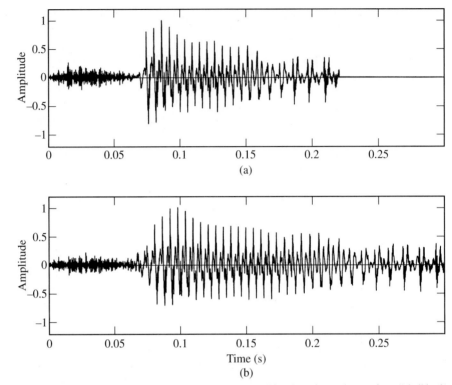

Figure 9.23 Example of sinewave-based time-scale modification of speech waveform "cha" in the word "<u>cha</u>nge": (a) original; (b) adaptive expansion with target $\rho = 1.7$.

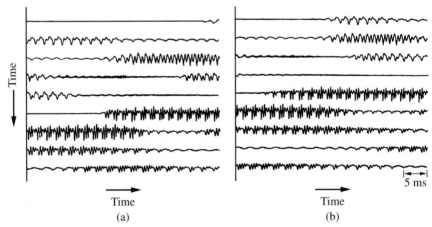

Figure 9.24 Example of time-varying time-scale modification of speech waveform using sinusoidal analysis/synthesis: (a) original; (b) time-scaled with rate change factor ρ changing from 0.5 (compression) to 1.5 (expansion).

SOURCE: T.F. Quatieri and R.J. McAulay, "Shape-Invariant Time-Scale and Pitch Modification of Speech" [39]. ©1992, IEEE. Used by permission.

Key-Fowle-Haggarty Phase Design: In the radar signal design problem, the signal is given as the output of a transmit filter whose input consists of impulses. The spectral magnitude of the filter's transfer function is specified and its phase is typically chosen so that the response over its duration is flat. This design allows the waveform to have maximum average power given a peak-power limit on the radar transmitter. The basic unit of the radar waveform is the impulse response $h[n]$ of the transmit filter. Following the development in [41],[42], it is expedient to view this response in the discrete time domain as an FM chirp signal with envelope $a[n]$, phase $\phi[n]$, and length N:

$$h[n] = a[n]\cos(\phi[n]), \quad 0 \le n < N \tag{9.39}$$

that has a Fourier transform $H(\omega)$ with magnitude $M(\omega)$ and phase $\Psi(\omega)$, i.e.,

$$H(\omega) = M(\omega)\exp[j\Psi(\omega)]. \tag{9.40}$$

By exploiting the analytic signal representation of $h[n]$, Key, Fowle, and Haggarty [20] have shown that, under a constraint of a large product of signal time duration and bandwidth, i.e., a large "time-bandwidth product," specifying the two amplitude components, $a[n]$ and $M(\omega)$, in Equation pair (9.39) and (9.40) is sufficient to approximately determine the remaining two phase components. How large the time-bandwidth product must be for these relations to hold accurately depends on the shape of the functions $a[n]$ and $M(\omega)$ [6],[10].

Ideally, for the minimum ratio of the signal peak to the square root of the signal average energy, referred to as the *peak-to-rms ratio* in the radar signal, the time envelope $a[n]$ should be flat over the duration N of the impulse response. With this and the additional constraint that the spectral magnitude is specified (a flat magnitude is usually used in the radar signal design problem), Key, Fowle, and Haggarty's (KFH) general relation among the envelope and phase components of $h[n]$ and its Fourier transform $H(\omega)$ reduces to an expression for the unknown phase $\Psi(\omega)$:

$$\Psi(\omega) = N\int_0^\omega \int_0^\beta \hat{M}^2(\alpha)d\alpha d\beta \tag{9.41}$$

where the "hat" indicates that the magnitude has been normalized by its energy, i.e.,

$$\hat{M}^2(\omega) = \frac{M^2(\omega)}{\int_0^\pi M^2(\alpha)d\alpha}.$$

The accuracy of the approximation in Equation (9.41) increases with increasing time-bandwidth product [42].

Equation (9.41) shows that the resulting phase $\Psi(\omega)$ depends only on the normalized spectral magnitude $\hat{M}(\omega)$ and the impulse response duration N. It can be shown that the envelope level of the resulting waveform can be determined, with the application of appropriate energy constraints, from the unnormalized spectrum and duration [42]. Specifically, if the envelope of $h[n]$ is constant over its duration N and zero elsewhere, the envelope constant has the value

$$A = \left[\frac{1}{2\pi N}\int_0^\pi M^2(\omega)d\omega\right]^{1/2}, \quad 0 \le n < N. \tag{9.42}$$

The amplitude and phase relation in Equations (9.41) and (9.42) will be used to develop the sinewave-based approach to peak-to-rms reduction.

Waveform Dispersion: In the above radar signal design context, the spectral magnitude was assumed known and a phase characteristic was estimated from the magnitude. Alternately, a filter impulse response with some arbitrary magnitude or phase is given and the objective is to disperse the impulse response to be maximally flat over some desired duration N. This requires first removing the phase of the filter and then replacing it with a phase characteristic from the KFH calculation. This same approach can be used to optimally disperse a voiced speech waveform.[18] The goal is to transform a system impulse response with arbitrary spectral magnitude and phase into an FM chirp response which is flat over the duration of a pitch period. The basis for this optimal dispersion is the sinewave source/filter phase model of Section 9.5.1. The sinewave analysis/synthesis system first separates the excitation and system phase components from the composite phase of the sinewaves that make up the waveform, as was done for time-scale modification. The system phase component is then removed and the new KFH phase replaces the natural phase dispersion to produce the optimally dispersed waveform.

Applying the KFH phase to dispersion requires estimation of the spectral magnitude $M(lL, \omega)$ of the system impulse response and the pitch period of the excitation P^l on each frame l. The duration of the synthetic impulse response is set close to the pitch period P^l so that the resulting waveform is as dense as possible. Estimation of the spectral envelope $M(lL, \omega)$ can be performed with a straight-line spectral smoothing technique that uses a linear interpolation across sinewave amplitudes (Chapter 10) [37]. The synthetic system phase on each frame is derived from $M(lL, \omega)$ using the KFH solution in Equation (9.41) and is denoted by $\Psi_{kfh}(lL, \omega)$ where "kfh" refers to the KFH phase.

Applying the KFH phase dispersion solution in the synthesis requires that the synthetic system phase, $\Psi_{kfh}(lL, \omega)$, a continuous function of frequency, be sampled at the time-varying sinewave frequencies ω_k^l. We write the resulting sampled phase function as

$$\Psi_{k,kfh}^l = \Psi_{kfh}(lL, \omega_k^l) \tag{9.43}$$

where the subscript "k, kfh" and superscript "l" denote the KFH phase sampled at the kth frequency on the lth analysis frame, respectively. The phase solution in Equation (9.43) is used only where an approximate periodicity assumption holds, whereas at aharmonic regions the original system phase is maintained. Therefore, the KFH phase is assigned only to those frequencies designated "harmonic." The assignment can be made using the same approach applied in the phase model of Equation (9.35) where a frequency cutoff ω_c adapts to a measure of the degree of voicing. Use of the system phase during aharmonic (noise or impulsive) regions does not change the preprocessor's effectiveness in reducing the peak-to-rms ratio since these regions contribute negligibly to this measure. Moreover, the preservation of as much of the original waveform as possible helps to preserve the original quality. Thus, the phase assignment for the kth sinewave on the lth frame is given by

$$\theta_k^l = \phi_k^l + b_k(\omega_c)\Phi_k^l + [1 - b_k(\omega_c)]\Psi_{k,kfh}^l, \tag{9.44}$$

---

[18] Schroeder has derived a different method to optimally flatten a harmonic series by appropriate selection of harmonic phases [48],[49]. This solution can be shown to approximately follow from the KFH dispersion formula.

where $b_k(\omega_c)$, given in Equation (9.36), takes on a value of zero for a harmonic frequency (below ω_c) and unity for an aharmonic frequency (above ω_c), and where ϕ_k^l is the excitation phase, Φ_k^l is the original system phase, and $\Psi_{k,kfh}^l$ is the synthetic system phase.

EXAMPLE 9.13   An example of dispersing a synthetic periodic waveform with a fixed pitch and fixed system spectral envelope is illustrated in Figure 9.25. Estimation of the spectral envelope uses a piecewise-linear spectral smoothing technique of Chapter 10 [37] and the pitch estimation is performed with a sinewave-based pitch estimator also described in Chapter 10. For the same peak level as the original waveform, the processed waveform has a larger rms value and so has a lower peak-to-rms ratio. The simulated vocal tract phase is modified significantly, as illustrated in Figures 9.25b and 9.25c. In Figure 9.25d the magnitude of the dispersed waveform is compared with the original magnitude and the agreement is very close, a property that is important to maintaining intelligibility and minimizing perceived distortion in speech.   ▲

With speech waveforms, a smoothed KFH phase solution (Exercise 9.18) that reduces the short-time peakiness within a pitch period can be combined with conventional dynamic range compression techniques that reduce long-time envelope fluctuations [4]. The result is a processed speech waveform with significant peak-to-rms reduction (as great as 8 dB) and good quality [42],[43]. The KFH solution introduces about a 3 dB contribution to this peak-to-rms reduction over that of dynamic range compression alone. The effect is a waveform with much greater average power but the same peak level, and thus a waveform that is louder than the original under the peak constraint. This property is useful in applications such as speech transmission (e.g., AM radio transmitters [43]) and speech enhancement (e.g., devices for the hearing-impaired).

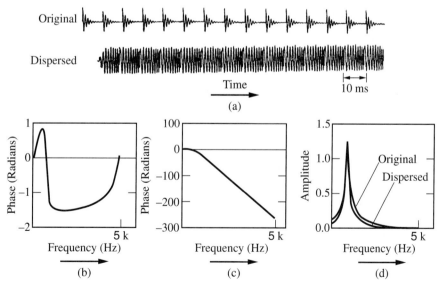

Figure 9.25   KFH dispersion using the sinewave preprocessor: (a) waveforms; (b) original phase; (c) modified phase; (d) spectral magnitudes.

SOURCE: T.F. Quatieri and R.J. McAulay, "Peak-to-rms Reduction of Speech Based on a Sinusoidal Model" [42]. ©1991, IEEE. Used by permission.

9.6 Additive Deterministic-Stochastic Model

We have seen in Chapter 3 that there are many aharmonic contributions to speech, including the numerous noise components (e.g., aspiration at the glottis and frication at an oral tract constriction) of unvoiced and voiced fricatives and plosives. Even during sustained vowels, aharmonicity can be present, both in distinct spectral bands and in harmonic spectral regions, due to aspiration (generated at the glottis) that is a subtle but essential part of the sound; this aharmonic component during sustained portions is different from the aharmonic sounds from unvoiced fricatives and plosives. Aharmonic contributions may also be due to sound transitions or from modulations and sources that arise from nonlinearities in the production mechanism such as vortical airflow. Speech analysis and synthesis are often deficient with regard to the accurate representation of these aharmonic components. In the context of sinewave analysis/synthesis, in particular, the harmonic and aharmonic components are difficult to distinguish and separate. One approach to separating these components, given in the previous section, assumes that they fall in two separate time-varying bands. Although the adaptive harmonicity measure is effective in specifying a split-band cutoff frequency (as well as generalizations to multi-bands as discussed in Chapter 10 [14]), it is, however, overly simple when the harmonic and aharmonic components are simultaneously present over the full band. An alternative representation, developed by Serra and Smith [50],[51], assumes the two components are *additive* over the full speech band. This approach, referred to as the *deterministic-stochastic* sinewave representation, is the focus of this section.

9.6.1 Signal Model

Following the formulation in [41], we express the Serra/Smith deterministic-stochastic model [50],[51] in continuous time as

$$s(t) = d(t) + e(t)$$

where $d(t)$ and $e(t)$ are the deterministic and stochastic components, respectively. The deterministic component $d(t)$ is of the form

$$d(t) = \sum_{k=1}^{K(t)} A_k(t) \cos[\theta_k(t)]$$

where the phase is given by the integral of the instantaneous frequency $\Omega_k(t)$:

$$\theta_k(t) = \int_0^t \Omega_k(\tau)d\tau$$

where the frequency trajectories $\Omega_k(t)$ are not necessarily harmonic and correspond to *sustained* sinusoidal components that are relatively long in duration with slowly varying amplitude and frequency. The deterministic component[19] is, therefore, defined in the same way as in the baseline sinusoidal model except with one caveat: The sinewaves are restricted to be "sustained." In the baseline sinusoidal model, the spectral peaks need not correspond to such long-term frequency trajectories. A mechanism for determining these sustained sinewave components is described in the following section.[20]

---

[19] Referring to sustained sinewave components as "deterministic" pertains more to our intuition rather than our strict mathematical sense.

[20] As an alternative deterministic-stochastic separation scheme, Therrien, Cristi, and Allison [54] have introduced an adaptive pole-zero model for sample-by-sample tracking of sinewave amplitudes and frequencies of the deterministic signal. This component is subtracted from the original signal and parameters of the resulting residual are also adaptively estimated using a pole-zero representation. This representation has been used to synthetically increase limited training data for a signal classification task [54].

The stochastic component, sometimes referred to as a "residual," is defined as the difference between the speech waveform and its deterministic part, i.e., $e(t) = s(t) - d(t)$, and can be thought of as anything not deterministic. It is modeled as the output of a linear time-varying system $h(t, \tau)$ with a white-noise input $u(t)$, i.e.,

$$e(t) = \int_0^t h(t, \tau)u(t - \tau)d\tau. \tag{9.45}$$

This is a different approach to modeling a stochastic component from that taken in the baseline sinewave model where noise is represented as a sum of sinewaves. A limitation with this stochastic representation, as discussed further below, is that not all aharmonic signals are appropriately modeled by this stochastic signal class; for example, a sharp attack at a vowel onset or a plosive may be better represented by a sum of short-duration coherent sinewaves or the output of an impulse-driven linear system, respectively. Another limitation of this representation is that harmonic and aharmonic components may be nonlinearly combined. Nevertheless, this simplification leads to a useful representation for a variety of applications.

9.6.2 Analysis/Synthesis

The analysis/synthesis system for the deterministic-stochastic model is similar to the baseline sinewave analysis/synthesis system. The primary differences lie in the frequency matching stage for extraction of the deterministic component, in the subtraction operation to obtain the residual (stochastic component), and in the synthesis of the stochastic component.

Although the matching algorithm of Section 9.3.4 can be used to determine frequency tracks (i.e., frequency trajectories from beginning to end of sinewaves), this algorithm does not necessarily extract the sustained sinewave components; rather, all sinwaves are extracted. In order to obtain these sustained components, Serra [50] developed a matching algorithm based on prediction of frequencies into the future, as well as extrapolation of past frequencies, over multiple frames. In this algorithm, frequency guides (which is a generalization of the frequency matching window of Section 9.4.3), advance in time through spectral peaks looking for slowly varying frequencies according to constraint rules. When the signal is known to be harmonic, the matcher is assisted by constraining each frequency guide to search for a specific harmonic number. A unique feature of the algorithm is the generalization of the birth and death process by allowing each frequency track to enter a "sleep" state and then reappear as part of a single frequency track. This *peak continuation algorithm* is described in detail in [50]. The algorithm helps to prevent the artificial breaking up of tracks, to eliminate spurious peaks, and to generate sustained sinewave trajectories, which is important in representing the time evolution of true frequency tracks.

With matched frequencies from the peak continuation algorithm, the deterministic component can be constructed using the linear amplitude and cubic phase interpolation of Section 9.4.1. The interpolators use the peak amplitudes from the peak continuation algorithm and the measured phases at the matched frequencies. The residual component can then be obtained by subtraction of the synthesized deterministic signal from the measured signal.

The residual is simplified by assuming it to be stochastic, represented by the output of a time-varying linear system with white-noise input as in Equation 9.45, a model useful in a variety of applications. In order to obtain a functional form for this stochastic process, the power spectrum of the residual must be estimated on each frame from the periodogram of the residual, i.e., the normalized squared STFT magnitude of the residual, $\frac{1}{N_w}|E(n,\omega)|^2$. This power spectrum estimate can be obtained, for example, with linear predictive (all-pole) modeling (Chapter 5) [50],[51]. We denote the power spectral estimate on each frame by $\hat{S}_e(lL, \omega)$. A synthetic version of the process is then obtained by passing a white-noise sequence into a time-varying linear filter with the square root of the residual power spectral estimate, $\hat{S}_e(lL, \omega)^{1/2}$, as the frequency response. A frame-based implementation of this time-varying linear filtering is to filter windowed blocks of white noise and overlap and add the outputs over consecutive frames. The time-varying impulse response of the linear system is given by

$$h[n, lL] = \frac{1}{2\pi} \int_{-\pi}^{\pi} \hat{S}_e(lL, \omega)^{1/2} \exp(jn\omega)d\omega,$$

which is a zero-phase response.[21] The synthetic stochastic signal over the lth frame is then given by

$$\hat{e}[n, lL] = h[n, lL] * (w[n - lL]u[n]),$$

where $u[n]$, a white-noise input, is multiplied by the sliding analysis window with a frame interval of L samples. Because the window $w[n]$ and frame interval L are designed so that $\sum_{l=-\infty}^{\infty} w[n - lL] = 1$, the overlapping sequences $\hat{e}[n, lL]$ can be summed to form the synthesized residual

$$\hat{e}[n] = \sum_{l=-\infty}^{\infty} \hat{e}[n, lL].$$

Although, as we will show in the application to speech modification, the stochastic model can provide an important alternative to a sinewave representation of noise-like sounds, it has its limitations. The residual which results from the deterministic-stochastic model generally contains everything which is not sustained sinewaves. One of the more challenging unsolved problems is the representation of transient events that reside in the residual; examples include plosives and transitions in speech and other sounds that are neither quasi-periodic nor random. Nevertheless, the deterministic-stochastic analysis/synthesis has the ability to separate out subtleties in the sound that the baseline sinewave representation may not reveal. For example, the residual can include low-level aspiration and transitions, as well as the stronger fricatives and plosives in speech [50]. The following example illustrates this property of the residual, as well as the generality of the technique, for a musical sound:

EXAMPLE 9.14   The deterministic-stochastic analysis/synthesis reveals the strong presence of non-tonal components in a piano sound [50]. The residual is an important component of the sound and includes the transient attack and noise produced by the piano action. An example of the decomposition of the attack (with noise) and initial sustained portion of a piano tone is illustrated in Figure 9.26. Synthesizing only the deterministic component, without the residual, results in a lifeless sound with a distorted attack and no noise component.   ▲

---

[21] It is left to the reader to show that a minimum-phase version of the filter can also be constructed through the homomorphic filtering methods of Chapter 6 and that its output has a power spectrum no different from the zero-phase counterpart response.

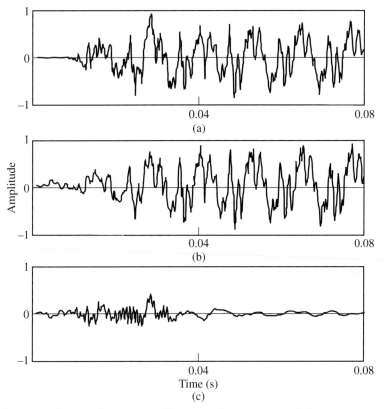

Figure 9.26   Deterministic-stochastic decomposition of a piano sound: (a) beginning segment of a piano tone; (b) deterministic component; (c) residual component, including the attack and noise.

SOURCE: X. Serra, *A System for Sound Analysis/Transformation/Synthesis Based on a Deterministic Plus Stochastic Decomposition* [50]. ©1989, X. Serra. Used by permission.

Generally, treating the residual as the output of a white-noise driven system when it contains transient events, as in the previous example, can alter the sound quality. An approach to improve the quality of a stochastic synthesis is to introduce a second layer of decomposition where transient events are separated from the residual and represented with an approach tailored to transient events. One method performs a wavelet analysis on the residual to estimate and remove transients in the signal [15]; the remainder is processed separately, an approach useful in applications such as speech coding or modification.

9.6.3 Application to Signal Modification

The decomposition approach has been applied successfully to speech and music modification where modification is performed differently on the two deterministic-stochastic components [50],[51]. Consider, for example, time-scale modification. With the deterministic component, the modification is performed as with the baseline system; using Equation (9.24), sustained sinewaves are compressed or stretched. For the aharmonic component, the white-noise input in the stochastic synthesis lingers over longer or shorter time intervals and is matched to the impulse responses (per frame) that vary slower or faster in time.

In one approach to implement synthesis of the modified stochastic component, the window length and frame interval are modified according to the rate change. A new window $w'[n]$ and frame interval L' are selected such that $\sum_{l=-\infty}^{\infty} w'[n - lL'] = 1$, and the factor L'/L equals the desired rate change factor ρ, which is assumed rational. The resulting time-scaled stochastic waveform is given by

$$e'[n] = \sum_{l=-\infty}^{\infty} h[n, lL'] * (w'[n - lL']u'[n]),$$

where $u'[n]$ is the white-noise input generated on the new time scale.

The advantage of separating out the additive stochastic component is that the character of a noise component is not modified; in particular, the noise may be stretched without the "tonality" that occurs in very large stretching of sinewaves. On the other hand, we noted earlier that the quality of transient aharmonic sounds may be altered. In addition, component separation and synthesis may suffer from a lack of "fusion," which is not a problem in sinewave modification because all components are modeled similarly. One approach to improve fusion of the two components is to exploit the property that for many sounds the stochastic component is in "synchrony" with the deterministic component. In speech, for example, the amplitude of the noise component is known to be modulated by the glottal air flow, as we saw in Example 3.4 of Chapter 3. Improved fusion can thus be obtained by temporal shaping of the stochastic component with the temporal envelope of the glottal air flow [21]. Indeed, more generally, further development of the deterministic-stochastic model will require accounting for harmonic/aharmonic components that are nonlinearly combined.

9.7 Summary

In this chapter, a sinusoidal model for the speech waveform has been developed. In sinewave analysis, the amplitudes, frequencies, and phases of the component sinewaves are extracted from the short-time Fourier transform of the speech. In order to account for effects due to a time-varying source and vocal tract events, the sinewaves are allowed to come and go in accordance with a birth-death frequency tracking algorithm. Once contiguous frequencies are matched, a cubic phase interpolation function is obtained that is consistent with all the frequency and phase measurements and that performs maximally smooth phase unwrapping. Synthesis is performed by applying this phase function to a sinewave generator that is amplitude-modulated by a linear interpolation of successive matched sinewave amplitudes. The process is repeated for all sinewave components, the final speech being the sum of the component sinewaves. In some respects, the basic model has similarities to the filter-bank representation used in the phase vocoder of Chapter 8 [13],[41]. Although the sinusoidal analysis/synthesis system is based on the discrete Fourier transform (DFT), which can be interpreted as a filter bank (Chapter 7), the use of the DFT in combination with peak-picking renders a highly *adaptive filter bank*, since only a subset of all of the DFT filters is used on any one frame. It is the use of the frequency tracker and cubic phase interpolator that allows the filter bank to move with highly resolved speech components.

The sinewave model for an arbitrary signal class results in a sinewave analysis/synthesis framework applicable to a variety of problems in speech and audio sound processing, including, for example, sound modification, morphing, and peak-to-rms reduction. Tailoring the sinewave representation, however, to specific signal classes can improve performance. A source/filter phase model for quasi-periodic signals leads to a means to preserve sinewave phase coherence through a model of onset times of the excitation impulses. This approach to phase coherence is similar in style to that introduced into the phase vocoder in Chapter 8. In addition, the sinewave analysis/synthesis was tailored to signals with additive harmonic and aharmonic components by introducing a deterministic-stochastic model. It was also shown that, in some applications, computational advantages can be achieved by performing sinewave synthesis using the FFT and an overlap-and-add procedure.

There are many extensions, refinements, and applications of the sinusoidal approach that we were not able to cover in this chapter. For example, we merely touched upon the many considerations of time-frequency resolution and led the reader to some important work in combining the baseline sinewave model, and its deterministic-stochastic extension, with a wavelet representation for multi-resolution analysis and synthesis. The problem of the separation of two voices from a single recording is another area that we introduced only briefly through Exercise 9.17. Although the sinewave paradigm has been useful in signal separation, the problem remains largely unsolved [41],[45]. Multi-speaker pitch estimation [41],[45] and the synchrony of movement of the time-varying sinewave parameters within a voice [5], may provide keys to solving this challenging separation problem. In addition, there remain a variety of other application areas not addressed within this chapter, including speech coding (Chapter 12) and enhancement for the hearing impaired [19],[47]. Other applications exploit the capability of sinewave analysis/synthesis to blend signal operations, an example being joint time-scale and pitch modification (Exercise 9.11) for prosody manipulation in concatenative speech synthesis [3],[24].

Appendix 9.A: Derivation of the Sinewave Model

Consider the speech production model in continuous time which is illustrated in Figure 9.2, where $h(t, \tau)$ is the time-varying vocal tract impulse response, i.e., the response at time t to an impulse applied τ samples earlier, at time $t - \tau$ (Chapter 2). The frequency response of the system at time t is given by the Fourier transform of $h(t, \tau)$ and can be written in polar form as

$$H(t, \Omega) = M(t, \Omega) \exp[j\Phi(t, \Omega)].$$

The source $u(t)$ is modeled as a sum of sinewaves (with, for simplicity, the number of sinewaves K assumed fixed):

$$u(t) = \text{Re} \sum_{k=1}^{K} a_k(t) \exp[j\phi_k(t)]$$

representing an arbitary source and thus is not constrained to a periodic, impulsive, or white-noise form. The functions $a_k(t)$ are the time-varying sinewave amplitudes and

$$\phi_k(t) = \int_0^t \Omega_k(\sigma)d\sigma + \phi_k$$

are the sinewave phase functions where, for convenience, we henceforth eliminate the "Re" notation.

The speech waveform $s(t)$ can be written as a time-varying convolution

$$s(t) = \int_0^t h(t, t - \tau) u(\tau) d\tau$$

which can be written as

$$s(t) = \int_0^t h(t, t - \tau) \sum_{k=1}^K a_k(\tau) \exp[j\phi_k(\tau)] d\tau.$$

Interchanging the $\int$ and $\sum$ above, we have

$$s(t) = \sum_{k=1}^K \int_{t'}^t h(t, t - \tau) a_k(\tau) \exp[j\phi_k(\tau)] d\tau$$

where t' is the effective starting time of $h(t, t - \tau)$. If we assume that the excitation amplitude and frequency are constant over the effective duration of $h(t, t - \tau)$ (Figure 9.27a), we have

$$\left. \begin{aligned} a_k(\tau) &= a_k \\ \Omega_k(\tau) &= \Omega_k \end{aligned} \right\} \text{ for } t' \leq \tau \leq t.$$

Then $\phi_k(\tau)$ can be written as

$$\begin{aligned} \phi_k(\tau) &= \int_0^\tau \Omega_k(\sigma) d\sigma + \phi_k \\ &= \int_{t'}^\tau \Omega_k d\sigma + \int_0^{t'} \Omega_k d\sigma + \phi_k \\ &= \Omega_k(\tau - t') + \phi_k(t'). \end{aligned}$$

Therefore, $s(t)$ can be written as

$$\begin{aligned} s(t) &= \sum_{k=1}^K a_k \int_{t'}^t h(t, t - \tau) \exp[j\Omega_k(\tau - t')] \exp[j\phi_k(t')] d\tau \\ &= \sum_{k=1}^K a_k \exp[j\phi_k(t')] \int_{t'}^t h(t, t - \tau) \exp[j\Omega_k(\tau - t')] d\tau \end{aligned}$$

where $\exp[j\Omega_k(t - t')]$ is considered as an eigenfunction of $h(t, \tau)$ at each time instant t. The above equation can therefore be further rewritten as

$$s(t) = \sum_{k=1}^K a_k \exp[j\phi_k(t')] H(t, \Omega_k) \exp[j\Omega_k(t - t')]$$

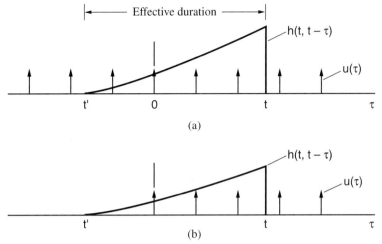

Figure 9.27 Effective duration of time-varying vocal tract impulse response $h(t, \tau)$ relative to the source $u(\tau)$: (a) an impulsive source overlapping $h(t, \tau)$; (b) a causal impulsive source not overlapping $h(t, \tau)$. In panel (a) the region of overlap of $u(\tau)$ and $h(t, \tau)$ in the time-varying convolution provides sinewave eigenfunctions as input to $h(t, \tau)$, but in panel (b) the eigenfunction assumption breaks down.

and thus

$$s(t) = \sum_{k=1}^{K} H(t, \Omega_k) a_k \exp[j\{\phi_k(t') + \Omega_k(t - t')\}]$$

$$= \sum_{k=1}^{K} H[t, \Omega_k(t)] a_k(t) \exp[j\phi_k(t)].$$

Expressing the system function in terms of its magnitude and phase, namely

$$H[t, \Omega_k(t)] = M[t, \Omega_k(t)] \exp[j\Phi(t, \Omega_k(t))],$$

we can express the speech waveform model as

$$s(t) = \sum_{k=1}^{K} M[t, \Omega_k(t)] a_k(t) \exp[j\{\phi_k(t) + \Phi[t, \Omega_k(t)]\}].$$

Combining amplitude and phase terms, we have

$$s(t) = \sum_{k=1}^{K} A_k(t) \exp[j\theta_k(t)]$$

where,

$$A_k(t) = a_k(t) M[t, \Omega_k(t)]$$
$$\theta_k(t) = \phi_k(t) + \Phi[t, \Omega_k(t)]$$

with

$$\phi_k(t) = \int_0^t \Omega_k(\sigma)d\sigma + \phi_k.$$

Finally, it is important to observe in this derivation that at rapid transitions the eigenfunction condition does not hold. Consider, for example, the onset of an initial vowel whose source is illustrated in Figure 9.27b as a causal impulse train. If we model this function $u(t)$ as a sum of sinewaves, each with a unit step function amplitude, $a_k(t) = u(t)$, then, as $h(t, \tau)$ slides over the beginning of the excitation function, the amplitudes of the excitation sinewaves are not constant over the duration of $h(t, \tau)$ (Problem 9.3).

Appendix 9.B: Derivation of Optimal Cubic Phase Parameters

The continuous-time cubic phase and frequency model for a sinewave over an analysis time interval $[0, T]$ is given by (where l is the frame index)

$$\theta(t) = \theta^l + \Omega^l t + \alpha t^2 + \beta t^3, \qquad \dot\theta(t) = \Omega^l + 2\alpha t + 3\beta t^2 \tag{9.46}$$

where we assume the initial phase $\theta(0) = \theta^l$ and frequency $\dot\theta(0) = \Omega^l$ values are known. We also assume a known endpoint phase (modulo 2π) and frequency so that at $t = T$

$$\theta(T) = \theta^l + \Omega^l T + \alpha T^2 + \beta T^3 = \theta^{l+1} + 2\pi M$$
$$\dot\theta(T) = \Omega^l + 2\alpha T + 3\beta T^2 = \Omega^{l+1}.$$

Therefore,

$$T^2\alpha + T^3\beta = \theta^{l+1} + 2\pi M - \theta^l - \Omega^l T$$
$$2T\alpha + 3T^2\beta = \Omega^{l+1} - \Omega^l$$

which can be written in matrix form as

$$\begin{bmatrix} T^2 & T^3 \\ 2T & 3T^2 \end{bmatrix} \begin{bmatrix} \alpha \\ \beta \end{bmatrix} = \begin{bmatrix} \theta^{l+1} - \theta^l - \Omega^l T + 2\pi M \\ \Omega^{l+1} - \Omega^l \end{bmatrix}$$

giving the solution for the unknown cubic parameters

$$\begin{bmatrix} \alpha(M) \\ \beta(M) \end{bmatrix} = \frac{\begin{bmatrix} 3T^2 & -T^3 \\ -2T & T^2 \end{bmatrix}}{3T^4 - 2T^4} \begin{bmatrix} \theta^{l+1} - \theta^l - \Omega^l T + 2\pi M \\ \Omega^{l+1} - \Omega^l \end{bmatrix}.$$

Now let

$$a_{11} = \frac{3}{T^2}, \qquad a_{12} = -\frac{1}{T}, \qquad b_1 = \theta^{l+1} - \theta^l - \Omega^l T,$$
$$a_{21} = -\frac{2}{T^3}, \qquad a_{22} = \frac{1}{T^2}, \qquad b_2 = \Omega^{l+1} - \Omega^l.$$

We can then write

$$\begin{bmatrix} \alpha(M) \\ \beta(M) \end{bmatrix} = \begin{bmatrix} a_{11} & a_{12} \\ a_{21} & a_{22} \end{bmatrix} \begin{bmatrix} b_1 + 2\pi M \\ b_2 \end{bmatrix}. \tag{9.47}$$

Suppose the frequency is constant. Then $\Omega^{l+1} = \Omega^l$ and

$$\dot{\theta}(t) = \Omega_l, \quad \ddot{\theta}(0) = 0, \quad t \in [0, T]$$

which motivates, for slowly changing frequencies, selection of the value of M that minimizes

$$\int_0^T [\ddot{\theta}(t)]^2 dt$$

which, using Equation (9.46), can be expressed as

$$\int_0^T \left[\ddot{\theta}(\tau)\right]^2 d\tau = 4\alpha^2 T + 12\alpha\beta T^2 + 12\beta^2 T^3.$$

Now, from Equation (9.47),

$$\alpha(M) = a_{11}b_1 + a_{12}b_2 + 2\pi a_{11} M$$
$$\beta(M) = a_{21}b_1 + a_{22}b_2 + 2\pi a_{21} M,$$

and letting

$$c_1 = a_{11}b_1 + a_{12}b_2, \quad d_1 = 2\pi a_{11}$$
$$c_2 = a_{21}b_1 + a_{22}b_2, \quad d_2 = 2\pi a_{21},$$

we have

$$\alpha(M) = c_1 + d_1 M, \quad \alpha'(M) = \frac{d\alpha(M)}{dM} = d_1$$

$$\beta(M) = c_2 + d_2 M, \quad \beta'(M) = \frac{d\beta(M)}{dM} = d_2. \tag{9.48}$$

Then

$$\int_0^T \left[\ddot{\theta}(t)\right]^2 dt = 4T \underbrace{\left[\alpha^2(M) + 3\alpha(M)\beta(M)T + 6\beta^2(M)T^2\right]}_{g(m)}. \tag{9.49}$$

Let $g(M)$ equal the bracketed expression in Equation (9.49). Then, to minimize $\int_0^T \left[\ddot{\theta}(t)\right]^2 dt$, we must minimize $g(M)$ with respect to the integer value M. To simplify this difficult nonlinear optimization problem, we assume a continuous argument x for $g(x)$, minimize $g(x)$ with respect to x, and then pick the closest integer to the optimizing solution. Therefore, we want

$$\frac{dg(x)}{dx} = 2\alpha(x)\alpha'(x) + 3T\left[\alpha(x)\beta'(x) + \alpha'(x)\beta(x)\right] + 3T^2\beta(x)\beta'(x) = 0. \tag{9.50}$$

Substituting Equation (9.48) into Equation (9.50), it can then be shown that the optimizing value x^* is given by

$$x^* = -\frac{c_1 d_1 + (\frac{3}{2})T(c_1 d_2 + c_2 d_1) + 3T^2 c_2 d_2}{d_1^2 + 3T(d_1 d_2) + 3T^2 d_2^2}.$$

To reduce x^* further, the denominator and numerator of x^* are written as, respectively (with appropriate substitutions from the above expressions),

$$
\begin{aligned}
\text{denom} &= d_1^2 + 3Td_1d_2 + 3T^2d_2^2 \\
&= (2\pi a_{11})^2 + 3T(2\pi a_{11}2\pi a_{21}) + 3T^2(2\pi a_{21})^2 \\
&= 4\pi^2 \frac{3}{T^4} = \frac{12\pi^2}{T^4} \\
\text{num} &= c_1d_1 + \frac{3}{2}T(d_1d_2 + c_2d_1) + 3T^2c_2d_2 \\
&= (a_{11}b_1 + a_{12}b_2)(2\pi a_{11}) \\
&\quad + \frac{3}{2}T\left[(a_{11}b_1 + a_{12}b_2)(2\pi a_{21}) + (a_{21}b_1 + a_{22}b_2)(2\pi a_{11})\right] \\
&\quad + 3T^2(a_{21}b_1 + a_{22}b_2)(2\pi a_{21}) \\
&= 2\pi \frac{3}{T^4}\left(b_1 - \frac{T}{2}b_2\right),
\end{aligned}
$$

from which we have

$$
\begin{aligned}
x^* &= -\frac{2\pi \frac{3}{T^4}\left(b_1 - \frac{T}{2}b_2\right)}{(2\pi)^2 \frac{3}{T^4}} \\
&= -\frac{1}{2\pi}\left(b_1 - \frac{T}{2}b_2\right).
\end{aligned}
$$

Finally, substituting for b_1 and b_2

$$
x^* = \hat{m} = -\frac{1}{2\pi}\left[(\theta^{l+1} - \theta^l - \Omega^l T) - \frac{T}{2}(\Omega^{l+1} - \Omega^l)\right],
$$

from which the nearest integer M^* is chosen.

EXERCISES

9.1 Consider analyzing with the STFT a sequence $x[n]$ consisting of a single sinewave, i.e.,

$$
x[n] = A\cos(\omega n).
$$

Show that normalizing the analysis window $w[n]$ used in the STFT according to

$$
\sum_{n=-\infty}^{\infty} w[n] = 1
$$

yields a spectral peak of value $\frac{A}{2}$, i.e., half the amplitude of the underlying sinewave. *Hint:* Use the windowing (multiplication) theorem that was reviewed in Chapter 2.

9.2 Consider reconstruction of a single sinewave in Example 9.3 from spectral peaks of the STFT.

(a) Describe an approach to short-time synthesis by estimating the phase at the spectral peaks over successive frames. What problem in sinewave frequency and phase estimation do you encounter with a DFT implementation of the STFT? *Hint:* Use the result that the STFT phase at the spectral peaks represents the phase offset of the sinewave relative to the center of the analysis window. Assume each short-time segment is shifted to the region $0 \leq n < N_w$, where N_w is the window length. Also consider that the sinewave frequency can be estimated no more accurately that half the distance between two consecutive DFT samples.

(b) Suppose now that sinewave parameters are time-varying. Explain why even more acute waveform discontinuity arises with your approach from part (b).

9.3 Propose a source/filter speech model that addresses the eigenfunction approximation problem at speech transitions described in Appendix 9.A.

9.4 One approach to phase interpolation, as described in Section 9.4.1, is through a cubic polynomial. An alternative is to use a piecewise quadratic interpolator, as illustrated in Figure 9.28, which shows the phase function in continuous time for one frame from $t = 0$ to $t = T$. In this interpolator, the frequency is linear over each half segment with the midpoint frequency being unknown. Assume the phase has been unwrapped up to time $t = 0$. The midpoint frequency is a free parameter which will be selected so the phase measurement $\theta(T)$, at the right boundary, is met. On two consecutive frames, the measurements θ^l (unwrapped), θ^{l+1} (modulo 2π), Ω^l, and Ω^{l+1} are known.

(a) Suppose for the moment that $2\pi M$ is also known. Derive an expression for $\theta(T)$ in terms of θ^l, Ω^l, Ω^{l+1}, and the unknown midpoint frequency $\tilde{\Omega}$. Then determine $\tilde{\Omega}$, assuming again that M is known.

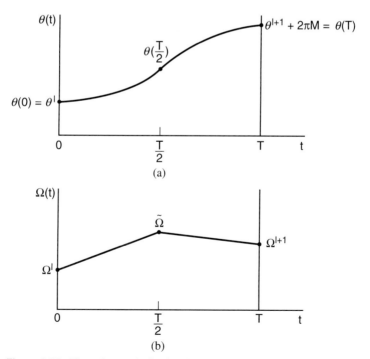

Figure 9.28 Piecewise quadratic phase interpolator over one synthesis frame:
(a) phase; (b) frequency [derivative of (a)].

(b) From your answer in part (a), $\theta(t)$ is now a function only of the unknown integer M. Propose an optimization criterion for selecting M such that the resulting unwrapped phase over the interval $[0, T]$ is "maximally smooth." Recall the derivation of the cubic phase interpolation. Do not solve.

9.5 An alternative to the cubic (Section 9.4.1) and quadratic (Exercise 9.4) phase interpolator used in sinewave synthesis is given in Figure 9.29, which shows the continuous-time phase over one synthesis frame from $t = 0$ to $t = T$. On two consecutive frames, the measurements θ^l, θ^{l+1}, Ω^l, and Ω^{l+1} are given. In this interpolator, the frequency is assumed piecewise constant over three equal intervals. Assume the phase $\theta(0) = \theta_l$ has been unwrapped up to time $t = 0$, and the measurement θ^{l+1} is made modulo 2π. The mid-frequency value $\tilde{\Omega}$ is a free parameter that will be selected so that the phase measurement $\theta(T)$ at the right boundary is met.

(a) Suppose for the moment that $2\pi M$ is known. Derive an expression for $\tilde{\Omega}$ in terms of the known quantities so that the phase $\theta(T) = \theta^{l+1} + 2\pi M$ is achieved.

(b) Give an expression for the phase derivative $\dot{\theta}(t, M)$ over the interval $[0, T]$ in terms of $\tilde{\Omega}$, Ω^k, and Ω^{k+1}. In order to select the unknown M, one can minimize the following criterion:

$$E(M) = \frac{1}{T} \int_0^T [\dot{\theta}(t, M) - f(t)]^2 dt$$

where $f(t)$ is a linearly interpolated frequency between the known frequency measurements Ω^l and Ω^{l+1}:

$$f(t) = \Omega^l + \left(\frac{\Omega^{l+1} - \Omega^l}{T} \right) t.$$

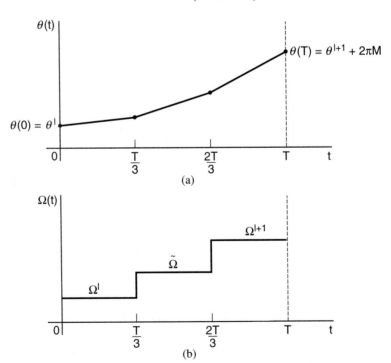

Figure 9.29   Piecewise linear-phase interpolator: (a) phase; (b) frequency [derivative of (a)].

Find the value of M (closest integer to) that minimizes $E(M)$. You are not asked to solve the equations that result from the minimization. Explain why this is a reasonable criterion for selecting M.

(c) Do you think the step discontinuities in frequency in Figure 9.29 would result in an annoying perceptual effect? Explain your reasoning.

9.6 In Section 9.4.2, we described an overlap-add approach to sinewave synthesis, one which does not give high-quality synthesis for a long frame interval. Consequently, overlap-add interpolation with the generation of an artificial mid-frame sinewave to reduce discontinuities in the resynthesized waveform was suggested. Consider the problem of finding an interpolated sinewave parameter set for a pair of matched sinewaves when a 20-ms frame interval is too low for high-quality synthesis. At frame l a sinewave is generated with amplitude A^l, frequency Ω^l, and phase θ^l, and a 20-ms triangular window is applied. The process is repeated 20 ms later for the sinewave at frame $l + 1$. The problem is to find the amplitude, frequency, and phase of a sinewave to apply at the midpoint between frames l and $l + 1$ such that the result of overlapping and adding it with the sinewave at frame l and the sinewave at frame $l + 1$ results in a "best fit." If $\bar{A}$, $\bar{\Omega}$ and $\bar{\theta}$ represent the mid-point amplitude, frequency, and phase, then a reasonable choice for the mid-point amplitude is simply the average amplitude

$$\bar{A} = \frac{A^l + A^{l+1}}{2}.$$

Suppose the time evolution of the phase is described by the cubic phase interpolator, Equation (9.21), i.e.,

$$\theta(t) = \theta^l + \Omega^l t + \alpha(M^*)t^2 + \beta(M^*)t^3$$

where $\alpha(M^*)$ and $\beta(M^*)$ are computed from Equation (9.20). Propose reasonable estimates for the midpoint frequency and phase, considering how well the new sinewaves match to their overlapping neighbors. Determine the computational complexity with the specific operations required in this scheme relative to decreasing the analysis frame interval to 10 ms.

9.7 Explain, conceptually, the differences between the baseline sinewave analysis/synthesis and the phase vocoder. Compare the two methods for time-scale modification with respect to waveform dispersion and sinewave frequency estimation and resolution. Compare also refinements to the two approaches that have been described in this and the previous chapter in achieving phase coherence with time-scale modification.

9.8 Consider the sinewave frequency matching algorithm in Section 9.3.4. Suppose that the frequency matching interval Δ is zero so that, in effect, no matching is performed.

(a) Over how many synthesis frames does each sinewave persist? *Hint:* The resulting sinewave frequencies are constant over this duration. Also, the sinewave birth and death process of Section 9.3.4 occurs very often.

(b) Consider a frequency set for two consecutive analysis frames l and $l + 1 : \{\omega_k\}^l$ and $\{\omega_k\}^{l+1}$. Show that when the matching interval Δ is zero, the sinewave synthesis over these two consecutive frames is equivalent to an overlap-and-add (OLA) procedure where the window weight is triangular and has duration equal to twice the synthesis frame interval.

(c) Argue why the overlap-add view of synthesis of parts (a) and (b) helps explain the ability of the sinewave analysis/synthesis to represent short-lived and rapidly varying speech events such as plosives and unvoiced-to-voiced transitions.

9.9 We saw in Chapter 7 that digital filtering of a sequence $x[n]$ could be performed with the short-time Fourier transform by way of the operation

$$Y(n, \omega) = X(n, \omega)H(n, \omega)$$

where $H(n, \omega)$ is a time-varying multiplicative modification. For FBS synthesis, the corresponding time-domain modification $h[n, m]$ is weighted by the analysis window with respect to the variable m prior to convolution with $x[n]$; in OLA synthesis, the time-domain modification is convolved with the window with respect to the variable n prior to convolution with $x[n]$. This problem asks you to consider the effect of spectral modifications in sinewave analysis/synthesis.

(a) Given the sinewave parameter sets $\{A_k\}$, $\{\omega_k\}$, and $\{\theta_k\}$ on frame l, consider the multiplicative modifier $H(lL, \omega_k)$. Letting

$$\tilde{A}_k = A_k|H(lL, \omega_k)|$$

$$\tilde{\theta}_k = \theta_k + \angle H(lL, \omega_k),$$

we do sinewave synthesis from the modified parameter sets $\{\tilde{A}_k\}$, $\{\tilde{\theta}_k\}$, and $\{\omega_k\}$. Suppose $H(lL, \omega_k)$ is time-invariant, i.e.,

$$H(lL, \omega_k) = H(\omega_k). \tag{9.51}$$

Write an approximate expression for the modified time-domain sequence in terms of the original sequence $x[n]$ and the time-domain filter $h[n]$. Assume we know the sinewave frequencies ω_k exactly.

(b) Now write an approximate expression for the modified time-domain sequence in terms of the original sequence $x[n]$ and the time-domain time-varying filter $h[n, m]$. Assume we know the sinewave frequencies ω_k exactly.

(c) Assume the sinewave frequencies are obtained by peak-picking the amplitude of the DFT samples. How does this affect your results from part (a) and part (b)?

9.10 Given a transmission channel bandwidth constraint, it is sometimes desirable to compress a signal's spectrum, a procedure referred to as "frequency compression." This problem explores a method of frequency compression based on sinewave analysis/synthesis. Suppose we have performed sinewave analysis in continuous time. We can then synthesize a signal as

$$s(t) = \sum_{k=0}^{K(t)} A_k(t) \cos[\theta_k(t)]$$

where the phase for the kth sine-wave, $\theta_k(t)$, is the integral of the kth sinewave's instantaneous frequency. Over each synthesis frame, the phase is assumed to have a cubic trajectory with some initial phase offset at the birth of a sinewave.

(a) Suppose that in sinewave synthesis, we scale the phase function by a scale factor B, i.e.,

$$\tilde{\theta}_k(t) = B\theta_k(t)$$

so that

$$\tilde{s}(t) = \sum_{k=-\infty}^{\infty} A_k(t) \cos[\tilde{\theta}_k(t)].$$

Show that the frequency trajectory of the resulting sinewaves are scaled by B. What is the resulting pitch if the original sinewaves are harmonic with fundamental frequency Ω_o?

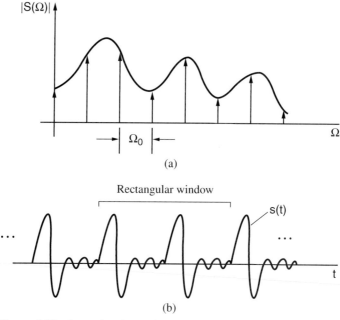

$|S(\Omega)|$

Ω_0

Ω

(a)

Rectangular window

s(t)

... ...

t

(b)

Figure 9.30 Spectral and waveform illustrations for Exercise 9.10: (a) vocal tract spectrum and harmonic lines; (b) periodic speech waveform with rectangular window.

(b) Suppose $s(t)$ is perfectly voiced with vocal tract spectrum and harmonic lines as shown in Figure 9.30a. Also, suppose a rectangular window, of duration twice the pitch period, is applied to $s(t)$ as illustrated in Figure 9.30b. Sketch the magnitude of the STFT of $s(t)$. Assume the scale factor $B = 1/2$. Then sketch the magnitude of the STFT of $\tilde{s}(t)$ [in contrast to $s(t)$], assuming the window duration is twice the new pitch period. Comment on the shape of $\tilde{s}(t)$ in comparison to $s(t)$, considering waveform dispersion over a single pitch period.

(c) Suppose we transmit the waveform $\tilde{s}(t)$ from part (b) over a channel with half the bandwidth of $s(t)$ and we want to reconstruct the original full-bandwidth waveform $s(t)$ at a receiver. One approach to recover $s(t)$ from $\tilde{s}(t)$ is to invert the frequency compression, i.e., perform "frequency expansion," using sinewave analysis/synthesis and modification with a scale factor of $B = 2$ on the cubic phase derived from $\tilde{s}(t)$. Assuming steady voiced speech, what is the duration of the rectangular analysis window (relative to a pitch period) required at the receiver to obtain sinewave amplitudes, phases, and frequencies from $\tilde{s}(t)$? Explain your reasoning.

(d) Do you think the method of recovering $s(t)$ from $\tilde{s}(t)$ in part (c) suffers from waveform dispersion? Briefly explain your reasoning. Given the time-varying nature of speech, what fundamental limitation might our bandwidth reduction scheme suffer from? Are we really getting "something for nothing"?

9.11 In pitch modification, it is desired to modify the pitch of a speaker while preserving the vocal tract transfer function. For voiced speech, the pitch is modified by multiplying the sinewave frequencies by a factor β prior to sinewave synthesis, i.e., $\Omega'_k = \beta\Omega_k$, with an associated change in the pitch period P given by

$$P' = \frac{P}{\beta}.$$

The change in pitch period corresponds to modification of the onset times of an assumed impulse train input (Section 9.5.1).

(a) Propose a sinewave model for the pitch-modified excitation function in terms of onset times and sinewave frequencies for voiced speech. *Hint:* This model is a simple variation of the excitation model in Equation (9.30) and involves a scaling of the original onset times and frequencies.

(b) Consider now the magnitude and phase of the system function $H(t, \Omega) = M(t, \Omega)$ $\exp[j\Phi(t, \Omega)]$. Propose a model for the system magnitude and phase contribution to a sinewave model for the pitch-modified voiced speech.

(c) Propose a sinewave model of pitch-modified unvoiced speech as a variation of the original sinewave model for unvoiced speech. *Hint:* Consider the changes that might occur in natural speech in generating noise-like and impulsive sources with pitch modification. Allow for adaptivity to the degree of voicing in your model and, in particular, assume a two-band spectrum with the upper band "voiced" and the lower band "unvoiced."

(d) In implementing sinewave analysis/synthesis for pitch modification we must estimate the amplitudes and phases of the excitation and system components at the scaled frequencies $\Omega'_k = \beta\Omega_k$ from the sinewave amplitudes and phases at the original sinewave frequencies; these parameters are then interpolated over consecutive frames prior to sinewave generation. The excitation and system amplitude functions are first considered. To preserve the shape of the magnitude of the vocal tract transfer function, the system amplitudes must be obtained at the new frequency locations, $\beta\Omega_k$. Propose a means to estimate the required excitation and system amplitude values from amplitudes at the measured sinewave frequencies. *Hint:* You are not allowed to simply shift the amplitude values of the original sinewaves obtained by peak-picking because this will modify the original formant structure and not simply the pitch.

(e) Consider now sinewave phases in implementing the pitch modification model. Propose first a method to implement the onset-time model from part (a) for the modified excitation, similar in spirit to that used for time-scale modification. Next, propose a means to estimate the system phase values from phases at the measured sinewave frequencies. As with the system amplitude, to preserve the shape of the phase of the vocal tract transfer function, the system phases must be computed at the new frequency locations, $\beta\Omega_k$. Again, you are not allowed to simply shift the phase values of the original sinewaves obtained by peak-picking because this will modify the original formant phase structure and not simply the pitch. Also, keep in mind that this is a more difficult problem than the corresponding one for the system amplitude because measured phase values are not unwrapped with respect to frequency.

(f) Using parts (a)-(e), propose a sinewave model and analysis/synthesis system for performing pitch and time-scale modification simultaneously.

9.12 Consider applying sinewave analysis/synthesis to slowing down the articulation rate of a speaker by a factor of two. The goal is to perform the modification without loss in phase coherence. This problem tests your understanding of the steps we have described in Section 9.5.2.

(a) Using the baseline analysis/synthesis system, if analysis is performed with a 10-ms frame interval, what is the corresponding synthesis frame interval over which amplitude and phase interpolation is performed?

(b) Explain your method of doing sinewave amplitude interpolation across the center of two consecutive frames.

(c) In generating the time-scaled phase, it is necessary to decompose the excitation and vocal tract phase functions. How do you perform this decomposition in the original time scale? The resulting system phase can be used in the new time scale. The excitation phase, however, must be obtained relative to the new time scale. Explain how, from the excitation onset times in the new time scale, you can obtain the linear excitation phase.

(d) Finally, give the composite phase at each frame center and describe the method of interpolation.

(e) Observe that we could have formed (in the modified time scale) an excitation phase $\rho\phi_k(n\rho^{-1})$ (why does this preserve pitch?), added this phase to the decomposed system phase at a frame center, and performed interpolation across frame boundaries in the new time scale. What is wrong with this approach for achieving phase coherence?

9.13 Suppose a 20-ms analysis window and a 10-ms frame interval are used in sinewave analysis/synthesis during unvoiced speech.

(a) Argue qualitatively that estimating sinewave amplitudes, frequencies, and phases by peak-picking the STFT results in a high-quality, noise-like synthesized waveform during unvoiced speech. Do not use the argument given in Section 9.3.2. Create your own argument.

(b) In doing time-scale expansion with sinewave analysis/synthesis, consider the preservation of the noise-like quality of unvoiced sounds. What problems arise in excessive stretching of sinewaves and how would you remedy these problems?

9.14 In the estimation of the sinewave model parameters (i.e., amplitudes, frequencies, and phases), the amplitudes and frequencies are held constant over the duration of the analysis frame. In synthesis, however, these parameters are assumed to vary over the synthesis interval.

(a) Suggest an approach to remove this inconsistency between analysis and synthesis.

(b) Consider now all-pole (linear prediction) analysis/synthesis of Chapter 5. Describe an inconsistency between all-pole analysis and synthesis similar in style to that of the sinewave analysis/synthesis. Propose a method to remove this inconsistency. *Hint:* See Exercise 5.13 in Chapter 5.

9.15 Observe that the overlap and add method of Section 9.4.2 can be interpreted as synthesizing a signal from a modified STFT corresponding to a sum of triangularly windowed sinewaves. Thus, we might consider the least-squared-error approach of Chapter 7 which results in a weighting of a short-time segment by the window and a normalization by the summed squared, shifted windows. Discuss the possible advantage of this method in the sinewave context. Note that you are implicitly comparing the overlap-and-add (OLA) and least-squared-error (LSE) methods of Chapter 7. *Hint:* Consider overlapping regions of low amplitude near the window edges.

9.16 We have seen that often it is desirable to do *time-varying* time-scale modification. Consider, for example, time-scale modification of a plosive followed by a vowel, where it may be desirable to maintain the original time-scale of the plosive and modify the time scale of the vowel. Another example is taken from music modification where it is often imperative to maintain the *attack* and *release* portion of the note, while lengthening or shortening the steady portion of the note; a change in the attack or release can radically alter the character of the sound.[22] An idealization of such

---

[22] For example, it has been said that replacing the attack of a flute with the attack of a clarinet can turn a flute into a clarinet.

a modification of a musical note is shown in Figure 9.31a. Suppose that the original tone is, as illustrated, a pure (complex) tone of the form

$$x(t) = a(t)e^{j\theta(t)}$$

where

$$a(t) = \text{rectangular envelope}$$
$$\theta(t) = \Omega t,$$

Ω being the frequency of the tone. Our goal is to stretch by a factor of two the steady portion of the tone, while maintaining phase and frequency continuity at its edges.

(a) One approach is to simply stretch out the amplitude and phase functions to form $a(t/2)$ and $\theta(t/2)$ during the steady portion and keep these functions intact during the attack and release. Stretching the phase in this way is shown in Figure 9.31b. Show that this operation on the phase lowers the frequency by a factor of two during the steady region.

(b) Show that one approach to preserve frequency during the steady portion is to scale $\theta(t/2)$ by 2. To preserve phase continuity at the attack, we also displace the resulting $2\theta(t/2)$ by a constant k, as illustrated in Figure 9.31c. (Assume you know the value of k.) Unfortunately, this implies, as shown, generally a phase discontinuity at the release. Explain why this discontinuity may occur. Heretoforth, we denote the three phase segments in the new time scale by $\phi_1(t)$, $\phi_2(t)$, and $\phi_3(t)$, as depicted in Figure 9.31c. One way to reduce the discontinuity between $\phi_2(t)$ and $\phi_3(t)$ is to add a 2π multiple to the third phase segment $\phi_3(t)$ so that

$$\phi_2(T) \approx \phi_3(T) + 2\pi M.$$

However, there is no guarantee of phase and frequency continuity at the boundary. We can, of course, shift $\phi_3(t)$ so that it matches $\phi_2(t)$ at the boundary, but this generally changes the principal phase value (i.e., the phase calculated modulo 2π) of the release which we want to preserve. Why is this principal phase value modified?

(c) An alternate approach to preserve both phase and frequency continuity at the boundaries of the attack and release, henceforth indicated by times $t = 0$ and $t = T$, respectively, is to add a "perturbation function," as in Figure 9.31d. Denote the perturbation function by $\phi_p(t)$ and also make it a *cubic* polynomial of the form

$$\phi_p(t) = a + bt + ct^2 + dt^3.$$

Then the new phase function over the stretched interval $[0, T]$ is given by

$$\theta_2(t) = \phi_2(t) + \phi_p(t).$$

For boundary phase and frequency continuity, there are four constraints that $\theta_2(t)$ must satisfy:

1. $\theta_2(0) = \phi_1(0)$
2. $\dot{\theta}_2(0) = \Omega$
3. $\theta_2(T) = \phi_3(T) + 2\pi M$
4. $\dot{\theta}_2(T) = \Omega$

Justify the four constraints.

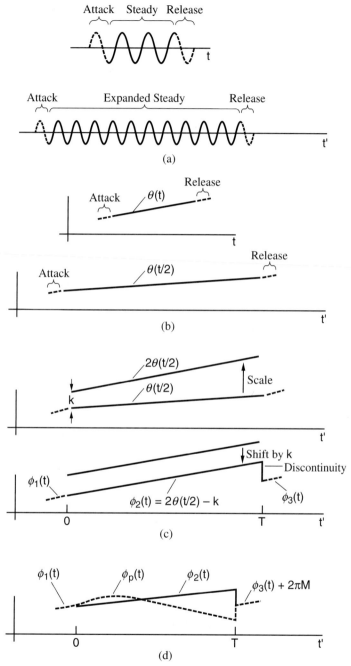

Figure 9.31 Illustrations of time-varying time-scale expansion in Exercise 9.16: (a) idealized stretching of musical tone; (b) simple stretching of phase of steady region only; (c) stretching with frequency preservation and phase continuity at the attack; (d) adding a phase perturbation function.

(d) Suppose for the moment that the integer M is known. Solve for the unknown cubic parameters a, b, c, and d such that all four constraints are satisfied. The solution will be a function of the known phase and frequency boundary values as well as the integer value M.

(e) We must now determine the value M. Propose an error criterion which is a function of M and which, when minimized with respect to M, gives a small frequency perturbation over the interval $[0, T]$. Do not solve the minimization.

9.17 An unsolved problem in speech processing is speaker separation, i.e., the separation of two speech waveforms that are additively combined, e.g.,in speech enhancement, a low-level speaker may be sought in the presence of a loud interfering talker. In this problem, you use a sinewave representation of speech to solve a simplified version of the speaker separation problem.

A speech waveform generated by two simultaneous talkers can be represented by a sum of two sets of sinewaves, each with time-varying amplitudes, frequencies, and phases. In the steady-state case where the source and vocal tract characteristics are assumed fixed over an analysis time interval, the sinewave model is given by

$$x[n] = x_a[n] + x_b[n] \tag{9.52}$$

with

$$x_a[n] = \sum_{k=1}^{K_a} a_k \cos(\omega_{a,k}n + \theta_{a,k})$$

$$x_b[n] = \sum_{k=1}^{K_b} b_k \cos(\omega_{b,k}n + \theta_{b,k})$$

where the sequences $x_a[n]$ and $x_b[n]$ denote the speech of Speaker A and the speech of Speaker B, respectively. The amplitudes and frequencies associated with Speaker A are denoted by a_k and $\omega_{a,k}$, and the phase offsets by $\theta_{a,k}$. A similar parameter set is associated with Speaker B.

(a) Let $s[n]$ represent a windowed speech segment extracted from the sum of two sequences

$$s[n] = w[n](x_a[n] + x_b[n])$$

where the analysis window $w[n]$ is non-zero over a duration N_w and is assumed centered and symmetric about the time origin. With the model Equation (9.52), show that the Fourier transform of the windowed summed waveforms, $S(\omega)$, for $\omega \geq 0$, is given by the summation of scaled and shifted versions of the transform of the analysis window $W(\omega)$:

$$S(\omega) = \sum_{k=1}^{K_a} a_k \exp(j\theta_{a,k}) W(\omega - \omega_{a,k}) + \sum_{k=1}^{K_b} b_k \exp(j\theta_{b,k}) W(\omega - \omega_{b,k}). \tag{9.53}$$

For the sake of simplicity, assume the negative frequency contribution is negligible and any scale factors have been embedded within the window.

(b) Consider the sum of two voiced segments. Figure 9.32 shows an example of the short-time spectral magnitude and phase of two actual voiced segments and of their sum. Assuming a Hamming window about two pitch periods in duration, justify the approximately piecewise flat phase functions in panels (a) and (b) of the figure (*Hint:* Recall the $w[n]$ is assumed to be centered at the time origin.) and justify the spectral magnitude and phase characteristics you see in panel (c).

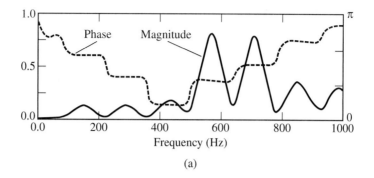

(a)

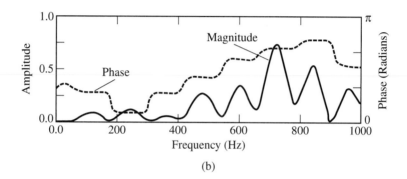

(b)

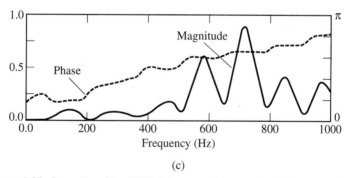

(c)

Figure 9.32 Properties of the STFT of the sum of two vowels of different speakers, $x[n] = x_a[n] + x_b[n]$: (a) STFT magnitude and phase of $x_a[n]$; (b) STFT magnitude and phase of $x_b[n]$; (c) STFT magnitude and phase of $x_a[n] + x_b[n]$. The vowels have roughly equal intensity and belong to two voices with dissimilar fundamental frequency. The duration of the short-time window is 25 ms.

SOURCE: T.F. Quatieri and R.J. McAulay, "Audio Signal Processing Based on Sinusoidal Analysis/Synthesis," chapter in *Applications of Digital Signal Processing to Audio and Acoustics* [41] (Figure 9.23). ©1998, Kluwer Academic Publishers. Used by permission.

(c) Assume now that you are given the frequencies $\omega_{a,k}$ and $\omega_{b,k}$ of Speakers A and B, respectively, and that you know which frequencies belong to each speaker. (The frequencies might be estimated by joint pitch estimation.) Explain why picking peaks of the spectral magnitude of the composite waveform in panel (c) of Figure 9.32 near the harmonics of each speaker is the basis for a flawed signal separation scheme. Suppose, on the other hand, that the spacing between the shifted versions of $W(\omega)$ in Equation (9.53) is such that the main lobes do not overlap. Briefly describe a sinewave-based analysis/synthesis scheme that allows you to separate the two waveforms $x_a[n]$ and $x_b[n]$.

(d) Because the frequencies of Speaker A and B may be arbitrarily close, and because the analysis window cannot be made arbitrarily long due to the time-varying nature of speech, the constraint of part (c) that window main lobes do not overlap is typically not met. One approach to address this problem is to find a least-squared error fit to the measured two-speaker waveform:

$$\text{minimize} \sum_{n=\infty}^{\infty} w[n](x_a[n] + x_b[n] - y[n])^2, \tag{9.54}$$

where $x_a[n]$ and $x_b[n]$ are the sinewave models in Equation (9.52), where $y[n]$ is the measured summed waveform, and where the minimization takes place with respect to the unknown sinewave parameters a_k, b_k, $\theta_{a,k}$, $\theta_{b,k}$, $\omega_{a,k}$, and $\omega_{b,k}$. We transform this nonlinear problem of forming a least-squared-error solution for the sinewave amplitudes, phases, and frequencies into a linear problem. We accomplish this by assuming, as above, that the sinewave frequencies are known, and by solving for the real and imaginary components of the quadrature representation of the sinewaves, rather than solving for the sinewave amplitudes and phases. Part (a) suggests that these parameters can be obtained by exploiting the linear dependence of $S(\omega)$ on scaled and shifted versions of the Fourier transform of the analysis window.

Suppose that Speaker A's waveform consists of a single (complex) sinewave with parameters A_1, ω_1, and θ_1, namely, $x_a[n] = A_1 e^{j\omega_1 n + \theta_1}$, and that Speaker B's waveform also consists of a single (complex) sinewave with parameters A_2, ω_2, and θ_2, namely, $x_b[n] = A_2 e^{j\omega_2 n + \theta_2}$. Figure 9.33 illustrates $S(\omega)$ of Equation (9.53) for this simple case as the sum of the Fourier transforms of the windowed $x_a[n]$ and $x_b[n]$, given by $X_a(\omega)$ and $X_b(\omega)$, respectively.

Using the spectral interpretation in Figure 9.33, solve the least-squared-error minimization of Equation (9.54) by solving for the samples of the Fourier transform of the separate waveforms at the known frequencies ω_1 and ω_2. Assume the Fourier transform of the analysis window, $W(\omega)$, is normalized so that $W(0) = 1$. Do not solve; just set up the equations in matrix form. (*Hint:* Consider the use of two linear equations in two unknowns.) From $X_a(\omega_1)$ and $X_b(\omega_2)$, describe a method for reconstruction of $x_a[n]$ and $x_b[n]$.

(e) Generalize the approach of part (d) to approximately solve the minimization problem in Equation (9.54) for the case of two arbitrary summed speech waveforms. Assume again that you know the frequencies of each speaker. Do not explicitly solve for the unknown sinewave parameters; rather, describe your approach and develop a set of linear equations in $K_a + K_b$ unknowns in matrix form. Describe the limitation of this approach as frequencies of voice A come arbitrarily close to those of voice B. Propose a scheme to address this problem by exploiting separation solutions from neighboring frames and frequencies.

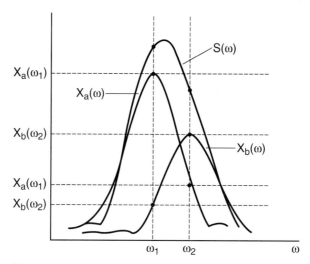

Figure 9.33 Least-squared-error solution for two sinewaves. $S(\omega)$ is the STFT of the composite waveform, while $X_a(\omega)$ and $X_b(\omega)$ are the STFTs of the component waveforms.

SOURCE: T.F. Quatieri and R.G. Danisewicz, "An Approach to Co-Channel Talker Interference Suppression Using a Sinusoidal Model for Speech" [45]. ©1990, IEEE. Used by permission.

9.18 As illustrated in Figure 9.25, the KFH phase typically traverses a very large range (e.g., from 0 to 300 radians over a bandwidth of 5000 Hz). This phase calculation can thus be sensitive to small measurement errors in pitch or spectrum.

    (a) For unity spectral magnitude and a one-sample error in the pitch period, derive the resulting change in the phase at $\omega = \pi$. Explain how this sensitivity might affect synthesis quality.

    (b) To reduce large frame-to-frame fluctuations in the KFH phase, both the pitch and the spectral envelope, used by the KFH solution, are smoothed in time over successive analysis frames. The strategy for adapting the degree of smoothing to signal characteristics is important for maintaining dispersion through rapidly changing speech events and for preserving the original phase in unvoiced regions where dispersion is unwanted. Suppose you are given a measure of the degree of voicing, and spectral and pitch derivatives which reflect the rate at which these parameters are changing in time. Propose a technique of smoothing the KFH phase over successive frames in time that reduces quality degradation due to the phase sensitivity shown in part (a).

9.19 Suppose you are given a set of sinewave frequencies $\omega_1, \omega_2, \ldots \omega_K$ and amplitudes $A_1, A_2, \ldots$ A_K and assume the corresponding phases are zero. Working *only in the frequency domain*, the objective of this problem is to generate a windowed sum of sinewaves of the form

$$x[n] = w[n] \sum_{k=1}^{K} A_k \cos(\omega_k n)$$

where $w[n]$ is an analysis window. The frequencies *do not* necessarily fall on DFT coefficients.

(a) Create a discrete Fourier transform (DFT), using the DFT of the window and the given frequencies and amplitudes, which when inverted yields the above windowed sum of sinewaves. Keep in mind that the frequencies *do not* necessarily fall on DFT coefficients, i.e., the frequencies do not necessarily fall at $\frac{2\pi}{N}m$ where N is the DFT length.

(b) Suppose the frequencies are multiples of a fundamental frequency. Describe the shape of the waveform $x[n]$. How will the shape change when the sinewave phases are made random, i.e., fall in the interval $[-\pi, \pi]$ with uniform probability density.

9.20 (MATLAB) In this MATLAB exercise, use workspace *ex9M1.mat* as well as the function *peak_pick.m* located in companion website directory Chap_exercises/chapter9. This exercise leads you through a (simplified) sinewave analysis/synthesis of a steady-state vowel.

(a) Plot the speech waveform *speech1_10k* (located in *ex9M1.mat*) and apply a 25-ms Hamming window to its center. (Think of this center point as the time origin $n = 0$.) Compute the DFT of the windowed waveform with a 1024-point FFT and display its log-magnitude spectrum (over 512 points).

(b) Do a *help* command on *peak_pick.m* and apply *peak_pick.m* to the FFT magnitude of part (a) (only the first 512 points). Sample the (complex) FFT at the peak locations, which is an output of *peak_pick.m*, and save these locations and corresponding FFT magnitudes and phases in a 3-D array. Superimpose the log-magnitude of the FFT peaks onto the original log-magnitude from part (a). One way to do this is to create a vector which is zero except at the peak locations at which the vector has values equal to the peak log-magnitudes. When plotted, this new vector will have a harmonic "line" structure.

(c) Write a MATLAB function to generate the sum of steady sinewaves with magnitudes and frequencies from part (b). The phase offsets of each sinewave should be such that at the time origin, $n = 0$ (center of original waveform), the sinewaves take on the measured phases from part (b), i.e., the synthesized waveform should be of the form

$$\hat{s}[n] = \sum_{k=1}^{L} \hat{a}_k \cos(\hat{\omega}_k n + \hat{\phi}_k),$$

where $\hat{a}_k$, $\hat{\omega}_k$, and $\hat{\phi}_k$ are the measured magnitudes, frequencies, and phases at the original waveform center ($n = 0$). Plot the synthesized waveform over a duration equal to the length of *speech1_10k*. How well does the synthesized waveform match the original? Explain any deviation from the original. Note that in analysis, from Section 9.3.3, you need to circularly shift the analysis window to the time origin (with respect to the DFT) to avoid large phase quantization errors. Otherwise, you will not achieve an adequate waveform match.

(d) Repeat part (c) with the phase offsets for all sinewaves set to zero, i.e.,

$$\hat{s}[n] = \sum_{k=1}^{K} \hat{a}_k \cos(\hat{\omega}_k n).$$

How does the waveform shape and its aural perception (using *sound.m*) change relative to your result in part (c)? Try also a random set of phase offsets, using the MATLAB function *rand.m* or *randn.m*, with normalization to the interval $[-\pi, \pi]$. In both cases, justify your observations.

BIBLIOGRAPHY

[1] L.B. Almeida and F.M. Silva, "Variable-Frequency Synthesis: An Improved Harmonic Coding Scheme," *Proc. IEEE Int. Conf. Acoustics, Speech and Signal Processing*, San Diego, CA, pp. 27.5.1–27.5.4, 1984.

[2] D. Anderson, "Speech Analysis and Coding Using a Multi-Resolution Sinusoidal Transform," *Proc. IEEE Conf. Acoustics, Speech, and Signal Processing*, Atlanta, GA, May 1996.

[3] E.R. Banga and C. Garcia-Mateo, "Shape-Invariant Pitch-Synchronous Text-to-Speech Conversion," *Proc. IEEE Int. Conf. Acoustics Speech, and Signal Processing*, Detroit, MI, vol. 4, pp. 656–659, April 1995.

[4] B.A. Blesser, "Audio Dynamic Range Compression for Minimum Perceived Distortion," *IEEE Trans. Audio and Electroacoustics*, vol. AU–17, no. 1, pp. 22–32, March 1969.

[5] A.S. Bregman, *Auditory Scene Analysis: The Perceptual Organization of Sound*, The MIT Press, Cambridge, MA, 1990.

[6] C.E. Cook and M. Bernfeld, *Radar Signals*, Academic Press, New York, NY, 1967.

[7] R.G. Danisewicz, "Speaker Separation of Steady State Vowels," Masters Thesis, Department of Electrical Engineering and Computer Science, Massachusetts Institute of Technology, June 1987.

[8] D.P. Ellis, *A Perceptual Representation of Sound*, Masters Thesis, Department of Electrical Engineering and Computer Science, Massachusetts Institute of Technology, Feb. 1992.

[9] D.P. Ellis, B.L. Vercoe, and T.F. Quatieri, "A Perceptual Representation of Audio for Co-Channel Source Separation," *Proc. 1991 Workshop on Applications of Signal Processing to Audio and Acoustics*, Mohonk Mountain House, New Paltz, NY, 1991.

[10] E.N. Fowle, "The Design of FM Pulse-Compression Signals," *IEEE Trans. Information Theory*, vol. IT–10, no. 10, pp. 61–67, Jan. 1964.

[11] E.B. George, *An Analysis-by-Synthesis Approach to Sinusoidal Modeling Applied to Speech and Music Signal Processing*, Ph.D. Thesis, Georgia Institute of Technology, Nov. 1991.

[12] O. Ghitza, "Speech Analysis/Synthesis Based on Matching the Synthesized and the Original Representations in the Auditory Nerve Level," *Proc. of IEEE Int. Conf. Acoustics, Speech, and Signal Processing*, Tokyo, Japan, pp. 1995–1998, 1986.

[13] M.M. Goodwin, *Adaptive Signal Models: Theory, Algorithms, and Audio Applications*, Kluwer Academic Publishers, Boston, MA, 1992.

[14] D. Griffin and J.S. Lim, "Multi-band Excitation Vocoder," *IEEE Trans. Acoustics, Speech, and Signal Processing*, vol. ASSP–36, no. 8, pp. 1223–1235, 1988.

[15] K.N. Hamdy, M. Ali, and A.H. Tewfik, "Low Bit Rate High Quality Audio Coding with Combined Harmonic and Wavelet Representations," *Proc. IEEE Int. Conf. Acoustics, Speech, and Signal Processing*, Atlanta, GA, vol. 2, pp. 1045–1048, May 1996.

[16] P. Hedelin, "A Tone-Oriented Voice-Excited Vocoder," *Proc. IEEE Int. Conf. Acoustics, Speech, and Signal Processing*, Atlanta, GA, pp. 205–208, 1981.

[17] H.L.F. von Helmholz, *On the Sensations of Tone*, translated by A.J. Ellis for Longmans & Co., 1885, Dover Publications, Inc., New York, NY, 1952.

[18] D.J. Hermes, "Synthesis of Breathy Vowels: Some Research Methods," *Speech Communications*, vol. 10, pp. 497–502, 1991.

[19] J.M. Kates, "Speech Enhancement Based on a Sinusoidal Model," *J. Speech and Hearing Research*, vol. 37, pp. 449–464, April 1994.

[20] E.L. Key, E.N. Fowle, and R.D. Haggarty, "A Method of Pulse Compression Employing Nonlinear Frequency Modulation," Technical Report 207, DDC 312903, Lincoln Laboratory, Massachusetts Institute of Technology, Aug. 1959.

[21] J. Laroche, Y. Stylianou, and E. Moulines, "Hns: Speech Modification Based on a Harmonic+noise Model," *Proc. IEEE Int. Conf. Acoustics, Speech, and Signal Processing*, Minneapolis, MN, vol. 2, pp. 550–553, April 1993.

[22] S.N. Levine, T.S. Verma, and J.O. Smith, "Alias-Free, Multiresolution Sinusoidal Modeling for Polyphonic, Wideband Audio," *Proc. 1997 Workshop on Applications of Signal Processing to Audio and Acoustics*, Mohonk Mountain House, New Paltz, NY, 1997.

[23] M.W. Macon and M.A. Clements, "Sinusoidal Modeling and Modification of Unvoiced Speech," *IEEE Trans. Speech and Audio Processing*, vol. 5, no. 6, pp. 557–560, Nov. 1997.

[24] M.W. Macon and M.A. Clements, "Speech Concatenation and Synthesis Using an Overlap-Add Sinusoidal Model," *Proc. IEEE Conf. Acoustics, Speech, and Signal Processing*, Atlanta, GA, vol. 1, pp. 361–364, May 1996.

[25] J.S. Marques and L.B. Almeida, "New Basis Functions for Sinusoidal Decomposition," *Proc. EUROCON*, Stockholm, Sweden, 1988.

[26] R. Maher and J. Beauchamp, "An Investigation of Vocal Vibrato for Synthesis," *Applied Acoustics*, vol. 30, pp. 219–245, 1990.

[27] J.S. Marques and L.B. Almeida, "Sinusoidal Modeling of Speech: Representation of Unvoiced Sounds with Narrowband Basis Functions," *Proc. EUSIPCO*, 1988.

[28] J.S. Marques and L.B. Almeida, "Frequency-Varying Sinusoidal Modeling of Speech," *IEEE Trans. Acoustics, Speech, and Signal Processing*, vol. 37, no. 5, pp. 763–765, May 1989.

[29] S. McAdams, "Spectral Fusion, Spectral Parsing, and the Formation of Auditory Images," Report no. STAN-M–22, CCRMA, Department of Music, Stanford, CA, May 1984.

[30] R.J. McAulay and T.F. Quatieri, "Speech Analysis-Synthesis Based on a Sinusoidal Representation," *IEEE Trans. Acoustics, Speech, and Signal Processing*, vol. ASSP–34, no. 4, pp. 744–754, Aug. 1986.

[31] R.J. McAulay and T.F. Quatieri, "Speech Analysis/Synthesis Based on a Sinusoidal Representation," Technical Report 693, Lincoln Laboratory, Massachusetts Institute of Technology, May 17, 1985.

[32] R.J. McAulay and T.F. Quatieri, "Pitch Estimation and Voicing Detection Based on a Sinusoidal Speech Model," *IEEE Proc. Int. Conf. Acoustics, Speech, and Signal Processing*, Albuquerque, NM, vol. 2, pp. 249–252, April 1990.

[33] R.J. McAulay and T.F. Quatieri, "Phase Modeling and Its Application to Sinusoidal Transform Coding," *Proc. IEEE Int. Conf. Acoustics, Speech, and Signal Processing*, Tokyo, Japan, pp. 1713–1715, April 1986.

[34] R.J. McAulay and T.F. Quatieri, "Low Rate Speech Coding Based on the Sinusoidal Speech Model," chapter in *Advances in Speech Signal Processing*, S. Furui and M.M. Sondhi, eds., Marcel Dekker, 1992.

[35] R.J. McAulay and T.F. Quatieri, "Computationally Efficient Sinewave Synthesis and its Application to Sinusoidal Transform Coding," *IEEE Int. Conf. Acoustics, Speech, and Signal Processing*, New York, NY, pp. 370–373, April 1988.

[36] J.A. Moorer, "Signal Processing Aspects of Computer Music," *Proc. IEEE*, vol. 65, no. 8, pp. 1108–1137, Aug. 1977.

[37] D. Paul, "The Spectral Envelope Estimation Vocoder," *IEEE Trans. Acoustics, Speech, and Signal Processing*, vol. ASSP–29, no. 4, pp. 786–794, Aug. 1981.

[38] T.F. Quatieri and R.J. McAulay, "Speech Transformations Based on a Sinusoidal Representation," *IEEE Trans. Acoustics, Speech, Signal Processing*, vol. ASSP–34, no. 6, pp. 1449–1464, Dec. 1986.

[39] T.F. Quatieri and R.J. McAulay, "Shape-Invariant Time-Scale and Pitch Modification of Speech," *IEEE Trans. Acoustics, Speech, and Signal Processing*, vol. 40, no. 3, pp. 497–510, March 1992.

[40] T.F. Quatieri and R.J. McAulay, "Phase Coherence in Speech Reconstruction for Enhancement and Coding Applications," *Proc. IEEE Int. Conf. Acoustics, Speech, and Signal Processing*, Glasgow, Scotland, May 1989.

[41] T.F. Quatieri and R.J. McAulay, "Audio Signal Processing Based on Sinusoidal Analysis/ Synthesis," chapter in *Applications of Digital Signal Processing to Audio and Acoustics*, M. Kahrs and K. Brandenburg, eds., Kluwer Academic Publishers, Boston, MA, 1998.

[42] T.F. Quatieri and R.J. McAulay, "Peak-to-rms Reduction of Speech Based on a Sinusoidal Model," *IEEE Trans. Signal Processing*, vol. 39, no. 2, pp. 273–288, Feb. 1991.

[43] T.F. Quatieri, J. Lynch, M.L. Malpass, R.J. McAulay, and C. Weinstein, "Speech Processing for AM Radio Broadcasting," Technical Report 681, Lincoln Laboratory, Massachusetts Institute of Technology, Nov. 1991.

[44] T.F. Quatieri, R.B. Dunn, and R.J. McAulay, "Signal Enhancement in AM–FM Interference," Technical Report 993, Lincoln Laboratory, Massachusetts Institute of Technology, May 1994.

[45] T.F. Quatieri and R.G. Danisewicz, "An Approach to Co-Channel Talker Interference Suppression using a Sinusoidal Model for Speech," *IEEE Trans. Acoustics, Speech, and Signal Processing*, vol. 38, no. 1, pp. 56–69, Jan. 1990.

[46] M.A. Ramalho, *The Pitch Mode Modulation Model with Applications in Speech Processing*, Ph.D. Thesis, Department of Electrical Engineering, Rutgers University, Jan. 1994.

[47] J.C. Rutledge, *Time-Varying, Frequency-Dependent Compensation for Recruitment of Loudness*, Ph.D. Thesis, Georgia Institute of Technology, Dec. 1989.

[48] M.R. Schroeder, "Synthesis of Low-Peak-Factor Signals and Binary Sequences with Low Autocorrelation," *IEEE Trans. Information Theory*, vol. IT–16, pp. 85–89, Jan. 1970.

[49] M.R. Schroeder, *Number Theory in Science and Communication*, Springer-Verlag, New York, NY, 2nd Enlarged Edition, 1986.

[50] X. Serra, *A System for Sound Analysis/Transformation/Synthesis Based on a Deterministic Plus Stochastic Decomposition*, Ph.D. Thesis, CCRMA, Department of Music, Stanford University, 1989.

[51] X. Serra and J.O. Smith, III, "Spectral Modeling Synthesis: A Sound Analysis/Synthesis System Based on a Deterministic Plus Stochastic Decomposition," *Computer Music Journal*, vol. 14, no. 4, pp. 12–24, Winter 1990.

[52] R. Smits and B. Yegnanarayana, "Determination of Instants of Significant Excitation in Speech Using Group Delay Function," *IEEE Trans. Speech and Audio Processing*, vol. 3, no. 5, pp. 325–333, Sept. 1995.

[53] E. Tellman, L. Haken, and B. Holloway, "Timbre Morphing of Sounds with Unequal Number of Features," *J. Audio Engineering Society*, vol. 43, no. 9, pp. 678–689, Sept. 1995.

[54] C.W. Therrien, R. Cristi, and D.E. Allison, "Methods for Acoustic Data Synthesis," *Proc. IEEE 1994 Digital Signal Processing Workshop*, Yosemite National Park, Oct. 1994.

[55] H. Van Trees, *Detection, Estimation and Modulation Theory, Part I*, John Wiley and Sons, New York, NY, 1968.

CHAPTER 10

Frequency-Domain Pitch Estimation

10.1 Introduction

We have seen throughout the text that different approaches to speech analysis/synthesis naturally lead to different methods of pitch and voicing estimation. For example, in homomorphic analysis, the location of quefrency peaks (of lack thereof) in the cepstrum provides a pitch and voicing estimate. Likewise, the distance between primary peaks in the linear prediction error yields an estimate of the pitch period. We also saw that the wavelet transform lends itself to pitch period estimation by way of the correlation of maxima in filter-bank outputs across different scales; this "parallel processing" approach is similar in style to that of an early successful pitch estimation method conceived by Gold and Rabiner that looks across a set of impulse trains generated by different peaks and valleys in the signal [1]. These latter methods based on linear prediction, the wavelet transform, and temporal peaks and valleys, provide not only a pitch-period estimate, but also an estimate of the time occurrence of the glottal pulse, an important parameter in its own right for a variety of applications. We can think of all the above methods as *time-domain* approaches to pitch, voicing, and glottal pulse time estimation. In this chapter, we take an alternative view to these estimation problems in the frequency domain motivated by the sinewave representation of Chapter 9.[1]

In Chapter 9, it was shown that it is possible to generate synthetic speech of very high quality using an analysis/synthesis system based on a sinusoidal speech model which, except in updating the average pitch to adjust the width of the analysis window, made no explicit use of pitch and voicing. Pitch and voicing, however, played an important role in accomplishing

---

[1] We cannot hope to cover the vast variety of all pitch, voicing, and glottal pulse time estimators in the time and frequency domain. By focusing on a few specific classes of estimators, however, we illustrate the goals and problems common to the many approaches.

sinewave-based modification in Chapter 9 and will also play an important role in reducing the bit rate in sinewave-based speech coding in Chapter 12, much as they do in the speech analysis/synthesis and coding based on linear prediction. In this chapter, we will see that the sinewave representation brings new insights to the problems of pitch estimation, voicing detection, and glottal pulse time estimation. Specifically, pitch estimation can be thought of as fitting a harmonic set of sinewaves to the measured set of sinewaves, and the accuracy of the harmonic fit is an indicator of the voicing state. It is the purpose of this chapter to explore this idea in detail. The result is a powerful pitch and voicing algorithm which has become a basic component in all of the applications of the sinewave system.

We begin with a simple pitch estimator based on the autocorrelation function, which becomes the basis for transiting into the frequency domain through a sinewave representation. A variety of sinewave-based pitch estimators of increasing complexity and accuracy are then derived. We end this chapter with an application of the sinewave model to glottal pulse time estimation, and finally a generalization of the sinewave model to multi-band pitch and voicing estimation [2].

10.2 A Correlation-Based Pitch Estimator

Consider a discrete-time short-time sequence given by

$$s_n[m] = s[m]w[n - m]$$

where $w[n]$ is an analysis window of duration N_w. The short-time autocorrelation function $r_n[\tau]$ is defined by

$$r_n[\tau] = s_n[\tau] * s_n[-\tau] = \sum_{m=-\infty}^{\infty} s_n[m]s_n[m + \tau].$$

When $s[m]$ is periodic with period P, $r_n[\tau]$ contains peaks at or near the pitch period, P. For unvoiced speech, no clear peak occurs near an expected pitch period. Typical sequences $r_n[\tau]$ for different window lengths were shown in Figure 5.6 of Chapter 5. We see that the location of a peak (or lack thereof) in the pitch period range provides a pitch estimate and voicing decision. This is similar to the strategy used in determining pitch and voicing from the cepstrum in homomorphic analysis.

It is interesting to observe that the above correlation pitch estimator can be obtained more formally by minimizing, over possible pitch periods ($P > 0$), the error criterion given by

$$E[P] = \sum_{m=-\infty}^{\infty} (s_n[m] - s_n[m + P])^2. \qquad (10.1)$$

Minimizing $E[P]$ with respect to P yields

$$\hat{P} = \max_P \left(\sum_{m=-\infty}^{\infty} s_n[m]s_n[m + P] \right) \qquad (10.2)$$

where $P > \epsilon$, i.e., P is sufficiently far from zero (Exercise 10.1). This alternate view of autocorrelation pitch estimation is used in the following section.

The autocorrelation function is a measure of "self-similarity," so we expect that it peaks near P for a periodic sequence. Partly due to the presence of the window, however, the peak at the pitch period does not always have the greatest amplitude. We saw in Chapter 5 that the envelope of the short-time autocorrelation function of a periodic waveform decreases roughly linearly with increasing P (Exercise 5.2). A longer window helps in assuring the peak at $\tau = P$ is largest, but a long window causes other problems when the sequence $s[m]$ is not exactly periodic. Peaks in the autocorrelation of the vocal tract impulse response, as well as peaks at multiple pitch periods,[2] may become larger than the peak at $\tau = P$ due to time-variations in the vocal tract and pitch. Another problem arises as a result of the interaction between the pitch and the first formant. If the formant bandwidth is narrow relative to the harmonic spacing (so that the vocal tract impulse response decays very slowly within a pitch period), the correlation function may reflect the formant frequency rather than the underlying pitch. Nonlinear time-domain processing techniques using various types of waveform *center-clipping* algorithms have been developed to alleviate this problem [11],[12],[13].

In the autocorrelation method of pitch estimation, Equation (10.2), the effective limits on the sum get smaller as the candidate pitch period P increases, i.e., as the windowed data segment shifts relative to itself, causing the autocorrelation to have the roughly linear envelope alluded to above. If we could extrapolate the periodic data segment so as to pretend that a longer periodic segment exists, we could avoid the effect of the window. Extrapolation has two advantages. First, we can use the full interval of duration N_w without assuming the data is zero outside the window duration and, second, we can make the interval long while maintaining stationarity (Figure 10.1). In the next section, we exploit this extrapolation concept together with the error criterion of Equation (10.1).

10.3 Pitch Estimation Based on a "Comb Filter"

One can imagine the design of a pitch estimator in the frequency domain by running the waveform through a *comb filter* with peaks at multiples of a hypothesized fundamental frequency (pitch) and selecting the comb with a pitch that matches the waveform harmonics and thus, hopefully, giving the largest energy. In this section, we derive a type of comb filter for pitch estimation, not by straightforward filtering, but based on the extrapolation and error criterion concepts of the previous section.

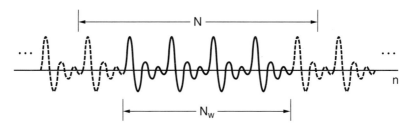

Figure 10.1 Extrapolation of a short-time segment from a periodic sequence.

---

[2] Two common problems in pitch estimation are referred to as *pitch-halving* and *pitch-doubling*, whereby the pitch estimate is half or double the true pitch.

One approach to extrapolating a segment of speech beyond its analysis window as in Figure 10.1 is through the sinewave model of Chapter 9. Suppose that, rather than using the short-time segment $s_n[m]$ (where the subscript n refers to the window location), we use its sinewave representation. Then we form a sequence $\tilde{s}[m]$ expressed as

$$\tilde{s}[m] = \sum_{k=1}^{K} A_k e^{jm\omega_k + \theta_k} \tag{10.3}$$

where the sinewave amplitudes, frequencies, and phases A_k, ω_k, and θ_k are obtained from the short-time segment[3] $s_n[m]$, but where $\tilde{s}[m]$ is thought of as infinite in duration. The sinewave representation, therefore, can extrapolate the signal beyond the analysis window duration. In this chapter, as in Equation (10.3), it is particularly convenient to use only the complex sinewave representation; hence the real part notation introduced in Chapter 9 has been omitted.

Consider now the error criterion in Equation (10.1) over an extrapolated interval of length N. Substituting Equation (10.3) into Equation (10.1), we have

$$E[P] = \frac{1}{N} \sum_{m=-(N-1)/2}^{(N-1)/2} \left| \sum_{k=1}^{K} A_k \left[e^{j(m\omega_k + \theta_k)} - e^{j((m+P)\omega_k + \theta_k)} \right] \right|^2$$

$$= \frac{1}{N} \sum_{m=-(N-1)/2}^{(N-1)/2} \left| \sum_{k=1}^{K} A_k e^{j(m\omega_k + \theta_k)} [1 - e^{jP\omega_k}] \right|^2 \tag{10.4}$$

where the extrapolation interval N is assumed odd and centered about the origin. Note that we could have begun with Equation (10.2), i.e., the autocorrelation perspective, rather than Equation (10.1). Observe also that, in Equation (10.4), the limits on the sum are $[-(N-1)/2, (N-1)/2]$ for all P and that data truncation does not occur. Rearranging terms in Equation (10.4), we obtain

$$E[P] = \frac{1}{N} \sum_{m=-(N-1)/2}^{(N-1)/2} \left\{ \sum_{k=1}^{K} A_k e^{j(m\omega_k + \theta_k)} \left[1 - e^{-jP\omega_k} \right] \right\}$$

$$\times \left\{ \sum_{\ell=1}^{K} A_\ell e^{-j(m\omega_\ell + \theta_\ell)} \left[1 - e^{jP\omega_\ell} \right] \right\}$$

$$= e^{j(\theta_k - \theta_\ell)} \sum_{k=1}^{K} \sum_{\ell=1}^{K} A_k A_\ell \left(1 - e^{jP\omega_k} \right) \left(1 - e^{-jP\omega_\ell} \right)$$

$$\times \underbrace{\frac{1}{N} \sum_{m=-(N-1)/2}^{(N-1)/2} e^{jm(\omega_k - \omega_\ell)}}_{q(\omega_k - \omega_\ell)},$$

---

[3] The sinewave amplitudes, frequencies, and phases A_k, ω_k, and θ_k are *fixed* parameters from one analysis window on one frame and not the interpolation functions used for synthesis in Chapter 9. Recall that θ_k is a phase offset relative to the analysis frame center, as described in Section 9.3.

for which we write

$$q(\omega_k - \omega_\ell) = \frac{1}{N} \sum_{m=-(N-1)/2}^{(N-1)/2} e^{jm(\omega_k - \omega_\ell)}$$

$$= \frac{1}{N} \frac{\sin\left[\frac{N}{2}(\omega_k - \omega_l)\right]}{\sin\left[\frac{(\omega_k - \omega_l)}{2}\right]}.$$

If we let the extrapolation interval N go to infinity, the function $q(\omega)$ approaches zero except at the origin so that

$$q(\omega_k - \omega_\ell) \approx \begin{cases} 0 & \text{for } \omega_k - \omega_\ell \neq 0 \\ 1 & \text{for } \omega_k = \omega_\ell \end{cases}$$

then

$$E[P] \approx \sum_{k=1}^{K} A_k^2 |1 - e^{jP\omega_k}|^2$$

$$= \sum_{k=1}^{K} A_k^2 \underbrace{\left(1 - e^{jP\omega_k}\right)\left(1 - e^{-jP\omega_k}\right)}_{2 - 2\cos(P\omega_k)}$$

$$= \sum_{k=1}^{K} 2A_k^2 [1 - \cos(P\omega_k)].$$

Let $P = \frac{2\pi}{\omega_o}$ where ω_o is the fundamental frequency. Then

$$E[P] = \sum_{k=1}^{K} A_k^2 \left[1 - \cos\left(\frac{2\pi}{\omega_o}\omega_k\right)\right] \tag{10.5}$$

where for convenience we have deleted the scale factor of two in Equation (10.5). We now want to minimize $E[P]$, as expressed in Equation (10.5), with respect to ω_o. To give this minimization an intuitive meaning, we rewrite Equation (10.5) as

$$E[P] = \sum_{k=1}^{K} A_k^2 - \sum_{k=1}^{K} A_k^2 \cos\left(\frac{2\pi}{\omega_o}\omega_k\right). \tag{10.6}$$

Minimization of $E[P]$ is then equivalent to maximizing with respect to ω_o the term

$$Q(\omega_o) = \sum_{k=1}^{K} A_k^2 \cos\left(\frac{2\pi}{\omega_o}\omega_k\right). \tag{10.7}$$

We refer to $Q(\omega)$ as a *likelihood function* because as the value of the function increases then so does the likelihood of the hypothesized fundamental frequency as the true value. One way

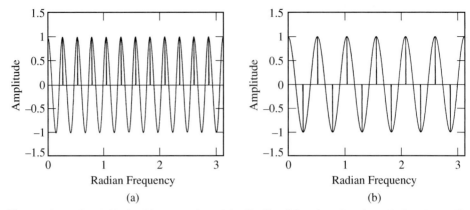

Figure 10.2 "Comb-filtering" interpretation of the likelihood function, Equation (10.7): (a) sampling by harmonic frequencies with the candidate pitch equals to the true pitch; (b) sampling by harmonic frequencies with the candidate pitch equal to twice the true pitch. For the latter case, the cancellation effect shown in panel (b) reduces the likelihood function. A candidate of half the true pitch, however, will not change the likelihood.

to view $Q(\omega_o)$ is to first replace ω_k in Equation (10.7) by a continuous ω. For each ω_o, the continuous function

$$F(\omega) \;=\; \cos\left(\frac{2\pi}{\omega_o}\omega\right)$$

is sampled at ω_k, $\forall_k$ and each $F(\omega_k)$ is weighted by A_k^2, and these weighted values are summed to form Equation (10.7). We can, loosely speaking, think of this as "comb-filtering" for each ω_o, and we want the ω_o whose comb filter has the maximum output. Figure 10.2a shows an example sampling of the function $F(\omega)$.

   If the ω_k's are multiples of a fundamental frequency $\tilde{\omega}_o$, and if $\omega_o = \tilde{\omega}_o$, we have $F(\omega_k) = 1$ (as in Figure 10.2a) and $E[P] = 0$ and minimization is achieved. Specifically,

$$\cos\left(\frac{2\pi}{\omega_o}\omega_k\right) \;=\; \cos\left(2\pi k\frac{\tilde{\omega}_o}{\tilde{\omega}_o}\right) \;=\; 1$$

and thus, from Equation (10.6) $E[P] = 0$. In Figure 10.2b we see that the estimator is insensitive to *pitch doubling* due to a cancellation effect. Notice, however, a disturbing feature of this pitch estimator. A fundamental frequency estimate of $\frac{\omega_o}{2}$, i.e., half the true pitch, will also yield zero error. Thus, the solution is *ambiguous*. These properties are illustrated in the following example:

EXAMPLE 10.1   Figure 10.3a,b,d shows the result of Equation (10.7) for a true pitch of $\tilde{\omega}_o = 50, 100$, and $200\,\mathrm{Hz}$ and where $A_k = 1$. A frequency increment of $0.5\,\mathrm{Hz}$ was used for the hypothesized pitch candidates (Exercise 10.12). One sees multiple peaks in the pitch likelihood function and thus the pitch estimate ambiguity. Nevertheless, the example also shows that the correct pitch is given by the last large peak, i.e., the peak of greatest frequency. There are no peaks beyond the true frequency because there is a cancellation effect for multiples of the true pitch (as in Figure 10.2b), and thus the likelihood function falls rapidly. Finally, Figure 10.3c shows the effect of white noise (added to the

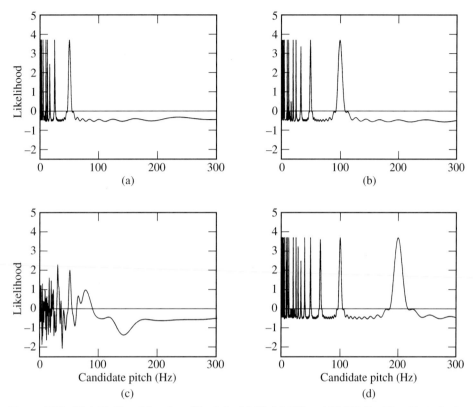

Figure 10.3   Pitch likelihood function $Q(\omega_o)$ for (a) 50, (b) 100, and (d) 200 Hz. The effect of noise on (a) is shown in (c).

measured frequencies) on the likelihood function for a true pitch of 50 Hz. In this case, although the last large peak of $Q(\omega_o)$ falls at roughly 50 Hz, there is little confidence in the estimate due to the multiple peaks and, in particular, a larger peak at about 25 Hz and 50 Hz.   ▲

This problem of *pitch-halving* (i.e., underestimating the pitch by a factor of two) is typical of many pitch estimators such as the homomorphic and autocorrelation pitch estimators. In fact, *explicit* comb filtering approaches have also been applied, but the tendency of these methods to suffer from pitch-halving has limited their use. Our goal is to design a pitch likelihood function that is characterized by a distinct peak at the true pitch. In the following Section 10.4, we propose an alternative sinewave-based pitch estimator which utilizes additional information. This new estimator is capable of resolving the above ambiguity.

10.4   Pitch Estimation Based on a Harmonic Sinewave Model

The previous sinewave-based correlation pitch estimator was derived by minimizing the mean-squared error between an estimated sinewave model and itself shifted by P samples. In this section, we *fit* a sinewave model, with unknown amplitudes, phases, and *harmonic* frequencies, to a waveform measurement [5],[8]. Although the resulting pitch estimator is prone to pitch doubling ambiguity, we show that, with *a priori* knowledge of the vocal tract spectral envelope, the pitch doubling problem is alleviated. This approach to pitch estimation leads naturally to a measure of the degree of voicing within a speech segment. Methods to evaluate the pitch estimator are also described.

10.4.1 Parameter Estimation for the Harmonic Sinewave Model

Consider a sinusoidal waveform model with unknown amplitudes and phases, and with harmonic frequencies:

$$\hat{s}(n; \omega_o, \underline{B}, \underline{\phi}) = \sum_{k=1}^{K(\omega_o)} B_k \exp[j(nk\omega_o + \phi_k)],$$

where ω_o is an unknown fundamental frequency, where $\underline{B}$ and $\underline{\phi}$ represent vectors of unknown amplitudes and phases, $\{B_k\}$ and $\{\phi_k\}$, respectively, and where $K(\omega_o)$ is the number of harmonics in the speech bandwidth. (For clarity, we have changed notation from $e^{j\phi}$ to $\exp[j\phi]$.) A reasonable estimation criterion is to seek the minimum mean-squared error between the harmonic model and the measured speech waveform:

$$E(\omega_o, \underline{B}, \underline{\phi}) = \frac{1}{N_w} \sum_{n=-(N_w-1)/2}^{(N_w-1)/2} |s[n] - \hat{s}[n; \omega_o, \underline{B}, \underline{\phi}]|^2 \qquad (10.8)$$

where we assume that the analysis window duration N_w is odd and the window is centered about the time origin. Then we can show that

$$E(\omega_o, \underline{B}, \underline{\phi}) \approx E_s - 2\text{Re} \sum_{k=1}^{K(\omega_o)} B_k \exp(-j\phi_k)S(k\omega_o) + \sum_{k=1}^{K(\omega_o)} B_k^2, \qquad (10.9)$$

where $S(\omega)$ represents one slice of the STFT of the speech waveform over the interval $[-(N_w - 1)/2, (N_w - 1)/2]$ and

$$E_s = \sum_{-(N_w-1)/2}^{(N_w-1)/2} |s[n]|^2. \qquad (10.10)$$

Minimizing Equation (10.9) with respect to $\underline{\phi}$, we see that the phase estimates $\hat{\phi}_k$ are

$$\hat{\phi}_k = \angle S(k\omega_o) \qquad (10.11)$$

so that

$$E(\omega_o, \underline{B}, \hat{\underline{\phi}}) \approx E_s - 2 \sum_{k=1}^{K(\omega_o)} B_k |S(k\omega_o)| + \sum_{k=1}^{K(\omega_o)} B_k^2, \qquad (10.12)$$

and minimizing Equation (10.12) with respect to $\underline{B}$:

$$\frac{\partial E}{\partial B_k} = -2|S(k\omega_o)| + 2B_k = 0, \qquad (10.13)$$

so that

$$B_k = |S(k\omega_o)|.$$

Therefore,

$$E(\omega_o, \underline{\hat{B}}, \underline{\hat{\phi}}) \approx E_s - \sum_{k=1}^{K(\omega_o)} |S(k\omega_o)|^2. \tag{10.14}$$

The reader is asked to work through the algebraic steps of this derivation in Exercise 10.2. Thus, the optimal ω_o is given by

$$\omega_o^* = \max_{\omega_o} \sum_{k=1}^{K(\omega_o)} |S(k\omega_o)|^2 \tag{10.15}$$

where we assume that $\omega_o > \epsilon$ (a small positive value close to zero) to avoid a bias toward a low fundamental frequency. As in the previous estimator, we see that this estimator acts like a comb filter, but here samples of the measured spectrum are at candidate harmonics. For a perfectly periodic waveform, ω_o would be chosen to correspond to the harmonic peaks in $|S(k\omega_o)|$. This criterion could lead, however, to a pitch-halving error similar to what we found in the sinewave-based correlation pitch estimator (Exercise 10.2).

Consider now giving ourselves some additional information; in particular, we assume a known vocal-tract spectral envelope,[4] $|H(\omega)|$. With this *a priori* information, we will show that the resulting error criterion can *resolve the pitch-halving ambiguity* [5],[8]. The following Section 10.4.2 derives the pitch estimator, while Section 10.4.4 gives a more intuitive view of the approach with respect to time-frequency resolution considerations.

10.4.2 Parameter Estimation for the Harmonic Sinewave Model with *a priori* Amplitude

The goal again is to represent the speech waveform by another for which all of the frequencies are harmonic, but now let us assume *a priori* knowledge of the vocal tract spectral envelope, $|H(\omega)|$. Under the assumption that the excitation amplitudes are unity, i.e., $a_k(t) = 1$ in Equation (9.5), $|H(\omega)|$ also provides an envelope for the sinewave amplitudes A_k. The harmonic sinewave model then becomes

$$\hat{s}[n; \omega_o, \underline{\phi}] = \sum_{k=1}^{K(\omega_o)} \bar{A}(k\omega_o) \exp[j(nk\omega_o + \phi_k)] \tag{10.16}$$

where ω_o is the fundamental frequency (pitch), $K(\omega_o)$ is the number of harmonics in the speech bandwidth, $\bar{A}(\omega) = |H(\omega)|$ is the vocal tract envelope, and $\underline{\phi} = \{\phi_1, \phi_2, \cdots, \phi_{K(\omega_o)}\}$ represents the phases of the harmonics. We would like to estimate the pitch ω_o and the phases $\{\phi_1, \phi_2, \cdots, \phi_{K(\omega_o)}\}$ so that $\hat{s}[n]$ is as close as possible to the speech measurement $s[n]$, according to some meaningful criterion. While a number of methods can be used for estimating the envelope $\bar{A}(\omega)$, for example, linear prediction or homomorphic estimation techniques, it is desirable to use a method that yields an envelope that passes through the measured sinewave

---

[4] We do not utilize the full capacity of the sinewave pitch estimator since we could have also supplied an *a priori* phase envelope, thus giving temporal alignment information (Exercise 10.2).

amplitudes. Such a technique has been developed in the Spectral Envelope Estimation Vocoder (SEEVOC) [10]. This estimator will also be used later in this chapter as the basis of the minimum-phase analysis for estimating the source excitation onset times and so we postpone its description (Section 10.5.3).[5]

A reasonable estimation criterion is to seek the minimum of the mean-squared error (MSE),

$$E(\omega_o, \underline{\phi}) = \frac{1}{N_w} \sum_{n=-(N_w-1)/2}^{(N_w-1)/2} |s[n] - \hat{s}[n; \omega_o, \underline{\phi}]|^2 \qquad (10.17)$$

over ω_o and $\underline{\phi}$. The MSE in Equation (10.17) can be expanded as

$$E(\omega_o, \underline{\phi}) = \frac{1}{N_w} \sum_{n=-(N_w-1)/2}^{(N_w-1)/2} \left\{ |s[n]|^2 - 2\mathrm{Re}(s[n]\hat{s}^*[n; \omega_o, \underline{\phi}]) + |\hat{s}[n; \omega_o, \underline{\phi}]|^2 \right\}. \qquad (10.18)$$

Observe that the first term of Equation (10.18) is the average energy (power) in the measured signal. We denote this average by P_s [an averaged version of the previously defined total energy in Equation (10.10)], i.e.,

$$P_s = \frac{1}{N_w} \sum_{n=-(N_w-1)/2}^{(N_w-1)/2} |s[n]|^2. \qquad (10.19)$$

Substituting Equation (10.16) in the second term of Equation (10.18) leads to the relation

$$\sum_{n=-(N_w-1)/2}^{(N_w-1)/2} s[n]\hat{s}^*[n; \omega_o, \underline{\phi}] = \sum_{k=1}^{K(\omega_o)} \bar{A}(k\omega_o) \exp(j\phi_k) \sum_{n=-(N_w-1)/2}^{(N_w-1)/2} s[n] \exp(-jnk\omega_o). \qquad (10.20)$$

Finally, substituting Equation (10.16) in the third term of Equation (10.18) leads to the relation

$$\frac{1}{N_w} \sum_{n=-(N_w-1)/2}^{(N_w-1)/2} |\hat{s}(n; \omega_o, \underline{\phi})|^2 \approx \sum_{k=1}^{K(\omega_o)} \bar{A}^2(k\omega_o),$$

where the approximation is valid provided the analysis window duration satisfies the condition $N_w \gg \frac{2\pi}{\omega_o}$, which is more or less assured by making the analysis window 2.5 times the average pitch period. Letting

$$S(\omega) = \frac{1}{N_w} \sum_{n=-(N_w-1)/2}^{(N_w-1)/2} s[n] \exp(-jn\omega) \qquad (10.21)$$

---

[5] We will see that, in the application of source onset time estimation, it is appropriate to linearly interpolate between the successive sinewave amplitudes. In the application to mean-squared-error pitch estimation in this section, however, the main purpose of the envelope is to eliminate pitch ambiguities. Since the linearly interpolated envelope could affect the fine structure of the mean-squared-error criterion through its interaction with the measured peaks in the correlation operation (to follow later in this section), better performance is obtained by using piecewise-constant interpolation between the SEEVOC peaks.

denote one slice of the short-time Fourier transform (STFT) of the speech signal and using this in Equation (10.20), then the MSE in Equation (10.18) becomes (Exercise 10.2)

$$E(\omega_o, \underline{\phi}) = P_s - 2\text{Re} \sum_{k=1}^{K(\omega_o)} \bar{A}(k\omega_o) \exp(-j\phi_k) S(k\omega_o) + \sum_{k=1}^{K(\omega_o)} \bar{A}^2(k\omega_o). \quad (10.22)$$

Since the phase parameters $\{\phi_k\}_{k=1}^{K(\omega_o)}$ affect only the second term in Equation (10.22), the MSE is minimized by choosing the phase estimates

$$\hat{\phi}_k = \angle S(k\omega_o)$$

and the resulting MSE is given by

$$E(\omega_o) = P_s - 2 \sum_{k=1}^{K(\omega_o)} \bar{A}(k\omega_0) |S(k\omega_o)| + \sum_{k=1}^{K(\omega_o)} \bar{A}^2(k\omega_o), \quad (10.23)$$

where the second term is reminiscent of a correlation function in the frequency domain. The unknown pitch affects only the second and third terms in Equation (10.23), and these can be combined by defining

$$\rho(\omega_o) = \sum_{k=1}^{K(\omega_o)} \bar{A}(k\omega_o)[|S(k\omega_o)| - \frac{1}{2}\bar{A}(k\omega_o)]. \quad (10.24)$$

The smooth weighting function $\bar{A}(k\omega_o)$ biases the method away from excessively low pitch estimates. The MSE can then be expressed as

$$E(\omega_o) = P_s - 2\rho(\omega_o). \quad (10.25)$$

Because the first term is a known constant, the minimum mean-squared error is obtained by maximizing $\rho(\omega_o)$ over ω_o.

It is useful to manipulate this metric further by making explicit use of the sinusoidal representation of the input speech waveform. Assume, as in Section 10.3, that a frame of the input speech waveform has been analyzed in terms of its sinusoidal components using the analysis system described in Chapter 9.[6] The measured speech data $s[n]$ is, therefore, represented as

$$s[n] = \sum_{k=1}^{K} A_k \exp[j(n\omega_k + \theta_k)] \quad (10.26)$$

---

[6] This mean-squared-error pitch extractor is predicated on the assumption that the input speech waveform has been represented in terms of the sinusoidal model. This implicitly assumes that the analysis has been performed using a Hamming window approximately two and one-half times the average pitch. It seems, therefore, that the pitch must be known in order to estimate the average pitch that is needed to estimate the pitch. This circular dilemma can be broken by using some other method to estimate the average pitch based on a fixed window. Since only an average pitch value is needed, the estimation technique does not have to be accurate on every frame; hence, any of the well-known techniques can be used.

where $\{A_k, \omega_k, \theta_k\}_{k=1}^{K}$ represents the amplitudes, frequencies, and phases of the K measured sinewaves. The sinewave representation allows us to extrapolate the speech measurement beyond the analysis window duration N_w to a larger interval N, as we described earlier. With a sinewave representation, it is straightforward to show that the signal power is given by the approximation

$$P_s \approx \sum_{k=1}^{K} A_k^2$$

and substituting the sinewave representation in Equation (10.26) in the short-time Fourier transform defined in Equation (10.21) leads to the expression

$$S(\omega) = \sum_{k=1}^{K} A_k \exp(j\theta_k) \operatorname{sinc}(\omega_k - \omega),$$

where

$$\operatorname{sinc}(x) = \frac{1}{N} \sum_{n=-(N-1)/2}^{(N-1)/2} \exp(jnx) = \frac{\sin(N\frac{x}{2})}{N \sin\frac{x}{2}}.$$

Because the sinewaves are well-resolved, the magnitude of the STFT can then be approximated by

$$|S(\omega)| \approx \sum_{k=1}^{K} A_k D(\omega_k - \omega)$$

where $D(x) = |\operatorname{sinc}(x)|$. The MSE criterion then becomes

$$\rho(\omega_o) = \sum_{k=1}^{K(\omega_o)} \bar{A}(k\omega_o) \left[\sum_{\ell=1}^{K} A_\ell D(\omega_\ell - k\omega_o) - \frac{1}{2}\bar{A}(k\omega_o) \right] \qquad (10.27)$$

where ω_ℓ are the frequencies of the sinewave representation in Equation (10.26) and ω_o is the candidate pitch.

To gain some insight into the meaning of this criterion, suppose that the input speech is periodic with pitch frequency ω^*. Then (barring measurement error) $\omega_\ell = \ell\omega^*$, $A_\ell = \bar{A}(\ell\omega^*)$, and

$$\rho(\omega^*) = \frac{1}{2} \sum_{k=1}^{K(\omega^*)} \bar{A}^2(k\omega^*).$$

When ω_o corresponds to sub-multiples of the pitch, the first term in Equation (10.27) remains unchanged, since $D(\omega_\ell - k\omega_o) = 0$ at the submultiples, but the second term, because it is an envelope and always non-zero, will increase at the submultiples of ω^*. As a consequence,

$$\rho\left(\frac{\omega^*}{m}\right) < \rho(\omega^*), \quad m = 2, 3, \cdots$$

which shows that *the MSE criterion leads to unambiguous pitch estimates*. To see this property more clearly, consider the example illustrated in Figure 10.4. In this example, the true pitch is

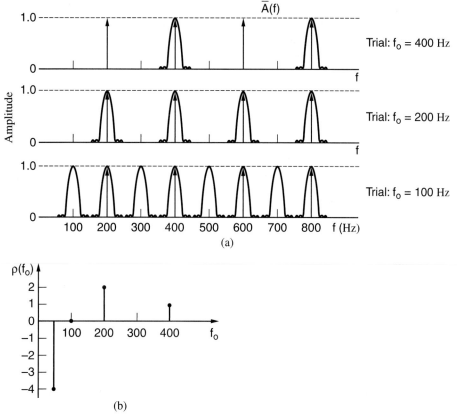

Figure 10.4   Illustration of sinewave-based pitch estimator for a periodic input: (a) the first term in Equation (10.28) with $D(f - kf_o)$ (a set of "comb filters") spaced by the candidate fundamental frequency f_o, where f denotes the frequency in Hertz. The comb is sampled at the measured frequencies $f_\ell = \ell f^*$; (b) values of the likelihood function $\rho(f_o)$ for pitch candidates 50, 100, 200, and 400 Hz. The true pitch is 200 Hz.

200 Hz and the sinewave envelope is constant at unity. One can then simplify Equation (10.27) as

$$\rho(\omega_o) = \sum_{k=1}^{K(\omega_o)} \sum_{\ell=1}^{K} D(\omega_\ell - k\omega_o) - \frac{1}{2} \sum_{\ell=1}^{K(\omega_o)} \bar{A}(k\omega_o)$$

$$= \sum_{\ell=1}^{K} \left[\sum_{k=1}^{K(\omega_o)} D(\omega_\ell - k\omega_o) \right] - \frac{1}{2} K(\omega_o) \qquad (10.28)$$

where $K(\omega_o)$ is the number of harmonics over a fixed bandwidth of 800 Hz. The first term in Equation (10.28) corresponds to laying down a set of "comb filters" $D(\omega - k\omega_o)$ (yet again different in nature from those previously described) spaced by the candidate fundamental frequency ω_o. The comb is then sampled at the measured frequencies $\omega_\ell = \ell\omega*$ and the samples are summed. Finally, the resulting value is reduced by half the number of harmonics over the fixed band.

For the candidate (trial) of f_0=200 Hz, $\rho(f_0)$=2, as illustrated in Figure 10.4b. For the candidate f_0=100 Hz (pitch-halving), the first term is the same (as for f_0=200 Hz), but the second term decreases (negatively) so that $\rho(f_0) = 0$. The remaining two cases in Figure 10.4b are straightforward to evaluate. This argument with constant $\bar{A}(\omega)$ holds more generally since we can write Equation (10.27) as

$$\rho(\omega_o) = \sum_{\ell=1}^{K} A_\ell \left[\sum_{k=1}^{K} \bar{A}(k\omega_o)D(\omega_\ell - k\omega_o) \right] - \frac{1}{2} \sum_{k=1}^{K(\omega_o)} \bar{A}^2(k\omega_o), \qquad (10.29)$$

where the first term is a correlation-like term [similar in style to the frequency-domain correlation-based pitch estimators, i.e., "comb filters," in Equations (10.7) and (10.15)] and the second term is the generalized negative compensation for low-frequency fundamental candidates. Possibly the most significant attribute of the sinewave-based pitch extractor is that the usual problems with pitch-halving and pitch-doubling do not occur with the new error criterion (Exercise 10.2). This pitch estimator has been further refined to improve its resolution, resolve problems with formant-pitch interaction (as alluded to in the context of the autocorrelation pitch estimator), and improve robustness in additive noise by exploiting the auditory masking principle that small tones are masked by neighboring high tones [8]. (This auditory masking principle is described in Chapter 13 in the context of speech enhancement.) The following example compares the sinewave-based pitch estimator for voiced and unvoiced speech:

EXAMPLE 10.2   In one implementation of the sinewave-based MSE pitch extractor, the speech is sampled at 10 kHz and analyzed using a 1024-point FFT. The sinewave amplitudes and frequencies are determined over a 1000-Hz bandwidth. In Figure 10.5(b), the measured amplitudes and frequencies are shown along with the piecewise-constant SEEVOC envelope for a voiced speech segment. Figure 10.5(c) is a plot of the first term in Equation (10.29) over a candidate pitch range from 38 Hz to 400 Hz and the inherent ambiguity of the correlator (comb-filter) is apparent. It should be noted that the peak at the correct pitch is largest, but during steady vowels the ambiguous behavior illustrated in the figure commonly occurs. Figure 10.5(d) is a plot of the complete likelihood function Equation (10.29) derived from the above MSE criterion and the manner in which the ambiguities are eliminated is clearly demonstrated. Figure 10.6 illustrates typical results for a segment of unvoiced fricative speech for which there is no distinct peak in the likelihood function.   ▲

10.4.3 Voicing Detection

In the context of the sinusoidal model, the degree to which a given frame of speech is voiced is determined by the degree to which the harmonic model fits the original sinewave data [5],[6]. The previous example indicated that the likelihood function is useful as a means to determine this degree of voicing. The accuracy of the harmonic fit can be related, in turn, to the *signal-to-noise ratio* (SNR) defined by

$$\text{SNR} = \frac{\sum_{n=-(N-1)/2}^{(N-1)/2} |s[n]|^2}{\sum_{n=-(N-1)/2}^{(N-1)/2} |s[n] - \hat{s}(n; \omega_o)|^2}$$

where $\hat{s}(n; \omega_o)$ is the sinewave harmonic model at the selected pitch ω_o. From Equation (10.25),

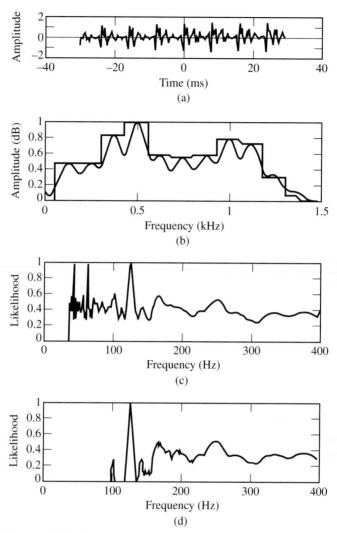

Figure 10.5   Sinewave pitch estimator performance for voiced speech: (a) input speech; (b) piecewise-constant SEEVOC envelope; (c) autocorrelation (comb-filter) component of the likelihood function; (d) complete likelihood function.

SOURCE: R.J. McAulay and T.F. Quatieri, "Sinusoidal Coding," chapter in *Speech Coding and Synthesis* [8]. ©1995, Elsevier Science. Reprinted with permission from Elsevier Science.

it follows that

$$ \text{SNR} = \frac{P_s}{P_s - 2\rho(\omega_o)} $$

where the input power P_s can be computed from the sinewave amplitudes. If the SNR is large, then the MSE is small and the harmonic fit is very good, which indicates that the input speech

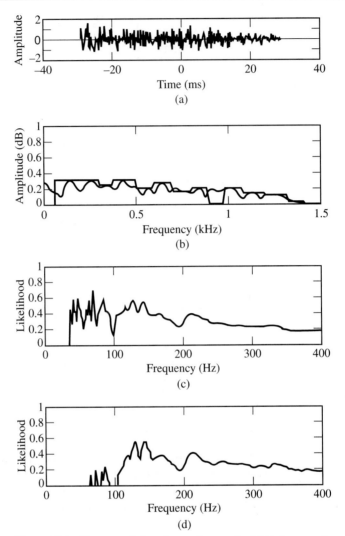

Figure 10.6 Sinewave pitch estimator for unvoiced fricative speech: (a) input speech; (b) piecewise-constant SEEVOC envelope; (c) autocorrelation (comb-filter) component of the likelihood function; (d) complete likelihood function.

SOURCE: R.J. McAulay and T.F. Quatieri, "Sinusoidal Coding," chapter in *Speech Coding and Synthesis* [8]. ©1995, Elsevier Science. Reprinted with permission from Elsevier Science.

is probably voiced. For small SNR, on the other hand, the MSE is large and the harmonic fit is poor, which indicates that the input speech is more likely to be unvoiced. Therefore, the degree of voicing is functionally dependent on the SNR. Although the determination of the exact functional form is difficult, a rule that has proven useful in several speech applications is the

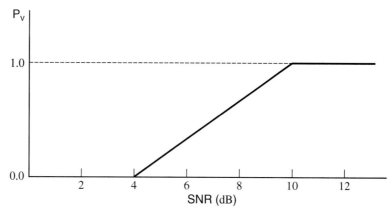

Figure 10.7 Voicing probability measure derived from the SNR associated with the MSE of the sinewave pitch estimator.

following (Figure 10.7):

$$P_v = \begin{cases} 1 & \text{SNR} > 10\,\text{dB} \\ \frac{1}{6}(\text{SNR} - 4) & 4\,\text{dB} \leq \text{SNR} \leq 10\,\text{dB} \\ 0 & \text{SNR} < 4\,\text{dB} \end{cases}$$

where P_v represents the probability that speech is voiced, and the SNR is expressed in dB. It is this quantity that was used to control the voicing-adaptive sinewave-based modification schemes in Chapter 9 and the voicing-adaptive frequency cutoff for the phase model to be used later in this chapter and in Chapter 12 for sinewave-based speech coding.

The next two examples illustrate the performance of the sinewave-based pitch estimator of Section 10.4.2 and the above voicing estimator for normal and "anomalous" voice types.

EXAMPLE 10.3   Figure 10.8 illustrates the sinewave pitch and voicing estimates for the utterance, "Which tea party did Baker go to?" from a female speaker. Although the utterance is a question, and thus, from our discussion in Chapter 3, we might expect a rising pitch at the termination of the passage, we see a rapid falling of pitch in the final word because the speaker is relaxing her vocal cords.   ▲

EXAMPLE 10.4   Figure 10.9 illustrates the sinewave pitch and voicing estimates for the utterance, "Jazz hour" from a very low-pitched male speaker. We see that the diplophonia in the utterance causes a sudden doubling of pitch during the second word "hour" where secondary pulses are large within a glottal cycle. Secondary pulses can result in an amplitude dip on every other harmonic (Exercise 3.3) and thus a doubling of the pitch estimate. We also see that diplophonia causes a severe raggedness in the voicing probability measure. The reader is asked to consider these contours further in Exercise 10.13.   ▲

10.4.4 Time-Frequency Resolution Perspective

We now summarize the sinewave pitch estimator strategy from a time-frequency resolution perspective. In Equation (10.17), we are attempting to fit the measurement $s[n]$ over an N_w-sample analysis window by a sum of sinewaves that are harmonically related, i.e., by a model

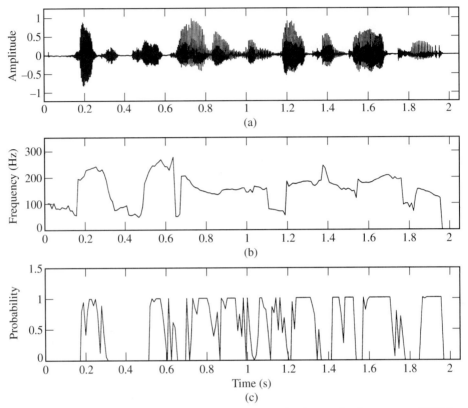

Figure 10.8 Pitch and voicing contours from sinewave-based estimators for the utterance, "Which tea party did Baker go to?" from female speaker: (a) waveform; (b) pitch contour; (c) voicing probability contour.

signal that is perfectly periodic. We have the spectral envelope of the sinewave amplitudes and thus the sinewave amplitudes for any fundamental frequency.

When we work through the algebra, we obtain the error function Equation (10.23) to be minimized over the unknown pitch ω_o and its equivalent form Equation (10.24) to be maximized. In this analysis, we have not yet made any assumptions about the measurement $s[n]$. We might simply stop at this point and perform the maximization to obtain a pitch estimate. For some conditions, this will give a reasonable pitch estimate and, in fact, the error function in Equation (10.24) benefits from the presence of the negative term on the right side of the equation in the sense that it helps to avoid pitch halving. But there is a problem with this approach.

Observe that Equation (10.24) contains the short-time Fourier transform magnitude of the measurement and recall that we make the analysis window as short as possible to obtain a time resolution as good as possible. Because the window is short, the main lobe of the window's Fourier transform is wide. For a low-pitch candidate ω_o, then, the first term in the expression for $\rho(\omega)$ in Equation (10.24) will be sampled many times within the main lobes of the window Fourier transforms embedded inside $|S(k\omega_o)|$. This may unreasonably bias the pitch to low estimates, even with the negative term on the right. Therefore, we are motivated to make the

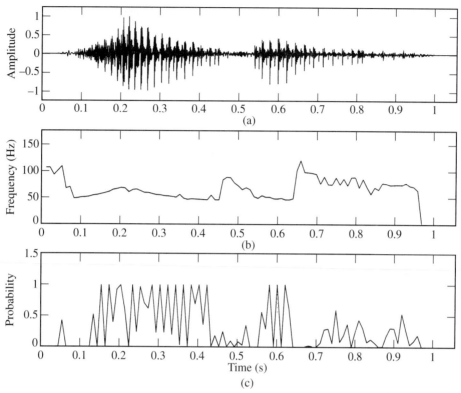

Figure 10.9 Pitch and voicing contours from sinewave-based estimators for the utterance, "Jazz hour" from a very low-pitched male speaker with diplophonia: (a) waveform; (b) pitch contour; (c) voicing probability contour.

analysis window very long to narrow the window's main lobe. This reduces time resolution, however, and can badly blur time variations and voiced/unvoiced transitions.

Our goal, then, is a representation of the measurement $s[n]$ that is derived from a short window but has the advantage of being able to represent a long interval in Equation (10.24). Recall that we were faced with this problem with the earlier autocorrelation pitch estimator of Equation (10.1). In that case, we used a sinewave representation of the measurement $s[n]$ to allow extrapolation beyond the analysis window interval. We can invoke the same procedure in this case. We modify the original error criterion in Equation (10.17) so that $s[n]$ is a sinewave representation derived from amplitude, frequencies, and phases at the STFT magnitude peaks. This leads to Equation (10.27) in which the main lobe of the function $D(\omega) = |\text{sinc}(\omega)|$ is controlled by the (pseudo) window length N. That is, we can make N long (and longer than the original window length N_w) to make the main lobe of $D(\omega)$ narrow, thus avoiding the above problem of biasing the estimate to low pitch. Observe we are using two sinewave representations: a harmonic sinewave model that we use to fit to the measurement, and a sinewave model of the measurement that is not necessarily harmonic since it is derived from peak-picking the STFT magnitude.

10.4.5 Evaluation by Harmonic Sinewave Reconstruction

Validating the performance of a pitch extractor can be a time-consuming and laborious procedure since it requires a comparison with hand-labeled data. An alternative approach is to reconstruct the speech using the harmonic sinewave model in Equation (10.16) and to listen for pitch errors. The procedure is not quite so straightforward as Equation (10.16) indicates, however, because, during unvoiced speech, meaningless pitch estimates are made that can lead to perceptual artifacts whenever the pitch estimate is greater than about 150 Hz. This is due to the fact that, in these cases, there are too few sinewaves to adequately synthesize a noiselike waveform. This problem has been eliminated by defaulting to a fixed low pitch ($\approx$ 100 Hz) during unvoiced speech whenever the pitch exceeds 100 Hz. The exact procedure for doing this is to first define a voicing-dependent cutoff frequency, ω_c (as we did in Chapter 9):

$$\omega_c(P_v) = \pi P_v \qquad (10.30)$$

which is constrained to be no smaller than $2\pi(1500 \text{ Hz}/f_s)$, where f_s is the continuous-to-discrete time sampling frequency. If the pitch estimate is ω_o, then the sinewave frequencies used in the reconstruction are

$$\omega_k = \begin{cases} k\omega_o & \text{for } k\omega_o \leq \omega_c(P_v) \\ k^*\omega_o + (k - k^*)\omega_u & \text{for } k\omega_o > \omega_c(P_v) \end{cases} \qquad (10.31)$$

where k^* is the largest value of k for which $k^*\omega_o \leq \omega_c(P_v)$, and where ω_u, the unvoiced pitch, corresponds to 100 Hz (i.e., $\omega_u = 2\pi(100 \text{ Hz}/f_s)$. If $\omega_o < \omega_u$, then we set $\omega_k = k\omega_o$ for all k. The harmonic reconstruction then becomes

$$\hat{s}[n; \omega_o] = \sum_{k=1}^{K} \bar{A}(\omega_k) \exp[j(n\omega_k + \phi_k)] \qquad (10.32)$$

where ϕ_k is the phase obtained by sampling a piecewise-constant phase function derived from the measured STFT phase using the same strategy to generate the SEEVOC envelope (Section 10.5.3). Strictly speaking, this procedure is harmonic only during strongly-voiced speech because if the speech is a voiced/unvoiced mixture the frequencies above the cutoff, although equally spaced by ω_u, are aharmonic, since they are themselves not multiples of the fundamental pitch.

The synthetic speech produced by this model is of very high quality, almost perceptually equivalent to the original. Not only does this validate the performance of the MSE pitch extractor, but it also shows that if the amplitudes and phases of the harmonic representation could be efficiently coded, then only the pitch and voicing would be needed to code the information in the sinewave frequencies. Moreover, when the measured phases are replaced by those defined by a voicing-adaptive phase model to be derived in the following section, the synthetic speech is also of good quality, although not equivalent to that obtained using the phase samples of the piecewise-flat phase envelope derived from the STFT phase measurements. However, it provides an excellent basis from which to derive a low bit-rate speech coder as described in Chapter 12.

10.5 Glottal Pulse Onset Estimation

In the previous section, the sinewave model was used as the basis for pitch estimation, which also led naturally to a harmonic sinewave model and associated harmonic analysis/synthesis system of high quality. This harmonic representation is important not only for the testing of the pitch estimation algorithm, but also, as stated, for the development of a complete parameterization for speech coding; an efficient parametric representation must be developed to reduce the size of the parameter set. That is, the raw sinewave amplitudes, phases, and frequencies cannot all be coded efficiently without some sort of parameterization, as we will discuss further in Chapter 12. The next step toward this goal is to develop a model for the sinewave phases by explicitly identifying the source linear phase component, i.e., the glottal pulse onset time, and the vocal tract phase component. In Chapter 9, in the application of time-scale modification, we side-stepped the *absolute* onset estimation problem with the use of a *relative* onset time derived by accumulating pitch periods. In this chapter, we propose one approach to absolute onset estimation [6],[7],[8]. Other methods include inverse filtering-based approaches described in Chapter 5 and a phase derivative-based approach [14].

10.5.1 A Phase Model Based on Onset Time

We first review the excitation onset time concept that was introduced in Chapter 9. For a periodic voiced speech waveform with period P, in discrete time the excitation is given by a train of periodically spaced impulses, i.e.,

$$u[n] = \sum_{r=-\infty}^{\infty} \delta[n - n_o - rP]$$

where n_o is a displacement of the pitch pulse train from the origin. The sequence of excitation pitch pulses can also be written in terms of complex sinewaves as

$$u[n] = \sum_{k=1}^{K} a_k \exp[j(n - n_o)\omega_k] \tag{10.33}$$

where $\omega_k = (2\pi/P)k$. We assume that the excitation amplitudes a_k are unity, implying that the measured sinewave amplitudes at spectral peaks are due solely to the vocal tract system function, the glottal flow function being embedded within the vocal tract system function. Generally, the sinewave excitation frequencies ω_k are assumed to be aharmonic but constant over an analysis frame. The parameter n_o corresponds to the time of occurrence of the pitch pulse nearest the center of the current analysis frame. The occurrence of this temporal event, called the *onset time* and introduced in Chapter 9, ensures that the underlying excitation sinewaves will be "in phase" at the time of the pitch pulse.

The amplitude and phase of the excitation sinewaves are altered by the vocal tract system function. Letting $H(\omega)$ denote the composite system function, the speech signal at its output becomes

$$\hat{s}[n] = \sum_{k=1}^{K} H(\omega_k) \exp[j(n - n_o)\omega_k], \tag{10.34}$$

which we write in terms of system and excitation components as

$$\hat{s}[n] = \sum_{k=1}^{K} M(\omega_k) \exp\{j(\Phi(\omega_k) + \phi_k[n])\},\tag{10.35}$$

where

$$M(\omega_k) = |H(\omega_k)|$$
$$\Phi(\omega_k) = \angle H(\omega_k)$$
$$\phi_k[n] = (n - n_o)\omega_k.$$

The time dependence of the system function, which was included in Chapter 9, is omitted here under the assumption that the vocal tract (and glottal flow function) is stationary over the duration of each synthesis frame. The excitation phase is linear with respect to frequency and time. As in Chapter 9, we assume that the system function has no linear phase so that all linear phase in $\hat{s}[n]$ is due to the excitation.

Let us now write the composite phase in Equation (10.35) as

$$\theta_k[n] = \Phi(\omega_k) + \phi_k[n].$$

At time $n = 0$ (i.e., the analysis and synthesis frame center), then

$$\theta_k[0] = \Phi(\omega_k) + \phi_k[0]$$
$$= \Phi(\omega_k) - n_o\omega_k.$$

The excitation and system phase components of the composite phase are illustrated in Figure 10.10 for continuous frequency ω. The phase value $\theta_k[0]$ is obtained from the STFT at frequencies ω_k at the center of an analysis frame, as described in Chapter 9. Likewise, the value $M(\omega_k)$ is the corresponding measured sinewave amplitude. Toward the development of

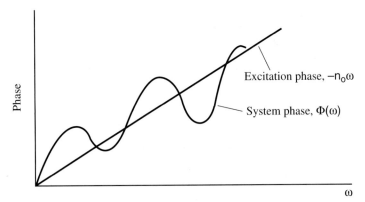

Figure 10.10 Vocal tract and excitation sinewave phase contributions. At the center of the analysis frame, i.e., at time $n = 0$, the excitation phase is linear in frequency and given by $-n_o\omega$.

a sinewave analysis/synthesis system based on the above phase decomposition, a method for estimating the onset time will now be described, given the measured sinewave phase $\theta_k[0]$ and amplitude $M(\omega_k)$ values [6],[7].

10.5.2 Onset Estimation

From Chapter 9, the speech waveform can also be represented in terms of measured sinewave amplitude, frequency, and phase parameters as

$$s[n] = \sum_{k=1}^{K} A_k \exp[j(n\omega_k + \theta_k)] \tag{10.36}$$

where θ_k denotes the measured phase $\theta_k[0]$ at the analysis frame center and where sinewave parameters are assumed to be fixed over the analysis frame. We now have two sinewave representations, i.e., $s[n]$ in Equation (10.36) obtained from parameter measurements and the sinewave model $\hat{s}[n]$ in Equation (10.35) with a linear excitation phase in terms of an unknown onset time. In order to determine the excitation phase, the onset time parameter n_o is estimated by choosing the value of n_o so that $\hat{s}[n]$ is as close as possible to $s[n]$, according to some meaningful criterion [6],[7],[8]. A reasonable criterion is to seek the minimum of the mean-squared error (MSE) over a time interval N (generally greater than the original analysis window duration N_w), i.e.,

$$E(n_o) = \frac{1}{N} \sum_{n=-(N-1)/2}^{(N-1)/2} |s[n] - \hat{s}[n; n_o]|^2 \tag{10.37}$$

over n_o and where we have added the argument n_o in $\hat{s}[n]$ to denote that this sequence is a function of the unknown onset time.

The MSE in Equation (10.37) can be expanded as

$$E(n_o) = \frac{1}{N} \sum_{n=-(N-1)/2}^{(N-1)/2} \{|s[n]|^2 - 2\mathrm{Re}(s[n]\hat{s}^*[n; n_o]) + |\hat{s}[n; n_o]|^2\}. \tag{10.38}$$

If the sinusoidal representation for $s[n]$ in Equation (10.36) is used in the first term of Equation (10.38), then, as before, the power in the measured signal can be defined as

$$P_s = \frac{1}{N} \sum_{n=-(N-1)/2}^{(N-1)/2} |s[n]|^2 \approx \sum_{k=1}^{K} A_k^2.$$

Letting the system transfer function be written in terms of its magnitude $M(\omega)$ and phase $\Phi(\omega)$, namely

$$H(\omega) = M(\omega) \exp[j\Phi(\omega)]$$

and using this as in Equation (10.35), the second term of Equation (10.38) can be written as

$$\frac{1}{N} \sum_{n=-(N-1)/2}^{(N-1)/2} s[n]\hat{s}^*[n; n_o] = \sum_{k=1}^{K} A_k M(\omega_k) \exp\{j[\theta_k + n_o\omega_k - \Phi(\omega_k)]\}.$$

Finally, the third term in Equation (10.38) can be written as

$$\frac{1}{N} \sum_{n=-(N-1)/2}^{(N-1)/2} |\hat{s}[n; n_o]|^2 = \sum_{k=1}^{K} M^2(\omega_k).$$

These relations are valid provided the sinewaves are well-resolved, a condition that is basically assured by making the interval N, two and one-half times the average pitch period, which was the condition assumed in the estimation of the sinewave parameters in the first place. [Indeed, we can make N as long as we like because we are working with the sinewave representation in Equation (10.36).] Combining the above manipulations leads to the following expression for the MSE:

$$E(n_o) = P_s - 2 \sum_{k=1}^{K} A_k M(\omega_k) \cos[\theta_k + n_o \omega_k - \Phi(\omega_k)] + \sum_{k=1}^{K} M^2(\omega_k). \quad (10.39)$$

Equation (10.39) was derived under the assumption that the system amplitude $M(\omega)$ and phase $\Phi(\omega)$ were known. In order to obtain the optimal onset time, therefore, the amplitude and phase of the system function must next be estimated from the data. We choose here the SEEVOC amplitude estimate $\hat{M}(\omega)$ to be described in Section 10.5.3. If the system function is assumed to be minimum phase, then, as we have seen in Chapter 6, we can obtain the corresponding phase function by applying a rightsided lifter to the real cepstrum associated with $\hat{M}(\omega)$, i.e.,

$$c[n] = \frac{1}{2\pi} \int_{-\pi}^{\pi} \log \hat{M}(\omega) d\omega, \qquad n = 0, 1, \cdots$$

from which the system minumum-phase estimate follows as

$$\hat{\Phi}(\omega) = \text{Im} \sum_{n=0}^{\infty} l[n]c[n]e^{jn\omega}, \qquad (10.40)$$

where $l[0] = 1$, $l[n] = 0$ for $n < 0$, and $l[n] = 2$ for $n > 0$. Use of Equation (10.40) is incomplete, however, because the minimum-phase analysis fails to account for the sign of the input speech waveform. This is due to the fact that the sinewave amplitudes from which the system amplitude and phase are derived are the same for $-s[n]$ and $s[n]$. This ambiguity can be accounted for by generalizing the system phase $\hat{\Phi}(\omega)$ in Equation (10.40) by

$$\hat{\Phi}(\omega) + \beta\pi, \quad \beta = 0 \text{ or } 1 \qquad (10.41)$$

and then choosing n_o and β to minimize the MSE simultaneously. Substituting Equation (10.40) and Equation (10.41) into Equation (10.39) leads to the following equation for the MSE:

$$E(n_o, \beta) = P_s - 2 \sum_{k=1}^{K} A_k^2 \cos[\theta_k + n_o \omega_k - \beta\pi - \hat{\Phi}(\omega_k)] + \sum_{k=1}^{K} A_k^2.$$

Since only the second term depends on the phase model, it suffices to choose n_o and β to maximize the "likelihood" function

$$\rho(n_o, \beta) = \sum_{k=1}^{K} A_k^2 \cos[\theta_k + n_o \omega_k - \beta\pi - \hat{\Phi}(\omega_k)].$$

However, since

$$\rho(n_o, \beta = 1) = -\rho(n_o, \beta = 0)$$

it suffices to maximize $|\rho(n_o)|$ where now

$$\rho(n_o) = \sum_{k=1}^{K} A_k^2 \cos[\theta_k + n_o \omega_k - \hat{\Phi}(\omega_k)] \tag{10.42}$$

and if $\hat{n}_0$ is the maximizing value, then choose $\beta = 0$ if $\rho(\hat{n}_0) > 0$ and $\beta = 1$, otherwise.

EXAMPLE 10.5   In this experiment, the sinewave analysis uses a 10-kHz sampling rate and a 1024-point FFT to compute the STFT [7]. The magnitude of the STFT is computed and the underlying sinewaves are identified by determining the frequencies at which a change in the slope of the STFT magnitude occurs. The SEEVOC algorithm was used to estimate the piecewise-linear envelope of the sinewave amplitudes. A typical set of results is shown in Figure 10.11. The logarithm of the STFT magnitude is shown in Figure 10.11d together with the SEEVOC envelope. The cepstral coefficients are used in Equation (10.40) to compute the system phase, from which the onset likelihood function is computed using Equation (10.42). The result is shown in Figure 10.11b. It is interesting to note that the peaks in the onset function occur at the points which seem to correspond to a sharp transition in the speech waveform, probably near the glottal pulse, rather than at a peak in the waveform. We also see in Figure 10.11c that the *phase residual* is close to zero below about 3500 Hz, indicating the accuracy of our phase model in this region. The phase residual is the difference between the measured phase and our modeled phase, interpolated across sinewave frequencies.

An example of the application of the onset estimator for an unvoiced fricative case is shown in Figure 10.12 [7]. The estimate of the onset time is meaningless in this case, and this is reflected in the relatively low value of the likelihood function. The onset time, however, can have meaning and importance for non-fricative consonants such as plosives where the phase and event timing is important for maintaining perceptual quality.   ▲

10.5.3 Sinewave Amplitude Envelope Estimation

The above results show that if the envelope of the sinewave amplitudes is known, then the MSE criterion can lead to a technique for estimating the glottal pulse onset time under the assumption that the glottal pulse and vocal tract response are minimum-phase. This latter assumption was necessary to derive an estimate of the system phase, hence the performance of the estimator depends on the system magnitude. An ad hoc estimator for the magnitude of the system function is simply to apply linear interpolation between successive sinewave peaks. This results in the function

$$M(\omega) = A_{k-1} + \frac{A_k - A_{k-1}}{\omega_k - \omega_{k-1}}(\omega - \omega_{k-1}), \quad \omega_{k-1} \leq \omega \leq \omega_k. \tag{10.43}$$

The problem with such a simple envelope estimator is that the system phase is sensitive to low-level peaks that can arise due to time variations in the system function or signal processing

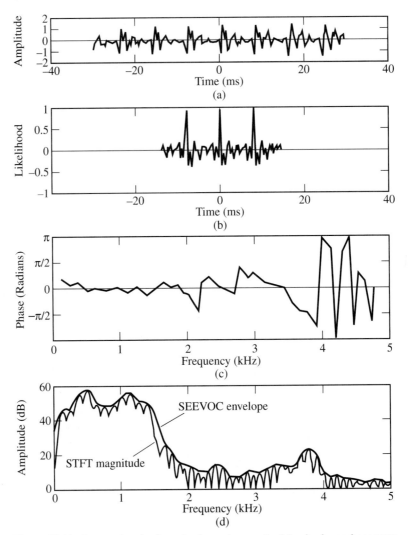

Figure 10.11   Onset estimation for voiced speech example: (a) voiced speech segment; (b) likelihood function; (c) phase residual; (d) STFT magnitude and superimposed piecewise-linear SEEVOC envelope.

SOURCE: R.J. McAulay and T.F. Quatieri, "Low-Rate Speech Coding Based on the Sinusoidal Model," chapter in *Advances in Speech Signal Processing* [7]. ©1992, Marcel Dekker, Inc. Courtesy of Marcel Dekker, Inc.

artifacts such as side-lobe leakage. Fortunately, this problem can be avoided using the technique proposed by Paul [8],[10] in the development of the Spectral Envelope Estimation Vocoder (SEEVOC).

The SEEVOC algorithm depends on having an estimate of the average pitch, denoted here by $\bar{\omega}_0$. The first step is to search for the largest sinewave amplitude in the interval $[\frac{\bar{\omega}_0}{2}, \frac{3\bar{\omega}_0}{2}]$. Having found the amplitude and frequency of that peak, labeled (A_1, ω_1), then one searches

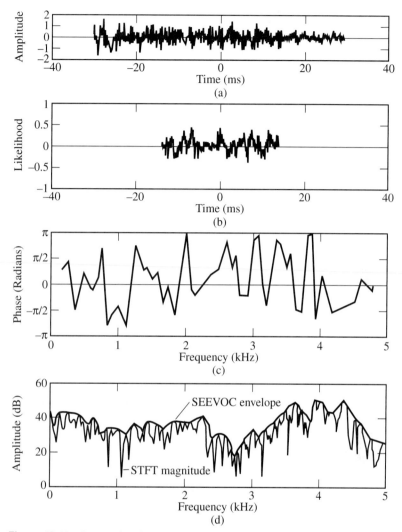

Figure 10.12 Onset estimation for unvoiced fricative speech example: (a) unvoiced speech segment; (b) likelihood function; (c) phase residual; (d) STFT magnitude and superimposed piecewise-linear SEEVOC envelope.

SOURCE: R.J. McAulay and T.F. Quatieri, "Low-Rate Speech Coding Based on the Sinusoidal Model," chapter in *Advances in Speech Signal Processing* [7]. ©1992, Marcel Dekker, Inc. Courtesy of Marcel Dekker, Inc.

the interval $[\omega_1 + \frac{\bar{\omega}_0}{2}, \omega_1 + \frac{3\bar{\omega}_0}{2}]$ for its largest peak, labeled (A_2, ω_2), as illustrated in Figure 10.13. The process is continued by searching the intervals $[\omega_{k-1} + \frac{\bar{\omega}_0}{2}, \omega_{k-1} + \frac{3\bar{\omega}_0}{2}]$ for the largest peaks (A_k, ω_k) until the edge of the speech bandwidth is reached. If no peak is found in a search bin, then the largest endpoint of the short-time Fourier transform magnitude is used and placed at the bin center, from which the search procedure is continued. The principle advantage of this pruning method is the fact that any low-level peaks within a pitch interval are masked

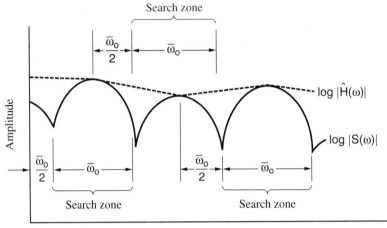

Figure 10.13 SEEVOC spectral envelope estimator. One spectral peak is selected for each harmonic bin. The envelope estimate is denoted by $|\hat{H}(\omega)|$ and the speech spectrum by $|S(\omega)|$.

by the largest peak, presumably a peak that is close to an underlying harmonic. Moreover, the procedure is not dependent on the peaks' being harmonic, nor on the exact value of the average pitch since the procedure resets itself after each peak has been found. The SEEVOC envelope, the envelope upon which the above minimum-phase analysis is based, is then obtained by applying the linear interpolation rule,[7] Equation (10.43), where now the sinewave amplitudes and frequencies are those obtained using the SEEVOC peak-picking routine.

10.5.4 Minimum-Phase Sinewave Reconstruction

Consider now a minimum-phase model which comes naturally from the discussion in Section 10.5.2. In order to evaluate the accuracy of a minimum-phase model, as well as the onset estimator, it is instructive to examine the behavior of the phase error (previously referred to as "phase residual") associated with this model that we saw earlier. Let

$$\tilde{\theta}(\omega) = -\hat{n}_0\omega + \hat{\Phi}(\omega) + \hat{\beta}\pi \tag{10.44}$$

be the model estimate of the sinewave phase at any frequency, ω where $\hat{\Phi}(\omega)$ denotes the minimum-phase estimate. Then, for the sinewave at frequency ω_k, for which the measured phase is θ_k, the phase error is

$$\epsilon(\omega_k) = \theta_k - \tilde{\theta}(\omega_k).$$

The phase error for the speech sample in Figure 10.11a is shown in Figure 10.11c. In this example, the phase error is small for frequencies below about 3.5 kHz. The larger structured error beyond 3.5 kHz probably indicates the inadequacy of the minimum-phase assumption and the possible presence of noise in the speech source or maximum-phase zeros in the transfer

---

[7] In the sinewave pitch estimation of the previous section, a piecewise-constant envelope was derived from the pruned peaks rather than a piecewise-linear envelope, having the effect of maintaining at harmonic frequencies the original peak amplitudes.

function $H(\omega)$ (Chapter 6). Nevertheless, as with the pitch estimator, it is instructive to evaluate the estimators by reconstruction [6],[7],[8].

One approach to reconstruction is to assume that the phase residual is negligible and form a phase function for each synthesis frame that is the sum of the minimum-phase function and a linear excitation phase and sign derived from the onset estimator. In other words, from Equation (10.44), the sinewave phases on each frame are given by

$$\tilde{\theta}(\omega_k) = -\hat{n}_0\omega_k + \hat{\Phi}(\omega_k) + \hat{\beta}\pi. \qquad (10.45)$$

When this phase is used, along with the measured sinewave amplitudes at ω_k, the resulting quality of voiced speech is quite natural but some slight hoarseness is introduced due to a few-sample inaccuracy in the onset estimator that results in a randomly changing pitch period (pitch jitter); it was found that even a sporadic one- or two-sample error in the onset estimator can introduce this distortion. This property was further confirmed by replacing the above *absolute* onset estimator by the *relative* onset estimator introduced in Chapter 9, whereby onset estimates are obtained by accumulating pitch periods. With this relative onset estimate, the hoarseness attributed to error in the absolute onset time is removed and the synthesized speech is free of artifacts. In spite of this limitation of the absolute onset estimator, it provides a means to obtain useful glottal flow timing information in applications where reduction in the linear phase is important (Exercise 10.5). Observe that we have described reconstruction of only voiced speech. When the phase function Equation (10.45) is applied to unvoiced speech, particularly fricatives and voicing with frication or strong aspiration, the reconstruction is "buzzy" because an unnatural waveform peakiness arises due to the sinewaves no longer be randomly displaced from one another (Exercise 10.6). An approach to remove this distortion when a minimum-phase vocal tract system is invoked is motivated by the phase-dithering model of Section 9.5.2 and will be described in Chapter 12 in the context of speech coding.

10.6 Multi-Band Pitch and Voicing Estimation

In this section, we describe a generalization of sinewave pitch and voicing estimation that yields a voicing decision in multiple frequency bands. A byproduct is sinewave amplitude estimates derived from a minimum-mean-squared error criterion. These pitch and voicing estimates were developed in the context of the Multi-Band Excitation (MBE) speech representation developed by Griffin and Lim [2],[8] where, as in Section 10.4, speech is represented as a sum of harmonic sinewaves.

10.6.1 Harmonic Sinewave Model

As above, the synthetic waveform for a harmonic set of sinewaves is written as

$$\hat{s}[n] = \sum_{k=1}^{K(\omega_o)} B_k \exp(jnk\omega_o + \phi_k). \qquad (10.46)$$

Whereas in Section 10.4.2 the sinewave amplitudes were assumed to be harmonic samples of an underlying vocal tract envelope, in MBE they are allowed to be unconstrained free variables and are chosen to render $\hat{s}[n]$ a minimum-mean-squared error fit to the measured speech signal $s[n]$. For analysis window $w[n]$ of length N_w, the short-time speech segment and its harmonic sinewave representation is denoted by $s_w[n] = w[n]s[n]$ and $\hat{s}_w[n] = w[n]\hat{s}[n]$ (thus, simplifying to one STFT slice), respectively. The mean-squared error between the two signals is given by

$$E(\omega_o, \underline{B}, \underline{\phi}) = \sum_{-(N_w-1)/2}^{(N_w-1)/2} |\, s_w[n] - \hat{s}_w[n]\, |^2 \qquad (10.47)$$

where $\underline{B} = (B_1, B_2, \cdots, B_{K(\omega_o)})$ and $\underline{\phi} = (\phi_1, \phi_2, \cdots, \phi_{K(\omega_o)})$ are the vectors of unknown amplitudes and phases at the sinewave harmonics. Following the development in [8],

$$S_w(\omega) = \int_{-\pi}^{\pi} s_w[n] \exp(-jn\omega)d\omega$$

denotes the discrete-time Fourier transform of $s_w[n]$ and, similarly, $\hat{S}_w(\omega)$ denotes the discrete-time Fourier Transform of $\hat{s}_w[n]$. Then using Parseval's Theorem, Equation (10.47) becomes

$$E(\omega_o, \underline{B}, \underline{\phi}) = \int_{-\pi}^{\pi} |\, S_w(\omega) - \hat{S}_w(\omega)\, |^2\, d\omega$$

$$= \int_{-\pi}^{\pi} \left\{ |\, S_w(\omega)\, |^2 - 2\mathrm{Re}\left[S_w(\omega)\hat{S}_w^*(\omega) \right] + |\, \hat{S}_w(\omega)\, |^2 \right\} d\omega. \quad (10.48)$$

The first term, which is independent of the pitch, amplitude, and phase parameters, is the energy in the windowed speech signal that we denote by E_w. Letting $B_k \exp(j\phi_k)$ represent the complex amplitude of the kth harmonic and using the sinewave decomposition in Equation (10.46), $\hat{S}_w(\omega)$ can be written as

$$\hat{S}_w(\omega) = \sum_{k=1}^{K(\omega_o)} \beta_k W(\omega - k\omega_o)$$

with

$$\beta_k = 2\pi B_k \exp(j\phi_k)$$

where $W(\omega)$ is the discrete-time Fourier transform of the analysis window $w[n]$. Substituting this relation into Equation (10.48), the mean-squared error can be written as

$$E[\omega_o, \underline{\beta}(\omega_o)] = E_w - 2\mathrm{Re}\left[\sum_{k=1}^{K(\omega_o)} \beta_k^* \int_{-\pi}^{\pi} S_w(\omega) W(\omega - k\omega_o)d\omega \right]$$

$$+ \sum_{k=1}^{K(\omega_o)} \sum_{m=1}^{K(\omega_o)} \beta_k \beta_m^* \int_{-\pi}^{\pi} W(\omega - k\omega_o)W^*(\omega - m\omega_o)d\omega.$$

For each value of ω_o this equation is quadratic in β_k. Therefore, each β_k is a function of ω_o and we denote this set of unknowns by $\underline{\beta}(\omega_o)$. It is straightforward to solve for the $\underline{\beta}(\omega_o)$ that results in the minimum-mean-squared error, $E[\omega_o, \underline{\beta}(\omega_o)]$. This process can be repeated for each value of ω_o so that the optimal minimum-mean-squared error estimate of the pitch can be determined. Although the quadratic optimization problem is straightforward to solve, it requires solution of a simultaneous set of linear equations for each candidate pitch value. This makes the resulting pitch estimator complicated to implement. However, following [2], we assume that $W(\omega)$ is essentially zero in the region $|\omega| > \frac{\omega_o}{2}$, which corresponds to the condition posed in Section 10.4 to insure that the sinewaves are well-resolved. We then define the frequency region about each harmonic frequency $k\omega_o$ as

$$\Delta_k = \{\omega : k\omega_o - \frac{\omega_o}{2} \le \omega \le k\omega_o + \frac{\omega_o}{2}\} \tag{10.49}$$

with which the mean-squared error can be approximated as

$$E[\omega_o, \underline{\beta}(\omega_o)] \approx E_w - 2\mathrm{Re}\left[\sum_{k=1}^{K(\omega_o)} \beta_k^* \int_{\Delta_k} S_w(\omega)W(\omega)d\omega\right]$$

$$+ \sum_{k=1}^{K(\omega_o)} |\beta_k|^2 \int_{\Delta_k} |W(\omega - k\omega_o)|^2 d\omega,$$

from which it follows that the values of the complex amplitudes that minimize the mean-squared error are

$$\hat{\beta}_k(\omega_o) \approx \frac{\int_{\Delta_k} S_w(\omega)W(\omega - k\omega_o)d\omega}{\int_{\Delta_k} |W(\omega - k\omega_o)|^2 d\omega}. \tag{10.50}$$

The best mean-squared error fit to the windowed speech data is therefore given by

$$\hat{S}_w(\omega; \omega_o) = \sum_{k=1}^{K(\omega_o)} \hat{\beta}_k(\omega_o)W(\omega - k\omega_o).$$

This expression is then used in Equation (10.47) to evaluate the mean-squared error for the given value of ω_o. This procedure is repeated for each value of ω_o in the pitch range of interest and the optimum estimate of the pitch is the value of ω_o that minimizes the mean-squared error. While the procedure is similar to that used in Section 10.4, there are important differences. The reader is asked to explore these differences and similarities in Exercise 10.8. Extensions of this algorithm by Griffin and Lim exploit pitch estimates from past and future frames in "forward-backward" pitch tracking to improve pitch estimates during regions in which the pitch and/or vocal tract are rapidly changing [2].

10.6.2 Multi-Band Voicing

As in Section 10.4.3, distinguishing between voiced and unvoiced spectral regions is based on how well the harmonic set of sinewaves fits the measured set of sinewaves. In Section 10.4.3, a signal-to-noise ratio (SNR) was defined in terms of the normalized mean-squared error, and this was mapped into a cutoff frequency below which the sinewaves were declared voiced and above which they were declared unvoiced. This idea, which originated with the work of Makhoul, Viswanathan, Schwartz, and Huggins [4], was generalized by Griffin and Lim [2] to allow for an arbitrary sequence of voiced and unvoiced bands with the measure of voicing in each of the bands determined by a normalized mean-squared error computed for the windowed speech signals. Letting

$$\gamma_m = \{\omega : \omega_{m-1} \le \omega \le \omega_m\}, \qquad m = 1, 2, \cdots, M$$

denote the mth band of M multi-bands over the speech bandwidth, then using Equation (10.48), the normalizing mean-squared error for each band can be written as

$$E_m(\hat{\omega}_0) = \frac{\int_{\gamma_m} \mid S_w(\omega) - \hat{S}_w(\omega; \hat{\omega}_0) \mid^2 d\omega}{\int_{\gamma_m} \mid S_w(\omega) \mid^2 d\omega}. \qquad (10.51)$$

Each of the M values of the normalized mean-squared error is compared with a threshold function to determine the binary voicing state of the sinewaves in each band. If $E_m(\hat{\omega}_0)$ is below the threshold, the mean-squared error is small, hence the harmonic sinewaves fit well to the input speech and the band is declared voiced. The setting of the threshold uses several heuristic rules to obtain the best performance [3].

It was observed that when the multi-band voicing decisions are combined into a two-band voicing-adaptive cutoff frequency, as was used in Section 10.4.3, no loss in quality was perceived in low-bandwidth (e.g., 3000–4000 Hz) synthesis [3],[8],[9]. Nevertheless, this scheme affords the possibility of a more accurate two-band cutoff than that in Section 10.4.3 and the multi-band extension may be useful in applications such as speech transformations (Exercise 10.9). An additional advantage of multi-band voicing is that it can make reliable voicing decisions when the speech signal has been corrupted by additive acoustical noise [2],[3],[9]. The reason for this lies in the fact that the normalized mean-squared error essentially removes the effect of the spectral tilt, which means that the sinewave amplitudes contribute more or less equally from band to band. When one wide-band voicing decision is made, as in Section 10.4.3, only the largest sinewave amplitudes will contribute to the mean-squared error, and if these have been corrupted due to noise, then the remaining sinewaves, although harmonic, may not contribute enough to the error measure to offset those that are corrupted. Finally, observe that although the multi-band voicing provides a refinement to the two-band voicing strategy, it does not account for the *additive* nature of noise components in the speech spectrum, as was addressed in the deterministic-stochastic model of Chapter 9.

10.7 Summary

In this chapter, we introduced a frequency-domain approach to estimating the pitch and voicing state of a speech waveform, in contrast to time-domain approaches, based on analysis/synthesis techniques of earlier chapters. Specifically, we saw that pitch estimation can be thought of as fitting a harmonic set of sinewaves to a measured set of sinewaves, and the accuracy of the harmonic fit is an indicator of the voicing state. A simple autocorrelation-based estimator, formulated in the frequency domain, was our starting point and led to a variety of sinewave-based pitch estimators of increasing complexity and accuracy. A generalization of the sinewave model to a multi-band pitch and voicing estimation method was also described. In addition, we applied the sinewave model to the problem of glottal pulse onset time estimation. Finally, we gave numerous mixed- and minimum-phase sinewave-based waveform reconstruction techniques for evaluating the pitch, voicing, and onset estimators. A spinoff of these evaluation techniques is analysis/synthesis structures based on harmonic sinewaves, a minimum-phase vocal tract, and a linear excitation phase that form the basis for frequency-domain sinewave-based speech coding methods in Chapter 12. Other pitch and voicing estimators will be described as needed in the context of speech coding, being more naturally developed for a particular coding structure.

We saw both the features and limitations of the various pitch estimators using examples of typical voice types and a diplophonic voice with secondary glottal pulses occurring regularly, and about midway, between primary glottal pulses within a glottal cycle. Many "anomalous" voice types that we have described earlier in this text, however, were not addressed. These include, for example, the creaky voice with erratically-spaced glottal pulses, and the case of every other vocal tract impulse response being amplitude modulated. These cases give difficulty to pitch and voicing estimators, generally, and are considered in Exercises 10.10, 10.11, 10.12, and 10.13. Such voice types, as well as voiced/unvoiced transitions, and rapidly-varying speech events, in spite of the large strides in improvements of pitch and voicing estimators, render the pitch and voicing problem in many ways still challenging and unsolved.

EXERCISES

10.1 Show that the autocorrelation-based pitch estimator in Equation (10.2) follows from minimizing the error criterion in Equation (10.1) with respect to the unknown pitch period P. Justify why you must constrain $P > \epsilon$ (a small positive value close to zero), i.e., P must be sufficiently far from zero.

10.2 In this problem you are asked to the complete missing steps in the harmonic sinewave model-based pitch estimator of Sections 10.4.1 and 10.4.2.

(a) Show that Equation (10.9) follows from Equation (10.8).

(b) Show that minimizing Equation (10.9) with respect to ϕ gives Equation (10.11), and thus gives Equation (10.12).

(c) Show that minimizing Equation (10.12) with respect to $\underline{B}$ gives Equation (10.13), and thus gives Equation (10.14).

(d) Explain why Equation (10.15) can lead to pitch-halving errors.

(e) Show how Equation (10.18) can be manipulated to obtain Equation (10.22). Fill in all of the missing steps in the text. Interpret Equation (10.29) in terms of its capability to avoid pitch halving relative to correlation-based pitch estimators. Argue that the pitch estimator also avoids pitch doubling.

(f) Explain how to obtain the results in Figure 10.4b for the cases $f_o = 400$ Hz and $f_o = 50$ Hz.

(g) Propose an extension of the sinewave-based pitch estimator where an *a priori* vocal tract system phase envelope is known, as well as an *a priori* system magnitude envelope. Qualitatively describe the steps in deriving the estimator and explain why this additional phase information might improve the pitch estimate.

10.3 In the context of homomorphic filtering, we saw in Chapter 6 one approach to determining voicing state (i.e., a speech segment is either voiced or unvoiced) which requires the use of the real cepstrum, and in this chapter we derived a voicing measure based on the degree of harmonicity of the short-time Fourier transform. In this problem, you consider some other simple voicing measurements. Justify the use of each of the following measurements as a voicing indicator. For the first two measures, use your knowledge of acoustic phonetics, i.e., the waveform and spectrogram characteristics of voiced and unvoiced phonemes. For the last two measurements use your knowledge of linear prediction analysis.

(a) The relative energy in the outputs of complementary highpass and lowpass filters of Figure 10.14.

(b) The number of zero crossings in the signal.

(c) The first reflection coefficient $k_1 = \alpha_1^{(1)}$ generated in the Levinson recursion.

(d) The linear prediction residual obtained by inverse filtering the speech waveform by the inverse filter $A(z)$.

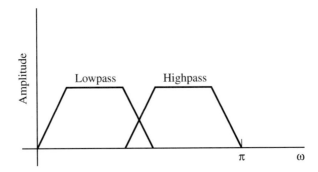

Figure 10.14 Highpass and lowpass filters.

10.4 Suppose, in the sinewave-based pitch estimator of Section 10.4.2, we do not replace the spectrum by its sinewave representation. What problems arise that the sinewave representation helps prevent? How does it help resolve problems inherent in the autocorrelation-based sinewave pitch estimators?

10.5 Show how the onset estimator of Section 10.5.2, which provides a linear excitation phase estimate, may be useful in obtaining a vocal tract phase estimate from sinewave phase samples. *Hint:* Recall from Chapter 9 that one approach to vocal tract system phase estimation involves interpolation of the real and imaginary parts of the complex STFT samples at the sinewave frequencies. When the vocal tract impulse response is displaced from the time origin, a large linear phase can be introduced.

10.6 It was observed that when the phase function in Equation (10.45), consisting of a minimum-phase system function and a linear excitation phase derived from the onset estimator of Section 10.5.2, is applied to unvoiced speech, particularly fricatives and voicing with frication or strong aspiration, the reconstruction is "buzzy." It was stated that this buzzy perceptual quality is due to an unnatural waveform peakiness. Give an explanation for this peakiness property, considering the characteristics of the phase residual in Figures 10.11 and 10.12, as well as the time-domain characteristics of a linear excitation phase.

10.7 In this problem, you investigate the "magnitude-only" counterpart to the multi-band pitch estimator of Section 10.6. Consider a perfectly periodic voiced signal of the form

$$x[n] = h[n] * p[n]$$

where

$$p[n] = \sum_{k=-\infty}^{\infty} \delta[n - kP],$$

P being the pitch period. The windowed signal $y[n] = w[n]x[n]$ (the window $w[n]$ being a few pitch periods in duration) is expressed by

$$y[n] = w[n]x[n]$$
$$= w[n](h[n] * p[n]).$$

(a) Show that the Fourier transform of $y[n]$ is given by

$$Y(\omega) = \frac{2\pi}{P} \sum_{k=0}^{N-1} H(k\omega_o)W(\omega - k\omega_o),$$

where $\omega_o = \frac{2\pi}{P}$ is the fundamental frequency (pitch) and N is the number of harmonics.

(b) A frequency-domain pitch estimator uses the error criterion:

$$E = \frac{1}{2\pi} \int_{-\pi}^{\pi} [|S(\omega)| - |Y(\omega)|]^2 d\omega$$

where $S(\omega)$ is the short-time spectral measurement and $Y(\omega)$ is the model from part (a). In the model $Y(\omega)$ there are two unknowns: the pitch ω_o and the vocal tract spectral values $H(k\omega_o)$. Suppose for the moment that we know the pitch ω_o. Consider the error E over a region around a harmonic and suppose that the window transform is narrow enough so that main window lobes are independent (non-overlapping). Then, the error around the kth harmonic can be written as

$$E(k) = \frac{1}{2\pi} \int_{k\omega_o-\Delta}^{k\omega_o+\Delta} [|S(\omega)| - |H(k\omega_o)W(\omega - k\omega_o)|]^2 d\omega$$

and the total error is approximately

$$E = \sum_{k=0}^{N-1} E(k).$$

Given ω_o, find an expression for $H(k\omega_o)$ that minimizes $E(k)$. With this solution, write an expression for E. Keep your expression in the frequency domain and do not necessarily simplify. It is possible with Parseval's Theorem to rewrite this expression in the time domain in terms of autocorrelation functions, which leads to an efficient implementation, but you are not asked to show this here.

(c) From part (b), propose a method for estimating the pitch ω_o that invokes minimization of the total error E. Do not attempt to find a closed-form solution, but rather describe your approach qualitatively. Discuss any disadvantages of your approach.

10.8 Consider similarities and differences in the sinewave-based pitch and voicing estimator developed in Sections 10.4.2 and 10.4.3 with the multi-band pitch and voicing estimators of Section 10.6 for the following:

1. Pitch ambiguity with pitch halving or pitch doubling. *Hint:* Consider the use of unconstrained amplitude estimates in the multi-band pitch estimator and the use of samples of a vocal tract envelope in the sinewave pitch estimator of Section 10.4.2.

2. Voicing estimation with voiced fricatives or voiced sounds with strong aspiration.

3. Computational complexity.

4. Dependence on the phase of the discrete-time Fourier transform over each harmonic lobe. *Hint:* The phase of the discrete-time Fourier transform is not always constant across every harmonic lobe. How might this changing phase affect the error criterion in Equation (10.48) and thus the amplitude estimation in the multi-band amplitude estimator of Equation (10.50)?

10.9 This problem considers the multi-band voicing measure described in Section 10.6.2.

(a) Propose a strategy for combining individual band decisions into a two-band voicing measure, similar to that described in Section 10.4.3, where the spectrum above a cutoff frequency is voiced and below the cutoff is unvoiced. As a point of interest, little quality difference has been observed between this type of reduced two-band voicing measure and the original multi-band voicing measure when used in low-bandwidth (e.g., 3000–4000 Hz) synthesis.

(b) Explain why the multi-band voicing measure of part (a), reduced to a two-band decision, gives the possibility of a more accurate two-band voicing cutoff than the sinewave-based method of Section 10.4.3.

(c) Explain how the multi-band voicing measure may be more useful in sinewave-based speech transformations than the reduced two-band decision, especially for wide-bandwidth (e.g., > 4000 Hz) synthesis. Consider, in particular, pitch modification.

10.10 In this problem, you investigate the different "anomalous" voice types of Figure 10.15 with diplophonic, creaky, and modulation (pitch periods with alternating gains) characteristics. Consider both the time-domain waveform and the short-time spectrum obtained from the center of each waveform segment.

(a) For the diplophonic voice, describe how secondary pulses generally affect the performance of both time- and frequency-domain pitch estimators.

(b) For the creaky voice, describe how erratic glottal pulses generally affect the performance of both time- and frequency-domain pitch estimators.

(c) For the modulated voice, explain why different spectral bands exhibit different pitch values. Consider, for example, time-domain properties of the signal. Propose a multi-band pitch estimator for such voice cases.

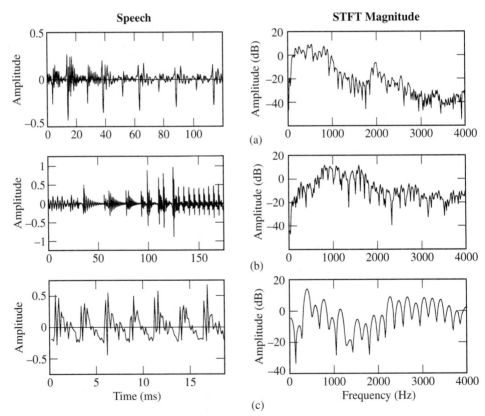

Figure 10.15 Examples of "anomalous" voice types: (a) diplophonic; (b) creaky; (c) alternating gain over successive pitch periods. The waveform and STFT magnitude are shown for each case.

10.11 (MATLAB) In this problem, you investigate the correlation-based pitch estimator of Section 10.2.

    (a) Implement in MATLAB the correlation-based pitch estimator in Equation (10.2). Write your function to loop through a speech waveform at a 10-ms frame interval and to plot the pitch contour.

    (b) Apply your estimator to the voiced speech waveform *speech1_8k* (at 8000 samples/s) in workspace *ex10M1.mat* located in companion website directory Chap_exercises/chapter10. Plot the short-time autocorrelation function for numerous frames. Describe possible problems with this approach for typical voiced speech.

    (c) Apply your estimator to the "anomalous" voice types in workspace *ex10M2.mat* located in companion website directory Chap_exercises/chapter10. Three voice types are given (also shown in Figure 10.15): *diplo1_8k* (diplophonic), *creaky1_8k* (creaky), and *modulated1_8k* (pitch periods with alternating gain). Describe problems that your pitch estimator encounters with these voice types.

10.12 (MATLAB) In this problem, you investigate the "comb-filtering" pitch estimator of Section 10.3.

    (a) Implement in MATLAB the comb-filtering pitch estimator of Equation (10.7) and used in Example 10.1 by selecting the last large peak in the pitch likelihood function $Q(\omega)$. Write the function to loop through a speech waveform at a 10-ms frame interval and to plot the pitch contour.

    (b) Apply your pitch estimator to the voiced speech waveform *speech1_8k* (at 8000 samples/s) in workspace *ex10M1.mat* located in companion website directory Chap_exercises/chapter10. Add white noise to the speech waveform and recompute your pitch estimate. Plot the pitch likelihood function for numerous frames. Describe possible problems with this approach.

    (c) Apply your estimator to the "anomalous" voice types in workspace *ex10M2.mat* located in companion website directory Chap_exercises/chapter10. Three voice types are given (and are shown in Figure 10.15): *diplo1_8k* (diplophonic), *creaky1_8k* (creaky), and *modulated1_8k* (pitch periods with alternating gain). Describe problems that your pitch estimator encounters with these voice types.

10.13 (MATLAB) In this problem you investigate the sinewave-based pitch estimators developed in Sections 10.4.1 and 10.4.2.

    (a) Implement in MATLAB the sinewave-based pitch estimator in Equation (10.15). Write the function to loop through a speech waveform at a 10-ms frame interval and to plot the pitch contour.

    (b) Apply your estimator to the voiced speech waveform *speech1_8k* (at 8000 samples/s) in workspace *ex10M1.mat* located in companion website directory Chap_exercises/chapter10. Discuss possible problems with this approach, especially the pitch-halving problem and the problem of a bias toward low pitch.

    (c) Apply your estimator from part (a) to the "anomalous" voice types in workspace *ex10M2.mat* located in companion website directory Chap_exercises/chapter10. Three voice types are given (Figure 10.15): *diplo1_8k* (diplophonic), *creaky1_8k* (creaky), and *modulated1_8k* (pitch periods with alternating gain). Describe problems that your pitch estimator encounters with these voice types.

    (d) Compare your results of parts (a) and (b) with the complete sinewave-based estimator in Equation (10.29) of Section 10.4.2. In companion website directory Chap_exercises/chapter10 you will find the scripts *ex10_13_speech1.m*, *ex10_13_diplo.m*, *ex10_13_creaky.m*, and *ex10_13_modulate.m*. By running each script, you will obtain a plot of the pitch and voicing contours from the complete sinewave-based estimator for the four cases: *speech1_8k*, *creaky1_8k*, *diplo1_8k*, and *modulated1_8k*, respectively. Using both time- and frequency-domain signal properties, explain your observations and compare your results with those from parts (a) and (b).

(e) Now consider the time-domain homomorphic pitch and voicing estimators of Chapter 6. Predict the behavior of the homomorphic estimators for the three waveform types of Figure 10.15. How might the behavior differ from the frequency-domain estimator of Section 10.4.2?

BIBLIOGRAPHY

[1] B. Gold and L.R. Rabiner, "Parallel Processing Techniques for Estimating Pitch Periods of Speech in the Time Domain," *J. Acoustical Society of America*, vol. 46, no. 2, pp. 442–448, 1969.

[2] D. Griffin and J.S. Lim, "Multiband Excitation Vocoder," *IEEE Trans. Acoustics, Speech and Signal Processing*, vol. ASSP–36, pp. 1223–1235, 1988.

[3] A. Kondoz, *Digital Speech: Coding for Low Bit Rate Communication Systems*, John Wiley & Sons, New York, NY, 1994.

[4] J. Makhoul, R. Viswanathan, R. Schwartz, and A.W.F. Huggins, "A Mixed-Source Model for Speech Compression and Synthesis," *Proc. IEEE Int. Conf. Acoustics, Speech, and Signal Processing*, Tulsa, OK, pp. 163–166, 1978.

[5] R.J. McAulay and T.F. Quatieri, "Pitch Estimation and Voicing Detection Based on a Sinusoidal Model," *Proc. Int. Conf. Acoustics, Speech, and Signal Processing*, Albuquerque, NM, pp. 249–252, 1990.

[6] R.J. McAulay and T.F. Quatieri, "Phase Modeling and its Application to Sinusoidal Transform Coding," *Proc. IEEE Int. Conf. Acoustics, Speech, and Signal Processing*, Tokyo, Japan, pp. 1713–1715, April 1986.

[7] R.J. McAulay and T.F. Quatieri, "Low-Rate Speech Coding Based on the Sinusoidal Speech Model," chapter in *Advances in Speech Signal Processing*, S. Furui and M.M. Sondhi, eds., Marcel Dekker, 1992.

[8] R.J. McAulay and T.F. Quatieri, "Sinusoidal Coding," chapter in *Speech Coding and Synthesis*, W.B. Kleijn and K.K. Paliwal, eds., Elsevier, 1995.

[9] M. Nishiguchi, J. Matsumoto, R. Wakatsuki, and S. Ono, "Vector Quantized MBE with Simplified V/UV Division at 3.0 kb/s," in *Proc. Int. Conf. on Acoustics, Speech, and Signal Processing*, Minneapolis, MN, vol. 2, pp. 151–154, April 1993.

[10] D.B. Paul, "The Spectral Envelope Estimation Vocoder," *IEEE Trans. Acoustics, Speech and Signal Processing*, vol. ASSP–29, pp. 786–794, 1981.

[11] L.R. Rabiner and R.W. Schafer, *Digital Processing of Speech Signals*, Prentice Hall, Englewood Cliffs, NJ, 1978.

[12] L.R. Rabiner, "On the Use of Autocorrelation Analysis for Pitch Detection," *IEEE Trans. Acoustics, Speech, and Signal Processing*, vol. ASSP–25, no. 1, pp. 24–33, 1977.

[13] M.M. Sondhi, "New Methods of Pitch Extraction," *IEEE Trans. Audio and Electroacoustics*, vol. AU–16, no. 2, pp. 262–266, June 1968.

[14] R. Smits and B. Yegnanarayana, "Determination of Instants of Significant Excitation in Speech Using Group Delay Function," *IEEE Trans. Speech and Audio Processing*, vol. 3, no. 5, pp. 325–333, Sept. 1995.

Nonlinear Measurement and Modeling Techniques

11.1 Introduction

Throughout the text, we have seen or alluded to examples of speech events that occur on a fine time scale, for example, on the order of about a pitch period or less. Such speech events, which we will loosely refer to as "fine structure" in the speech waveform, arise from either time-varying linear or nonlinear elements of the speech production mechanism and have importance in distinguishing sounds and voice types. In the context of the linear model, rapid transitions occur across speech phones such as unvoiced-to-voiced transitions, abrupt onsets of sounds such as the attack of a plosive or a voiced sound, and events that are short-lived, closely spaced, or even overlapping such as the burst and aspiration components of a plosive. In this linear context, rapid formant and pitch modulation also occurs, as with formant movement through a dipthong and with vibratto in the singing voice. We introduced mathematical models for many of these events in previous chapters.

Nonlinear aspects of speech production also contribute to speech fine structure. We saw in Chapter 5 that nonlinear coupling between the vocal tract pressure and glottal flow velocity is responsible for rapid formant and bandwidth modulation within a pitch period; such coupling is assumed to be nonexistent in the linear source/filter model. In particular, first formant modulation within a pitch period is manifested by a rapidly increasing bandwidth and the *truncation effect* that can characterize the speech waveform. Also, nonlinear modes of vocal cord vibration can occur that are responsible for secondary glottal pulses within a glottal cycle (the diplophonic voice), erratic glottal pulses across successive pitch periods (the creaky voice) (Figure 10.15), and small glottal pulse time variations, referred to as pitch jitter (the hoarse voice). Modulations and secondary sources may also arise from the interaction of *vortices* with vocal tract boundaries such as the epiglottis (false vocal folds), teeth, or occlusions in the oral tract. As an example, a "vortex ring" might modulate the effective area of the vocal tract and thus modulate the response pattern. In a second example, a vortex, traveling along the oral tract, may be converted to a secondary acoustic sound source. In these latter nonlinear models, vortices are generated by a fast-moving air jet from the glottis, quite different from the small compression and rarefraction perturbations associated with the static (i.e., without net displacement) acoustic sound wave in the tract. We alluded to this nonacoustic phenomenon in Chapters 3 and 4 and will more formally describe it in this chapter.

Speech fine structure is often difficult to measure with signal processing techniques such as the short-time Fourier transform and the wavelet transform that are characterized by time-frequency resolution tradeoffs. That is, according to the uncertainty principle reviewed in Chapter 2, we cannot obtain unlimited time and frequency resolution simultaneously, which (to some extent) is often required in the representation of speech fine structure. A purpose of this chapter is to describe alternative high-resolution measurement techniques that aim to "beat" the uncertainty principle and reveal fine structure not seen by the standard approaches.

In Section 11.2, we first review the limitation of the short-time Fourier transform and wavelet transform associated with the uncertainty principle, and discuss some generalizations of these transforms in addressing this limitation. We next move beyond the STFT and wavelet transform in Section 11.3 to more advanced time-frequency analysis methods, including the *Wigner distribution* and its variations referred to as *bilinear time-frequency distributions*. These distributions, which seek to undermine the uncertainty principle, have historically been refined and applied in estimating fine structure speech events, largely in the context of linear speech models. In the latter half of this chapter, we introduce a second approach to analysis of fine structure motivated by nonlinear aeroacoustic models for spatially distributed sound sources and modulations induced by nonacoustic fluid motion. Section 11.4 presents the aeroacoustic modeling approach which provides the impetus for the high-resolution *Teager energy operator* described in Section 11.5. This operator is characterized by a time resolution that can track rapid signal energy changes within a glottal cycle and is the basis for a method of high-resolution estimation of amplitude and frequency modulation in speech formants.

11.2 The STFT and Wavelet Transform Revisited

In Chapter 7, we saw conditions under which an STFT analysis window, typically 20–30 ms in duration, can result in a poor representation of nonstationary resonances or harmonics.[1] Formal studies also have been made of the inadequacy of the STFT in representing speech dynamics. In a study by Smits [45], for example, a 25-ms analysis window was shown for the STFT to result in many undesirable effects, including staircase formant tracks of rapidly varying formants and flattening-off of formants close to voicing onset. Likewise, in Chapter 8 (also Exercise 8.12), we saw that the wavelet transform, being constrained to good time resolution for high frequencies *or* good frequency resolution for low frequencies, is inadequate for the representation of many speech sounds.

It behooves us, therefore, to seek alternative analysis methods that provide a time-frequency resolution with which fine structure speech parameters can be represented. This objective takes

---

[1] In all analysis methods considered thus far, speech parameters are considered stationary under the analysis window, typically about 25 ms in duration. Only in certain special cases have we considered or alluded to explicit nonstationary models. These include a polynomial model for linear prediction coefficients (Exercise 5.13) and a Gaussian model for time-varying harmonic components (Section 9.4.5). The time-varying linear prediction analysis has been shown to lead to formant transitions that tightly cluster for the three plosive categories /p/, /t/, and /k/; with a stationary assumption, a more diffuse clustering occurs [35].

us to an approach based on *nonlinear* signal analysis methods that we refer to as *bilinear time-frequency distributions*. Before giving this more advanced time-frequency analysis, however, we present a common perspective on the STFT and wavelet transform analysis methods using a basis representation, thus setting the stage for more advanced time-frequency analysis.

11.2.1 Basis Representations

We saw in Chapter 8 that in the mathematical development of the wavelet transform, in continuous time, we define a set of functions as the time-scaled and shifted versions of a prototype $h(t)$, i.e.,

$$h_{\tau,a}(t) = \frac{1}{\sqrt{a}} h\left(\frac{t-\tau}{a}\right)$$

where $h(t)$ is the *basic wavelet*, $h_{\tau,a}(t)$ are the associated *wavelets*, τ is the time shift, and a is the scaling factor. We then define the continuous wavelet transform of a signal $x(t)$ as

$$X_w(\tau, a) = \frac{1}{\sqrt{a}} \int_{-\infty}^{\infty} x(t) h^*\left(\frac{t-\tau}{a}\right) dt \qquad (11.1)$$

which we can think of as a measure of "similarity" or correlation of $x(t)$ with $h(t)$ at different scales and time shifts, as was illustrated in Figure 8.15. We think of the functions $h_{\tau,a}(t)$ as basis functions in our representation that are concentrated at time τ and scale a. We saw in Chapter 8 that the wavelet transform is invertible under the admissibility condition in Equation (8.33). Also under this condition, we can show a relation analogous to that of Parseval's Theorem:

$$\int_{-\infty}^{\infty} |x(t)|^2 dt = \frac{1}{C_h} \int_{-\infty}^{\infty} \int_{-\infty}^{\infty} |X_w(\tau, a)|^2 d\tau \frac{da}{a^2} \qquad (11.2)$$

[with C_h defined in Equation (8.33)] where we think of $|X_w(\tau, a)|^2$ as an *energy density* in time and scale (Exercise 11.1). Equation (11.2) states that the total energy in the signal is the integral of the energy density over time and scale with the appropriate scale factor.

A basis representation can also be formulated for the STFT. In continuous time, we define the STFT basis function as

$$h_{\tau,\Omega}(t) = w(t-\tau)e^{j\Omega t}$$

which is concentrated near time τ and frequency Ω. Then the continuous-time STFT is written as

$$X(\tau, \Omega) = \int_{-\infty}^{\infty} x(t) h_{\tau,\Omega}^*(t) dt. \qquad (11.3)$$

In this case, the total energy in the signal is the integral of the energy density over time and frequency, provided that the energy in the analysis window is unity (Exercise 11.1). Both basis representations can be discretized, as we have seen with the discrete STFT and the discrete wavelet transform in Chapters 7 and 8, respectively.

11.2.2 Minimum Uncertainty

For the STFT and the wavelet transform, the basis functions define the time-frequency resolving power of the transform. In both cases, this resolution is limited by the uncertainty principle, requiring, as in Chapter 2, the variance about the mean for each dimension, i.e., frequency and time or scale and time. For frequency and time, we have the variances of a short-time segment governed by those of the short-time analysis window $w(t)$. Specifically, we are interested in the variances

$$\sigma_\Omega^2 = \int_{-\infty}^{\infty} (\Omega - \bar{\Omega})^2 |W(\Omega)|^2 d\Omega$$

$$\sigma_t^2 = \int_{-\infty}^{\infty} (t - \bar{t})^2 |w(t)|^2 dt \qquad (11.4)$$

where the mean values are

$$\bar{\Omega} = \int_{-\infty}^{\infty} \Omega |W(\Omega)|^2 d\Omega$$

$$\bar{t} = \int_{-\infty}^{\infty} t |w(t)|^2 dt. \qquad (11.5)$$

The product of the variances must not be less than the lower limit of the uncertainty principle:

$$\sigma_t^2 \sigma_\Omega^2 \geq \frac{1}{4}.$$

For the STFT, the variances, as well as their product, remain fixed as $h_{\tau,\Omega}(t) = w(t - \tau)e^{j\Omega t}$ moves about in the time-frequency plane. Because $h_{\tau,\Omega}(t)$ defines the resolving power of the STFT, we refer to $h_{\tau,\Omega}(t)$ as a "cell" or "atom" in the time-frequency plane. It is of interest to consider the window that gives the minimum uncertainty. For the STFT, this optimal signal can be shown equal to the Gaussian function modulated by a complex sinusoid. In other words,

$$\sigma_t^2 \sigma_\Omega^2 = \frac{1}{4} \qquad (11.6)$$

for the basis functions of the form

$$h_{\tau,\Omega}(t) = \alpha e^{-b(t-\tau)^2} e^{j\Omega t}$$

which are concentrated at time τ and frequency Ω.

An analogous result holds for time and scale with the wavelet transform. The uncertainty principle for time and scale is given by

$$\sigma_t^2 \sigma_a^2 \geq \frac{1}{2}|\bar{t}| \qquad (11.7)$$

where $|\bar{t}|$ equals the mean time of the wavelet $h_{\tau,a}(t)$ as defined in Equation (11.5) [10],[17]. The intuition behind the changing limit $\frac{1}{2}|\bar{t}|$ is as follows. Suppose that we stretch out a basic wavelet in time to form a wavelet, thus increasing its time variance σ_t^2. Because scale can be thought of as the reciprocal of frequency (Example 8.3), the scale variance σ_a^2 must also increase and so does the product $\sigma_t^2 \sigma_a^2$. In stretching out the basic wavelet, the mean time (i.e., location) of the resulting wavelet likewise increases. Equation (11.7) thus provides a constraint between these three increasing quantities.[2]

For the wavelet transform, the modulated Gaussian is not the function to achieve the lower bound $\frac{1}{2}|\bar{t}|$; rather, for the wavelet transform the *gammachirp function* achieves this lower bound [17]. The gammachirp function is of the form

$$g(t) = \alpha t^b e^{-ct} e^{j[\Omega t + d\log(t)]}$$

where the envelope $\alpha t^b e^{-ct}$ is a gamma distribution function and where the complex exponential $e^{j[\Omega t + d\log(t)]}$ has instantaneous frequency decreasing with time, $\Omega + \frac{d}{t}$, and thus the name "gammachirp." The difference in the functions that achieve the Heisenberg limit for the STFT and the wavelet transform, corresponding to different basis representations, further illustrates that the formulation of an uncertainty principle depends on the particular framework in which it is derived.

In the frequency domain, the gammachirp filter has an asymmetric spectral magnitude characteristic. Moreover, it is fascinating that its particular shape has been found to provide a good fit to a variety of auditory filters, including the cochlear filters along the basilar membrane that we studied in Chapter 8 and critical band masking filters that we will study in Chapter 13 [17],[27]. It is also fascinating to observe that the optimality (in the sense of giving minimum uncertainty) for gammachirp filters in aural perception is analogous to that of modulated Gaussian functions in visual perception. Indeed, measurements in a cat's visual cortex have found the presence of modulated Gaussian filters with respect to spatial resolution [11]. In these biological systems, the modulated Gaussian and the gammachirp filters represent a highly redundant and overcomplete basis.[3] This redundancy, however, may be desirable in that the loss of certain filters still allows a representation from which a signal may be recovered.

In both the STFT and the wavelet transform, we have enforced an atom (or "cell") with a particular time-frequency (scale) resolution on the signal. We can, on the other hand, consider *adapting* the time-frequency resolution of each atom to account for local signal properties [28]. Adaptive bases can be formulated to meet certain desired conditions such as completeness and invertibility. One adaptive basis is the *local cosine basis* which nonuniformly divides the time axis while uniformly segmenting frequency to adapt to local temporal properties. The dual to the local cosine basis is the *wavelet packet basis* that uniformly shifts the analysis window in time and nonuniformly divides the frequency dimension, adapting to the local frequency properties of the signal. A comparison of the "atomic tilings" of a wavelet packet basis and local cosine basis is shown in Figure 11.1. The particular cosine or wavelet packet basis can be selected with algorithms referred to as "best basis" searches that adapt to the particular time-frequency characteristics of the signal [28]. The basis is chosen to represent a signal, in a mean-squared-error sense, with a set number of time-frequency atoms. These techniques result

---

[2] Nevertheless, the uncertainty relations for frequency and scale are essentially the same because normalizing the scale variance by the mean time in the uncertainty relation for time and scale, we obtain for Equation (11.7) $\sigma_t^2(\sigma_a^2/|\bar{t}|) \geq \frac{1}{2}$. This is consistent with Figure 8.14 which shows that, although the time-frequency atom of the wavelet transform changes with frequency, in contrast to the fixed time-frequency resolution of the STFT, the Heisenberg constant remains fixed in the time-frequency plane.

[3] This is in contrast to a complete representation where there are as many representation coefficients as there are sample values of a signal or an undercomplete representation for which some signals cannot be recovered.

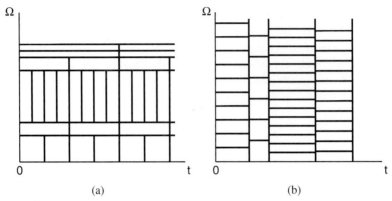

(a)   (b)

Figure 11.1   Comparison of (a) wavelet packet and (b) local cosine bases.   With
the wavelet packet basis, for each nonuniform frequency division there is one time
resolution. With the local cosine basis, for each nonuniform time division there is one
frequency resolution.

SOURCE: S. Mallat, *A Wavelet Tour of Signal Processing* [28] (Figures 1.4 and 1.5 of Chapter
1). ©1998, Academic Press. Used by permission of the publisher.

in basis coefficients that highlight certain properties of a signal and have found usefulness in
speech coding and noise reduction. Recall that we saw an example in Chapter 8 of how wavelet
coefficients (albeit, not a "best basis") can represent signal singularities from which the signal
can be reconstructed.

11.2.3 Tracking Instantaneous Frequency

An important illustration of the properties of the STFT and the wavelet transform is through
the problem of estimating a time-varying frequency in a complex exponential signal of the
form $x(t) = a(t)e^{j\phi(t)}$. For this signal, we have defined the *instantaneous frequency* as the
derivative of the phase, i.e., $\dot{\phi}(t) = \frac{d\phi(t)}{dt}$. Recall also that we have considered the STFT
magnitude squared of a signal, $|X(\Omega)|^2$, as an energy density in frequency and the magnitude
squared of the signal, $|x(t)|^2$, as an energy density in time. With these signal characteristics, it
can then be shown that average and instantaneous frequency are related by (Exercise 11.2)

$$\bar{\Omega} = \int_{-\infty}^{\infty} \Omega |X(\Omega)|^2 d\Omega$$

$$= \int_{-\infty}^{\infty} \dot{\phi}(t)|x(t)|^2 dt = \int_{-\infty}^{\infty} \dot{\phi}(t)a^2(t)dt. \tag{11.8}$$

This is an intriguing result relating the average frequency, as defined in the frequency domain,
to average instantaneous frequency, as defined in the time domain using our definition of instantaneous frequency. Consider now the average bandwidth of a signal as defined in Equation
(11.4). We saw in Chapter 2 (Exercise 2.8) that this average bandwidth for a continuous-time
signal of the form $x(t) = a(t)e^{\phi(t)}$ can be expressed as

$$B = \int_{-\infty}^{\infty} \left(\frac{da(t)}{dt}\right)^2 dt + \int_{-\infty}^{\infty} \left(\frac{d\phi(t)}{dt} - \bar{\Omega}\right)^2 a^2(t)\, dt. \tag{11.9}$$

The first term in Equation (11.9) is the average rate of change of the signal envelope $a(t)$ and thus reflects how fast the general signal shape is changing, as well as the signal duration in that a shorter signal gives a more rapidly changing envelope. We will see shortly that the quantity $\frac{\dot{a}(t)}{a(t)}$ can be interpreted as an *instantaneous bandwidth* of the signal. The second term is the average deviation of the instantaneous frequency $\frac{d\phi(t)}{dt}$ around the average frequency $\bar{\Omega}$ and thus is a measure of frequency spread.

One application of the STFT and wavelet transform is the estimation of the instantaneous frequency of a complex sinewave $x(t) = a(t)e^{j\phi(t)}$. Indeed, we have seen this application in Chapter 9 in sinewave analysis where peak-picking the STFT magnitude was applied to estimate sinewave amplitudes, frequencies, and phases. It can be shown that the sinewave frequency $\Omega(t) = \dot{\phi}(t)$ corresponds to the location of the maximum of the spectrogram $|X(t, \Omega)|^2$, provided that the instantaneous bandwidth $\frac{\dot{a}(t)}{a(t)}$ and instantaneous frequency $\dot{\phi}(t)$ have small variation over the duration of the analysis window [28]. This "ridge tracking" [28] method of instantaneous frequency estimation is expressed formally as

$$\dot{\phi}(t) \approx \max_{\Omega} |X(t, \Omega)|^2$$

and

$$a(t) \approx |X(t, \dot{\phi}(t))|^2$$

provided the instantaneous frequency and bandwidth are slowly varying over the analysis window.

In ridge tracking, a problem arises with rapidly changing frequency, as occurs, for example, with rapidly changing pitch; in this case, our assumption of slow variation in instantaneous frequency and bandwidth breaks down and the ridges become less distinct and meaningful. We saw in Figure 7.8 an example of this signal characteristic for a single frequency-modulated sinewave. An example of this property for a harmonic complex of sinewaves, comparing constant pitch and linearly changing pitch, is shown in Figure 11.2. We see that the spectral slice in the latter case is ragged at high frequencies, which across time gives no distinct ridge or harmonic structure. Exercise 11.3 asks the reader to investigate this condition for ridge structure in more detail.

The wavelet transform also provides instantaneous frequency estimation by using ridges of the scalogram, defined in Chapter 8 as $|X_w(t, a)|^2$. In particular, with time-varying frequency, a scalogram ridge analysis can more accurately track rapid changes at high frequencies by virtue of its increased temporal resolution. As with the STFT, under certain conditions on how fast the sinewave amplitude and frequency changes, the wavelet ridges correspond to the sinewave instantaneous frequency, i.e.,

$$\frac{c}{\dot{\phi}(t)} \approx \max_{a} |X_w(t, a)|^2$$

and

$$a(t) \approx \left| X_w\left(t, \frac{c}{\dot{\phi}(t)}\right) \right|^2$$

where c is an appropriate constant that relates scale to frequency. As an example, for a single sinewave, an improvement in frequency estimation over the STFT has been demonstrated for a hyperbolic chirp frequency, a signal well-suited to wavelet analysis because the rate of frequency

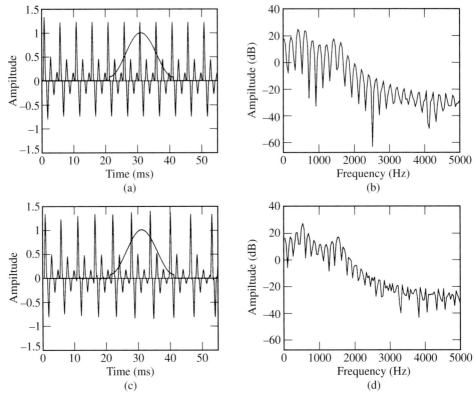

Figure 11.2　Short-time spectral comparison of two voiced synthetic waveforms: (a) waveform with constant pitch; (b) log-spectral magnitude of (a); (c) waveform with linearly changing pitch; (d) log-spectral magnitude of (c). The waveforms are obtained by exciting a linear system with a pulse train. The linear system consists of resonances at 500 Hz and 1500 Hz. The period of the pulse train is fixed at 5 ms in (a), while it is increasing by 0.2 ms per period in (c). The Hamming window used in the short-time analysis is superimposed on the waveforms.

change increases with frequency [28]. In this case, a staircase frequency track arises using the spectrogram, while the scalogram ridges approximately follow the hyperbolic chirp [28].

　　The wavelet transform, as well as its extension to adaptive bases, can also help in the representation of speech signal components, for example, transient and rapidly varying sinewaves, as was noted in Chapter 9. Nevertheless, this resolution gain is limited. With harmonic spectra, for example, there is a limit to frequency resolution at high frequencies, especially when frequencies are closely spaced (Exercises 8.12 and 11.3). The problem is that, regardless of the configuration of time-frequency atoms, these approaches are constrained by the time-frequency uncertainty principle, both in the representation and in the estimation of speech features; the transforms are obtained by correlating the signal with a time-frequency atom that is characterized by a limited time-frequency resolution. As pointed out by Mallat [28], ideally, we want a time-frequency distribution that does *not spread energy in time and frequency*. Our next section moves in this direction with a nonlinear signal representation, in contrast to our analysis thus far that has been linear.

11.3 Bilinear Time-Frequency Distributions

In this section, we describe a number of time-frequency representations that possess many of the properties lacking in the spectrogram and scalogram. We find, however, that these advantages may be traded off for a loss in some of the desirable properties of the conventional representations.

11.3.1 Properties of a Proper Time-Frequency Distribution

The spectrogram $|X(t, \Omega)|^2$ and the scalogram $|X_w(t, a)|^2$ are two examples of an *energy density* that describes how energy is distributed in time and frequency (or scale); hence we use the nomenclature "time-frequency distribution" interchangeably with the term "energy density." Often the energy density is viewed similarly to that of a probability density function. As such, means, variances, and integrals along one or two dimensions should behave similarly. Based on this relation to a probability function, we first define what is referred to as a *proper* time-frequency distribution [10].

One property that must be satisfied by a proper time-frequency distribution $P(t, \Omega)$ is positivity, i.e.,

$$P(t, \Omega) > 0$$

because we want $P(t, \Omega)$ to represent energy density. A second property is that the *marginals* must be satisfied. By this we mean

$$\int_{-\infty}^{\infty} P(t, \Omega)d\Omega = |x(t)|^2 = P(t)$$

$$\int_{-\infty}^{\infty} P(t, \Omega)dt = |X(\Omega)|^2 = P(\Omega). \tag{11.10}$$

That is, integration of the distribution over all frequencies at a particular time yields the energy density at that time, while integration of the distribution over all time at a particular frequency yields the energy density at that frequency. As shown above, we denote these purely time and frequency energy densities by $P(t)$ and $P(\Omega)$, respectively. A third desired property is that the total energy of the distribution sum to one:

$$E = \int_{-\infty}^{\infty} \int_{-\infty}^{\infty} P(t, \Omega)d\Omega\, dt = 1.$$

Then, from the marginal property in Equation (11.10), the total energy E equals the energy in the signal, i.e.,

$$E = \int_{-\infty}^{\infty} |x(t)|^2 dt = \int_{-\infty}^{\infty} |X(\Omega|^2 d\Omega$$

which equals unity. This unity energy constraint simply implies that we have properly normalized the signal in order that the notion of *density* be valid. Another property of a proper density function is that the density function has finite support in one of the variables, given that the signal itself or its spectrum has finite support in that variable. That is, if a signal is zero outside the range $[t_1, t_2]$, then we want the distribution $P(t, \Omega)$ to be zero outside this range. We call this a *weak finite support* condition [10]. A similar condition can be made along the frequency variable, i.e., for a bandlimited signal. Observe that the energy density cannot have weak finite support in both time and frequency because this would violate the uncertainty principle. An even stronger condition, the *strong finite support* condition, is that strips of zero energy over specific time or frequency intervals map to strips of zero energy in the time-frequency distribution. This finite support constraint is loosely related to a property that preserves the number of components; for example, the distribution of a two-component signal should reflect only these two components and not additional unwanted terms, sometimes referred to as *cross terms*, due to nonlinearity of a distribution operator. This notion of cross terms will become clearer as we proceed.

Based on this density concept and the above properties, we can now give certain global and conditional averages and variances of the time and frequency variables. We have already encountered the global averages and variances in our earlier discussion of the uncertainty principle. These are given in Equations (11.4) and (11.5), respectively, that represent the spread of time (duration) and the spread of frequency (bandwidth) and mean time and frequency. We are also interested in the conditional quantities of the mean frequency at a certain time and the dual quantity of the mean time at a certain frequency. We often also refer to these conditionals as "local" quantities. Motivated by our analogy with a probability density function, we write $P(\Omega|t) = \frac{P(t,\Omega)}{P(t)}$ and, likewise, $P(t|\Omega) = \frac{P(t,\Omega)}{P(\Omega)}$, and thus define conditional mean frequency and time, respectively, as

$$\bar{\Omega}_t = \int_{-\infty}^{\infty} \Omega P(\Omega|t) d\Omega = \frac{1}{P(t)} \int_{-\infty}^{\infty} \Omega P(t, \Omega) d\Omega$$

$$\bar{t}_\Omega = \int_{-\infty}^{\infty} t P(t|\Omega) dt = \frac{1}{P(\Omega)} \int_{-\infty}^{\infty} t P(t, \Omega) dt. \tag{11.11}$$

Furthermore, when the marginals are satisfied, we have from Equation (11.10)

$$\bar{\Omega}_t = \frac{1}{|x(t)|^2} \int_{-\infty}^{\infty} \Omega P(t, \Omega) d\Omega$$

$$\bar{t}_\Omega = \frac{1}{|X(\Omega)|^2} \int_{-\infty}^{\infty} t P(t, \Omega) dt.$$

Likewise, the corresponding conditional variances are given by

$$\sigma_{\Omega|t}^2 = \frac{1}{P(t)} \int_{-\infty}^{\infty} (\Omega - \bar{\Omega})^2 P(t, \Omega) d\Omega$$

$$\sigma_{t|\Omega}^2 = \frac{1}{P(\Omega)} \int_{-\infty}^{\infty} (t - \bar{t})^2 P(t, \Omega) dt \tag{11.12}$$

where, when the marginals are satisfied by $P(t, \Omega)$, then, as with the mean expressions, we can replace $P(t)$ and $P(\Omega)$ by $|x(t)|^2$ and $|X(\Omega)|^2$, respectively.

Finally, we are interested in the relation between the global and local means and variances. Here again, we are motivated by our probability density function analogy. For the means, we can show that

$$\bar{\Omega} = \int_{-\infty}^{\infty} \bar{\Omega}_t P(t) dt$$

$$\bar{t} = \int_{-\infty}^{\infty} \bar{t}_\Omega P(\Omega) d\Omega. \tag{11.13}$$

For the variances, Cohen [10] has shown that for an arbitrary proper time-frequency distribution, the global variances of frequency and time are given by

$$
\sigma_\Omega^2 = \int_{-\infty}^{\infty} \sigma_{\Omega|t}^2 P(t)dt + \int_{-\infty}^{\infty} (\bar{\Omega}_t - \bar{\Omega})^2 P(t)dt
$$

$$
\sigma_t^2 = \int_{-\infty}^{\infty} \sigma_{t|\Omega}^2 P(\Omega)d\Omega + \int_{-\infty}^{\infty} (\bar{t}_\Omega - \bar{t})^2 P(\Omega)d\Omega. \tag{11.14}
$$

We see then that there are present two contributions to the global variances. For the frequency variable, for example, we have (1) the average over all time of the conditional variance in frequency and (2) the average over all time of the conditional mean frequency about its global mean. An analogous result holds for the time variable. Exercise 11.4 asks you to work through the derivation of these results.

To gain intuition for these concepts, we consider the specific case of a complex sinewave of the form $x(t) = a(t)e^{j\phi(t)}$. For this signal, we saw in Equation (11.8) the relation between the global frequency average and the instantaneous frequency. This relation supports the notion of the phase derivative $\dot{\phi}(t)$ as the *instantaneous frequency*, i.e., the frequency at each time of the complex sinewave $x(t) = a(t)e^{j\phi(t)}$, because the relation reveals the global frequency average as the average of $\dot{\phi}(t)$ weighted by $a^2(t)$ at each time. Moreover, when the marginals are satisfied, it follows from Equation (11.13) that the (conditional) mean frequency at a particular time is the instantaneous frequency, i.e., $\bar{\Omega}_t = \dot{\phi}(t)$ (Exercise 11.5). The dual result for the global time average is $\bar{t} = \int_{-\infty}^{\infty} \dot{\theta}(\Omega)|X(\Omega)|^2 d\Omega$, where $\theta(\Omega) = \angle X(\Omega)$. We call $\dot{\theta}(\Omega)$ the *group delay* which, likewise, when the marginals are satisfied can be shown to equal the (conditional) mean time at a particular frequency. When the marginals are satisfied, we thus have the equation pair:

$$
\bar{\Omega}_t = \dot{\phi}(t)
$$

$$
\bar{t}_\Omega = \dot{\theta}(\Omega). \tag{11.15}
$$

It is important to note that the marginals are not necessarily satisfied for an arbitrary time-frequency distribution $P(t, \Omega)$, and thus instantaneous frequency and group delay cannot always be computed as above from the distribution. We will see shortly that the spectrogram is one such distribution.

Consider now the global and conditional variances for the complex sinewave $x(t) = a(t)e^{j\phi(t)}$. We have seen above that the instantaneous frequency is the (conditional) *average* of frequencies that exist at a particular time. Therefore, we can think of the conditional variance as the spread in frequency about this average at each time. We have also seen that the global variance of frequency for the complex sinewave is given by Equation (11.9). To gain further insight into global variance, we rewrite Equation (11.9) as

$$
\sigma_\Omega^2 = \int_{-\infty}^{\infty} \left(\frac{\dot{a}(t)}{a(t)}\right)^2 a^2(t)dt + \int_{-\infty}^{\infty} [\dot{\phi}(t) - \bar{\Omega}]^2 a^2(t)dt. \tag{11.16}
$$

A dual expression can be obtained for the global variance of time. By comparing Equations (11.14) and (11.16), Cohen has argued that under the condition that the marginals are satisfied and that the conditional variance is positive (a plausible constraint), then the conditional variance of frequency at a particular time is given by

$$\sigma^2_{\Omega|t} = \left| \frac{\dot{a}(t)}{a(t)} \right|^2. \tag{11.17}$$

Because $\sigma^2_{\Omega|t}$ is the spread of frequency at a particular time, it can be thought of as an "instantaneous bandwidth" [10], depending only on the amplitude and not the phase of the signal. For a constant amplitude, frequency-modulated sinewave, the instantaneous bandwidth is zero, there being one frequency at each time instant, i.e., the derivative of the phase [10]. In addition, in this case, we see in Equation (11.16) that the only contribution to global bandwidth is the spread of instantaneous frequency about the global mean frequency.

Consider now applying the above concepts to signals with more than one complex sinewave component, as, for example, voiced speech signals. The value of the instantaneous bandwidth of each component determines whether a time-frequency distribution is capable of revealing the components. For example, for a two-component signal $x(t) = a_1(t)e^{j\phi_1(t)} + a_2(t)e^{j\phi_2(t)}$, the two instantaneous frequencies at a time instant t are revealed as separate ridges in the distribution, provided that (Exercise 11.7) [10]

$$\left| \frac{\dot{a}_1(t)}{a_1(t)} \right|^2, \left| \frac{\dot{a}_2(t)}{a_2(t)} \right|^2 \ll |\dot{\phi}_1(t) - \dot{\phi}_2(t)|^2. \tag{11.18}$$

Likewise, the separation of signal components in time at a particular frequency is determined by the time spread of each component at that frequency relative to its group delay (Exercise 11.7). These constraints are meaningful only when the marginals of a time-frequency distribution are satisfied and thus Equations (11.15) and (11.17) hold.

The properties of a proper time-frequency distribution are summarized as:

P1: The distribution is positive.

P2: The total energy integrates to one.

P3: The marginals are satisfied.

P4: The conditional mean frequency and its variance are positive and, for a complex sinewave $x(t) = a(t)e^{j\phi(t)}$, give the instantaneous frequency $\dot{\phi}(t)$ and bandwidth $\left| \frac{\dot{a}(t)}{a(t)} \right|^2$, respectively (when property **P3** holds). The dual relations for conditional mean time and its variance also hold.

P5: The distribution has strong finite support and there are no cross terms.

We now look at the properties of a number of specific distributions.

11.3.2 Spectrogram as a Time-Frequency Distribution

Two energy densities that we have studied thus far are the spectrogram, based on the STFT, and the scalogram, based on the wavelet transform. In general, these energy densities do not satisfy many of the properties of our proper time-frequency distribution. In fact, typically, an arbitrary energy density violates some of our desired properties. For illustration, we focus in this section on the spectrogram. Analogous properties of the scalogram can be found in [10],[28].

We begin with the total energy for the spectrogram. Denoting the analysis window by $w(t)$, it is possible to show that the total energy E is given by (Exercise 11.8)

$$E = \int_{-\infty}^{\infty} \int_{-\infty}^{\infty} |X(t,\Omega)|^2 dt\, d\Omega = \int_{-\infty}^{\infty} |x(t)|^2 dt \int_{-\infty}^{\infty} |w(t)|^2 dt.$$

Therefore, if the energy of the window function is unity, then the energy in the spectrogram equals the energy in the signal. The marginals, i.e., integrating out time and frequency from the spectrogram, however, are not the desired time and frequency densities, $|x(t)|^2$ and $|X(\Omega)|^2$, respectively (Exercise 11.8). A consequence of this is that the conditional mean frequency and its variance, although positive, do not in general equal the instantaneous frequency and instantaneous bandwidth, respectively, for a complex sinewave of the form $x(t) = a(t)e^{j\phi(t)}$. Specifically,

$$\bar{\Omega}_t \neq \dot{\phi}(t)$$

$$\sigma^2_{\Omega|t} \neq \left|\frac{\dot{a}(t)}{a(t)}\right|^2.$$

Nevertheless, it can be shown that as the analysis window duration goes to zero, approaching an impulse, then the conditional mean approaches the instantaneous frequency [10]. In this case, however, the conditional variance $\sigma^2_{\Omega|t}$ goes to infinity, making difficult the measurement of instantaneous frequency [10]. The dual relations for conditional mean time and its variance also do not generally hold in the measurement of group delay.

Another violation of a desired property is that of finite support. A signal with finite support does not maintain this property in the spectrogram, as with, for example, a sinewave of finite length; the onset and offset are smeared in time in the time-frequency distribution. Moreover, the spectrogram has neither weak nor strong finite support. These limitations are a reflection of the analysis window being "entangled" with the signal within the distribution [10]; we see in Equation (11.3) that the time-frequency resolution of the spectrogram is limited by the window's time-frequency resolution that is constrained by the uncertainty principle. Finally, we note that the spectrogram can introduce *cross terms* when more than one signal component is present. We earlier defined cross terms to mean terms that are not present in the original signal and that are created because the distribution is nonlinear in the signal. In the spectrogram, however, these cross terms are often hidden by the original components in the distribution [10].

In the next section, we explore the Wigner distribution that preserves many of the properties that the spectrogram violates. We will see, however, that certain desirable properties of the spectrogram are not held by the Wigner distribution. We then investigate variations of the Wigner distribution that merge the good properties of the spectrogram and the Wigner distribution. Finally, we illustrate the use of such distributions in speech analysis.

11.3.3 Wigner Distribution

The Wigner distribution was originally designed in the context of quantum mechanics by Wigner [53] and later in 1948 was introduced into signal processing by Ville [52], who was motivated by the measurement of instantaneous frequency. Since the work of Ville, further development of the Wigner distribution in the context of signal processing has been made, most notably in a series of papers by Claasen and Mecklenbräuker [7],[8],[9]. In this and the following few sections, we describe the Wigner distribution and its variations in continuous time. The extensions of the Wigner distribution to discrete time have been developed in [8],[18].

The Wigner distribution is given by

$$W(t, \Omega) = \frac{1}{2\pi} \int_{-\infty}^{\infty} x\left(t + \frac{\tau}{2}\right) x^*\left(t - \frac{\tau}{2}\right) e^{-j\Omega\tau} d\tau \qquad (11.19)$$

which can also be written in terms of the signal spectrum, i.e., (Exercise 11.11)

$$W(t, \Omega) = \frac{1}{2\pi} \int_{-\infty}^{\infty} X\left(\Omega + \frac{\theta}{2}\right) X^*\left(\Omega - \frac{\theta}{2}\right) e^{-jt\theta} d\theta. \qquad (11.20)$$

The Wigner distribution is one of a class of *bilinear* distributions, where the signal $x(t)$ enters twice into the calculation of the transform. We can view the Wigner distribution as correlating a signal with itself translated both in time and in frequency. A signal that is time- or frequency-limited remains so over the same time-frequency intervals in the Wigner representation, but zero gaps in a signal or its spectrum are not necessarily maintained in the distribution (Exercise 11.11). The Wigner distribution thus has weak, but not strong finite support. A consequence of bilinearity, however, is that the Wigner distribution typically goes negative and thus its meaning as an energy density is sometimes questioned.

Another important property of the Wigner distribution is that, unlike the spectrogram, it satisfies the time-frequency marginals (Exercise 11.11):

$$\int_{-\infty}^{\infty} W(t, \Omega) d\Omega = |x(t)|^2$$

$$\int_{-\infty}^{\infty} W(t, \Omega) dt = |X(\Omega)|^2 \qquad (11.21)$$

and because the marginals are satisfied, the desired energy relation follows:

$$E = \int_{-\infty}^{\infty} \int_{-\infty}^{\infty} W(t, \Omega) d\Omega \, dt = \int_{-\infty}^{\infty} |x(t)|^2 \, dt. \qquad (11.22)$$

In addition, when the marginals are satisfied, the conditional mean frequency and time are correctly calculated [10],[28], i.e.,

$$\bar{\Omega}_t = \frac{1}{|x(t)|^2} \int \Omega W(t, \Omega) d\Omega = \dot{\phi}(t)$$

$$\bar{t}_\Omega = \frac{1}{|X(\Omega)|^2} \int t W(t, \Omega) dt = -\dot{\theta}(\Omega).$$

The conditional variances, however, generally give less plausible results and can even go negative [10].

For certain signals, the Wigner distribution provides a representation more discernible than the spectrogram. For a chirp signal, i.e., a signal of the form $x(t) = a(t)e^{j\phi(t)}$ with constant amplitude and linear frequency, over all time, the *Wigner distribution is a two-dimensional delta function* concentrated along the instantaneous frequency, i.e., the Wigner distribution is non-zero only along the trajectory $\Omega(t) = \dot{\phi}(t)$. As such, for this signal the conditional variance (and hence instantaneous bandwidth) is zero:

$$\sigma_{\Omega|t}^2 = \left| \frac{\dot{a}(t)}{a(t)} \right| = 0.$$

We see then that for this particular signal class, because the distribution is a two-dimensional delta function, we have, *in this conditional framework*, "beaten" the uncertainty principle: the conditional, or what we also think of as the *local*, time-bandwidth product is zero, i.e., [25]

$$\sigma_{\Omega|t}^2 \sigma_{t|\Omega}^2 = 0.$$

Observe, however, that the *global* time-bandwidth product $\sigma_\Omega^2 \sigma_t^2$ is still constrained by the uncertainty principle.

On the other hand, because of the presence of the analysis window, the "beating" of the uncertainty principle by the local time-bandwidth product can never occur with the spectrogram. (Exercise 11.12 asks you to derive bounds on the conditional time-bandwidth product for an arbitrary distribution, compare differences in the local and global time-bandwidth products, and consider the conditional time-bandwidth product for the spectrogram.) For a finite-duration chirp, likewise, the Wigner distribution is spread in frequency [10]. For a cubic phase, i.e., quadratic instantaneous frequency, the Wigner distribution is concentrated along $\Omega(t) = \dot{\phi}(t)$, but it is also not a delta function. Higher-order instantaneous frequency functions can yield quite complicated Wigner distributions, even though the conditional mean frequency always gives the instantaneous frequency function; indeed, the spread about this function can make the time-frequency presentation nebulous.

In spite of certain features of the Wigner distribution for signals consisting of one amplitude- and frequency-modulated sinewave, its major difficulty arises with more realistic signals such as those in speech consisting of sums of amplitude- and frequency-modulated resonances or harmonics. Let us consider a signal with two additive components:

$$x(t) = x_1(t) + x_2(t).$$

It is straightforward to show that the Wigner distribution of $x(t)$ consists of the Wigner distribution of the signal components, but also some additional terms:

$$W(t, \Omega) = W_{11}(t, \Omega) + W_{22}(t, \Omega) + W_{12}(t, \Omega) + W_{21}(t, \Omega) \tag{11.23}$$

where

$$W_{12}(t, \Omega) = \frac{1}{2\pi} \int_{-\infty}^{\infty} x_1^* \left(t - \frac{\tau}{2} \right) x_2 \left(t + \frac{\tau}{2} \right) e^{-j\Omega\tau} d\tau$$

and similarly for $W_{21}(t, \Omega)$. The components $W_{12}(t, \Omega)$ and $W_{21}(t, \Omega)$ are *cross terms* we alluded to earlier. These cross terms come about due to the nonlinearity of the distribution, i.e., the signal is multiplied by itself shifted in time. One must be careful, however, in thinking that nonlinearity, and thus cross terms, are peculiar to the Wigner distribution. The spectrogram can also be considered as nonlinear in the signal $x(t)$ due to the magnitude squaring operation. The difference, as we see in the example below, is that the cross terms introduced by the spectrogram are often hidden under the self terms, while the cross terms from the Wigner distribution are typically distinct from the self terms.

EXAMPLE 11.1   An example illustrating the effect of cross terms placed between signal components in time and frequency is the sum of a complex sinewave and an impulse [10] expressed as

$$x(t) = e^{j\Omega_o t} + \sqrt{2\pi}\,\delta(t - t_o).$$

The Wigner distribution of this signal consists of three components: an impulse in frequency (the Wigner distribution of $e^{j\Omega_o t}$), an impulse in time [the Wigner distribution of $\sqrt{2\pi}\,\delta(t - t_o)$], and a sinusoidal cross term, i.e.,

$$W(t, \Omega) = \delta(\Omega - \Omega_o) + \delta(t - t_o) + \frac{1}{\sqrt{2\pi}}\cos[2(\Omega - \Omega_o)(t - t_o) - \Omega_o t_o].$$

Figure 11.3a shows the Wigner distribution of the sum of the two signals. The cross-term contribution from the Wigner distribution is clearly distinct (a mild case in contrast to the cross terms from other multi-component signals). For comparison, Figure 11.3b shows the corresponding spectrogram that gives a smeared representation of the two components, in contrast to the ideal resolution from the Wigner distribution. The cross terms of the spectrogram, however, are hidden under its self-terms [10].

▲

As with the STFT and the spectrogram (Chapter 7), the Wigner distribution is invertible, but up to a scale factor. To show this, we can use an approach similar to that used in showing the invertibility of the STFT. Because the Wigner distribution is the Fourier transform of $x(t + \frac{\tau}{2})x^*(t - \frac{\tau}{2})$, we can invert it to obtain

$$x\left(t + \frac{\tau}{2}\right)x^*\left(t - \frac{\tau}{2}\right) = \int_{-\infty}^{\infty} W(t, \Omega)e^{j\tau\Omega}d\Omega.$$

We then evaluate this expression at time $t = \frac{\tau}{2}$ and set $\tau = t$ giving

$$x(t) = \frac{1}{x^*(0)}\int_{-\infty}^{\infty} W\left(\frac{t}{2}, \Omega\right)e^{t\Omega}d\Omega$$

which is known up to the constant $\frac{1}{x^*(0)}$. Also, as with the STFT and spectrogram, not every two-dimensional function of time and frequency is a Wigner distribution [10]. In particular, when a legitimate Wigner distribution is modified in a desired way, e.g., to reduce interference or noise, the resulting time-frequency function is no longer necessarily a legitimate Wigner distribution. In this case, one typically uses least-squared-error methods for signal recovery, similar to those we developed in Chapter 7 for recovering a signal from a modified STFT or STFT magnitude [5].

An interesting property of the Wigner distribution is its relation to the spectrogram. One can show that for a window function $w(t)$, the spectrogram is the convolution of the Wigner distribution of the signal, denoted here by $W_x(t, \Omega)$, and the Wigner distribution of the window, $W_h(t, \Omega)$ [10]:

$$|X(t, \Omega)|^2 = \int_{-\infty}^{\infty}\int_{-\infty}^{\infty} W_x(t', \Omega')W_h(t - t', \Omega - \Omega')dt'd\Omega' \tag{11.24}$$

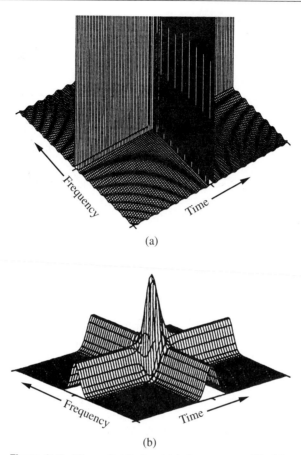

(a)

(b)

Figure 11.3 Wigner distribution (a) and spectrogram (b) of the sum of a complex sinewave and impulse. The ideal resolution of the Wigner distribution for these components is compromised by its cross terms. The spectrogram, on the other hand, has a relative loss in resolution, but has cross terms that are hidden.

SOURCE: L. Cohen, *Time-Frequency Analysis* [10]. ©1995, Pearson Education, Inc. Used by permission.

revealing a loss of time-frequency resolution in the spectrogram relative to that of the Wigner distribution. Nevertheless, this smoothing can have the positive effect of reducing the presence of unwanted cross terms in the Wigner distribution.

Ultimately, we want a time-frequency distribution that is useful in the visualization and estimation of speech events that are hidden in the spectrogram and scalogram energy densities. Cross terms and negativity in the Wigner distribution can prevent our reaching this objective. One approach to reduce these unwanted characteristics is to apply a window function to the product $x(t + \frac{\tau}{2})x^*(t - \frac{\tau}{2})$ [10]. This operation, however, also reduces time-frequency resolution. In the next section, we look at variations of the Wigner distribution that combine the desirable resolution of the Wigner distribution with the positivity and reduced cross terms of the spectrogram.

11.3.4 Variations on the Wigner Distribution

Cohen has proposed a class of time-frequency distributions that unite the spectrogram and Wigner distributions and, in addition, provide for the creation of new distributions [10]. This broad class of time-frequency distributions is written as

$$C(t, \Omega) = \frac{1}{4\pi^2} \int_{-\infty}^{\infty} \int_{-\infty}^{\infty} \int_{-\infty}^{\infty} x\left(t + \frac{\tau}{2}\right) x^*\left(t - \frac{\tau}{2}\right) k(\theta, \tau) e^{-j\theta t - j\Omega\tau + j\theta u} \, du \, d\tau \, d\theta$$

(11.25)

which can also be written in terms of the signal spectrum, in a fashion analogous to that of the frequency version of the Wigner distribution. The two-dimensional function $k(\theta, \tau)$ is referred to as the *kernel*. The kernel is selected to obtain certain time-frequency distributions with desired properties. For example, a kernel of unity gives the Wigner distribution, while the kernel for the spectrogram is given by[4] $k(\theta, \tau) = \int_{-\infty}^{\infty} w(t + \frac{\tau}{2}) w^*(t - \frac{\tau}{2}) e^{-j\theta t} dt$, where $w(t)$ is the analysis window [10]. The time-frequency marginals for $C(t, \Omega)$ are satisfied if (Exercise 11.14)

$$k(\theta, 0) = 1, \quad \text{for the time marginal}$$

$$k(0, \tau) = 1, \quad \text{for the frequency marginal.}$$

Other conditions on the kernel guarantee properties such as total energy conservation, preservation of local and global averages, and (important for speech analysis) reduced interference from cross terms with multi-component signals [10]. For low-level cross terms in $C(t, \Omega)$, the kernel must be peaked at its origin, i.e.,

$$k(\theta, \tau) \ll 1, \quad \text{for} \quad \theta\tau \gg 0.$$

The kernel for the Wigner distribution does not satisfy this property. The kernel for the spectrogram, on the other hand, satisfies this property for particular windows; e.g., a Hamming or Gaussian window results in good cross-term behavior [10]. Conditions for weak and strong finite support, important again in the speech context, can also be imposed through the kernel [10],[26].

An important distribution derived from the above canonic formulation is the *Choi-Williams distribution*, which has good cross-term behavior. The kernel for the Choi-Williams distribution is given by $k(\theta, \tau) = e^{-\frac{\theta^2\tau^2}{2}}$. This kernel gives not only reduced cross terms, but also a distribution that satisfies the marginals; thus, local averages result in instantaneous frequency and group delay for a single sinewave. An example provided by Williams and presented by Cohen [10] illustrating how the Choi-Williams distribution reduces cross terms, relative to that of the Wigner distribution, is shown in Figure 11.4.

11.3.5 Application to Speech Analysis

We return now to the topic at hand: speech analysis. One motivation for investigating time-frequency distributions is to exceed the time and frequency resolution capabilities of the spectrogram for the representation of speech fine structure. In particular, we desire a distribution that

---

[4] The function $k(-\theta, \tau) = \int_{-\infty}^{\infty} w(t + \frac{\tau}{2}) w^*(t - \frac{\tau}{2}) e^{j\theta t} dt$ is called the *ambiguity function* and is often used in radar signal processing.

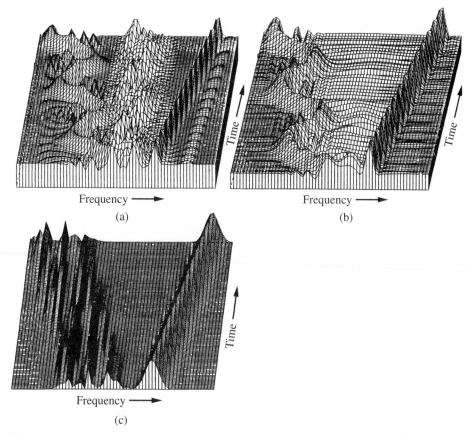

Figure 11.4 Comparison of the (a) Wigner distribution, (b) Choi-Williams distribution, and (c) spectrogram for a signal consisting of the sum of a linear FM sinewave and a sinusoidal FM sinewave [10]. The Choi-Williams distribution suppresses cross terms of the Wigner distribution and improves on the FM tracking by the spectrogram.

SOURCE: L. Cohen, *Time-Frequency Analysis* [10]. ©1995, W.J. Williams. Used by permission of W.J. Williams.

can show simultaneously transient events such as plosives with good temporal resolution and tonal events such as harmonics with good frequency resolution; furthermore, we hope that such a distribution will uncover speech events not revealed by conventional approaches. The Wigner distribution and its extensions, such as the Choi-Williams distribution for reducing cross terms, are attempts to obtain this joint time-frequency resolution. A limitation of these methods is that, although they have better resolution properties than the spectrogram, they can result in negative values and residual cross terms that make their interpretation as an energy density difficult [10], particularly for speech signals with multiple components, closely spaced in time and frequency.

Nevertheless, positive time-frequency distributions have been developed that hold promise in speech representations in that they can simultaneously give good temporal and spectral resolution while maintaining many desirable properties of a proper distribution. Development of these methods is beyond our scope, so we give the reader merely a flavor for the approach.

These positive time-frequency distributions have been developed by Pitton, Atlas, and Loughlin by working with a class of distributions of the form [37].

$$P(t, \Omega) = |x(t)|^2 |X(\Omega)|^2 \Lambda(t, \Omega, x(t)).$$

This relation shows that the energy density is low when the signal is low in either time or frequency, thus ensuring strong finite support in time and frequency. The function $\Lambda(t, \Omega, x(t))$ is constrained to be positive and to be such that $P(t, \Omega)$ satisfies the marginals. Under these constraints, a function $\Lambda(t, \Omega, s(t))$ is then selected that "maximizes the entropy" of the resulting distribution by using a nonlinear iterative estimation scheme [37]. Based on this formulation, Pitton, Atlas, and Loughlin have developed a number of positive time-frequency distributions that satisfy the marginals, have low cross-term interference, and good time-frequency resolution. The next two examples show first a synthetic and then a real speech signal case.

> **EXAMPLE 11.2**   Consider a simulated speech resonance with linearly changing center frequency, driven by a periodic impulse train.[5] Although the driving function is periodic, the signal is not periodic due to the time-varying filter coefficients. Figure 11.5 gives, for the synthetic speech signal, a comparison of a wideband spectrogram, a narrowband spectrogram, and a positive time-frequency distribution by Pitton, Atlas, and Loughlin [37]. One sees that, unlike either spectrogram, the positive time-frequency distribution shows simultaneously the harmonics of the excitation, resonances of the time-varying filter, and onset times of the excitation. An additional, interesting signal property, revealed by the positive time-frequency distribution, is noted by Pitton [37]: ". . . although the spectral harmonics of the positive time-frequency distribution are spaced 125 Hz about the center chirping frequency, the harmonics are not located at integer multiples of 125 Hz because the signal is not periodic and cannot be described by a Fourier series with fundamental frequency 125 Hz."   ▲

> **EXAMPLE 11.3**   Figure 11.6 gives a comparison of the wideband spectrogram, narrowband spectrogram, and a positive time-frequency distribution by Pitton, Atlas, and Loughlin [37] for a real speech waveform. As with the synthetic case, one sees that, unlike either spectrogram, the positive time-frequency distribution shows simultaneously the harmonics of the excitation, resonances of the time-varying filter, and onset times of the excitation, such as at the start of plosives.   ▲

Positive time-frequency distributions of the type illustrated in the previous examples present some important directions for speech analysis. Nevertheless, it has been argued by Pitton, Wang, and Juang [38] that these distributions have not yet "fundamentally altered" approaches to speech processing or speech and speaker recognition tasks. The expectation here is that such time-frequency distributions would replace the spectrogram in providing a means to a clearer picture of speech production and its associated parameters. The problem, however, is that with greater resolution come greater variability of speech parameters and a larger number of parameters required in the representation. It is important in many speech processing applications that we attain a small set of reliable parameters that preserve the essential dynamics of the speech signal [38]. For example, in the speaker recognition task that we study in Chapter 14, an increase in the feature vector dimension implies the need for more training data, and greater variability with increased parameter resolution can blur speaker differences. *A challenge is to reduce*

---

[5] This time-varying resonance is reminiscent of that given in Example 4.2 and Exercise 4.5 where an empty bottle is filled with running water.

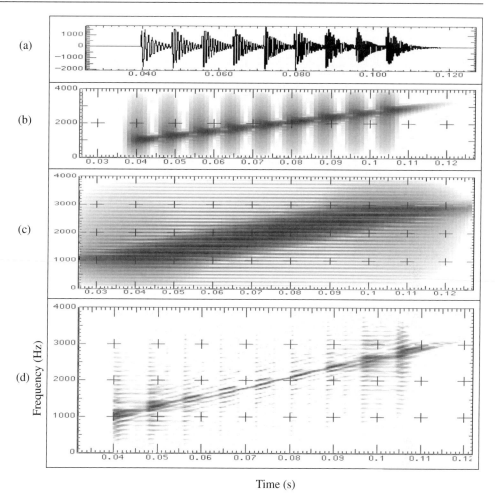

(a)

(b)

(c)

(d)

Frequency (Hz)

Time (s)

Figure 11.5 Comparison of the wideband/narrowband spectrograms and a positive time-frequency distribution for a simulated time-varying speech resonance: (a) waveform; (b) wideband spectrogram; (c) narrowband spectrogram; (d) positive time-frequency distribution. The positive time-frequency distribution reveals the harmonics moving along the path of the time-varying resonance.

SOURCE: J.W. Pitton, L.E. Atlas, and P.J. Loughlin, "Applications of Positive Time-Frequency Distributions to Speech Processing" [37]. ©1994, IEEE. Used by permission.

statistical variability while preserving time-frequency resolution. Nevertheless, the progress made in time-frequency distributions for speech analysis is an important step toward improved signal processing representations of the fine structure and dynamics of the speech waveform.

In the second half of this chapter, we look at alternative approaches to measuring speech fine structure and dynamics. As we noted in this chapter's introduction, these approaches have been motivated by a nonlinear, aeroacoustic view of speech production. We therefore take a short but fascinating and important detour before returning to methods of high-resolution speech analysis.

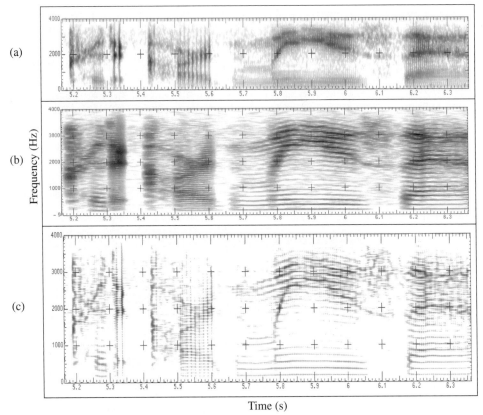

Figure 11.6 Comparison of the wideband/narrowband spectrograms and a positive time-frequency distribution for the speech utterance, "... credit card, and we use that ..." from a female speaker: (a) wideband spectrogram; (b) narrowband spectrogram; (c) positive time-frequency distribution. The positive time-frequency distribution simultaneously resolves the plosives (e.g., at 5.3 s), harmonics, and the time-varying formants (e.g, at 5.8 s).

SOURCE: J.W. Pitton, L.E. Atlas, and P.J. Loughlin, "Applications of Positive Time-Frequency Distributions to Speech Processing" [37]. ©1994, IEEE. Used by permission.

11.4 Aeroacoustic Flow in the Vocal Tract

The classical approach in speech science has been to reduce the problem of sound generation to one of purely acoustic motion where still air in the vocal tract is perturbed only by small one-dimensional planar compressions and expansions that comprise the sound field. In Chapter 4, this assumption led to the linear wave equation for describing the pressure/volume velocity relation within and at the input and output of the vocal tract. In this model, the source is independent from the vocal tract system. We have referred to this as the linear source/filter theory. While these approximations have allowed a great deal of progress to be made in understanding how speech sounds are produced and how to analyze, modify, synthesize, and recognize sounds, the approximations have led to limitations. In reality, acoustic motion is not the only kind of air motion involved. The air in the vocal system is not static, but moves from the lungs out of the mouth, carrying the sound field along with it, i.e., it contains a *nonacoustic* component. This nonacoustic phenomena, yielding a difference from the linear source/filter theory, can have an impact on the fine structure in the speech waveform and thus how speech is processed.

In this section, we outline some of the history and current thinking on *aeroacoustic* contributions to sound generation from the vocal tract, i.e., contributions to sound due to nonacoustic fluid motion. We noted in this chapter's introduction that these contributions may take the form of secondary sources and modulations that arise from *vortices* traveling along the vocal tract. The multiple sources we refer to here are not a result of the conventional linear paradigm as with, for example, voiced fricatives where a periodic glottal source occurs simultaneously with a noise source at a vocal tract constriction. Rather, we refer to the distributed sources due to a nonlinear exchange of kinetic energy in nonacoustic fluid motion into propagating sound. Likewise, the modulations we refer to are not a result of nonlinear coupling between the glottal source and the vocal tract system that we described in Chapter 5, but rather are due to aeroacoustic phenomenon.

11.4.1 Preliminaries

Conventional theories of speech production are based on a linearization of pressure and volume velocity relations. Furthermore, these variables are assumed constant within a given cross-section of the vocal tract, i.e., a one-dimensional planar wave assumption. The linear assumption neglects the influence of any nonacoustic motion of the fluid medium, or "flow," and leads to the classical wave equation for describing the pressure/volume velocity relation. In this model, the output acoustic pressure wave at the lips is due solely to energy from an injection of air mass at the glottis. It is known that in this process, only a small fraction (on the order of 1%) of the kinetic energy in the *flow* at the glottis[6] is converted to acoustic energy propagated by compression and rarefaction waves [15],[19]. The vocal tract acts as a passive acoustic filter, selectively amplifying some bands while attenuating others.

This linear, one-dimensional acoustic model has been suggested to be too tightly constrained to accurately model many characteristics of human speech. In particular, there is an increasing collection of evidence suggesting that nonacoustic fluid motion can significantly influence the sound field. For example, measurements by the Teagers [46],[47],[48] reveal the presence of *separated flow* within the vocal tract. Separated flow occurs when a region of fast moving fluid—a jet—detaches from regions of relatively stagnant fluid. When this occurs, viscous forces (neglected by linear models) create a tendency for the fluid to "roll up" into rotational fluid structures commonly referred to as *vortices* (Figure 11.7). The vortices can convect downstream at speeds much slower (90% slower or more) than acoustic propagation speed. Jet flow and associated vortices thus fall in the category of nonacoustic behavior.

There has been much effort in the confirmation of the existence and importance of these vortices for sound generation, encompassing theories, measurements, mechanical models, numerical solutions to the Navier Stokes equations (describing the complete nonlinear fluid dynamics), and more practical computational models [2],[15],[19],[33],[44],[48],[49]. Here we focus on three representative works: (1) the original motivating measurements of the Teagers [47] and their insightful interpretation by Kaiser [19], (2) the mechanical model of Barney, Shadle, and Davies [2],[43], and (3) the computational model of Sinder [44]. In all three cases, speech

---

[6] Thus far, we have used the term "flow" or "airflow," e.g., *the glottal flow waveform*, in an acoustic context of *local* velocity perturbations. In this chapter, the term "flow" takes on a different meaning when used in a nonacoustic context of net movement of air.

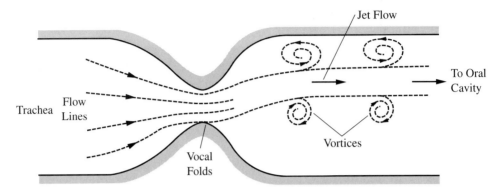

Figure 11.7   Jet flow through the glottis and the formation of a sequence of vortices.

spectral measurements are explained by the presence of both an acoustic source *and* a source generated from nonacoustic motion.

11.4.2 Early Measurements and Hypotheses of Aeroacoustic Flow in the Vocal Tract

The Teagers [47] speculated on the presence of nonacoustic phenomenon in the vocal tract, with a "flow not traveling by compression." They supported this speculation with extensive measurements made in the oral cavity with a hot-wire anemometer (a device that measures airflow velocity from changes in temperature in the surface of a material). With a steadily held voiced sound, it was found that the airflow velocity is not uniform across the cross section. In one experiment, the flow velocity at the tongue differed significantly from the flow velocity at the roof of the mouth (the palate) at the same location along the oral tract. Flow velocity measurements in this experiment were made simultaneously using an array of hot-wire anemometers. The measurements also indicated that the jet from the glottis attached itself to the cavity walls, but switching back and forth ("flapping"), at approximately the first formant frequency, between the tongue and the palate. With hot-wire anemometers that could measure normal flows, as well as radial and axial flows, the Teagers also observed vortices of both an axial and a radial type. These measurements are clearly inconsistent with planar one-dimensional acoustic flow velocity as predicted from the linear wave equation. A schematic comparison of the acoustic and nonacoustic models is illustrated in Figure 11.8. Note that the hot-wire anemometer measurement captures total flow velocity. Therefore, because the acoustic velocity is very small relative to the nonacoustic flow velocity, the Teagers' measurements are probably dominated by the nonacoustic contribution. An implication is that the measurements do not rule out the acoustic propagation and nonacoustic flow both present simultaneously. We return to this paradigm in our discussions of the work of Barney, Shadle, and Davies [2].

In order to understand where a discrepancy from conventional linear theory might arise, we return to our derivation of the wave equation in Chapter 4. The reader may recall a sleight-of-hand in arriving at the final form in Equation (4.1); the total derivative $\frac{dv}{dt}$ in Newton's 2nd Law was replaced with the partial derivative $\frac{\partial v}{\partial t}$. Because the particle velocity $v(x, t)$ is a function

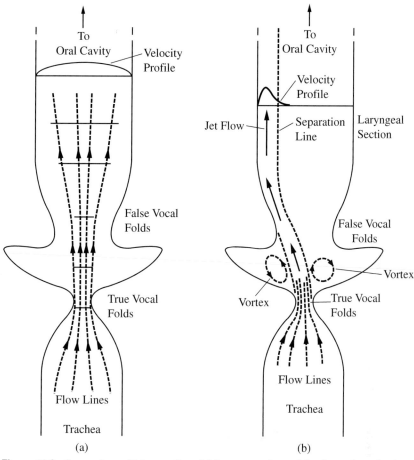

Figure 11.8 Comparison of (a) acoustic and (b) aeroacoustic models of speech production.

SOURCE: J.F. Kaiser, "Some Observations on Vocal Tract Operation from a Fluid Flow Point of View" [19]. ©1983, The Denver Center for the Performing Arts, Inc. Used by permission.

of space x as well as time t, we noted that the true acceleration of the air particles is given by [34],[48]

$$\frac{dv}{dt} = \frac{\partial v}{\partial t} + v \frac{\partial v}{\partial x}$$

and therefore Equation (4.1) is correctly written as

$$-\frac{\partial p}{\partial x} = \rho \left(\frac{\partial v}{\partial t} + v \frac{\partial v}{\partial x} \right) \tag{11.26}$$

which is a *nonlinear* equation because the fluid particle velocity v multiplies $\frac{\partial v}{\partial x}$, consequently making it difficult to determine a general solution. The linear approximation to Equation (11.26) is accurate when the correction term $\rho v \frac{\partial v}{\partial x}$ is small relative to $\rho \frac{\partial v}{\partial t}$. For sound waves, the corrective term is removed because there is no net flow, only very small velocity changes from local compressions and rarefactions. When the air is moving as a fast-moving jet, however, the air velocity may not be small and thus the nonlinear correction term can grow in importance [34]. This is one of numerous contributions that would arise from solution to the nonlinear Navier Stokes equation that describes the full complexity of velocity/pressure relations in the vocal tract and glottis.

The Teagers studied a variety of vowel sounds and found a unique flow pattern for each sound [46],[47],[48]. These flow patterns included a jet attached along the inside wall of the vocal tract and vortices "entrapped" in small cavities along the vocal tract such as due to the epiglottis (false vocal folds) (Figure 11.8b). They called the latter pattern the "whistle" because the jet velocity was observed to be modulated by the vortex that expands and contracts within the small cavity.[7] Another observed pattern is a fast-moving jet with large enough flow velocity to cause a train of vortices to be *shed* downstream (as we saw in Figure 11.7). The Teagers suggested that the presence of these traveling "smoke ring" vortices could result in additional acoustic sources throughout the vocal tract through the conversion of the nonacoustic flow to an acoustic propagation. Yet another observed pattern is a jet flapping across wall cavities, as noted earlier. For the various sounds, the Teagers found combinations of these flow patterns and a variety of interactions between them.

Motivated by the measurements of the Teagers, Kaiser re-examined the source/filter theory in light of aeroacoustic theories, giving further credence to a "jet-cavity flow paradigm" in the vocal tract [19]. Kaiser hypothesized that the interaction of the jet flow and vortices with the vocal tract cavity is responsible for much of the speech fine structure, particularly at high formant frequencies. This contribution is *in addition* to the glottal flow associated with the conventional linear source/filter theory, and the fine structure from the jet-cavity interactions is time-synchronized with this bulk glottal flow. He also hypothesized that the relative timing of the different flow interaction events and their effect on instantaneous formant frequencies within a glottal cycle are more important in speech processing and recognition than absolute spectral shape. Furthermore, these complex interactions may cause the appearance and disappearance of higher formants in the spectra over successive glottal cycles, with different excitations exciting different formants. Kaiser also outlined tasks needed to take the jet-cavity paradigm from the qualitative to the quantitative and functional realm, including flow measurements in mechanical models of the vocal tract, a mathematical model of acoustic wave generation from energy exchange in flow interaction, and computational models. Finally, he proposed the need for time-frequency analysis methods with greater resolution than the STFT for measuring fine structure within a glottal cycle due to the jet-cavity interaction paradigm, especially rapidly appearing and disappearing frequency components; the methods ideally should be insensitive to nonlinear transformations of the speech spectrum.[8]

Since the early work of the Teagers and Kaiser, aeroacousticans such as McGowan [33] and Hirschberg [15] have looked at more theoretical aspects of the nonacoustic velocity component within the vocal tract that result in acoustic sound sources due to interaction with the vocal tract boundaries at area discontinuities. In addition, two-dimensional numerical simulations of the Navier Stokes equation describing fluid flow in the vocal tract with appropriate boundary

---

[7] In the policeman's whistle, a jet is deflected into the whistle cavity and a vortex builds up inside the cavity. This vortex amplitude modulates the jet at a rate determined by the vortex's precession rate.

[8] Kaiser argues that because speech that is passed through a no-memory nonlinearity loses little intelligibility, the timing of jet-cavity interaction events and instantaneous formant frequencies may be more important than the absolute spectral shape [19].

conditions further suggest the existence of an unsteady jet at the glottal exit from which shedding vortices downstream from the glottis can evolve [23],[49]. More recently, Barney, Shadle, and Davies [2] have taken a different approach using a dynamical mechanical model, and Sinder [44] has developed computational models of unvoiced speech, both showing that a significant and measurable portion of sound at the lips can be explained by the presence of nonacoustic vortical flow. We now briefly describe these two developments.

11.4.3 Aeroacoustic Mechanical Model

Barney, Shadle, and Davies [2] have measured the acoustic contribution from vortices using a dynamical mechanical model (DMM) with simple geometry consistent with typical average male vocal tract measurements for voiced speech. The model consists of a duct 28 cm in length and 1.7 cm in diameter. A vibration generator with moving shutters gave a sinusoidal input driving function of 80 Hz that modulated three flow volume velocities of 200, 300, and 400 cm^3/s at the entrance to the duct.[9]

Using the linear acoustic theory of Chapter 4, Barney, Shadle, and Davies predicted the acoustic spectral power at a 45° angle and 60 cm from the duct exit for all three flow velocities. The prediction also invoked a model of a piston in an infinite baffle (Chapter 5) and accounted for a directivity factor because of the 45° off axis measurement. In addition, a spectral measurement of the duct output pressure, using a pressure-sensitive electret microphone, was made by averaging 13 short-time power spectra using a 4096-point FFT with a Hanning window overlap of 15%. Spectral predictions and measurements were plotted for harmonics of the shutter speed. The result is shown in Figure 11.9a. Except for the 4th, 5th, and 6th harmonics, the measured acoustic power is underpredicted by at least 10 dB, indicating the presence of a nonacoustic contribution.

To explore this inconsistency, Barney, Shadle, and Davies made hot-wire anemometer velocity flow measurements in the duct. Hot wire anemometers were inserted along the duct through a small slot. Their measurements revealed a jet flow at the opening with shear layers that roll up to form a succession of regularly-spaced vortices downstream from the shutter opening. The flow velocity measurements indicated symmetric "smoke ring" vortices traveling down the duct from about 4 cm downstream from the moving shutters. It was also found that the mean velocity was never zero, even when the shutters were closed. Their conclusion is that the velocity field in the duct consists of an acoustic standing wave due to the fluctuating mass at the exit to the shutters and a train of vortices, one generated during each shutter cycle, drifting down the duct at the mean flow speed. A model for the total velocity disturbance in the duct is therefore given by

$$v(t) = \bar{v} + v_a(t) + v_n(t)$$

where $\bar{v}$ is the mean velocity of the air particles, $v_a(t)$ is the acoustic particle velocity, and $v_n(t)$ is the nonacoustic particle velocity due to the vortex transport. The number of vortices in the duct can be calculated at any particular time from the shutter speed and mean flow velocity (Exercise 11.17).

---

[9] The self-noise component of the mechanism due to the shutter movement was estimated by measuring sound at the duct exit without airflow at the shutter entrance. With the proper relative waveform phase, this estimate was then subtracted from measurements.

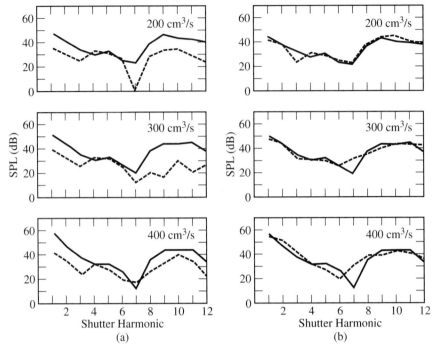

Figure 11.9 Comparison of far-field acoustic spectral power measurements (60 cm from the duct exit plane at a 45° angle to the duct center axis) and predictions for three volume velocity inputs: (a) acoustic model; (b) combined acoustic and nonacoustic models. The solid line represents the measurement and the dashed line represents the prediction. SPL denotes the sound pressure level with respect to the ambient pressure.

SOURCE: A. Barney, C.H. Shadle, and P.O.A.L. Davies, "Fluid Flow in a Dynamical Mechanical Model of the Vocal Folds and Tract, 1: Measurements and Theory" [2]. ©1999, *Acoustical Society of America.* Used by permission.

Although vortices themselves are not effective sound radiators, when interacting with changing solid boundaries, they can provide significant acoustic sources that can excite resonators. For the DMM model, the open end of the duct provides such a boundary; the nonacoustic flow is transferred to pressure and velocity fluctuations that can generate radiated sound at the duct exit. Davies has developed a model by which this transfer can occur and has found an expression for the generated acoustic power associated with the vortices under the infinite baffle assumption at the duct output [12]. This model was used to predict the spectral component from nonacoustic sources outside the DMM at the 45° angle and 60 cm from the duct exit for the three flow velocities given above. Under the assumption that the acoustic and nonacoustic contributions are independent, the *sum of the two terms far more accurately predicts the sound pressure level of each harmonic* than the acoustic contribution alone (Figure 11.9b). Discrepancies may be due to certain simplifying assumptions including a symmetric jet from the duct opening (thus not allowing wall attachment of the jet), the nondispersive nature of the vortices, and independence of the acoustic and nonacoustic components.

The question arises whether the DMM model extends to actual vocal tract geometries, i.e., whether vortices exist in the vocal tract, and, if so, whether the vortical train provides additional acoustic sources at boundary discontinuities. To answer this question, Shadle, Barney, and Davies [43] compared hot-wire anemometer measurements for the DMM with those of an actual vocal tract during phonation of the vowel /A/. The vowel /A/ was selected because it is most similar to the geometry of the DMM. Velocity field measurements in the vocal tract within a glottal cycle were found to be similar in shape, but often different in magnitude, to those of the DMM. The flow velocity typically consisted of a set of primary pulses at the pitch rate and secondary pulses between the primary pulses. Differences from the vocal tract were probably due to the sinusoidal shutter movement, in contrast to the skewed sawtooth-like glottal flow of humans with unequal open and closed phases, and to the interaction of the vortex train with a multitude of tract boundaries, unlike the single DMM duct exit. From the similarities and explained differences, it was concluded that the DMM acts qualitatively like a voiced vocal tract and thus suggests sound generation by a vortex train in actual vocal tracts. As with the DMM, the vortex train interacts with boundaries of the vocal cavity, producing acoustic sources.

Shadle, Barney, and Davies [43] also investigated implications of the presence of the nonacoustic, vortical source for glottal waveform estimation by inverse filtering, and noted inconsistencies with inverse filtering based on the conventional linear source/filter theory. For example, at the shutter opening *different* glottal flow waveforms are measured for different hot-wire positions within the cross-sectional plane. Perhaps the most serious inconsistency, however, is that multiple acoustic and nonacoustic sources may exist in the vocal tract cavity, not all located at the opening of the glottis, but where the vortical train meets boundary discontinuities. Acoustic and nonacoustic sources have different travel times and excite different regions of the vocal tract cavity, i.e., *different sources can excite different formants, thus extending the notion of a single inverse filter*. This notion is consistent with Kaiser's speculation of formants that disappear and reappear over successive glottal cycles [19]. The following example illustrates a speech measurement consistent with this phenomenon, but also shows that care must be taken in how we interpret our observations:

EXAMPLE 11.4   Multi-source, multi-response models of production within a glottal cycle suggests as a simplification a convolutional model of speech at the lip output. This model involves more than one source with corresponding different resonant system functions. We have observed a variety of speech signals where different vocal tract impulse responses appear within a glottal cycle. It is easiest to see these events for very low-pitched speakers. An example of this phenomenon is shown in Figure 11.10a, which shows secondary responses about two thirds into the glottal cycle. Figure 11.10b zooms in on one period, illustrating that the secondary response is different in shape from the primary response, a property further confirmed by the spectra in Figure 11.10c. A 14th-order all-pole spectral fit (obtained using the autocorrelation method of linear prediction) is shown for each response. We see that, in the secondary component, the first formant (as well as the last formant) is noticeably absent, and so the two responses have different spectral characteristics. We can also see this property in time in the isolated glottal cycle (Figure 11.10b) where the low- and high-frequency components of the primary response are not present in the secondary response. Without further analysis, however, we cannot definitely give the cause of this change; different formants may arise, for example, because the secondary glottal source has a different shape from the primary pulse or because a secondary flow-induced source occurs downstream from the glottis as, for example, when a sound source is created by energy exchange from a vortex at a location along the vocal tract such as at the epiglottis. An important consequence of this difference in primary and secondary components is that

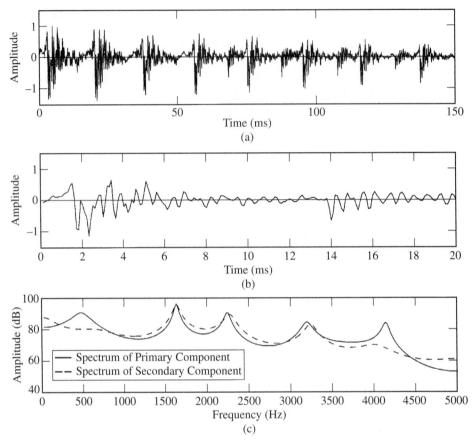

Figure 11.10 Example of secondary pulses with different spectrum from that of the primary pulses: (a) speech waveform; (b) one glottal cycle; (c) spectra (14th-order all-pole fit) of primary and secondary responses.

the low-order autocorrelation coefficients used typically in speech analysis, e.g., in linear prediction analysis, consist of the *sum of the autocorrelation coefficients from two different impulse responses* (Exercise 11.18).   ▲

We seek signal processing tools that can reveal the type of fine structure in the previous example, even when such fine structure is hidden, for example, when secondary sources occur with high-pitched speakers. Ideally, our signal processing tools should also help in the very difficult task of distinguishing between acoustic and nonacoustic sources.

11.4.4 Aeroacoustic Computational Model

A functional model that incorporates the nonlinear effects of nonacoustic motion allows for improved prediction and analysis of speech signals. In this paradigm, vortices in the flow lead to additional acoustic sources within the vocal tract. Speech output at the lips is computed as the sum of vocal-tract-filtered responses to the glottal source and aeroacoustic flow sources. We have seen that, although vortices themselves are not efficient sound radiators, when interacting with a non-uniform vocal tract they can be a source of sound. In this manner the vocal tract resonator can be excited, for example, at the epiglottis, velum, palate, teeth, and lips, as well as at other locations with a changing cross section. A computational model requires that we know the spectrum, strength (gain), and location of these secondary sound sources.

Sinder [44] has applied aeroacoustic theory to produce a computationally efficient model of vortex sound generation. In this model, the characteristics of the source—spectrum, strength, and spatial distribution—are automatically computed as a function of vocal tract shape and lung pressure. The essential result from aeroacoustic theory incorporated into this work is that of Howe [16]. Howe's result relates the motion of vortices through a duct of changing cross-section *to the plane wave acoustic sound field generated by that motion.* This relation is used to compute the strength and location of acoustic pressure sources in the duct due to traveling vortices generated by jet flow. The model is applicable to either voiced or unvoiced speech with vortices originating from a jet formed at or downstream from the glottis. The computational model used by Sinder determines automatically the location of jet formation (i.e., flow separation boundaries), the time of vortex creation, and vortex motion based simply upon the vocal tract geometry and the local air flow speed. For voiced speech, the jet is created at the glottal exit and can contribute a distributed aspiration source during voicing. For unvoiced fricatives, the jet forms near the tongue-palate constriction. For unvoiced plosives, a glottal jet follows, in time, a jet formed at a tongue-palate constriction, thus allowing for the aspiration component of the plosive. A jet is formed only when local particle velocity exceeds a threshold. For voiced fricatives and plosives, jet formation occurs in a way similar to that of unvoiced sounds, except that in these cases the valving of the glottis modulates the jet at a constriction and thus modulates the vortex formation; the noise source due to the vortices is therefore pitch synchronous. Sinder has applied this computational model for improved distributed source creation of unvoiced sounds in speech synthesis [44] for which point (non-distributed) sources are conventionally used.

11.5   Instantaneous Teager Energy Operator

We have noted that speech fine structure, as, for example, secondary sources and modulations within a glottal cycle, is often not "seen" by conventional Fourier analysis methods that look over multiple glottal cycles. This limitation motivated the Teagers to propose an operator characterized by a time resolution that can track rapid signal energy changes *within* a glottal cycle [20],[21]. The *Teager energy operator* is based on a definition of energy that accounts for the energy in the system that generated the signal. An extension of this operator also allows for a simple algorithm for separating out from the Teager energy sinewave amplitude modulation (AM) and frequency modulation (FM) with excellent time resolution [29],[30],[31]. We refer to this algorithm as the *energy separation algorithm* and show its application to estimating AM and FM in speech formants.

11.5.1 Motivation

As a means to motivate the Teager energy operator and the energy separation algorithm for AM-FM estimation, we first consider the different perspective on energy introduced by the Teagers and extended by Kaiser [20],[21]. In the conventional view of "energy," we compute the sum of squared signal elements, i.e., an average of energy density. This means that tones at 10 Hz and at 1000 Hz with the same amplitude have the same energy. Teager observed, however, that the energy required to *generate* the signal at 1000 Hz is much greater than that at 10 Hz. This alternative notion of energy can be understood with the sinusoidal oscillation that occurs with a simple harmonic oscillator. In this case, the total energy in the system, i.e., the sum of the potential and kinetic energy, is proportional to the *product* of frequency and amplitude squared.

For a harmonic oscillator, applying Newton's law of motion to the motion of a mass m suspended by a spring with force constant k, gives the 2nd-order differential equation

$$\frac{d^2x}{dt^2} + \frac{k}{m}x = 0$$

whose solution is the harmonic motion $x(t) = A\cos(\Omega t + \phi)$, where A is the amplitude and Ω the frequency of oscillation. The total energy is the sum of the potential energy (in the spring) and the kinetic energy (of the mass) and is

$$E = \frac{1}{2}kx^2 + \frac{1}{2}m\dot{x}^2. \tag{11.27}$$

Substituting for $x(t)$ our harmonic solution, we obtain

$$E = \frac{1}{2}mA^2\Omega^2$$

or

$$E \propto A^2\Omega^2$$

which is the "true" energy in the harmonic system, i.e., the energy required to generate the signal.

11.5.2 Energy Measurement

Consider the *energy operator* introduced by Kaiser [21] and defined by

$$\Psi_c[x(t)] = \left(\frac{d}{dt}x(t)\right)^2 - x(t)\frac{d^2}{dt^2}x(t) = [\dot{x}(t)]^2 - x(t)\ddot{x}(t) \tag{11.28}$$

where $\dot{x} = \frac{d}{dt}x$. When Ψ_c is applied to signals produced by simple harmonic oscillators, without explicit requirement of values of mass m and spring constant k, the operator can track the oscillator's energy (per half unit mass), which is equal to the squared product of the oscillation amplitude and frequency (Exercise 11.25):

$$\Psi_c[x(t)] = A^2\Omega^2$$

for a signal of the form

$$x(t) = A\cos(\Omega t + \theta).$$

To discretize this continuous-time operator, we can select a variety of sampling strategies. One possibility is to replace t with nT (T is the sampling period), $x(t)$ by $x(nT)$ or simply $x[n]$,

$\dot{x}(t)$ by its first backward difference $y[n] = \frac{x[n]-x[n-1]}{T}$, and $\ddot{x}(t)$ by $\frac{y[n]-y[n-1]}{T}$. Then, we have (Exercise 11.25)

$$\dot{x}(t) \;\rightarrow\; \frac{x[n] - x[n-1]}{T}$$

$$\ddot{x}(t) \;\rightarrow\; \frac{x[n] - 2x[n-1] + x[n-2]}{T^2}$$

$$\Psi_c[x(t)] \;\rightarrow\; \frac{\Psi_d(x[n-1]}{T^2} \tag{11.29}$$

where $\Psi_d(x[n])$ is the *discrete-time energy operator* (the counterpart to the continuous-time energy operator $\Psi_c[x(t)]$) defined as

$$\Psi_d(x[n]) \;=\; x^2[n] - x[n-1]x[n+1] \tag{11.30}$$

for discrete-time signals $x[n]$. Analogous to the case of a continuous-time sinewave, we can show (Exercise 11.25)

$$\Psi_d(x[n]) \;=\; A^2 \sin^2(\omega)$$

for a signal of the form

$$x[n] \;=\; A\cos(\omega n + \theta).$$

This result of applying $\Psi_d(x[n])$ to a discrete-time sinewave is exact, provided that $\omega < \frac{\pi}{2}$. Furthermore, for small values of ω, because $\sin(\omega) \approx \omega$, it follows that

$$\Psi_d(x[n]) \;=\; x^2[n] - x[n-1]x[n+1] \approx A^2\omega^2$$

and so, as with the continuous-time operator, the discrete-time operator can track the energy in a discrete-time sinewave. These energy operators were developed by the Teagers in their work on modeling speech production [47],[48] and were first introduced by Kaiser [20],[21].

The energy operator is also very useful for analyzing oscillatory signals with time-varying amplitude and frequency. In continuous time, such real-valued AM–FM signals of the form

$$x(t) \;=\; a(t)\cos[\phi(t)] \;=\; a(t)\cos\left(\Omega_c t + \Omega_m \int_0^t q(\tau)d\tau + \theta\right) \tag{11.31}$$

can model time-varying amplitude and frequency patterns in speech resonances [29]. The signal $x(t)$ is a cosine of carrier frequency Ω_c with a time-varying amplitude signal $a(t)$ and a time-varying instantaneous frequency signal

$$\Omega_i(t) \;=\; \frac{d}{dt}\phi(t) \;=\; \Omega_c + \Omega_m q(t),$$

where $|q(t)| \leq 1$, $\Omega_m \in [0, \Omega_c]$ is the maximum frequency deviation, and θ is a constant phase offset. Specifically, it has been shown [29],[30] that Ψ_c applied to an AM–FM signal can approximately estimate the squared product of the amplitude $a(t)$ and instantaneous frequency $\Omega_i(t)$ signals, i.e.,

$$\Psi_c\left[a(t)\cos\left(\int_0^t \Omega_i(\tau)d\tau + \theta\right)\right] \;\approx\; [a(t)\Omega_i(t)]^2 \tag{11.32}$$

assuming that the signals $a(t)$ and $\Omega_i(t)$ do not vary too fast (time rate of change of value) or too greatly (range of value) in time compared to the carrier frequency Ω_c. Likewise, a discrete-time version of this result can be obtained using the discrete-time energy operator $\Psi_d(x[n]) = x^2[n] - x[n-1]x[n+1]$. Specifically, consider an AM-FM discrete-time sinewave of the form

$$x[n] = a[n]\cos(\phi[n]) = a[n]\cos\left(\omega_c n + \omega_m \int_0^n q[m]dm + \theta\right) \qquad (11.33)$$

with instantaneous frequency

$$\omega_i[n] = \frac{d}{dn}\phi[n] = \omega_c + \omega_m q[n]$$

where $|q[n]| \leq 1$, $\omega_m \in [0, \omega_c]$ is the maximum frequency deviation, and θ is a constant phase offset. Note that the continuous-time frequencies Ω_c, Ω_m, and Ω_i have been replaced by their discrete-time counterparts ω_c, ω_m, and ω_i. All discrete-time frequencies are assumed to be in the interval $[0, \pi]$. Then, we can show [29],[30]

$$\Psi_d\left(a[n]\cos\left\{\int_0^n \omega_i[m]dm + \theta\right\}\right) \approx a^2[n]\sin^2(\omega_i[n])$$

again provided that the signals $a[n]$ and $\omega_i[n]$ do not vary too fast (time rate of change of value) or too greatly (range of value) in time compared to the carrier frequency ω_c. More quantitative constraints for these approximations to hold in both continuous- and discrete-time can be found in [29],[30]. Specific cases of an exponetially decaying sinewave and a linear chirp signal are explored in Exercise 11.25.

As with other time-frequency representations we have studied, the energy operator warrants a close look with the presence of more than one sinewave component. Consider, for example, a sequence with two components

$$x[n] = x_1[n] + x_2[n]$$

The output of the Teager energy operator can be shown to be (Exercise 11.21)

$$\Psi_d(x[n]) = \Psi_d^1(x[n]) + \Psi_d^2(x[n]) + \Psi_d^{12}(x[n]) + \Psi_d^{21}(x[n]) \qquad (11.34)$$

where $\Psi_d^{12}(x[n])$ and $\Psi_d^{21}(x[n])$ are "cross Teager energies" which can be thought of as cross-correlation-type terms between the two sequences. The following example by Kaiser [20] gives some insight into the use of the Teager energy on a two-component signal:

EXAMPLE 11.5   Consider a discrete-time signal consisting of two sinewaves

$$x[n] = A_1\cos(\omega_1 n + \theta_1) + A_2\cos(\omega_2 n + \theta_2).$$

For this case, it can be shown that the Teager energy is the sum not only of the individual energy components but also an additional cross term at the difference frequency of the two frequency components (Exercise 11.21). Figure 11.11 illustrates the energy measure for the two-component signal where the effect of the cross term is similar to that of beating between two sinewaves. For this example $A_1 = A_2 = 1.0$, $\omega_1 = \frac{\pi}{4}$, and $\omega_2 = \frac{\pi}{6}$. Kaiser observes [20] "... the output of the algorithm is maximum when the composite signal is at its peak and minimum when at zero; it is as if the algorithm is able to extract the envelope function of the signal."   ▲

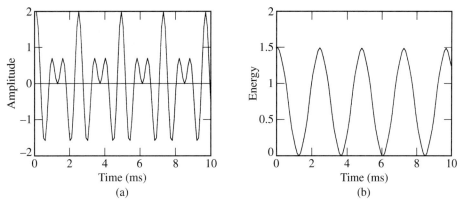

Figure 11.11   Teager energy measure for the two-component signal of Example 11.5: (a) sum of two sinewaves; (b) Teager energy of (a).

SOURCE: J.F. Kaiser, "On a Simple Algorithm to Calculate the 'Energy' of a Signal," [20]. ©1990, IEEE. Used by permission.

We see then that the cross-correlation may contain useful information (as with other time-frequency representations). Nevertheless, we must separate the components prior to using the energy operator when only one component is of interest.

An important property of the Teager energy operator in discrete time, Equation (11.30), is that it is nearly *instantaneous* given that only three samples are required in the energy computation at each time instant: $x[n-1]$, $x[n]$, and $x[n+1]$. This excellent time resolution provided the Teagers with the capability to capture energy fluctuations (in the sense of squared product of amplitude and frequency) *within a glottal cycle*. In the context of the above harmonic oscillator formulation, the Teagers loosely referred to a "resonance" as an oscillator system formed by local cavities of the vocal tract emphasizing certain frequencies and de-emphasizing others during speech production. We saw earlier that the Teagers' experimental work provided evidence that speech resonances can change rapidly within a single pitch period,[10] possibly due to the rapidly varying and separated speech airflow and modulating vortices in the vocal tract [46],[47],[48]. In addition, vortices that shed from the jet and move downstream can interact with the vocal tract cavity, forming distributed sources due to a nonlinear exchange of kinetic energy in nonacoustic fluid motion into propagating sound. We noted that these distributed sources may reveal themselves as secondary sources within a glottal cycle. We also noted that Kaiser hypothesized that these complex interactions may cause the appearance and disappearance of higher formants in the spectra over successive glottal cycles with different excitations exciting different formants.

The Teagers' approach to revealing the result of these complex interactions from a speech pressure measurement is the following [48]. First, after locating the resonance center frequencies of the vocal tract cavity, each resonance was bandpass-filtered (with a modulated Gaussian filter) around the resonance center frequency. This filtering approximately separates out the resonant components. Each bandpass-filter output was then processed with the discrete energy operator of Equation (11.30). The energy traces for six consecutive pitch periods from the vowel

---

[10] Time variations of the elements of simple harmonic oscillators can result in amplitude or frequency modulation of the oscillator's cosine response [51].

sound /i/ of the Teagers' original work is shown in Figure 11.12. The Teagers reasoned that if the formants were simply damped sinewaves, as predicted by linear acoustic theory, then the energy traces would be exponentially decaying over each pitch period (Exercise 11.25). The Teagers hypothesized that the bumps observed in Figure 11.12 within a glottal cycle, on the other hand, reflect the aeroacoustic behavior, in particular "pulsatile vortical flow interactions" with boundaries of the vocal tract; these interactions are dependent on the physical locations and types of vortices present in the vocal tract, perhaps "in three separate locations, working from the glottis to the tongue" [48]. Alternatively, as we will see shortly, one can interpret these observations as the result of amplitude and frequency modulations of cavity resonances.

Finally, given that our framework in this chapter is high resolution time-frequency analysis, it is of interest to consider the Teager energy as the basis or component of a time-frequency distribution. A detailed study is beyond the scope of this chapter. It suffices to say that a time-frequency distribution has been defined based on the Teager energy function [24]. Specifically, it has been shown that a time-frequency distribution can be defined as a modified Wigner distribution, with the property that its frequency marginal yields the Teager energy. A connection between the Teager energy operator and the ambiguity function, as well as other modifications of the Wigner distribution, has also been established for a generalized Teager operator for complex-valued signals [14]. [The ambiguity function, as we noted earlier in this chapter (footnote number 4), is related to the Wigner and spectrogram distributions.] This has led to the use of the Teager energy operator in measuring various moments of a signal and its spectrum [14].

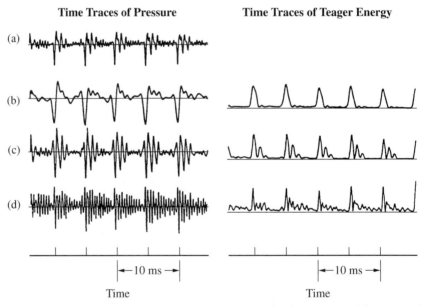

Figure 11.12 Traces of the Teager energy operator output for the vowel sound /i/: (a) speech waveform; (b)–(d) bandpass-filtered first, second, and third formants and their energy traces.

Source: H.M. Teager and S.M. Teager, "Evidence for Nonlinear Sound Production Mechanisms in the Vocal Tract" [48] (Figure 5). ©1990, Kluwer Academic Publishers, The Netherlands. Used by kind permission from Kluwer Academic Publishers.

11.5.3 Energy Separation

We have seen that the Teager energy operator, when applied to an AM-FM sinewave, yields the squared product of the AM and FM components (more strictly speaking, amplitude envelope and instantaneous frequency). We now describe an approach to *separate* the time-varying amplitude envelope $a(t)$ and instantaneous frequency $\Omega_i(t)$ of an arbitrary AM-FM signal, based on the energy operator. First, we present a continuous-time solution for exact estimation of the constant amplitude and frequency of a sinewave and then show that the same equations approximately apply to an AM-FM signal with time-varying amplitude and frequency. We call these algorithms *energy separation algorithms* because an oscillator's energy depends on the product of amplitude and frequency [31]. Next, we develop counterpart separation algorithms for discrete-time signals. We then apply discrete-time energy separation to search for modulations in speech formants that we model using AM-FM sinewaves. To simplify notation, we henceforth drop the subscripts from the continuous and discrete energy operator symbols and use Ψ for both.

Consider a cosine $x(t)$ with constant amplitude A and frequency Ω_c and its derivative

$$x(t) = A\cos(\Omega_c t + \theta); \quad \dot{x}(t) = -A\Omega_c \sin(\Omega_c t + \theta).$$

Then applying Ψ yields

$$\Psi[x(t)] = A^2\Omega_c^2; \quad \Psi[\dot{x}(t)] = A^2\Omega_c^4.$$

Hence, the constant frequency and the amplitude of the cosine can be obtained from the equations

$$\Omega_c = \sqrt{\frac{\Psi[\dot{x}(t)]}{\Psi[x(t)]}}; \quad A = \frac{\Psi[x(t)]}{\sqrt{\Psi[\dot{x}(t)]}}.$$

Now consider the more general AM-FM sinewave of the form $x(t) = a(t)\cos[\phi(t)]$ in Equation (11.31). Its derivative is

$$\dot{x}(t) = \underbrace{\dot{a}(t)\cos[\phi(t)]}_{y_1(t)} - \underbrace{a(t)\Omega_i(t)\sin[\phi(t)]}_{y_2(t)}.$$

To make Equation (11.32) a valid approximation, we henceforth assume the two constraints:

C1: The functions $a(t)$ and $q(t)$ are bandlimited with highest frequencies Ω_a and Ω_q, respectively, and $\Omega_a, \Omega_q \ll \Omega_c$.
C2: $\Omega_a^2 + \Omega_m\Omega_q \ll (\Omega_c + \Omega_m)^2$.

Further, if we define the *order of magnitude of a signal* $z(t)$ to be $O(z) = O(z_{max})$ where $z_{max} = \max_t |z(t)|$, then $O(y_1) \approx O(a\Omega_a)$ and $O(y_2) \approx O(a\Omega_i)$ (See [30],[31] for details and proofs). Since $O(\Omega_a) \ll O(\Omega_i)$, then we can obtain the approximation [30],[31]

$$\Psi[\dot{x}(t)] \approx \Psi[a(t)\Omega_i(t)\sin\phi(t)] \approx a^2(t)\,\Omega_i^4(t). \tag{11.35}$$

By combining Equations (11.32) and (11.35), we obtain

$$\Omega_i(t) \approx \sqrt{\frac{\Psi[\dot{x}(t)]}{\Psi[x(t)]}}\;; \quad a(t) \approx \frac{\Psi[x(t)]}{\sqrt{\Psi[\dot{x}(t)]}}$$

At each time instant, this algorithm estimates the instantaneous frequency and the amplitude envelope by using only the two instantaneous output values from the energy operator applied to the signal and its derivative. Although we derived this algorithm by assuming bandlimited modulating signals, there are also other special cases of AM–FM signals where the algorithm yields approximately correct solutions. Examples include chirp signals with linear FM, i.e., $A \cos(\Omega_c t + \frac{\Omega_m t^2}{2L} + \theta)$, $0 \leq t \leq L$, whose amplitude and linear instantaneous frequency can be estimated via the above continuous-time energy separation algorithm, provided that $\frac{\Omega_m}{L} \ll (\Omega_c + \Omega_m)^2$.

For the validity of the continuous-time energy separation algorithm, it is assumed that $\Psi(x)$, $\Psi(\dot{x}) \geq 0$. There are large classes of signals satisfying this condition [30], e.g., all AM–FM signals whose modulation amounts do not exceed 50% and Ω_a, $\Omega_f < \frac{\Omega_c}{10}$. In discrete simulations of the continuous energy separation algorithm on noise-free AM–FM and bandpass filtered speech signals, negative $\Psi(x)$ values are rarely encountered; when they do, they appear to be due to round-off errors. Note also that at times t_o when $\Psi[x(t_o)] = 0$, we have $a(t_o) \approx 0$ if $\Omega_i(t)$ is assumed to be always positive. At such rare time instants, we need additional information to estimate $\Omega_i(t_o)$. For example, in discrete simulations, we can interpolate $\Omega_i(t_o)$ from its immediate neighbors.

Consider now the discrete-time counterpart to the continuous energy separation algorithm for AM–FM sequences of the form of Equation (11.33). Our objective is to estimate their instantaneous frequency

$$\omega_i[n] = \frac{d}{dn}\phi[n] = \omega_c + \omega_m q[n]$$

and to estimate their amplitude envelope $a[n]$ from the discrete-time energy operator output, $\Psi(x[n]) = x^2[n] - x[n-1]x[n+1]$, and from the operator applied to differences (that approximate the signal derivative). A number of different discrete-time energy separation algorithms have been derived using different approximating derivatives and different assumptions on the form of the AM and FM functions [31]. One algorithm is based on symmetric differences of the form

$$y[n] = \frac{x[n+1] - x[n-1]}{2}$$

that looks one sample ahead and one sample behind in time.[11] Assuming that the instantaneous frequency $\omega_i[n]$ is a finite sum of cosine functions and assuming bandwidth constraints on $a[n]$ and $q[n]$ of the types **C1** and **C2** above, then using $\Psi(x[n])$ and $\Psi(y[n])$ we obtain the following formulas for estimating the time-varying frequency and amplitude envelope:

$$\omega_i[n] \approx \arcsin\left(\sqrt{\frac{\Psi(x[n+1] - x[n-1])}{4\Psi(x[n])}}\right)$$

$$a[n] \approx \frac{2\Psi(x[n])}{\sqrt{\Psi(x[n+1] - x[n-1])}}. \tag{11.36}$$

---

[11] Alternative discrete algorithms result if we replace derivatives with backward and forward differences $x[n] - x[n-1]$ and $x[n+1] - x[n]$, respectively.

The frequency estimation part assumes that $0 < \omega_i[n] < \frac{\pi}{2}$ because the computer's implementation of $\arcsin(u)$ function assumes that $|u| \leq \frac{\pi}{2}$. Thus, this discrete energy separation algorithm can be used to estimate instantaneous frequencies $\leq \frac{1}{4}$ the sampling frequency. Therefore, doubling the original sampling frequency f_s allows the algorithm to estimate frequencies up to $\frac{f_s}{2}$. Note that, if $x[n] = A\cos(\omega_c n + \theta)$, then the AM–FM separation formulas yield the exact constant frequency and amplitude (Exercise 11.22). In [31] the (mean absolute and rms) errors of this and other discrete energy separation algorithms are obtained in estimating the amplitude and frequency of synthetic AM–FM signals. On the average (for AM and FM amounts of 5%–50%), the algorithms yield very small errors in the order of 1% or less. In the presence of added white Gaussian noise at a 30-dB SNR, the algorithm yields errors in the order of 10% or less; this low error is probably achieved because the two estimation equations above are the quotients of two functions, each of which is approximately a low-bandwidth function.

EXAMPLE 11.6   Consider the AM–FM sinewave with a sinusoidal FM and decaying sinusoidal AM:

$$x[n] = a[n]\cos\left[\frac{\pi}{5}n + \sin\left(\frac{\pi}{50}n\right)\right]$$

where

$$a[n] = (0.998)^n\left[1 + 0.8\cos\left(\frac{\pi}{100}n\right)\right].$$

Figure 11.13 shows the Teager energy and the estimated amplitude envelope and instantaneous frequency using the discrete energy separation algorithm. Because the AM is much larger than the FM, the FM is hidden in the energy function.   ▲

An important property of the discrete energy separation algorithm in Equation (11.36) is that the short "window" (five samples) required by the algorithm implies excellent temporal resolution (Exercises 11.19 and 11.26) and thus it is useful in instantaneously adapting during speech transitions and within a glottal cycle. This time resolution is often not achieved by other AM–FM estimation methods such as the Hilbert transform[12] (Exercise 11.19).

With speech fine structure due to secondary sources and modulations of the nonlinear models described earlier as motivation, a model of a speech resonance within a pitch period by an exponentially-damped AM–FM signal $r^n a[n]\cos(\phi[n])$ was proposed [29],[31]. In this model, the instantaneous frequency $\omega_i[n] = \omega_c + \omega_m q[n]$ models the deviation of a time-varying formant from its center value ω_c and $r \in (0, 1)$ is related to the rate of energy dissipation. By assuming that the time-varying amplitude and frequency modulating signals can be modeled by a sum of slowly varying sinusoids within a pitch period, i.e., the AM and FM do not vary too fast in time or too greatly compared to the carrier ω_c, it follows that the energy separation algorithm can be applied to estimate the AM and FM variations within a glottal cycle. The following example illustrates the energy separation approach on a speech signal:

---

[12] Although experiments on speech signals show that the energy separation algorithm yields similar AM-FM estimates, in a mean-squared error sense, to that of the Hilbert transform, the extremely short window allows the separation algorithm to instantaneously adapt. In addition, the separation algorithm requires less computation [39].

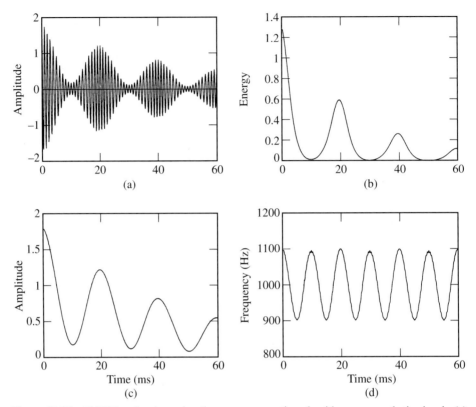

Figure 11.13   AM-FM estimation using the energy separation algorithm on a synthetic signal: (a) AM–FM signal of Example 11.6; (b) Teager energy; (c) estimated amplitude envelope; (d) estimated instantaneous frequency. A 10000 samples/s sampling rate is assumed.

EXAMPLE 11.7   Figure 11.14a shows a filtered segment of a speech vowel /e/ sampled at 30 kHz, extracted around a formant at $f_c = 3400$ Hz by convolving the speech signal with a bandpass Gaussian filter's impulse response $h(t) = \exp(-\alpha^2 t^2) \cos(2\pi f_c t)$, properly discretized and truncated [29],[30] with $\alpha = 1000$. The corresponding Teager energy is shown in Figure 11.14b. The observed multiple pulses are similar to the "energy pulses" [48] found by the Teagers (Figure 11.12). The observations provide evidence that speech resonances exhibit a structure that does not originate from a linear time-invariant resonator model because, as we noted earlier, applying Ψ to an impulse response $x[n] = Ar^n \cos(\omega_c n + \theta)$ yields a decaying exponential $\Psi(x[n]) = A^2 r^{2n} \sin^2(\omega_c)$ without any multi-pulse structure. Figure 11.14c shows the estimated (via the energy separation algorithm) amplitude signal, where we see a strong AM modulation and two pulses per pitch period. The estimated instantaneous frequency in Figure 11.14d oscillates around its center value with a deviation that reaches 200 Hz. It contains some isolated narrow spikes, which are usually caused either by amplitude valleys or by the onset of a new pitch pulse. These spikes can be eliminated by post-smoothing the frequency signal using a median filter [31]. Excluding these narrow spikes, in vowels the instantaneous frequency and amplitude envelope profiles follow simple oscillatory, roughly sinusoidal, patterns. Typically, two-to-four amplitude pulses and formant frequency oscillations are seen within a pitch period in experiments with signals from speech vowels, particularly in higher formants. Consistent with the

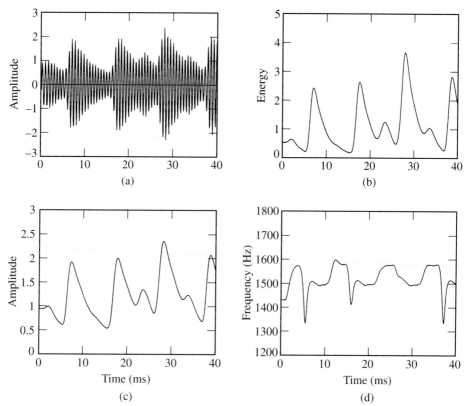

Figure 11.14 Estimation of AM–FM in a speech resonance (around 1500 Hz) using the energy separation algorithm: (a) signal from filtered speech vowel; (b) Teager energy of (a); (c) estimated amplitude envelope; (d) estimated instantaneous frequency.

Teagers' original observations, for signals from low formants of vowels it is typical to observe one energy pulse per pitch period. This may be partially explained by a large amount of damping, or by a low amount of modulation for the specific speaker/sound/formant combination.   ▲

Finally, we note that the AM–FM model of a *single* resonance does not explicitly take into consideration that actual speech vowels are quasi-periodic and usually consist of multiple resonances. Both of these phenomena may affect the AM-FM estimates. We mentioned above that the pitch periodicity induces narrow spikes in the amplitude and (mainly) the frequency signal around the onset of each glottal pulse. In addition, the effect of neighboring formants that have not been completely rejected by the bandpass filter is to cause "parasitic" FM and AM modulations, which have a smaller order of magnitude than the main observed modulations [31]. Various approaches have been proposed for jointly estimating the multiple AM–FM sinewave components of speech signals [22],[50].

11.6 Summary

We have seen that speech fine structure in the speech waveform can arise from either time-varying linear or nonlinear elements of the speech production mechanism. These speech events typically occur on the order of a glottal cycle and are thus difficult to reveal and measure with signal processing techniques such as the short-time Fourier transform and the wavelet transform that are characterized by time-frequency resolution tradeoffs. In this chapter, we described alternative high-resolution measurement techniques that aim to undermine the uncertainty principle and reveal fine structure not seen by the standard approaches. These more advanced time-frequency analysis methods included the Wigner distribution and its variations, referred to as bilinear time-frequency distributions, that have historically been developed in the context of estimating fine structure speech events associated largely with linear models. We also introduced a second approach to the analysis of fine structure motivated by the nonlinear aeroacoustic models for spatially distributed sound sources and modulations induced by nonacoustic fluid motion. We described qualitatively this modeling approach which then provided the motivation for the high-resolution Teager energy operator. Although significant progress has been made in nonlinear measurement and modeling techniques for speech analysis and synthesis, as we indicated throughout this chapter, the speech research community has only brushed the surface of these challenging and important areas.

An underlying assumption in this chapter is that a human listener perceives the fine structure that we are attempting to measure. It behooves us then to close with a few words on the time-frequency resolution of the auditory system. We do this briefly, and speculatively, from the perspectives of the two primary analysis tools of this chapter: (1) advanced time-frequency representations and (2) the Teager energy operator. From the perspective of advanced time-frequency analysis, we saw numerous speculations in Chapter 8. For example, an interesting possibility that arose from the auditory model of Yang, Wang, and Shamma [54] is that the auditory system performs jointly wideband and narrowband analysis. In this model, good joint time-frequency resolution is obtained from the lateral inhibition that occurs *across* adjacent auditory filters, e.g., in the form of differences across adjacent channels.

The second analysis tool of this chapter, the Teager energy operator, is consistent with the hypothesis that the auditory system exploits an FM *transduction* mechanism. Psychoacoustic experiments by Saberi and Hafter [41] provide evidence that FM and AM sounds must be transformed into a common neural code before binaural convergence in the brain stem. They showed that listeners can accurately determine if the phase of an FM signal presented to one ear is leading or lagging the phase of an AM signal presented to the other ear, with peak performance occurring when the signals are $180°$ out of phase. Saberi and Hafter's explanation of their experiments is based on FM-to-AM transduction in which the instantaneous frequency of an FM signal sweeps through the passband of a cochlear filter and the change in frequency is transformed into a change in amplitude. Motivated by this hypothesis, an approach to AM–FM estimation was developed based on the transduction of FM into AM by linear filters and on using the quotient of the amplitude envelopes of two overlapping filters (Exercise 11.23) [32],[40]. An auditory neural model was then proposed for how AM-FM separation may be achieved from pairs of overlapping auditory filter envelope outputs [3]. Specifically, it was shown that the transduction approach can be realized as a bank of auditory-like bandpass filters followed by envelope detectors and laterally inhibiting shunting neural networks that perform ratio processing, i.e., quotients between signal levels.[13] The resulting nonlinear dynamical system is capable of robust AM-FM estimation in noisy environments. The relation of this auditory processing model with the Teager energy operator lies in the observation that the transduced auditory filter output envelope is to a first approximation the *product* of AM and FM in the output (more strictly, the product of instantaneous amplitude and frequency) [40]. (This interpretation can be argued when the transduction filters have linear shape.) An interesting speculation is that the amplitude envelope of an auditory filter's output, being thus related to the "energy" required to generate the input sinewave, may provide a robust representation for further auditory stages.

In spite of these intriguing possibilities, it is important to emphasize that such auditory models are speculative. Even if this sort of signal processing is exploited, the mechanism may not be a linear one. For example, the models assume that the filter shapes are constant. In the cochlea, however, there is a fast-acting automatic gain control (AGC) that provides compression in the main part of the filter passband (the tip of the filter), while leaving the gain in the low-frequency portion of the filter (the tail) unaffected. The compression may be the result of an almost instantaneous saturating nonlinearity in the cochlear mechanics [42], and could both introduce fluctuations in the system output for a low-frequency sinusoid with constant amplitude and reduce the fluctuations in an AM-FM modulated tone. Incorporating the influence of such a nonlinearity and other mechanisms, such as those described in Chapter 8, is essential in understanding the full complexity of signal processing in the auditory system.

EXERCISES

11.1 Show in continuous time for the wavelet transform that the total energy in a signal equals the scaled energy in the scalogram, i.e., the relation in Equation (11.2) is satisfied. Then show in continuous time for the STFT that the total energy in a signal equals the energy in the spectrogram:

$$\int_{-\infty}^{\infty} |x(t)|^2 dt = \frac{1}{2\pi} \int_{-\infty}^{\infty} |X(t, \Omega)|^2 dt \, d\Omega$$

provided that the energy in the analysis window is unity.

11.2 Consider the complex exponential signal $x(t) = a(t)e^{j\phi(t)}$. For this signal we have defined the instantaneous frequency as the derivative of the phase, i.e., $\dot{\phi}(t) = \frac{d\phi(t)}{dt}$. Show that the average and instantaneous frequency are related by

$$\bar{\Omega} = \int_{-\infty}^{\infty} \Omega |X(\Omega)|^2 d\Omega$$

$$= \int_{-\infty}^{\infty} \dot{\phi}(t)|x(t)|^2 dt = \int \dot{\phi}(t) a^2(t) dt.$$

11.3 Figure 11.2 shows an example of short-time spectra of two synthetic speech waveforms, both from a linear time-invariant (steady) vocal tract with a transfer function $H(\Omega)$. The difference in the two

---

[13] Shunting neural networks have been used to explain a range of biological phenomena through ratio processing. For example, these networks have been shown to provide dynamic range compression and contrast enhancement in human vision [13]. Since shunting neural networks provide ratio processing, it is natural to implement the transduction approach with a feedforward, two-neuron, shunting circuit in which one neuron provides an FM estimate and the other provides an AM estimate. The coupled equations for neuronal activity can be written as ordinary differential equations with time-varying coefficients [3].

cases lies in the excitation. For Figure 11.2a, the underlying excitation is a pure impulse train with pitch Ω_o, while for Figure 11.2b, the pitch changes linearly as

$$\Omega_o(t) = \Omega_o + \alpha t.$$

(a) Write the expression for a sinewave model for the speech waveform corresponding to each spectrum in Figure 11.2 in terms of the vocal tract transfer function $H(\Omega)$ and pitch (Ω_o or $\Omega_o + \alpha t$).

(b) Using Figure 7.8, explain the observed spectra in terms of short-time Fourier analysis. What is the implication of your reasoning for the use of spectral peak-picking in Chapter 9 for accurate estimation of sinewave parameters?

(c) Under the condition that the pitch is changing rapidly under the short-time analysis window, propose an analysis scheme that may result in more accurate estimates of sinewave amplitude and frequency than simple peak-picking. Consider methods such as the wavelet transform and the Teager energy-based AM-FM separation method. If there appears to be no simple way to improve performance with these methods, then explain your reasoning.

11.4 Derive the expressions in Equations (11.11) and (11.12) for the conditional mean and variance of time and frequency. Derive also the expressions in Equations (11.13) and (11.14) for the relation between conditional and global quantities. You will need the relations $P(\Omega|t) = \frac{P(t,\Omega)}{P(t)}$ and, likewise, $P(t|\Omega) = \frac{P(t,\Omega)}{P(\Omega)}$.

11.5 From the expressions in Equation (11.13) for the global frequency average and its time-domain dual, argue that the instantaneous frequency and group delay are given by the conditional mean frequency $\bar{\Omega}_t$ and conditional mean time $\bar{t}_\Omega$, i.e., Equation (11.15), provided the marginals are satisfied.

11.6 Consider a signal consisting of a complex sinewave of the form $x(t) = a(t)e^{j\Omega t}$. Compute the instantaneous bandwidth of the following special cases:

(a) $x(t) = Ae^{j\Omega t}$

(b) $x(t) = Ae^{j\Omega t}u(t)$

(c) $x(t) = Ae^{-\alpha t}e^{j\Omega t}$

(d) $x(t) = Ae^{-\alpha t}e^{j\Omega t}u(t)$

where $u(t)$ is the unit step. What special characteristics arise with the unit step signal $u(t)$?

11.7 In this problem, you consider the separability of signal components for time-frequency distributions.

(a) Suppose that for a particular time-frequency distribution, the instantaneous frequency and bandwidth are given by Equations (11.15) and (11.17). Argue that Equation (11.18) must hold for the "ridges" of the distribution to reveal the components in frequency at a particular time instant.

(b) Consider now the dual problem of the separation of signal components in time at a particular frequency as revealed in a time-frequency distribution. Argue that for a two-component signal spectrum of the form

$$X(\Omega) = M_1(\Omega)e^{\theta_1(\Omega)} + M_2(\Omega)e^{\theta_2(\Omega)}$$

that the spectral magnitudes $M_1(\Omega)$, $M_2(\Omega)$ and group delays $\theta_1(\Omega)$, $\theta_2(\Omega)$ are constrained by

$$\left|\frac{\dot{M}_1(\Omega)}{M_1(\Omega)}\right|, \left|\frac{\dot{M}_2(\Omega)}{M_2(\Omega)}\right| \ll |\dot{\theta}_1(\Omega) - \dot{\theta}_2(\Omega)|,$$

when the marginals of the distribution are satisfied.

11.8 In this problem, you consider properties of the spectrogram $|X(t, \Omega)|^2$ of a signal $x(t)$, i.e., the squared STFT magnitude, as a time-frequency distribution.

(a) Show that with analysis window $w(t)$

$$E = \int_{-\infty}^{\infty} \int_{-\infty}^{\infty} |X(t, \Omega)|^2 dt d\Omega = \int_{-\infty}^{\infty} |x(t)|^2 dt \int_{-\infty}^{\infty} |w(t)|^2 dt$$

and thus that if the energy of the window function is unity, then the energy in the spectrogram equals the energy in the signal.

(b) Show that integrating the spectrogram over frequency gives

$$E = \int_{-\infty}^{\infty} |X(t, \Omega)|^2 d\Omega$$

$$= \int_{-\infty}^{\infty} |x(\tau)|^2 w(t - \tau) d\tau$$

and thus the time marginal is not satisfied. Show the dual relation when integrating over time.

11.9 Define the time spread of a signal, $x(t)$, as

$$\sigma_t^2 = \int_{-\infty}^{\infty} (t - \bar{t})^2 |x(t)|^2 dt$$

where $\bar{t}$ is the temporal mean of the signal. Similarly, define the frequency spread or bandwidth of $x(t)$ as

$$\sigma_\Omega^2 = \int_{-\infty}^{\infty} (\Omega - \bar{\Omega})^2 |X(\Omega)|^2 d\Omega.$$

(a) Compute the time-bandwidth product, $\sigma_t^2 \sigma_\Omega^2$, for the following signal:

$$x_1(t) = e^{-a(t-t_1)^2} \cos \Omega_1 t.$$

(b) Compute the time-bandwidth product for the following signal:

$$x_2(t) = e^{-b(t-t_2)^2} e^{j[ct^2 + m \sin(\Omega_m t) + \Omega_2 t]}.$$

(c) Can $x_2(t)$ be designed to have the same time-bandwidth product as $x_1(t)$? If so, what are the constraints on the signal parameters b, c, m, t_2, Ω_m, and Ω_2?

11.10 One approach to achieve good joint time-frequency resolution is through a geometric or additive mean of the narrowband and wideband spectrograms, given by $[|X_1(n, \omega)|^2 |X_2(n, \omega)|^2]^{1/2}$ and $[|X_1(n, \omega)|^2 + |X_2(n, \omega)|^2]/2$, respectively. $[|X_1(n, \omega)|^2$ denotes the narrowband spectrogram and $|X_2(n, \omega)|^2$ the wideband spectrogram. Describe which of the properties **P1–P5** of Section 11.3.1, that characterize a proper time-frequency distribution, holds for each of these mean distributions.

11.11 In this problem you investigate various properties of the Wigner distribution.

(a) Show that the Wigner distribution can be written in terms of its spectrum as in Equation (11.20).

(b) Show, using Equations (11.19) and (11.20), that a signal compact in time or in frequency remains compact in these dimensions within the Wigner distribution, thus avoiding loss in time-frequency resolution. Show, however, that zero gaps in a signal or its spectrum are not preserved in the distribution. Therefore, the Wigner distribution has weak, but not strong finite support.

(c) Show that the marginals, Equation (11.21), and thus the energy equivalence relation, Equation (11.22), are satisfied for the Wigner distribution. Why are the marginals for the Wigner distribution more desirable than those for the spectrogram?

11.12 In this problem, you consider the local and global time-bandwidth products for a distribution.

(a) Starting with the relation between a joint time-frequency density $P(t, \Omega)$ and the conditional densities $P(\Omega|t)$ and $P(t|\Omega)$, derive expressions for the conditional (local) time and frequency variances $\sigma^2_{\Omega|t}$ and $\sigma^2_{t|\Omega}$. How is the product $\sigma^2_{\Omega|t}\sigma^2_{t|\Omega}$ bounded?

(b) Compare differences in the local and global time-bandwidth products, i.e., between $\sigma^2_{\Omega|t}\sigma^2_{t|\Omega}$ and $\sigma^2_{\Omega}\sigma^2_{t}$.

(c) Because of the presence of the analysis window, explain why "beating" of the uncertainty principle by the local time-bandwidth product can never occur with the spectrogram. This property is therefore in contrast to that of the Wigner distribution.

11.13 Determine the Wigner distribution, $W(t, \Omega)$, for the signal

$$x(t) = e^{j(at^2+bt+c)}.$$

How is the conditional mean frequency for the Wigner distribution related to the instantaneous frequency for this signal?

11.14 Show that if

$$k(\theta, 0) = 1, \quad \text{for the time marginal}$$

$$k(0, \tau) = 1, \quad \text{for the frequency marginal}$$

then the time-frequency marginals of the canonic distribution in Equation (11.25) are satisfied.

11.15 Consider a voiced speech waveform where the pitch is linearly changing and the vocal tract is steady. Design a waveform time-warping procedure resulting in a steady pitch. What is the effect of time-warping on the vocal tract impulse response? Discuss the implications of your time-warping operation on the STFT time-frequency resolution when the pitch is rapidly varying. How might you exploit the modified resolution in sinusoidal analysis/synthesis? *Hint:* Consider the stationarity of the resulting sinewave amplitudes and frequencies.

11.16 Consider the following psychoacoustic experiment. Suppose we add two synthetically generated vowels, both having steady pitch and steady vocal tract. Human listening experiments indicate that the two vowels are not perceived as perceptually separate, i.e., we do not hear two distinct vowels. However, when we add vibrato, i.e., sinusoidal modulation to the pitch [i.e., $\Omega_o + \alpha \sin(\Omega_m t)$, Ω_o being the pitch] of one vowel, then a listener is more likely to perceive two separate vowel sounds, indicating the importance of modulation in human perception. Given the constant-Q filter-bank model of auditory cochlear front-end filters of Chapter 8 and the filter transduction approach to AM-FM separation described in Section 11.6, propose an auditory mechanism for sound separation based on the presence of modulation.

11.17 Consider the dynamical mechanical model of Barney, Shadle, and Davies [2], described in Section 11.4.3. Suppose that the shutter speed is 80 Hz, the mean flow rate $\bar{v}$ is 80 cm/s, and the duct length is 17 cm. Assuming a train of equally-spaced vortices moving at the mean flow rate, compute the number of vortices in the duct at any time.

11.18 We have observed that multiple acoustic and nonacoustic sources may exist in the vocal tract cavity, not all located at the opening of the glottis, especially where the vortical train meets boundary discontinuities. Consider now determining a solution approach for estimating the frequency response of each contribution from a measured correlation function of a very low-pitched speaker.

(a) Consider an example where a primary source pulse occurs at the glottis due to a conventional acoustic source and a secondary source pulse occurs downstream, e.g., behind the teeth due to a traveling vortex, i.e., the speech waveform is given by

$$s[n] = \delta[n] * h_1[n] + \delta[n - n_o] * h_2[n]$$

where $h_1[n]$ and $h_2[n]$ are responses to the primary and secondary source pulses, respectively. Show that the autocorrelation function of $s[n]$ consists of two components, one near the origin and the second near $n = n_o$, i.e.,

$$r_s[\tau] = r_{11}[\tau] + r_{22}[\tau] + r_{12}[\tau + n_o] + r_{12}[-\tau - n_o]$$

where

$$r_{11}[\tau] = h_1[n] * h_1[-n]$$
$$r_{22}[\tau] = h_2[n] * h_2[-n]$$
$$r_{12}[\tau] = h_1[n] * h_2[-n].$$

An important observation here is that the low-order correlation coefficients that are typically used in speech analysis, e.g., linear prediction, consist of the *sum of correlation coefficients from two different impulse responses*. The higher-order coefficients near $n = n_o$ are due primarily to the cross-correlation between these two responses. Argue why the correlation coefficients around $n = n_o$ are not symmetric for the conditions of this problem.

(b) To do accurate modeling, our goal becomes to separate the correlation functions $r_{11}[\tau]$ and $r_{22}[\tau]$. In general, this is a difficult problem. To simplify, we constrain n_o to be large enough to separate the sum near the origin, $r_{11}[\tau] + r_{22}[\tau]$, and the cross-correlation function, $r_{12}[\tau + n_o]$. Show that we can then formulate the following quadratic equation

$$X^2(\omega) + A(\omega)X(\omega) + B(\omega) = 0$$

with

$$A(\omega) = -(|S_{11}(\omega)|^2 + |S_{22}(\omega)|^2)$$
$$B(\omega) = |S_{12}(\omega)|^2.$$

where $|S_{11}(\omega)|^2$, $|S_{22}(\omega)|^2$, and $|S_{12}(\omega)|^2$ are the Fourier transforms of $r_{11}[\tau]$, $r_{22}[\tau]$, and $r_{12}[\tau]$, respectively. Show that the two solutions to this quadratic equation are $|S_{11}(\omega)|^2$ and $|S_{22}(\omega)|^2$.

11.19 Figure 11.15 shows (a) a voiced speech segment, (b) its bandpass-filtered version (with a modulated Gaussian filter centered at the first formant), and (c) the envelope of the bandpass-filtered waveform. The envelope is the amplitude of the bandpass-filtered waveform obtained from the energy separation algorithm of Section 11.5. Assume that the bandpass filter separates the first formant exactly (i.e., there is no leakage from neighboring formants) and assume that the vocal tract and vocal cords are in an approximate steady-state.

(a) Explain why the envelope measurement of Figure 11.15c violates the simple voiced speech model of a linear system driven by a periodic pulse train. Suggest possible "anomalies" in speech production that may describe the observed envelope.

(b) How well do you think the STFT (computed with a window of duration 2.5 times a pitch period) would capture the fine structure in the observed envelope?

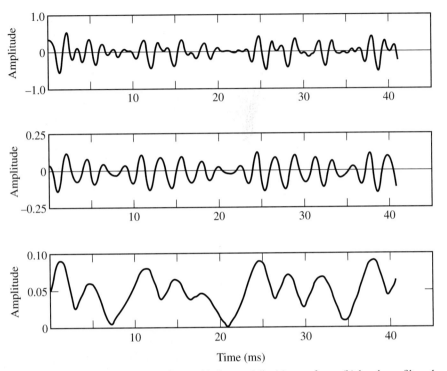

Figure 11.15 Voiced speech waveform with "anomaly": (a) waveform; (b) bandpass-filtered waveform; (c) envelope of bandpass-filtered waveform.

    (c) Suppose that you attempt to estimate the AM and FM of a resonance where more than a cycle of modulation occurs over a very short-time duration, say, less than a pitch period. Comment on problems you might have in estimating these modulations by ridge tracking with the STFT (the spectrogram). Does the "instantaneous" nature of the Teager energy-based methods help avoid these problems? Explain your reasoning.

    (d) Compare the temporal resolution of AM-FM estimation based on the Teager operator and based on the Hilbert transform for a finite data segment with rapidly varying modulation. *Hint:* Consider the Hilbert transform response of Equation (2.10).

11.20 We showed in Section 11.5 that the Teager energy operator $\Psi(x)$ output for a sinewave of the form

$$x(t) = A\cos(\Omega t)$$

is given by

$$\Psi(x) = A^2\Omega^2.$$

Also we derived in Section 11.5 a method to separate the amplitude A and frequency Ω, given both the Teager energy output of the signal and its derivative, i.e., $\Psi(x)$ and $\Psi(\frac{dx}{dt})$. Suppose, on the other hand, you are given $x(t)$ and its integral $y(t) = \int_0^t x(\tau)d\tau$. Derive a method to separate the amplitude A and frequency Ω from the Teager energy of $x(t)$ and $y(t)$, i.e., from $\Psi(x)$ and $\Psi(y)$. Why might this algorithm have better frequency resolution than the derivative form for very low-frequency sinewaves?

11.21 In this problem, we consider the Teager energy for the sum of two sequences:

$$x[n] = x_1[n] + x_2[n]$$

(a) Show that the output of the discrete-time Teager energy operator can be expressed as

$$\Psi_d(x[n]) = \Psi_d^1(x[n]) + \Psi_d^2(x[n]) + \Psi_d^{12}(x[n]) + \Psi_d^{21}(x[n]),$$

where $\Psi_d^{12}(x[n])$ and $\Psi_d^{21}(x[n])$ are the Teager energy cross terms.

(b) For the sum of two sinewaves

$$x[n] = A_1 \cos(\omega_1 n + \theta_1) + A_2 \cos(\omega_2 n + \theta_2)$$

show that the Teager energy is the sum not only of the individual energy components but also an additional cross term at the difference frequency of the two frequency components.

11.22 Show that when $x[n] = A \cos(\omega_c n + \theta)$, then the AM–FM separation formula of Equation (11.36) yields the exact constant frequency and amplitude.

11.23 In this problem, you consider the FM-to-AM transduction approach to AM-FM sinewave estimation. Consider an AM-FM signal of the form $x(t) = A(t) \cos \phi(t)$. The AM component is $A(t)$. The FM component is considered as the instantaneous frequency, $\Omega(t) = \frac{d}{dt}\phi(t)$. Suppose the output of a front-end auditory filter, $y(t)$, in response to the input signal, is the convolution of the impulse response of the filter, $h(t)$, with the input signal, i.e. $y(t) = x(t) * h(t)$. We define the amplitude envelope of the filter output as the rectified signal $|y(t)|$. (We assume $h(t)$ is in analytic form.) Transduction consists of sweeping an FM signal through a nonuniform bandpass filter; the change in frequency induces a corresponding change in the amplitude envelope of the filter output.

(a) Show that under a quasi-stationary assumption on the sinewave AM and FM, an FM-to-AM transduction approximation holds such that the amplitude envelope of the filter output is given by $|y(t)| \approx A(t)|H[\Omega(t)]|$, where $H[\Omega]$ is the frequency response of the filter.

(b) The AM-FM estimation problem can be solved by taking the ratio of the output of two overlapping auditory filters, $H_1[\Omega(t)]$ and $H_2[\Omega(t)]$. Specifically, under the transduction approximation, show that

$$\frac{|y_2(t)|}{|y_1(t)|} = \frac{|H_2[\Omega(t)]|}{|H_1[\Omega(t)]|}$$

so that when the filters have a Gaussian shape, one can solve for $\Omega(t)$ and recover the FM component under certain mild constraints. Then show that the recovered FM component can be used to recover the AM component. What general class of overlapping bandpass filters allow AM and FM estimation through transduction?

11.24 (MATLAB) In this problem, you explore the analysis of frequency-modulated sinewaves using two different time-frequency distributions: the short-time Fourier transform (STFT) magnitude and the Wigner distribution. There are three functions you will use: *cross.m*, *stftm.m*, and *wigner.m* located in companion website directory Chap_exercises/chapter11.

Run *cross.m* in your MATLAB workspace by typing *cross*. This will generate two (among other items) signals x and $x1$ that you will use. $x1$ is a single FM sinewave and x consists of the sum of two crossing FM sinewaves.

(a) Run *stftm.m* (with a 256-point window) on $x1$ and also run *wigner.m* on $x1$. Compare and explain your results. The function *stftm.m*, in addition to a spectrogram, also displays an estimate of instantaneous frequency equal to the location of the spectrogram ridge corresponding to the peak in the STFT magnitude at each time instant.

(b) Repeat part (a) for the signal x with crossing frequency trajectories. Include discussion of the phenomena at the trajectory intersection. Also explain any peculiarities of the Wigner distribution. What difficulties might you encounter in doing frequency estimation by ridge tracking on the Wigner distribution?

(c) One way to achieve better temporal resolution with the STFT magnitude is to shorten the analysis window. Run *stftm.m* with window lengths 128, 64, and 32, achieving increasingly better time resolution. Explain your observations.

(d) Load the speech segment *speech1_10k* (at 10000 samples/s) of workspace *ex11M1.mat* located in companion website directory Chap_exercises/chapter11 and run *wigner.m* on it. Compare with the spectrogram with regard to the difficulty in interpreting spurious vertical and horizontal striations that come about from cross terms. How does this affect your ability to perceive speech fine structure?

Note: To run *wigner.m* successfully, you must truncate the input speech signal length to the nearest power of 2. (This is why the signals in *cross.m* are of length 1024.)

11.25 (MATLAB) In this problem we explore the Teager energy operator in continuous and discrete time.

(a) Show that for a simple continuous-time harmonic oscillation of the form

$$x(t) = A \cos(\Omega t + \theta)$$

the continuous-time energy operator Equation (11.28) gives the energy (within a scale factor) required to generate the signal, i.e., $A^2 \Omega^2$.

(b) Show that discretizing $\dot{x}(t)$ and $\ddot{x}(t)$ by first backward differences in the continuous-time energy operator leads to the time-shifted and scaled discrete-time energy operator in Equation (11.29).

(c) Show that for a simple discrete-time harmonic oscillation of the form

$$x[n] = A \cos(\omega n + \theta),$$

the discrete-time energy operator Equation (11.30) gives $\Psi_d(x[n]) = A^2 \sin^2(\omega)$. Explain why your result is exact provided that the frequency $\Omega < \pi/2$, i.e., one quarter of the sampling frequency. Then show that the operator gives approximately the energy as the squared product of frequency and amplitude, $\omega^2 A^2$, for small values of ω. Suppose that you limit the frequency $\omega < \pi/8$. What percent relative error do you entail in measuring the energy $A^2 \omega^2$?

(d) Show that the discrete-time Teager energy for an exponentially-damped sinewave of the form

$$x[n] = A r^n \cos(\omega n + \phi)$$

is given by

$$\Psi_d(x[n]) = A^2 r^{2n} \sin^2(\omega).$$

How can you approximate the resulting expression for small frequency? Write MATLAB code for the discrete-time energy operator Equation (11.30) and apply it to a damped sinewave with $r = 1, r = .99$, and $\phi = 0$. Let the frequency change from 100 Hz to 2000 Hz in steps of 500 Hz (for a sampling rate of 10000 Hz). What observations can you make?

(e) Using the discrete-time energy operator Equation (11.30), derive the Teager energy for a chirp signal with linear frequency sweep

$$\omega[n] = \omega_o - \frac{\Delta\omega}{N} n$$

i.e., for a signal of the form

$$x[n] = A \cos\left(\omega_o n - \frac{\Delta\omega}{2N}n^2\right).$$

Assume that the rate of change in frequency is small enough so that $\cos(\frac{\Delta\omega}{N}) \approx 1$ and $\sin(\frac{\Delta\omega}{N}) \approx 0$. Use your MATLAB code for the discrete-time energy operator from part (d) and apply it to the chirped sinewave with $\omega_o = 500$ Hz. Let the sweep rate change from $\Delta\omega = 100$ Hz to 2000 Hz in steps of 500 Hz (for a sampling rate of 10000 Hz). What observations can you make?

11.26 (MATLAB) In this problem you investigate the temporal resolution of the discrete energy separation algorithm of Equation (11.36).

    (a) Explain why the algorithm in Equation (11.36) requires only five samples to perform AM–FM estimation.

    (b) Concatenate two sinewaves of the form $A \cos(2\pi f t)$, each of 0.5 s duration, the first with $A = 1$ and $f = 500$ Hz, and the second with $A = 1.5$ and $f = 700$ Hz. (In discrete form, assume a sampling rate of 10000 samples/s.) Use the provided energy separation function (*amfm_sep.m* in companion website directory Chap_exercises/chapter11) to estimate the AM and FM of the signal. Comment on the temporal resolution in comparison with other AM-FM estimators (from time-frequency distributions) we have studied in this chapter.

11.27 (MATLAB) You will find in the companion website directory Chap_exercises/chapter11 functions *amfm_sep.m*, *gaussian_filt.m*, *specgram_speech.m* and a workspace *ex11M2.mat*. In this problem, you are asked to analyze a speech waveform to determine certain underlying modulations of the speech.

    (a) Estimate the center frequency of the second formant of the waveform *speech_10k* (at 10000 samples/s) in *ex11M2.mat* and apply a Gaussian bandpass filter (using *gaussian_filt.m*) to approximately filter out this second formant. Plot the resulting amplitude- (AM) and frequency-modulated (FM) filtered signal over the signal range [501:1000]. (Use any technique you have learned thus far to estimate the center frequency of the second formant.)

    (b) Now apply *amfm_sep.m* to the filtered signal from part (a) and plot the AM and FM estimates and Teager energy estimate over the signal range [501:1000]. The function *amfm_sep.m* implements the AM-FM energy separation algorithm. Plot also the conventional energy measure (signal squared); how does it differ from the Teager energy?

    (c) Use the function *specgram_speech.m* to find the spectrogram of the entire signal *speech_10k* and observe the global trend on the second formant. Then redo parts (a) and (b) to find the AM and FM estimates over the signal sample range [501:3000] (in contrast to [501:1000]). In the former, you will see the detail in the AM and FM over a long time interval, as well as the general trends. In the latter, you will see the detail over a short time interval.

    (d) Propose an explanation for your observations in parts (b) and (c). Think about the AM and FM *within* a glottal cycle; speculate on how the speech source and the vocal tract activity might contribute to what you observe. Finally, might the AM–FM energy separation algorithm have certain limitations? Can you propose an alternative AM–FM estimator that might address these limitations?

BIBLIOGRAPHY

[1] J.D. Anderson, Jr., *Modern Compressible Flow*, McGraw-Hill, New York, NY, 1982.

[2] A. Barney, C.H. Shadle, and P.O.A.L. Davies, "Fluid Flow in a Dynamical Mechanical Model of the Vocal Folds and Tract, 1: Measurements and Theory," *J. Acoust. Soc. Am.*, vol. 105, no. 1, pp. 444–455, Jan. 1999.

[3] R. Baxter and T.F. Quatieri, "Shunting Networks for Multi-Band AM-FM Decomposition," *Proc. of IEEE 1999 Workshop on Applications of Signal Processing to Audio and Acoustics*, New Paltz, NY, Oct. 1999.

[4] B. Boashash, "Estimating and Interpreting the Instantaneous Frequency of a Signal," *Proc. IEEE*, vol. 80, pp. 520–538, April 1992.

[5] G.F. Boudreaux-Bartels and T.W. Parks, "Time-Varying Filtering and Signal Estimation Using Wigner Distribution Synthesis Techniques," *IEEE Trans. Acoustics, Speech, and Signal Processing*, vol. 34, pp. 442–451, 1986.

[6] A. Bovik, P. Maragos, and T.F. Quatieri, "AM-FM Energy Detection and Separation in Noise Using Multiband Energy Operators," *IEEE Trans. Signal Processing*, special issue on wavelets, vol. 41, no. 12, pp. 3245–3265, Dec. 1993.

[7] T.A.C.M. Claasen and W.F.G. Mecklenbräuker, "The Wigner Distribution—A Tool for Time-Frequency Analysis—Part 1: Continuous Time Signals," *Philips Jour. Research*, vol. 35, pp. 217–250, 1980.

[8] T.A.C.M. Claasen and W.F.G. Mecklenbräuker, "The Wigner Distribution—A Tool for Time-Frequency Analysis—Part 2: Discrete-Time Signals," *Philips Jour. Research*, vol. 35, pp. 276–300, 1980.

[9] T.A.C.M. Claasen and W.F.G. Mecklenbräuker, "The Wigner Distribution—A Tool for Time-Frequency Analysis—Part 3: Relations with Other Time-Frequency Signal Transformations," *Philips Jour. Research*, vol. 35, pp. 372–389, 1980.

[10] L. Cohen, *Time-Frequency Analysis*, Prentice Hall, Englewood Cliffs, NJ, 1995.

[11] J.G. Daugman, "Uncertainty Relation for Resolution in Space, Spatial Frequency, and Orientation Optimized by Two-Dimensional Visual Cortical Filters," *J. Opt. Soc. Am. A*, vol. 2, no. 7, pp. 1160–1169, July 1985.

[12] P.O.A.L. Davies, "Aeroacoustics and Time-Varying Systems," *J. Sound Vibration*, vol. 190, pp. 345–362, 1996.

[13] S. Grossberg, "The Quantized Geometry of Visual Space: The Coherent Computation of Depth, Form, and Lightness," *The Brain and Behavioral Sciences*, vol. 6, pp. 625–692, 1983.

[14] R. Hamila, J. Astola, F. Alaya Cheikh, M. Gabbouj, and M. Renfors, "Teager Energy and the Ambiguity Function," *IEEE Trans. on Signal Processing*, vol. 47, no. 1, pp. 260–262, Jan. 1999.

[15] A. Hirschberg, "Some Fluid Dynamic Aspects of Speech," *Bulletin de la Communication Parlee*, vol. 2, pp. 7–30, 1992.

[16] M.S. Howe, "Contributions to the Theory of Aerodynamic Sound, with Application to Excess Jet Noise and the Theory of the Flute," *J. Fluid Mechanics*, vol. 67, part 3, pp. 625–673, 1975.

[17] T. Irino and R.D. Patterson, "A Time-Domain, Level-Dependent Auditory Filter: The Gammachirp," *J. Acoustical Society of America*, vol. 101, no. 1, pp. 412–419, Jan. 1997.

[18] J. Jeong and W.J. Williams. "Alias-Free Generalized Discrete-Time Time-Frequency Distributions," *IEEE Trans. on Signal Processing*, vol. 40, no. 11, pp. 2757–2765, Nov. 1992.

[19] J.F. Kaiser, "Some Observations on Vocal Tract Operation from a Fluid Flow Point of View," chapter in *Vocal Fold Physiology: Biomechanics, Acoustics, and Phonatoy Control*, R. Titze and R.C. Scherer, eds., The Denver Center for the Performing Arts, pp. 358–386, May 1983.

[20] J.F. Kaiser, "On a Simple Algorithm to Calculate the 'Energy' of a Signal," in *Proc. IEEE Int. Conf. Acoustics, Speech, and Signal Processing*, vol. 1, Albuquerque, NM, pp. 381–384, April 1990.

[21] J.F. Kaiser, "On Teager's Energy Algorithm and its Generalization to Continuous Signals," in *Proc. 4th IEEE Digital Signal Processing Workshop*, Mohonk (New Paltz), NY, Sept. 1990.

[22] R. Kumaresan, A.G. Sadasiv, C.S. Ramalingam, and J.F. Kaiser, "Instantaneous Non-Linear Operators for Tracking Multicomponent Signal Parameters," *Proceedings Sixth SP Workshop on Statistical Signal & Array Processing*, Victoria, British Columbia, Canada, pp. 404–407, Oct. 1992.

[23] J. Liljencrants, "Numerical Simulations of Glottal Flow," chapter in *Vocal Fold Physiology*, J. Gauffin and B. Fritzell, eds., Raven Press, 1990.

[24] W. Lin and P. Chitrapu, "Time-Frequency Distributions Based on Teager-Kaiser Energy Function," *Proc. Int. Conf. Acoustics, Speech, and Signal Processing*, Atlanta, GA, vol. 3, pp. 1818–1821, May 1996.

[25] P.J. Loughlin and K.L. Davidson, "Positive Local Variances of Time-Frequency Distributions and Local Uncertainty," *Proc. IEEE-SP Int. Symp. on Time-Frequency and Time-Scale Analysis*, pp. 541–544, Oct. 1998.

[26] P.J. Loughlin, J. Pitton, and L.E. Atlas, "Bilinear Time-Frequency Representations: New Insights and Properties," *IEEE Trans. Signal Processing*, vol. 41, pp. 750–767, no. 2, Feb. 1993.

[27] R.F. Lyon, "The All-Pole Gammatone Filter and Auditory Models," chapter in *Proc. Forum Acusticum 96*, Antwerp, Belgium.

[28] S. Mallat, *A Wavelet Tour of Signal Processing*, Academic Press, New York, NY, 1998.

[29] P. Maragos, T. F. Quatieri, and J. F. Kaiser, "Speech Nonlinearities, Modulations, and Energy Operators," *Proc. Int. Conf. Acoustics, Speech, and Signal Processing*, Toronto, Canada, pp. 421–424, May 1991.

[30] P. Maragos, J. F. Kaiser, and T. F. Quatieri, "On Amplitude and Frequency Demodulation Using Energy Operators," *IEEE Trans. Signal Processing*, vol. 41, no. 4, pp. 1532–1550, April 1993.

[31] P. Maragos, J. F. Kaiser, and T. F. Quatieri, "Energy Separation in Signal Modulations with Application to Speech Analysis," *IEEE Trans. Signal Processing*, vol. 41, no. 10, pp. 3024–3051, Oct. 1993.

[32] R. McEachern, "How the Ear Really Works," *Proc. IEEE-SP Int. Symp. on Time-Frequency and Time-Scale Analysis*, pp. 437–440, Oct. 1992.

[33] R.S. McGowan, "An Aeroacoustic Approach to Phonation," *J. Acoustical Society of America*, vol. 83, pp. 696–704, 1988.

[34] P.M. Morse and K.U. Ingard, *Theoretical Acoustics*, Princeton University Press, Princeton, NJ, 1986.

[35] K. Nathan and H. Silverman, "Time-Varying Feature Selection and Classification of Unvoiced Stop Consonants," *IEEE Trans. Speech and Audio Processing*, vol. 2, no. 3, pp. 395–405, 1991.

[36] A. Papoulis, *The Fourier Integral and its Applications*, McGraw-Hill, New York, NY, 1962.

[37] J.W. Pitton, L.E. Atlas, and P.J. Loughlin, "Applications of Positive Time-Frequency Distributions to Speech Processing," *IEEE Trans. Speech and Audio Processing*, vol. 2, no. 4, pp. 554–566, Oct. 1994.

[38] J.W. Pitton, K. Wang, and B.H. Juang, "Time-Frequency Analysis and Auditory Modeling for Automatic Recognition of Speech," *Proc. IEEE*, vol. 84, no. 9, pp. 1199–1215, Sept. 1996.

[39] A. Potamianos and P. Maragos, "A Comparison of the Energy Operator and the Hilbert Transform Approach to Signal and Speech Demodulation," *Signal Processing*, vol. 37, pp. 95–120, May 1994.

[40] T.F. Quatieri, T.E. Hanna, and G.C. O'Leary, "AM-FM Separation Using Auditory-Motivated Filters," *IEEE Trans. Speech and Audio Processing*, vol. 5, no. 5, pp. 465–480, Sept. 1997.

[41] K. Saberi and E.R. Hafter, "A Common Neural Code for Frequency- and Amplitude-Modulated Sounds," *Nature*, vol. 374, pp. 537–539, April 1995.

[42] P.M. Sellick, R. Patuzzi, and B.M. Johnstone, "Measurement of Basilar Membrane Motion in the Guinea Pig Using the Mössbauer Technique," *J. Acoustical Society of America*, vol. 72, no. 1, pp. 1788–1803, July 1982.

[43] C.H. Shadle, A. Barney, and P.O.A.L. Davies, "Fluid Flow in a Dynamical Mechanical Model of the Vocal Folds and Tract, 2: Implications for Speech Production Studies," *J. Acoustical Society of America*, vol. 105, no. 1, pp. 456–466, Jan. 1999.

[44] D.J. Sinder, *Speech Synthesis Using an Aeroacoustic Fricative Model*, Ph.D. Thesis, Rutgers University, New Brunswick, NJ, Oct. 1999.

[45] R. Smits, "Accuracy of Quasistationary Analysis of Highly Dynamic Speech Signals," *J. Acoustical Society of America*, vol. 96, no. 6, pp. 3401–3415, Dec. 1994.

[46] H.M. Teager, "Some Observations on Oral Air Flow During Phonation," *IEEE Trans. Acoustics, Speech, Signal Processing*, vol. ASSP–28, no. 5, pp. 599–601, Oct. 1980.

[47] H.M. Teager and S.M. Teager, "A Phenomenological Model for Vowel Production in the Vocal Tract," chapter in *Speech Science: Recent Advances*, R.G. Daniloff, ed., College-Hill Press, San Diego, CA, pp. 73–109, 1985.

[48] H.M. Teager and S.M. Teager, "Evidence for Nonlinear Sound Production Mechanisms in the Vocal Tract," chapter in *Speech Production and Speech Modelling*, W.J. Hardcastle and A. Marchal, eds., NATO Adv. Study Inst. Series D, vol. 55, Bonas, France; Kluwer Academic Publishers, Boston, MA, pp. 241–262, 1990.

[49] T.J. Thomas, "A Finite Element Model of Fluid Flow in the Vocal Tract," *Computer Speech and Language*, vol. 1, pp. 131–151, 1986.

[50] W.P. Torres and T.F. Quatieri, "Estimation of Modulation Based on FM-to-AM Transduction: Two-Sinusoid Case," *IEEE Trans. Signal Processing*, vol. 47, no. 11, pp. 3084–3097, Nov. 1999.

[51] B. van der Pol, "Frequency Modulation," *Proc. IRE*, vol. 18, pp. 1194–1205, July 1930.

[52] J. Ville, "Theorie et Applications de la Notion de Signal Analytique," *Cables et Transmissions*, vol. 2A, pp. 61–74, 1948. Translated from French by I. Selin, "Theory and Applications of the Notion of Complex Signal," RAND Corporation Technical Report T–92, Santa Monica, CA, 1958.

[53] E.P. Wigner, "On the Quantum Correction for Thermodynamic Equilibrium," *Physical Review*, vol. 40, pp. 749–759, 1932.

[54] X. Yang, K. Wang, and S.A. Shamma, "Auditory Representations of Acoustic Signals," *IEEE Trans. Information Theory*, vol. 38, no. 2, pp. 824–839, March 1992.

Speech Coding

12.1 Introduction

We define speech coding as any process that leads to the representation of analog waveforms by sequences of binary digits (i.e., 0's and 1's) or *bits*. Although high-bandwidth channels and networks are becoming more viable (as with fiber optics), speech coding for bit-rate reduction has retained its importance. This is due to the need for low rates with, for example, cellular and Internet communications over constrained-bandwidth channels, and with voice storage playback systems. In addition, coded speech, even at high bit rates, is less sensitive than analog signals to transmission noise and is easier to error-protect, encrypt, multiplex, and packetize. An example of speech coding in a digital telephone communication scenario is illustrated in Figure 12.1. An analog waveform is first digitized with an A/D converter and then analyzed with one of the speech analysis techniques of the previous chapters. The resulting speech parameters are quantized and then encoded into a bit pattern that may also be encrypted. Next, the bits are converted back to an analog communication signal, i.e., they are modulated using, for example, phase or frequency shift keying [65], and transmitted over an analog telephone channel. With a digital link, e.g., a digital satellite channel, this step is not performed. Because the channel may introduce distortion, the encrypted bits are error-protected before modulation. Finally, at the receiver the inverse operations are performed.

Speech coders are typically categorized in three groups: *waveform coders, hybrid coders, and vocoders*. Waveform coders quantize the speech samples directly and operate at high bit rates in the range 16–64 kbps (bps, denoting bits per second). Hybrid coders are partly waveform-based and partly speech model-based and operate in the 2.4–16 kbps range. Finally, vocoders

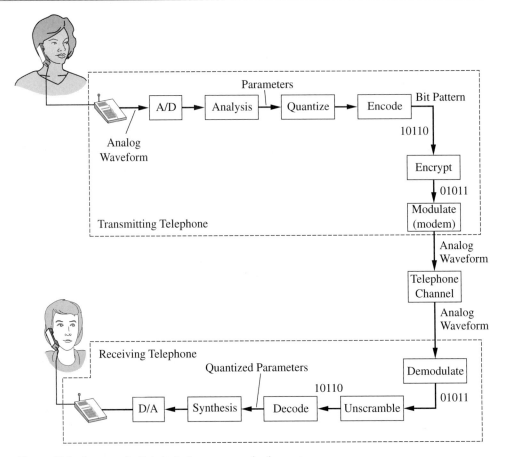

Figure 12.1 An example digital telephone communication system.

are largely model-based, operate at the low bit rate range of 1.2–4.8 kbps, and tend to be of lower quality than waveform and hybrid coders. In this chapter, a representative set of coders from each class is developed, using speech modeling principles and analysis/synthesis methods described throughout the text.

Observe that we have used the word "quality" without definition. We can think of quality in terms of the closeness of the processed speech (e.g., coded, enhanced, or modified) to the original speech or to some other desired speech waveform (e.g., a time-scaled waveform). Quality has many dimensions. For example, a processed speech waveform can be rated in terms of naturalness (e.g., without a machine-like characteristic), background artifacts (e.g., without interfering noise or glitches), intelligibility (e.g., understandability of words or content), and speaker identifiability (i.e., fidelity of speaker identity). Different subjective and objective tests have been developed to measure these different attributes of quality. For example, in subjective testing, based on opinions formed from comparative listening tests, the *diagnostic rhyme test* (DRT) measures intelligibility, while the *diagnostic acceptability measure* (DAM) and *mean opinion score* (MOS) tests provide a more complete quality judgment. Examples of objective tests include the *segmental signal-to-noise ratio* (SNR), where the average of SNR over short-time segments is computed, and the *articulation index* that relies on an average SNR across frequency bands. Although useful in waveform coders, SNR in the time domain is not always meaningful, however, for model-based coders because the phase of the original speech may be modified, disallowing waveform differencing (Exercise 12.19). Alternative objective measures have thus been developed that are based on the short-time Fourier transform magnitude only. A more complete study of subjective and objective measures can be found in the literature, as in [17],[70], and a number of these tests will be further described as they are needed. Our purpose here is to give the reader a qualitative understanding of the term "quality" to be used in this chapter in describing the output of speech coders (as well as the output of speech processing systems in other parts of the text); a careful assessment of individual quality attributes for each coder is beyond our scope.

We begin in Section 12.2 with statistical models of speech that form the backbone of coding principles and lead naturally to the waveform coding and decoding techniques of Section 12.3. These methods are based on one-dimensional *scalar quantization* where each sample value is coded essentially independently. An optimal method of scalar quantization, referred to as the *Max quantizer*, is derived. Extensions of this basic approach include companding, adaptive quantization, and differential quantization, and they form the basis for waveform coding standards at high rates such as 32–64 kbps. In Section 12.4, scalar quantization is then generalized to *vector quantization* that quantizes a set of waveform or speech parameters rather than single scalars, exploiting the correlation across values. Vector quantization is an important tool in the development of mid-rate and low-rate hybrid and vocoding strategies. In Section 12.5, we move to the frequency domain and describe two important hybrid coding techniques: *subband coding* and *sinusoidal coding* that are based on the STFT and sinewave analysis/synthesis techniques in Chapters 8 and 9, respectively. These coding strategies provide the basis for 2.4–16 kbps coding.

Section 12.6 then develops a number of all-pole model-based coding methods using the linear prediction analysis/synthesis of Chapter 5. The section opens with one of the early classic vocoding techniques at 2.4 kbps, referred to as *linear prediction coding* (LPC) that is based on a simple binary impulse/noise source model. To design higher-quality linear-prediction-based speech coders at such low rates or at higher rates, generalizations of the binary excitation model are required. One such approach is the *mixed excitation linear prediction* (MELP) *vocoder* that uses different blends of impulses and noise in different frequency bands, as well as exploits many excitation properties which we have seen in earlier chapters such as jittery and irregular pitch pulses and waveform properties that reflect the nonlinear coupling between the source and vocal tract. Another such approach is *multipulse LPC* which uses more than one impulse per glottal cycle to model the voiced speech source and a possibly random set of impulses to model the unvoiced speech source. An important extension of multipulse LPC is *code excited linear prediction* (CELP) that models the excitation as one of a number of random sequences, or "codewords," together with periodic impulses. Although these later two approaches, studied in the final Section 12.7 of this chapter, seek a more accurate representation of the speech excitation, they also address the inadequacies in the basic linear all-pole source/filter model; as such, these methods are referred to as linear prediction *residual coding* schemes.

12.2 Statistical Models

When applying statistical notions to speech signals, it is necessary to estimate probability densities and averages (e.g., the mean, variance, and autocorrelation) from the speech waveform which is viewed as a random process. One approach to estimate a probability density function (pdf) of $x[n]$ is through the *histogram*. Assuming the speech waveform is an ergodic process (Appendix 5.A), in obtaining the histogram, we count up the number of occurrences of the value of each speech sample in different ranges, $x_o - \frac{\Delta}{2} \leq x \leq x_o + \frac{\Delta}{2}$ (covering all speech values). We do this for many speech samples over a long time duration, and then normalize the area of the resulting curve to unity.

A histogram of the speech waveform was obtained by Davenport [13] and also by Paez and Glisson [66]. The histogram was shown to approximate a *gamma density* which is of the form [71]

$$p_x(x) = \left(\frac{\sqrt{3}}{8\pi \sigma_x |x|} \right)^{\frac{1}{2}} e^{-\frac{\sqrt{3}|x|}{2\sigma_x}},$$

where σ_x is the standard deviation of the pdf. An even simpler approximation is given by the Laplacian pdf of the form

$$p_x(x) = \frac{1}{\sqrt{2}\sigma_x} e^{-\frac{\sqrt{2}|x|}{\sigma_x}}.$$

(In the above pdfs, we have simplified $x[n]$ notation as x.) The two density approximations, as well as the speech waveform histogram, are shown in Figure 12.2. Features of the speech pdf include a distinct peak at zero which is the result of pauses and low-level speech sounds. In spite of this characteristic, there is a significant probability of high amplitudes, i.e., out to $\pm 3\sigma_x$. Histogram analysis is useful not only in determining the statistical properties of the speech waveform, but also of speech model parameters such as linear prediction and cepstral coefficients or of non-parametric representations such as filter-bank outputs. The mean, variance, and autocorrelation of the speech pressure waveform can be obtained from its pdf or from long-time averages of the waveform itself when ergodicity holds (Appendix 5.A). Alternatively, given the nonstationarity of speech, a counterpart pdf over a short time duration, and associated short-time measures of the mean, variance and autocorrelation (as we saw in Chapter 5), may be a more appropriate characterization for certain coding strategies.

12.3 Scalar Quantization

Assume that a speech waveform has been lowpass-filtered and sampled at a suitable rate giving a sequence $x[n]$ and that the sampling is performed with infinite amplitude precision, i.e., the A/D is ideal. We can view $x[n]$ as a sample sequence of a discrete-time random process. In the first step of the waveform coding process, the samples are quantized to a finite set of amplitudes which are denoted by $\hat{x}[n]$. Associated with the quantizer is a quantization step size Δ that we will specify shortly. This quantization allows the amplitudes to be represented by a finite set of bit patterns, or *symbols*, which is the second step in the coding process. The mapping of $\hat{x}[n]$ to a finite set of symbols is called *encoding* the quantized values and this yields a sequence of *codewords*, denoted by $c[n]$, as illustrated in Figure 12.3a. Likewise, a *decoder* takes a sequence of codewords $c'[n]$, the prime denoting a codeword that may be altered in the transmission process, and transforms them back to a sequence of quantized samples, as illustrated

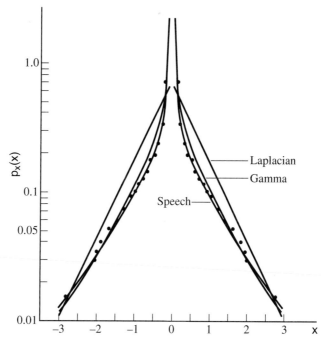

Figure 12.2 Comparison of histograms from real speech and gamma and Laplacian probability density fits to real speech. The densities are normalized to have mean $m_x = 0$ and variance $\sigma_x^2 = 1$. Dots (and the corresponding fitted curve) denote the histogram of the speech.

SOURCE: M.D. Paez and T.H. Glisson, "Minimum Mean-Squared Error Quantization in Speech" [66]. ©1972, IEEE. Used by permission.

in Figure 12.3b. In this section, we look at the basic principles of quantizing individual sample values of a waveform, i.e., the technique of scalar quantization, and then extend these principles to adaptive and differential quantization techniques that provide the basis of many waveform coders used in practice.

12.3.1 Fundamentals

Suppose we quantize a signal amplitude into M levels. (We use the term "amplitude" to mean signed signal value.) We denote the quantizer operator by $Q(x)$; specifically,

$$\hat{x}[n] = \hat{x}_i = Q(x[n]), \quad x_{i-1} < x[n] \leq x_i$$

where $\hat{x}_i$ denotes M possible *reconstruction levels*, also referred to as *quantization levels*, with $1 \leq i \leq M$ and where x_i denotes $M+1$ possible *decision levels* with $0 \leq i \leq M$. Therefore, if $x_{i-1} < x[n] \leq x_i$, then $x[n]$ is quantized (or mapped) to the reconstruction level $\hat{x}_i$. As above, we denote the quantized $x[n]$ by $\hat{x}[n]$. We call this *scalar quantization* because each value (i.e., scalar) of the sequence is individually quantized, in contrast to *vector quantization* where a group of values of $x[n]$ are coded as a vector. Scalar quantization is best illustrated through an example.

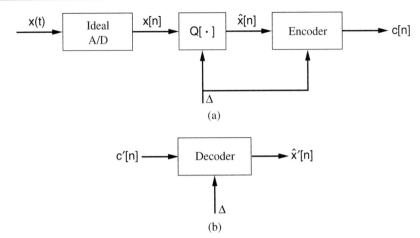

Figure 12.3 Waveform (a) coding and (b) decoding. Coding involves quantizing $x[n]$ to obtain a sequence of samples $\hat{x}[n]$ and encoding them to codewords $c[n]$. Decoding takes a sequence of codewords $c'[n]$ back to a sequence of quantized samples $\hat{x}'[n]$. $c'[n]$ denotes the codeword $c[n]$ that may be distorted by a channel and Δ is the quantization step size.

SOURCE: L.R. Rabiner and R.W. Schafer, *Digital Processing of Speech Signals* [71]. ©1978, Pearson Education, Inc. Used by permission.

EXAMPLE 12.1   Suppose that the number of reconstruction levels $M = 4$ and assume that the amplitude of the input $x[n]$ falls in the range $[0, 1]$. In addition, assume that the decision and reconstruction levels are equally spaced. Then the decision levels are $[0, \frac{1}{4}, \frac{1}{2}, \frac{3}{4}, 1]$, as shown in Figure 12.4. There are many possibilities for equally spaced reconstruction levels, one of which is given in Figure 12.4 as $[0, \frac{1}{8}, \frac{3}{8}, \frac{5}{8}, \frac{7}{8}]$.   ▲

Example 12.1 illustrates the notion of *uniform quantization*. Simply stated, a uniform quantizer is one whose decision and reconstruction levels are uniformly spaced. Specifically, we write the uniform quantizer as

$$x_i - x_{i-1} = \Delta, \quad 1 \le i \le M$$
$$\hat{x}_i = \frac{x_i + x_{i-1}}{2}, \quad 1 \le i \le M \tag{12.1}$$

where Δ is the step size equal to the spacing between two consecutive decision levels, which is the same as the spacing between two consecutive reconstruction levels (Exercise 12.1).

Having selected the reconstruction levels, we then attach to each reconstruction level a symbol, i.e., the codeword. The collection of codewords is called the *codebook*. In most cases, it is convenient to use binary numbers to represent the quantized samples. Figure 12.4 illustrates the codeword assignment for the 4-level uniform quantizer of Example 12.1, given by binary codewords $[00, 01, 10, 11]$. We think of this set as a 2-bit binary codebook because for each of the two places in the binary number we can select a 0 or 1, so that there exists $2^2 = 4$ possible binary numbers. More generally, it is possible with a B-bit binary codebook to represent 2^B different quantization (or reconstruction) levels; that is, B bits gives 2^B quantization levels. *Bit rate* in a coded representation of a signal is defined as the number of bits B per sample multiplied

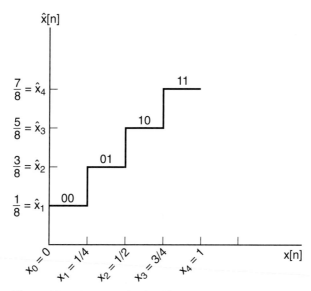

Figure 12.4 An example of uniform 2-bit quantization where the reconstruction and decision levels are uniformly spaced. The number of reconstruction levels $M = 4$ and the input falls in the range $[0, 1]$.

by the number of samples per second f_s, i.e.,

$$I = Bf_s$$

which is sometime referred as the "information capacity" required to transmit or store the digital representation. Finally, the decoder inverts the coder operation, taking the codeword back to a quantized amplitude value, e.g., $01 \rightarrow \hat{x}_2$. Often the goal in speech coding/decoding is to maintain the bit rate as low as possible while maintaining a required level of quality. Because the speech sampling rate is fixed in most applications, this goal implies that the bit rate be reduced by decreasing the number of bits per sample.

In designing the decision regions for a scalar quantizer, we must consider the maximum value of the sequence. In particular, it is typical to assume that the range of the speech signal is proportional to the standard deviation of the signal. Specifically, in designing decision regions, we often assume that $-4\sigma_x \leq x[n] \leq 4\sigma_x$, where σ_x is the signal's standard deviation. There will, however, occur values outside of this region. For example, if σ_x is derived under the assumption that speech values are characterized by a Laplacian pdf, then approximately 0.35% of speech samples fall outside of the range $-4\sigma_x \leq x[n] \leq 4\sigma_x$. Nevertheless, under this assumed range we select parameters for the uniform quantizer. If we want to use a B-bit binary codebook, then we want the number of quantization (reconstruction) levels to be 2^B. Denoting the (almost) maximum signal value $x_{\max} = 4\sigma_x$, then with a uniform quantization step size by Δ, we have

$$\frac{2x_{\max}}{\Delta} = 2^B$$

or

$$2x_{\max} = \Delta 2^B$$

so that the quantization step size is given by

$$\Delta = \frac{2x_{\max}}{2^B}. \tag{12.2}$$

The size of Δ is related to the notion of *quantization noise*.

12.3.2 Quantization Noise

There are two classes of quantization noise, also referred to as quantization error. The first is *granular distortion*. Let

$$\hat{x}[n] = x[n] + e[n]$$

where $x[n]$ is the unquantized signal and $e[n]$ is the quantization noise. Consider a quantization step size Δ. For this step size, the magnitude of the quantization noise $e[n]$ can be no greater than $\frac{\Delta}{2}$, i.e.,

$$-\frac{\Delta}{2} \leq e[n] \leq \frac{\Delta}{2}.$$

This can be seen by plotting the error $e[n] = \hat{x}[n] - x[n]$, as illustrated in Figure 12.5.

The second form of quantization noise is *overload distortion*. From our maximum-value constraint, $x_{\max} = 4\sigma_x$ (with $-4\sigma_x \leq x \leq 4\sigma_x$). Assuming a Laplacian pdf, we noted that 0.35% of the speech samples fall outside of the range of the quantizer. These *clipped* samples incur a quantization error in excess of $\pm\frac{\Delta}{2}$. Nevertheless, the number of clipped samples is so small that it is common to neglect the infrequent large errors in theoretical calculations.

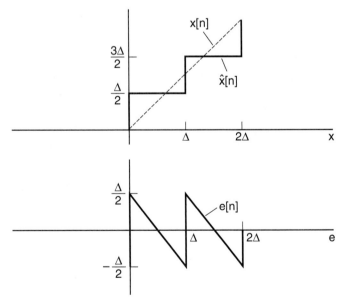

Figure 12.5 Quantization noise for a linearly changing input $x[n] = n$. The quantization noise is given by $e[n] = x[n] - \hat{x}[n]$, where $\hat{x}[n] = Q(x[n])$, with Q the quantization operator. For this case of a linearly-changing input, the quantization noise is seen to be signal-dependent.

In some applications, it is useful to work with a statistical model of quantization noise. In doing so, we assume that the quantization error is an ergodic white-noise random process. The autocorrelation function of such a process is expressed as

$$r_e[m] = E(e[n]e[n + m])$$
$$= \sigma_e^2, \quad m = 0$$
$$= 0, \quad m \neq 0$$

where the operator E denotes expected value. That is, the process is uncorrelated (Appendix 5.A). We also assume that the quantization noise and the input signal are uncorrelated, i.e.,

$$E(x[n]e[n + m]) = 0, \quad \text{for all } m.$$

Finally, we assume that the pdf of the quantization noise is uniform over the quantization interval, i.e.,

$$p_e(e) = \frac{1}{\Delta}, \quad -\frac{\Delta}{2} \leq e \leq \frac{\Delta}{2}$$
$$= 0, \quad \text{otherwise.}$$

We must keep in mind, however, that these assumptions are not always valid. Consider for example a signal that is constant, linearly changing, or, more generally, slowly varying. In particular, when the signal is linearly changing, then $e[n]$ also changes linearly and is signal-dependent, thus violating being uncorrelated with itself as well as with the input. An example of this signal type was illustrated in Figure 12.5. Correlated quantization noise can be aurally quite annoying. On the other hand, our assumptions for the noise being uncorrelated with itself and the signal are roughly valid when the signal fluctuates rapidly among all quantization levels, and when Δ is small. Then the signal traverses many quantization steps over a short-time interval and the quantization error approaches a white-noise process with an impulsive autocorrelation and flat spectrum [33],[71].

It is interesting to observe that one can force $e[n]$ to be white and uncorrelated with $x[n]$ by the deliberate addition of white noise to $x[n]$ prior to quantization. This is called *dithering* or the Roberts' pseudo noise technique [72]. Such decorrelation can be useful in improving the performance of the quantizer and in improving the perceptual quality of the quantization noise. This technique has been used not only with speech and audio, but also with image signals.[1]

In order to quantify the severity of the quantization noise, we define *signal-to-noise ratio* (SNR) by relating the strength of the signal to the strength of the quantization noise. As such, we define SNR as

$$\text{SNR} = \frac{\sigma_x^2}{\sigma_e^2}$$
$$= \frac{E(x^2[n])}{E(e^2[n])} \approx \frac{\frac{1}{N}\sum_{n=0}^{N-1} x^2[n]}{\frac{1}{N}\sum_{n=0}^{N-1} e^2[n]}$$

---

[1] Dithering is used in image coding to reduce the signal-dependent quantization noise that can give an image a "contouring" distortion, which is the result of abrupt changes of output level in a slowly changing gray area of the input scene. The use of dithering breaks up these contours [41].

where the latter estimates, over a time duration N, are based on a zero-mean ergodic assumption. Given our assumed quantizer range $2x_{max}$ and a quantization interval $\Delta = \frac{2x_{max}}{2^B}$ for a B-bit quantizer, and the uniform pdf, it is possible to show that (Exercise 12.2)

$$\sigma_e^2 = \frac{\Delta^2}{12}$$

$$= \frac{(2x_{max}/2^B)^2}{12} = \frac{x_{max}^2}{(3)2^{2B}}. \tag{12.3}$$

We can then express the SNR as

$$SNR = \frac{\sigma_x^2}{\sigma_e^2}$$

$$= \sigma_x^2 \left(\frac{(3)2^{2B}}{x_{max}^2} \right) = \frac{(3)2^{2B}}{(x_{max}/\sigma_x)^2}$$

or in decibels (dB) as

$$SNR(dB) = 10 \log_{10} \left(\frac{\sigma_x^2}{\sigma_e^2} \right)$$

$$= 10 \left(\log_{10} 3 + 2B \log_{10} 2 \right) - 20 \log_{10} \left(\frac{x_{max}}{\sigma_x} \right)$$

$$\approx 6B + 4.77 - 20 \log_{10} \left(\frac{x_{max}}{\sigma_x} \right). \tag{12.4}$$

Because $x_{max} = 4\sigma_x$, then

$$SNR(dB) \approx 6B - 7.2$$

and thus each bit contributes 6 dB to the SNR.

This simple uniform quantization scheme is called *pulse code modulation* (PCM) [33],[71]. Here B bits of information per sample are transmitted as a codeword. The advantages of the scheme are that it is *instantaneous*, i.e., there is no coding delay, and it is not signal-specific, e.g., it does not distinguish between speech and music. A disadvantage is that at least 11 bits are required for "toll quality," i.e., equivalent to typical telephone quality. For a sampling rate of 10000 samples/s, for example, the required bit rate is $B = (11 \text{ bits}) \times (10000 \text{ samples/s}) = 110,000$ bps in transmission systems.

EXAMPLE 12.2   Consider a compact disc (CD) player that uses 16-bit PCM. This gives a SNR $= 96 - 7.2$ dB $= 88.8$ dB for a bit rate of 320,000 bps. This high bit rate is not of concern because space is not a limitation in this medium.   ▲

Although uniform quantization is quite straightforward and appears to be a natural approach, it may not be optimal, i.e., the SNR may not be as small as we could obtain for a certain number of decision and reconstruction levels. To understand this limitation, suppose that the amplitude of $x[n]$ is much more likely to be in one particular region than in another, e.g., low values occurring much more often than high values. This certainly is the case for a speech signal, given the speech pdf of Figure 12.2. Large values do not occur relatively often, corresponding to a very large peak-to-rms ratio of about 15 dB. (Recall the definition of peak-to-rms value in Chapter 9, Section 9.5.2.) Thus it seems that we are not effectively utilizing decision and reconstruction levels with uniform intervals over $\pm x_{max}$. Rather, for a random variable $x[n]$ with such a pdf, intuition tells us to select small intervals where the probability of occurrence is high and select large intervals when the probability of occurrence is low. Quantization in which reconstruction and decision levels do not have equal spacing is called *nonuniform quantization*. A nonuniform quantization that is optimal (in a least-squared-error sense) for a particular pdf is referred to as the *Max quantizer*; an example is illustrated in Figure 12.6 for a Laplacian pdf, where spacing is seen to be finer for low signal values than for high values. Scalar quantization of speech based

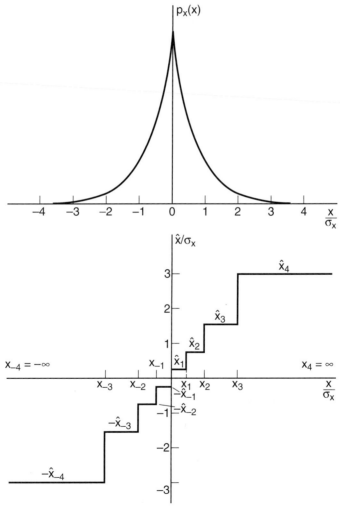

Figure 12.6   3-bit nonuniform quantizer: (a) Laplacian pdf; (b) decision and reconstruction levels.

SOURCE: L.R. Rabiner and R.W. Schafer, *Digital Processing of Speech Signals* [71]. ©1978, Pearson Education, Inc. Used by permission.

on this method can achieve toll quality at 64 kbps and 32 kbps. We study the optimal Max quantizer in the following section.

12.3.3 Derivation of the Max Quantizer

Max in 1960 [48] solved the following problem in scalar quantization:

> For a random variable x with a known pdf, find the set of M quantizer levels (both decision boundaries and reconstruction levels) that minimizes the quantization error variance. In other words, find the decision and boundary levels x_i and $\hat{x}_i$, respectively, that minimize the mean-squared error (MSE) distortion measure
>
> $$D = E[(\hat{x} - x)^2]$$
>
> where E denotes expected value and $\hat{x}$ is the quantized version of x.

To determine the optimal (generally nonuniform) quantizer in this sense for a speech sequence, suppose x is a random variable denoting a value of the sequence $x[n]$ with pdf, $p_x(x)$. For simplicity, we have removed time notation. Then, using the minimum MSE criterion, we determine x_i and $\hat{x}_i$ by minimizing

$$D = E[(\hat{x} - x)^2]$$
$$= \int_{-\infty}^{\infty} p_x(x)(\hat{x} - x)^2. \tag{12.5}$$

Noting that $\hat{x}$ is one of the M reconstruction levels $\hat{x} = Q[x]$, then we can write D as

$$D = \sum_{i=1}^{M} \int_{x_{i-1}}^{x_i} p_x(x)(\hat{x}_i - x)^2 dx$$

where the integral is divided into contributions to D over each decision interval (Figure 12.7). To minimize D, the optimal decision and reconstruction levels must satisfy

$$\frac{\partial D}{\partial \hat{x}_k} = 0, \quad 1 \leq k \leq M \tag{12.6}$$

$$\frac{\partial D}{\partial x_k} = 0, \quad 1 \leq k \leq M - 1 \tag{12.7}$$

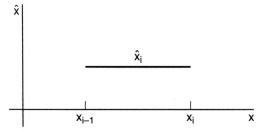

Figure 12.7 Contribution of the MSE over each decision interval in the derivation of the Max quantizer.

where we assume (Figure 12.8)

$$x_0 = -\infty, \quad x_M = +\infty.$$

In order to differentiate in Equations (12.6) and (12.7), we note that there are two contributions to the sum for each k. For Equation (12.7), we thus have

$$\frac{\partial D}{\partial x_k} = \frac{\partial}{\partial x_k} \int_{x_{k-1}}^{x_k} p_x(x)(\hat{x}_k - x)^2 dx$$

$$+ \frac{\partial}{\partial x_k} \int_{x_k}^{x_{k+1}} p_x(x)(\hat{x}_{k+1} - x)^2 dx$$

and flipping the limits on the second term,

$$\frac{\partial D}{\partial x_k} = \frac{\partial}{\partial x_k} \int_{x_{k-1}}^{x_k} p_x(x)(\hat{x}_k - x)^2 dx$$

$$- \frac{\partial}{\partial x_k} \int_{x_{k+1}}^{x_k} p_x(x)(\hat{x}_{k+1} - x)^2 dx.$$

Now, recall the *Fundamental Theory of Calculus*:

$$\frac{d}{dx} \int_{x_o}^{x} g(\sigma) d\sigma = g(x).$$

Therefore,

$$\frac{\partial D}{\partial x_k} = p_x(x_k)(\hat{x}_k - x_k)^2 - p_x(x_k)(\hat{x}_{k+1} - x_k)^2 = 0$$

and thus with some algebraic manipulation, we obtain

$$x_k = \frac{\hat{x}_{k+1} + \hat{x}_k}{2}, \quad 1 \le k \le M - 1. \tag{12.8}$$

The optimal decision level x_k is then the average of the reconstruction levels x_k and $\hat{x}_{k+1}$.

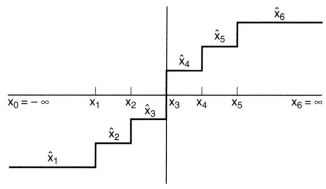

Figure 12.8 Nonuniform quantization example with number of reconstruction levels $M = 6$.

To solve (12.6), we have

$$\frac{\partial D}{\partial \hat{x}_k} = \sum_{i=1}^{M} \frac{\partial}{\partial \hat{x}_k} \int_{x_{i-1}}^{x_i} p_x(x)(\hat{x}_i - x)^2 dx$$

$$= \frac{\partial}{\partial \hat{x}_k} \int_{x_{k-1}}^{x_k} p_x(x)(\hat{x}_k - x)^2 dx$$

$$= \int_{x_{k-1}}^{x_k} p_x(x) \frac{\partial}{\partial \hat{x}_k} (\hat{x}_k - x)^2 dx$$

and setting this result to zero

$$\int_{x_{k-1}}^{x_k} p_x(x) 2(\hat{x}_k - x) dx = 0$$

or

$$\int_{x_{k-1}}^{x_k} p_x(x) \hat{x}_k dx = \int_{x_{k-1}}^{x_k} p_x(x) x dx$$

and thus

$$\hat{x}_k = \frac{\int_{x_{k-1}}^{x_k} p_x(x) x dx}{\int_{x_{k-1}}^{x_k} p_x(x) dx}, \quad 1 \le k \le M.$$

The optimal reconstruction level $\hat{x}_k$ is the *centroid* of $p_x(x)$ over the interval $x_{k-1} \le x \le x_k$ (Figure 12.9). Alternately,

$$\hat{x}_k = \int_{x_{k-1}}^{x_k} \left[\frac{p_x(x)}{\int_{x_{k-1}}^{x_k} p_x(x') dx'} \right] x dx$$

$$= \int_{x_{k-1}}^{x_k} \tilde{p}_x(x) x dx \qquad (12.9)$$

which is interpreted as the mean value of x over interval $x_{k-1} \le x \le x_k$ for the normalized pdf $\tilde{p}(x)$.

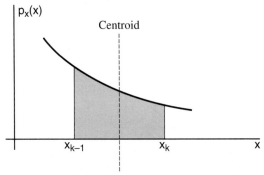

Figure 12.9 Centroid calculation in determining the optimal MSE reconstruction level.

Solving Equations (12.8) and (12.9) simultaneously for x_k an $\hat{x}_k$ is a *nonlinear* problem in these two variables. For a known pdf, Max proposed an iterative technique involving a numerical solution to the required integrations [48]. This technique, however, is cumbersome and, furthermore, requires obtaining the pdf, which can be difficult. An alternative approach proposed by Lloyd [43], not requiring an explicit pdf, is described in a following section where we generalize the optimal scalar quantizer to more than one dimension.

We now look at an example of the Max quantizer designed using the iterative solution as proposed by Max and which compares quantization schemes that result from uniform and Laplacian pdfs [48],[66],[71].

EXAMPLE 12.3   For a uniform pdf, a set of uniform quantization levels result using the Max quantizer, as expected, because all values of the signal are equally likely (Exercise 12.5). Consider, however, a nonuniform Laplacian pdf with mean zero and variance $\sigma^2 = 1$, as illustrated in Figure 12.6a, and let the number of quantization levels $M = 8$. The resulting optimal quantization levels were shown in Figure 12.6b, where we see smaller decision and reconstruction levels for low values of $x[n]$ because with a Laplacian pdf of zero mean, low signal values occur most often.   ▲

For a certain number of bits, the variance of the quantization noise for the uniform quantizer is always greater than the optimal design (or equal if the pdf is uniform). Generally, the more the pdf deviates from being uniform, the higher the gain in this sense from a nonuniform quantizer; the performance gain also increases with an increasing number of bits [33].

In this section, we have designed an optimal quantizer for a given signal. An alternative quantizer design strategy is to achieve the effect of nonuniform quantization by designing the signal to match a uniform quantizer, rather than designing the quantizer to match the signal. Companding is one such approach described in the following section.

12.3.4 Companding

An alternative to the nonuniform quantizer is *companding*, a method that is suggested by the fact that the uniform quantizer is optimal for a uniform pdf. The approach is illustrated in Figure 12.10. In the coding stage, a nonlinearity is applied to the waveform $x[n]$ to form a new sequence $g[n]$ whose pdf is uniform. A uniform quantizer is then applied, giving $\hat{g}[n]$, which at the decoder is inverted with the inverse nonlinearity. One can show that the following nonlinear

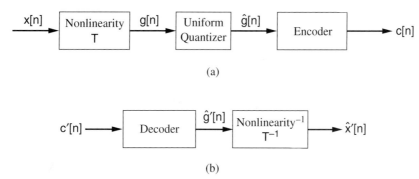

(a)

(b)

Figure 12.10   The method of companding in coding and decoding: (a) coding stage consisting of a nonlinearity followed by uniform quantization and encoding; (b) an inverse nonlinearity occurring after decoding.

transformation T results in the desired uniform density:

$$g[n] = T(x[n]) = \int_{-\infty}^{x[n]} p_x(\beta)d\beta - \frac{1}{2}, \quad \frac{-1}{2} \leq g[n] \leq \frac{1}{2}$$

$$= 0, \qquad\qquad\qquad\qquad \text{elsewhere} \qquad (12.10)$$

which acts like a dynamic range compression, bringing small values up and large values down (Exercise 12.6). This procedure of transforming from a random variable with an arbitrary pdf to one with a uniform pdf is easier to solve than the Max quantizer nonlinear equations; it is not, however, optimal in a mean-squared-error sense [41]. This is because the procedure minimizes the distortion measure

$$D' = E[(\hat{g}[n] - g[n])^2]$$

and does not minimize the optimal criterion

$$D = E[(\hat{x}[n] - x[n])^2].$$

In addition, the method still requires a pdf measurement. Nevertheless, the transformation has the effect of a nonuniform quantizer.

Other nonlinearities approximate the companding operation, but are easier to implement and do not require a pdf measurement. These transformations also give the effect of a nonuniform quantizer, and tend to make the SNR resistant to signal characteristics. The logarithm, for example, makes the SNR approximately independent of signal variance and dependent only upon quantization step size (Exercise 12.6). The μ-law transformation approximates the logarithm operation, but avoids the infinite dynamic range of the logarithm. The μ-law companding, ubiquitous in waveform coding, is given by [71]

$$T(x[n]) = x_{\max} \frac{\log\left(1 + \mu \frac{|x[n]|}{x_{\max}}\right)}{\log(1 + \mu)} \operatorname{sign}(x[n]).$$

Together with uniform quantization, this nonlinearity for large values of μ yields an SNR approximately independent of the ratio of the signal peak $x_{\max}$ and rms σ_x over a large range of signal input, unlike the SNR of the uniform quantizer in Equation (12.4) [71],[80]. An example of the use of the μ-law companding is the international (CCITT) standard coder at 64 kbps [37]. Here, a μ-law transformation is applied followed by 7-bit uniform quantization (some bits remaining for channel bit-error protection), giving toll quality speech. This quality is equivalent to that which would be produced by a uniform 11-bit quantization on the original speech.

12.3.5 Adaptive Quantization

The optimal nonuniform quantizer helps in addressing the problem of large values of a speech signal occurring less often than small values, i.e., it accounts for a nonuniform pdf. We might ask at this point, however, given the time-varying nature of speech, whether it is reasonable to assume a *single* pdf derived from a long-time speech waveform, as we did with a Laplacian or gamma density at the beginning of this chapter. There can be significant changes in the speech waveform as in transitioning from unvoiced to voiced speech. These changes occur in the waveform temporal shape and spectrum, but can also occur in less complex ways such as with slow, as well as rapid, changes in volume. Motivated by these observations, rather than estimate a long-time pdf, we may, on the other hand, choose to estimate a short-time pdf. Pdfs derived over short-time intervals, e.g., 20–40 ms, are also typically single-peaked, as is the gamma or Laplacian pdf, but are more accurately described by a Gaussian pdf regardless of the speech class [33]. A pdf derived from short-time speech segments, therefore, more accurately represents the speech *nonstationarity*, not blending many different sounds. One approach to characterizing the short-time pdf is to assume a pdf of a specific *shape*, in particular a Gaussian, with an unknown variance σ^2. If we can measure the local variance σ^2, then we can *adapt* a nonuniform quantizer to the resulting local pdf. This quantization method is referred to as *adaptive quantization*.

The basic idea of adaptive quantization then is to assume a known pdf, but with unknown variance that varies with time. For a Gaussian, we have the form

$$p_x(x) = \frac{1}{\sqrt{2\pi\sigma_x^2}} e^{\frac{-x^2}{2\sigma_x^2}}.$$

The variance σ_x^2 of values for a sequence $x[n]$ is measured as a function of time, and the resulting pdf is used to design the optimal max quantizer. Observe that a change in the variance simply scales the time signal, i.e., if $E(x^2[n]) = \sigma_x^2$, then $E[(\beta x[n])^2] = \beta^2 \sigma_x^2$, so that one needs to design only one nonuniform quantizer with unity variance and then scale the decision and reconstruction levels according to a particular variance. Alternatively, one can fix the quantizer and apply a time-varying gain to the signal[2] according to the estimated variance, i.e., we scale the

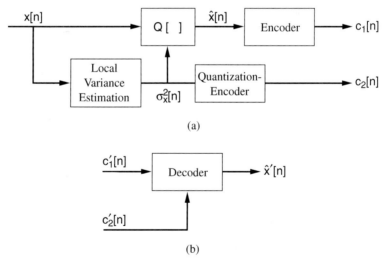

(a)

(b)

Figure 12.11   Adapting a nonuniform quantizer to a local pdf. By measuring the local variance $\sigma_x^2[n]$, we characterize the assumed Gaussian pdf. $c_1[n]$ and $c_2[n]$ are codewords for the quantized signal $\hat{x}[n]$ and time-varying variance $\hat{\sigma}^2[n]$, respectively. This feed-forward structure is one of a number of adaptive quantizers that exploit local variance.

---

[2] Motivation for a time-varying gain is similar to that for applying dynamic range compression prior to audio transmission, as we did in Chapter 9. In coding or in audio transmission, we preprocess the signal so that the SNR is improved in the face of quantization or background noise, respectively.

signal to match the quantizer. When adapting to a changing variance, as shown in Figure 12.11, in a coder/decoder we now need to transmit both the quantizer signal and the variance. This implies, therefore, that the variance itself must be quantized and transmitted; such a parameter is sometimes referred to as *side information* because it is used "on the side" to define the design of the decoder at the receiver. Observe in Figure 12.11 that errors in decoding can now occur due to errors in both the signal codewords $c_1'[n]$ and variance codewords $c_2'[n]$.

Two approaches for estimation of a time-varying variance $\sigma^2[n]$ are based on a feed-forward method (shown in Figure 12.11) where the variance (or gain) estimate is obtained from the input, and a feedback method where the estimate is obtained from a quantizer output. The feedback method has the advantage of not needing to transmit extra side information; however, it has the disadvantage of additional sensitivity to transmission errors in codewords [71]. Because our intent here is not to be exhaustive, we describe the feed-forward approach only, and we refer the reader to [33],[71] for more complete expositions. In one feed-forward estimation method, the variance is assumed proportional to the short-time energy in the signal, i.e.,

$$\sigma_x^2[n] = \beta \sum_{m=-\infty}^{\infty} x^2[n]h[n-m] \tag{12.11}$$

where $h[n]$ is a lowpass filter and β is a constant. The bandwidth of the time-varying variance $\sigma_x^2[n]$ is controlled by the time width of the filter $h[n]$. There is thus a tradeoff between tracking the true variance of a signal and the amenability of the time-varying variance estimate for adaptive quantization (Exercise 12.8). Because $\sigma_x^2[n]$ must also be *sampled* and transmitted, we want to make this slowly varying, i.e., with small bandwidth, but without sacrificing the time resolution of the estimate. In general, adaptive quantizers can have a large advantage over their nonadaptive counterparts. A comparison for a feed-forward quantizer by Noll [60],[61] is given in the following example. This example also illustrates the importance of assuming a Gaussian function for the local pdf, in contrast to a pdf tailored to long-term statistics.

EXAMPLE 12.4   Noll considered a feed-forward scheme in which the variance estimate is [60],[62]

$$\sigma_x^2[n] = \frac{1}{M} \sum_{m=n}^{n+M-1} x^2[m]$$

and where the variance is evaluated and transmitted every M samples. Table 12.1 shows that the adaptive quantizer can achieve as high as 8 dB better SNR for the speech material used by Noll. Both a Gaussian and a Laplacian pdf are compared, as well as a $\mu-law$ companding scheme which provides an additional reference. Observe that the faster smoothing with $M = 128$, although requiring more samples/s, gives a better SNR than the slower smoothing with $M = 1024$. In addition, with $M = 128$ the Gaussian pdf provides a higher SNR than the Laplacian pdf.   ▲

Although we see in the above example that the optimal adaptive quantizer can achieve higher SNR than the use of μ-law companding, μ-law companding is generally preferred for high-rate waveform coding because of its lower background noise when the transmission channel is idle [37]. Nevertheless, the methodology of optimal adaptive quantization is useful in a variety of other coding schemes.

Table 12.1 Comparison of 3-bit adaptive and nonadaptive quantization schemes [60]. Adaptive schemes use feed-forward adaptation.

SOURCE: Table from L.R. Rabiner and R.W. Schafer, *Digital Processing of Speech Signals* [71]. ©1978, Pearson Education, Inc. Used by permission. Data by Noll [60].

Nonuniform Quantizers	*Nonadaptive* SNR *(dB)*	*Adaptive* ($M = 128$) SNR *(dB)*	*Adaptive* ($M = 1024$) SNR *(dB)*
μ-law ($\mu = 100$, $x_{\max} = 8\sigma_x$)	9.5	–	–
Gaussian	7.3	15.0	12.1
Laplacian	9.9	13.3	12.8
Uniform Quantizers			
Gaussian	6.7	14.7	11.3
Laplacian	7.4	13.4	11.5

12.3.6 Differential and Residual Quantization

Up to now we have investigated instantaneous quantization, where *individual* samples are quantized. We have seen, however, that speech is highly correlated both on a short-time basis (e.g., on the order of 10–15 samples) and on a long-time basis (e.g., a pitch period). In this section, we exploit the short-time correlation to improve coding performance; later in this chapter, we will exploit the long-time correlation.

The meaning of short-time correlation is that neighboring samples are "self-similar," not changing too rapidly from one another. The *difference* between adjacent samples should, therefore, have a lower variance than the variance of the signal itself, thus making more effective use of quantization levels. With a lower variance, we can achieve an improved SNR for a fixed number of quantization levels [33],[71]. More generally, we can consider *predicting* the next sample from previous ones and finding the best prediction coefficients to yield a minimum mean-squared prediction error, just as we did in Chapter 5. We can use in the coding scheme a *fixed* prediction filter to reflect the average correlation of a signal, or we can allow the predictor to short-time *adapt* to the signal's local correlation. In the latter case, we need to transmit the quantized prediction coefficients as well as the prediction error. One particular prediction error encoding scheme is illustrated in Figure 12.12 where the following sequences are required:

$\tilde{x}[n]$ = prediction of the input sample $x[n]$; this is the output of the predictor $P(z)$ whose input is a quantized version of the input signal $x[n]$, i.e., $\hat{x}[n]$.

$r[n]$ = prediction error signal.

$\hat{r}[n]$ = quantized prediction error signal.

The prediction error signal $r[n]$ is also referred to as the *residual* and thus this quantization approach is sometimes referred to as *residual coding*. The quantizer can be of any type, e.g., fixed, adaptive, uniform, or nonuniform. In any case, its parameters are adjusted to match the variance of $r[n]$. Observe that this differential quantization approach can be applied not only to the speech signal itself, but also to parameters that represent the speech, e.g., linear prediction coefficients, cepstral coefficients from homomorphic filtering, or sinewave parameters. In fact, our focus in the later part of this chapter will be on quantization of such representations.

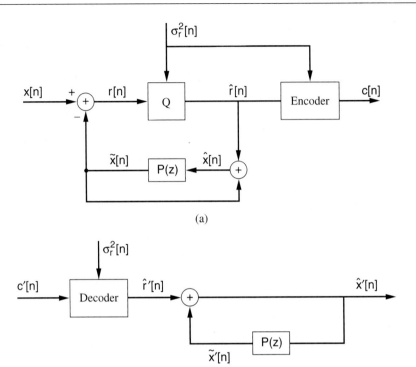

(a)

(b)

Figure 12.12 Differential coding (a) and decoding (b) schemes. The coefficients of the predictor $P(z)$, as well as the variance of the residual $r[n]$, must be quantized, coded, and decoded (not shown in the figure).

SOURCE: L.R. Rabiner and R.W. Schafer, *Digital Processing of Speech Signals* [71]. ©1978, Pearson Education, Inc. Used by permission.

In the differential quantization scheme of Figure 12.12, we are interested in properties of the quantized residual and, in particular, its quantization error. As such, we write the quantized residual as

$$\hat{r}[n] = r[n] + e[n]$$

where $e[n]$ is the quantization error. Figure 12.12 shows that the quantized residual $\hat{r}[n]$ is added to the predicted value $\tilde{x}[n]$ to give the quantized input $\hat{x}[n]$, i.e.,

$$\hat{x}[n] = \tilde{x}[n] + \hat{r}[n]$$
$$= \tilde{x}[n] + r[n] + e[n]$$
$$= \tilde{x}[n] + x[n] - \tilde{x}[n] + e[n]$$
$$= x[n] + e[n].$$

Therefore, the quantized signal samples differ from the input only by the quantization error $e[n]$, which is the quantization error of the residual. If the prediction of the signal is accurate, the variance of $r[n]$ will be smaller than the variance of $x[n]$ so that a quantizer with a given number of levels can be adjusted to give a smaller quantization error than would be possible when quantizing the signal directly [71]. Observe that the procedure for reconstructing the quantized input from the codewords associated with the quantized residual $\hat{r}[n]$ is *implicit in the encoding scheme*, i.e., if $c'[n] = c[n]$, i.e., no channel errors occur, then the quantized signal at the encoder is simply $\hat{x}'[n] = \hat{x}[n]$, as illustrated in Figure 12.12.

The differential coder of Figure 12.12, when using a fixed predictor and fixed quantization, is referred to as *differential PCM* (DPCM). Although this scheme can improve SNR, i.e., improve the variance in the signal $x[n]$ relative to the variance in quantization noise $e[n]$ over what we achieve by a direct quantization scheme, its improvement is not dramatic [71]. Rather, DPCM with both adaptive prediction (i.e., adapting the predictor to the local correlation), and adaptive quantization (i.e., adapting the quantizer to the local variance of $r[n]$), referred to as ADPCM, yields the greatest gains in SNR for a fixed bit rate. The international coding standard, CCITT G.721, with toll quality speech at 32 kbps (8000 samples/s $\times$ 4 bits/sample) has been designed based on ADPCM techniques. Both the predictor parameters and the quantizer step size are calculated using only the coded quantizer output (feedback adaptation), which means that there is no need to transmit this information separately [7]. The interested reader is referred to [33],[37],[71] for a thorough comparison of DPCM and ADPCM approaches, in addition to other classic differential waveform coding schemes such as delta modulation and continuous variable slope delta modulation where the sampling rate is increased to many times the Nyquist rate, thus increasing sample-to-sample correlation, and exploiting one-bit quantization. These methods are also used for high bit-rate waveform coding. To achieve lower rates with high quality requires further dependence on speech model-based techniques and the exploiting of long-time prediction, as well as short-time prediction, to be described later in this chapter.

Before closing this section, we describe an important variation of the differential quantization scheme of Figure 12.12. Observe that our prediction has assumed an all-pole, or autoregressive, model. Because in this model a signal value is predicted from its past samples, any error in a codeword (due, for example, to bit errors over a degrading channel) propagate over considerable time during decoding. Such error propagation is particularly severe when the signal values represent speech model parameters computed frame-by-frame, rather than sample-by-sample. An alternative is a finite-order *moving-average* predictor derived from the residual. One common approach to the use of a moving-average predictor is illustrated in Figure 12.13 [75]. At the coder stage, for a predictor parameter time sequence $a[n]$, we write the residual as the difference of the true value and the value predicted from the moving average of K quantized residuals, i.e.,

$$r[n] = a[n] - \sum_{k=1}^{K} p[k]\hat{r}[n - k]$$

where $p[k]$ represents the coefficients of $P(z)$. Then the predicted value at the decoder, also embedded within the coder, is given by

$$\hat{a}[n] = \hat{r}[n] + \sum_{k=1}^{K} p[k]\hat{r}[n - k]$$

for which one can see that an error propagation is limited to only K samples (or K analysis frames for the case of model parameters).

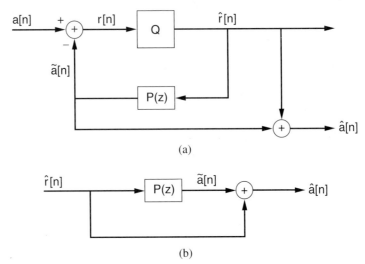

Figure 12.13 Differential coding (a) and decoding (b) schemes with moving average predictor.

12.4 Vector Quantization (VQ)

We have in the previous section investigated *scalar quantization*, which is the basis for high-rate waveform coders. In this section, we describe a generalization of scalar quantization, referred to as *vector quantization*, in which a block of scalars are coded as a vector, rather than individually. As with scalar quantization, an optimal quantization strategy can be derived based on a mean-squared error distortion metric. Vector quantization is essential in achieving low bit rates in model-based and hybrid coders.

12.4.1 Approach

We motivate vector quantization with a simple example in which the vocal tract transfer function is characterized by only two poles, thus requiring four reflection coefficients. Furthermore, suppose that the vocal tract can take on only one of four possible shapes. This implies that there exist only four possible sets of the four reflection coefficients, as illustrated in Figure 12.14. Now consider scalar quantizing each of the reflection coefficients individually. Because each reflection coefficient can take on 4 different values, 2 bits are required to encode each coefficient. Since there are 4 reflection coefficients, we need $2 \times 4 = 8$ bits per analysis frame to code the vocal tract transfer function. On the other hand, we know that there are only four possible positions of the vocal tract corresponding to *only four possible vectors* of reflection coefficients. The scalar values of each vector are, therefore, highly correlated in the sense that a particular value of one reflection coefficient allows a limited choice of the remaining coefficients. In this case, we need only 2 bits to encode the 4 reflection coefficients. If the scalars were independent of each other, treating them as a vector would have no advantage over treating them individually. In vector quantization, we are exploiting correlation in the data to reduce the required bit rate.

We can think of vector quantization (VQ) as first grouping scalars into blocks, viewing each block as a unit, and then quantizing each unit. This scheme is, therefore, also sometimes called *block quantization*. We now look at VQ more formally. Consider a vector of N continuous

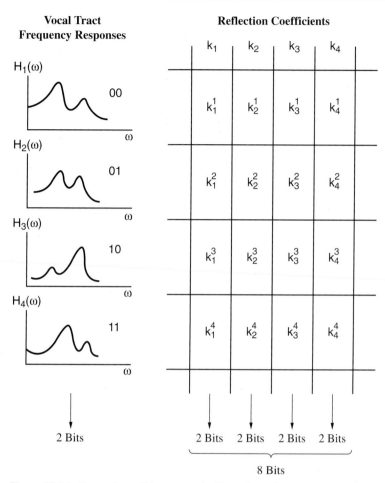

Figure 12.14 Comparison of bit rates required by scalar and vector representations of reflection coefficients for four two-pole vocal tract frequency responses. The four required reflection coefficients are highly correlated, the vocal tract being limited to only four configurations.

scalars

$$\underline{x} = [x^1, x^2, x^3, \ldots x^N]^T$$

where T denotes matrix transpose. With quantization, the vector $\underline{x}$ is mapped to another N-dimensional vector

$$\underline{\hat{x}} = [\hat{x}^1, \hat{x}^2, \hat{x}^3, \ldots \hat{x}^N]^T.$$

The vector $\underline{\hat{x}}$ is chosen from M possible reconstruction (quantization) levels, thus representing a generalization of the scalar case, i.e.,

$$\underline{\hat{x}} = VQ[\underline{x}] = \underline{r}_i, \qquad \text{for} \quad \underline{x} \in C_i$$

Max quantizer (1–D)

$$\hat{x} = Q\,[x]$$

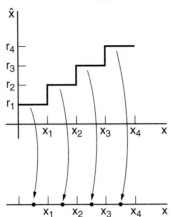

Vector quantizer (2–D)

$$\underline{\hat{x}} = VQ\,[\underline{x}]$$

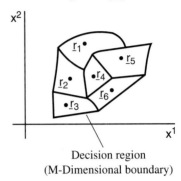

Decision region
(M-Dimensional boundary)

• = Centroid over the decision interval   • = Centroid over the decision region

$$D = E\,[(\hat{x} - x)^2]$$ $$D = E\,[(\underline{\hat{x}} - \underline{x})^2(\underline{\hat{x}} - \underline{x})]$$

Figure 12.15   Comparison of scalar and vector quantization.

where

$$VQ \; = \; \text{vector quantization operator,}$$

$$\underline{r}_i \; = \; M \text{ possible reconstruction levels for } 1 \; \leq \; i \; < \; M,$$

$$C_i \; = \; i\text{th "cell" or cell boundary,}$$

and where if $\underline{x}$ is in the cell C_i, then $\underline{x}$ is mapped to $\underline{r}_i$. As in the scalar case, we call $\underline{r}_i$ the *codeword* and the complete set of codewords $\{\underline{r}_i\}$ the *codebook*. An example of one particular cell configuration is given in Figure 12.15 in which the order of the vector $N = 2$ and the number of reconstruction levels is $M = 6$. The dots represent the reconstruction levels and the solid lines the cell boundaries. Also in Figure 12.15, we summarize the comparison between scalar and vector quantization. In this figure, we see the following properties that we elaborate on shortly:

P1: In vector quantization a cell can have an arbitrary size and shape. By contrast, in scalar quantization a "cell" (region between two decision levels) can have an arbitrary size, but its shape is fixed.

P2: As in the scalar case, we define a distortion measure $D(\underline{x}, \underline{\hat{x}})$, which is a measure of dissimilarity or error between $\underline{x}$ and $\underline{\hat{x}}$.

12.4.2 VQ Distortion Measure

Vector quantization noise is represented by the vector $\underline{e} = \underline{\hat{x}} - \underline{x}$ so that the distortion is the average of the sum of squares of scalar components given by $D = E[\underline{e}^T \underline{e}]$ where E denotes expected value. We then can write for a multi-dimensional pdf $p_{\underline{x}}(\underline{x})$

$$
\begin{aligned}
D &= E[(\hat{\underline{x}} - \underline{x})^T (\hat{\underline{x}} - \underline{x})] \\
&= \int_{-\infty}^{\infty} \int_{-\infty}^{\infty} \cdots \int_{-\infty}^{\infty} (\hat{\underline{x}} - \underline{x})^T (\hat{\underline{x}} - \underline{x}) p_{\underline{x}}(\underline{x}) d\underline{x} \\
&= \sum_{i=1}^{M} \int \int_{\underline{x} \in C_i} \cdots \int (\underline{r}_i - \underline{x})^T (\underline{r}_i - \underline{x}) p_{\underline{x}}(\underline{x}) d\underline{x}
\end{aligned}
\tag{12.12}
$$

where C_i are the cell boundaries. We can think of this distortion as a generalization of the 1-D (scalar) case in Equation (12.5). Note the change of notation for the reconstruction levels from the formulation of the 1-D case: $\hat{x}_i \rightarrow r_i$. Our next goal is to minimize $D = E[(\hat{\underline{x}} - \underline{x})^T (\hat{\underline{x}} - \underline{x})]$ with respect to unknown reconstruction levels $\underline{r}_i$ and cell boundaries C_i (similar to the unknown decision levels for the 1-D scalar case). The solution to this optimization problem is nonlinear in the unknown reconstruction levels and cell boundaries. Following an approach similar to that used in deriving the Max quantizer for scalars results in two necessary constraints for optimization. These two constraints have been formulated by Lim as [41]:

C1: A vector $\underline{x}$ must be quantized to a reconstruction level $\underline{r}_i$ that gives the *smallest distortion* between $\underline{x}$ and $\underline{r}_i$. This is because each of the elements of the sum in Equations (12.5) and (12.12) must be minimized.

C2: Each reconstruction level $\underline{r}_i$ must be the *centroid* of the corresponding decision region, i.e., of the cell C_i.

Observe that the first condition implies that, having the reconstruction levels, we can vector-quantize without explicit need of the cell boundaries; i.e., to quantize a given vector, we find the reconstruction level that minimizes its distortion. This requires a very large search, but in theory the search is possible. The second condition gives a way to determine a reconstruction level from a cell boundary.

The two conditions motivate the following iterative solution [41]. Assume that we are given an initial estimate of $\underline{r}_i$. Then from the first condition, we can determine all vectors from an ensemble that quantize to each $\underline{r}_i$, and thus it follows that we have the corresponding cell boundaries for the ensemble. Having the cell boundaries, we use the second condition to obtain a new estimate of the reconstruction levels, i.e., the centroid of each cell. We continue the iteration until the change in the reconstruction levels is small. A problem with this algorithm is the requirement of using an ensemble of all vectors $\underline{x}$ and their corresponding joint pdfs, the latter required in the determination of the distortion measure and the multi-dimensional centroid. An alternative algorithm proposed by Lloyd for 1-D [43] and extended to multi-D by Forgy [18] uses a finite set of input vectors as *training vectors* to replace difficult-to-obtain joint pdfs and the ensemble requirement. The algorithm is referred to as the *k-means algorithm*, where k is the number of reconstruction levels and (nearly) minimizes D. The k-means algorithm is given by the following steps:

S1: Replace the ensemble average D with an average denoted by $D' = \frac{1}{N} \sum_{k=0}^{N-1} (\hat{\underline{x}}_k - \underline{x}_k)^T (\hat{\underline{x}}_k - \underline{x}_k)$, where $\underline{x}_k$ are the training vectors and $\hat{\underline{x}}_k$ are the quantized vectors that are functions of cell boundaries and reconstruction levels. (We can think of the unknown pdf as having been replaced by the set of training vectors.) Our new goal is to minimize D' with respect to cell boundaries and reconstruction levels.

(a)   (b)

Figure 12.16 Two steps in the k-means algorithm: (a) cluster formation; (b) mean (centroid) formation.

S2: Pick an initial guess at the reconstruction levels $\{\underline{r}_i\}$.

S3: For each $\underline{x}_k$ select an $\underline{r}_i$ closest to $\underline{x}_k$. The set of all $\underline{x}_k$ nearest $\underline{r}_i$ forms a *cluster* (Figure 12.16a), which is why this iteration is sometimes referred to as a "clustering algorithm."

S4: Find the mean of $\underline{x}_k$ in each cluster which gives a new $\underline{r}_i$, approximating a centroid of the cluster from **S3**, denoted by $\underline{r}_i'$ (Figure 12.16b). Calculate D'.

S5: Stop when the change in D' over two consecutive iterations is insignificant.

The iteration can be shown to converge to a local minimum of D'. Note that minimizing D' does not necessarily minimize D, but it is close. Because a local minimum may be found, we can try a number of different initial conditions and pick the solution with the smallest mean-squared error.

An advantage of VQ over scalar quantization is that it can reduce the number of reconstruction levels when the distortion D' is held constant, and thus lower the number of required bits (Exercise 12.10). Likewise, when the number of reconstruction levels is held fixed, it can lower the distortion D'. As indicated, these advantages are obtained by exploiting the correlation (i.e., linear dependence) among scalars of the vector.[3] Therefore, any transformation[4] of the vector that reduces this correlation also reduces the advantage of vector quantization [41]. Nevertheless, there are conditions under which VQ has an advantage even without correlation across vector elements because VQ gives more flexibility than scalar quantization in partitioning an M-dimensional space [44]; this advantage is clarified and illustrated later in this chapter in the description of code-excited linear prediction coding. A quantitative discussion of these VQ properties is beyond our scope here; the interested reader can find further tutorials and examples in [20],[41],[44].

12.4.3 Use of VQ in Speech Transmission

A generic data transmitter/receiver based on VQ is described as follows. At the transmitter, the first step is to generate the codebook of vector codewords. To do so, we begin with a set of training vectors, and then use the k-means algorithm to obtain a set of reconstruction levels $\underline{r}_i$ and possibly cell boundaries C_i. Attached to each reconstruction level is a codeword with an associated index in a codebook. Our second step is to quantize the vectors to be transmitted. As mentioned above, we can vector quantize without explicit need of the cell boundaries, i.e., to quantize a given vector, we find the reconstruction level that minimizes its distortion. Therefore,

---

[3] Scalars may be linearly independent yet statistically dependent, corresponding to *nonlinear dependence* between scalars. VQ can also exploit such nonlinear dependence [41],[44].

[4] For example, the elements of a waveform vector may be linearly dependent, but its DFT or cepstral coefficients may be uncorrelated.

we do not need to store the cell boundaries. We then transmit the selected codeword. At the receiver, the codeword is decoded in terms of a codebook index and a table lookup procedure is used to extract the desired reconstruction level. This VQ strategy has been particularly useful in achieving high-quality coded speech at middle and low bit rates, as we will see in the following sections.

12.5 Frequency-Domain Coding

We have seen that with basic PCM waveform coding and adaptive and differential refinements of PCM, we can obtain good quality at about 32 kbps for telephone-bandwidth speech. These are purely time-domain approaches. To attain lower rates, as required, for example, over low-bandwidth channels, we take a different strategy. One approach is to exploit the frequency-domain structure of the signal using the short-time Fourier transform (STFT). From Chapter 7, we saw that the STFT can be interpreted by either a filter-bank or Fourier-transform perspective. In this section, we first describe speech coding techniques based, in particular, on the filter-bank view and using the scalar quantization schemes of the previous section. This strategy is called *subband coding* and leads naturally into *sinusoidal coding* based on sinewave analysis/synthesis and its multi-band variations of Chapter 9; these methods exploit both scalar and vector quantization. Sinusoidal coders are *hybrid* coders in the sense of imposing speech model-based structure on the spectrum, in contrast to subband coders.

12.5.1 Subband Coding

Subband coding is based on the generalized filter-bank summation method introduced in Chapter 7, which is essentially the FBS method discretized in time. In this analysis/synthesis method, to reduce the number of samples used in coding, the output of the kth complex[5] analysis filter, $h_k[n] = w[n]e^{j2\pi nk/N}$, is first decimated by the time-decimation factor L and modulated down to DC by $e^{-j2\pi nk/N}$ to form the subband sequence $X(nL, k)$. At the receiving end of the coder, the decimated filter outputs are interpolated by the synthesis filter $f[n]$ and the resulting interpolated sequences are modulated and summed, thus giving Equation (7.17), i.e.,

$$y[n] = \frac{L}{N} \sum_{r=-\infty}^{\infty} \sum_{k=0}^{N-1} f[n - rL]X(rL, k)e^{-j\frac{2\pi}{N}kn}.$$

We saw in Chapter 7 that the condition for perfect reconstruction of $x[n]$ is given by Equation (7.18). In coding, the time-decimated subband outputs are quantized and encoded, then are decoded at the receiver.

In subband coding, a small number of filters (≤ 16) with wide and overlapping bandwidths[6] are chosen and each output is quantized using scalar quantization techniques described in the previous section. Typically, each bandpass filter output is quantized individually. The subband coder derives its advantage over (fullband) waveform coding by limiting the quantization noise from the coding/decoding operation largely to the band in which it is generated

---

[5] In practice, the subbands are implemented as a lowpass translation of a frequency band to DC in a manner similar to single-sideband modulation, giving real signals rather than the complex signals [81].

[6] In contrast, ideal and nonoverlapping bandpass filters imply a long time-domain filter impulse response and thus difficulty in achieving the perfect reconstruction filter-bank property.

and by taking advantage of known properties of aural perception. To satisfy the first condition, although the bandpass filters are wide and overlapping, careful design of the filters results in a *cancellation* of quantization noise that leaks across bands, thus keeping the bandpass quantization noise independent across channels from the perspective of the synthesis. *Quadrature mirror filters* are one such filter class; for speech coding, these filters have the additional important property of, under certain conditions, canceling aliasing at the output of each decimated filter when the Nyquist criterion is not satisfied [81]. Figure 12.17 shows an example of a two-band subband coder using two overlapping quadrature mirror filters [81]. Quadrature mirror filters can be further subdivided from high to low filters by splitting the fullband into two (as in Figure 12.17), then the resulting lower band into two, and so on. Because of the octave-band nature of the resulting bandpass filters, repeated decimation is invoked. This octave-band splitting, together with the iterative decimation, can be shown to yield a perfect reconstruction filter bank (i.e., its output equals its input) [77]. In fact, such octave-band filter banks, and their conditions for perfect reconstruction, are closely related to wavelet analysis/synthesis structures and their associated invertibility condition as described in Chapter 8 [12],[77].

We also saw in Chapter 8 that the octave-band filters, having a constant-Q property, provide an auditory front-end-like analysis. In the context of speech coding, we are interested in the effect of quantization noise in each auditory-like subband on speech perception. In particular, it is known that the SNR in each subband is correlated with speech intelligibility, and that for high intelligibility, as evaluated in speech-perception tests, it is desirable to maintain a constant SNR

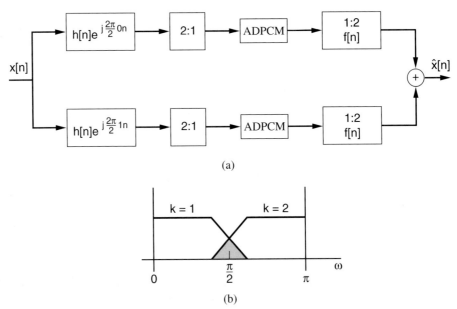

(a)

(b)

Figure 12.17 Example of a two-band subband coder using two overlapping quadrature mirror filters and ADPCM coding of the filter outputs: (a) complete analysis, coding/decoding, and synthesis configuration; (b) overlapping quadrature mirror filters.

SOURCE: J.M. Tribolet and R.E. Crochiere, "Frequency Domain Coding of Speech" [81]. ©1979, IEEE. Used by permission.

across the near constant-Q subbands that span the speech spectrum[7] [39]. In addition, when the SNR in a subband is high enough, then the quantization noise in a subband can be perceptually *masked*, or "hidden," by the signal in that subband.[8]

As in fullband waveform coding, the variance of each subband signal, denoted by $\sigma_x^2[nL, k]$ for the kth subband on the nth frame, is determined and used to control the decision and reconstruction levels of the quantization of each subband signal.[9] For example, the adaptive quantization scheme of Section 12.3.5 can be used, where the subband variance estimate controls the quantizer step size in each band. Given that we have a fixed number of bits, one must decide on the bit allocation across the bands. Suppose that each band is given the same number of bits. Then, bands with lower signal variance have smaller step sizes and contribute less quantization noise. Subbands with larger signal variance have larger step sizes and, therefore, contribute more quantizing noise. The quantization noise, by following the input speech spectrum energy, results in a constant SNR across subbands.

It is of interest to ask whether this uniform bit allocation is "optimal" according to minimizing an average measure of quantization noise by allowing the number of bits assigned to each frequency band to vary according to the time-varying variance of each band. Suppose we are given a total of B bits for N channels; then

$$B = \sum_{k=0}^{N-1} b(nL, k) \tag{12.13}$$

where $b(nL, k)$ denotes the number of bits for the kth subband at time nL. In order to determine an optimal bit assignment, we first define the quantization noise over all bands as

$$D = \frac{1}{N} \sum_{k=0}^{N-1} E\{[\hat{X}(nL, k) - X(nL, k)]^2\}$$

$$= \frac{1}{N} \sum_{k=0}^{N-1} \sigma_e^2(nL, k) \tag{12.14}$$

where $\hat{X}(nL, k)$ is the quantized version of the STFT $X(nL, k)$ and where $\sigma_e^2(nL, k)$ denotes the variance of the quantization noise on the kth channel at time nL. Suppose, for example, that $x[n]$ is a Gaussian random process. Then each subband signal has a Gaussian pdf because a linear transformation of a Gaussian random process remains a Gaussian random process [63]. Then it can be shown (Exercise 12.11) that the minimum-mean-squared error results in the

---

[7] As mentioned in this chapter's introduction, speech intelligibility is sometimes represented by the *articulation index*, which uses a weighted sum of the SNR in each auditory-like subband [39].

[8] Other principles of auditory masking that involve hiding of small spectral components by large *adjacent* components have also been exploited in speech coding, as well as in speech enhancement. We study one such approach in Chapter 13.

[9] The variance of each subband signal can be obtained by local averaging of the spectrum in frequency, and thus it increases with increasing spectral energy [81]. This variance is sent as side information by the coder.

optimal bit assignment given approximately by [14],[27]

$$b(nL, k) \approx \frac{1}{2} \log_2 \left[\frac{\sigma_x^2(nL, k)}{D^*} \right] \qquad (12.15)$$

where $\sigma_x^2(nL, k)$ are the variances of the channel values and where D^* denotes the minimum-mean-squared error. The number of bits $b(nL, k)$, therefore, increases with increasing variance. It can be shown that this bit assignment rule, based on the minimum-mean-squared error criterion, leads to a flat noise variance (power spectrum) in frequency, i.e., $\sigma_e^2(nL, k) = D^*$ (Exercise 12.11). A flat noise variance, however, is not the most desirable because the SNR decreases with decreasing subband signal energy and thus a loss in intelligibility is incurred due to subbands with relatively low energy (assuming the importance of equal SNR in all bands).

We can modify the shape of the quantization noise variance across subbands by allowing a positive weighting factor $Q(k)$ that weights the importance of the noise in different frequency bands. The new distortion measure becomes

$$D = \frac{1}{N} \sum_{k=0}^{N-1} E\{Q(k)[\hat{X}(nL, k) - X(nL, k)]^2\}$$

$$= \frac{1}{N} \sum_{k=0}^{N-1} Q(k)\sigma_e^2(nL, k) \qquad (12.16)$$

which, when minimized, gives the resulting noise spectrum

$$\sigma_e^2(nL, k) \approx \alpha \frac{D^*}{Q(k)}$$

where α is a constant. The quantization noise variance, therefore, is inversely proportional to the weighting. The corresponding optimal bit assignment is given by (Exercise 12.11).

$$b(nL, k) \approx \frac{1}{2} \log_2 \left[\frac{Q(k)\sigma_x^2(nL, k)}{D^*} \right]. \qquad (12.17)$$

When the weighting function is given by the reciprocal of the subband signal variance, i.e., $Q(k) = \sigma_x^{-2}(nL, k)$, giving less weight to high-energy regions, then the optimal bit assignment is constant. In this case, the quantization noise spectrum follows the signal spectrum (with more error in the high-energy regions) and the SNR is constant as a function of frequency. Thus the uniform bit assignment that we had described previously is not only perceptually desirable, but also approximately satisfies an optimal error criterion with a reasonable perceptual weighting. This subband coding strategy using ADPCM techniques, along with variations of these approaches, has been applied to speech with good quality in the range of 12–16 kbps [10],[11].

The coding strategies of this section have also been applied to the STFT viewed from the Fourier transform perspective [81]. Typically, the number of frequency channels (e.g., transform sizes of 64 to 512) used in this approach is much greater than in subband coding in order to capitalize on the spectral harmonic structure of the signal, as well as the general spectral formant shape. This technique, referred to as *adaptive transform coding*, can be thought of as narrowband analysis/synthesis, in contrast to subband coding (with far fewer and wider bands) classified as wideband analysis/synthesis, analogous to the difference between narrowband and wideband spectrogram analysis.

Although subband and transform coding methods can attain good quality speech in the range of 12–16 kbps, they are hard-pressed to attain such quality at lower bit rates. One approach to attaining these lower bit rates with frequency-domain approaches is to incorporate more speech-specific knowledge, as we did in the framework of the phase vocoder (which can also be thought of as falling in the class of subband coders). One of the earliest such coders, even prior to the introduction of the phase vocoder, was the *channel vocoder* first introduced by Dudley [15],[23] and extended by Gold and Radar [22]. This coder uses a representation of the speech source, as well as a subband spectral representation (Exercise 12.14). In the remainder of this chapter, we investigate two current approaches to explicitly using knowledge of the speech production mechanism in the frequency domain. These coders, based on a sinewave representation of speech, are capable of attaining high quality at lower bit rates by speech-dependent frequency-domain modeling.

12.5.2 Sinusoidal Coding

In Chapter 10, it was shown that synthetic speech of high quality could be synthesized using a harmonic set of sinewaves, provided the amplitudes and phases are the harmonic samples of a magnitude and phase function derived from the short-time Fourier transform at the sinewave frequencies. Specifically, the harmonic sinewave synthesis is of the form

$$\hat{s}[n; \omega_o] = \sum_{k=1}^{K} \bar{A}(k\omega_o) \exp[j(nk\omega_o + \phi_k)] \tag{12.18}$$

where ω_o is the fundamental frequency (pitch) estimate [but where ω_o is fixed above an adaptive voicing frequency cutoff, as in Equation (10.31)], where $\bar{A}(\omega_k)$ are samples of the vocal tract piecewise-linear spectral envelope, derived from the SEEVOC estimation strategy of Chapter 10, and where ϕ_k is the phase obtained by sampling a piecewise-flat phase function derived from the measured STFT phase, using the same strategy to generate the SEEVOC envelope. (Recall from Chapter 10 that use of a piecewise-flat phase, in contrast to a piecewise-linear function, avoids the difficult problem of phase interpolation across frequency.)

In the context of speech coding, even though the harmonic synthesis eliminates the need to code the sinewave frequencies, the amplitudes and phases must be quantized, and, in general, there remain too many parameters to encode and achieve operation at low bit rates. A further reduction of required sinewave parameters was obtained in Chapter 10 for voiced speech by using speech production to model the measured phase samples. It was shown that, using the properties of the speech production mechanism, a minimum-phase model of the vocal tract and an onset model of the source allows a simplified representation during voicing which was said to provide a reduced parameter set for coding. In this section, this model for the glottal excitation and vocal tract transfer function is generalized to both voiced and unvoiced speech. Then a coding scheme based on the resulting reduced parameter set is developed.

Minimum-phase harmonic sinewave speech model —
Voiced speech sinewave model: In Chapter 9, the excitation impulse train during voicing is represented by a sum of sinewaves "in phase" at glottal pulse times, and the amplitude and phase of the excitation sinewaves are altered by the glottal airflow and vocal tract filter. Letting

$H(\omega) = |H(\omega)| \exp[j\Phi(\omega)]$ denote the composite transfer function for these latter effects, which we have also been referring to as the system function, then the speech signal at its output due to the excitation impulse train at its input can be written as

$$\hat{s}[n] = \sum_{k=1}^{K} |H(\omega_k)| \exp[j(n - n_o)\omega_k + \Phi(\omega_k)]$$

where the onset time n_o corresponds to the time of occurrence of the impulse nearest the center of the current analysis frame. (In this section, we avoid frame index notation unless necessary.) We saw that $|H(\omega)|$ can be estimated as the SEEVOC envelope, denoted by $\bar{A}(\omega)$. Furthermore, if we assume that the vocal tract is minimum-phase, then (as we saw in Chapter 10), a system phase can be obtained through the cepstral representation of $|H(\omega)|$ that we denote by $\bar{\Phi}(\omega)$. Then for harmonic frequencies, these minimim-phase amplitude and phase functions can be sampled at $\omega_k = k\omega_o$.

Since the function of the onset time is to bring the sinewaves into phase at times corresponding to the occurrence of a glottal pulse, then, rather than attempt to estimate the *absolute* onset time from the data, we saw in Chapters 9 and 10 that it is possible to achieve the same perceptual effect simply by keeping track of successive onset times generated by a succession of pitch periods that are available at the synthesizer. This *relative* onset time was given by Equation (9.38). Another way to compute the onset times while accounting for the effects of time-varying pitch is to define the phase of the fundamental frequency to be the integral of the instantaneous frequency. For the lth frame, we express this phase as

$$\phi_o[n] = \phi_o[(l-1)L] + \int_{(l-1)T}^{nT} \Omega_o(\sigma)d\sigma \tag{12.19}$$

where $\Omega_o(t)$ is the time-varying pitch frequency in continuous time t, T is the sampling time, l denotes the analysis frame index, and L is the frame interval in discrete time. Because this phase is monotonically increasing with time n, a sequence of onset times can be found at the values of n for which $\phi_o[n_o] = 2\pi M$ for integer values of M. If ω_o^{l-1} and ω_o^l denote the estimated pitch frequencies on frames $l-1$ and l, respectively, then a reasonable model for the frequency variation in going from frame $l-1$ to frame l is

$$\omega_o[n] = \omega_o^{l-1} + \frac{\omega_o^l - \omega_o^{l-1}}{L} n$$

which can be used to numerically compute the phase in Equation (12.19) and subsequently the onset time. If all of the sinewaves are harmonically related, then the phase of the kth sinewave is simply k times the phase of the fundamental, which means that the excitation sinewaves are in phase for every point in time. This leads to a phase model for which it is unnecessary to compute the onset time explicitly (i.e., the absolute onset time). Therefore, using the minimum-phase system phase representation, denoted above by $\bar{\Phi}(\omega)$, and the excitation phase onset representation, we have with harmonic frequencies the sinewave voiced speech model

$$\hat{s}[n] = \sum_{k=1}^{K} \bar{A}(k\omega_o) \exp[jk\phi_o[n] + \bar{\Phi}(k\omega_o)] \tag{12.20}$$

where $\phi_o[n]$ is given in Equation (12.19). This shows that for voiced speech the sinewave reconstruction depends only on the pitch, and, through the SEEVOC envelope, the sinewave amplitudes, the system phase being derived from the amplitude envelope under a minimum-phase assumption. In synthesis, where cubic phase interpolation (or overlap-add) is invoked, we define the excitation phase offset for the kth sinewave as k times the phase of the fundamental [Equation (12.19)], evaluated at the center of the current synthesis frame [50].

Unvoiced speech sinewave model: If the above phase model is used in place of the measured sinewave phases, the synthetic speech is quite natural during voiced speech, but "buzzy" during the unvoiced segments. As noted in Chapter 10, an important difference between the measured and synthetic phase is a phase residual which typically contains a noise component that is particularly important during unvoiced speech and during voicing with a high aspiration component. On the other hand, if the phases are replaced by uniformly distributed random variables on $[-\pi, \pi]$, then the speech is quite natural during unvoiced speech but sounds like whispered speech during the voiced segments (look ahead to Figure 14.12). This suggests that the phase model in Equation (12.20) can be generalized by adding a voicing-dependent component which would be zero for voiced speech and random on $[-\pi, \pi]$ for unvoiced speech. However, it should be expected that such a binary voiced/unvoiced phase model would render the sinewave system overly dependent on the voicing decision, causing artifacts to occur in the synthetic speech when this decision was made erroneously. The deleterious effects of the binary decision can be reduced significantly by using a more general mixed excitation model based on a voicing-dependent frequency cutoff described in Chapters 9 and 10. In this model, a voicing transition frequency is estimated below which voiced speech is synthesized and above which unvoiced speech is synthesized. Letting ω_c denote the voicing-dependent cutoff frequency, then the unvoiced phase component can be modeled by

$$\hat{\epsilon}(\omega) = \begin{cases} 0 & \text{if } \omega \leq \omega_c \\ U[-\pi, \pi] & \text{if } \omega > \omega_c \end{cases} \tag{12.21}$$

representing an estimate to the actual phase residual $\epsilon(\omega)$ introduced in Section 10.5.4 and where $U[-\pi, \pi]$ denotes a uniformly distributed random variable on $[-\pi, \pi]$. If this is added to the voiced-speech phase model in Equation (12.20), the complete sinewave model becomes

$$\hat{s}[n] = \sum_{k=1}^{K} \bar{A}(k\omega_o) \exp[jk\phi_o[n] + \bar{\Phi}(k\omega_o) + \hat{\epsilon}(k\omega_o)] \tag{12.22}$$

where the residual $\hat{\epsilon}(\omega)$ is given in Equation (12.21). As in the basic sinewave reconstruction system described in Chapter 9, speech is synthesized over contiguous frames using the linear amplitude and cubic phase interpolation algorithms, the amplitude and phase functions in Equation (12.22) being given at each frame center prior to interpolation. Using this interpolation, a frame rate of 10 ms was found adequate; above 10 ms perceptual artifacts become apparent. Approaches to estimate the voicing-adaptive cutoff frequency, based on the degree of harmonicity in frequency bands, were described in Chapter 10.

Postfiltering: While the synthetic speech produced by this system is of good quality, a "muffling" effect can be detected particularly, for certain low-pitched speakers. Such a quality loss has also been found in code-excited linear prediction systems (which we study later in this chapter), where it has been argued that the muffling is due to coder noise in the formant nulls. Because the synthetic speech produced by the minimum-phase harmonic sinewave system has not yet been quantized, the muffling cannot be attributed to quantization noise, but to the front-end analysis that led to the sinewave representation.[10] The origin of this muffling effect is not completely understood. Speculations include sidelobe leakage of the window transform filling in the formant nulls, or harmonic sampling of the vocal tract SEEVOC estimate that broadens the formant bandwidths. These speculations, however, are in question because muffling is effectively removed when the synthetic phase function of Equation (12.22) is replaced by the piecewise-flat phase derived from the measured phase at sinewave frequencies. This reduced muffling is due, perhaps, to the *auditory system's exploiting of the original phase to obtain a high-resolution formant estimate*, as we saw in Chapter 8. In addition, the original phase recovers proper timing of speech events, such as plosives and voiced onset times.

Techniques have been developed for filtering out quantization noise in formant nulls and sharpening formant bandwidths by passing the synthesized speech through a *postfilter* [9]. A variant of a code-excited linear prediction postfilter design technique, that uses a frequency-domain design approach, has been developed for sinewave systems [49],[52]. Essentially, the postfilter is a normalized, compressed version of the spectrally-flattened vocal tract envelope, which, when applied to the vocal tract envelope, results in formants having deeper nulls and sharper bandwidths that, in turn, result in synthetic speech that is less muffled. The first step in the design of the postfiltering procedure is to remove the spectral tilt from the log-spectrum. (Spectral tilt was described in Chapter 3, Example 3.2.) One approach to estimating the spectral tilt is to use the first two of the real cepstrum coefficients. These coefficients are computed using the equation

$$c[m] = \frac{1}{\pi} \int_0^\pi \log[\bar{A}(\omega)] \cos(m\omega) d\omega, \quad m = 0, 1$$

where $\bar{A}(\omega)$ is the SEEVOC envelope. The spectral tilt is then given by

$$\log T(\omega) = c[0] + 2c[1] \cos \omega$$

and this is removed from the measured speech spectral envelope to give the residual envelope

$$\log R(\omega) = \log \bar{A}(\omega) - \log T(\omega). \tag{12.23}$$

This is then normalized to have unity gain, and compressed using a root-γ compression rule ($\gamma \approx 0.2$). That is, if $R_{\max}$ is the maximum value of the residual envelope, then the postfilter is taken to be

$$P(\omega) = \left[\frac{R(\omega)}{R_{\max}} \right]^\gamma, \quad 0 \le \gamma \le 1.$$

The idea is that at the formant peaks the normalized residual envelope has unity gain and is not altered by the compressor (Figure 12.18). In the formant nulls, the compressor reduces the fractional values so that overall, a Wiener-like filter characteristic will result (see Chapter 13). In order to insure that excessive spectral shaping is not applied, giving an unnatural tonal characteristic, a clipping rule [32] is introduced such that the final postfilter value cannot fall less than 0.5. The resulting compressed envelope then becomes the postfilter and is applied to the measured envelope to give

$$\hat{A}(\omega) = P(\omega)\bar{A}(\omega) \tag{12.24}$$

---

[10] Because some additional muffling can be introduced with quantization, the postfilter should be applied at the decoding stage.

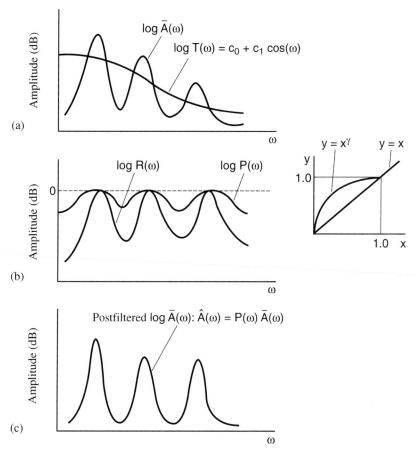

Figure 12.18   Frequency-domain postfilter design: (a) spectral tilt, $\log T(\omega)$, as provided by the order-two cepstrum; (b) spectral tilt removal to obtain flattened spectral envelope, $\log R(\omega)$, and spectral compression to obtain postfilter, $\log P(\omega)$; (c) postfiltered log-spectral envelope.

SOURCE: R.J. McAulay, T.M. Parks, T.F. Quatieri, and M. Sabin, "Sinewave Amplitude Coding at Low Data Rates" [52]. ©1989, IEEE. Used by permission.

which cause the formants to narrow and the formant nulls to deepen, thereby reducing the effects of the analysis and coder noise. Unfortunately, the spectral tilt does not always adequately track the formant peaks and the resulting postfilter can introduce significant spectral imbalance. The effect of this imbalance can be reduced somewhat by renormalizing the postfiltered envelope so that the energy after postfiltering is equal to the energy before postfiltering. Other methods of estimating spectral tilt have also been developed based on all-pole modeling [9],[50].

Experimental results: When the voicing-dependent synthetic-phase model in Equation (12.22) is used to replace the measured phases in the harmonic sinewave synthesizer and the above postfilter is applied, the speech is found to be of very high quality. It is particularly notable that the synthetic speech does not have a "reverberant" characteristic, an effect which, as we saw in Chapter 9, arises in sinewave synthesis when the component sinewaves are not forced to be phase locked. The minimum-phase model preserves a (controlled) dispersive characteristic in the voiced and unvoiced sinewave phases, helping the synthetic speech to sound natural. Moreover, the effect of the postfilter is to make the synthetic speech sound more crisp, considerably reducing the muffled quality often perceived in low-rate speech coders.

Sinewave parameter coding at low data rates — We saw in the previous section that a synthetic minimum-phase function can be obtained from the SEEVOC magnitude estimate and that the excitation sinewave phases can be modeled in terms of a linear phase due to the onset times and voicing-adaptive random phases due to a noise component. Because the pitch can be scalar-quantized using ≈ 8 bits and the voicing probability using ≈ 3 bits, then good quality speech at low data rates appears to be achievable provided the sinewave amplitudes can be coded efficiently. A number of approaches can be applied to code the sinewave amplitudes. To obtain a flavor for the required coding issues, we describe a cepstral modeling approach and then briefly allude to other techniques.

The cepstral model: The first step in coding the sinewave amplitudes is to develop a parametric model for the SEEVOC envelope[11] using the cepstral representation. Because the SEEVOC estimate is real and even, such a model can be written as

$$\log \bar{A}(\omega) \;=\; c[0] + 2 \sum_{m=1}^{\infty} c[m] \cos(m\omega)$$

$$c[m] \;=\; \frac{1}{\pi} \int_0^{\pi} \log \bar{A}(\omega) \cos(m\omega) d\omega, \quad m = 0, 1, 2, \cdots \qquad (12.25)$$

The real cepstral $c[m]$ is then truncated to M points ($M \le 40$ has been found to be adequate) and the cepstral length becomes a design parameter that is varied depending on the coder rate. This representation provides a functional form for *warping* the frequency axis prior to quantization, which can exploit perceptual properties of the ear to reduce the number of required cepstral coefficients [49].

Spectral warping: We saw in Chapter 8 that, above about 1000 Hz, the front-end human auditory system consists of filters whose center frequency increases roughly logarithmically and whose bandwidth is near constant-Q with increasing frequency. This increasing decimation and smoothing with frequency, said to make the ear less sensitive to details in the spectral envelope at higher frequencies, can be exploited by a nonlinear sampling at the higher frequencies, thereby providing a more efficient allocation of the available bits. Let $\omega' = W(\omega)$ and $\omega = W^{-1}(\omega')$ represent the warping function and its inverse. An example of a warping function is shown in Figure 12.19. Then the warped envelope is computed from the SEEVOC envelope through the cepstral representation in Equation (12.25) using

$$\log \bar{A}_w(\omega') \;=\; \log \bar{A}[W^{-1}(\omega')]$$

where the mapping occurs on uniform DFT frequency samples. In order to simulate the effect of auditory filters, including uniformly-spaced filters at low frequency, we select a mapping that is

---

[11] The estimate of the log-magnitude of the vocal tract transfer function can also be taken as the cubic spline fit to the logarithm of the sinewave amplitudes at the frequencies obtained using the SEEVOC peak-picking routine, providing a smoother function than the SEEVOC estimate [50].

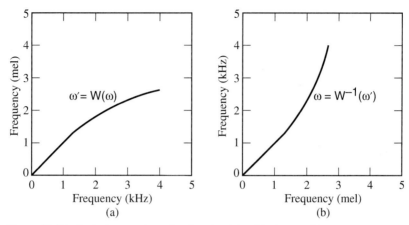

Figure 12.19   Typical spectral warping function (a) and its inverse (b).

SOURCE: R.J. McAulay and T.F. Quatieri, "Low-Rate Speech Coding Based on the Sinusoidal Model," chapter in *Advances in Speech Signal Processing* [49]. ©1992, Marcel Dekker, Inc. Courtesy of Marcel Dekker, Inc.

linear in the low-frequency region (≤ 1000 Hz) and exponential in the high-frequency region, a relationship which is described parametrically as

$$W(\omega) = \begin{cases} \omega, & 0 \leq \omega \leq \omega_L \\ \omega_L(1 + \alpha)^{\omega - \omega_L}, & \omega_L \leq \omega. \end{cases}$$

If M cepstral coefficients are used in the representation, then the amplitude envelope on the original frequency scale is approximately recovered using the relation

$$\log \bar{A}(\omega) \approx c[m] + 2\sum_{m=1}^{M} c[m]\cos[mW(\omega)].$$

The value of M, as well as the warping parameters (α, ω_L), are varied to give the best performance at a given bit rate. Some typical designs which have been found to give reasonable performance are given in [49],[50].

Cepstral transform coding: A potential problem in quantizing the speech envelope using the cepstral coefficients is their large dynamic range (Exercise 12.24) (see also Chapter 6). To avoid this problem, the M cepstral coefficients (derived from the warped spectral envelope) are transformed back into the frequency domain to result in another set of coefficients satisfying the cosine transform pair [49]:

$$g_k = c[0] + 2\sum_{m=1}^{M-1} c[m]\cos\left(2\pi\frac{mk}{2M-1}\right)$$

$$c[m] = g_0 + 2\sum_{k=1}^{M-1} g_k\cos\left(2\pi\frac{km}{2M-1}\right).$$

The advantage in using the transform coefficients g_k instead of the cepstral coefficients $c[m]$ is that the transform coefficients correspond to equally-spaced samples on the warped frequency scale (and thus nonlinearly-spaced on the original frequency scale) of the log-magnitude and, hence, can be quantized using the subband coding principles described earlier in Section 12.5.1. In addition, because of the spectral warping, the transform coefficients can be coded in accordance with the perceptual properties of the ear. Finally, the lowpass filtering incurred in truncation to M coefficients implies that the coefficients are correlated and thus differential PCM techniques in frequency can be exploited[12] [49]. The overall spectral level is set using 4–5 bits to code the first channel gain. Depending on the data rate, the available bits are partitioned among the $M - 1$ remaining channel gains using step-sizes that range from 1–6 dB with the number of bits decreasing with increasing frequency. Bit rates from 1.2–4.8 kbps have been obtained with a variety of bit allocation schemes. An example of a set of coded channel gains is shown in Figure 12.20 for 2.4 kbps operation. The reconstructed (decoded) envelope is also shown in Figure 12.20. Coding strategies include methods for *frame-fill interpolation* with decreasing frame rate [49],[58]. This is a simple method to reduce bit rate by transmitting the speech parameters every second frame, using control information to instruct the synthesizer how to reconstruct or "fill in" the missing data [58]. The frame-fill method has also proved useful at higher rates because it adds temporal resolution to the synthetic waveform [49].

This section has described a cepstral approach to parameterizing and quantizing the sinewave amplitudes. An advantage of the cepstral representation is that it assumes no con-

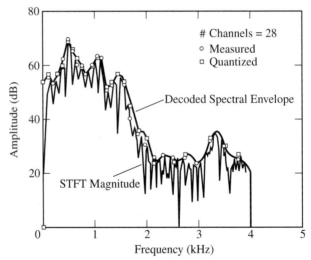

Figure 12.20 Typical unquantized and quantized channel gains derived from the cepstral transform coefficients. (Only one gain is shown when the two visually coincide.) The dewarped spectral envelope after decoding is also illustrated.

SOURCE: R.J. McAulay and T.F. Quatieri, "Low-Rate Speech Coding Based on the Sinusoidal Model," chapter in *Advances in Speech Signal Processing* [49]. ©1992, Marcel Dekker, Inc. Courtesy of Marcel Dekker, Inc.

---

[12] Techniques have been invoked that protect against a positive slope-overload condition in which the coder fails to track the leading edge of a sharply-peaked format [49],[74].

straining model shape, except that the SEEVOC envelope must represent a vocal tract transfer function that is minimum-phase. Other methods further constrain the vocal tract envelope to be *all-pole* [50],[53]. This more constrained model leads to quantization rules that are more bit-rate efficient than those obtained with cepstral modeling methods, leading to higher quality at lower bit rates such as 2.4 kbps and 1.2 kbps [50]. In these schemes, the SEEVOC envelope itself is all-pole-modeled, rather than the speech waveform. Using the frequency-domain view in Chapter 5 of linear prediction analysis, we can derive an all-pole fit to the SEEVOC envelope simply by replacing the autocorrelation coefficients of the speech by those associated with the squared SEEVOC envelope (Exercise 12.17). An advantage of the all-pole fit to the SEEVOC (rather than the waveform) is that it removes dependency on pitch and allows variable order models which in the limit fit the SEEVOC exactly, i.e., there is no problem with autocorrelation aliasing, as illustrated in Figure 5.7. In addition, spectral warping can be applied and further coding efficiencies can be gained [50] by applying frequency-differential methods to a transformation of the all-pole predictor coefficients, referred to as *line spectral frequencies* (LSFs) which we study later in the chapter.

Finally, we note that higher rates than 4.8 kbps (e.g., 6.4 kbps) can be achieved with improved quality by utilizing measured sinewave phases, rather than the synthetic minimum phase used at the lower rates. In one scheme, the measured sinewave phases for the eight lowest harmonics are coded and the remaining high-frequency sinewave phases are obtained from the synthetic phase model of Equation (12.22) [1]. However, because the linear phase component of the measured phases differs from that of the synthetic phase model, a linear phase from an absolute onset time must be estimated and subtracted from the measured phases. The measured phases are then aligned with the synthetic phases by adding to them the synthetic linear phase. Other methods of incorporating phase have also led to improved synthesis quality [2].

Multi-Band Excitation Vocoder — One of the many applications of the sinewave modeling technique to low-rate speech coding is the *multi-band excitation* (MBE) *speech coder* developed by Griffin, Lim, and Hardwick [24],[25]. The starting point for MBE is to represent speech as a sum of harmonic sinewaves. This framework and the resulting pitch and multi-band voicing estimates were described in Chapter 10 (Section 10.6). Versions of this coder have been chosen as standards for the INMARSAT-M system in 1990 [28], the APCO/NASTD/Fed Project 25 in 1992 [32], and the INMARSAT-Mini-M system in 1994 [16], demonstrating good quality in a multitude of conditions, including acoustic noise and channel errors. The purpose of this section is to show the similarities and differences between MBE and the generic sinewave coding methods described in previous sections.

Sinewave analysis and synthesis: The amplitudes that were computed in Chapter 10 as a byproduct during the MBE pitch estimation process have proven not to be reliable estimates of the underlying sinewave amplitudes (Exercise 10.8). Consequently, the MBE coder uses a different method to estimate sinewave amplitudes for synthesis and coding. Because the pitch has been determined, the discrete-time Fourier Transform of the windowed harmonic sinewave speech model, $\hat{s}[n]$ [Equation (10.46)], within the region Δ_k of the kth harmonic [Equation (10.49)] can be written explicitly as (ignoring a 2π scaling)

$$\hat{S}_w(\omega) = B_k \exp(j\phi_k) W(\omega - k\omega_0), \qquad \text{for } \omega \text{ in } \Delta_k$$

where ω_o is the fundamental frequency estimate and where we have used the notation of Chapter 10 in which $\hat{S}_w(\omega)$ denotes the discrete-time Fourier Transform of $\hat{s}_w[n] = w[n]\hat{s}[n]$ with

$w[n]$ the analysis window. In the MBE coder, the amplitude is chosen such that over each harmonic region, Δ_k, the energy in the model matches the measured energy. This leads to the amplitude estimator

$$\hat{B}_k = \left[\frac{\int_{\Delta_k} \mid \hat{S}_w(\omega) \mid^2 d\omega}{\int_{\Delta_0} W^2(\omega) d\omega} \right]^{\frac{1}{2}}.$$

An additional processing step that is required before the sinewave amplitude data is presented to the synthesizer is the postfiltering operation based on the frequency-domain design principles introduced in Section 12.5.2. The phase function in MBE uses a synthetic excitation phase model described in Section 12.5.2 in which the excitation phase of the kth sinewave is k times the phase of the fundamental.

In MBE, two distinct methods are used to synthesize voiced and unvoiced speech (Figure 12.21). Voiced speech is synthesized using either sinewave generation similar to that in Chapter 9 with frequency matching and amplitude and phase interpolation, or using the overlap-add method of Section 9.4.2. Details of the conditions under which different harmonics use each synthesis method are given in [37].

For those speech bands for which the sinewaves have been declared unvoiced, MBE synthesis is performed using filtered white noise. Care must be taken to insure that the effect of the analysis window has been removed so that the correct synthesis noise level is achieved. The details of the normalization procedure are given in [37]. This approach to unvoiced synthesis is in contrast to the sinewave synthesis of Section 12.5.2 which uses random phases in the unvoiced regions. The advantage of using random sinewave phases is that the synthesizer is simpler to implement and more natural "fusing" of the two components may occur, as the same operations are performed for voiced and unvoiced speech.

Sinewave parameter coding: In order to operate MBE as a speech coder, the pitch, voicing, and sinewave amplitudes must be quantized. For low-rate coding, measured phase is not coded

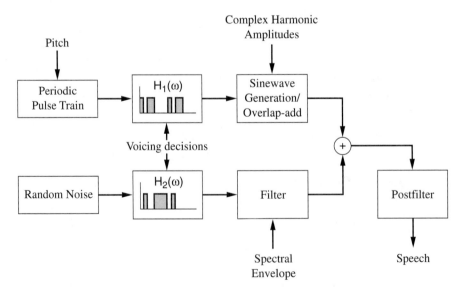

Figure 12.21 Schematic of the MBE synthesis as a basis for speech coding.

and the above synthetic phase model is derived from pitch and a sinewave amplitude envelope when minimum-phase dispersion is invoked. Consider, for example, the particular MBE coding scheme used in the INMARSAT-M system. Here 8 bits are allocated to pitch quantization. Because allowing all voiced/unvoiced multi-band decisions for each harmonic requires an excessive number of bits, the decisions are made on groups of adjacent harmonics. A maximum of 12 bands (and thus 12 bits for one bit per binary decision) is set; if this maximum is exceeded, as with low-pitched speakers, then all remaining high-frequency harmonics are designated unvoiced. For the INMARSAT-M system, 83 bits per frame are allocated. The bits allocated for the sinewave amplitudes per frame are then given by the required total number of bits per frame reduced by the sum of the pitch and voicing bits, i.e., $83 - 8 - b$, where b denotes the bits (≤ 12) for multi-band voicing decisions. For a 20-ms frame interval, 83 bits result in 4.15 kbps. The gross bit rate in the INMARSAT-M system is 6.4 kbps with 45 bits per frame allocated for bit transmission error correction [37]. As with the basic sinewave coder of Section 12.5.2, lower rates (e.g., 1.2 and 2.4 kbps) and more efficient quantization can be obtained using an all-pole model of the sinewave amplitude estimates [37],[83]. In higher-rate MBE coding, time-differential coding of both the sinewave amplitudes and measured phases is used [24].

12.6 Model-Based Coding

We have already introduced the notion of model-based coding in using an all-pole model to represent the sinewave-based SEEVOC spectral envelope for an efficient coding of sinewave amplitudes; this can be considered a hybrid speech coder, blending a frequency-domain representation with the all-pole model. The purpose of model-based approaches to coding is to increase the bit efficiency of waveform-based or frequency-domain coding techniques either to achieve higher quality for the same bit rate or to achieve a lower bit rate for the same quality. In this section, we move back in history and trace through a variety of model-based and hybrid approaches in speech coding using, in particular, the all-pole speech representation. We begin with a coder that uses the basic all-pole linear prediction analysis/synthesis of the speech waveform. The deficiencies of scalar quantization in a simple binary excitation impulse/noise-driven linear-prediction coder lead to the need for applying vector quantization. Deficiencies in the binary source representation itself lead to a hybrid vocoder, the *mixed excitation linear prediction* (MELP) coder. In this coder, the impulse/noise excitation is generalized to a multi-band representation, and other important source characteristics are incorporated, such as pitch jitter, glottal flow velocity, and time-varying formant bandwidths from nonlinear interaction of the source with the vocal tract. A different approach to source modeling, not requiring explicit multi-band decisions and source characterization, uses a *multipulse* source model, ultimately, resulting in the widely used *code-excited linear prediction* (CELP) coder.

12.6.1 Basic Linear Prediction Coder (LPC)

Recall the basic speech production model in which the vocal tract system function, incorporating the glottal flow velocity waveform during voicing, is assumed to be all-pole and is given by

$$H(z) = \frac{A}{A(z)}$$

$$= \frac{A}{1 - P(z)}$$

where the predictor polynomial

$$P(z) = \sum_{k=1}^{p} a_k z^{-k}$$

and whose input is the binary impulse/noise excitation. Suppose that linear prediction analysis is performed at 100 frames/s and 13 parameters (10 all-pole spectrum parameters, pitch, voicing decision, and gain) are extracted on each frame, resulting in 1300 parameters/s to be coded per frame. For telephone bandwidth speech of 4000 Hz, 1300 parameters/s is clearly a marked reduction over 8000 samples/s required by strict waveform coding. Hence we see the potential for low bit rate coding using model-based approaches. We now look at this linear prediction coder (LPC), first introduced by Atal and Hanauer [5], in more detail.

Rather than use the prediction coefficients a_i, we use the corresponding poles b_i, the partial correlation (PARCOR) coefficients k_i (equivalently, from Chapter 5, the reflection coefficients), or some other transformation on the prediction coefficients. This is because, as we noted in Chapter 5, the behavior of the prediction coefficients is difficult to characterize, having a large dynamic range (and thus a large variance); in addition, their quantization can lead to an unstable system function (poles outside the unit circle) at synthesis. On the other hand, the poles b_i or PARCOR coefficients k_i, for example, have a limited dynamic range and can be enforced to give stability because $|b_i| < 1$ and $|k_i| < 1$. There are many ways to code the linear prediction parameters. Ideally, the optimal quantization uses the Max quantizer based on known or estimated pdfs of each parameter. One early scenario at 7200 bps involves the following bit allocation per frame [5]:

1. Voiced/unvoiced decision: 1 bit (on or off)
2. Pitch (if voiced): 6 bits (uniform)
3. Gain: 5 bits (nonuniform)
4. Poles b_i: 10 bits (nonuniform), including 5 bits for bandwidth and 5 bits for center frequency, for each of 6 poles giving a total of 60 bits

At 100 frames/s with (1+6+5+60) bits per frame, we then have the desired 7200 bps. This gives roughly the same quality as the uncoded synthesis, although a mediocre quality by current standards, limited by the simple impulse/noise excitation model.

Refinements to this basic coding scheme involve, for example, companding in the form of a logarithmic operator on pitch and gain, typical values being 5-bit or 6-bit logarithmic coding of pitch and 5-bit logarithmic coding of gain. Another improvement uses the PARCOR coefficients that are more amenable to coding than the pole locations. As noted earlier, the stability condition on the PARCOR coefficients k_i is $|k_i| < 1$ and is simple to preserve under quantization; therefore, interpolation between PARCOR coefficients of stable filters guarantees stable filters.

It has been shown empirically (using histogram analysis) that the first few PARCOR coefficients, k_1 and k_2, have an asymmetric pdf for many voiced sounds, k_1 being near -1 and k_2 being near $+1$ [46]. The higher-order coefficients have a pdf closer to Gaussian, centered around zero. Therefore, nonuniform quantization is desirable. Alternatively, using companding, the PARCOR coefficients can be transformed to a new set of coefficients with a pdf close to uniform. In addition, there is a second reason for the companding transformation: The PARCOR coefficients do not have as good a spectral sensitivity as one would like. By *spectral sensitivity*, we mean the change in the spectrum with a change in the spectral parameters, a characteristic that we aim to minimize. It has been found that a more desirable transformation in this sense is the logarithm of the vocal tract area function ratio $\frac{A_{i+1}}{A_i} = \frac{1-k_i}{1+k_i}$ (Chapter 5), i.e.,

$$g_i = T[k_i]$$
$$= \log\left(\frac{1-k_i}{1+k_i}\right)$$
$$= \log\left(\frac{A_{i+1}}{A_i}\right)$$

where the parameters g_i have a pdf close to uniform and a smaller spectral sensitivity than the PARCOR coefficients, i.e., the all-pole spectrum changes less with a change in g_i than with a change in k_i [13] (and less than with a change in pole positions). Typically, these parameters can be coded at 5–6 bits each, a marked improvement over the 10-bit requirement of pole parameters. Therefore, with 100 frames/s and an order 6 predictor (6 poles), we require $(1+6+5+36)100 = 4800$ bps for about the same quality at 7200 bps attained by coding pole positions for telephone bandwidth speech (4000 Hz) [46].

Observe that by reducing the frame rate by a factor of two to 50 frames/s, we attain a bit rate of 2.4 kbps. This basic LPC structure was used as a government standard for secure communications at 2.4 kbps for about a decade. Although the quality of this standard allowed a usable system, in time, quality judgments became stricter and the need for a new generation of speech coders arose. This opened up research on two primary problems with speech coders based on all-pole linear prediction analysis: (1) The inadequacy of the basic source/filter speech production model, and (2) The unaccounting for possible parameter correlation with the use of one-dimensional scalar quantization techniques. We first explore vector quantization methods to exploit correlation across prediction parameters.

12.6.2 A VQ LPC Coder

One of the first speech coders that exploited linear dependence across scalars applied vector quantization (VQ) to the LPC PARCOR coefficients. The components of this coder are illustrated in Figure 12.22. A set of PARCOR coefficients provides the training vectors from which reconstruction levels $\underline{r}_i$ are derived with the k-means algorithm. For given PARCOR coefficients $\underline{k}$ on each analysis frame, the closest reconstruction level is found to give quantized value $\hat{k}$ and a table index is transmitted as a codeword. At the receiver, a binary codeword is decoded as an index for finding the reconstruction level in a table lookup. The pitch and voicing decision are also coded and decoded.

There are two cases that have been investigated with the VQ LPC structure shown in Figure 12.22. First, we try to achieve the same quality as obtained with scalar quantization of the PARCOR coefficients, but at a lower bit rate. Wong, Juang, and Gray [82] experimented with a 10-bit codebook (1024 codewords) and found they could achieve with roughly a 800 bps vector quantizer a quality comparable to a 2400-bps scalar quantizer. At 44.4 frames/s, 440 bits were required to code the PARCOR coefficients each second; 8 bits were used for pitch, voicing, and gain on each frame, and 1 bit for frame synchronization each second, which

---

[13] The log-area ratio represents a step improvement over the PARCOR coefficients in quantization properties. An even greater advantage is gained by *line spectral frequencies* that we introduce in a following section.

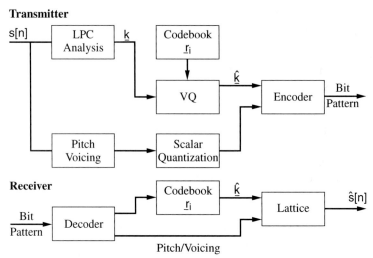

Figure 12.22 A VQ LPC vocoder. In this example, the synthesizer uses an all-pole lattice structure.

roughly makes up the remaining bits. In the second case, the goal is to maintain the bit rate and achieve a higher quality. A higher-quality 2400 bps coder was achieved with a 22-bit codebook, corresponding to $2^{22} = 4200000$ codewords that need be searched. These systems, which were developed in the early 1980s, were not pursued for a number of reasons. First, they were intractable, with respect to computation and storage. Second, the quality of vector-quantized LPC was characterized by a "wobble" due to the LPC-based spectrum's being quantized. When a spectral vector representation is near a VQ cell boundary, the quantized spectra "wobble" to and from two cell centroids from frame to frame; perceptually, the quantization never seems fine enough. Therefore, in the mid 1980s the emphasis changed from improved VQ, but with a focus on the spectrum and to better excitation models and ultimately to a return to VQ on the excitation. We now look at some of these approaches, beginning with a refinement of the simple binary impulse/noise excitation model.

12.6.3 Mixed Excitation LPC (MELP)

A version of LPC vocoding developed by McCree and Barnwell [55] exploits the multi-band voicing decision concept introduced in Section 12.5.2 and is referred to as Mixed Excitation (multi-band) LPC (MELP). MELP addresses shortcomings of conventional linear prediction analysis/synthesis primarily through a realistic excitation signal, time-varying vocal tract formant bandwidths due to the nonlinear coupling of the glottal flow velocity and vocal tract pressure, and production principles of the "anomalous" voice that we introduced in Chapters 3, 4, and 5. Aspects of this improved linear prediction excitation were also explored by Kang and Everett [34].

Model — As with the foundation for MBE, in the model on which MELP is based, different mixtures of impulses and noise are generated in different frequency bands (4–10 bands). The impulse train and noise in the MELP model are each passed through time-varying spectral shaping filters and are added together to form a fullband signal. Important unique components of MELP are summarized as:

1. An auditory-based approach to multi-band voicing estimation for the mixed impulse/noise excitation.

2. Aperiodic impulses due to pitch jitter, the creaky voice, and the diplophonic voice.

3. Time-varying resonance bandwidth within a pitch period accounting for nonlinear source/system interaction and introducing the truncation effect.

4. More accurate shape of the glottal flow velocity source.

We look briefly at how each component is incorporated within the LPC framework.

To estimate the degree of voicing (i.e., the degree of harmonicity) within each band, a normalized correlation coefficient of each output $x[n]$ from a bank of constant-Q bandpass filters is calculated over a duration N at the pitch period lag P as

$$r[P] = \frac{\sum_{n=0}^{N-1} x[n]x[P+n]}{\sqrt{\sum_{n=0}^{N-1} x[n]x[n] \sum_{n=0}^{N-1} x[P+n]x[P+n]}}.$$

A problem with this approach is its sensitivity to varying pitch, the value of $r[P]$ being significantly reduced, resulting in a slightly whispered quality to the synthetic speech. On the other hand, motivated by auditory signal processing principles, McGree and Barnwell [55] found that the *envelope* of the bandpass signal output is less sensitive to nonstationarity, characterized by a broad peak at the pitch period. Therefore, the maximum of the normalized correlation at the pitch lag of the waveform *and* of the envelope is selected as a measure of the degree of voicing, and this determines how much noise is added in each band.

To reduce buzziness due to a pitch contour that is unnaturally stationary (characteristic of conventional LPC), pitch periods are varied with a pulse position jitter uniformly distributed up to $\pm 25\%$, mimicking the erratic behavior of the glottal pulses that can occur in regions such as during transitions, creakiness, or diplophonia. However, to avoid with jitter excessive hoarseness (recall from Chapter 3 that the hoarse voice is characterized by irregularity in the pitch period), the "jittery state" is detected using both correlation and peakiness in the speech waveform as given by a peak-to-rms value, i.e., the jitter increases with decreasing correlation (at the pitch period) and increasing peak-to-rms. Each speech frame is classified as either voiced, jittery voiced, or unvoiced. In both voiced states, the synthesizer uses a mixed impulse/noise excitation, but in the jittery voiced state the synthesizer uses controlled aperiodic pulses to account for irregular movement of glottal pulses. Allowing irregular pulses also allows better representation of plosives, detected through the peakiness measure for the jittery voiced state, without the need for a separate plosive class. This finding is consistent with the suggestion by Kang and Everett [34] of improving the representation of plosives in linear prediction synthesis with the use of randomly spaced pulses.

The synthesizer also includes a mechanism for introducing a time-varying bandwidth into the vocal tract response that we saw in Chapters 4 and 5 can occur *within* a pitch period. Specifically, a formant bandwidth can increase as the glottis opens, corresponding to an increasing decay rate of the response into the glottal cycle. This effect can be mimicked in linear prediction synthesis by replacing z^{-1} in the all-pole transfer function by αz^{-1} and letting α decrease, thus moving the filter poles away from the unit circle; this and other pole bandwidth modulation techniques, as well as an approach to their estimation, were considered in [55]. Finally, the simple impulsive source during voicing is dispersed with an FIR filter to better match an actual glottal flow input.

MELP Coding — A 2.4-kbps coder has been implemented based on the MELP model and has been selected as a government standard for secure telephone communications [56]. In the original version of MELP, 34 bits are allocated to scalar quantization of the LPC coefficients [specifically, the line spectral frequencies (LSFs) described below], 8 bits for gain, 7 bits for pitch and overall voicing (estimated using an autocorrelation technique on the lowpass filtered LPC residual), 5 bits to multi-band voicing, and 1 bit for the jittery state (aperiodic) flag. This gives a total of 54 bits per 22.5-ms frame equal to 2.4 kbps. In the actual 2.4-kbps standard, greater bit efficiency is achieved with vector quantization of the LSF coefficients [56]. Further quality improvement was obtained at 2.4 kbps by combining the MELP model with sinusoidal modeling of the low-frequency portion of the linear prediction residual, formed as the output of the vector-quantized LPC inverse filter [55],[57]. Similar approaches were also used to achieve a higher rate 4.8-kbps MELP coder.

We have indicated that in MELP, as well as in almost all linear prediction-based coders, a more efficient parameter set for coding the all-pole model is the line spectral frequencies (LSFs) [78],[79]. We saw that the LSF parameter set is also used in the sinewave coder to represent the all-pole model of the sinewave amplitudes for coding at 2.4–4.8 kbps. The LSFs for a pth order all-pole model are defined as follows. Two polynomials of order $p+1$ are created from the pth order inverse filter $A(z)$ according to

$$P(z) = A(z) + z^{-(p+1)}A(z^{-1})$$
$$Q(z) = A(z) - z^{-(p+1)}A(z^{-1}). \qquad (12.26)$$

The line spectral frequencies ω_i correspond to the roots of $P(z)$ and $Q(z)$ which are on the unit circle (i.e., at $z = e^{j\omega_i}$), where the trivial roots that always occur at $\omega_i = \pi$ and $\omega_i = 0$ are ignored. Substantial research has shown that the LSFs can be coded efficiently and the stability of the resulting synthesis filter can be guaranteed when they are quantized. This parameter set has the advantage of better quantization and interpolation properties than the corresponding PARCOR coefficients [79]. However, it has the disadvantage that solving for the roots of $P(z)$ and $Q(z)$ can be more computationally intensive than computing the PARCOR coefficients. Finally, the polynomial $A(z)$ is easily recovered from the LSFs (Exercise 12.18).

12.7 LPC Residual Coding

In linear prediction analysis/synthesis based on the all-pole model $H(z) = \frac{A}{A(z)}$, we first estimate the parameters of $A(z) = \sum_{k=1}^{p} a_k z^{-k}$ (and the gain A) using, for example, the autocorrelation method of linear prediction. A *residual* waveform $u[n]$ is obtained by inverse filtering the speech waveform $s[n]$ by $A(z)$, i.e,

$$u[n] \leftrightarrow U(z) = S(z)A(z).$$

In the binary impulse/noise excitation model, this residual is approximated during voicing by a quasi-periodic impulse train and during unvoicing by a white noise sequence. We denote this approximation by $\hat{u}[n]$. We then pass $\hat{u}[n]$ through the filter $\frac{1}{A(z)}$ and compare the result to the original speech signal. During voicing, if the residual does indeed resemble a quasi-periodic impulse train, then the reconstructed signal is close to the original. More often, however, the residual during voicing is far from a periodic impulse train, as illustrated through examples in Chapter 5 (Figure 5.11). As with MELP, the motivation for the two coding schemes in this section is the inability of basic linear prediction analysis/synthesis to accurately model the speech waveform due, in part, to the complexity of the excitation. Unlike MELP, the methods of this section view the inverse filter output without necessarily explicit recourse to the underlying physics of source production, hence the use of the term "residual."

12.7.1 Multi-Pulse Linear Prediction

Multi-pulse linear prediction analysis/synthesis can be considered the basis of a hybrid speech coder that blends waveform-based and all-pole model-based approaches to coding. During voicing, the basic goal of the multi-pulse coder is to represent the excitation waveform, which passes into $H(z) = \frac{1}{A(z)}$, with additional impulses between the primary pitch impulses in order to make the resulting reconstruction $\hat{s}[n]$ a closer fit to $s[n]$ than can be achieved with pitch impulses only. There are numerous justifications for such additional impulses, including the inadequacy of the all-pole model in representing the vocal tract spectrum and glottal flow derivative, and including "secondary" impulses due, for example, to vocal fry and diplophonia, as well as aspiration from turbulent air flow and air displacements resulting from small movements of the surface vocal folds [26]. Other events that help explain the complexity of the residual are due to nonlinear phenomena, including, for example, the glottal flow ripple we encountered in Chapter 5 (Section 5.7.2) and the vortical shedding proposed in Chapter 11 (Section 11.4). In addition to an improved representation of speech during voicing, multi-pulse prediction provides a more accurate representation of speech during unvoicing and in regions where the binary voiced/unvoiced decision is ambiguous, such as in voiced/unvoiced transitions, voiced fricatives, and plosives. The beauty of this approach is that a voicing decision is not required, the analysis being freed of a binary decision, over the full speech band or its multi-bands.

Analysis-by-Synthesis — The objective in the analysis stage of multi-pulse linear prediction is to estimate the source pulse positions and amplitudes. To do so, we define an error criterion between the reconstructed waveform and its original over the lth synthesis frame as

$$E[lL] = \sum_{n=lL}^{(l+1)L-1} (s[n] - \hat{s}[n])^2 \tag{12.27}$$

where the sequence $\hat{s}[n]$ is the output of the filter $H(z) = \frac{1}{A(z)}$ (where we assume here the gain is embedded in the excitation) with a multi-pulse input $\hat{e}[n] = \sum_{k=1}^{M} A_k \delta[n - n_k]$, i.e.,

$$\hat{s}[n] = h[n] * \sum_{k=1}^{M} A_k \delta[n - n_k] \tag{12.28}$$

where each input impulse has amplitude A_k and location n_k. For each speech frame, we minimize $E[lL]$ with respect to the unknown impulse positions and amplitudes. Because we need to synthesize the speech waveform for different candidate impulse parameters in doing the minimization, this approach falls under the class of *analysis-by-synthesis* methods whereby analysis is accomplished via synthesis, as illustrated in Figure 12.23. Finding A_k and n_k for $k = 1, 2, \ldots, M$ by this minimization problem is difficult, given that the impulse positions are nonlinear functions of $\hat{s}[n]$. Before addressing this general problem, we simplify the optimization in the following example:

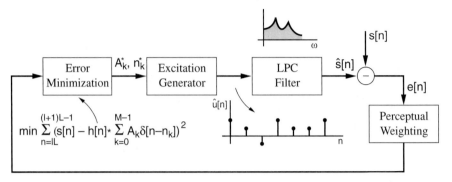

Figure 12.23 Multi-pulse linear prediction as an analysis-by-synthesis approach.

EXAMPLE 12.5   Consider one impulse $A\delta[n-N]$ so that the error in Equation (12.27) becomes

$$E(lL; N, A) = \sum_{n=lL}^{(l+1)L-1} (s[n] - Ah[n - N])^2$$

with $h[n]$ the all-pole impulse response and with impulse position N and amplitude A unknown. Differentiating with respect to A, we have

$$\frac{\partial E}{\partial A} \sum_{n=lL}^{(l+1)L-1} 2(s[n] - Ah[n - N])h[n - N] = 0$$

resulting in the optimal value of A:

$$A^* = \frac{\sum_{n=lL}^{(l+1)L-1} s[n]h[n - N]}{\sum_{n=lL}^{(l+1)L-1} h^2[n - N]}.$$

To obtain the optimal value of N, we substitute the expression for A^* into $E(lL; N, A)$ which becomes a function of only the unknown position N. We then have a nonlinear minimization problem in solving for the optimal position N^*. One solution is to perform an exhaustive search over the frame interval $[lL, (l + 1)L - 1]$ to obtain N^* from which the value for A^* follows.   ▲

Example 12.5 motivates a suboptimal solution to the general nonlinear multi-pulse minimization problem. For the general multi-pulse problem, an efficient but suboptimal solution is obtained by determining the location and amplitude of impulses, one impulse at a time. Thus an optimization problem with many unknowns is reduced to two unknowns.[14] A closed-form solution for an impulse amplitude is given as above, followed by determination of the impulse location by an exhaustive search. We then subtract out the effect of the impulse from the speech waveform and repeat the process. More specifically, we have the following steps for a particular frame [6]:

---

[14] The optimal solution requires consideration of the interaction among all impulses. One iterative approach first assumes an initial pulse positioning and solves the resulting linear optimization problem for the unknown amplitudes (Exercise 12.21). Search methods can then be used to re-estimate the impulse positions and the iteration proceeds until a small enough error is achieved. Although this approach can give greater bit efficiency, it does so at great computational expense [37].

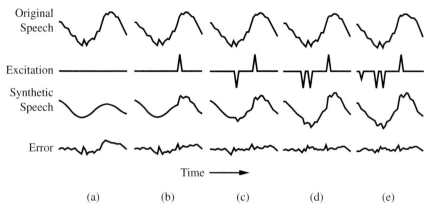

Original
Speech

Excitation

Synthetic
Speech

Error

Time ⟶

(a)   (b)   (c)   (d)   (e)

Figure 12.24 Illustration of the suboptimal (iterative) impulse-by-impulse multi-pulse solution. Each panel shows the waveform of a 5-ms frame interval, the excitation, the memory hangover from the previous frame, and the error resulting from estimating and subtracting successive impulses, and shows the error decreasing with each additional impulse.

SOURCE: B.S. Atal and J.R. Remde, "A New Model of LPC Excitation of Producing Natural-Sounding Speech at Low Bit Rates," [6]. ©1982, IEEE. Used by permission.

S1: We begin with no excitation and an estimate of the all-pole vocal tract impulse response. Contribution from the past all-pole filter memory is subtracted from the current frame.

S2: The amplitude and location from one single impulse, which minimizes the mean-squared error, is determined.

S3: A new error is determined by subtracting out the contribution from step **S2**.

S4: Repeat steps **S2–S3** until a desired minimum error is obtained.

Figure 12.24 illustrates the method [6]. Panel (a) shows the waveform of a 5-ms frame interval, an initial zero excitation, the memory hangover from the previous frame, and the error resulting from subtracting the memory hangover. Panels (b)-(e) show the result of estimating and subtracting successive impulses, and show the error decreasing with each additional impulse. It has been found that negligible error decrease occurs after eight impulses over a 10-ms interval and the resulting speech is perceptually close to the original. The synthetic speech is not characterized by the unnatural or buzzy quality of basic linear prediction analysis/synthesis.

EXAMPLE 12.6   An all-pole filter with 16 poles is estimated every 10 ms using the covariance method of linear prediction with a 20-ms analysis window. Impulse locations and amplitudes are determined with the method of the above steps **S1–S4** over successive 5-ms frame intervals, i.e., the 10-ms all-pole response estimation time interval was split into two sub-frames. Two examples from this procedure are illustrated in Figure 12.25 [6]. As seen, the multi-pulse excitation is able to follow rapid changes in the speech, as during transitions. It is also interesting to observe various secondary pulses, including pulses in the excitation very close to the primary pulse, contributed perhaps from secondary glottal pulses, secondary nonacoustic sources, or zeros in the vocal tract system function that are not adequately modeled by the all-pole transfer function (Exercise 12.20).   ▲

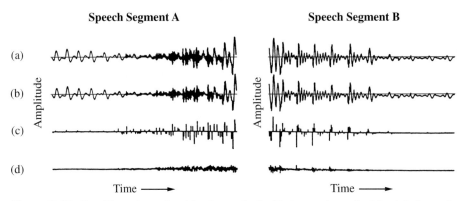

Figure 12.25 Two illustrations of multi-pulse synthesis. Each case shows the (a) original speech, (b) synthetic speech, (c) excitation signal, and (d) error.

SOURCE: B.S. Atal and J.R. Remde, "A New Model of LPC Excitation of Producing Natural-Sounding Speech at Low Bit Rates," [6]. ©1982, IEEE. Used by permission.

We have omitted up to now an important component of the error minimization. This component is illustrated in Figure 12.23 as a *perceptual weighting filter*, accounting for the inadequacy of the mean-squared error criterion for human perception. This weighting de-emphasizes error in the formant regions by applying a linear filter that attenuates the error energy in these regions, i.e., we give less weight to the error near formant peaks, much as we did in selecting quantization noise in subband speech coders. The result is a synthetic speech spectrum with more error near formant peaks, roughly preserving the signal-to-noise ratio across frequency. In the frequency domain, the error criterion becomes

$$E = \int_{-\pi}^{\pi} Q(\omega)|S(\omega) - \hat{S}(\omega)|^2 d\omega$$

where $Q(\omega)$ is the weighting function chosen to de-emphasize the error near formant peaks, and where $S(\omega)$ and $\hat{S}(\omega)$ refer to the Fourier transforms of $s[n]$ and $\hat{s}[n]$, respectively, over a frame duration. Thus, qualitatively, we want $Q(\omega)$ to take an inverse-like relation to $S(\omega)$ as we did in the subband coding context. Let $P(z)$ be the prediction error filter corresponding to an all-pole vocal tract transfer function. Then the inverse filter $A(z) = 1 - P(z)$. One choice of $Q(z)$ is given by

$$Q(z) = \frac{1 - P(z)}{1 - P(z/\alpha)}$$

$$= \frac{1 - \sum_{k=1}^{p} a_k z^{-k}}{1 - \sum_{k=1}^{p} a_k \alpha^k z^{-k}}$$

where $0 \le \alpha \le 1$ [6]. The filter changes from $Q(z) = 1$ for $\alpha = 1$ to $Q(z) = 1 - P(z)$ (the inverse filter) for $\alpha = 0$. The particular choice of α, and thus the degree to which error is de-emphasized near formant peaks, is determined by perceptual listening tests. Figure 12.26 shows an example for the typical value $\alpha = 0.8$ used in Example 12.6 [6].

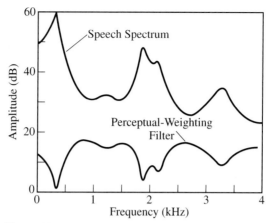

Figure 12.26   Example of perceptual weighting filter.

SOURCE: B.S. Atal and J.R. Remde, "A New Model of LPC
Excitation of Producing Natural-Sounding Speech at Low Bit
Rates," [6]. ©1982, IEEE. Used by permission.

Multi-Pulse Parameter Coding — In multi-pulse speech coding, the predictor coefficients (or transformations thereof such as the log area ratios or LSFs), pulse locations, and pulse amplitudes are quantized. In one 9.6-kbps coder, for example, 8 impulses/10-ms frame are used or 800 impulses/s. In this particular scheme, the corresponding bit allocation is 2.4 kbps for the all-pole system function $A(z)$ and 7.2 kbps for the impulses (or 7200/800 = 9 bits/impulse) [37].

Given only 9 bits/impulse for impulse amplitudes and locations, the quantization strategy must be carefully selected. The impulse locations are coded differentially to reduce dynamic range, i.e., the difference $\Delta_i = n_i - n_{i-1}$ between impulse locations is quantized rather than the absolute locations. The amplitudes are normalized to unity to push the dynamic range into one gain parameter, making all amplitudes lie in the range [0, 1]. The result is good quality for 9.6 kbps, but quality that degrades rapidly with bit distributions for bit rates < 9.6 kbps. This is particularly true for high-pitched voices where a large portion of the 8 impulses/frame is used in representing the primary pitch pulses, leaving few impulses for the remaining excitation. For good quality below 9.6 kbps, as well as greater bit efficiency at 9.6 kbps, a different approach is taken invoking long-time periodicity prediction and vector quantization of the excitation.

12.7.2 Multi-Pulse Modeling with Long-Term Prediction

Primary pitch pulses, i.e., from the periodic glottal flow, can waste many of the available bits for the above 8 impulses/frame in multi-pulse speech coding. Therefore, one would like to represent the periodic correlation in the speech with a few parameters per frame, freeing up impulses in the remaining excitation. One strategy that avoids the coding of individual pitch impulses first performs pitch estimation and then introduces pitch impulses in the excitation. This eliminates the need for primary pitch impulses in the multi-pulse representation. The problem with this approach is that synchrony of the primary impulses with the original speech waveform periodicity is required; in addition, primary pitch pulses in speech vary in amplitude from frame-to-frame. An alternative is to introduce *long-term prediction*.

Long-term prediction is based on the observation that primary pitch pulses due to glottal pulses are correlated and *predictable* over consecutive pitch periods, so that

$$s[n] \approx bs[n - P]$$

where P is the pitch period and b is a scale factor. In fact, we can consider the speech waveform as having a *short-term and long-term correlation*. As illustrated in Figure 12.27, the short-term correlation (with which we are already familiar from our linear prediction analysis of Chapter 5) occurs over the duration of the vocal tract response *within* a pitch period, while the long-term correlation occurs *across* consecutive pitch periods. The approach that we take, therefore, is to first remove short-term correlation by short-term prediction followed by removing long-term correlation by long-term prediction.

The short-term prediction-error filter is the pth-order polynomial $A(z) = 1 - P(z) = \sum_{k=1}^{p} a_k z^{-1}$, where p is typically in the range 10–16. The result of the short-term prediction error is a residual function $u[n]$ that includes primary pitch pulses (long-term correlation). The long-term prediction-error filter is of the form

$$B(z) = 1 - bz^{-P}$$

where bz^{-P} is the long-term predictor in the z-domain. The output of the long-term prediction-error filter is a residual

$$v[n] = u[n] - bu[n - p]$$

with fewer large (long-term correlated) pulses to code than in $u[n]$ (Figure 12.28). After removing the long-term prediction contribution, the residual $v[n]$ forms the basis for an efficient coding of a multi-pulse excitation. Having the short-term predictor and long-term predictor, we can then invert the process and recover the original speech waveform as shown in Figure 12.29 where we assume knowledge of the residual $v[n]$ as well as inverse filters $A(z)$ and $B(z)$. In synthesis, with a frame-by-frame implementation, the memory hangover from a previous frame is added into the result of filtering with $\frac{1}{A(z)}$ and $\frac{1}{B(z)}$ on the current frame.

In estimating the long-term predictor, we must estimate both the pitch period P and the scale factor b. The pitch period can be estimated independently with any pitch estimator. However, it is preferred to tie the estimation of P to the prediction problem because our goal is to remove pulses correlated over consecutive periods, reducing the prediction error. In the time domain, the long-term prediction error filter $B(z) = 1 - bz^{-P}$ is expressed by

$$b[n] = \delta[n] - b\delta[n - P].$$

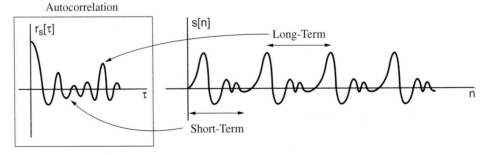

Figure 12.27 Illustration of short- and long-term correlation in the speech waveform.

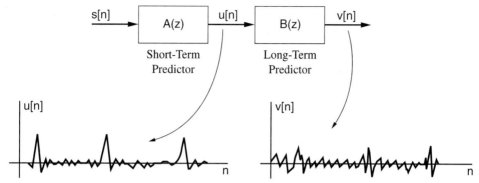

Figure 12.28   Illustration of short- and long-term prediction. The residual sequence $u[n]$ is the short-term prediction error and the residual $v[n]$ is the long-term prediction error.

We then define the error criterion for the lth frame as

$$E[lL; P, b] = \sum_{n=lL}^{(l+1)L-1} v^2[n]$$

$$= \sum_{n=lL}^{(l+1)L-1} (u[n] * b[n])^2$$

$$= \sum_{n=lL}^{(l+1)L-1} (u[n] - bu[n - P])^2$$

where L is a frame duration. The objective is to minimize $E[lL; P, b]$ with respect to P and b.

Suppose that we know the pitch period P. Then we differentiate with respect to the unknown b, i.e.,

$$\frac{\partial}{\partial b} E[lL; P, b] = 0$$

leading to

$$b^* = \frac{\sum_{n=lL}^{(l+1)L-1} u[n]u[n - P]}{\sum_{n=lL}^{(l+1)L-1} u^2[n - P]}. \tag{12.29}$$

Observe that this procedure corresponds to the *covariance method of linear prediction* because the fixed limits of summation imply that we have not truncated the data. Due to the long delay P, it is important to use the covariance method rather than the autocorrelation method; with the autocorrelation at long lags, we would obtain poor estimates of the correlation (Exercise 12.22).

Figure 12.29   Illustration of speech reconstruction from short- and long-term prediction filter outputs.

Now substituting b^* in the expression for $E[lL; P, b]$, we obtain (Exercise 12.23)

$$E[lL; P] = \sum_{n=lL}^{(l+1)L-1} u^2[n] - E'[lL; P] \qquad (12.30)$$

where

$$E'[lL; P] = \frac{(\sum_{n=lL}^{(l+1)L-1} u[n]u[n-P])^2}{\sum_{n=lL}^{(l+1)L-1} u^2[n-P]}.$$

We now want to maximize $E'[lL; P]$ (which minimizes $E[lL; P]$). This maximization is similar to the operation required in the autocorrelation pitch estimator that we derived in Chapter 10. To do so, we compute $E'[lL; P]$ for all integer values over some expected pitch period range. When speech is voiced, this results in local maxima at delays that correspond to the pitch period and its multiples.

The analysis and synthesis of Figures 12.28 and 12.29, respectively, form the basis for the *multi-pulse* residual estimation algorithm in Figure 12.30. Observe in Figure 12.30 that multi-pulse estimation is performed to attain a more accurate waveform match at the receiver. As in the previous section (12.7.1), we select impulse amplitudes and locations to minimize the mean-squared error between the original and synthetic speech where the frequency-domain error function is perceptually weighted. Now, however, impulses are selected to represent $v[n]$, the output of the long-term predictor, rather than $u[n]$, the output of the short-term predictor. It is interesting to observe that, although long-term prediction intends to remove quasi-periodicity in the signal, multi-pulse modeling of the residual still results in impulses that are placed at or near the primary pitch pulses, i.e., some of the time multi-pulses lie near pitch pulses [37], as we indicated schematically by $v[n]$ in Figure 12.28. This is probably due to the inadequacy of the linear-prediction analysis in modeling a time-varying vocal tract that contains (among other contributions) zeros that can be modeled as impulses near the primary pitch pulses, as we alluded to earlier. Another interesting observation is the nature of the long-term predictor filter estimate used in synthesis, i.e.,

$$\frac{1}{\hat{B}(z)} = \frac{1}{1 - be^{-jPz}}$$

which, when magnitude-squared, along the unit circle, results in

$$\frac{1}{|\hat{B}(\omega)|^2} = \frac{1}{(1+b)^2 - 2b\cos(P\omega)}.$$

This filter has a "comb" structure that imparts a quasi-harmonicity onto an estimate of the multi-pulse residual $\hat{v}[n]$ even when the residual $v[n]$ is not reproduced exactly [73].

Observe, as indicated in Figure 12.30, that we have solved for $A(z)$ and $B(z)$ "open-loop" in the sense that we are not doing waveform matching at these steps. In obtaining $B(z)$, for example, the matching is performed on the predicted residual, rather than on the speech waveform itself. In contrast, we solve for the multi-pulse model of the residual $v[n]$ "closed-loop" because we are waveform-matching at this final stage of the optimization. Thus, we have performed the open-loop and closed-loop analysis sequentially; the multi-pulse approach (as is the CELP coder in the following system) is considered "analysis-by-synthesis," but this is not strictly true, due to this sequential nature of the processing. On the other hand, we could have optimized all system components in closed-loop, but this would be computationally infeasible.

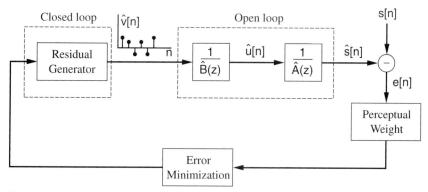

Figure 12.30 Closed-loop/open-loop multi-pulse analysis/synthesis with short- and long-term prediction. The polynomials $\hat{A}(z)$ and $\hat{B}(z)$ represent the short- and long-term predictor estimates with possibly quantized coefficients.

With the analysis/synthesis scheme of Figure 12.30, we can obtain higher quality in the bit range 8–9.6 kbps than with the basic multi-pulse approach. In fact, a multi-pulse coder based on the configuration of Figure 12.30, including coding of the long-term predictor gain and delay, short-term predictor coefficients, and locations and amplitudes of residual impulses, forms the essence of a 9.6 kbps coder that was selected for public communication between commercial in-flight aircraft and ground, as well as a government standard secure telephone unit (STU 3) and an international standard for aeronautical mobile satellite telecommunications [37]. A variant of this multi-pulse scheme has also been selected for the Pan-European Digital Cellular Mobile Radio System (GSM) at 12.2 kbps. In this coder, referred to as *regular pulse-excited LPC* (RPELPC), the residual from the long-term predictor is constrained to consist of equally spaced impulses of different amplitudes [19],[38]. Because the impulse locations are constrained, optimal selection of the impulse amplitudes becomes a linear problem (Exercise 12.21).

Although the multi-pulse approach with long-term prediction provides the basis for satisfactory coders near 9.6 kbps, achieving high-quality coding much below this rate requires a different way to represent and quantize the residual $v[n]$, i.e., the coding of individual impulse amplitudes and locations is too taxing for lower rate coders. In the next section, we present the fundamentals of the code-excited linear prediction (CELP) coder designed to more efficiently represent the residual based on vector quantization.

12.7.3 Code-Excited Linear Prediction (CELP)

Concept — The basic idea of code-excited linear prediction (CELP) is to represent the residual from long-term prediction on each frame by codewords from a VQ-generated codebook, rather than by multi-pulses. This codeword approach can be conceptualized by replacing the residual generator in Figure 12.30 by a codeword generator; on each frame, a codeword is chosen from a codebook of residuals such as to minimize the mean-squared error between the synthesized and original speech waveforms. The length of a codeword sequence is determined by the analysis frame length. For example, for a 10-ms frame interval split into 2 inner frames of 5 ms each, a codeword sequence is 40 samples in duration for an 8000-Hz sampling rate. The residual and long-term predictor is estimated with twice the time resolution (a 5-ms frame) of the short-term predictor (a 10-ms frame) because the excitation is more nonstationary than the vocal tract. As with multi-pulse linear prediction, a perceptual weighting is used in the selection of the codewords (Figure 12.30). The weighting function is chosen similarly to that in multi-pulse, and thus again yields a quantization error that is roughly proportional to the speech spectrum. In addition, a postfilter, similar to that described in previous sections, is introduced in synthesis for formant bandwidth sharpening. Two approaches to formation of the codebook have been considered: "deterministic" and "stochastic" codebooks. In either case, the number of codewords in the codebook is determined by the required bit rate and the portion of bits allocated to the residual $v[n]$. A 10-bit codebook, for example, allows $2^{10} = 1024$ codewords in the codebook.

A *deterministic* codebook is formed by applying the k-means clustering algorithm to a large set of residual training vectors. We saw earlier in this chapter that k-means clustering requires a distortion measure in training, as well as in selection of reconstruction levels in coding. For CELP, the natural distortion measure used for clustering in training is the waveform quantization error over each inner frame, i.e., $\sum_{n=lL}^{(l+1)L-1} (s[n] - \hat{s}[n])^2$, where L here denotes the inner frame length and $\hat{s}[n]$ is the quantized speech waveform. Successful codebooks of this type can be constructed using a large, representative set of speech training data, and have yielded good-quality speech [38],[76]. Nevertheless, CELP coders typically do not apply k-means training [38]. One reason is that often the training data and the speech being coded (the "test data") are recorded over different channels, thus resulting in a *channel mismatch* condition when the VQ selection process occurs. Other reasons for not using trained codebooks are that trained codebooks may limit the coverage of the signal space and their lack of structure may make them unsuitable for fast codebook search algorithms.

One of the first alternative *stochastic* codebooks was motivated by the observation that the histogram of the residual from the long-term predictor follows roughly a Gaussian probability density function (pdf) [76]. This is roughly valid except at plosives and voiced/unvoiced transitions. In addition, the cumulative distributions (integral of the pdf) derived from histograms for actual residuals are nearly identical to those for white Gaussian random variables. As such, an alternative codebook is constructed of white Gaussian random numbers with unit variance.[15] A gain must thus also be estimated for each codeword selection. Although suboptimal (in theory), this stochastic codebook offers a marked reduction in computation for codebook generation, with speech quality equivalent to that of trained codebooks [38],[76]. Codebooks are also used that consist of codeword vectors representing residual impulses of ± 1 gain and with certain predetermined positions [75]. These are called *algebraic codebooks* because they are obtained at a receiver from the transmitted index using linear algebra rather than a table lookup as described earlier in Section 12.4.3 for a typical VQ speech coding scenario. Algebraic codebooks have an advantage with respect to coding efficiency and search speed.

There have evolved many variations of the basic CELP concept, not only with respect to codebook design and search (lookup), but also with respect to the short- and long-term predictors. For example, in the multi-pulse scheme, we noted that the short-term and long-term predictors

---

[15] It appears that there should be no gain in coding efficiency with VQ used with vectors consisting of white Gaussian random variables because VQ was designed to exploit correlation among the vector elements. This dilemma is resolved by observing that for a fixed number of bits per vector, VQ allows *fractional bits*. Consider, for example, a 100-element vector at a 10-ms frame interval. Scalar quantization allows a minimum of one bit per vector element for 100 bits per frame or 10000 bits/s. On the other hand, VQ allows any number of bits per frame and thus fractional bits per element, e.g., 50 bits per frame correspond to $\frac{1}{2}$ bit per vector element for 5000 bits/s. In addition, VQ gives flexibility in partitioning the vector space not possible with scalar quantization [44].

were estimated open-loop while the residual was estimated close-loop. In various renditions of CELP, the short-term predictor is estimated in open-loop form, while the long-term predictor is estimated either in open-loop or closed-loop form and sometimes in both [8],[37],[75]. The open-loop long-term predictor estimation is performed, as in our previous discussion, whereby the long-term prediction $bu[n - P]$ attempts to match the short-term prediction residual $u[n]$, i.e., we minimized the mean-square of the error

$$v[n] = u[n] - bu[n - P].$$

This minimization resulted in a solution for the long-term gain and delay ("pitch") given by Equations (12.29) and (12.30), respectively. In the closed-loop estimation of the long-term predictor, rather than matching the short-term prediction error waveform, we match the speech waveform itself [37]. Suppose that we have estimated and quantized the short-term predictor polynomial $\hat{P}(z)$; then we denote by $\hat{h}[n]$ the impulse response of $\frac{1}{1-\hat{P}(z)}$. In order to determine the long-term predictor delay and gain closed-loop, one needs to simultaneously determine the optimal excitation codewords. Here the prediction error to be minimized seeks to match the speech waveform and is formed as

$$q[n] = \hat{s}[n] - b\hat{s}[n - P]$$

where the synthetic waveform $\hat{s}[n]$ is given by

$$\hat{s}[n] = \hat{h}[n] * \hat{u}[n]$$

where $\hat{u}[n]$ is the quantized excitation. Because solution by exhaustive search is computationally infeasible, various two-step (sub-optimal) procedures have been developed. The interested reader is referred to [37],[38]. The advantage of this closed-loop strategy is that the parameter estimation is performed with respect to the desired signal, rather than to an intermediary residual. As such, it can attain higher quality than the original CELP coder that performs estimation of the long-term predictor open-loop [37],[38].

CELP Coders — The CELP strategy has been widely used in a variety of government and international standard coders. One example is that of a CELP coder used in an early 1990s government standard for secure communications at 4.8 kbps at a 4000-Hz bandwidth (Fed-Std–1016) [8], one of three coders in a secure STU–3 telephone that invokes three bit rates: 9.6 kbps (multi-pulse), 4.8 kbps (CELP), and 2.4 kbps (LPC). In this 4.8-kbps standard, the short-time predictor is determined at a 30-ms frame interval and is coded with 34 bits per frame. The prediction coefficients of a 10th-order vocal tract spectrum are transformed to LSFs, a parameter set that we saw is more amenable to quantization than other all-pole representations. The 10 LSFs are coded using nonuniform quantization. The short-term and long-term predictors are estimated in open-loop, while the residual codewords are determined in closed-loop form. The residual vectors and long-term pitch and gain are updated at a 7.5-ms inner frame and both are coded with VQ schemes (long-term pitch and gain as a separate vector). This entails a 512-element stochastic (algebraic) codebook for the residual, requiring 9 bits per inner frame, and an associated gain coded with 5-bit nonuniform scalar quantization per inner frame. The long-term pitch and gain use 256 codewords, requiring 28 bits per outer frame. The total number of bits/s is thus 4600 bps. The remaining 200 bps are used for frame synchronization at the receiver, for error protection, and future expansion. Further details of this bit allocation are given in [8].

The CELP concept also forms the basis for many current international standards, for cellular, satellite, and Internet communications of limited bandwidth, with bit rates in about the range of 5–13 kbps. As does the 4.8-kbps STU–3 coder, these coders apply the residual models of this section and the quantization principles described throughout the chapter. Two such coders are G.729 and G.723.1. The two coders are based on a residual/LSF/postfilter analysis/synthesis, with the primary difference being the manner of coding the excitation residual. The G.729 coder runs at 8 kbps and was standardized by ITU (International Telecommunication Union) for personal communication and satellite systems, and is based on an algebraic CELP residual coding scheme [30]. The gain and pitch from an open-loop estimation are refined with estimates from a closed-loop predictor. This coder also exploits time-differential quantization of the LSF coefficients, in particular, the finite-order moving average predictor approach of Figure 12.13 for reducing propagation over time of channel bit errors. The G.723.1 coder is also based on a variation of the basic CELP concept, and is a current multi-media standard coder at 5.3 and 6.3 kbps [31].

12.8 Summary

In this chapter, we have described the principles of one-dimensional scalar quantization and its generalization to vector quantization that form the basis for waveform, model-based, and hybrid coding techniques. Not having the space to cover the wide array of all speech coders, we gave a representative set from each of the three major classes. The area of hybrid coders, in particular, is rapidly evolving and, itself, requires a complete text to give it justice. One important new class of hybrid coders, for example, is *waveform-interpolation coders* that view the speech waveform as a slowly evolving series of glottal cycles; this coder class combines a sinusoidal model with waveform representation methods [35]. This, in turn, has led to the development of a further decomposition of the sinewave representation of evolving glottal cycles into slowly varying and rapidly varying components [36]. By computing sinewave parameters at a relatively high frame rate ($\approx$ 5 ms), matching the sinewave parameters from frame-to-frame, and applying complementary highpass and lowpass filters to the real and imaginary parts along each of the sinewave tracks, the rapidly varying and slowly varying components of the speech signal can be isolated. If the rapidly varying components are quantized crudely but often, and the slowly varying components are quantized accurately but infrequently, high-quality synthetic speech can be obtained at 2.4 kbps. This quantization technique exploits a form of perceptual masking where noise is hidden by spectrally dynamic (the rapidly varying) regions, and which we will again encounter in Chapter 13 for speech enhancement in noise.

Perceptual masking in speech coding, more generally, is an important area that we have only touched upon in this chapter. We have mentioned that, by working in the frequency domain, we are able to exploit perceptual masking principles to "hide" quantization noise under the speech signal in a particular frequency band. However, more general masking principles based on the masking of one tone by adjacent tones can also be exploited. Such principles have been applied successfully in wideband coding at high bit rates [29] and have also been investigated for use in sinewave coding [4],[21]. Finally, an important direction is the evolving area of speech coding based on nonlinear speech modeling. A number of new techniques are being developed based on the acoustic/nonacoustic modeling approaches of Chapter 11 in which parametric representations of speech modulations are exploited [67], as well as on nonlinear dynamical modeling approaches [40],[69].

EXERCISES

12.1 Show for the uniform scalar quantizer in Equation (12.1) that the quantizer step size Δ, which is the step size equal to the spacing between two consecutive decision levels, is the same as the spacing between two consecutive reconstruction levels.

12.2 Assume quantization noise has a uniform probability density function (pdf) of the form $p_e(e) = \frac{1}{\Delta}$ for $-\frac{\Delta}{2} \le e \le \frac{\Delta}{2}$ and zero elsewhere. Show that the variance of the quantization noise is given by $\sigma_e^2 = \frac{\Delta^2}{12}$, and thus that the noise variance of a uniform quantizer with step size Δ is given as in Equation (12.3).

12.3 Consider a 4-level quantizer. Suppose that values of a sequence $x[n]$ fall within the range [0, 1] but rarely fall between $x_3 = \frac{1}{2}$ and $x_4 = 1$. Propose a nonuniform quantization scheme, not necessarily optimal in a least-squared-error sense, that reduces the least-squared error relative to that of a uniform quantization scheme.

12.4 Let x denote the signal sample whose pdf $p_x(x)$ is given by

$$p_x(x) = \begin{cases} 1 & \frac{-1}{2} \le x \le \frac{1}{2} \\ 0 & \text{otherwise.} \end{cases}$$

(a) Suppose we assign only one reconstruction level to x. We denote the reconstruction level by $\hat{x}$. We want to minimize $E\{(x - \hat{x})^2\}$. Determine $\hat{x}$. How many bits are required to represent the reconstruction level?

(b) Suppose again we assign only one reconstruction level to x, but now the reconstruction level is set to unity, i.e., $\hat{x} = 1$. Compute the signal-to-noise ratio (SNR) defined as

$$\text{SNR} = \frac{\sigma_x^2}{\sigma_e^2}$$

i.e., the ratio of signal and quantization noise variances, where the quantization noise is given by

$$e = \hat{x} - x.$$

(c) Suppose a uniform quantizer is applied to x with four reconstruction levels. Determine the reconstruction levels that minimize $E\{(x - \hat{x})^2\}$. Assign a codeword to each of the reconstruction levels.

12.5 For a random variable with a uniform pdf, show that the Max quantizer results in a set of uniform quantization levels.

12.6 In this problem, you consider the method of companding introduced in Section 12.3.4.

(a) Show that the transformation in Equation (12.10) results in the random process $g[n]$ whose elements have a uniform pdf. *Hint:* First note that the transformation T gives the probability distribution of x (x denoting the value of signal sample $x[n]$), i.e.,

$$\text{Prob}\{x \le x_o\} = F_x[x_o]$$

where F_x denotes probability distribution. Then consider the probability distribution of $g[n]$ itself.

(b) Show that logarithmic companding makes the SNR (i.e., signal-to-quantization noise ratio) resistant to signal characteristics. Specifically, show that the logarithm makes the SNR approximately independent of signal variance and dependent only upon quantization step size.

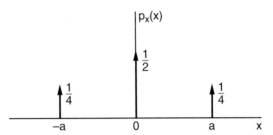

Figure 12.31 Symmetric probability density function of random variable x.

12.7 Let x denote the random variable whose pdf $p_x(x)$ is given in Figure 12.31.

    (a) A symmetric quantizer is defined such that if it has a reconstruction level of r, then it also has a reconstruction level of $-r$. Given 1 bit to quantize the random variable x, determine the *symmetric* minimum mean-squared-error quantizer. What is the corresponding mean-squared error?

    (b) It has been suggested that using a *non-symmetric* quantizer would result in lower mean-squared error than that of part (a). If you think that the above statement is true, then determine the non-symmetric quantizer and its corresponding mean-squared error. If it is false, then justify your answer.

12.8 Show that the expected value of the signal variance estimate in Equation (12.11) is equal to the true variance, σ_x^2, for a stationary $x[n]$ and for an appropriate choice of the constant β. Discuss the time-frequency resolution tradeoffs in the estimate for adaptive quantization with a nonstationary signal.

12.9 Consider a signal $x[n]$ that takes on values according to the pdf given in Figure 12.32.

    (a) Derive the signal-to-noise ratio (SNR) defined as

$$\text{SNR} = \frac{\sigma_x^2}{\sigma_e^2}$$

    for a 2-bit quantizer where the reconstruction levels are uniformly-spaced over the interval [0, 1], as shown in Figure 12.32c. Assume the pdf of the quantization noise is uniform. Note that you should first find the range of the quantization error as a function of the quantization step size Δ.

    (b) Now suppose we design a nonuniform quantizer that is illustrated in Figure 12.32d. Derive the SNR for this quantizer. Assume the pdf of the quantization error $e[n]$ is uniform for $x[n]$ in the interval [0, 1/2], and also uniform for $x[n]$ in the interval [1/2, 1], but within each interval a different quantization step size is applied, as shown. *Hint:* Use the relation:

$$p_e(e[n]) = \text{Prob}\{x \in [0, 1/2]\} p_e(e[n]|x \in [0, 1/2])$$
$$+ \text{Prob}\{x \in [1/2, 1]\} p_e(e[n]|x \in [1/2, 1]).$$

    (c) Is the nonuniform quantizer of part (b) an optimal Max quantizer? Explain your reasoning.

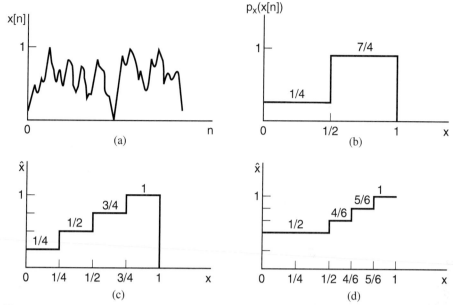

Figure 12.32 Quantization conditions for Exercise 12.9: (a) sample function; (b) pdf of $x[n]$; (c) uniform quantizer; (d) nonuniform quantizer.

12.10 Consider two random variables x_1 and x_2 with joint pdf $p_{x_1,x_2}(x_1, x_2)$ where the separate pdf's $p_{x_1}(x_1)$ and $p_{x_2}(x_2)$ are triangular, as illustrated in Figure 12.33, with range $[-a, a]$ and with peak amplitude $1/a$. In this problem, you investigate the advantage of vector quantization for correlated scalars.

   (a) From Figure 12.33, we can deduce that the two scalars are linearly dependent, i.e., correlated. Show this property analytically for the variables, i.e., show that

$$E[x_1x_2] \neq E[x_1]E[x_2].$$

   (b) Consider quantizing x_1 and x_2 separately using scalar quantization with the Max quantizer (invoking the minimum-mean-squared-error criterion). Suppose, in particular, that we use two reconstruction levels for each scalar. Then find the optimal reconstruction levels. Sketch for the two quantized scalars, the four (2×2) resulting reconstruction levels in the two-dimensional space $[x_1, x_2]$, corresponding to a 2-bit representation.

   (c) From the joint pdf in Figure 12.33, we know two of the four reconstruction levels in part (b) are not possible; we have wasted two bits. Suppose you use only these two possible reconstruction levels in the 2-D space, requiring only a 1-bit codebook representation. Argue that the mean-squared error (MSE) in the two representations is the same. Therefore, by exploiting linear dependence, for the same MSE, we have saved 1 bit.

   (d) Rotate the pdf in Figure 12.33 45° clockwise and argue that you have eliminated the linear dependence of the two new scalars y_1 and y_2. In particular, show that

$$E[y_1y_2] = E[y_1]E[y_2]$$

   so that the advantage of VQ is removed. What might be an implication of such a decorrelating transformation for speech coding?

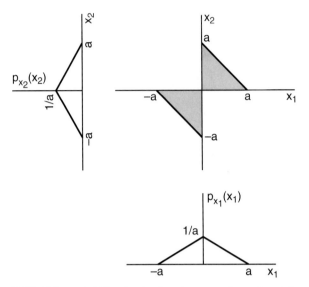

Figure 12.33 Joint probability density function $p_{x_1,x_2}(x_1, x_2)$ and its marginals $p_{x_1}(x_1)$ and $p_{x_2}(x_2)$.

12.11 Show that minimizing the subband coding distortion measure in Equation (12.14) results in the optimal bit assignment of Equation (12.15) and thus the flat quantization noise variance $\sigma_e^2(nL, k) = D^*$. Then derive the optimal frequency-weighted bit assignment rule of Equation (12.17).

12.12 We saw with subband coding that when the weight in the error function in Equation (12.16) across bands is inversely proportional to the variance of the subband signal, the optimal rule is to give each band the same number of bits, and this results in the same SNR for each band. Suppose that an octave band structure is invoked, e.g., quadrature mirror filter bandwidths decreasing by a factor of two with decreasing frequency. Then, as we have seen in Chapter 8, wider bands imply the need for a faster sampling rate (less decimation) to maintain the invertibility of the filter bank. Discuss the implications of this requirement of band-dependent decimation for bit allocation, given the above optimal uniform bit allocation per band.

12.13 Argue that a reasonable model of the spectral tilt used in postfiltering is a first-order all-pole function:

$$T(\omega) = \frac{\sigma}{1 - \rho \exp(-j\omega)},$$

and hence the spectrally flattened sinewave amplitude for the kth harmonic in Section 12.5.2, denoted by $F(k\omega_0)$, can be written as

$$F(k\omega_0) = \frac{\bar{A}(k\omega_0)}{|T(k\omega_0)|} = \bar{A}(k\omega_0)\left[\frac{(1 + \rho^2) - 2\rho \cos(k\omega_0)}{\sigma^2}\right]^{\frac{1}{2}}.$$

In the context of sinusoidal coding, then determine the prediction coefficient ρ by applying linear prediction analysis to the synthetic speech waveform in Equation (12.20). Specifically, show that

$\rho = r_1/r_0$, where r_0 and r_1 are the first two correlation coefficients of $\hat{s}[n]$ and that

$$r_0 = \sum_{k=1}^{K} \bar{A}(k\omega_o)^2$$

$$r_1 = \sum_{k=1}^{K} \bar{A}(k\omega_o)^2 \cos(k\omega_0).$$

12.14 The oldest form of speech coding, the channel vocoder, invented by Dudley [15], was illustrated in Figures 8.31 and 8.32. In this exercise, you are asked to consider the scalar quantization of channel parameter outputs determined in Exercise 8.11.

    (a) Suppose each channel parameter b_k spans a range of 0 to 8, and a 5-bit uniform quantizer is used. What is the mean-squared error (i.e., variance of the quantization noise) associated with the quantizer? Assume a uniform pdf for quantization noise.

    (b) If 10 bits are needed for the pitch and voicing decision, what is the total bit rate in bits per second? Assume that the pitch and voicing are transmitted at the decimation rate you selected in part (b) of Exercise 8.11.

12.15 An alternative to the cubic phase interpolator used in sinewave synthesis was given in Figure 9.29 of Exercise 9.5, that shows one synthesis frame from $t = 0$ to $t = T$. The measurements θ^l, θ^{l+1}, Ω^l, and Ω^{l+1} are given. In this interpolator, the frequency is assumed piecewise-constant over three equal intervals. The mid-frequency value $\tilde{\Omega}$ is a free parameter that was selected so that the phase measurement $\theta(T)$ at the right boundary is met. In this problem, you are asked to design a sinewave phase coding scheme based on your results from parts (a) and (b) of Exercise 9.5.

    (a) Suppose sinewave analysis/synthesis is performed every 5 ms. We want to determine a quantization scheme and corresponding bit rate for transmitting phase. Suppose we have transmitted up to frame l so the receiver has the values θ^l (assume unwrapped) and Ω^l. On the $(l+1)$st frame there are two parameters to be transmitted: θ^{l+1} and Ω^{l+1}, from which we use the results in parts (a) and (b) of Exercise 9.5 to obtain the interpolated phase on frame $l+1$. Note that θ^{l+1} is wrapped, thus falling in the interval $[-\pi, \pi]$ and Ω^{l+1} falls in the interval $[0, \pi]$. Consider a uniform quantization scheme with 16 levels for each parameter. Calculate the bit rate required to transmit phase for 50 sinewaves per 5-ms frame. Discuss the feasibility of this scheme for low-rate speech coding.

    (b) Comment on the effect of the phase quantization scheme of part (a) in maintaining the speech waveform shape in synthesis. That is, argue briefly either for or against the method's capability to avoid waveform dispersion in synthesis. Suppose you choose the phase difference $\theta^{l+1} - \theta^l$ and frequency difference $\Omega^{l+1} - \Omega^l$ to quantize and code, rather than the phase and frequency themselves. Would such a quantization scheme help or hurt your ability to (1) reduce quantization noise and/or (2) maintain speech waveform shape at the receiver? A qualitative argument is sufficient.

12.16 Consider a signal $x[n]$ consisting of two sinewaves with frequency ω_1 and ω_2, phase ϕ_1 and ϕ_2, and amplitude 2.0 and 1.0, respectively, i.e.,

$$x[n] = 2\cos(\omega_1 n + \phi_1) + \cos(\omega_2 n + \phi_2).$$

Suppose that the pdf of $x[n]$ is uniform, as given in Figure 12.34a.

    (a) Suppose you quantize $x[n]$ with a 9-bit optimal (in the mean-squared error sense of Max) uniform quantizer. What is the signal-to-noise (SNR) ratio in dB?

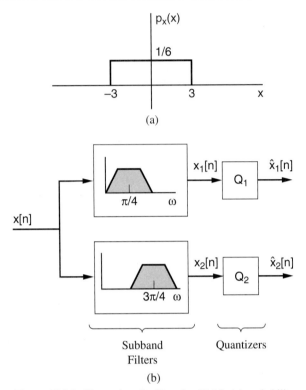

Figure 12.34 Illustrations for Exercise 12.16: (a) probability density function of $x[n]$; (b) two-band quantizer.

(b) Suppose now you (subband) filter $x[n]$ with two bandpass filters split roughly over the full-band interval $[0, \pi]$, one centered at $\omega = \frac{\pi}{4}$ and the second at $\omega = \frac{3\pi}{4}$, as illustrated in Figure 12.5b. Assume $H_1(\omega)$ passes only the sinewave at frequency ω_1 exactly, while $H_2(\omega)$ passes only the sinewave at frequency ω_2 exactly. Assume the pdfs of $x_1[n]$ and $x_2[n]$ are uniform over their respective ranges. Given that you have 9 bits to allocate over the two bands (i.e., you can use only a total of 9 bits, so that $B_1 + B_2 = 9$ where B_1 denotes the number of bits for the first band and B_2 denotes the number of bits for the second band), design an optimal uniform quantizer for each band such that the SNR is the same in each band. In your design, specify the number of bits you allocate for each band, as well as the quantization step size. What is the SNR in each band? (Assume quantization noise has uniform pdfs. Also, the number of bits may be fractional.)

(c) To reconstruct the signal, you add the two quantized outputs. What is the SNR of the reconstructed signal? (Assume quantization errors and signals are independent.)

(d) Suppose that the absolute phase relation between two sinewave components of a signal is perceptually unimportant. (Generally, this is not true, but let's assume so.) Suppose also that you are given *a priori* knowledge that only two sinewaves are present, each falling within a different subband of the system from part (b). Finally, suppose also that the magnitude and frequency of the sinewaves are varying slowly enough so that they can be sampled at 1/8 of the original input sampling rate. Modify the subband encoder of Figure 12.34b so that only the sinewave amplitude and the frequency per subband are required for encoding. If each parameter is allocated the same number of bits, given that you have 9 bits/sample at the input sampling rate, how many bits does each amplitude and frequency parameter receive? Explain qualitatively the steps involved in the design of the optimal nonuniform Max quantizer for the amplitude and frequency parameters. (Assume you have many sample functions and that you don't know the pdfs and need to measure them.) Keep in mind that pdfs of the amplitude and frequency parameters are not generally uniform, even though we are assuming the pdfs of $x_1[n]$ and $x_2[n]$ are uniform. *Hint:* Use the principles of the phase vocoder.

12.17 From the frequency-domain view in Chapter 5 of linear prediction analysis, show that we can derive an all-pole fit to the SEEVOC envelope simply by replacing the autocorrelation coefficients of the short-time speech by the autocorrelation coefficients associated with the squared SEEVOC envelope.

12.18 Determine a means to recover the inverse filter $A(z)$ from the roots of $P(z)$ and $Q(z)$ (i.e., the LSF coefficients) in Equation (12.26).

12.19 Model-based vocoders that do not reproduce frequency-domain phase typically do not allow a time-domain minimum-mean-squared error performance measure. Give reasoning for this limitation. Then propose for such coders an alternative frequency-domain objective measure. Suppose now that a class of vocoders generates a minimum-phase reconstruction. Propose an objective performance measure as a function of the amplitudes, frequencies, and phases of sinusoidal representations of the original and coded waveforms.

12.20 Show that a zero in the vocal tract transfer function manifests itself as a "multi-pulse" close to the primary pitch impulses of the excitation when the original speech is filtered by its all-pole transfer function component.

12.21 In multi-pulse linear prediction analysis/synthesis, assume that the impulse positions n_k in the excitation model [Equation (12.28)] are known and solve the minimum mean-squared error [Equation (12.27)] estimation of the unknown impulse amplitudes A_k. *Hint:* The solution is linear in the unknown amplitudes.

12.22 Consider estimation of the long-time predictor lag P in the long-term predictor $b[n] = \delta[n] - b\delta[n - P]$ of Section 12.7.2. Argue why at long lags, we might obtain poor estimates of the autocorrelation and, consequently, obtain a poor basis for the use of the autocorrelation method in obtaining the gain b required in the long-term predictor.

12.23 Derive the error expression in Equation (12.30) for the long-term predictor error.

12.24 (MATLAB) In this MATLAB exercise, use the speech waveform *speech1_10k* (at 10000 samples/s) in workspace *ex12M1.mat* located in companion website directory Chap_exercises/chapter12. This problem asks you to design a speech coder based on both homomorphic filtering and sinewave analysis/synthesis, and is quite open-ended, i.e., there is no one solution.

    (a) Window *speech1_10k* with a 25-ms Hamming window and compute the real cepstrum of the short-time segment using a 1024-point FFT.

    (b) Estimate the pitch (in DFT samples) using the real cepstrum. Quantize the pitch using a 7-bit uniform quantizer. Assume that the pitch of typical speakers can range between 50 Hz and 300 Hz, and map this range to DFT samples by rounding to the nearest integer.

    (c) Quantize the first 28 cepstral coefficients $c[n]$ for $n = 0, 1, 2, 3, \ldots 27$. Observe that there is a very large dynamic range in the cepstral coefficients; i.e., the low-quefrency coefficients are much larger than the high-quefrency coefficients, implying that the quantizer should adapt to the cepstral number. Since the gain term $c[0]$ swamps the cepstrum, remove this term and quantize it separately with a 5-bit quantizer. Assume that $c[0]$ can range over twice the measured value. Then compute an average value of $c[n]$ for $n = 1, 2, 3, \ldots 27$ and assume that these coefficients can range over twice this average value. Select some reasonable quantization step size and quantize the 27 cepstral coefficients. Alternatively, you can assume the range of *each* $c[n]$ is twice the given $c[n]$. This latter choice is more consistent with letting the quantizer adapt to the variance of the cepstral value. This ends the "transmitter" stage of the speech coder.

(d) At the receiver, form a minimum-phase reconstruction of the waveform from your quantized values. To do this, first apply the appropriate right-sided cepstral lifter to the quantized 28 cepstral coefficients and Fourier transform the result using a 1024-point FFT. The lifter multiplies $c[0]$ by 1 and $c[n]$ for $n = 1, 2 \ldots 27$ by 2 for a minimum-phase reconstruction. These operations provide a log-magnitude and phase estimate of the vocal tract frequency response.

(e) Sample the log-magnitude and phase functions from part (d) at the harmonic frequencies. The harmonic frequencies can be generated from the quantized pitch. Note that you will need to select harmonics closest to DFT samples.

(f) Exponentiate the log-magnitude and phase harmonic samples. Then use the resulting amplitudes, frequencies, and phases to form a sinewave-based reconstruction of the speech waveform using the MATLAB function designed in Exercise 9.20. The synthesized waveform should be the same length as the original (i.e., 1000 samples). How does your reconstruction compare to the original waveform visually and aurally? You have now synthesized a minimum-phase, harmonic waveform at your "receiver."

(g) Assuming that you code, reconstruct, and concatenate 1000-point short-time segments (as in this problem), i.e., the time decimation factor $L = 1000$, and that the waveform sampling rate is 10000 samples/s (in practice a $L = 100$ for 10000 samples/s is more realistic), what is the bit rate, i.e., number of bits/s, of your speech coder? In answering this question, you need to consider the bits required in coding the pitch, gain ($c[0]$), and cepstral coefficients ($c[n]$, $1 \le n \le 27$).

BIBLIOGRAPHY

[1] G. Aguilar, J. Chen, R.B. Dunn, R.J. McAulay, X. Sun, W. Wang, and R. Zopf, "An Embedded Sinusoidal Transform Codec with Measured Phases and Sampling Rate Scalability," *Proc. IEEE Int. Conf. Acoustics, Speech, and Signal Proc.*, Istanbul, Turkey, vol. 2, pp. 1141–1144, June 2000.

[2] S. Ahmadi and A.S. Spanias, "A New Phase Model for Sinusoidal Transform Coding of Speech," *IEEE Trans. on Speech and Audio Processing*, vol. 6, no. 5, pp. 495–501, Sept. 1998.

[3] L.B. Almeida and F.M. Silva, "Variable-Frequency Synthesis: An Improved Harmonic Coding Scheme," *Proc. IEEE Int. Conf. Acoustics, Speech, and Signal Processing*, San Diego, CA, pp. 27.5.1–27.5.4, May 1984.

[4] D. Anderson, "Speech Analysis and Coding Using a Multi-Resolution Sinusoidal Transform," *Proc. IEEE Conf. on Acoustics, Speech, and Signal Processing*, Atlanta, GA, vol. 2, pp. 1037–1040, May 1996.

[5] B.S. Atal and S.L. Hanauer, "Speech Analysis and Synthesis by Linear Prediction of the Speech Waveform," *J. Acoustical Society of America*, vol. 50, pp. 637–655, 1971.

[6] B.S. Atal and J.R. Remde, "A New Model of LPC Excitation for Producing Natural-Sounding Speech at Low Bit Rates," *Proc. IEEE Int. Conf. Acoustics, Speech, and Signal Processing*, Paris, France, vol. 1, pp. 614–617, April 1982.

[7] N. Benvenuto *et al.*, "The 32-kb/s ADPCM Coding Standard," *AT&T Technical J.*, vol. 65, pp. 12–22, Sept./Oct. 1990.

[8] J.P. Campbell, Jr., T.E. Tremain, and V.C. Welch, "The Federal Standard 1016 4800 bps CELP Voice Coder," *Digital Signal Processing*, Academic Press, vol. 1, no. 3, pp. 145–155, 1991.

[9] J.H. Chen and A. Gersho, "Real-Time Vector APC Speech Coding at 4800 b/s with Adaptive Postfiltering," *Proc. IEEE Int. Conf. Acoustics, Speech, and Signal Processing*, Dallas, TX, vol. 4, pp. 2185–2188, May 1987.

[10] R.V. Cox, S.L. Gay, Y. Shoham, S.R. Quackenbush, N. Seshadri, and N. Jayant, "New Directions in Subband Coding," *IEEE J. Selected Areas in Communications*, vol. 6. no. 2, pp. 391–409, Feb. 1988.

[11] R.E. Crochiere, S.A. Webber, and J.L. Flanagan, "Digital Coding of Speech in Sub-Bands," *Bell System Technical J.*, vol. 55, no. 8, pp. 1069–1085, Oct. 1976.

[12] I. Daubechies, *Ten Lectures on Wavelets*, SIAM, 1992.

[13] W.B. Davenport, "An Experimental Study of Speech-Wave Probability Distributions," *J. Acoustical Society of America*, vol. 24, pp. 390–399, July 1952.

[14] L.D. Davisson, "Rate-Distortion Theory and Application," *Proc. IEEE*, vol. 60, no. 7, pp. 800–808, July 1972.

[15] H. Dudley, R. Riesz, and S. Watkins, "A Synthetic Speaker," *J. Franklin Inst.*, vol. 227, no. 739, 1939.

[16] S. Dimolitsas, F. L. Corcoran, C. Ravishankar, R.S. Skaland, and A. Wong. "Evaluation of Voice Codec Performance for the INMARSAT Mini-M System," *Proc. 10th Int. Digital Satellite Conf.*, Brighton, England, May 1995.

[17] J.R. Deller, J.G. Proakis, and J.H.L. Hansen, *Discrete-Time Processing of Speech*, Macmillan Publishing Co., New York, NY, 1993.

[18] E.W. Forgy, "Cluster Analysis of Multivariate Data: Efficiency vs. Interpretability of Classifications," *Biometrics*, abstract, vol. 21, pp. 768–769, 1965.

[19] European Telecommunication Standards Institute, "European Digital Telecommunications System (Phase2); Full Rate Speech Processing Functions (GSM 06.01)," ETSI, 1994.

[20] A. Gersho and R. Gray, *Vector Quantization and Signal Compression*, Kluwer Academic Publishers, Dordrecht, Holland, 1991.

[21] O. Ghitza, "Speech Analysis/Synthesis Based on Matching the Synthesized and Original Representations in the Auditory Nerve Level," *Proc. Int. Conf. Acoustics, Speech, and Signal Processing*, pp. 1995–1998, Tokyo, Japan, 1986.

[22] B. Gold and C.M. Radar, "The Channel Vocoder," *IEEE Trans. on Audio and Electroacoustics*, vol. AU–15, no. 4, Dec. 1967.

[23] B. Gold and N. Morgan, *Speech and Audio Signal Processing*, John Wiley and Sons, New York, NY, 2000.

[24] D. Griffin and J.S. Lim, "Multiband Excitation Vocoder," *IEEE Trans. Acoustics, Speech, and Signal Processing*, vol. ASSP–36, no. 8, pp. 1223–1235, 1988.

[25] J.C. Hardwick and J.S. Lim, "A 4800 bps Improved Multi-Band Excitation Speech Coder," *Proc. IEEE Workshop on Speech Coding for Telecommunications*, Vancouver, B.C., Canada, Sept. 5–8, 1989.

[26] J.N. Holmes, "Formant Excitation Before and After Glottal Closure," *Proc. IEEE Int. Conf. Acoustics, Speech, and Signal Processing*, Philadelphia, PA, pp. 39–42, April 1976.

[27] J. Huang and F. Schultheiss, "Block Quantization of Correlated Gaussian Random Variables," *IEEE Trans. Communications Systems*, vol. CS–11, pp. 289–296, Sept. 1963.

[28] "INMARSAT-M Voice Codec," *Proc. Thirty-Sixth Inmarsat Council Meeting*, Appendix I, July 1990.

[29] J.D. Johnson, "Transform Coding of Audio Signals Using Perceptual Noise Criteria," *IEEE J. Selected Areas in Communications*, vol. 6, no. 2, pp. 314–323, Feb. 1988.

[30] ITU-T Recommendation G.729, "Coding of Speech at 8 kb/s Using Conjugate-Structure Algebraic-Code-Excited Linear Prediction," June 1995.

[31] ITU-T Recommendation G.723.1, "Dual Rate Speech Coder for Multimedia Communications Transmitting at 5.3 and 6.3 kb/s," March 1996.

[32] "APCO/NASTD/Fed. Project 25 Vocoder Description," Telecommunications Industry Association Specifications, 1992.

[33] N.S. Jayant and P. Noll, *Digital Coding of Waveforms: Principles and Applications to Speech and Video*, Prentice Hall, Englewood Cliffs, NJ, 1984.

[34] G.S. Kang and S.S. Everett, "Improvement of the Excitation Source in the Narrowband LPC Analysis," *IEEE Trans. Acoustics, Speech, and Signal Processing*, vol. ASSP–33, no. 2, pp. 377–386, April 1985.

[35] W.B. Kleijn, "Encoding Speech Using Prototype Waveforms," *IEEE Trans. on Speech and Audio Processing*, vol. 1, no. 4, pp. 386–399, Oct. 1993.

[36] W.B. Kleijn and J. Haagen, "A Speech Coder Based on Decomposition of Characteristic Waveforms," *Proc. Int. Conf. Acoustics, Speech, and Signal Processing*, Detroit, MI, pp. 508–511, May 1995.

[37] A. Kondoz, *Digital Speech: Coding for Low Bit Rate Communication Systems*, John Wiley & Sons, NY, 1994.

[38] P. Kroon and W.B. Kleijn, "Linear-Prediction Analysis-by-Synthesis Coding," chapter in *Speech Coding and Synthesis*, W.B. Kleijn and K.K. Paliwal, eds., Elsevier, Amsterdam, the Netherlands, 1995.

[39] K.D. Kryter, "Methods for the Calculation and Use of the Articulation Index," *J. Acoustical Society of America*, vol. 34, pp. 1689–1697, 1962.

[40] G. Kubin, "Nonlinear Processing of Speech," chapter in *Speech Coding and Synthesis*, W.B. Kleijn and K.K. Paliwal, eds., Amsterdam, the Netherlands, Elsevier, 1995.

[41] J.S. Lim, *Two-Dimensional Signal and Image Processing*, Prentice Hall, Englewood Cliffs, NJ, 1990.

[42] Y. Linde, A. Buzo, and R.M. Gray, "An Algorithm for Vector Quantizer Design," *IEEE Trans. Communications*, vol. COM–28, no. 1, pp. 84–95, Jan. 1980.

[43] S.P. Lloyd, "Least-Squares Quantization in PCM," *IEEE Trans. Information Theory*, vol. IT–28, pp. 129–137, March 1982.

[44] J. Makhoul, S. Roucos, and H. Gish, "Vector Quantization in Speech Coding," *Proc. IEEE*, vol. 73, pp. 1551–1588, Nov. 1985.

[45] J. Makhoul, R. Viswanathan, R. Schwartz, and A.W.F. Huggins, "A Mixed-Source Model for Speech Compression and Synthesis," *Proc. IEEE Int. Conf. Acoustics, Speech, and Signal Processing*, Tulsa, OK, pp. 163–166, 1978.

[46] J.D. Markel and A.H. Gray, *Linear Prediction of Speech*, Springer-Verlag, New York, NY, 1976.

[47] J.S. Marques and L.B. Almeida, "New Basis Functions for Sinusoidal Decomposition," *Proc. EUROCON*, Stockholm, Sweden, 1988.

[48] J. Max, "Quantizing for Minimum Distortion," *IRE Trans. Information Theory*, vol. IT–6, pp. 7–12, March 1960.

[49] R.J. McAulay and T.F. Quatieri, "Low-Rate Speech Coding Based on the Sinusoidal Model," chapter in *Advances in Speech Signal Processing*, S. Furui and M.M. Sondhi, eds., Marcel Dekker, New York, NY, 1992.

[50] R.J. McAulay and T.F. Quatieri, "Sinusoidal Coding," chapter in *Speech Coding and Synthesis*, W.B. Kleijn and K.K. Paliwal, eds., Amsterdam, the Netherlands, Elsevier, 1995.

[51] R.J. McAulay and T.F. Quatieri, "Speech Analysis-Synthesis Based on a Sinusoidal Representation," *IEEE Trans. Acoustics, Speech, and Signal Processing*, vol. ASSP–34, no. 4, pp. 744–754, 1986.

[52] R.J. McAulay, T.M. Parks, T.F. Quatieri, and M. Sabin, "Sinewave Amplitude Coding at Low Data Rates," *Proc. IEEE Workshop on Speech Coding*, Vancouver, B.C., Canada, 1989.

[53] R.J. McAulay, T.F. Quatieri, and T.G. Champion, "Sinewave Amplitude Coding Using High-Order All-Pole Models," *Proc. EUSIPCO–94*, Edinburgh, Scotland, U.K., pp. 395–398, Sept. 1994.

[54] R.J. McAulay and T.F. Quatieri, "Multirate Sinusoidal Transform Coding at Rates from 2.4 kb/s to 8 kb/s," *Proc. IEEE Int. Conf. Acoustics, Speech, and Signal Processing*, Dallas, TX, vol. 3, pp. 1645–1648, May 1987.

[55] A.V. McCree and T.P. Barnwell, "A Mixed Excitation LPC Vocoder Model for Low Bit Rate Speech Coding," *IEEE Trans. on Speech and Audio Processing*, vol. 3, no. 4, pp. 242–250, July 1995.

[56] A.V. McCree, K. Truong, E.B. George, T.P. Barnwell, and V. Viswanathan, "A 2.4 kbit/s MELP Coder Candidate for the New US Federal Standard," *Proc. IEEE Int. Conf. Acoustics, Speech and Signal Processing*, Atlanta, GA, vol. 1, pp. 200–203, May 1996.

[57] A.V. McCree, "A 4.8 kbit/s MELP Coder Candidate with Phase Alignment," *Proc. IEEE Int. Conf. Acoustics, Speech, and Signal Processing*, Istanbul, Turkey, vol. 3, pp. 1379–1382, June 2000.

[58] E. McLarnon, "A Method for Reducing the Frame Rate of a Channel Vocoder by Using Frame Interpolation," *Proc. IEEE Int. Conf. Acoustics, Speech, and Signal Processing*, Washington, D.C., pp. 458–461, 1978.

[59] M. Nishiguchi, J. Matsumoto, R. Wakatsuki, and S. Ono, "Vector Quantized MBE with Simplified V/UV Division at 3.0 kb/s," *Proc. Int. Conf. Acoustics, Speech and Signal Processing*, Minneapolis, MN, vol. 2, pp. 151–154, April 1993.

[60] P. Noll, "A Comparative Study of Various Schemes for Speech Encoding," *Bell System Tech. J.*, vol. 54, no. 9, pp. 1597–1614, Nov. 1975.

[61] P. Noll, "Adaptive Quantizing in Speech Coding Systems," *Proc. 1974 Zurich Seminar on Digital Communications*, Zurich, Switzerland, March 1974.

[62] A.V. Oppenheim and R.W. Schafer, *Discrete-Time Signal Processing*, Prentice Hall, Englewood Cliffs, NJ, 1989.

[63] A. Papoulis, *Probability, Random Variables, and Stochastic Processes*, McGraw-Hill, New York, NY, 1965.

[64] D. Paul, "The Spectral Envelope Estimation Vocoder," *IEEE Trans. Acoustics, Speech, and Signal Processing*, vol. ASSP–29, no. 4, pp. 786–794, Aug. 1981.

[65] J.G. Proakis, *Digital Communications*, McGraw-Hill, New York, NY, 1983.

[66] M.D. Paez and T.H. Glisson, "Minimum Mean Squared-Error Quantization in Speech," *IEEE Trans. Communications*, vol. COM–20, pp. 225–230, April 1972.

[67] A. Potamianos and P. Maragos, "Applications of Speech Processing Using an AM-FM Modulation Model and Energy Operators," chapter in *Signal Processing VII: Theories and Applications*, M. Holt, C.F. Cowan, P.M. Grant, and W.A. Sandham, eds., vol. 3, pp. 1669–1672, Amsterdam, the Netherlands, Elsevier, Sept. 1994.

[68] T.F. Quatieri and R.J. McAulay, "Phase Coherence in Speech Reconstruction for Enhancement and Coding Applications," *Proc. IEEE Int. Conf. Acoustics, Speech, and Signal Processing*, Glasgow, Scotland, vol. 1, pp. 207–209, May 1989.

[69] T.F. Quatieri and E.M. Hofstetter, "Short-Time Signal Representation by Nonlinear Difference Equations," *Proc. IEEE Int. Conf. Acoustics, Speech, and Signal Processing*, Albuquerque, NM, vol. 3, pp. 1551–1554, April 1990.

[70] S.R. Quackenbush, T.P. Barnwell, and M.A. Clements, *Objective Measures of Speech Quality*, Prentice Hall, Englewood Cliffs, NJ, 1988.

[71] L.R. Rabiner and R.W. Schafer, *Digital Processing of Speech Signals*, Prentice Hall, Englewood Cliffs, NJ, 1978.

[72] L.G. Roberts, "Picture Coding Using Pseudo-Random Noise," *IRE Trans. on Information Theory*, vol. 8, pp. 145–154, Feb. 1962.

[73] R.C. Rose, *The Design and Performance of an Analysis-by-Synthesis Class of Predictive Speech Coders*, Ph.D. Thesis, Georgia Institute of Technology, School of Electrical Engineering, April 1988.

[74] M.J. Sabin, "DPCM Coding of Spectral Amplitudes without Positive Slope Overload," *IEEE Trans. Signal Processing*, vol. 39. no. 3, pp. 756–758, 1991.

[75] R. Salami, C. Laflamme, J.P. Adoul, A. Kataoka, S. Hayashi, T. Moriya, C. Lamblin, D. Massaloux, S. Proust, P. Kroon, and Y. Shoham, "Design and Description of CS-ACELP: A Toll Quality 8 kb/s Speech Coder," *IEEE Trans. on Speech and Audio Processing*, vol. 6, no. 2, pp. 116–130, March 1998.

[76] M.R. Schroeder and B.S. Atal, "Code-Excited Linear Prediction (CELP): High-Quality Speech at Very Low Bit Rates," *Proc. IEEE Int. Conf. Acoustics, Speech, and Signal Processing*, vol. 3, pp. 937–940, April 1985.

[77] M.J.T. Smith and T.P. Barnwell, "Exact Reconstruction Techniques for Tree-Structured Subband Coders," *IEEE Trans. Acoustics, Speech, and Signal Processing*, vol. ASSP–34, no. 3, pp. 434–441, June 1986.

[78] F.K. Soong and B.H. Juang, "Line Spectral Pair and Speech Compression," *Proc. IEEE Int. Conf. Acoustics, Speech, and Signal Processing*, San Diego, CA, vol. 1, pp. 1.10.1–1.10.4, 1984.

[79] N. Sugamura and F. Itakura, "Speech Data Compression by LSP Analysis/Synthesis Technique," *Trans. of the Institute of Electronics, Information, and Computer Engineers*, vol. J64-A, pp. 599–606, 1981.

[80] B. Smith, "Instantaneous Companding of Quantized Signals," *Bell System Technical J.*, vol. 36, no. 3, pp. 653–709, May 1957.

[81] J.M. Tribolet and R.E. Crochiere, "Frequency Domain Coding of Speech," *IEEE Trans. Acoustics, Speech, and Signal Processing*, vol. ASSP–27, no. 5, pp. 512–530, Oct. 1979.

[82] D.Y. Wong, B.H. Juang, and A.H. Gray, Sr., "An 800 bit/s Vector Quantization LPC Vocoder," *IEEE Trans. Acoustics, Speech, and Signal Processing*, vol. ASSP–30, no. 5, pp. 770–779, Oct. 1982.

[83] S. Yeldener, A.M. Kondoz, and B.G. Evans, "High-Quality Multi-Band LPC Coding of Speech at 2.4 kb/s," *IEE Electronics*, vol. 27, no. 14, pp. 1287–1289, July 1991.

CHAPTER 13

Speech Enhancement

13.1 Introduction

Throughout the text, we have introduced a number of speech enhancement techniques, including homomorphic deconvolution in Chapter 6 for removing convolutional distortion and a spectral magnitude-only reconstruction approach in Chapter 7 to reducing additive noise. In this chapter, we further develop speech enhancement methods for addressing these two types of distortion. We cannot hope to cover all such methods; rather, we illustrate how certain analysis/synthesis schemes described in the text are used as a basis for enhancement, focusing on signal processing principles.

In this chapter, we judge enhancement techniques in part by the extent to which they reduce additive noise or convolutional distortion, by the severity of artifacts in the remaining (or "residual") disturbance, and by the degree of distortion in the desired speech signal. Because of the typical requirement of average speech characteristics, such as average short-time spectra, enhancement algorithms tend to degrade transient and dynamic speech components—for example, plosive fine structure and formant modulation—that contribute significantly to the quality of the speech signal. Therefore, with regard to speech distortion, we are interested in preserving not only slowly varying short-time spectral characteristics, but also instantaneous temporal properties such as signal attack, decay, and modulation. In addition, we judge the enhanced speech by using the subjective and objective measures of quality that were outlined in the introduction of Chapter 12. In subjective quality evaluation, which is ultimately the deciding evaluation in human listening, we consider speech naturalness, intelligibility, and speaker identifiability. Speech enhancement, however, is not always performed for the human listener, but often is performed for the machine recognizer. As an example, in Chapter 14, we will see how many of the techniques in this chapter are applied to improve automatic speaker recognition performance when convolutional and additive disturbances are present.

We begin in Section 13.2 with the Fourier transform and filtering perspectives of the short-time Fourier transform (STFT) as bases for approaches to the problems of reducing additive noise and convolutional distortion. We describe, as a foundation, *spectral subtraction* that operates on STFT short-time sections for additive noise reduction, and *cepstral mean subtraction* that operates on STFT bandpass filter outputs for removing stationary convolutional distortion. In Section 13.3, the *Wiener filter* and its adaptive renditions for additive noise removal are then developed. A variety of Wiener-filter-based approaches are given, including a method that adapts to spectral change in a signal to help preserve signal nonstationarity while exploiting auditory temporal masking, and also a stochastic-theoretical approach to obtain a mean-squared error estimate of the desired spectral magnitude. Sections 13.4 and 13.5 then develop all-pole model-based and further auditory-based approaches to additive noise reduction, respectively. The auditory-based methods use frequency-domain perceptual masking principles to conceal annoying residual noise under the spectral components of interest. Section 13.6 next generalizes cepstral mean subtraction (CMS), introduced in Section 13.2, to reduce *time-varying* convolutional distortion. This generalization, commonly referred to as *RASTA*, (along with CMS) can be viewed as homomorphic filtering along the time dimension of the STFT, rather than with respect to its frequency dimension. These approaches represent a fascinating application of homomorphic filtering theory in a domain different from that studied in Chapter 6. Moreover, we will see that CMS and RASTA can be viewed as members of a larger class of enhancement algorithms that filter *nonlinearly* transformed temporal envelopes of STFT filter outputs.

13.2 Preliminaries

In this section, we first formulate the additive noise and convolutional distortion problems in the context of the STFT, from both the Fourier transform and filtering viewpoints introduced in Chapter 7. With this framework, we then develop the method of spectral subtraction for additive noise suppression and cepstral mean subtraction for reducing a stationary convolutional distortion.

13.2.1 Problem Formulation

Additive Noise — Let $y[n]$ be a discrete-time noisy sequence

$$y[n] = x[n] + b[n] \tag{13.1}$$

where $x[n]$ is the desired signal, which we also refer to as the "object," and $b[n]$ is the unwanted background noise. For the moment, we assume $x[n]$ and $b[n]$ to be wide-sense stationary, uncorrelated random processes with power spectral density functions (Appendix 5.A) denoted by $S_x(\omega)$ and $S_b(\omega)$, respectively. One approach to recovering the desired signal $x[n]$ relies on the additivity of the power spectra, i.e.,

$$S_y(\omega) = S_x(\omega) + S_b(\omega). \tag{13.2}$$

With STFT analysis, however, we work with the short-time segments given by

$$y_{pL}[n] = w[pL - n](x[n] + b[n])$$

where L is the frame length and p is an integer, which in the frequency domain is expressed as

$$Y(pL, \omega) = X(pL, \omega) + B(pL, \omega)$$

where $X(pL, \omega)$, $B(pL, \omega)$, and $Y(pL, \omega)$ are the STFTs of the object $x[n]$, the background noise $b[n]$, and the measurement $y[n]$, respectively, computed at frame interval L. The STFT magnitude squared of $y[n]$ is thus given by

$$|Y(pL, \omega)|^2 = |X(pL, \omega)|^2 + |B(pL, \omega)|^2$$
$$+ X^*(pL, \omega)B(pL, \omega) + X(pL, \omega)B^*(pL, \omega) \qquad (13.3)$$

from which our objective is to obtain an estimate of $|X(pL, \omega)|^2$. We think of the relation in Equation (13.3) as the "instantaneous" counterpart to the stochastic Equation (13.2).

In the above approach to signal estimation, we do not estimate the STFT phase.[1] Therefore, the best we can do for each short-time segment is an estimate of the form

$$\hat{X}(pL, \omega) = |X(pL, \omega)|e^{j\angle Y(pL, \omega)}, \qquad (13.4)$$

i.e., the ideal STFT estimate consists of the clean STFT magnitude and noisy measured STFT phase. We refer to this as the *theoretical limit* in estimating the original STFT when only the STFT magnitude is estimated [53]. By considering the threshold of perception of phase deviation due to additive noise, it has been shown that speech degradation is not perceived with an average short-time "segmental" signal-to-noise ratio[2] (SNR) > 6 dB for the theoretical limit in Equation (13.4). With this SNR considerably below 6 dB, however, a roughness of the reconstruction is perceived [53].

Convolutional Distortion — Consider now a sequence $x[n]$ that has passed through a linear time-invariant distortion $g[n]$ resulting in a sequence $y[n] = x[n] * g[n]$. Our objective is to recover $x[n]$ from $y[n]$ without *a priori* knowledge of $g[n]$. This problem is sometimes referred to as *blind deconvolution*. We saw an example of blind deconvolution in Exercise 6.20 where homomorphic filtering reduced the effect of the distorting impulse response $g[n]$ of an acoustic recording horn. As in Exercise 6.20, we assume that in short-time analysis the window $w[n]$ is long and smooth relative to the distortion $g[n]$, so that a short-time segment of $y[n]$ can be written as

$$y_{pL}[m] = w[pL - m](x[m] * g[m])$$
$$\approx (x[m]w[pL - m]) * g[m].$$

Then we can write the STFT of the degraded signal as [2] (Exercise 13.9 gives a formal argument)

$$Y(pL, \omega) = \sum_{m=-\infty}^{\infty} w[pL - m](x[m] * g[m])e^{-j\omega m}$$

$$\approx \sum_{m=-\infty}^{\infty} (w[pL - m]x[m]) * g[m]e^{-j\omega m}$$

$$= X(pL, \omega)G(\omega). \qquad (13.5)$$

---

[1] We noted in Chapter 7 (Section 7.4) that STFT phase estimation for speech signals in noise is a more difficult problem than STFT magnitude estimation. This is in part due to the difficulty in characterizing phase in low-energy regions of the spectrum, and in part due to the use of only second-order, statistical averages, e.g., the autocorrelation function and its corresponding power spectrum, in standard noise reduction algorithms. One approach to estimate phase is through the STFT magnitude as described in Sections 7.5.3 and 7.6.2. A different approach is proposed in Exercise 13.4.

[2] The average short-time SNR is the ratio (in dB) of the energy in the short-time clean speech and short-time noise disturbance averaged over all frames. This ratio is sometimes referred to as the *segmental signal-to-noise ratio*, a nomenclature that was introduced in Chapter 12.

Because the Fourier transform of the distortion, $G(\omega)$, is a multiplicative modification of the desired signal's STFT, one is tempted to apply the homomorphic filtering described in Chapter 6 whereby $g[m]$ is removed from each short-time section $y_{pL}[m]$ using cepstral liftering. Our objective, however, is to obtain $x[n]$ even when the cepstra of the signal and the distortion are not necessarily disjoint in the quefrency domain.

13.2.2 Spectral Subtraction

We return now to the problem of recovering an object sequence $x[n]$ from the noisy sequence $y[n]$ of Equation (13.1). We assume that we are given an estimate of the power spectrum of the noise, denoted by $\hat{S}_b(\omega)$, that is typically obtained by averaging over multiple frames of a known noise segment. We also assume that the noise and object sequences are uncorrelated. Then with short-time analysis, an estimate of the object's short-time squared spectral magnitude is suggested from Equation (13.2) as [5]

$$
\begin{aligned}
|\hat{X}(pL, \omega)|^2 &= |Y(pL, \omega)|^2 - \hat{S}_b(\omega), \quad \text{if } |Y(pL, \omega)|^2 - \hat{S}_b(\omega) \geq 0 \\
&= 0, \qquad\qquad\qquad\qquad\quad \text{otherwise.} \qquad\qquad\qquad (13.6)
\end{aligned}
$$

When we combine this magnitude estimate with the measured phase, we then have the STFT estimate

$$
\hat{X}(pL, \omega) = |\hat{X}(pL, \omega)|e^{j\angle Y(pL, \omega)}.
$$

An object signal estimate can then be formed with any of the synthesis techniques described in Chapter 7 including the overlap-add (OLA), filter-bank summation (FBS), or least-squared-error (LSE) synthesis.[3] This noise reduction method is a specific case of a more general technique given by Weiss, Aschkenasy, and Parsons [57] and extended by Berouti, Schwartz, and Makhoul [4] that we introduce in Section 13.5.3. This generalization, Equation (13.19), allows in Equation (13.6) a compression of the spectrum and over- (or under-) estimation of the noise contribution.

It is interesting to observe that spectral subtraction can be viewed as a filtering operation where high SNR regions of the measured spectrum are attenuated less than low SNR regions. This formulation can be given in terms of an "instantaneous" SNR defined as

$$
R(pL, \omega) = \frac{|X(pL, \omega)|^2}{\hat{S}_b(\omega)} \qquad\qquad (13.7)
$$

---

[3] In spectral subtraction, as well as in any noise reduction scheme based on a modified STFT, synthesis occurs from a discrete STFT with N uniformly spaced frequencies $\omega_k = \frac{2\pi}{N}k$. The DFT length N, therefore, must be sufficiently long to account for the inverse Fourier transform of $\hat{X}(pL, \omega)$ being possibly longer than that of the original short-time segment.

resulting in a spectral magnitude estimate

$$|\hat{X}(pL,\omega)|^2 = |Y(pL,\omega)|^2 - \hat{S}_b(\omega)$$

$$= |Y(pL,\omega)|^2 \left[1 - \frac{\hat{S}_b(\omega)}{|Y(pL,\omega)|^2} \right]$$

$$\approx |Y(pL,\omega)|^2 \left[1 + \frac{1}{R(pL,\omega)} \right]^{-1}$$

where we have used the approximation $|Y(pL,\omega)|^2 \approx |X(pL,\omega)|^2 + \hat{S}_b(\omega)$. The time-varying *suppression filter* applied to the STFT measurement is therefore given approximately by

$$H_s(pL,\omega) = \left[1 + \frac{1}{R(pL,\omega)} \right]^{-1/2}. \tag{13.8}$$

The filter attenuation is given in Figure 13.1 as a function of $R(pL,\omega)$, illustrating that low SNR signals are attenuated more than high SNR signals.

STFT magnitude estimation has also been more formally posed in the context of stochastic estimation theory that leads to a modification of spectral subtraction. Specifically, a maximum-likelihood (ML) estimation (Appendix 13.A) of the desired STFT magnitude was proposed by McAulay and Malpass [37]. The resulting ML estimate of $|X(pL,\omega)|$ is expressed as

$$|\hat{X}(pL,\omega)| = \frac{1}{2}|Y(pL,\omega)| + \frac{1}{2}[|Y(pL,\omega)|^2 - \hat{S}_b(\omega)]^{1/2}$$

where it is assumed that the noise, mapped to the frequency domain from the time domain, is Gaussian at each frequency. As with spectral subtraction, the ML solution can be formulated as a suppression filter with the instantaneous SNR as a variable and has been shown to give a more gradual attenuation than spectral subtraction [37]. A further extension by McAulay and Malpass [37] modifies the ML estimate by the *a priori* probability of the presence of speech; when the probability of speech being present is estimated to be small, then noise attenuation in the above ML estimate is further increased.

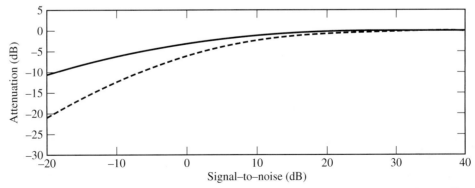

Figure 13.1 Comparison of suppression curves for spectral subtraction (solid line) and the Wiener filter (dashed line) as a function of the instantaneous SNR.

An important property of noise suppression by spectral subtraction, as well as other STFT-based suppression techniques, is that attenuation characteristics change with the length of the analysis window. Because we are ultimately interested in speech signals, we look next at an example of enhancing a sinewave in noise (adapted from Cappe and Laroche [8]), a basic building block of the speech signal. The missing steps in this example are carried through in Exercise 13.1.

EXAMPLE 13.1   Consider a sinewave $x[n] = A\cos(\omega_o n)$ in stationary white noise $b[n]$ with variance σ^2 and analyzed by a short-time window $w[n]$ of length N_w. When the sinewave frequency ω_o is larger than the width of the main lobe of the Fourier transform of the analysis window, $W(\omega)$, then it follows that the STFT magnitude of $x[n]$ at frequency ω_o is given approximately by $|X(pL, \omega_o)| \approx \frac{A}{2}|W(0)|$ where $W(0) = \sum_{n=-\infty}^{\infty} w[n]$. Then, denoting by E, the expectation operator, the average short-time signal power at frequency ω_o is given by

$$\hat{S}_x(pL, \omega_o) = E[|X(pL, \omega_o)|^2] \approx \frac{A^2}{4}\left|\sum_{n=-\infty}^{\infty} w[n]\right|^2.$$

Likewise, it is possible to show that the average power of the windowed noise is constant over all ω and given by

$$\hat{S}_b(pL, \omega) = E[|B(pL, \omega)|^2] = \sigma^2 \sum_{n=-\infty}^{\infty} w^2[n].$$

In the above expressions, $\hat{S}_x(pL, \omega)$ and $\hat{S}_b(pL, \omega)$ denote estimates of the underlying power spectra $S_x(\omega)$ and $S_b(\omega)$, respectively (which in this example are unchanging in time). Using the property that $x[n]$ and $b[n]$ are uncorrelated, we then form the following ratio at the frequency ω_o:

$$\frac{E\left[|Y(pL, \omega_o)|^2\right]}{\hat{S}_b(pL, \omega_o)} = 1 + \frac{\hat{S}_x(pL, \omega_o)}{\hat{S}_b(pL, \omega_o)}$$

$$= 1 + \frac{A^2}{4}[S_b(\omega)\Delta_w]^{-1} \tag{13.9}$$

where $S_b(\omega) = \sigma^2$ and

$$\Delta_w = \frac{\sum_{n=-\infty}^{\infty} w^2[n]}{\left|\sum_{n=-\infty}^{\infty} w[n]\right|^2}$$

which can be shown to be the 3-dB bandwidth of the analysis window main lobe. It follows that $S_b(\omega)\Delta_w$ is the approximate power in a band of noise in the STFT centered at frequency ω_o. Consequently, the second term in Equation (13.9) is the ratio of the (half) power in the sinewave, i.e., $A^2/4$, to the power in the noise band (equal to the window bandwidth) centered at frequency ω_o, i.e., $S_b(\omega)\Delta_w$. As we increase the window length, we increase this SNR at frequency ω_o because the window bandwidth decreases, thus decreasing the power in the noise relative to that of the signal.

As the window length is decreased, we see from our result that, for a sinewave at frequency ω_o, the SNR decreases and thus, in spectral subtraction, the sinewave is more likely to be attenuated and perhaps removed because of the thresholding operation in Equation (13.6). Therefore, to preserve a sinewave after suppression, we must use an analysis filter of adequate length. However, a long window conflicts with the need to preserve transient and changing components of a signal such as plosives, fast attacks, and modulations.   ▲

The previous example illustrates the time-frequency tradeoff encountered in enhancing a simple signal in noise using spectral subtraction. In addition, we must consider the *perceptual consequences* of what appears to be a benign filtering scheme. In particular, a limitation of spectral subtraction is the aural artifact of "musicality" that results from the rapid coming and going of sinewaves over successive frames [32]. We can understand how musicality arises in spectral subtraction when we consider that the random fluctuations of the periodogram of noise, as well as possibly the desired sinewave itself (as in the previous example), rise above and below the spectral subtraction threshold level over time at different frequencies. Consequently, numerous smoothing techniques have been applied to reduce such annoying fluctuations.[4] We discuss some of these smoothing techniques in a following section in the context of Wiener filtering, where spectral smoothing also helps reduce musicality.

13.2.3 Cepstral Mean Subtraction

Consider now the problem of recovering a sequence $x[n]$ from the convolution $y[n] = x[n] * g[n]$. Motivated by Equation (13.5), we apply the nonlinear logarithm operator to the STFT of $y[n]$ to obtain

$$Y(pL, \omega) \approx \log[X(pL, \omega)] + \log[G(\omega)].$$

Because the distortion $g[n]$ is time-invariant, the STFT views $\log[G(\omega)]$ at each frequency as fixed along the time index variable p. If we assume that the speech component $\log[X(pL, \omega)]$ has zero mean in the time dimension, then we can remove the convolutional distortion $g[n]$ while keeping the speech contribution intact. This can be accomplished in the quefrency domain by computing cepstra, along each STFT time trajectory, of the form

$$\hat{y}[n, \omega] \approx F_p^{-1}(\log[X(pL, \omega)]) + F_p^{-1}(\log[G(\omega)])$$
$$= \hat{x}[n, \omega] + \hat{g}[n, \omega]$$
$$= \hat{x}[n, \omega] + \hat{g}[0, \omega]\delta[n]$$

where F_p^{-1} denotes the inverse Fourier transform of sequences along the time dimension p. Applying a cepstral lifter, we then have:

$$\hat{x}[n, \omega] \approx l[n]\hat{y}[n, \omega]$$

where $l[n] = 0$ at $n = 0$ and unity elsewhere. Because the 0th value of the cepstrum equals the mean of $\log[Y(pL, \omega)]$ (along the time dimension) for each ω, the method is called *cepstral mean subtraction* (CMS). Although this approach is limited due to the strictness of the assumption of a zero-mean speech contribution, it has significant advantages in feature estimation for recognition applications (Chapter 14). (Historically, CMS was first developed in the context of speech recognition as described in [23].) In these applications, often the mean

---

[4] To date, no useful *intelligibility* improvements (i.e., a better understanding of words or content) have been reported with current noise reduction systems, even with smoothing techniques that remove musicality and give perceived noise reduction. Such "discrepancies" are essentially always found in formal evaluation of these noise reduction systems: the perceived noise level or an objective error is reduced but intelligibility is not improved or is reduced [34]. This is probably because intelligibility is dependent on short, transient speech components, such as subtle differences between voiced and unvoiced consonants and formant modulations, which are not enhanced or are degraded in the noise reduction process. Nevertheless, the processed speech can be more pleasing and less fatiguing to the listener, as well as more easily transcribed [34].

of the $\log[Y(pL, \omega)]$ is computed and subtracted directly.[5] Because, in practice, the mean is computed over a finite number of frames, we can think of CMS as a highpass, non-causal FIR filtering operation [2],[23].

13.3 Wiener Filtering

An alternative to spectral subtraction for recovering an object sequence $x[n]$ corrupted by additive noise $b[n]$, i.e., from a sequence $y[n] = x[n] + b[n]$, is to find a linear filter $h[n]$ such that the sequence $\hat{x}[n] = y[n]*h[n]$ minimizes the expected value of $(\hat{x}[n] - x[n])^2$. Under the condition that the signals $x[n]$ and $b[n]$ are uncorrelated and stationary, the frequency-domain solution to this stochastic optimization problem is given by the suppression filter (Exercise 13.2)

$$H_s(\omega) = \frac{S_x(\omega)}{S_x(\omega) + S_b(\omega)} \tag{13.10}$$

which is referred to as the *Wiener filter* [32]. When the signals $x[n]$ and $b[n]$ meet the conditions under which the Wiener filter is derived, i.e., uncorrelated and stationary object and background, the Wiener filter provides noise suppression without considerable distortion in the object estimate and background residual. The required power spectra, $S_x(\omega)$ and $S_b(\omega)$, can be estimated by averaging over multiple frames when sample functions of $x[n]$ and $b[n]$ are provided. Typically, however, the desired signal and background are nonstationary in the sense that their power spectra change over time, i.e., they can be expressed as time-varying functions $S_x(n, \omega)$ and $S_b(n, \omega)$. Thus, ideally, each frame of the STFT is processed by a different Wiener filter. For the simplifying case of a stationary background, we can express the *time-varying* Wiener filter as

$$H_s(pL, \omega) = \frac{\hat{S}_x(pL, \omega)}{\hat{S}_x(pL, \omega) + \hat{S}_b(\omega)}$$

where $\hat{S}_x(pL, \omega)$ is an estimate of the time-varying power spectrum of $x[n]$, $S_x(n, \omega)$, on each frame, and $\hat{S}_b(\omega)$ is an estimate of the power spectrum of a stationary background, $S_b(\omega)$. The time-varying Wiener filter can also be expressed as (Exercise 13.2)

$$H_s(pL, \omega) = \left[1 + \frac{1}{R(pL, \omega)}\right]^{-1} \tag{13.11}$$

with a signal-to-noise ratio $R(pL, \omega) = \hat{S}_x(pL, \omega)/\hat{S}_b(\omega)$. A comparison of the suppression curves for spectral subtraction and Wiener filtering is shown in Figure 13.1, where we see the attenuation of low SNR regions relative to the high SNR regions to be somewhat stronger for the Wiener filter, consistent with the filter in Equation (13.8) being a compressed (square-rooted) form of that in Equation (13.11). A second important difference from spectral subtraction is that the Wiener filter does not invoke an absolute thresholding. Finally,

---

[5] In practice, only the STFT magnitude is used in recognition applications and the 0th cepstral value is computed by computing the mean of $\log|X(pL, \omega)|$ along p, rather than computing an explicit inverse Fourier transform. Often, however, in recognition applications, the cepstrum of $\log|X(pL, \omega)|$ is computed with respect to ω to obtain a cepstral feature vector for each frame. Equivalently, one can then subtract the mean cepstrum (across frames) to remove the distortion component, and hence we have an alternative and, perhaps, more legitimate motivation for the nomenclature *cepstral mean subtraction*.

as with spectral subtraction, an enhanced waveform is recovered from the modified STFT, $\hat{X}(pL, \omega) = Y(pL, \omega)H_s(pL, \omega)$, by any of the synthesis methods of Chapter 7. Observe that the Wiener filter is zero-phase so that the original phase of $Y(pL, \omega)$ is again used in synthesis.

In forming an estimate $\hat{S}_x(pL, \omega)$ of the time-varying object power spectrum, we typically must use very short-time and local measurements. This is because when the desired signal is on the order of a few milliseconds in duration, and when its change is rapid, as with some plosives, its spectrum is difficult to measure, requiring an estimate to be made essentially "instantaneously." In the remainder of this section, we study a variety of spectral estimation methods and then apply these methods to speech enhancement. Representations for binaural presentation are also considered, indicating that further enhancement can be obtained with stereo aural displays of combinations of the object estimate and the signal not passed by the Wiener filter, i.e., the result of subtracting the object estimate from the original signal.

13.3.1 Basic Approaches to Estimating the Object Spectrum

Suppose a signal $y[n]$ is short-time processed at frame interval L samples and we have available an estimate of the Wiener filter on frame $p - 1$, denoted by $\hat{H}_s((p-1)L, \omega)$. We assume, as before, that the background $b[n]$ is stationary and that its power spectrum, $S_b(\omega)$, is estimated by averaging spectra over a known background region. For a nonstationary object signal $x[n]$, one approach to obtain an estimate of its time-varying power spectrum on the pth frame uses the past Wiener filter $H_s((p-1)L, \omega)$ to enhance the current frame [32]. This operation yields an enhanced STFT on the pth frame:

$$\hat{X}(pL, \omega) = Y(pL, \omega)H_s((p-1)L, \omega) \tag{13.12}$$

which is then used to update the Wiener filter:

$$H_s(pL, \omega) = \frac{|\hat{X}(pL, \omega)|^2}{|\hat{X}(pL, \omega)|^2 + \hat{S}_b(\omega)}. \tag{13.13}$$

The estimate of the time-varying object power spectrum $\hat{S}_x(pL, \omega) = |\hat{X}(pL, \omega)|^2$ may be initialized with, for example, the raw spectral measurement or a spectrum derived from spectral subtraction. There is, however, little control over how rapidly the object power spectrum estimate changes[6] in Equation (13.13). Because the filter in Equation (13.13) can vary rapidly from frame to frame, as with spectral subtraction, the result is a noise *residual* with fluctuating artifacts. These fluctuations again are perceived as annoying musicality because of peaks in the periodogram $|Y(pL, \omega)|^2$ that influence the object estimate $\hat{X}(pL, \omega)$ in Equation (13.12) and thus the filter estimate in Equation (13.13).

---

[6] In the examples to follow in this section, we use overlap-add (OLA) synthesis. We saw in Chapter 7 (Section 7.5.1) that the effect of OLA synthesis, from a multiplicatively modified STFT (as occurs with Wiener filtering), is a time-varying linear filter smoothed in time by the window. Thus, the window bandwidth constrains how fast the time-domain Wiener filter can change. For a very short window, however, the resulting large bandwidth implies little smoothing.

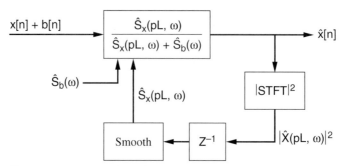

Figure 13.2 Classic Wiener filter with smoothing of the object spectrum estimate.

One approach to slow down the rapid frame-to-frame movement of the object power spectrum estimate, and thus reduce annoying fluctuations in the residual, is to apply temporal smoothing to the object spectrum of Equation (13.12). Denote the object power spectrum estimate on the pth frame by $\hat{S}_x(pL, \omega) = |\hat{X}(pL, \omega)|^2$. Then the smooth power spectrum estimate is obtained as

$$\tilde{S}_x(pL, \omega) = \tau \tilde{S}_x((p - 1)L, \omega) + (1 - \tau)\hat{S}_x(pL, \omega) \tag{13.14}$$

where τ is the smoothing constant. $\tilde{S}_x(pL, \omega)$ then replaces $|\hat{X}(pL, \omega)|^2$ within the Wiener filter of Equation (13.13) (Figure 13.2). The smoothing constant of Equation (13.14) controls how fast we adapt to a nonstationary object spectrum. A fast adaptation, with a small smoothing constant, implies improved time resolution, but more noise in the spectral estimate, and thus more musicality in the synthesis. A large smoothing constant improves the spectral estimate in regions of stationarity, but it smears onsets and other rapid events. We now look at an example of the time-frequency resolution tradeoff inherent in this approach.

EXAMPLE 13.2   Figure 13.3b shows an example of a synthetic train of rapidly-decaying sinewaves in white Gaussian noise; the original 1000-Hz sinewave pulses of Figure 13.3a are uniformly spaced by 2 ms. In noise reduction with a Wiener filter, a 4-ms triangular analysis window, a 1-ms frame interval, and overlap-add (OLA) synthesis are applied. The particular analysis window and frame interval ensure that the OLA constraint is satisfied.[7] A Wiener filter was derived using the spectral smoothing in Equation (13.14) with $\tau = 0.85$. The background power spectrum estimate was obtained by averaging the squared STFT magnitude over the first 0.08 seconds of $y[n]$, and the initial object power spectrum estimate $|\hat{X}(0, \omega)|^2$ was obtained by applying spectral subtraction to $|Y(0, \omega)|^2$. Panel (c) illustrates the result of applying the Wiener filter. An advantage of the spectral smoothing is that it has removed musicality in the noise residual. However, the initial attack of the signal is reduced, resulting in an aural "dulling" of the sound, i.e., it is perceived as less "crisp" than that of its original noiseless counterpart. In addition, although the Wiener filter adapts to the object spectrum, the effect of this adaptation lingers beyond the object, thus preventing noise reduction for some time thereafter and resulting in a perceived "hiss." In this example, the smoothing constant τ is selected to give substantial noise reduction while reducing musicality, but at the expense of slowness in the filter adaptation, resulting in the smeared object attack and the trailing hiss. This sluggish adaptivity

---

[7] The short window corresponds to a large window bandwidth and thus ensures that OLA synthesis does not inherently impose a strong temporal smoothing of the Wiener filter, as seen in Chapter 7 (Section 7.5.1).

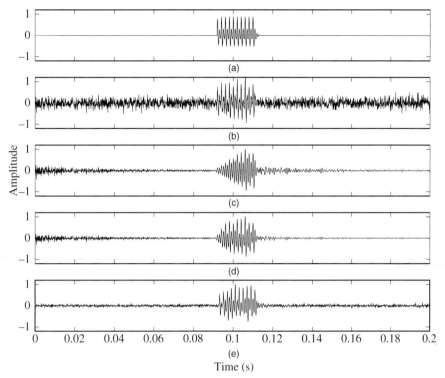

Figure 13.3 Enhancement by adaptive Wiener filtering of a train of closely-spaced decaying sinewaves in 10 dB of additive white Gaussian noise: (a) original clean object signal; (b) original noisy signal; (c) enhanced signal without use of spectral change; (d) enhanced signal with use of spectral change; (e) enhanced signal using spectral change, the iterative filter estimate (2 iterations), and background adaptation.

can be avoided by reducing the smoothing constant, but at the expense of less noise suppression and the return of residual musicality due to a fluctuating object power spectrum estimate. ▲

13.3.2 Adaptive Smoothing Based on Spectral Change

In this section, we present an adaptive approach to smoothing the object spectrum estimate with emphasis on preserving the object nonstationarity, while avoiding perceived musicality and hiss in the noise residual. The essence of the enhancement technique is a Wiener filter that uses an object signal power spectrum whose estimator adapts to the "degree of stationarity" of the measured signal [46]. The degree of stationarity is derived from a short-time spectral derivative measurement that is motivated by the sensitivity of biological systems to spectral change, as we have seen in Chapter 8 in the phasic/tonic auditory principle, but also by evidence that noise is perceptually masked by rapid spectral changes,[8] as has been demonstrated with quantization noise in the context of speech coding [30]. The approach, therefore, preserves dynamic regions

---

[8] This aural property is analogous to a visual property. With noise reduction in images, noise in a stationary region, such as a table top, is more perceptible than noise in a nonstationary region, such as a table edge [35].

important for perception, as we described in Chapter 8 (Section 8.6.3), but also *temporally shapes the noise residual* to coincide with these dynamic regions according to this particular perceptual masking criterion. Additional evidence for the perceptual importance of signal dynamics has been given by Moore who, in his book on the psychology of hearing [41] (p. 191), states: "The auditory system seems particularly well-suited to the analysis of changes in sensory input. The perceptual effect of a change in a stimulus can be roughly described by saying that the preceding stimulus is subtracted from the present one, so what remains is the change. The changed aspect stands out perceptually from the rest . . . a powerful demonstration of this effect may be obtained by listening to a stimulus with a particular spectral structure and then switching rapidly to a white noise stimulus. . . . The noise sounds colored, and the coloration corresponds to the inverse of the spectrum of the preceding sound."

Our goal is to make the adaptive Wiener filter more responsive to the presence of the desired signal without sacrificing the filter's capability to suppress noise. We can accomplish this by making the smoothing constant of the recursive smoother in Equation (13.14) adapt to the spectrum of the measurement. In particular, a time-varying smoothing constant is selected to reflect the degree of stationarity of the waveform whereby, when the spectrum is changing rapidly, little temporal smoothing is introduced resulting in a near instantaneous object spectrum estimate used in the Wiener filter. On the other hand, when the measurement spectrum is stationary, as in background or in steady object regions, an increased smoothing improves the object spectral estimate. Although this filter adaptation results in relatively more noise in nonstationary regions, as observed earlier, there is evidence that, perceptually, noise is masked by rapid spectral changes and accentuated in stationary regions.

One measure of the degree of stationarity is obtained through a spectral derivative measure defined for the pth frame as

$$\Delta Y(pL) = \left[\frac{1}{\pi} \int_0^\pi |Y(pL, \omega) - Y((p-1)L, \omega)|^2 d\omega \right]^{1/2}. \qquad (13.15)$$

Because this measure is itself erratic across successive frames, it, too, is temporally smoothed as $\Delta \tilde{Y}(pL) = f_\Delta[p] * \Delta Y(pL)$ where $f_\Delta[p]$ is a noncausal linear filter. The smooth spectral derivative measure is then mapped to a time-varying smoothing constant as

$$\tau(p) = Q[1 - 2(\Delta \tilde{Y}(pL) - \Delta \bar{Y})]$$

where

$$Q(x) = \begin{cases} x, & 0 \le x \le 1 \\ 0, & x < 0 \\ 1, & x > 1 \end{cases}$$

and where $\Delta \bar{Y}$ is the average spectral derivative over the known background region. Subtraction of $\Delta \bar{Y}$ and multiplication by 2 in the argument of Q are found empirically to normalize $\tau(p)$ to fall roughly between zero and unity [46]. The resulting smooth object spectrum is given by

$$\tilde{S}_x(pL, \omega) = \tau(p)\tilde{S}_x((p-1)L, \omega) + [1 - \tau(p)]\hat{S}_x(pL, \omega). \qquad (13.16)$$

The resulting enhancement system is illustrated in Figure 13.4 where the time-varying, rather than fixed, smoothing constant controls the estimation of the object power spectrum estimate, and is derived from the spectral derivative. As we will see shortly through an example, this refined Wiener filter improves the object attack and reduces the residual hiss artifact in synthesis.

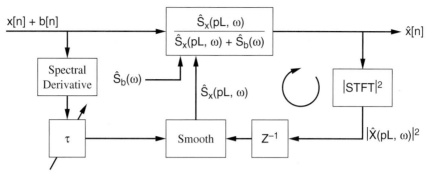

Figure 13.4 Noise-reduction Wiener filtering based on spectral change. The time-varying smoothing constant $\tau(p)$ controls the estimation of the object power spectrum, and is derived from the spectral derivative.

One approach to further recover the initial attack is to iterate the Wiener filtering on each frame; the iteration attempts to progressively improve the Wiener filter by looping back the filter output, i.e., iteratively updating the Wiener filter with a new object spectrum estimate derived from the enhanced signal. This iterative approach is indicated in Figure 13.4 by the clockwise arrow. Such an iterative process better captures the initial attack through a refined object spectrum estimate obtained from the enhanced signal and by reducing the effect of smoothing delay. Only a few iterations, however, are possible on each frame because with an increasing number of iterations, resonant bandwidths of the object spectral estimate are found empirically to become unnaturally narrow.

EXAMPLE 13.3   In this example, we improve on the object signal enhancement of Example 13.2 using the adaptive and iterative Wiener filter described above. As before, a 4-ms triangular analysis window, a 1-ms frame interval, and OLA synthesis are applied. In addition to providing good temporal resolution, the short 4-ms analysis window prevents the inherent smoothing by the OLA process from controlling the Wiener filter dynamics. A 2-ms rectangular filter $f_\Delta[p]$ is selected for smoothing the spectral derivative. Figure 13.3d shows that the use of spectral change in the Wiener filter adaptation [Equation (13.16)] helps to reduce both the residual hiss artifact and improve attack fidelity, giving a cleaner and crisper synthesis. Nevertheless, the first few object components are still reduced in amplitude because, although, indeed, the spectral derivative rises, and the resulting smoothing constant falls in the object region, they do not possess the resolution nor the predictive capability necessary to track the individual object components. The iterative Wiener filter (with 2 iterations) helps to further improve the attack, as illustrated in Figure 13.3e. In addition, the uniform background residual was achieved in Figure 13.3e by allowing the background spectrum to adapt during frames declared to be background, based on an energy-based detection of the presence of speech in an analysis frame [46]. ▲

EXAMPLE 13.4   Figure 13.5 shows a frequency-domain perspective of the performance of the adaptive Wiener filter based on spectral change and with iterative refinement (2 iterations). In this example, a synthetic signal consists of two FM chirps crossing one another and repeated, and the background is white Gaussian noise. As in the previous example, the analysis window duration is 4 ms, the frame interval is 1 ms, and OLA synthesis is applied. Likewise, a 2-ms rectangular filter $f_\Delta[p]$ is selected for smoothing the spectral derivative. The ability of the Wiener filter to track the FM is illustrated by spectrographic views, as well as by snapshots of the adaptive filter at three different signal transition time instants. ▲

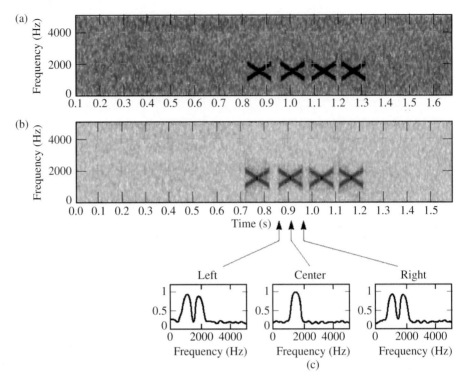

Figure 13.5 Frequency-domain illustration of adaptive Wiener filter for crossing chirp signals: (a) spectrogram of original noisy signal; (b) spectrogram of enhanced signal; (c) adaptive Wiener filters at three signal transition time instants.

13.3.3 Application to Speech

In the speech enhancement application, a short analysis window and frame interval, e.g., a 4-ms triangular window and a 1-ms frame interval, and OLA synthesis, as in the previous examples, provide good temporal resolution of sharp attacks, transitions, and modulations. This temporal resolution is obtained without the loss in frequency resolution and the loss of low-level speech components (Example 13.1), which is typical of processing with a short window. (We cannot necessarily project results from simple signals, such as a sinewave in white noise, to more complex signals.) Some insight into this property is found in revisiting the difference between the narrowband and wideband spectrograms as described in Chapter 3. Although, the reader is left to further ponder and to question this intriguing observation.

For voiced speech, we approximate the speech waveform $x[n]$ as the output of a linear time-invariant system with impulse response $h[n]$ (with an embedded glottal airflow) and with an impulse train input $p[n] = \sum_{k=-\infty}^{\infty} \delta[n - kP]$, i.e., $x[n] = p[n] * h[n]$. This results in the windowed speech waveform

$$x_n[m] = w[n - m](p[m] * h[m]).$$

The difference between the narrowband and wideband spectrograms is the length of the window $w[n]$. For the narrowband spectrogram, we use a long window with a duration of at least two pitch periods; as seen in Chapter 3 (Section 3.3), the short-time section $x_n[m]$ maps in the frequency domain to

$$|X(pL, \omega)|^2 \approx 4\pi^2 \sum_{k=-\infty}^{\infty} |H(\omega_k)|^2 |W(\omega - \omega_k)|^2.$$

where $H(\omega)$ is the glottal flow spectrum/vocal tract frequency response and where $\omega_k = 2\pi k/P$. Thus spectral slices in the narrowband spectrogram of voiced speech consist of a set of narrow "harmonic lines," whose width is determined by the Fourier transform of the window, shaped by the magnitude of the product of the glottal airflow spectrum and vocal tract frequency response. With the use of a long window, the Wiener filter would appear as a "comb" following the underlying harmonic speech structure,[9] *reducing noise in harmonic nulls*. A consequence of a long window, however, is that rapid events are smeared by the Wiener filtering process.

For the wideband spectrogram, we assume a short window with a duration of less than a single pitch period (e.g., 4 ms), and this provides a different view of the speech waveform. Because the window length is less than a pitch period, as the window slides in time, it essentially "sees" pieces of the periodically occurring vocal tract/glottal flow response $h[n]$ (assuming tails of previous responses have died away), and, as in Chapter 3 (Section 3.3), this can be expressed as

$$|X(pL, \omega)|^2 \approx |H(\omega)|^2 E[pL]$$

where $E[pL]$ is the energy in the waveform under the sliding window at time pL. In this case, the spectrogram shows the formant frequencies of the vocal tract along the frequency dimension, but also gives vertical striations at the pitch rate in time, rather than the harmonic horizontal striations as in the narrowband spectrogram because, for a sufficiently small frame interval L, the short window is sliding through fluctuating energy regions of the speech waveform. The Wiener filter, therefore, follows the resonant structure of the waveform, rather than the harmonic structure, providing *reduction of noise in formant nulls* rather than harmonic nulls. Nevertheless, noise reduction does occur at the fundamental frequency rate, i.e., within a glottal cycle, via the short window and frame interval.

In our description of the narrowband and wideband spectrograms, we have used the example of voiced speech. For unvoiced sound classes (e.g., fricatives and plosives), either spectrogram shows greater intensity at formants of the vocal tract; neither shows horizontal or vertical pitch-related striations because periodicity is not present except when the vocal cords are vibrating simultaneously with unvoiced noise-like or impulse-like sounds. For plosive sounds in particular, the wideband spectrogram is preferred because it gives better temporal resolution of the sound's fine structure and when the plosive is closely followed by a vowel. A short 4-ms window used in the Wiener filter estimation is thus consistent with this property. Informal listening shows the approach to yield considerable noise reduction and good speech quality without residual musicality. Formal evaluations, however, have not been performed. We demonstrate in Example 13.5 the performance of the adaptive Wiener filter with an example of speech in a white noise background.

---

[9] In a voiced speech segment, if we make the analysis window duration multiple pitch periods, the Wiener filter becomes a comb, accentuating the harmonics. Alternatively, given a pitch estimate, it is straightforward to explicitly design a comb filter [32]. Many problems are encountered with this approach, however, including the need for very accurate pitch (the effect of pitch error increasing with increasing harmonic frequency), the presence of mixed voicing states, and the changing of pitch over a frame duration, as well as the lack of a single pitch over the speech bandwidth due to nonlinear production phenomena.

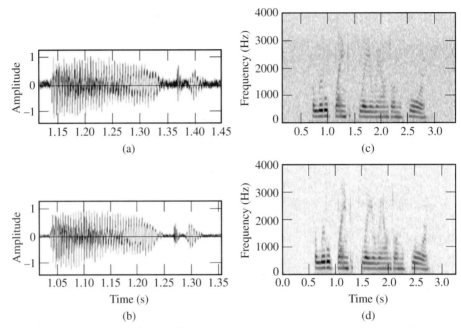

Figure 13.6 Reduction of additive white noise corrupting a speech waveform from a female speaker: (a) excerpt of original waveform; (b) enhancement of (a) by adaptive Wiener filtering; (c)-(d) spectrograms of full waveforms corresponding to panels (a) and (b).

EXAMPLE 13.5 In this example, the speech of a female speaker is corrupted by white Gaussian noise at a 9-dB SNR. As in the previous examples, a 4-ms triangular window, 1-ms frame interval, and OLA synthesis are used. A Wiener filter was designed using adaptation to the spectral derivative, two iterative updates, and background adaptation, as described in the previous section. Likewise, a 2-ms rectangular filter $f_\Delta[p]$ was selected for smoothing the spectral derivative. Figure 13.6 illustrates the algorithm's performance in enhancing the speech waveform. Good temporal resolution is achieved by the algorithm, while maintaining formant and harmonic trajectories. The speech synthesis is subjectively judged informally to be of high quality and without background residual musicality. It is important to note, however, that the speech quality and degree of residual musicality can be controlled by the length of the noncausal smoothing filter $f_\Delta[p]$ used to achieve a smooth spectral derivative measure, i.e., $\Delta\tilde{Y}(pL) = f_\Delta[p] * \Delta Y(pL)$. For example, an excessively long $f_\Delta[p]$ can smear speech dynamics, while an $f_\Delta[p]$ too short causes some musicality to return.   ▲

13.3.4 Optimal Spectral Magnitude Estimation

Motivated by the maximum-likelihood (ML) estimation method of McAulay and Malpass [37], Ephraim and Malah [13] proposed a different stochastic-theoretic approach to an optimal spectral magnitude estimator. Specifically, they obtained the least-squared-error estimate of $|X(pL, \omega)|$, given the noisy observation $y[n] = x[n] + b[n]$, which is its expected value given $y[n]$, i.e., $E\{|X(pL, \omega)| \,|\, y[n]\}$. In contrast, neither spectral subtraction, ML estimation, nor Wiener filtering solves this optimization problem. Its solution (with its derivation beyond the scope of our presentation) results in significant noise suppression with negligible musical residual noise. For each analysis frame, the suppression filter involves measures of both *a priori* and *a posteriori* SNRs that we denote by $\gamma_{pr}(pL, \omega)$ and $\gamma_{po}(pL, \omega)$, respectively [7],[13],[17]:

$$H_s(pL, \omega) = \sqrt{\frac{\pi}{2}} \sqrt{\left(\frac{1}{1 + \gamma_{po}(pL, \omega)}\right)\left(\frac{\gamma_{pr}(pL, \omega)}{1 + \gamma_{pr}(pL, \omega)}\right)}$$
$$\times G\left[\frac{\gamma_{pr}(pL, \omega) + \gamma_{po}(pL, \omega)\gamma_{pr}(pL, \omega)}{1 + \gamma_{pr}(pL, \omega)}\right]$$

with

$$G[x] = e^{-x/2}[(1 + x)I_0(x/2) + xI_1(x/2)]$$

where $I_o(x)$ and $I_1(x)$ are the modified Bessel functions of the 0th and 1st order, respectively. The term *a priori* is used because it is an estimate of the SNR on the current frame based in part on an object estimate from the previous frame. The term *a posteriori* is used because it is an estimate of the SNR in the current frame based on an object estimate from the current frame. The *a priori* and *a posteriori* SNRs are written specifically in the form [7],[13],[17]

$$\gamma_{po}(pL, \omega) = \frac{P[|Y(pL, \omega)|^2 - \hat{S}_b(\omega)]}{\hat{S}_b(\omega)}$$

$$\gamma_{pr}(pL, \omega) = (1 - \alpha)P[\gamma_{po}(pL, \omega)] + \alpha\frac{|H_s((p - 1)L, \omega)Y((p - 1)L, \omega)|^2}{\hat{S}_b(\omega)}$$

where $P(x) = x$ for $x \geq 0$ and $P(x) = 0$ for $x < 0$, which ensures that the SNRs are positive. The constant α is a weighting factor satisfying $|\alpha| < 1$ that is set to a value close to 1. We see that the *a priori* SNR is, then, a combination of the (current) *a posteriori* SNR and an estimate of the (previous) instantaneous SNR with the object power spectrum estimated from filtering by $H_s((p - 1)L, \omega)$.

The above expression for $H_s(pL, \omega)$ is quite complicated, but it can be argued [7],[13], [17] that the *a priori* SNR $\gamma_{pr}(pL, \omega)$ is the dominant factor. $\gamma_{pr}(pL, \omega)$ can be interpreted as a heavily smoothed version of $\gamma_{po}(pL, \omega)$ when $\gamma_{po}(pL, \omega)$ is small, and as a delayed version of $\gamma_{po}(pL, \omega)$ when $\gamma_{po}(pL, \omega)$ is large [7],[13],[17]. Thus, for low SNR cases, $\gamma_{pr}(pL, \omega)$ is heavily smoothed, which results in a smooth $H_s(pL, \omega)$ in the low regions of the spectrum. Because musical residual noise occurs in the low SNR regions of the spectrum, the smooth suppression function gives reduced musicality. In the high SNR regions, $\gamma_{pr}(pL, \omega)$ roughly tracks $\gamma_{po}(pL, \omega)$, which is an estimate of the instantaneous SNR with the object power spectrum estimated from spectral subtraction. This results in an SNR-based suppression filter similar to the Wiener filter of Section 13.3.1 for high SNR. Thus, the Ephraim and Malah algorithm can be interpreted as using a fast-tracking SNR estimate for high-SNR frequency components and a highly-smoothed, slow-tracking SNR estimate for low SNR components [7],[17].

This algorithm significantly reduces the amount of musical noise compared to the spectral subtraction and basic Wiener filtering methods [7],[13],[17]. As with other approaches to enhancement, the speech quality and background artifacts can be controlled by the degree of spectral smoothing. Some musical residual noise can be perceived, for example, if the smoothing of the *a posteriori* SNR estimate is not sufficient. On the other hand, when the smoothing is excessive, the beginnings and ends of sounds that are low in SNR are distorted due to the *a posteriori* SNR estimate's being too slow in catching up with the transient speech [7],[17].

13.3.5 Binaural Representations

Consider now the output of the filter, $1 - H_s(pL, \omega)$, with $H_s(pL, \omega)$ being a suppression filter. This filter output, that we refer to as the output *complement*, serves two purposes. First, it gives a means for assessing performance, showing pieces of the object signal, as well as the noise background, that were eliminated by the suppression filter. Because the complement and object estimate together form an identity, the complement contains anything that is not captured by the suppression filter.

The filter output complement also opens the possibility of forming a binaural presentation for the listener. For example, we can send the object estimate and its complement into separate ears. In experiments using the Wiener filter of Section 13.3.2, this stereo presentation appears to give the illusion that the object and its complement emanate from different directions, and thus there is further enhancement [46]. A further advantage of this presentation is that, because the complement plus object estimate are an identity, no object component is lost by separating the complement. A disadvantage is possible confusion in having components of one signal appear to come from different directions.

There are also other binaural presentations of interest for enhancement. For example, with the object estimate plus its complement (i.e., the original noisy signal) in one ear and the complement only in the second ear, one perceives the noise coming from directly ahead, while the object appears to come from an angle. Alternately, one can present the object estimate in both ears and its complement in one ear. Other variations include passing the complement signal in both ears and the object estimate in one ear and the inverted (negative of) the object in the other ear. It is interesting to note the following early experiment described by van Bergeijk, Pierce, and David [3] (pp. 164–165): "Imagine that we listen to a noise reaching both ears simultaneously, as it might from a source dead ahead, together with a low-frequency tone reaching one ear direct and the other ear inverted, as it might from a source to one side. In such circumstances we can hear a tone that is around 10 dB weaker than if it reached both ears in the same manner, uninverted, undelayed. . . .This is another way of saying that we are using our power of directional discrimination in separating the tone from the noise."

13.4 Model-Based Processing

Heretofore, the additive noise reduction methods of this chapter have not relied on a speech model. On the other hand, we can design a noise reduction filter that exploits estimated speech model parameters. For example, the Wiener filter can be constructed with an object power spectrum estimate that is based on an all-pole vocal tract transfer function. This filter can then be applied to enhance speech, just as we did with the nonparametric Wiener filter in the previous sections. Such a filter could be obtained by simply applying the deconvolution methods that we have studied such as the all-pole vocal tract (correlation and covariance linear prediction) estimation methods of Chapter 5 or homomorphic filtering methods of Chapter 6. The problem with this approach, however, is that such estimation methods that work for clean speech often degrade for noisy speech. A more formal approach in additive noise is to apply stochastic estimation methods such as maximum likelihood (ML), maximum *a posteriori* (MAP), or minimum-mean-squared error (MMSE) estimation (Appendix 13.A).

MAP estimation of all-pole parameters[10] has been applied by Lim and Oppenheim [32], maximizing the *a posteriori* probability density of the linear prediction coefficients $\underline{a}$ (in vector form), given a noisy speech vector $\underline{y}$ (for each speech frame), i.e., maximizing $p_{\underline{A}|\underline{Y}}(\underline{a}|\underline{y})$ with respect to $\underline{a}$. For the speech-in-noise problem, solution to the MAP problem requires solving a set of nonlinear equations. Reformulating the MAP problem, however, leads to an iterative approach that requires a linear solution on each iteration and thus avoids the nonlinear equations [32]. (For convenience, we henceforth drop the $\underline{A}|\underline{Y}$ notation.) Specifically, we maximize $p(\underline{a}, \underline{x}|\underline{y})$ where $\underline{x}$ represents the clean speech, so we are estimating the all-pole parameters and the desired speech *simultaneously*. The iterative algorithm, referred to as linearized MAP (LMAP), begins with an initial guess $\hat{\underline{a}}^0$ and estimates the speech as the conditional mean $E[\underline{x}|\hat{\underline{a}}^0, \underline{y}]$, which is a linear problem. Then, having a speech estimate, a new parameter vector $\hat{\underline{a}}^1$ is estimated using the autocorrelation method of linear prediction and the procedure is repeated to obtain a series of parameter vectors $\hat{\underline{a}}^i$ that increases $p(\underline{a}, \underline{x}|\underline{y})$ on each iteration. For a stationary stochastic speech process, it can be shown that when an infinitely-long signal is available, estimating the clean speech as $E[\underline{x}|\hat{\underline{a}}^i, \underline{y}]$ (on each iteration i) is equivalent to applying a zero-phase Wiener filter with frequency response

$$H^i(\omega) = \frac{\hat{S}_x^i(\omega)}{\hat{S}_x^i(\omega) + \hat{S}_b(\omega)}$$

and where the power spectrum estimate of the speech on the ith iteration is given by

$$\hat{S}_x^i(\omega) = \frac{A^2}{|1 - \sum_{k=1}^{p} \hat{a}_k^i e^{-j\omega k}|^2}$$

where a_k^i for $k = 1, 2, \ldots, p$ are the predictor coefficients estimated using the autocorrelation method and A is the linear prediction gain as determined in Chapter 5 (Section 5.3.4). The LMAP algorithm is illustrated in Figure 13.7, where it is emphasized that the LMAP algorithm estimates not only the all-pole parameter vector, but also the clean speech from Wiener filtering on each iteration.[11]

The LMAP algorithm was evaluated by Lim and Oppenheim [32] and shown to improve speech quality using the subjective diagnostic acceptability measure (DAM) [Chapter 12 (Section 12.1)], in the sense that perceived noise is reduced. In addition, an objective mean-squared error in the all-pole envelope was reduced for a variety of SNRs. Nevertheless, based on the subjective diagnostic rhyme test (DRT) (Chapter 12), intelligibility does not increase.

Numerous limitations of the LMAP algorithm were improved upon by Hansen and Clements [21]. These limitations include decreasing formant bandwidth with increasing iteration (as occurs with the iterative Wiener filter described in Section 13.3.2), frame-to-frame pole jitter in stationary regions, and lack of a formal convergence criterion. In order to address these limitations, Hansen and Clements [21] introduced a number of spectral constraints within the LMAP iteration steps. Specifically, spectral constraints were imposed on all-pole parame-

---

[10] Without noise, all-pole MAP estimation can be shown to reduce to the autocorrelation method of linear prediction.

[11] When the speech is modeled with vocal tract poles and zeros, LMAP can be generalized to estimate both the poles and zeros, and speech simultaneously (Exercise 13.5).

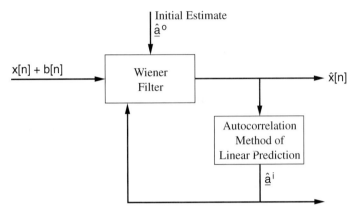

Figure 13.7 Linearized MAP (LMAP) algorithm for the estimation of both the all-pole parameters and speech simultaneously from a noisy speech signal.

SOURCE: J.S. Lim and A.V. Oppenheim, "Enhancement and Bandwidth Compression of Noisy Speech" [32]. ©1979, IEEE. Used by permission.

ters across time and, within a frame, across iterations so that poles do not fall too close to the unit circle, thus preventing excessively narrow bandwidths, and so that poles do not have large fluctuations from frame to frame.

Finally, we end this section by providing a glimpse into a companion problem to vocal tract parameter estimation in noise: estimation of the speech source, e.g., pitch and voicing. This is an important problem not only for its own sake, but also for an alternative means of noise reduction through model-based synthesis. Therefore, reliable source estimation, in addition, serves to improve model- and synthesis-based speech signal processing in noise, such as various classes of speech coders and modification techniques. In Chapter 10, we described numerous classes of pitch and voicing estimators. Although some of these estimators have been shown empirically to have certain immunity to noise disturbances, additive noise was not made explicit in the pitch and voicing model and in the resulting estimation algorithms. A number of speech researchers have brought more statistical decision-theoretic formalisms to this problem. McAulay [38] was one of the first to introduce optimum speech classification to source estimation. This approach is based on principles of decision theory in which voiced/unvoiced hypotheses are formalized for voicing estimation. McAulay [39], as well as Wise, Caprio, and Parks [58], have also introduced maximum-likelihood approaches to pitch estimation in the presence of noise. These efforts represent a small subset of the many possibilities for source estimation under degrading conditions.

13.5 Enhancement Based on Auditory Masking

In the phenomenon of auditory masking, one sound component is concealed by the presence of another sound component. Heretofore in this chapter and throughout the text, we have sporadically made use of this auditory masking principle in reducing the perception of noise. In Chapter 12, we exploited masking of quantization noise by a signal, both noise and signal occurring within a particular frequency band. In Section 13.3.2 of this chapter, we exploited the masking of additive noise by rapid change in a signal, both noise and signal change occurring at a particular time instant. These two different psychoacoustic phenomena are referred to as *frequency and temporal masking*, respectively. Research in psychoacoustics has also shown that we can have difficulty hearing weak signals that fall in the frequency or time *vicinity* of stronger signals (as well as those superimposed in time or frequency on the masking signal, as in the above two cases). A small spectral component may be masked by a stronger nearby spectral component. A similar masking can occur in time for two closely-spaced sounds. In this section, this principle of masking is exploited for noise reduction in the frequency-domain. While temporal masking by adjacent sounds has proven useful, particularly in wideband audio coding [28], it has been less widely used in speech processing because it is more difficult to quantify.

In this section, we begin with a further look at frequency-domain masking that is based on the concept of a *critical band*. Then using the critical band paradigm, we describe an approach to determine the *masking threshold* for complex signals such as speech. The speech masking threshold is the spectral level (determined from the speech spectrum) below which non-speech components are masked by speech components in frequency. Finally, we illustrate the use of the masking threshold in two different noise reduction systems that are based on generalizing spectral subtraction.

13.5.1 Frequency-Domain Masking Principles

We saw in Chapter 8 that the basilar membrane, located at the front-end of the human auditory system, can be modeled as a bank of about 10,000 overlapping bandpass filters, each tuned to a specific frequency (the characteristic frequency) and with bandwidths that increase roughly logarithmically with increasing characteristic frequency. These physiologically-based filters thus perform a spectral analysis of sound pressure level appearing at the ear drum. In contrast, there also exist psychoacoustically-based filters that relate to a human's ability to *perceptually* resolve sound with respect to frequency. The bandwidths of these filters are known as the *critical bands* of hearing and are similar in nature to the physiologically-based filters.

Frequency analysis by a human has been studied by using perceptual masking. Consider a tone at some intensity that we are trying to perceive; we call this tone the *maskee*. A second tone, adjacent in frequency, attempts to drown out the presence of the maskee; we call this adjacent tone the *masker*. Our goal is to determine the intensity level of the maskee (relative to the absolute level of hearing) at which it is not audible in the presence of the masker. This intensity level is called the *masking threshold* of the maskee. The general shape[12] of the masking curve for a masking tone at frequency Ω_o with a particular sound pressure level (SPL) in decibels was first established by Wegel and Lane [56] and is shown in Figure 13.8. Adjacent tones that have an SPL below the solid lines are not audible in the presence of the tone at Ω_o. We see then that there is a range of frequencies about the masker whose audibility is affected.

We see in Figure 13.8 that maskee tones above the masking frequency are more easily masked than tones below this frequency. The masking threshold is therefore asymmetric, the masking threshold curve for frequencies higher than Ω_o having a milder slope, as we see in Figure 13.8. Furthermore, the steepness of this slope in the higher frequencies is dependent on the level of the masking tone at frequency Ω_o, with a milder slope as the level of the masking tone increases. On the other hand, for frequencies lower than Ω_o, the masking curve is modeled with a fixed slope [18],[56].

---

[12] The precise shape is more complicated due to the generation of harmonics of the tonal masker by non-linearities in the auditory system; the shape in Figure 13.8 more closely corresponds to a narrow band of noise centered at Ω_o acting as the masker [18].

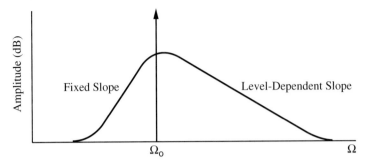

Figure 13.8 General shape of the masking threshold curve for a masking tone at frequency Ω_o. Tones with intensity below the masking threshold curve are masked (i.e., made inaudible) by the masking tone.

Another important property of masking curves is that the bandwidth of these curves increases roughly logarithmically as the frequency of the masker increases. In other words, the range of frequencies that are affected by the masker increases as the frequency of the masking tone increases. This range of frequencies in which the masker and maskee interact was quantified by Fletcher [15],[16],[18] through a different experiment. In Fletcher's experiment, a tone (the maskee) is masked by a band of noise centered at the maskee frequency. The level of the tone was set so that the tone was not audible in the presence of wideband white noise. The bandwidth of the noise was decreased until the tone became audible. This experiment was repeated at different frequencies and the resulting bandwidths were dubbed by Fletcher the *critical bands*. The critical band also relates to the effective bandwidth of the general masking curve in Figure 13.8. Critical bands reflect the frequency range in which two sounds are not experienced independently but are affected by each other in the human perception of sound, thus also relating to our ability to perceptually resolve frequency components of a signal.

Given the roughly logarithmically increasing width of the *critical band filters*, this suggests that about 24 critical band filters cover our maximum frequency range of 15000 Hz for human perception. A means of mapping linear frequency to this perceptual representation is through the *bark scale*. In this mapping, one bark covers one critical band with the functional relation of frequency f to bark z given by [44]

$$z = 13 \tan^{-1}(0.76 f) + 3.5 \tan^{-1}(f/7500). \qquad (13.17)$$

In the low end of the bark scale (< 1000 Hz), the bandwidths of the critical band filters are found to be about 100 Hz and in higher frequencies the bandwidths reach up to about 3000 Hz [18]. A similar mapping (which we apply in Chapter 14), uses the *mel scale*. The mel scale is approximately linear up to 1000 Hz and logarithmic thereafter [44]:

$$m = 2595 \log_{10}(1 + f/700). \qquad (13.18)$$

Although Equation (13.17) provides a continuous mapping from linear to bark scale, most perceptually motivated speech processing algorithms use quantized bark numbers of $1, 2, 3 \ldots 24$ that correspond approximately to the upper band edges of the 24 critical bands that cover our range of hearing. We must keep in mind that although these bark frequencies cover our hearing frequency range, physiologically, there exist about 10,000 overlapping cochlear filters along the basilar membrane. Nevertheless, this reduced bark representation (as well as a quantized mel scale) allows us to exploit perceptual masking properties with feasible computation in speech signal processing, and also provides a perceptual-based framework for feature extraction in recognition applications, as we will see in Chapter 14.

13.5.2 Calculation of the Masking Threshold

For complex signals such as speech, the effects of individual masking components are additive; the overall masking at a frequency component due to all the other frequency components is given by the sum of the masking due to the individual frequency components, giving a single masking threshold [47],[49],[52]. This threshold tells us what is or is not perceptible across the spectrum. For a background noise disturbance (the maskee) in the presence of speech (the masker) we want to determine the masking threshold curve, as determined from the speech spectrum, below which the background noise would be inaudible. For the speech threshold calculation, however, we must consider that the masking ability of tonal and noise components of speech (in masking background noise) is different [22].

Based on the above masking properties, a common approach to calculating the background noise masking threshold on each speech frame, denoted by $T(pL, \omega)$, was developed by Johnston [29], who does the analysis on a critical-band basis. This approach approximates the masking threshold and reduces computation as would be required on a linear-frequency basis. The method can be stated in the following four steps [29],[54]:

S1: The masking threshold is obtained on each analysis frame from the clean speech by first finding spectral energies (by summing squared magnitude values of the discrete STFT), denoted by E_k with k the bark number, within the above 24 critical bands; as we have seen, the critical band edges have logarithmically increasing frequency spacing. This step accounts approximately for the frequency selectivity of a masking curve associated with a single tone at bark number k with energy E_k. Because only noisy speech is available in practice, an approximate estimate of the clean speech spectrum is computed with spectral subtraction.

S2: To account for masking among neighboring critical bands, the critical band energies E_k from step **S1** are convolved with a "spreading function" [47]. This spreading function has an asymmetric shape similar to that in Figure 13.8, but with fixed slopes, [13] and has a range of about 15 on a bark scale [29],[54]. If we denote the spreading function by h_k on the bark scale, then the resulting masking threshold curve is given by $T_k = E_k * h_k$.

S3: We next subtract a threshold offset that depends on the noise-like or tone-like nature of the masker. One approach to determine this threshold offset uses the method of Sinha and Tewfik, based on speech being typically tone-like in low frequencies and noise-like in high frequencies [50],[54].

S4: We then map the masking threshold T_k resulting from step **S3** from the bark scale back to a linear frequency scale to obtain $T(pL, \omega)$ where ω is sampled as DFT frequencies [29].

An example by Virag [54] of masking curves for clean and noisy speech is shown in Figure 13.9 for a voiced speech segment over a 8000 Hz band. The above critical-band approach gives a step-like masking curve, as is shown by Figure 13.9, because we are mapping a bark scale back to a linear frequency scale. A comparison of the masking thresholds derived from both the clean and noisy speech (enhanced with spectral subtraction) shows little difference in the masking curves.

---

[13] In other methods [49],[52], the masking threshold has a slope dependent on the masker level for frequencies higher than the maskers.

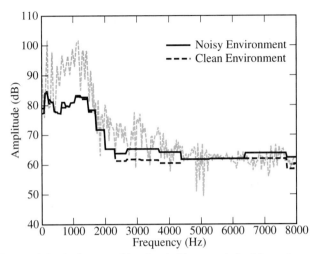

Figure 13.9 Auditory masking threshold curves derived from clean and noisy speech. The light dashed curve is the STFT magnitude in dB.

SOURCE: N. Virag, "Single Channel Speech Enhancement Based on Masking Properties of the Human Auditory System" [54]. ©1999, IEEE. Used by permission.

13.5.3 Exploiting Frequency Masking in Noise Reduction

In exploiting frequency masking, the basic approach is to attempt to make inaudible spectral components of annoying background residual (from an enhancement process) by forcing them to fall below a masking threshold curve as derived from a measured speech spectrum.[14] We are interested in masking this annoying (often musical) residual while maximizing noise reduction and minimizing speech distortion. There are a variety of psychoacoustically motivated speech enhancement algorithms that seek to achieve this goal by using suppression filters similar to those from spectral subtraction and Wiener filtering [10],[17],[49],[54]. Each algorithm establishes a different optimal perceptual tradeoff between the noise reduction, background residual (musical) artifacts, and speech distortion. In this section, for illustration, we describe two particular suppression algorithms that exploit masking in different ways. The first approach by Virag [54] applies less attenuation when noise is heavily masked so as to limit speech distortion. The second approach by Gustafsson, Jax, and Vary [17],[19] seeks residual noise that is perceptually equivalent to an attenuated version of the input noise without explicit consideration of speech distortion.

In the approach of Virag, a masking threshold curve is used to modify parameters of a suppression filter that is a generalization of spectral subtraction. The suppression filter in this method, originally proposed by Berouti, Schwartz, and Makhoul [4], can be written as

$$
\begin{aligned}
H_s(pL, \omega) &= [1 - \alpha Q(pL, \omega)^{\gamma_1}]^{\gamma_2}, \quad \text{if } Q(pL, \omega)^{\gamma_1} < \frac{1}{\alpha + \beta} \\
&= [\beta Q(pL, \omega)^{\gamma_1}]^{\gamma_2}, \qquad \text{otherwise,}
\end{aligned}
\tag{13.19}
$$

---

[14] An interesting question arises as to whether the masking phenomenon of the auditory system provides its own mechanism for speech enhancement. In fact, masking is associated with lateral inhibition, which we discussed in Chapter 8, and which led Wang and Shamma [55] to propose the "auditory spectrum" that, in the presence of noise, represents an enhanced version of spectral components important for perception.

where $Q(pL, \omega)$ is the ratio of the estimated background power spectrum to the measured STFT magnitude:

$$Q(pL, \omega) = \left[\frac{\hat{S}_b(\omega)}{|Y(pL, \omega)|^2} \right]^{1/2}.$$

An advantage of this filtering scheme over basic spectral subtraction is that it provides a tradeoff between noise reduction and speech and background residual distortion. In Virag's algorithm, the parameters of the generalized spectral subtraction filter are adapted to the masking threshold curve on each frame [54]. The factor α controls the extent of noise reduction. Typically, for $\alpha > 1$ noise reduction is obtained at the expense of speech distortion. The additional factor β gives the minimum noise floor and provides a means to add background noise to mask the perceived residual (musical) noise but at the expense of added background noise. The exponent $\gamma_1 = 1/\gamma_2$ controls the sharpness of the transition in the suppression curve associated with $H_s(pL, \omega)$ (similar to that in Figure 13.1).

The steps in Virag's noise reduction algorithm are stated as follows:

S1: Estimation of the background noise power spectrum required in the spectral subtraction filter by averaging over non-speech frames.

S2: Calculation, from the short-time speech spectrum, of the masking threshold curve on each frame, denoted by $T(pL, \omega)$, using the approach by Johnston described in the previous section.

S3: Adaptation of the parameters α and β of the spectral subtraction filter to the masking curves $T(pL, \omega)$. Ideally, we want the residual (musical) noise to fall below the masking curves, thus making it inaudible. When the masking threshold is high, the background noise is already masked and does not need to be reduced to avoid speech distortion. When the masking threshold is low, we reduce the residual noise to avoid the residual appearing above the masking threshold. Maximum and minimum parameter values are defined such that α_{min} and β_{min} map to $T(pL, \omega)_{max}$ corresponding to the least noise reduction, and α_{max} and β_{max} map to $T(pL, \omega)_{min}$ corresponding to the largest noise reduction. Interpolation is performed across these extrema.

S4: Application of the noise suppression filter from step **S3** and overlap-add (OLA) synthesis.

In determining the performance of the algorithm, Virag used both objective and subjective measurements of speech quality. Two objective measures used by Virag are the articulation index and a perceptually-weighted all-pole spectral distance measure (giving less weight to regions of the spectrum with greater energy), both of which were briefly described in Chapter 12. For the above enhancement algorithm these methods were found to correlate with subjective testing, unlike other objective measures such as segmental signal-to-noise ratio.[15] Although not achieving the theoretical limit (defined in Equation (13.4) as having the clean STFT magnitude and noisy phase), Virag found that the proposed spectral subtraction scheme that adapts to auditory masking outperformed the more classical spectral subtraction approaches, according to the above two objective measures. Finally, Virag used the subjective Mean Opinion Score (MOS)[16] test to show that the auditory-based algorithm also outperforms other subtractive-type

---

[15] The standard segmental signal-to-noise ratio weights equally all spectral components and does not impart a perceptual weighting, thus not correlating well with subjective measures.

[16] The Mean Opinion Score (MOS), alluded to in Chapter 12, is one standardized subjective test. In this test, the listener is asked to rank a test utterance between 1 (least desirable) and 5 (most desirable). The score reflects the listener's opinion of the speech distortion, noise reduction, and residual noise.

noise suppression algorithms with respect to human perception; the algorithm was judged to reduce musical artifacts and give acceptable speech distortion [54].

In an alternative suppression algorithm based on auditory masking, rather than using the masking threshold curve to modify a standard suppression filter, Gustafsson, Jax, and Vary [19] use the masking threshold to derive a new suppression filter that results in perceived noise which is an attenuated version of the original background noise. In this formulation, given an original noisy signal $y[n] = x[n] + b[n]$, the desired signal can be written as $d[n] = x[n] + \alpha b[n]$, where α is a noise suppression scale factor. With $h_s[n]$ denoting the impulse response of the suppression filter, an estimate of the short-time power spectrum of the noise error, $\alpha b[n] - h_s[n] * b[n]$, can be shown to be (Exercise 13.6):

$$\hat{S}_e(pL, \omega) = |H_s(pL, \omega) - \alpha|^2 \hat{S}_b(\omega) \qquad (13.20)$$

where $H_s(pL, \omega)$ is the frequency response of $h_s[n]$ on the pth frame and $\hat{S}_b(\omega)$ is an estimate of the background power spectrum. If this error falls below the speech masking threshold curve, then only the attenuated noise is perceived. Thus we form the constraint:

$$|H_s(pL, \omega) - \alpha|^2 \hat{S}_b(\omega) < T(pL, \omega)$$

so that

$$\alpha - \sqrt{\frac{T(pL, \omega)}{\hat{S}_b(\omega)}} < H_s(pL, \omega) < \alpha + \sqrt{\frac{T(pL, \omega)}{\hat{S}_b(\omega)}}. \qquad (13.21)$$

This gives a range of values on $H_s(pL, \omega)$ for which the output noise is perceived as the desired attenuated original noise, so that no musicality occurs in the residual noise. Selecting the upper limit of Equation (13.21) (and constraining $H(pL, \omega) \leq 1$) gives the minimum attenuation (and thus distortion) of the speech signal. As expected, this algorithm gives a noise output that is perceptually equivalent to an attenuated version of the noise input and thus contains no musical artifacts, while it gives speech distortion similar to that in conventional spectral subtraction [17],[19]. Tradeoffs in the degree of noise attenuation and the speech distortion can be obtained through the parameter α. The reader is asked to explore this tradeoff, as well as a comparison with the basic Wiener filter in Exercise 13.6. An extension of the suppression algorithm by Gustafsson, Jax, and Vary that reduces speech distortion has been introduced by Govindasamy [17]. This method uses frequency-domain masking to explicitly seek to hide speech distortion (defined as $\beta x[n] - h_s[n] * x[n]$, β a constant scale factor) simultaneously with the noise distortion ($\alpha b[n] - h_s[n] * b[n]$).

13.6 Temporal Processing in a Time-Frequency Space

In this chapter, we have thus far performed speech enhancement by applying various forms of spectral subtraction and Wiener filtering on short-time speech segments, holding the time variable fixed in the STFT. We now take a different approach in which we hold the frequency variable fixed and filter along *time-trajectories* of STFT filter-bank outputs.

13.6.1 Formulation

Recall from Chapter 7 the filter bank interpretation of the STFT:

$$X(n, \omega) = \sum_{m=-\infty}^{\infty} w[n - m]x[m]e^{-j\omega m}$$
$$= e^{-j\omega n}(x[n] * w[n]e^{j\omega n}) \qquad (13.22)$$

where $w[n]$ is the analysis window, also referred to as the *analysis filter*. We refer to the demodulated output of each filter as the *time-trajectory* at frequency ω. Unless needed, for simplicity throughout this section, we assume no time decimation of the STFT by the frame interval L.

Suppose we denote a short-time segment at time n, $x_n[m] = w[n-m]x[m]$, by the two-dimensional function $f[n, m]$ in the time index variables n and m. Then the corresponding STFT can be expressed as

$$X(n, \omega) = \sum_{m=-\infty}^{\infty} x[n, m]e^{-j\omega m}.$$

Considering the Fourier transform of $X(n, \omega)$ along the time axis, at specific frequencies, we then have the function [2]

$$\tilde{X}(\theta, \omega) = \sum_{n=-\infty}^{\infty} X(n, \omega)e^{-j\theta n}$$

$$= \sum_{n=-\infty}^{\infty} \sum_{m=-\infty}^{\infty} x[n, m]e^{-j(\theta n + \omega m)} \tag{13.23}$$

which is a two-dimensional (2-D) Fourier transform of the 2-D sequence $x[n, m]$. The 2-D transform $\tilde{X}(\theta, \omega)$ can be interpreted as a frequency analysis of the filter-bank outputs. The frequency composition of the time-trajectory of each channel is referred to as the *modulation spectrum* with *modulation frequency* as the frequency variable θ [25].

13.6.2 Temporal Filtering

The temporal processing of interest to us is the filtering of time-trajectories to remove distortions incurred by a sequence $x[n]$. The blind homomorphic deconvolution method of cepstral mean subtraction (CMS) introduced in Section 13.2.3 aims at this same objective and is related to the temporal filtering of this section. We formalize this temporal processing with a multiplicative modification $\tilde{P}(\theta, \omega)$:

$$\tilde{Y}(\theta, \omega) = \tilde{P}(\theta, \omega)\tilde{X}(\theta, \omega)$$

which can be written as a filtering operation in the time-domain variable n [2]; i.e., for each ω, we invert $\tilde{Y}(\theta, \omega)$ with respect to the variable θ to obtain

$$Y(n, \omega) = \sum_{m=-\infty}^{\infty} P(n-m, \omega)X(m, \omega)$$

$$= P(n, \omega) * X(n, \omega) \tag{13.24}$$

where $P(n, \omega)$ denotes the time-trajectory filter at frequency ω. We see then that the STFT is convolved along the time dimension, while it is multiplied along the frequency dimension.

We now consider the problem of obtaining a sequence from the modified 2-D function $Y(n, \omega)$. In practice, we replace continuous frequency ω by $\omega_k = \frac{2\pi}{N}k$ corresponding to N uniformly spaced filters in the discrete STFT that we denote by $Y(n, k)$. One approach to synthesis[17] is to apply the filter-bank summation (FBS) method given in Equation (7.9) of Chapter 7. For the modified discrete STFT, $Y(n, k)$, the FBS method gives

$$y[n] = \frac{1}{Nw[0]} \sum_{k=0}^{N-1} Y(n, k) e^{j\frac{2\pi}{N}kn}. \tag{13.25}$$

We saw in Chapter 7 that when no modification is applied, the original sequence $x[n]$ is recovered uniquely with this approach under the condition that the STFT bandpass filters sum to a constant (or, more strictly, that $w[n]$ has length less than N). On the other hand, with modification by time-trajectory filters $P(n, \frac{2\pi}{N}k)$ (representing a different filter for each uniformly spaced frequency $\frac{2\pi}{N}k$), we can show that (Exercise 13.7)

$$y[n] = x[n] * \left[\frac{1}{Nw[0]} \sum_{k=0}^{N-1} \{w[n] * P\left(n, \frac{2\pi}{N}k\right)\} e^{j\frac{2\pi}{N}kn} \right]. \tag{13.26}$$

Filtering along the N discrete STFT time-trajectories (with possibly different filters) is thus equivalent to a *single linear time-invariant* filtering operation on the original time sequence $x[n]$ [2]. The following example considers the simplifying case where the same filter is used for each channel:

EXAMPLE 13.6   Consider a convolutional modification along time-trajectories of the discrete STFT with the same causal sequence $p[n]$ for each discrete frequency $\frac{2\pi}{N}k$. The resulting discrete STFT is thus

$$Y(n, k) = X(n, k) * p[n], \qquad k = 0, 1, \ldots N - 1$$

i.e., we filter by $p[n]$ along each time-trajectory of the discrete STFT. For the FBS output we have Equation (13.26). We are interested in finding the condition on the filter $p[n]$ and the analysis window $w[n]$ under which $y[n] = x[n]$.

Using Equation (13.26), we can show that $y[n]$ can be written as

$$y[n] = \frac{1}{w[0]} \sum_{m=-\infty}^{\infty} x[m] \frac{1}{N} \sum_{k=0}^{N-1} (w[n - m] * p[m]) e^{j\frac{2\pi}{N}kn}.$$

Rearranging this expression, we have

$$y[n] = \frac{1}{w[0]} \sum_{m=-\infty}^{\infty} x[m]w[n - m] * p[m] \frac{1}{N} \sum_{k=0}^{N-1} e^{j\frac{2\pi}{N}kn}$$

$$= \frac{1}{w[0]} \sum_{m=-\infty}^{\infty} x[m]\tilde{w}[n - m] \sum_{r=-\infty}^{\infty} \delta[n - m - rN] \tag{13.27}$$

---

[17] Throughout this section, the filter-bank summation (FBS) method is used for synthesis. Other methods of synthesis from a modified STFT, such as the overlap-add (OLA) or least-squared-error (LSE) method of Chapter 7, however, can also be applied. As in Chapter 7, each method has a different impact on the synthesized signal. The effect of OLA synthesis, for example, from a temporally processed STFT has been investigated by Avendano [2].

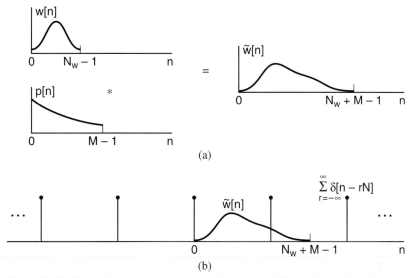

Figure 13.10   Illustration of the constraint on $w[n]$ and $p[n]$ for $y[n] = x[n]$ in Example 13.6: (a) the sequence $\tilde{w}[n] = w[n] * p[n]$ with $p[n]$ assumed causal; (b) the constraint given by $\frac{1}{w[0]} \tilde{w}[n] \sum_{r=-\infty}^{\infty} \delta[n - rN] = \delta[n]$.

where $\tilde{w}[n] = w[n] * p[n]$ and where we have used the result that a sum of harmonically related complex exponentials equals an impulse train. We can then simplify Equation (13.27) as

$$y[n] = x[n] * v[n]$$

where

$$v[n] = \frac{1}{w[0]} \tilde{w}[n] \sum_{r=-\infty}^{\infty} \delta[n - rN].$$

Then the constraint on $p[n]$ and $w[n]$ for $y[n] = x[n]$, illustrated pictorially in Figure 13.10, is given by

$$\frac{1}{w[0]} \tilde{w}[n] \sum_{r=-\infty}^{\infty} \delta[n - rN] = \delta[n]$$

or, assuming no zeros within the duration of $\tilde{w}[n]$, the constraint for $y[n] = x[n]$ becomes

$$N_w + M - 1 < N$$
$$\frac{w[0]p[0]}{w[0]} = p[0] = 1$$

where N_w and M are the lengths of the window $w[n]$ and filter $p[n]$, respectively.   ▲

We see from the previous example that we can recover $x[n]$ when $X(n, \omega)$ is temporally filtered along the n dimension by a single filter. We have also seen in this section that linear filtering along time-trajectories is equivalent to applying a *single* linear time-invariant filter to the sequence $x[n]$. Although these results are of academic interest, our ultimate objective is to use temporal filtering along channels to remove distortion. We will now see that one approach to realizing the advantage of temporal processing lies with *nonlinear transformations of the trajectories prior to filtering*, for which there does not always exist an equivalent time-domain filtering of $x[n]$ [2].

13.6.3 Nonlinear Transformations of Time-Trajectories

We have seen in Chapter 6 examples of nonlinear processing in which the logarithm and spectral root operators were applied to the STFT. Although the resulting homomorphic filtering was applied to spectral slices of the nonlinearly transformed STFT, and not along STFT filter-bank output trajectories, the concept motivates the temporal processing of this section.

Magnitude-Only Processing — We begin with "magnitude-only" processing where the STFT magnitude only of a sequence $x[n]$ is processed along time-trajectories to yield a new STFT magnitude:

$$|Y(n, \omega)| = \sum_{m=-\infty}^{\infty} P(n - m, \omega)|X(m, \omega)|$$

where we assume that the filter $P(n, \omega)$ is such that $|Y(n, \omega)|$ is a positive 2-D function. As with other processing methods of this chapter that utilize only the STFT magnitude, we attach the phase of the original STFT to the processed STFT magnitude,[18] resulting in a modified 2-D function of the form

$$Y(n, \omega) = |Y(n, \omega)|e^{j\phi(n,\omega)}$$

where $\phi(n, \omega)$ denotes the phase of $X(n, \omega)$. Then, using the modified discrete STFT, $Y(n, k)$, we apply FBS synthesis to obtain a sequence. This nonlinear process can be shown to be equivalent to a time-varying filter that consists of a sum of time-varying bandpass filters corresponding to the discretized frequencies, specifically (Exercise 13.8),

$$y[n] = \sum_{m=-\infty}^{\infty} x[n - m] \left[\sum_{k=-\infty}^{\infty} g_k[n, m]e^{j\omega_k n} \right] \tag{13.28}$$

where $\omega_k = \frac{2\pi}{N}k$ and where the time-varying filters

$$g_k[n, m] = \sum_{r=-\infty}^{\infty} p_k[n, r]w[m - r]$$

with

$$p_k[n, r] = P(r, \omega_k)e^{j[\phi(n,\omega_k)-\phi(n-r,\omega_k)]}.$$

Denoting the composite filter $\sum_{k=-\infty}^{\infty} g_k[n, m]e^{j\omega_k n}$ by $q[n, m]$, according to our interpretation in Chapter 2, $q[n, m]$ is a time-varying filter impulse response at time n to a unit sample applied m samples earlier. In contrast to the time-varying filter of Equation (13.26) [derived from temporal processing of $X(n, \omega)$], this time-varying filter is quite complicated, nonlinear, and *signal-dependent* in the sense that it requires the phase of the STFT of $x[n]$. This is also in contrast to the time-varying multiplicative modification of Chapter 7 for which the FBS method

---

[18] We can also consider a spectral magnitude-only reconstruction, e.g., the iterative solution in Section 7.5.3, whereby a signal phase is obtained through the modified STFT magnitude.

results in an equivalent time-varying linear filter, Equation (7.23), that is not signal-dependent. For example, such an equivalent time-domain linear filter can be found for the zero-phase spectral subtraction and Wiener filters developed in the earlier sections of this chapter.

The magnitude function is one example of a nonlinear operation on the STFT prior to temporal filtering along time-trajectories. More generally, we can write

$$Y(n, \omega) = O^{-1}\left\{ \sum_{m=-\infty}^{\infty} P(n - m, \omega)O[X(m, \omega)] \right\} \qquad (13.29)$$

where O is a nonlinear operator, and where we invoke the inverse operator O^{-1}, having in mind the objective of signal synthesis. We now look at two other methods of nonlinear temporal processing, for which there is no equivalent time-domain linear filter (not even time-varying and signal-dependent) [2], and which have effective application with convolutional and additive disturbances. In these applications, we do not always seek to synthesize a signal; rather, the modified STFT may be replaced by some other feature set. We touch briefly upon these feature sets here and discuss them in more detail in Chapter 14 in the specific application domain of speaker recognition.

RASTA Processing — A generalization of cepstral mean subtraction (CMS), that we introduced in Section 13.2.3, is RelAtive SpecTrAl processing (RASTA) of temporal trajectories. RASTA, proposed by Hermansky and Morgan [23], addresses the problem of a slowly time-varying linear channel $g[n, m]$ (i.e., convolutional distortion) in contrast to the time-invariant channel $g[n]$ removed by CMS. The essence of RASTA is a cepstral lifter that removes low and high modulation frequencies and not simply the DC component, as does CMS.

In addition to being motivated by a generalization of CMS, RASTA is also motivated by certain auditory principles. This auditory-based motivation is, in part, similar to that for adaptivity in the Wiener filter of Section 13.3.2: The auditory system is particularly sensitive to change in a signal. However, there is apparent evidence that auditory channels have preference for modulation frequencies near 4 Hz [23]. This peak modulation frequency is sometimes called the *syllabic rate* because it corresponds roughly to the rate at which syllables occur. RASTA exploits this modulation frequency preference. With slowly varying (rather than fixed) channel degradation, and given our insensitivity to low modulation frequencies,[19] in RASTA a filter that notches out frequency components at *and near* DC is applied to each channel. In addition, the RASTA filter suppresses high modulation frequencies to account for the human's preference for signal change at a 4 Hz rate. (We let the reader explore the possible inconsistency of this high-frequency removal with our auditory-based motivation for RASTA.)

In RASTA, the nonlinear operator O in the general nonlinear processing scheme of Equation (13.29) becomes the magnitude followed by the logarithm operator. The RASTA filtering scheme then attenuates the slow and fast changes in the temporal trajectories of the logarithm of the STFT magnitude. Using the signal processing framework in Equation (13.29), we thus write the modified STFT magnitude used in RASTA enhancement as:

$$|\hat{X}(n, \omega)| = \exp\left\{ \sum_{m=-\infty}^{\infty} p[n - m] \log |Y(m, \omega)| \right\}$$

---

[19] It is known, for example, that humans become relatively insensitive to stationary background noises over long time durations.

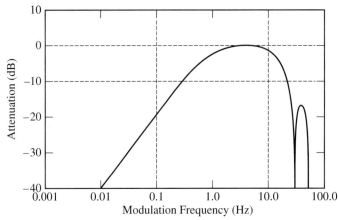

Figure 13.11 Frequency response of the RASTA bandpass filter.

SOURCE: H. Hermansky and N. Morgan, "RASTA Processing of Speech" [23].
©1994, IEEE. Used by permission.

where a single filter $p[n]$ is used along each temporal trajectory and where here $Y(n, \omega)$ denotes the STFT of a convolutionally distorted sequence $x[n]$. The particular discrete-time filter used in one formulation of RASTA is an IIR filter given by [23]

$$P(z) = \frac{2 + z^{-1} - z^{-3} - 2z^{-4}}{1 - 0.98z^{-1}}$$

where the denominator provides a lowpass effect and the numerator a highpass effect. The sampling frequency of this RASTA filter is 100 Hz, i.e., the frame interval L corresponds to 10 ms.[20]

The frequency response of the resulting bandpass RASTA filter is shown in Figure 13.11. The RASTA filter is seen to peak at about 4 Hz. As does CMS, RASTA reduces slowly varying signal components, but, in addition, suppresses variations above about 16 Hz. The complete RASTA temporal processing for blind deconvolution is illustrated in Figure 13.12. In this figure, a slowly varying distortion $\log |G(n, \omega)|$, due to a convolutional distortion $g[n]$ to be removed by the RASTA filter $p[n]$, is added to the rapidly varying speech contribution $\log |X(n, \omega)|$.

Although we have formulated RASTA as applied to the STFT, the primary application for RASTA has been in speech and speaker recognition where filtering is performed on temporal trajectories of critical band energies used as features in these applications. In fact, the characteristics of the RASTA filter shown in Figure 13.11 were obtained systematically by optimizing performance of a speech recognition experiment in a degrading and changing telephone environment [23]. In Chapter 14, we describe another common use of RASTA in the particular application of speaker recognition.

RASTA-Like Additive Noise Reduction — In addition to reducing convolutional distortion, RASTA can also be used to reduce additive noise. The temporal processing is applied to the

---

[20] The RASTA filter was originally designed in the context of speech recognition where speech features are not obtained at every time sample [23].

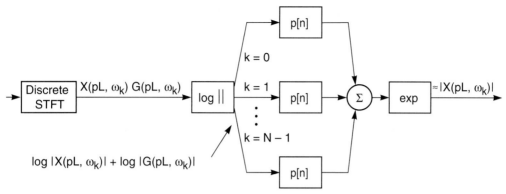

Figure 13.12   Complete flow diagram of RASTA processing for blind deconvolution. A linear filter $p[n]$ is applied to each nonlinearly-processed STFT temporal trajectory consisting of the rapidly varying object component $\log |X(pL, \omega_k)|$ and the slowly varying distortion component $\log |G(pL, \omega_k)|$ where $\omega_k = 2\pi k/N$.

STFT magnitude and the original (noisy) phase along each temporal trajectory is kept intact. In performing noise reduction along STFT temporal trajectories,[21] we assume that the noise background changes slowly relative to the rate of change of speech, which is concentrated in the 1–16 Hz range.

A nonlinear operator used in Equation (13.29) such as the logarithm, however, does not preserve the additive noise property and thus linear trajectory filtering is not strictly appropriate. Nevertheless, a cubic-root nonlinear operator followed by RASTA filtering has led to noise reduction similar in effect to spectral subtraction, including the characteristic musicality [2],[24].

An alternative approach taken by Avendano [2] is to design a Wiener-like optimal filter that gives a best estimate of each desired temporal trajectory. In his design, Avendano chose a power-law modification of the magnitude trajectories, i.e.,

$$|\hat{X}(n, \omega)| = \left\{ \sum_{m=-\infty}^{\infty} P(n - m, \omega)|Y(m, \omega)|^{1/\gamma} \right\}^{\gamma}$$

and, with the original phase trajectories, an enhanced signal is obtained using FBS synthesis. Here $Y(m, \omega)$ denotes the STFT of a sequence $x[n]$ degraded by additive noise. Motivation for the power-law, as well as the earlier cube-root modification, is evidence that the auditory system exploits this form of nonlinearity on envelope trajectories [2],[24]. For the discretized frequencies $\omega_k = \frac{2\pi}{N}k$ used in FBS, the objective is to find for each trajectory an optimal filter $p_k[n] = P(n, \omega_k)$ that, when applied to the noisy trajectory, $y_k[n] = |Y(n, \omega_k)|^{1/\gamma}$, results in a modified trajectory that meets a desired trajectory, $d_k[n]$, which is the trajectory of the clean speech. This filter is designed to satisfy a least-squared-error criterion, i.e., we seek to minimize

$$E_k = \sum_{n=-\infty}^{\infty} (y_k[n] * p_k[n] - d_k[n])^2. \tag{13.30}$$

---

[21] Noise reduction has also been performed by temporally filtering along sinewave amplitude, frequency, and phase trajectories [48]. Given the continuity of these functions by the interpolation methods of Chapter 9, it is natural to consider this form of temporal filtering.

In forming the error E_k, a clean speech reference is used to obtain the desired trajectories $d_k[n]$. Assuming a non-causal filter $p_k[n]$ that is non-zero over the interval $[-L/2, L/2]$ (L assumed even), Avendano then solved this mean-squared error optimization problem for the $2L + 1$ unknown values of $p_k[n]$ (Exercise 13.11) and found that the resulting filters have frequency responses that resemble the RASTA filter of Figure 13.11 but with somewhat milder rolloffs. The filters are Wiener-like in the sense of preserving the measured modulation spectrum in regions of high SNR and attenuating this spectrum in low SNR regions. Interestingly, it was observed that the non-causal filter impulse responses are nearly symmetric and thus nearly zero-phase. With this filter design technique, the value of $\gamma = 1.5$ for the power-law nonlinearity was found in informal listening to give preferred perceptual quality.

For various noise conditions, the optimal RASTA-like filtering was also found to give a reduction in mean-squared error in critical band energies relative to conventional spectral subtraction and Wiener filtering [2]. This reduction in mean-squared error, however, was obtained at the expense of more annoying residual "level fluctuations." Nevertheless, an important aspect of this comparison is that because of the nonlinear power operation, there is no equivalent single time-domain filtering process as with conventional short-time spectral subtraction and Wiener filtering [2]. Although this optimal filtering of temporal trajectories suffers from residual fluctuations, it has not exploited any form of adaptive smoothing as in the Wiener filtering of time slices of Sections 13.3.2 and 13.3.4. Moreover, the approach is based on auditory principles that give promise for further improvement.

13.7 Summary

In this chapter, we have studied approaches to reduce additive noise and convolutional distortion in a speech signal over a single channel.[22] For each disturbance, we first investigated techniques of filtering one spectral slice at a particular time instant, motivated by the Fourier transform view of the STFT. These techniques include spectral subtraction, Wiener filtering, and optimal spectral magnitude estimation. For each disturbance, we also investigated filtering a multitude of time slices at a particular frequency, motivated by the bandpass filtering view of the STFT. We observed that one approach to realizing the advantage of the later temporal processing lies with nonlinear transformations of the temporal trajectories, prior to filtering, thus mimicking, in part, early stages of auditory front-end processing. These techniques include cepstral mean subtraction and RASTA processing. We also introduced the principle of auditory masking both in time, with adaptive Wiener filtering for concealing noise under dynamically changing components of speech and, in frequency, with spectral subtraction for concealing small spectral components by nearby large spectral components. In addition, we exploited auditory phenomena in creating binaural presentations of the enhanced signal and its complement for possible further enhancement. A characteristic of all the techniques studied is that the original noisy STFT phase is preserved. No attempt is made to estimate the phase of the desired signal; rather, the noisy measurement phase is returned to the estimate. We saw that the work of Vary [53], however, indicates that phase can be important perceptually at low SNR, pointing to the need for the development of phase estimation algorithms in noise (one possibility given in Exercise 13.4). Finally, in this chapter, we have observed or implied that, in spite of significant improvements

---

[22] This is in contrast to enhancement with multiple recordings of the degraded speech as from multiple microphones. A useful survey of such techniques is given by Deller, Proakis, and Hansen [11].

in reducing signal distortion and residual artifacts, improved intelligibility of the enhanced waveform for the human listener remains elusive.

Given the space constraints of this chapter, there remain many developing areas in speech enhancement that we did not cover. Three of these areas involve multi-resolution analysis, nonlinear filtering, and advanced auditory processing models. Here we mention a few representative examples in each area. In the first area of multi-resolution analysis, Anderson and Clements [1] have exploited auditory masking principles in the context of multi-resolution sinusoidal analysis/synthesis, similar to that illustrated in Figure 9.18. Irino [26] has developed a multi-resolution analysis/synthesis scheme based explicitly on level-adaptive auditory filters with impulse responses modeled as gammachirps [Chapter 11 (Section 11.2.2)], and Hansen and Nandkumar [20] have introduced an auditory-based multi-resolution Wiener filter that exploits lateral inhibition across channels. Wavelet-based noise reduction systems, whose bases approximate critical band responses of the human auditory system, bridge multi-resolution analysis and nonlinear filtering by applying nonlinear thresholding techniques to wavelet coefficients [12],[45]. Other nonlinear filtering methods include Occam filters based on data compression [42], and Teager-based and related quadratic-energy-based estimation algorithms [14],[27]. All of these methods, both multi-resolution and nonlinear, provide different approaches to seek the preserving of fine temporal structure while suppressing additive noise. In the third developing area, advanced auditory processing models, we have only begun to explore the possibilities. In the growing field of *auditory scene analysis* [6], for example, components of sounds that originate from different sources (e.g., speech and additive background noise) are grouped according to models of auditory perception of mixtures of sounds. Component groups for signal separation include harmonically related sounds, common onset and offset times, and common amplitude or frequency modulation[23] [9].

Although these emerging areas are in their infancy, they are providing the basis for new directions not only for the reduction of additive noise and convolutional distortions, but also for the mitigation of other common distortions not covered in this chapter. These disturbances include reverberation, nonlinear degradations that we introduce in Chapter 14, and interfering speakers that we touched on briefly in Chapter 9 (Exercise 9.17).

Appendix 13.A: Stochastic-Theoretic Parameter Estimation

In this appendix, we consider the estimation of a parameter vector $\underline{a}$ given a measurement vector $\underline{y}$ that is a probabilistic mapping of $\underline{a}$, e.g., $\underline{a}$ are linear prediction coefficients and $\underline{y}$ are noisy speech measurements. Three estimation methods of interest are the maximum likelihood (ML), maximum *a posteriori* (MAP), and minimum-mean-squared error (MMSE) estimation [32].

Maximum Likelihood (ML): Suppose that the parameter vector $\underline{a}$ is deterministic and the probability density of $\underline{y}$ given $\underline{a}$ is known. In ML estimation, the desired parameter vector is selected that most likely resulted in the observation vector $\underline{y}$. This corresponds to maximizing the conditional probability function $p_{\underline{Y}|\underline{A}}(\underline{y}|\underline{a})$ over all $\underline{a}$ in the parameter space $\underline{A}$ and where $\underline{y}$ falls in the space $\underline{Y}$.

---

[23] A fascinating set of experiments by McAdams [36] shows the importance of modulation in perceptually separating two summed synthetically-generated vowels. In one experiment, with fixed pitch and vocal tract, the vowels were not perceived as distinct. When frequency modulation is added to the pitch of one vowel, aural separation of the vowels significantly improves.

Maximum *a posteriori* (MAP): Suppose that the parameter vector $\underline{a}$ is random and the *a posteriori* probability density of $\underline{a}$ given $\underline{y}$, $p_{\underline{A}|\underline{Y}}(\underline{a}|\underline{y})$, is known. In MAP estimation, the parameter vector is selected to maximize $p_{\underline{A}|\underline{Y}}(\underline{a}|\underline{y})$ over the space of parameter vectors $\underline{A}$. When the *a priori* probability density $p_{\underline{A}}(\underline{a})$ is flat over the range of $\underline{a}$, ML and MAP yield the same parameter estimation.

Minimum-Mean-Squared Error (MMSE): Suppose again that the parameter vector $\underline{a}$ is random and the *a posteriori* probability density of $\underline{a}$ given $\underline{y}$, $p_{\underline{A}|\underline{Y}}(\underline{a}|\underline{y})$, is known. In MMSE estimation, the parameter vector is selected by maximizing the mean-squared error $E[(\hat{\underline{a}} - \underline{a})^2]$ which can be shown to result in the conditional *a posteriori* mean $E[\underline{a}|\underline{y}]$; thus, when the maximum of $p_{\underline{A}|\underline{Y}}(\underline{a}|\underline{y})$ equals its mean, the MAP and MMSE estimates are equal.

EXERCISES

13.1 Consider a signal $y[n]$ of the form $y[n] = x[n] + b[n]$ where $x[n]$ is a sinewave, i.e., $x[n] = A\cos(\omega_o n)$, and $b[n]$ is a white noise background disturbance with variance σ^2. In this problem, you investigate the signal properties used in Example 13.1 for $y[n]$, analyzed by a short-time window $w[n]$ of length N_w.

(a) Show that when the sinewave frequency ω_o is larger than the width of the main lobe of the analysis window, it follows that

$$|X(pL, \omega_o)| \approx \left|\frac{A}{2}W(0)\right|$$

where $W(0) = \sum_{n=-\infty}^{\infty} w[n]$ and thus

$$E[|X(pL, \omega_o)|^2] \approx \frac{A^2}{4}\left|\sum_{n=-\infty}^{\infty} w[n]\right|^2$$

where E denotes the expectation operator.

(b) Show that the average power of the windowed noise is constant with frequency, i.e.,

$$E[|B(pL, \omega)|^2] = \sigma^2 \sum_{n=-\infty}^{\infty} w^2[n].$$

(c) Using the property that $x[n]$ and $b[n]$ are uncorrelated, show that the SNR ratio in Equation (13.9) follows. Argue that Δ_w in Equation (13.9) represents the 3-dB bandwidth of the window main lobe.

13.2 For the signal $y[n] = x[n] + b[n]$ with the object random process $x[n]$ uncorrelated with the background noise random process $b[n]$, derive the Wiener filter in Equation (13.10). Then show its time-varying counterpart in Equation (13.11) in terms of signal-to-noise ratio $R(n, \omega) = \hat{S}_x(n, \omega)/\hat{S}_b(\omega)$.

13.3 Consider a filter bank $h_k[n] = w[n]e^{j\frac{2\pi}{N}kn}$ that meets the FBS constraint. In this problem, you develop a single noise suppression filter, applied to all N channel outputs for a noisy input $y[n] = x[n] + b[n]$. Assume the object random process $x[n]$ uncorrelated with the background noise random process $b[n]$. Specifically, find the optimal noise suppression filter $h_s[n]$ that minimizes the error criterion

$$E\left[\sum_{k=0}^{N-1}\{h_s[n] * (h_k[n] * y[n]) - (h_k[n] * x[n])\}^2\right].$$

Express your solution in terms of the object and background spectra $S_x(\omega)$ and $S_b(\omega)$, respectively. Explain intuitively the difference between your solution and the standard Wiener filter.

13.4 All enhancement methods in this chapter avoid phase estimation, whether in processing STFT slices along frequency or temporal trajectories. In this problem you are asked to develop a method of estimating the phase of the Fourier transform of a sequence $x[n]$, i.e., the phase of $X(\omega)$, from the noisy sequence $y[n] = x[n] + b[n]$. We work here with the entire sequences, rather than the STFT, although the approach can be generalized to the STFT.

    Suppose that we switch the roles of time and frequency. Then one approach to estimating the phase function $\angle X(\omega)$ is to estimate the complex Fourier transform $X(\omega)$ in the presence of the disturbance $B(\omega)$ (i.e., the Fourier transform of $b[n]$) where we view the complex functions $X(\omega)$, $B(\omega)$, and $Y(\omega)$ as "time signals."

(a) Find the best frequency-domain linear smoothing filter $H_s(\omega)$ such that

$$\hat{X}(\omega) = H_s(\omega) * Y(\omega)$$

gives the minimum-mean-squared error between the desired $X(\omega)$ and its estimate $\hat{X}(\omega)$, i.e., minimize the error

$$E = \int_{-\pi}^{\pi} |H_s(\omega) * Y(\omega) - X(\omega)|^2 d\omega.$$

Using Parseval's Theorem, express your result in the time domain as an operation on $y[n]$ and assume you know $|x[n]|^2$ and $|b[n]|^2$. The procedure yields both a magnitude and phase estimate of $X(\omega)$ through *smoothing* the real and imaginary parts of $Y(\omega)$. The magnitude may not be as accurate an estimate as that derived from the conventional Wiener filter, but the phase estimate may be an improvement. *Hint:* Given that we have switched the roles of time and frequency, we expect the result to invoke a *multiplication* of $y[n]$ by a function, corresponding to a smoothing of $Y(\omega)$. An example of a phase estimate from this method, for an exponentially-decaying sinewave $x[n]$ and white Gaussian noise $b[n]$, is shown in Figure 13.13b. In this example, $|x[n]|^2$ is assumed to be known and $|b[n]|^2$ is replaced by the variance of $b[n]$.

(b) Suppose we now combine the two estimates from the frequency-domain filter of part (a) and a conventional time-domain Wiener filter. Propose an iterative estimation scheme that uses both estimates and that may provide a phase estimate that improves on each iteration. *Hint:* Apply the conventional time-domain Wiener filter, but "deterministic" version where $|X(\omega)|^2$ and $|B(\omega)|^2$ are assumed known, followed by the frequency-domain filter from part (a), first estimating magnitude and then estimating phase. Heuristically, the time-domain method de-emphasizes regions of the complex spectrum at low-energy regions where the phase is known to bear little resemblance to the phase of the actual signal. The frequency-domain filter then smooths the complex spectrum. The resulting smooth phase estimate for the example of an exponentially-decaying sinewave in white noise is shown in Figure 13.13c after three iterations. For the frequency-domain filter, as in part (a), $|x[n]|^2$ is assumed to be known and $|b[n]|^2$ is replaced by the variance of $b[n]$. For the time-domain filter, $|B(\omega)|^2$ is replaced by the variance of $b[n]$.

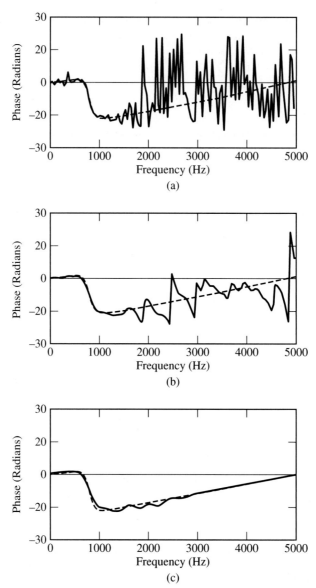

Figure 13.13 Iterative phase estimation based on time-frequency Wiener filtering of an exponentially-decaying sinewave in white noise in Exercise 13.4: (a) original noisy phase; (b) phase estimate from the frequency-domain Wiener filter; (c) phase estimate from the combined time- and frequency-domain Wiener filters after three iterations. In each panel, the phase of the decaying sinewave is shown as a dashed line and the phase estimate by a solid line.

13.5 Suppose a speech waveform is modeled with vocal tract poles and zeros, and consider the problem of estimating the speech in the presence of additive noise. Propose a (linear) iterative method, as a generalization of the LMAP algorithm for all-pole estimation of Section 13.4, that estimates both the poles and zeros, as well as the speech, simultaneously. *Hint:* Maximize the *a posteriori* probability density $p(\underline{a}, \underline{b}, \underline{x}/\underline{y})$. The vectors $\underline{a}$ and $\underline{b}$ represent the pole and zero polynomial coefficients, respectively, and the vectors $\underline{x}$ and $\underline{y}$ represent (for each analysis frame) the clean and noisy speech, respectively.

13.6 In this problem, you explore the use of a masking threshold in a suppression filter $h_s[n]$ that results in perceived noise that is an attenuated version of the original noise. In this formulation, given an original noisy signal $y[n] = x[n] + b[n]$, the desired signal can be written as $d[n] = x[n] + \alpha b[n]$ where α is the noise suppression scale factor.

    (a) Show that an estimate of the short-time power spectrum of the noise error $\alpha b[n] - h_s[n] * b[n]$ is given by Equation (13.20).

    (b) Derive the suppression filter range, Equation (13.21), for which the short-time noise error power spectrum estimate falls below the masking threshold $T(pL, \omega)$.

    (c) Discuss tradeoffs in the degree of noise attenuation and the speech distortion that can be obtained though the parameter α. Compare this suppression tradeoff with that of the standard Wiener filter.

13.7 Show that with FBS synthesis in Section 13.6.2, the operation equivalent to filtering an STFT by $P\left(n, \frac{2\pi}{N}k\right)$ along temporal trajectories is a filtering by a single time-invariant linear filter consisting of a sum of bandpass filters, i.e.,

$$y[n] = x[n] * \left[\frac{1}{Nw[0]} \sum_{k=0}^{N-1} \left\{ w[n] * P\left(n, \frac{2\pi}{N}k\right) \right\} e^{j\frac{2\pi}{N}kn} \right].$$

13.8 Suppose that we apply the filter $P(n, \omega)$ along each time-trajectory of the STFT magnitude of a sequence $x[n]$, as in Section 13.6.3. Show that applying the filter-bank summation (FBS) method with discretized frequencies $\omega_k = \frac{2\pi}{N}k$ results in the input sequence $x[n]$ being modified by the time-varying filter of Equation (13.28).

13.9 Consider a signal $y[n]$ of the form $y[n] = x[n] * g[n]$ where $g[n]$ represents a linear time-invariant distortion of a desired signal $x[n]$. In this problem you explore different formulations of the STFT of $y[n]$.

    (a) Given $y[n] = x[n] * g[n]$, show that

$$Y(n, \omega) = (g[n]e^{-j\omega n}) * X(n, \omega)$$

    where the above convolution is performed with respect to the time variable n. Then argue that the two block diagrams in Figure 13.14 are equivalent.

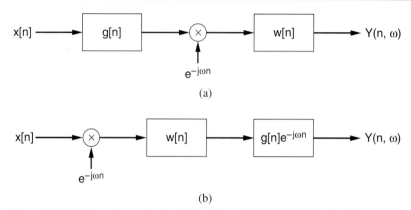

Figure 13.14 Effect of convolutional distortion on the STFT: (a) filter-bank interpretation; (b) equivalence to (a).

SOURCE: C. Avendano, *Temporal Processing of Speech in a Time-Feature Space* [2]. ©1997, C. Avendano. Used by permission.

(b) Rewrite the STFT of $y[n]$ as

$$Y(n, \omega) = \sum_{m=-\infty}^{\infty} x[m]e^{-j\omega m} \sum_{r=-\infty}^{\infty} w[n - m - r]g[r]e^{-j\omega r}$$

and argue that if the window $w[n]$ is long and smooth relative to the impulse response $g[n]$ so that $w[n]$ is approximately constant over the duration of $g[n]$, then $w[n - m]g[m] \approx w[n]g[m]$, from which it follows that

$$Y(n, \omega) \approx X(n, \omega)G(\omega)$$

i.e., the convolutional distortion results in approximately a multiplicative modification $G(\omega)$ to the STFT of $x[n]$. Discuss practical conditions under which this approximation may not be valid.

13.10 Suppose we compute the complex cepstrum of the STFT of a sequence $x[n]$ at each frame p, i.e.,

$$c[p, n] = \frac{1}{2\pi} \int_{-\pi}^{\pi} \log[X(p, \omega)]e^{j\omega n}d\omega$$

where we have assumed a frame interval $L = 1$. Show that applying a linear time-invariant filter $h[p]$ to $\log[X(p, \omega)]$ along the time dimension for each frequency ω is equivalent to applying the same filter to $c[p, n]$ with respect to the time variable p for each time n (i.e., for each cepstral coefficient). That is, $h[p] * \log[X(p, \omega)]$ gives a complex cepstrum $h[p] * c[p, n]$. Show that this relation is also valid for an arbitrary frame length L. Filtering the temporal trajectories of the logarithm of the STFT is, therefore, equivalent to filtering the temporal trajectories of the corresponding cepstral coefficients.

13.11 This problem addresses the use of RASTA-like filtering to reduce additive noise, as described in Section 13.6.3. Consider the design of a distinct optimal filter $p_k[n]$ along each power-law-modified temporal trajectory $y_k[n]$ of the STFT magnitude for discretized frequencies $\omega_k = \frac{2\pi}{N}k$ and assume a known desired trajectory $d_k[n]$. Also assume a non-causal filter $p_k[n]$, non-zero over the interval $[-L/2, L/2]$ (L assumed even). Minimize the error function [Equation (13.30)]

$$E_k = \sum_{n=-\infty}^{\infty} (y_k[n] * p_k[n] - d_k[n])^2$$

with respect to the unknown $p_k[n]$. Discuss the relation of your solution for $p_k[n]$ with that of the conventional Wiener filter applied to short-time signal segments.

13.12 (MATLAB) Design in MATLAB a noise reduction system based on the spectral subtraction rule in Equation (13.6) and OLA synthesis. Use a 20-ms analysis window and a 10 ms frame interval. Apply your function to the noisy signal *speech_noisy_8k* (at 8000 samples/s and a 9 db SNR) given in workspace *ex13M1.mat* located in companion website directory Chap_exercises/chapter13. (This signal is the one used in Example 13.5.) You will need to obtain an estimate of the power spectrum of the background noise, $\hat{S}_b(\omega)$, from an initial waveform segment. The clean signal version, *speech_clean_8k*, and the signal enhanced by the adaptive Wiener filter of Section 13.3.2 (Figure 13.4), *speech_wiener_8k*, are also given in the workspace. Comment on the musicality artifact from spectral subtraction and compare your result to *speech_wiener_8k*. In your spectral subtraction design, apply a scale factor α to your background power estimate, $\hat{S}_b(\omega)$, in Equation (13.6). Then comment on the tradeoff between speech distortion, noise reduction, and musicality in the residual noise as you vary α below and above unity, corresponding to an under-and over-estimation of $\hat{S}_b(\omega)$, respectively.

BIBLIOGRAPHY

[1] D.V. Anderson and M.A. Clements, "Audio Signal Noise Reduction Using Multi-Resolution Sinusoidal Modeling," *Proc. IEEE Conf. Acoustics, Speech, and Signal Processing*, vol. 2, pp. 805–808, Phoenix, AZ, March 1999.

[2] C. Avendano, *Temporal Processing of Speech in a Time-Feature Space*, Ph.D. Thesis, Oregon Graduate Institute of Science and Technology, April 1997.

[3] W.A. van Bergeijk, J.R. Pierce, and E.E. David, *Waves and the Ear*, Anchor Books, Doubleday & Company, Garden City, NY, 1960.

[4] M. Berouti, R. Schwartz, and J. Makhoul, "Enhancement of Speech Corrupted by Additive Noise," *Proc. IEEE Int. Conf. Acoustics, Speech, and Signal Processing*, pp. 208–211, April 1979.

[5] S.F. Boll, "Suppression of Acoustic Noise in Speech Using Spectral Subtraction," *IEEE Trans. Acoustics, Speech, and Signal Processing*, vol. ASSP–29, no. 2, pp. 113–120, April 1979.

[6] A.S. Bregman, *Auditory Scene Analysis*, The MIT Press, Cambridge, MA, 1990.

[7] O. Cappe, "Elimination of the Musical Noise Phenomenon with the Ephraim and Malah Noise Suppressor," *IEEE Trans. Speech and Audio Processing*, vol. 2, no. 1, pp. 345–349, April 1994.

[8] O. Cappe and J. Laroche, "Evaluation of Short-Time Attenuation Techniques for Restoration of Musical Recordings," *IEEE Trans. Speech and Audio Processing*, vol. 3, no. 1, pp. 84–93, Jan. 1995.

[9] M. Cooke, *Modeling Auditory Processing and Organization*, Cambridge University Press, Cambridge, England, 1993.

[10] A. Czyzewski and R. Krolikowski, "Noise Reduction in Audio Signals Based on the Perceptual Coding Approach," *Proc. 1999 IEEE Workshop on Applications of Signal Processing to Audio and Acoustics*, New Paltz, NY, Oct. 1999.

[11] J.R. Deller, J.G. Proakis, and J.H.L. Hansen, *Discrete-Time Processing of Speech*, Macmillan Publishing Co., New York, NY, 1993.

[12] D. Donahue and I. Johnson, "Ideal Denoising in an Orthonormal Basis Chosen from a Library of Bases," *C.R. Academy of Science*, Paris, France, vol. 1, no. 319, pp. 1317–1322, 1994.

[13] Y. Ephraim and D. Malah, "Speech Enhancement Using a Minimum Mean-Square Error Short-Time Amplitude Estimator," *IEEE Trans. Acoustics, Speech, and Signal Processing*, vol. ASSP–32, no. 6, pp. 1109–1121, Dec. 1984.

[14] J. Fang and L.E. Atlas, "Quadratic Detectors for Energy Estimation," *IEEE Trans. Signal Processing*, vol. 43, no. 11, pp. 2582–2594, Nov. 1995.

[15] H. Fletcher, "Auditory Patterns," *Rev. Mod. Phys.*, vol. 12, pp. 47–65, 1940.

[16] B. Gold and N. Morgan, *Speech and Audio Signal Processing*, John Wiley and Sons, Inc., New York, NY, 2000.

[17] S. Govindasamy, *A Psychoacoustically Motivated Speech Enhancement System*, S.M. Thesis, Massachusetts Institute of Technology, Dept. Electrical Engineering and Computer Science, Jan. 2000.

[18] D.M. Green, *An Introduction to Hearing*, John Wiley and Sons, New York, NY, 1976.

[19] S. Gustafsson, P. Jax, and P. Vary, "A Novel Psychoacoustically Motivated Audio Enhancement Algorithm Preserving Background Noise Characteristics," *Proc. IEEE Conf. Acoustics, Speech, and Signal Processing*, vol. 1, pp. 397–400, Seattle, WA, May 1998.

[20] J.H. Hansen and S. Nandkumar, "Robust Estimation of Speech in Noisy Backgrounds Based on Aspects of the Auditory Process," *J. Acoustical Society of America*, vol. 97, no. 6, pp. 3833–3849, June 1995.

[21] J.H. Hansen and M.A. Clements, "Constrained Iterative Speech Enhancement with Application to Automatic Speech Recognition," *IEEE Trans. Signal Processing*, vol. 39, no. 4, pp. 795–805, April 1991.

[22] R.P. Hellman, "Asymmetry of Masking Between Noise and Tone," *Perception and Psychophysics*, vol. 11, pp. 241–246, 1972.

[23] H. Hermansky and N. Morgan, "RASTA Processing of Speech," *IEEE Trans. Speech and Audio Processing*, vol. 2, no. 4, pp. 578–589, Oct. 1994.

[24] H. Hermansky, N. Morgan, and H.G. Hirsch, "Recognition of Speech in Additive and Convolutional Noise Based on RASTA Spectral Processing," *Proc. IEEE Conf. Acoustics, Speech, and Signal Processing*, vol. 2, pp. 83–86, Minneapolis, MN, April 1993.

[25] T. Houtgast and H.J.M. Steeneken, "A Review of the MTF Concept in Room Acoustics and its Use for Estimating Speech Intelligibility in Auditoria," *J. Acoustical Society of America*, vol. 77, no. 3, pp. 1069–1077, March 1985.

[26] T. Irino, "Noise Suppression Using a Time-Varying, Analysis/Synthesis Gammachip Filter Bank," *Proc. IEEE Int. Conf. Acoustics, Speech, and Signal Processing*, vol. 1, pp. 97–100, Phoenix, AZ, March 1999.

[27] F. Jabloun and A.E. Cetin, "The Teager Energy Based Feature Parameters for Robust Speech Recognition in Noise," *Proc. IEEE Conf. Acoustics, Speech, and Signal Processing*, vol. 1, pp. 273–276, Phoenix, AZ, March 1999.

[28] N. Jayant, J. Johnston, and R. Safranek, "Signal Compression Based on Models of Human Perception," *Proc. IEEE*, vol. 81, no. 10, pp. 1385–1422, Oct. 1993.

[29] J.D. Johnston, "Transform Coding of Audio Signals Using Perceptual Noise Criteria," *IEEE J. Selected Areas Communication*, vol. 6, no. 2, pp. 314–323, Feb. 1988.

[30] H.P. Knagenhjelm and W. B. Kleijn, "Spectral Dynamics is More Important than Spectral Distortion," *Proc. IEEE Int. Conf. Acoustics, Speech, and Signal Processing*, vol. 1, pp. 732–735, Detroit, MI, May 1995.

[31] K.D. Kryter, "Methods for the Calculation and the Use of the Articulation Index," *J. Acoustical Society of America*, vol. 34, pp. 1689–1697, Nov. 1962.

[32] J.S. Lim and A.V. Oppenheim, "Enhancement and Bandwidth Compression of Noisy Speech," *Proc. of the IEEE*, vol. 67, no. 12, pp. 1586–1604, Dec. 1979.

[33] P. Lockwood and J. Boudy, "Experiments with a Nonlinear Spectral Subtractor (NSS), Hidden Markov Models and Projection, for Robust Recognition in Cars," *Speech Comm.*, vol. 11, pp. 215–228, June 1992.

[34] J. Makhoul, T.H. Crystal, D.M. Green, D. Hogan, R.J. McAulay, D.B. Pisoni, R.D. Sorkin, and T.G. Stockham, "Removal of Noise from Noise-Degraded Speech," Panel on Removal of Noise from a Speech/Noise Signal, National Academy Press, Washington, D.C. 1989.

[35] D. Marr, *Vision: A Computational Investigation into the Human Representation of Visual Information*, W.H. Freeman and Company, New York, NY, 1982.

[36] S. McAdams, *Spectral Fusion, Spectral Parsing, and Formation of Auditory Images*, Ph.D. Thesis, CCRMA, Stanford University, Dept. of Music, May 1984.

[37] R.J. McAulay and M.L. Malpass, "Speech Enhancement Using a Soft-Decision Maximum Likelihood Noise Suppression Filter," *IEEE Trans. Acoustics, Speech, and Signal Processing*, vol. ASSP–28, no. 2, pp. 137–145, April 1980.

[38] R.J. McAulay, "Optimum Speech Classification and Its Application to Adaptive Noise Classification," *IEEE Proc. Int. Conf. Acoustics, Speech, and Signal Processing*, pp. 425–428, Hartford, CT, April 1977.

[39] R.J. McAulay, "Design of a Robust Maximum Likelihood Pitch Estimator in Additive Noise," Technical Note 1979–28, Massachusetts Institute of Technology, Lincoln Laboratory, June 11, 1979.

[40] R.J. McAulay and T.F. Quatieri, "Speech Analysis-Synthesis Based on a Sinusoidal Representation," *IEEE Trans. Acoustics, Speech, and Signal Processing*, vol. ASSP–34, no. 4, pp. 744–754, Aug. 1986.

[41] B.C.J. Moore, *An Introduction to the Psychology of Hearing*, Academic Press, 2nd Edition, New York, NY, 1988.

[42] B.K. Natarajan, "Filtering Random Noise from Deterministic Signals via Data Compression," *IEEE Trans. Signal Processing*, vol. 43, no. 11, pp. 2595–2605, Nov. 1995.

[43] S.H. Nawab, T.F. Quatieri, and J.S. Lim, "Signal Reconstruction from Short-Time Fourier Transform Magnitude," *IEEE Trans. Acoustics, Speech, and Signal Processing*, vol. ASSP–31, no. 4, pp. 986–998, Aug. 1983.

[44] D. O'Shaughnessy, *Speech Communication: Human and Machine*, Addison-Wesley, Reading, MA, 1987.

[45] I. Pinter, "Perceptual Wavelet-Representation of Speech Signals and its Application to Speech Enhancement," *Computer Speech and Language*, vol. 10, pp. 1–22, 1996.

[46] T.F. Quatieri and R. Baxter, "Noise Reduction Based on Spectral Change," *Proc. IEEE Workshop on Applications of Signal Processing to Audio and Acoustics*, pp. 8.2.1–8.2.4, New Paltz, NY, Oct. 1997.

[47] M.R. Schroeder, B.S. Atal, and J.L. Hall, "Optimizing Digital Speech Coders by Exploiting Masking Properties of the Human Ear," *J. Acoustical Society of America*, vol. 66, pp. 1647–1652, Dec. 1979.

[48] A. Seefeldt, *Enhancement of Noise-Corrupted Speech Using Sinusoidal Analysis/Synthesis*, Masters Thesis, Massachusetts Institute of Technology, Dept. Electrical Engineering and Computer Science, May 1987.

[49] D. Sen, D.H. Irving, W.H. Holmes, "Use of an Auditory Model to Improve Speech Coders," *Proc. IEEE Int. Conf. Acoustics, Speech, and Signal Processing*, vol. 2, pp. 411–414, Minneapolis, MN, April 1993.

[50] D. Sinha and A.H. Tewfik, "Low Bit Rate Transparent Audio Compression Using Adapted Wavelets," *IEEE Trans. Signal Processing*, vol. 41, no. 2, pp. 3463–3479, Dec. 1993.

[51] T. Stockham, T. Cannon, and R. Ingebretsen, "Blind Deconvolution Through Digital Signal Processing," *Proc. IEEE*, vol. 63, pp. 678–692, April 1975.

[52] E. Terhardt, "Calculating Virtual Pitch," *Hearing Research*, vol. 1, pp. 155–199, 1979.

[53] P. Vary, "Noise Suppression by Spectral Magnitude Estimation—Mechanism and Theoretical Limits," *Signal Processing*, vol. 8, pp. 387–400, 1985.

[54] N. Virag, "Single Channel Speech Enhancement Based on Masking Properties of the Human Auditory System," *IEEE Trans. Speech and Audio Processing*, vol. 7, no. 2, pp. 126–137, March 1999.

[55] K. Wang and S.A. Shamma, "Self-Normalization and Noise-Robustness in Early Auditory Representations," *IEEE Trans. Speech and Audio Processing*, vol. 2, no. 3, pp. 421–435, July 1994.

[56] R.L. Wegel and C.E. Lane, "The Auditory Masking of One Pure Tone by Another and Its Probable Relation to the Dynamics of the Inner Ear," *Physical Review*, vol. 23, no. 2, pp. 266–285, 1924.

[57] M.R. Weiss, E. Aschkenasy, and T.W. Parsons, "Study and Development of the INTEL Technique for Improving Speech Intelligibility," Nicolet Scientific Corp, Final Rep. NSC-FR/4023, Dec. 1974.

[58] J.D. Wise, J.R. Caprio, and T.W. Parks, "Maximum Likelihood Pitch Estimation," *IEEE Trans. Acoustics, Speech, and Signal Processing*, vol. ASSP–24, no. 5, pp. 418–423, Oct. 1976.

C H A P T E R **14**

Speaker Recognition

14.1 Introduction

One objective in automatic speaker recognition is to decide which voice model from a known set of voice models best characterizes a speaker; this task is referred to as *speaker identification* [10]. In the different task of *speaker verification*, the goal is to decide whether a speaker corresponds to a particular known voice or to some other unknown voice. A speaker known to a speaker recognition system who is correctly claiming his/her identity is labeled a *claimant* and a speaker unknown to the system who is posing as a known speaker is labeled an *imposter*. A known speaker is also referred to as a *target speaker*, while an imposter is alternately called a *background speaker*. There are two types of errors in speaker recognition systems: false acceptances, where an imposter is accepted as a claimant, and false rejections, where claimants are rejected as imposters.

This chapter focuses on the signal processing components of speaker identification and verification algorithms, illustrating how signal processing principles developed in the text are applied in speech-related recognition problems. In addition, we apply methods of speech modification, coding, and enhancement in a variety of speaker recognition scenarios, and introduce some new signal processing tools such as methods to compensate for nonlinear, as well as linear, signal distortion. Such compensation is particularly important when differences arise over channels from which data is collected and is then used for speaker recognition. Other recognition tasks such as speech and language recognition could also have been used to illustrate the use of signal processing in recognition, but these require a more statistical framework and would take us far beyond the signal processing theme of this book; numerous tutorials exist in these areas [29],[32],[59],[85].

The first step in a speaker recognition system, whether for identification or verification, is to build a *model* of the voice of each target speaker, as well as a model of a collection of background speakers, using speaker-dependent *features* extracted from the speech waveform. For example, the oral and nasal tract length and cross-section during different sounds, the vocal fold mass and shape, and the location and size of the false vocal folds, if accurately measured from the speech waveform, could be used as features in an anatomical speaker model. We call this the *training* stage of the recognition system, and the associated speech data used in building a speaker model is called the *training data*. During the recognition or *testing* stage, we then match (in some sense) the features measured from the waveform of a test utterance, i.e., the *test data* of a speaker, against speaker models obtained during training. The particular speaker models we match against, i.e., from target and background, depends on the recognition task. An overview of these components of a speaker recognition system for the verification task is given in Figure 14.1.

In practice, it is difficult to derive speech anatomy from the speech waveform.[1] Rather, it is typical to use features derived from the waveform based on the various speech production and perception models that we have introduced in the text. The most common features characterize the magnitude of the vocal tract frequency response as viewed by the front-end stage of the human auditory system, assumed to consist of a linear (nearly) constant-Q filter bank. In Section 14.2, we give examples of these spectral-based features and discuss time- and frequency-resolution considerations in their extraction from the speech waveform. In Section 14.3, we next describe particular approaches to training and testing in speaker recognition systems, first minimum-distance and vector-quantization (VQ) methods, and then a more statistical pattern-recognition approach based on maximum-likelihood classification. Also in this section, we illustrate the loss in recognition accuracy that can occur when training and test data suffer from various forms of degradation. In Section 14.4, we investigate a feature class based on the source to the vocal tract, rather than its spectrum, using the glottal flow derivative estimates and parameterization of Chapter 5. In this section, we also present examples of speaker recognition performance that give insight into the relative importance of the vocal tract spectrum, source characteristics, and prosody (i.e., pitch intonation and articulation rate) in speaker recognition, both by machine

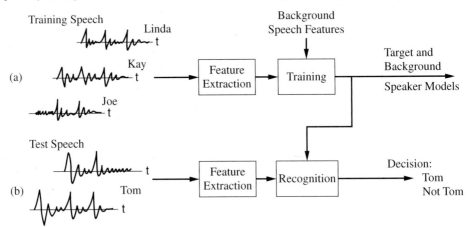

Figure 14.1 Overview of speaker recognition system for speaker verification. In this task, features extracted from test speech are matched against both target and background models.

---

[1] Nevertheless, there has been considerable effort in deriving physiological models of speech production, as with application to speech coding [74], which also holds promise for speaker recognition.

and by human. In this development, we alter speech characteristics using the sinewave-based modification of Chapter 9, and evaluate recognition performance from the modified speech.

In Section 4.5, we introduce the *channel mismatch* problem in speaker recognition that occurs when training data is collected under a different condition (or "channel") from the test data. For example, the training data may be essentially undistorted, being recorded with a high-quality microphone, while the test data may be severely degraded, being recorded in a harsh cellular environment. Section 14.5 provides different approaches to dealing with this challenging problem, specifically, signal processing for channel compensation, calling upon enhancement methods of Chapter 13 that address additive and convolutional distortion, including spectral subtraction, cepstral mean substraction (CMS), and RelAtive SpecTrAl processing (RASTA). Also in this section, we introduce a compensation approach for *nonlinear* distortion, an algorithm for removing badly corrupted features, i.e., "missing features," and the development of features that hold promise in being invariant under different degrading conditions. Finally, Section 14.6 investigates the impact of the various speech coding algorithms of Chapter 12 on speaker recognition performance, an increasingly important issue due to the broad range of digital communications from which recognition will be performed.

14.2 Spectral Features for Speaker Recognition

We have seen in Chapter 3 a variety of voice attributes that characterize a speaker. In viewing these attributes, both from the perspective of the human and the machine for recognition, we categorize speaker-dependent voice characteristics as "high-level" and "low-level." High-level voice attributes include "clarity," "roughness," "magnitude," and "animation" [60],[82]. Other high-level attributes are prosody, i.e., pitch intonation and articulation rate, and dialect. Voiers found that such high-level characteristics are perceptual cues in determining speaker identifiability [82]. On the other hand, these attributes can be difficult to extract by machine for automatic speaker recognition. In contrast, low-level attributes, being of an acoustic nature, are more measurable. In this chapter, we are interested in low-level attributes that contain speaker identifiability for the machine and, perhaps, as well as for the human. These attributes include primarily the vocal tract spectrum and, to a lesser extent, instantaneous pitch and glottal flow excitation, as well as temporal properties such as source event onset times and modulations in formant trajectories. In this section, we focus on features for speaker recognition derived from spectral measurements, and then later in the chapter branch out to a non-spectral characterization of a speaker.

14.2.1 Formulation

In selecting acoustic spectral features, we want our feature set to reflect the unique characteristics of a speaker. The short-time Fourier transform (STFT) is one basis for such features. The STFT can be written in polar form as

$$X(n, \omega) = \sum_{m=-\infty}^{\infty} x[m]w[n - m]e^{-j\omega m}$$

$$= |X(n, \omega)|e^{j\angle X(n,\omega)}.$$

In speaker recognition, only the magnitude component $|X(n, \omega)|$ has been used because features corresponding to the phase component are difficult to measure and are susceptible to channel distortion.[2] We have seen in Chapter 5 that the *envelope* of the speech STFT magnitude is characterized in part by the vocal tract resonances, i.e., the poles, that are speaker-dependent (as well as phoneme-dependent), determined by the vocal tract length and spatially-varying cross-section. In addition to the vocal tract resonances, its anti-resonances, i.e., its zeros, also contribute to the STFT magnitude envelope. Consider, as an example, nasalized vowels for which anti-resonances arise when energy is absorbed by a closed nasal cavity and open velum. Because the nasal cavity is fixed, the zeros during nasalized vowels introduced by this cavity may be a particularly important contribution to the envelope for speaker identifiability. A similar argument holds for nasal sounds for which the fixed nasal cavity is open and the oral cavity closed, together coupled by an open velum. Indeed, the resonances of the nasal passage, being fixed, were found by Sambur to possess considerable speaker identifiability, relative to other phoneme spectral resonances, for a certain group of speakers and utterances [69]. (A complete study of the relative importance of resonances and anti-resonances of all sounds has yet to be made.) In addition to the resonant and anti-resonant envelope contributions, the general trend of the envelope of the STFT magnitude (referred to in Example 3.2 of Chapter 3 as the "spectral tilt") is influenced by the coarse component of the glottal flow derivative. Finally, the STFT magnitude is also characterized by speaker-dependent fine structure including pitch, glottal flow components, and the distributed aeroacoustic effects of Chapter 11.

Most speaker recognition systems derive their features from a *smooth* representation of the STFT magnitude, assumed to reflect primarily vocal tract resonances and spectral tilt. We have seen a number of smooth spectral representations in this text, including ones obtained from linear prediction all-pole modeling (Chapter 5), homomorphic filtering (Chapter 6), and the SEEVOC representation (Chapter 10). For the linear prediction method, we can use the linear prediction coefficients or some transformation thereof as a feature vector. For the homomorphic filtering estimate, a liftered real cepstrum can serve as a feature vector. Another possibility is to combine approaches. For example, the first p values of the real cepstrum of a pth-order linear predictor uniquely characterize the predictor (Exercise 6.13) and serve as a feature vector. Conversely, one can obtain an all-pole spectrum of the homomorphically deconvolved vocal tract spectrum, a method referred to in Chapter 6 as "homomorphic prediction." Similarly, the SEEVOC representation serves as a basis for an all-pole feature set. Although such features have had some success in speaker recognition algorithms [4], they tend to be inferior to those that are derived using *auditory-based* principles of speech processing.

In this section, we study two different spectral-based feature sets for speaker recognition that exploit front-end auditory filter-bank models. These features are referred to as the *mel-cepstrum* and the *sub-cepstrum* and provide not only a means to study useful speech features for speaker recognition, but also provide an illustrative comparison of time-frequency tradeoffs in feature selection.

14.2.2 Mel-Cepstrum

The *mel-cepstrum*, introduced by Davies and Mermelstein [7], exploits auditory principles, as well as the decorrelating property of the cepstrum. In addition, the mel-cepstrum is amenable to compensation for convolutional channel distortion. As such, the mel-cepstrum has proven to

---

[2] We saw in Chapter 13 that speech enhancement techniques typically do not reduce phase distortion. Nonetheless, phase may have a role in speaker identifiability. Indeed, we have seen in earlier chapters that phase reflects the temporal nature of the glottal flow and the vocal tract impulse response shape. These temporal properties may be associated with unique, speaker-dependent anatomical characteristics.

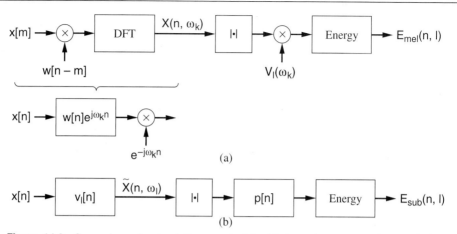

Figure 14.2 Comparison of computations required by (a) the mel-cepstrum and an equivalent structure with (b) the sub-cepstrum.

be one of the most successful feature representations in speech-related recognition tasks. In this section, and others to follow, we study these important properties of the mel-cepstrum.

The mel-cepstrum computation is illustrated in Figure 14.2a. The speech waveform is first windowed with analysis window $w[n]$ and the discrete STFT, $X(n, \omega_k)$, is computed:

$$X(n, \omega_k) = \sum_{m=-\infty}^{\infty} x[m]w[n - m]e^{-j\omega_k m}$$

where $\omega_k = \frac{2\pi}{N}k$ with N the DFT length. The magnitude of $X(n, \omega_k)$ is then weighted by a series of filter frequency responses whose center frequencies and bandwidths roughly match those of the auditory critical band filters. We saw in Chapter 13 that these filters follow the *mel-scale* whereby band edges and center frequencies of the filters are linear for low frequency (< 1000 Hz) and logarithmically increase with increasing frequency. We thus call these filters *mel-scale filters* and collectively a *mel-scale filter bank*. An example of a mel-scale filter bank used by Davies and Mermelstein [7] (in a speech recognition task) is illustrated in Figure 14.3; this filter bank, with 24 triangularly-shaped frequency responses, is a rough approximation to actual auditory critical-band filters covering a 4000 Hz range.

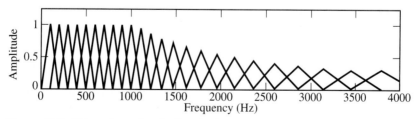

Figure 14.3 Triangular mel-scale filter bank used by Davies and Mermelstein [7] in determining spectral log-energy features for speech recognition. The 24 filters follow the *mel-scale* whereby band edges and center frequencies of these filters are linear for low frequency and logarithmically increase with increasing frequency, mimicking characteristics of the auditory critical bands. Filters are normalized according to their varying bandwidth.

The next step in determining the mel-cepstrum is to compute the energy in the STFT weighted by each mel-scale filter frequency response. Denote the frequency response of the lth mel-scale filter as $V_l(\omega)$. The resulting energies are given for each speech frame at time n and for the lth mel-scale filter as

$$E_{mel}(n, l) = \frac{1}{A_l} \sum_{k=L_l}^{U_l} |V_l(\omega_k)X(n, \omega_k)|^2 \qquad (14.1)$$

where L_l and U_l denote the lower and upper frequency indices over which each filter is nonzero and where

$$A_l = \sum_{k=L_l}^{U_l} |V_l(\omega_k)|^2$$

which normalizes the filters according to their varying bandwidths so as to give equal energy for a flat input spectrum [7],[60]. For simplicity of notation, we have (as we will throughout this chapter unless needed) removed reference to the time decimation factor L.

The real cepstrum associated with $E_{mel}(n, l)$ is referred to as the *mel-cepstrum* and is computed for the speech frame at time n as

$$C_{mel}[n, m] = \frac{1}{R} \sum_{l=0}^{R-1} \log\{E_{mel}(n, l)\} \cos\left(\frac{2\pi}{R}lm\right)$$

where R is the number of filters and where we have used the even property of the real cepstrum to write the inverse transform in terms of the cosine basis, sometimes referred to as the *discrete cosine transform*.[3] In the context of recognition algorithms, an advantage of the discrete cosine transform is that it is close to the Karhunen-Loeve transform [84] and thus it tends to decorrelate the original mel-scale filter log-energies. That is, the cosine basis has a close resemblence to the basis of the Karhunen-Loeve transform that results in decorrelated coefficients. Decorrelated coefficients are often more amenable to probabilistic modeling than are correlated coefficients, particularly to Gaussian mixture models that we will study shortly; indeed, speaker recognition with the Gaussian mixture model approach typically gives better performance with the mel-cepstrum than with the companion mel-scale filter log-energies [60].

Observe that we are not accurate in calling the STFT magnitude weighting by $|V_l(\omega_k)|$, as given in Equation (14.1), a "filtering" operation on the signal $x[n]$. This perspective has an important implication for time resolution in our analysis. To understand this point, recall from Chapter 7 that the STFT can be viewed as a filter bank:

$$X(n, \omega_k) = e^{-j\omega_k n}(x[n] * h_k[n]) \qquad (14.2)$$

where $h_k[n] = w[n]e^{j\omega_k n}$, i.e., for each frequency ω_k the STFT is the output of a filter whose response is the analysis window modulated to frequency ω_k (Figure 14.2a). It follows from the filtering view of the STFT in Equation (14.2) that the speech signal $x[n]$ is first filtered by

---

[3] In the original work of Davies and Mermelstein [7], the discrete cosine transform was written as $C_{mel}[n, m] = \frac{1}{R} \sum_{l=0}^{R-1} \log\{E_{mel}(n, l)\} \cos[\frac{2\pi}{R}(l + \frac{1}{2})m]$.

$h_k[n]$ that has a *fixed, narrow bandwidth*. The length of the filter $h_k[n]$ equals the length of the analysis window $w[n]$, typically about 20 ms in duration. The temporal resolution of the mel-scale filter energy computation in Equation (14.1), therefore, is limited by the STFT filters $h_k[n]$, together with the STFT magnitude operation, *in spite of the logarithmically increasing bandwidths* of the mel-scale filter frequency responses $V_l(\omega)$.[4]

14.2.3 Sub-Cepstrum

An alternate method to compute spectral features, addressing the limited temporal resolution of the mel-scale filter energies and better exploiting of auditory principles, is to convolve the mel-scale filter impulse responses *directly* with the waveform $x[n]$ [14], rather than applying the mel-scale frequency response as a weighting to the STFT magnitude. The result of this convolution can be expressed as

$$\tilde{X}(n, \omega_l) = x[n] * v_l[n] \tag{14.3}$$

where $v_l[n]$ denotes the impulse response corresponding to the frequency response,[5] $V_l(\omega)$, of the lth mel-scale filter centered at frequency ω_l. The mel-scale filter impulse responses are constructed in *analytic signal* form [Chapter 2 (Section 2.5)] by requiring $V_l(\omega) = 0$ for $-\pi \le \omega < 0$ so that $|\tilde{X}(n, \omega_l)|$ provides a temporal *envelope* for each mel-scale filter output. When invoking the convolution in Equation (14.3), we refer to $v_l[n]$ as a *subband filter*. The energy of the output of the lth subband filter can be taken as simply $|\tilde{X}(n, \omega_l)|^2$ or as a smoothed version of $|\tilde{X}(n, \omega_l)|^2$ over time [14], i.e.,

$$E_{sub}(n, l) = \sum_{m=-N/2}^{N/2} p[n - m]|\tilde{X}(m, \omega_l)|^2$$

using a smoothing filter $p[n]$, as illustrated in Figure 14.2, and where time is typically downsampled at the analysis frame rate, i.e., $n = pL$. The real cepstrum of the energies $E_{sub}(n, l)$ for $l = 0, 1, 2 \ldots R - 1$, R equal to the number of filters, is referred to as the *subband cepstrum* [14] and is written as

$$C_{sub}[n, m] = \frac{1}{R} \sum_{l=0}^{R-1} \log\{E_{sub}(n, l)\} \cos\left(\frac{2\pi}{R}lm\right)$$

where, as with the mel-cepstrum, we have exploited the symmetry of the real cepstrum.

The energy trajectories of the subband filters, $E_{sub}(n, l)$, can capture more temporal characteristics of the signal than the mel-scale filter energies, particularly for high frequencies. This is because, unlike in the mel-scale filtering operation, the short-duration, high-frequency subband filters are applied directly to the signal $x[n]$. Therefore, subband filters operate closer to front-end auditory filters, than do mel-scale filters, and likewise give more auditory-like

---

[4] Viewing the mel-scale filter weighting as a multiplicative modification to the STFT, i.e., $V_l(\omega_k)X(n, \omega_k)$, and using the filter bank summation method of Chapter 7, we obtain a sequence $x[n] * (w[n]v_l[n])$. In synthesis, then, assuming $v_l[n]$ is causal with length less than the window length, we recover temporal resolution through the product $w[n]v_l[n]$ (corresponding to summing over multiple filters $h_k[n] = w[n]e^{j\omega_k n}$). Nevertheless, this resolution is not realized in mel-filter energies because the STFT phase is discarded in the magnitude operation.

[5] The phase is selected to mimic that of auditory filters, either physiological or psychoacoustic, that are typically considered to be minimum-phase [47].

temporal characteristics. Although the smoothing filter $p[n]$ gives a loss of temporal resolution, its duration can be chosen such that, together with the duration of $v_l[n]$, more auditory-like temporal resolution results than can be achieved with a typical analysis window required in the first stage of computing the mel-scale filter energies. The difference in temporal resolution of the mel-scale filter and sub-filter energies is illustrated in the following example:

> **EXAMPLE 14.1**   Figure 14.4 shows the output energies from the mel-scale and subband filters for a speech segment that includes a voiced region and a voiced/unvoiced transition. For each filter bank, the energy functions are shown for two filters, one centered at about 200 Hz and the other at about 3200 Hz. The time-domain subband filter impulse responses are minimum phase and are derived from $|V_l(\omega)|$ through the minimum-phase construction method introduced in Chapter 6 (Section 6.7.1). The Hamming window duration used in the STFT of the mel-scale filter bank is 20 ms and is about equal to the length of the subband filters in the low 1000-Hz band. The length of the subband filters above 1000 Hz decreases with increasing filter center frequency. In comparing the time-frequency resolution properties of the mel-scale and subband filter bank energy trajectories [Figure 14.4(b-c) versus Figure 14.4(d-e)], we see that subband filter energies more clearly show speech transitions, periodicity, and short-time events, particularly in the high-frequency region. In this example, the subband filter energies have not been smoothed in time. Exercise 14.14 further explores the relative limits of the temporal resolution for the two filter-bank configurations.   ▲

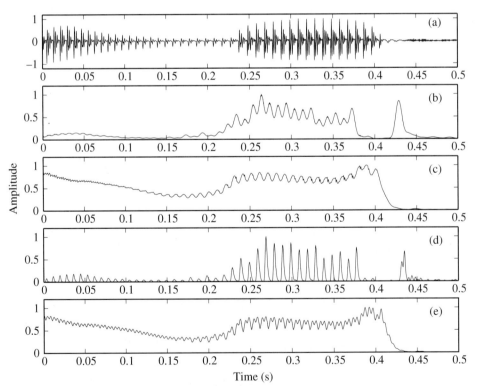

Figure 14.4   Energies from mel-scale and subband filter banks: (a) speech waveform; (b)–(c) mel-scale filter energy from filter number 22 ($\approx$ 3200 Hz) and filter number 2 ($\approx$ 200 Hz); (d)–(e) subband filter energy from filter number 22 ($\approx$ 3200 Hz) and filter number 2 ($\approx$ 200 Hz).

Other variations of the mel-cepstrum and sub-cepstrum based on the mel-scale filter and subband filter energy representations are possible. For example, a cepstral feature set is based on the Teager energy operator [30],[36],[37] [see Chapter 11 (Section 11.5)] applied to the subband-filter outputs [25]. The primary difference between the sub-cepstrum and the Teager-energy-based sub-cepstrum is the replacement of the conventional mean-squared amplitude energy with that of the Teager energy computed with the 3-point operator $\Psi\{x[n]\} = x^2[n] - x[n-1]x[n+1]$. The Teager energy, being the product of amplitude and frequency for an AM-FM sinewave input, may provide a more "information-rich" feature representation than that of conventional energy that involves only amplitude modulation [26],[27].

14.3   Speaker Recognition Algorithms

In this section, we describe three approaches to speaker recognition building in their complexity and accuracy [5],[60].

14.3.1   Minimum-Distance Classifier

Suppose in a speaker recognizer, we obtain a set of features on each analysis frame from training and test data. We refer to the feature set on each frame as a *feature vector*. One of the simplest approaches to speaker recognition is to compute the average of feature vectors over multiple analysis frames for speakers from testing and training data and then find the distance (in some sense) between these average test and training vectors [5],[44],[60]. In speaker verification, we set a distance threshold below which we "detect" the claimant speaker; in identification, we pick the target speaker with the smallest distance from that of the test speaker.

As an example of a feature vector, consider the average of mel-cepstral features for the test and training data:

$$\bar{C}_{mel}^{ts}[n] = \frac{1}{M}\sum_{m=1}^{M} C_{mel}^{ts}[mL, n]$$

and

$$\bar{C}_{mel}^{tr}[n] = \frac{1}{M}\sum_{m=1}^{M} C_{mel}^{tr}[mL, n]$$

where the superscripts ts and tr denote test and training data, respectively, M is the number of analysis frames which differs in training and testing, and L is the frame length. We can then form, as a distance measure, the mean-squared difference between the average testing and training feature vectors expressed as

$$D = \frac{1}{R-1}\sum_{n=1}^{R-1}(\bar{C}_{mel}^{ts}[n] - \bar{C}_{mel}^{tr}[n])^2$$

where R is the number of mel-cepstral coefficients, i.e., the length of our feature vector, and where the 0th value of the cepstral difference is not included due to its sensitivity to scale changes. In the speaker verification task, the speaker is detected when D exceeds some threshold, i.e.,

$$\text{if } D > T, \text{ then speaker present.}$$

We call this recognition algorithm the *minimum distance classifier*, a nomenclature that more appropriately applies to the alternate speaker identification task in which a speaker is chosen from a set of target speakers to have minimum distance to the average feature vector of the test speaker.

14.3.2 Vector Quantization

A problem with the minimum-distance classifier is that it does not distinguish between acoustic speech classes, i.e., it uses an average of feature vectors per speaker computed over all sound classes. Individual speech events are blurred. It seems reasonable then that we could do better if we average feature vectors over distinct sound classes, e.g., quasi-periodic, noise-like, and impulse-like sounds, or even finer phonetic categorization within these sound classes, compute a distance with respect to each sound class, and then average the distances over classes. This would reduce, for example, the phonetic differences in the feature vectors, to which we alluded earlier, and help focus on speaker differences. We illustrate this categorization, again using the example of mel-cepstral features.

Suppose we know *a priori* the speech segments corresponding to the sound classes in both the training and test data. We then form averages for the ith class as

$$\bar{C}_i^{ts}[n] = \frac{1}{M} \sum_{m=1}^{M} C_i^{ts}[mL, n]$$

and

$$\bar{C}_i^{tr}[n] = \frac{1}{M} \sum_{m=1}^{M} C_i^{tr}[mL, n]$$

where for convenience we have removed the "mel" notation. We then compute a Euclidean distance with respect to each class as

$$D(i) = \frac{1}{R-1} \sum_{n=1}^{R-1} (\bar{C}_i^{ts}[n] - \bar{C}_i^{tr}[n])^2.$$

Finally, we average over all classes as

$$D(I) = \frac{1}{I} \sum_{i=1}^{I} D_i$$

where I is the number of classes. To form this distance measure, we must identify class distinctions for the training and test data.

One approach to segmenting a speech signal in terms of sound classes is through speech recognition on a phoneme or word level. A Hidden Markov Model (HMM) speech recognizer,[6] for example, yields a segmentation of the training and testing utterances in terms of desired acoustic phonetic or word classes [8],[22],[29],[59], and can achieve good speaker recognition performance [71]. An alternative approach is to obtain acoustic classes without segmentation

---

[6] The highly popular and successful HMM speech recognizer is based on probabilistic modeling of "hidden" acoustic states (classes) and transitions between states through Markovian class constraints [8],[22],[29],[59]. The states are hidden because only measured feature vectors associated with underlying states are observed. Unlike a Gaussian Mixture Model (GMM) to be described in Section 14.3.3, the HMM imposes a temporal order on acoustic classes.

of an utterance and without connecting sound classes to specific phoneme or word categories; there is no identifying or labeling of acoustic classes. One technique to achieve such a sound categorization is through the vector-quantization (VQ) method described in Chapter 12, using the k-nearest neighbor clustering algorithm [5],[21],[60],[77]. Each centroid in the clustering is derived from training data and represents an acoustic class, but without identification or labeling. The distance measure used in the clustering is given by the Euclidean distance (as above) between a feature vector and a centroid. In recognition, i.e., in testing, we pick a class for each feature vector by finding the minimum distance with respect to the various centroids from the training stage. We then compute the average of the minimum distances over all test feature vectors. In speaker identification, for example, we do this for each known speaker and then pick a speaker with the smallest average minimum distance. The training (clustering) and testing (recognition) stages of the VQ approach are illustrated in Figure 14.5.

We can think of this VQ approach as making "hard" decisions in that a single class is selected for each feature vector in testing. An alternative is to make "soft" decisions by introducing probabilistic models using a multi-dimensional probability density function (pdf) of feature vectors. The components of the pdf, together sometimes referred to as a pdf *mixture model*, represent the possible classes. Each feature vector is given a "soft" decision with respect to each mixture component. In the next section, we look at the Gaussian pdf mixture model commonly used in a maximum-likelihood approach to recognition.

14.3.3 Gaussian Mixture Model (GMM)

Speech production is not "deterministic" in that a particular sound (e.g., a phone) is never produced by a speaker with exactly the same vocal tract shape and glottal flow, due to context, coarticulation, and anatomical and fluid dynamical variations. One way to represent this variability is probabilistically through a multi-dimensional Gaussian pdf [60],[61]. The Gaussian pdf is *state-dependent* in that there is assigned a different Gaussian pdf for each acoustic sound class. We can think of these states at a very broad level such as quasi-periodic, noise-like, and impulse-like sounds or on a very fine level such as individual phonemes. The Gaussian pdf of a feature vector $\underline{x}$ for the ith state is written as

$$b_i(\underline{x}) = \frac{1}{(2\pi)^{\frac{R}{2}} |\boldsymbol{\Sigma}_i|^{\frac{1}{2}}} e^{-\frac{1}{2}(\underline{x}-\underline{\mu}_i)^T \boldsymbol{\Sigma}_i^{-1}(\underline{x}-\underline{\mu}_i)}$$

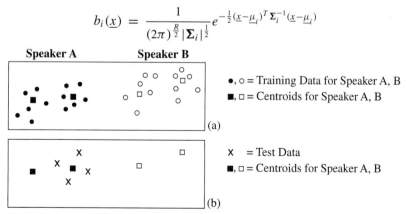

Figure 14.5 Pictorial representation of (a) training and (b) testing in speaker recognition by VQ. In this speaker identification example, in testing, the average minimum distance of test vectors to centroids is lower for speaker A than for speaker B.

where $\underline{\mu}_i$ is the state mean vector, Σ_i is the state covariance matrix, and R is the dimension of the feature vector. The vector $(\underline{x} - \underline{\mu}_i)^T$ denotes the matrix transpose of $\underline{x} - \underline{\mu}_i$ and $|\Sigma_i|$ and Σ_i^{-1} indicate the determinant and inverse, respectively, of matrix Σ_i. The mean vector $\underline{\mu}_i$ is the expected value of elements of the feature vector $\underline{x}$, while the covariance matrix Σ_i represents the cross-correlations (off-diagonal terms) and the variance (diagonal terms) of the elements of the feature vector.

The probability of a feature vector being in any one of I states (or acoustic classes), for a particular speaker model, denoted by λ, is represented by the union, or *mixture*, of different Gaussian pdfs:

$$p(\underline{x}|\lambda) = \sum_{i=1}^{I} p_i b_i(\underline{x}) \tag{14.4}$$

where the $b_i(\underline{x})$ are the component mixture densities and p_i are the mixture weights, as illustrated in Figure 14.6. Given that each individual Gaussian pdf integrates to unity, then we constrain $\sum_{i=1}^{I} p_i = 1$ to ensure that the mixture density represents a true pdf (Exercise 14.1). The speaker model λ then represents the set of GMM mean, covariance, and weight parameters, i.e.,

$$\lambda = \{p_i, \underline{\mu}_i, \Sigma_i\}.$$

We can interpret the GMM as a "soft" representation of the various acoustic classes that make up the sounds of the speaker; each component density can be thought of as the distribution of possible feature vectors associated with each of the I acoustic classes, each class representing possibly one speech sound (e.g., a particular phoneme) or a set of speech sounds (e.g., voiced or unvoiced). Because only the measured feature vectors are available, we can think of the acoustic classes as being "hidden" processes,[7] each feature vector being generated from a particular class i with probability p_i on each analysis frame [60]. However, for some specified number of mixtures in the pdf model, generally one cannot make a strict relation between component densities and specific acoustic classes [60].

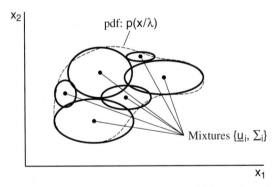

Figure 14.6 The Gaussian mixture model is a union of Gaussian pdfs assigned to each acoustic state.

---

[7] Therefore, conceptually, the GMM is similar to an HMM, except that there is no constrained temporal relation of acoustic classes in time through a Markovian model of the process [59]. In other words, the GMM does not account for temporal ordering of feature vectors.

Regardless of an acoustic class association, however, a second interpretation of the GMM is a *functional* representation of a pdf. The GMM, being a linear combination of Gaussian pdfs, has the capability to form an approximation to an arbitrary pdf for a large enough number of mixture components. For speech features, typically having smooth pdfs, a finite number of Gaussians (e.g., 8–64) is sufficient to form a smooth approximation to the pdf. The modeling of a pdf takes place through appropriate selection of the means, (co)variances, and probability weights of the GMM. Diagonal covariance matrices Σ_i are often sufficient for good approximations and significantly reduce the number of unknown variables to be estimated.

Now that we have a probabilistic model for the feature vectors, we must *train*, i.e., estimate, parameters for each speaker model, and then classify the test utterances (for speaker verification or identification). As we did with VQ classification, we form clusters within the training data, but now we go a step further and represent each cluster with a Gaussian pdf, the union of Gaussian pdfs being the GMM. One approach to estimate the GMM model parameters is by maximum-likelihood estimation (Appendix 13.A), i.e., maximizing, with respect to λ, the conditional probability $p(X|\lambda)$ where the vector $X = \{\underline{x}_0, \underline{x}_1, \ldots \underline{x}_{M-1}\}$ is the collection of all feature vectors for a particular speaker. An important property of maximum-likelihood estimation is that for a large enough set of training feature vectors, the model estimate converges (as the data length grows) to the true model parameters; however, there is no closed-form solution for the GMM representation, thus requiring an iterative approach to solution [40],[60].

This solution is performed using the *expectation-maximization* (EM) algorithm [9] (Appendix 14.A). The EM algorithm iteratively improves on the GMM parameter estimates by increasing (on each iteration) the probability that the model estimate λ matches the observed feature vectors, i.e., on each iteration $p(X|\lambda^{k+1}) > p(X|\lambda^k)$, k being the iteration number. The EM algorithm is similar, at least in spirit, to other iterative estimation algorithms we have seen throughout the text, such as spectral magnitude-only signal reconstruction [Chapter 7 (Section 7.5.3)], where the estimate is iteratively refined to best match the observed data. Having trained the GMM model, we can now perform speaker recognition.

Speaker Identification — GMM-based speaker verification and identification systems are shown in Figure 14.7. We begin with the identification task. Suppose we have estimated S target speaker models λ_j for $j = 1, 2 \ldots S$. Then for each test utterance, features at frame time n, $\underline{x}_n$, are calculated. One common approach to identification is to compute the probability of each speaker model given the features, i.e., $P(\lambda_j|\underline{x}_n)$, and then choose the speaker with the highest probability. This technique is called *maximum a posterior probability (MAP) classification* (Appendix 13.A). To express $P(\lambda_j|\underline{x}_n)$ in terms of a known pdf, for example, the GMM model, by Bayes' rule [11] we write

$$P(\lambda_j|\underline{x}_n) = \frac{p(\underline{x}_n|\lambda_j)P(\lambda_j)}{P(\underline{x}_n)}.$$

Because $P(\underline{x}_n)$ is constant, it is sufficient to maximize the quantity $p(\underline{x}_n|\lambda_j)P(\lambda_j)$ where $P(\lambda_j)$ is the *a priori* probability, also referred to as a *prior probability*, of speaker λ_j being the source of $\underline{x}_n$. If the prior probabilities $P(\lambda_j)$ are assumed equal, which is customary, the problem becomes finding the λ_j that maximizes $p(\underline{x}_n|\lambda_j)$, which is simply the GMM speaker model derived in the training procedure, evaluated at the input feature vector $\underline{x}_n$.

It is important to note that, in practice, there is not one $\underline{x}_n$ for a given testing utterance, but instead a stream of $\underline{x}_n$ generated at the front end of the recognizer at a frame interval L. Thus we must maximize $p(\{\underline{x}_0, \underline{x}_1, \ldots \underline{x}_{M-1}\}|\lambda_j)$, where M is the number of feature vectors for

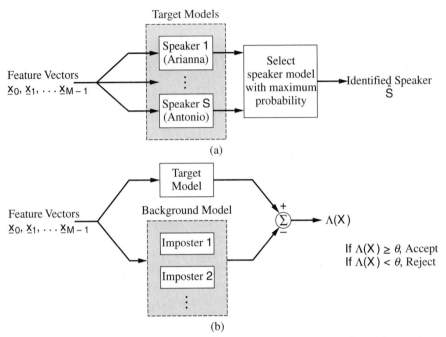

Figure 14.7 GMM-based speaker recognition systems: (a) speaker identification in which a speaker target model is selected with maximum probability; (b) speaker verification in which the speaker claimant is detected when the log-likelihood ratio between the target and the background pdf exceeds a threshold.

SOURCE: D.A. Reynolds, "Automatic Speaker Recognition Using Gaussian Mixture Speaker Models" [68]. ©1995, *MIT Lincoln Laboratory Journal.* Used by permission.

the utterance. (Observe that the vector index here and to follow denotes the frame number rather than absolute time.) This calculation is typically made by assuming that frames are independent; therefore, the likelihood for an utterance is simply the product of likelihoods for each frame:

$$p(\{\underline{x}_0, \underline{x}_1, \ldots \underline{x}_{M-1}\}|\lambda_j) = \prod_{m=0}^{M-1} p(\underline{x}_m|\lambda_j).$$

The assumption of frame independence is a very strong one; an implication is that both the speaker models λ_j and the likelihoods calculated above do not depend on the *order* of the feature vectors. Dynamics of feature vectors over time are thus not considered in the GMM, as we alluded to earlier. By applying the logarithm, we can write the speaker identification solution as

$$\hat{S} = \max_{1 \leq j \leq S} \sum_{m=0}^{M-1} \log[p(\underline{x}_m|\lambda_j)]$$

which is illustrated in Figure 14.7.

Reynolds has evaluated the performance of GMM models for speaker identification, using several speech databases [63],[64]. Speaker identification results with the TIMIT and NTIMIT databases are particularly interesting because they show differences in recognition performance in the face of channel distortion. The TIMIT speech database, a standard in recognition experiments, consists of 8-kHz bandwidth read (not conversational) speech recorded in a quiet environment without channel distortion using a high-quality Sennheiser microphone [17]. TIMIT has 630 speakers (438 males and 192 females) with 10 utterances per speaker, each 3 seconds long on average. It has balanced coverage of speech phonemes. The NTIMIT database [28] has the same properties of TIMIT except that NTIMIT was created by transmitting all TIMIT utterances over actual telephone channels. A primary difference in NTIMIT from TIMIT is that the recording microphone is of the carbon-button type that introduces a variety of distortions, most notably nonlinear distortion. Additional distortions in the NTIMIT channel include bandlimiting (3300 Hz), linear (convolutional) channel transmission, and additive noise. Because NTIMIT utterances are *identical* to TIMIT utterances except for the effects of the telephone channel (and because the two databases are time-aligned), we can compare speech processing algorithms between "clean" speech and "dirty" telephone speech, as illustrated in the following example:

EXAMPLE 14.2   In this example, a GMM recognition system is developed using the TIMIT and NTIMIT databases. For each database, 8 of the 10 available utterances are used for training (about 24 s) and 2 for testing (about 6 s). Non-speech regions were removed (as in all other recognition experiments of this chapter) using an energy-based speech activity detector [61]. The GMM speaker recognition system is trained with a 19-element mel-cepstral feature vector (the 0th value being removed) obtained at a 10-ms frame interval. Because telephone speech is typically bandlimited to 300–3300 Hz, we remove the first 2 and last 2 mel-scale filter energies that are derived from the 24-element filter bank in Figure 14.3 and that fall outside this telephone passband. Each Gaussian pdf in the mixture is characterized by a diagonal covariance matrix, and parameters λ of an 8-component GMM are obtained using the EM algorithm for each target speaker. In this and other GMM experiments of this chapter, ten iterations of the EM algorithm are sufficient for parameter convergence. Figure 14.8 shows the performance of

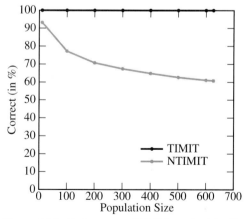

Figure 14.8 Performance of GMM speaker identification system on TIMIT and NTIMIT databases, as a function of number of speakers. Performance is measured as percent correct.

SOURCE: D. A. Reynolds, "Automatic Speaker Recognition Using Gaussian Mixture Speaker Models" [68]. ©1995, *MIT Lincoln Laboratory Journal*. Used by permission.

the GMM speaker identification system on both TIMIT and NTIMIT as a function of the number of speakers in the evaluation [63],[68]. Two interesting results appear. First, the speaker identification performance for clean speech (i.e., TIMIT) is near 100%, up to a population size of 630 speakers. This result suggests that the speakers are not inherently confusable with the use of mel-cepstral features; given features generated from speech produced in the best possible acoustic environment, a GMM classifier can perform almost error-free with these features. Second, speaker identification performance with telephone speech drops off considerably as population size increases. Recall that NTIMIT is the same speech as TIMIT, except that it is transmitted over a telephone network. Thus there is significant performance loss due to the sensitivity of the mel-cepstral features to telephone channel characteristics introduced by NTIMIT, in spite of both training and test data being from the same database. We will see shortly that even greater performance loss occurs when training and test data are collected under different recording conditions.   ▲

Reynolds has also compared the GMM method against VQ and the minimum-distance classifiers in a different speaker identification task [60]. The experiments were run with the KING database that consists of clean conversational utterances recorded locally with a high-quality Sennheiser microphone and also consists of conversational utterances recorded remotely after being transmitted over a telephone channel, including carbon-button microphones, as well as higher-quality electret microphones [33]. With the clean KING database, it was found that the GMM approach outperforms the VQ method, especially for small test speech duration, with the minimum-distance classifier being much inferior to both methods. In addition, the results of a second experiment using KING telephone speech only, as with the NTIMIT database, show a general decline in performance of all three methods while maintaining their relative performance levels.

Speaker Verification — Recall that the speaker verification problem requires that we make a binary decision (detection) based on two hypotheses, i.e., the test utterance belongs to the target speaker, hypothesis H_0, or comes from an imposter, hypothesis H_1. As noted in this chapter's introduction, a speaker known to the system who is correctly claiming his or her identity is labeled as a *claimant* and a speaker unknown to the system who is posing as a known speaker is labeled an *imposter*. Suppose that we have a GMM of the target speaker and a GMM for a collection of imposters that we call the *background model*; we then formulate a likelihood ratio test that decides between H_0 and H_1. This ratio is the quotient between the probability that the collection of feature vectors $X = \{\underline{x}_0, \underline{x}_1, \ldots \underline{x}_{M-1}\}$ is from the claimed speaker, $P(\lambda_C|X)$, and the probability that the collection of feature vectors X is not from the claimed speaker, $P(\lambda_{\bar{C}}|X)$, i.e., from the background. Using Bayes' rule, we can write this ratio as

$$\frac{P(\lambda_C|X)}{P(\lambda_{\bar{C}}|X)} = \frac{p(X|\lambda_C)P(\lambda_C)/P(X)}{p(X|\lambda_{\bar{C}})P(\lambda_{\bar{C}})/P(X)} \tag{14.5}$$

where $P(X)$ denotes the probability of the vector stream X. Discarding the constant probability terms and applying the logarithm, we have the *log-likelihood ratio* (Exercise 14.3)

$$\Lambda(X) = \log[p(X|\lambda_C)] - \log[p(X|\lambda_{\bar{C}})] \tag{14.6}$$

that we compare with a threshold to accept or reject whether the utterance belongs to the claimed speaker, i.e.,

$$\Lambda(X) \geq \theta, \quad \text{accept}$$
$$\Lambda(X) < \theta, \quad \text{reject.} \tag{14.7}$$

One approach to generating a background pdf $p(X|\lambda_{\bar{C}})$ is through models of a variety of background (imposter) speakers (Exercise 14.3). Other methods also exist [61]. We will see numerous examples of speaker verification later in this chapter in different feature vector and application contexts.

14.4 Non-Spectral Features in Speaker Recognition

Up to now we have focused on spectral-based vocal tract features for speaker recognition, specifically, the mel-cepstrum derived from a smoothed STFT magnitude. Although these spectral features are likely to contain some source information, e.g., the spectral tilt of the STFT magnitude influenced by the glottal flow derivative, we have assumed that the primary contribution to the mel-cepstrum is from the vocal tract system function. In this section, in contrast, we explore the explicit use of various aspects of the speech source in speaker recognition. First, we investigate the voiced glottal flow derivative (introduced in Chapters 3 through 5) and, then, briefly source event onset times (introduced in Chapters 10 and 11). In this section, we also provide some insight into the relative importance of source, system, and prosodic (i.e., pitch intonation and articulation rate) features by controlled speech modification using the sinusoidal speech transformation system presented in Chapter 9.

14.4.1 Glottal Flow Derivative

We saw in Chapters 3 through 5 that the glottal flow derivative appears to be speaker-dependent. For example, the flow can be smooth, as when the folds never close completely, corresponding perhaps to a "soft" voice, or discontinuous, as when they close rapidly, giving perhaps a "hard" voice. The flow at the glottis may be turbulent, as when air passes near a small portion of the folds that remains partly open. When this turbulence, referred to as aspiration, occurs often during vocal cord vibration, it results in a "breathy" voice. In order to determine quantitatively whether such glottal characteristics contain speaker dependence, we want to extract features such as the general shape of the glottal flow derivative, the timing of vocal fold opening and closing, and the extent and timing of turbulence at the vocal folds.

Feature Extraction — In Chapter 5 (Sections 5.7 and 5.8), we introduced a method for extracting and characterizing the glottal flow derivative during voicing by pitch-synchronous inverse filtering and temporal parameterization of the flow under the assumption that the time interval of glottal closure is known. The *coarse structure* of the flow derivative was represented by seven parameters, describing the shape and timing of the components of the piecewise-functional Liljencrants-Fant (LF) model [16]. A descriptive summary of the seven parameters of the LF model was given in Table 5.1. These parameters are obtained by a nonlinear estimation method of the Newton-Gauss type. The coarse structure was then subtracted from the glottal flow derivative estimate to give its *fine-structure* component; this component has characteristics not captured by the general flow shape such as aspiration and a perturbation in the flow referred to (in Chapter 5) as "ripple." Ripple is associated with first-formant modulation and is due to the time-varying and nonlinear coupling of the source and vocal tract cavity [2].

In determining fine-structure features, we defined five time intervals within a glottal cycle, introduced in Section 5.7.2 and illustrated in Figure 5.23, over which we make energy measurements on fine structure. The first three intervals correspond to the timing of the open, closed, and return glottal phase based on the LF model of coarse structure, while the last two intervals come from open and closed phase glottal timings, but timings based on formant modulation. The latter is motivated by the observation that when the vocal folds are not fully shut during the closed phase, as determined by the LF model, ripple can begin prior to the end of this closed phase estimate. The open- and closed-phase estimates using formant frequency modulation thus allow additional temporal resolution in the characterization of fine structure. Time-domain energy measures are calculated over these five time intervals for each glottal cycle and normalized by the total energy in the estimated glottal flow derivative waveform. The coarse- and fine-structure features can then be applied to speaker recognition.

Speaker Identification Experiments — In this section, we illustrate the speaker dependence of the glottal source features with a Gaussian mixture model (GMM) speaker identification system.[8] As we have done previously with the GMM, each Gaussian mixture component is assumed to be characterized by a diagonal covariance matrix, and maximum-likelihood speaker model parameters are estimated using the iterative Expectation-Maximization (EM) algorithm. Although for recognition, the GMM has been used most typically with the mel-cepstrum, as noted earlier, if the number of component densities in the mixture model is not limited, we can approximate virtually any smooth density. We select 16 Gaussians in the mixture model for LF and energy source features based on experiments showing that increasing the number of Gaussians beyond 16 does not improve performance and decreasing the number below 16 hurts performance of the classifier.

Before describing the speaker identification experiments, to obtain further insight into the nature of the glottal source parameters for this application, we give an example of an enlightening statistical analysis [48].

> **EXAMPLE 14.3**   Figure 14.9 compares the histograms of parameters from the two classes of glottal features: (1) the coarse structure of the glottal flow derivative using the glottal flow shape parameter α and the open quotient equal to the ratio of $T_e - T_o$ to the pitch period as determined from the LF model (Table 5.1), and (2) the fine-structure energy over the closed-phase interval $[0, T_o)$ also as determined from the LF model. The experiment used about 20 s of TIMIT utterances for each of two male speakers, and feature values were divided across 40 histogram bins. We see in Figure 14.9 that there is a separation of distributions of glottal features across the two speakers, particularly with the shape parameter α and the open quotient parameter. We also see in Figure 14.9 generally smooth distributions with specific energy concentrations, indicating their amenity to a GMM model. On the other hand, we see a strong asymmetry in certain histograms such as with the open quotient of Speaker B. This strong asymmetry has also been observed in the histograms of other glottal flow features (not shown), particularly with the return phase as determined from the LF model and the open quotient as determined from formant modulation. Consequently, these features may be more efficiently modeled by a non-GMM pdf such as sums of Rayleigh or Maxwell functions [46], being characterized by a sharp attack and slow decay. This sharp asymmetry may explain the need for a 16-Gaussian mixture, in contrast to the 8-Gaussian mixture used with mel-cepstra in Example 14.2.   ▲

---

[8] Source information has also been shown to possess speaker identifiability in other speaker recognition systems [35],[81]. These source features, however, are not based on an explicit temporal model of the glottal flow derivative, but rather a linear prediction residual. Furthermore, the residual estimation was not pitch synchronous, i.e., it did not use glottal open- and closed-phase timing and voiced speech identification. As such, in these recognition systems, the residual is a representation of the source, primarily in the form of pitch and voicing information, and not of a glottal flow derivative measure.

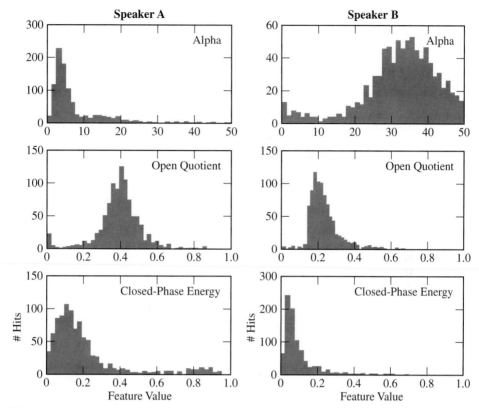

Figure 14.9   A comparison of the histograms of two coarse glottal flow features, the glottal flow shape parameter α and the open quotient (ratio of $T_e - T_o$ to the pitch period) determined from the LF model, and one fine-structure feature, the energy over the closed-phase interval $(0, T_o)$ determined from the LF model. The histograms are shown for two different male TIMIT speakers using about 20 s of data for each speaker and feature values are divided across 40 histogram bins.

SOURCE: M. Plumpe, T.F. Quatieri, and D.A. Reynolds, "Modeling of the Glottal Flow Derivative Waveform with Application to Speaker Identification" [48]. ©1999, IEEE. Used by permission.

Speaker identification experiments are performed with a subset of the TIMIT database [48],[49]. The male subset contains 112 speakers, while the female subset contains 56 speakers. As before, for each speaker, eight of the available ten sentences are used for training and two are used for testing. Male and female speaker recognitions are performed separately. Speaker identification results are obtained for coarse, fine, and combined coarse and fine glottal flow derivative features. The results in Table 14.1 show that the two categories of source parameters, although giving a marked decrease from performance with mel-cepstra, contain significant speaker-dependent information, the combined feature set giving about 70% accuracy.

   As an alternative to explicit use of the glottal features, one can derive features from glottal flow derivative waveforms. Specifically, as in Example 14.2, we consider a length 19 (the 0th value being removed) mel-cepstrum as a feature vector for speaker identification over a 4000-Hz bandwidth from both the glottal flow derivative waveform estimate, obtained from all-pole

Table 14.1 Speaker identification performance (percent correct) for various combinations of the source parameters for a subset of the TIMIT database [48].

Features		Male	Female
Coarse:	7 LF	58.3%	68.2%
Fine:	5 energy	39.5%	41.8%
Source:	12 LF & energy	69.1%	73.6%

(pitch-synchronous) inverse filtering, and its counterpart modeled waveform, i.e., the waveform synthesized using the LF-modeled glottal flow derivative during voicing. The results are shown in Table 14.2. We observe that the seven LF parameters shown in the first row of Table 14.1 better represent the modeled glottal flow derivative than the order–19 mel-cepstral parameter representation of the corresponding waveforem. This is probably because the signal processing to obtain the mel-cepstrum smears the spectrum of the glottal flow derivative and discards its phase. On the other hand, Table 14.2 also indicates that the mel-cepstra of the seven-parameter LF-modeled glottal flow derivative waveform contains significantly less information than that of the estimated glottal flow derivative waveform; this is not surprising given that the synthesized modeled glottal flow derivative is a much reduced version of the glottal flow derivative estimate that contains additional fine structure such as aspiration and ripple. Further study of these results is given in [48],[49].

Although the above results are fundamental for speech science, their potential practical importance lies in part with speaker identification in degrading environments, such as telephone speech. To test performance of the source features on degraded speech, we use the counterpart 120 male speakers and 56 female speakers from the telephone-channel NTIMIT database [48],[49]. The order–19 mel-cepstral representation of the synthesized LF-modeled source waveform, rather than the LF-model parameters themselves, is used in order to provide frame synchrony and similar feature sets for spectrum and source. While the LF-modeled source performs poorly on its own, i.e., 12.5% on males and 27.5% on females, when combined with the mel-cepstra of the speech waveform (with mel-cepstra of the speech and source concatenated into one feature vector), improvement of roughly 3.0% speaker identification for both males and females is obtained, representing a 5% error reduction from using only the mel-cepstra of the speech waveform. Nevertheless, the usefulness of this approach will require further confirmation and understanding, with larger databases and more general degrading conditions.

Table 14.2 Speaker identification performance (percent correct) for mel-cepstral representations of the glottal flow derivative (GFD) waveform and the modeled GFD waveform for a subset of the TIMIT database [48].

Features	Male	Female
Modeled GFD	41.1%	51.8%
GFD	95.1%	95.5%

14.4.2 Source Onset Timing

The glottal flow features used in the previous section were chosen based on common properties of the glottal source. Chapters 5 and 11, on the other hand, described numerous atypical cases. One example is multiple points of excitation within a glottal cycle (e.g., Figure 5.21). Such secondary pulses were observed in Chapter 11 (Section 11.4) to occasionally "excite" formants different from those corresponding to the primary excitation. Different hypotheses exist for explaining this phenomenon, including non-acoustic energy sources originating from a series of traveling vortices along the oral tract. The presence of such secondary source pulses may in part explain the improved speaker identification scores achieved by measuring pulse onset times in different formant bands and appending these onset times as features to mel-cepstral features [52]. This improvement was noted specifically for a limited (but confusable) subset of the telephone-channel NTIMIT database, the premise being that source pulse timings are robust to such degradations. These examples, as well as other atypical cases of glottal flow [49], point to the need for a more complete understanding of such features and their importance under adverse environmental conditions in automatic (and human) speaker recognition.

14.4.3 Relative Influence of Source, Spectrum, and Prosody

In speaker recognition, both by machine and human, we are interested in understanding the relative importance of the source, the vocal tract, and prosody (i.e., pitch intonation and articulation rate). One approach to gain insight into this relation is to *modify* these speech characteristics in the speech waveform and measure recognition performance from the modified speech. For example, little performance change was found in an HMM-based speaker recognizer when using synthetic speech from an all-pole noise-driven synthesis to produce the effect of a whispered speech source [39]. The sinusoidal modification system [51], introduced in Chapter 9, has also been used to study the relative importance of source, vocal tract, and prosody, both for a GMM recognizer and for the human listener [53]. In this section, we investigate the effect on speaker verification of sinusoidally modified speech, including change in articulation rate, pitch, and vocal tract spectrum, as well as a transformation to whispering.

Preliminaries — We saw in Chapter 9 that with sinewave analysis/synthesis as a basis, we can modify the articulation rate and change the pitch (see Exercise 9.11 for a development of pitch modification) of a speaker by stretching and compressing sinewave frequency trajectories in time and frequency, respectively. In addition, other transformations can also be simulated through sinewave analysis/synthesis. The effect of whispering, i.e., an aspirated excitation, can be created by replacing the excitation phases $\phi_k(t)$ in Equation (9.5), along the kth sinewave frequency trajectory $\Omega_k(t)$, by uniformly distributed random variables on the interval $[-\pi, \pi]$. The effect of lengthening or shortening the vocal tract is obtained by stretching or compressing along the frequency axis ω the amplitude and phase $M(t, \Omega)$ and $\Phi(t, \Omega)$, respectively, of the vocal tract system function in Equation (9.5).

The consequence of these transformations on GMM speaker verification is studied with the use of the 2000 National Institute of Standards and Technology (NIST) evaluation database [42]. The series of NIST evaluation databases consist of large amounts of spontaneous 4000-Hz telephone speech from hundreds of speakers, collected under home, office, and college campus acoustic conditions with a wide range of telephone handsets from high-quality electret to low-quality carbon-button. The evaluation data comes from the Switchboard Corpa collected by the Linguistic Data Consortium (LDC) [33]. For the female gender, 504 target speakers are used with 3135 target test trials and 30127 imposter test trials; for the male gender, 422 target speakers are used with 2961 target test trials and 30349 imposter test trials. The target training utterances are 30 s in duration and the test utterances 1 s in duration. Training and testing use the 19 mel-cepstral coefficients at a 10-ms frame interval as described in the previous identification experiments (along with the companion delta-cepstrum and CMS and RASTA channel compensation to be described in the following Section 14.5.1).

Because we are performing speaker verification, rather than speaker identification, a background (i.e., imposter) model is needed for the pdf $p(X|\lambda_{\bar{C}})$ of Equation (14.6). Using the EM algorithm, this background model is trained using a large collection of utterances from the 2000 NIST evaluation database. The background GMM consists of 2048 mixtures to account for a very large array of possible imposters, hence the terminology "universal background model" (UBM) [61],[62]. The target model training again uses the EM algorithm but, in addition, each target speaker's model is derived by adapting its 2048 mixtures from the UBM. This adaptation approach has been found to provide better performance than the more standard application of the EM algorithm for the speaker verification task. Details and motivation for the GMM recognizer using the adaptation-based target training, henceforth referred to as the GMM-UBM recognizer, can be found in [61],[62].

In evaluating recognition performance with modified speech, target and background training data are left intact and the test data (target and imposter) are modified with the sinusoidal modification system. To obtain a performance reference for these experiments, however, we first obtain the recognition performance with sinewave analysis/synthesis without modification. Performance for the speaker verification task is reported as Detection Error Tradeoff (DET) curves [38], produced by sweeping out the speaker-independent threshold of Equation (14.7) over all verification test scores and plotting estimated miss and false alarm probabilities at each threshold level. More specifically, performance is computed by separately pooling all target trial scores (cases where the speaker of the model and the test file are the same) and all nontarget (imposter) trial scores (cases where the speaker of the model and the test file are different) and sweeping out a speaker-independent threshold over each set to compute the system's miss and false alarm probabilities. The tradeoff in these errors is then plotted as a DET curve. Figure 14.10 shows that for male speakers, sinewave analysis/synthesis (thick dashed) does not change DET performance of the recognizer relative to the original data (thick solid). A similar characteristic is also seen for females speakers, but with a slight decrease in performance.

Transformation Experiments —

Articulation Rate: In modifying articulation rate, we expect little change in recognition performance because the stretching or compressing of the sinewave amplitude and phase functions affects only the duration of sinewave amplitude trajectories, and not their value. Thus, the mel-cepstrum features, derived from the STFT magnitude, should be negligibly affected. Indeed, Figure 14.10 (thin solid) shows that articulation rate change gives a small loss in performance for both male and female speakers for the case of time-scale expansion by a factor of two. A similar small loss occurs for an equivalent compression (not shown). Nevertheless, the loss, although small, indicates that articulation rate is reflected somewhat in recognition performance. This may be due to time-scale modification imparting greater or lesser stationarity to the synthetic speech than in the original time scale so that speech features are altered, particularly during event transitions and formant movement. In informal listening, however, the loss in automatic recognition performance, does not apparently correspond to a loss in aural speaker identifiability.

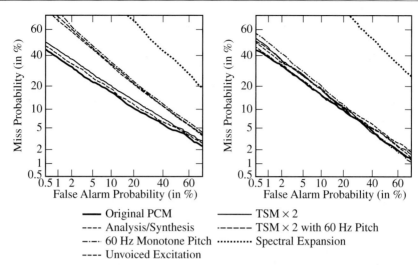

Figure 14.10   Speaker recognition performance on female (left panel) and male (right panel) speakers using various sinewave-based modifications on all test trial utterances [53]. Each panel shows the Detection Error Tradeoff (DET) curves (miss probability vs false alarm probability) for the original speech waveform (thick solid), sinewaves analysis/synthesis without modification (thick dashed), 60-Hz monotone pitch (thick dashed-dotted), unvoiced (aspirated) excitation (thin dashed), articulation rate slowed down by a factor of two (thin solid), 60-Hz monotone pitch and articulation rate slowed down by a factor of two (thin dashed-dotted), and spectral expansion by 20% (thick dotted). *TSM* denotes time-scale modification.

Pitch: Common knowledge has been that a change in pitch should not affect automatic recognition performance using the mel-cepstrum because pitch does not measurably alter the mel-scale filter energies. To test this hypothesis, we first perform a pitch change by modifying the time-varying pitch contour to take on a fixed, monotone pitch. For the case of 60-Hz monotone pitch, Figure 14.10 shows the unimportance of pitch for male speakers, but for female speakers, a significant loss in performance occurs (thick dashed-dotted). A similar result is found when the original pitch is replaced by a monotone 100-Hz pitch, as well as when pitch contours of the test data are shifted and scaled by numerous values to impart a change in the pitch mean and variance (not shown). In all cases of pitch modification, the male recognition performance changes negligibly, while female performance changes significantly. In informal listening, aural speaker identifiability is essentially lost with these modifications for both genders; for male speakers, this loss by human listeners is not consistent with the negligible performance loss by automatic recognition. The result of combining the transformations of rate and pitch change (time-scale expansion by two and a 60-Hz monotone pitch) is also shown in Figure 14.10 (thin dashed-dotted). Here we see that performance is essentially equivalent to that of the monotone pitch transformation, the change in mel-cepstral coefficients being dominated by the pitch change.

To understand the gender-dependent performance loss with pitch modification, we need to understand how the mel-cepstral features change with a change in pitch. Toward this end, we compare mel-scale filter energies (from which the mel-cepstrum is derived) from a high-pitched female and a low-pitched male. Figure 14.11 shows the speech log-spectrum and mel-scale filter log-energies for one frame from each speaker. It is seen that there is clear harmonic structure,

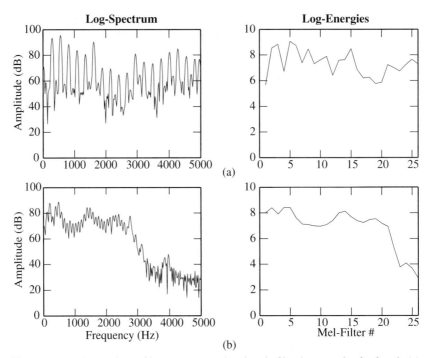

Figure 14.11 Comparison of log-spectrum and mel-scale filter log-energies for female (a) and male (b) speakers.

SOURCE: T.F. Quatieri, R.B. Dunn, D.A. Reynolds, J.P. Campbell, and E. Singer, "Speaker Recognition Using G. 729 Speech Codec Parameters" [57]. ©2000, IEEE. Used by permission.

most notably in the low frequencies of the mel-scale filter energies of the female speech, due to alternating (narrow) mel-scale filters weighting peaks and nulls of the harmonic structure. In practice, we see a continuum in the degree of harmonic structure in going from high-pitched to low-pitched speakers. Mel-scale filter energies do, therefore, contain pitch information. For high-pitched speakers, a change in pitch of the test data modifies the mel-cepstrum and thus creates a "mismatch" between the training and test features.

Whispering: The next transformation is that of converting the excitation to white noise to produce the effect of whispered speech. It was stated earlier that this effect can be created with sinewave analysis/synthesis by replacing the excitation sinewave phases by uniformly distributed random phases on the interval $[-\pi, \pi]$. Figure 14.12 illustrates that formant information remains essentially intact with this sinewave phase randomization. The recognition result for this transformation is given in Figure 14.10 (thin dashed). We see that the performance change is very similar to that with pitch modification: a large loss for female speakers and a negligible loss for male speakers; indeed, for each gender, the DET curves are close for the two very different transformations. We speculate that, as with pitch modification, this gender-dependent performance loss is due to the increasing importance of fine structure in the mel-scale filter energies with increasing pitch; as seen in the narrowband spectrograms of Figure 14.12, pitch is completely removed with phase randomization, while gross (formant) spectral structure is

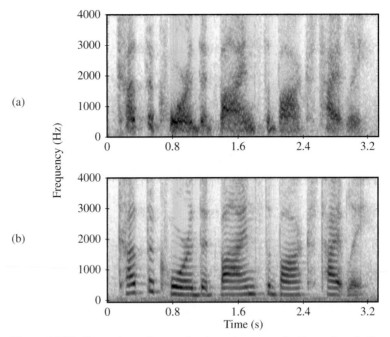

Figure 14.12 Comparison of narrowband spectrograms of whispered synthetic speech (a) and its original counterpart (b) [53].

preserved. For this transformation, there is a large loss in aural speaker identifiability, as there occurs with naturally whispered speech; yet there is little performance loss for male speakers in automatic speaker recognition.

Spectral Warping: The final transformation invokes a warping of the vocal tract spectral magnitude $M(t, \omega)$ and phase $\Phi(t, \omega)$ along the frequency axis, which corresponds to a change in length of the vocal tract. With this transformation, we expect a large loss in recognition performance because we are displacing the speech formants and modifying their bandwidths, resulting in a severe mismatch between the training and test features. Figure 14.10 (thick dotted) confirms this expectation, showing a very large loss in performance for both males and females using a 20% expansion; in fact, the performance is nearly that of a system with random likelihood ratio scores. For this transformation, there also occurs a significant loss in aural speaker identifiability. A similar result was obtained with a 20% spectral compression.

14.5 Signal Enhancement for the Mismatched Condition

We saw in Section 14.3.3 examples using NTIMIT and KING databases of how speaker recognition performance can degrade in adverse environments. In these examples, the training and test data were *matched* in the sense of experiencing similar telephone channel distortions. Training and test data, however, may experience different distortions, a scenario we refer to as the *mismatched* condition, and this can lead to far greater performance loss than for the matched condition in degrading environments. In this section, we address the mismatch problem by calling upon enhancement methods of Chapter 13 that reduce additive and convolutional distortion, including spectral subtraction, cepstral mean substraction (CMS), and RelAtive SpecTrAl processing (RASTA). Also in this section, we introduce a compensation approach for nonlinear distortion, an algorithm for removing badly corrupted features, i.e., "missing features," and the development of features that hold promise in being invariant under different degrading conditions.

14.5.1 Linear Channel Distortion

We saw in Chapter 13 that the methods of CMS and RASTA can reduce the effects of linear distortion on the STFT magnitude. In this section, we look at how these methods can improve speaker recognition performance. We begin, however, by first extending the use of CMS and RASTA to operate on *features*, in particular the mel-scale filter energies, rather than on the STFT magnitude.

Feature Enhancement — Recall from Chapter 13 (Section 13.2) that when a sequence $x[n]$ is passed through a linear time-invariant channel distortion $g[n]$ resulting in a sequence $y[n] = x[n] * g[n]$, the logarithm of the STFT magnitude can be approximated as

$$\log |Y(n, \omega)| \approx \log |X(n, \omega)| + \log |G(\omega)| \tag{14.8}$$

where we assume the STFT analysis window $w[n]$ is long and smooth relative to the channel distortion $g[n]$. As a function of time at each frequency ω, the channel frequency response $G(\omega)$ is seen as a constant disturbance. In CMS, assuming that the speech component $\log |X(n, \omega)|$ has zero mean in the time dimension, we can estimate and remove the channel disturbance while keeping the speech contribution intact. We recall from Chapter 13 that RASTA is similar to CMS in removing this constant disturbance, but it also attenuates a small band of low modulation frequencies. In addition, RASTA, having a passband between 1 and 16 Hz, attenuates high modulation frequencies.

For speaker recognition, however, we are interested in features derived from the STFT magnitude. We focus here specifically on the mel-scale filter energies from which the mel-cepstral feature vector is derived. After being modified by CMS or the RASTA filter, denoted by $p[n]$, the mel-scale filter log-energies become

$$\log[\tilde{E}_{mel}(n, l)] = \sum_{m=-\infty}^{\infty} p[n - m] \log[E_{mel}(m, l)]$$

where

$$E_{mel}(n, l) = \sum_{k=L_l}^{U_l} |V_l(\omega_k) X(n, \omega_k) G(\omega_k)|^2$$

and where $[L_l, U_l]$ is the discrete frequency range for the lth mel-scale filter. Analogous to Equation (14.8), we would like to state that the logarithm of the mel-scale filter energies of the speech and channel distortion are additive so that the CMS and RASTA operations are meaningful. This additivity is approximately valid under the assumption that the channel frequency response is "smooth" over each mel-scale filter. Specifically, we assume

$$G(\omega_k) \approx G_l, \qquad \text{for} \quad k = [L_l, U_l]$$

where G_l represents a constant level for the lth mel-scale filter over its bandwidth. Then we can write

$$E_{mel}(n, l) = \sum_{k=L_l}^{U_l} |V_l(\omega_k) X(n, \omega_k) G(\omega_k)|^2$$

$$\approx |G_l|^2 \sum_{k=L_l}^{U_l} |V_l(\omega_k) X(n, \omega_k)|^2$$

and so

$$\log[E_{mel}(n, l)] \approx \log |G_l|^2 + \log \left[\sum_{k=L_l}^{U_l} |V_l(\omega_k) X(n, \omega_k)|^2 \right] \qquad (14.9)$$

where the second term on the right of the above expression is the desired log-energy. CMS and RASTA can now view the channel contribution to the mel-scale filter log-energies as an additive disturbance to the mel-scale filter log-energies of the speech component. We must keep in mind, however, that *this additivity property holds only under the smoothness assumption of the channel frequency response*, thus barring, for example, channels with sharp resonances. Before describing the use of the mel-cepstrum features, derived from the enhanced mel-scale filter log-energies,[9] in speaker recognition, we introduce another feature representation that is widely used in improving recognition systems in the face of channel corruption, namely the *delta cepstrum*.

An important objective in recognition systems is to find features that are *channel invariant*. Features that are instantaneous, in the sense of being measured at a single time instant or frame, such as the mel-cepstrum, often do not have this channel invariance property. Dynamic features that reflect change over time, however, can have this property. The *delta cepstrum* is one such feature set when linear stationary channel distortion is present [19],[78]. The delta cepstrum also serves as a means to obtain dynamic, temporal speech characteristics which can possess speaker identifiability. One formulation of the delta cepstrum invokes the difference between two mel-cepstral feature vectors over two consecutive analysis frames. For the mel-cepstrum, due to the linearity of the Fourier transform, this difference corresponds to the difference of mel-scale filter log-energies over two consecutive frames, i.e.,

$$\Delta \log[E_{mel}(pL, l)] = \log[E_{mel}(pL, l)] - \log[E_{mel}((p - 1)L, l)] \qquad (14.10)$$

which can be considered as a first-backward difference approximation to the continuous-time differential of the log-energy in time. For a time-invariant channel, the channel contribution is

---

[9] Observe that we can also apply CMS and RASTA in time (across frames) to the mel-cepstral coefficients directly. That is, due to the linearity of the Fourier transform operator used to obtain the mel-cepstrum from the mel-scale filter log-energies, enhancing the mel-cepstrum with CMS or RASTA linear filtering is equivalent to the same operation on the mel-scale filter log-energies.

removed by the difference operation in Equation (14.10) (Exercise 14.4). The delta cepstrum has been found to contain speaker identifiability, with or without channel distortion, but does not perform as well as the mel-cepstrum in speaker recognition [78]. Nevertheless, the mel-cepstrum and delta cepstrum are often combined as one feature set in order to improve recognition performance. In GMM recognizers, for example, the two feature sets have been concatenated into one long feature vector [60], while in minimum-distance and VQ recognizers, distance measures from each feature vector have been combined with a weighted average [78].

A number of variations of the delta cepstrum, proven even more useful than a simple feature difference over two consecutive frames, form a polynomial fit to the mel-scale filter log-energies (or mel-cepstrum) over multiple successive frames and use the parameters of the derivative of the polynomial as features. It is common to use a first-order polynomial that results in just the slope of a linear fit across this time span. For example, a first-order polynomial typically used in speaker recognition fits a straight line over five 10-ms frames (i.e., 50 ms); this time span represents the tradeoff between good temporal resolution and a reliable representation of temporal dynamics. Experiments in both speaker and speech recognition have shown that a smoothed but more reliable estimate of the trend of spectral dynamics gives better recognition performance than a higher resolution but less reliable estimate [78].

Application to Speaker Recognition   We now look at an example of speaker recognition that illustrates use of the above channel compensation techniques in a degrading environment and under a mismatched condition. In this example we investigate speaker verification using the Tactical Speaker Identification Database (TSID) [33]. TSID consists of 35 speakers reading sentences, digits, and directions over a variety of very degraded and low bandwidth wireless radio channels, including cellular and push-to-talk. These channels are characterized by additive noise and linear and nonlinear distortions. For each such "dirty" recording, a low noise and high bandwidth "clean" reference recording was made simultaneously at the location of the transmitter. In this next example, we give performance in terms of the equal error rate (EER). The EER is the point along the DET where the % of false acceptance and % of false rejection errors are equal, and thus provides a concise means of performance comparison.

> **EXAMPLE 14.4**   The purpose of this example is to illustrate the effect of the mismatched condition, training on clean data and testing on dirty data, using TSID, and to investigate ways of improving performance by channel compensation [45]. The speaker recognition system uses the previously described 8-mixture GMM and 19 mel-cepstral coefficients over a 4000-Hz bandwidth. Details of the training and testing scenario are given in [45].
>
> As a means of comparison to what is possible in a matched scenario of training on clean and testing on clean data, referred to as the "baseline" case, the EER is obtained at approximately 4% (Table 14.3). In contrast, the matched case of training on dirty and testing on dirty data gives an ERR of about 9%. On the other hand, the mismatched case of training on clean and testing on dirty data has an EER of approximately 50%—performance equivalent to flipping a coin. To improve this result, a variety of compensation techniques are employed, individually and in combination. In particular, the result of applying delta cepstral coefficients (DCC) based on the above 1st-order polynomial, cepstral mean subtraction (CMS), and their combination (CMS + DCC) is shown in Table 14.3 for the mismatched case. CMS alone gives greater performance than DCC alone, while the combination of CMS and DCC yields greater performance than either sole compensation method. An analogous set of experiments was performed with RASTA. While maintaining the relative performance of CMS, RASTA gave a somewhat inferior performance to CMS. The combination of CMS and RASTA gained little over CMS alone.   ▲

Table 14.3 Speaker verification results with various channel compensation techniques using the TSID database [45]. Results using CMS, DCC, or combined CMS and DCC show ERR performance improvements in the Clean/Dirty mismatched case.

SOURCE: M. Padilla, *Applications of Missing Theory to Speaker Recognition* [45]. ©2000, M. Padilla and Massachusetts Institute of Technology. Used by permission.

Equalization	ERR(approx)
Clean/Dirty + CMS + DCC	23%
Clean/Dirty + CMS	28%
Clean/Dirty + DCC	43%
Clean/Dirty	49%
Dirty/Dirty	9%
Clean/Clean	4%

We see in this example that RASTA performs somewhat inferior to CMS and offers little gain when combined with CMS. This is because the degrading channels are roughly stationary over the recording duration. The relative importance of these compensation techniques, however, can change with the nature of the channel. Extensive studies of the relative importance of the CMS and RASTA techniques for speaker recognition have been performed by van Vuurem and Hermansky [83] for a variety of both matched and mismatched conditions.

14.5.2 Nonlinear Channel Distortion

Although linear compensation techniques can improve recognition performance, such methods address only part of the problem, not accounting for a nonlinear contribution to distortion [65]. Nonlinear distortion cannot be removed by the linear channel compensation techniques of CMS and RASTA and can cause a major source of performance loss in speaker recognition systems, particularly when they are responsible for mismatch between training and testing data. For example, in telephone transmission, we have seen that carbon-button microphones induce nonlinear distortion, while electret microphones are more linear in their response. In this section, we develop a nonlinear channel compensation technique in the context of this particular telephone handset mismatch problem. A method is described for estimating a system nonlinearity by matching the *spectral magnitude* of the distorted signal to the output of a nonlinear channel model, driven by an undistorted reference [55]. This magnitude-only approach allows the model to directly match unwanted speech resonances that arise over nonlinear channels.

Theory of Polynomial Nonlinear Distortion on Speech — Telephone handset nonlinearity often introduces resonances that are not present in the original speech spectrum. An example showing such added resonances, which we refer to as "phantom formants," is given in Figure 14.13 where a comparison of all-pole spectra from a TIMIT waveform and its counterpart carbon-button microphone version from HTIMIT [33],[66] is shown. HTIMIT is a handset-dependent corpus derived by playing a subset of TIMIT through known carbon-button and electret handsets.

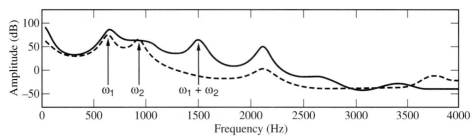

Figure 14.13 Illustration of phantom formants, comparing all-pole spectra from wideband TIMIT (dashed) and carbon-button HTIMIT (solid) recordings. The location of the first phantom formant ($\omega_1 + \omega_2$) is roughly equal to the sum of the locations of the first two original formants (ω_1 and ω_2). The (14th-order) all-pole spectra are derived using the autocorrelation method of linear prediction.

SOURCE: T.F. Quatieri, D.A. Reynolds, and G.C. O'Leary, "Estimation of Handset Nonlinearity with Application to Speaker Recognition" [55]. ©2000, IEEE. Used by permission.

The phantom formants occur at peaks in the carbon-button spectral envelope that are not seen in that of the high-quality microphone, and, as illustrated by the all-pole spectra in Figure 14.13, can appear at multiples, sums, and differences of original formants (Exercise 14.8). Phantom formants, as well as two other spectral distortions of bandwidth widening and spectral flattening, also seen in Figure 14.13, have been consistently observed not only in HTIMIT, but also in other databases with dual wideband/telephone recordings such as the wideband/narrowband KING and the TIMIT/NTIMIT databases that were introduced in earlier examples [26],[27].

To further understand the generation of phantom formants, we investigate the effect of polynomial nonlinearities first on a single vocal tract impulse response, and then on a quasi-periodic speech waveform. The motivation is to show that finite-order polynomials can generate phantom formants that occur at multiples, sums, and differences of the original formants, consistent with measurement of distorted handset spectra.

EXAMPLE 14.5   Consider a single-resonant impulse response of the vocal tract, $x[n] = r^n \cos(\omega n)u[n]$, with $u[n]$ the unit step, passed through a cubic nonlinearity, i.e., $y[n] = x^3[n]$. The output of the distortion element consists of two terms and is given by

$$y[n] = \frac{3}{4}r^{3n}\cos(\omega n)u[n] + \frac{1}{4}r^{3n}\cos(3\omega n)u[n].$$

The first term is a resonance at the original formant frequency ω, and the second term is a phantom formant at three times the original frequency. We observe that the resulting resonances have larger bandwidth than the original resonance, corresponding to a faster decay. The bandwidth widening and additional term contribute to the appearance of smaller spectral tilt. More generally, for a multi-formant vocal tract, the impulse response consists of a sum of damped complex sinewaves. In this case, a polynomial nonlinearity applied in the time domain maps to convolutions of the original vocal tract spectrum with itself, thus introducing sums and differences of the original formant frequencies with wider bandwidths and with decreased spectral tilt. The presence of additional resonances and the change in bandwidth can alternatively be derived using trigonometric expansions.   ▲

In selecting a polynomial distortion to represent measured phantom formants, it is important to realize that both odd and even polynomial powers are necessary to match a spectrum of typical observed measurements. For polynomial distortion, odd-power terms give (anti-) symmetric nonlinearities and alone are not capable of matching even-harmonic spectral distortion; even-power terms introduce asymmetry into the nonlinearity and the necessary even harmonics. A comparison of this (anti-) symmetric/asymmetric distinction is given in Figure 14.14. The (anti-) symmetric nonlinearity of Figure 14.14a gives phantom formants beyond the highest second formant, but never between the first and second formants as would any odd-powered polynomial such as a cubic polynomial $(x + x^3)$. The polynomial of Figure 14.14b is given by $y = x^3 + 1.7x^2 + x$, the even power responsible for asymmetry and a phantom formant between the first and second formants. Asymmetric nonlinearities are important because in dual wideband/telephone recordings, a phantom formant is often observed between the first and second formants. Moreover, asymmetry is characteristic of input/output relations of carbon-button microphones derived from physical principles (Exercise 4.15) [1],[43].

   We see then that phantom formants are introduced by polynomial nonlinearities on a single impulse response. Because a speech waveform is quasi-periodic, however, an analysis based on a single vocal tract impulse response holds only approximately for its periodic counterpart; in the

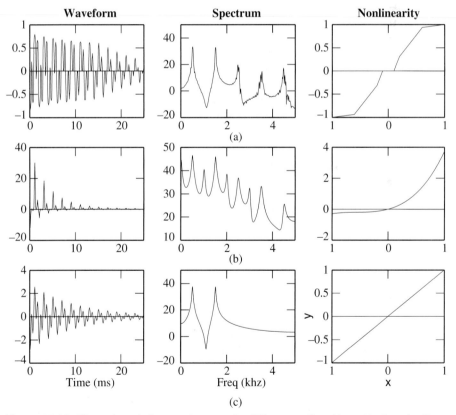

Figure 14.14   Illustration of phantom formants for different nonlinearities: (a) piecewise-linear (anti-) symmetric; (b) polynomial asymmetric; (c) no distortion, corresponding to the original two-resonant signal.

SOURCE: T.F. Quatieri, D.A. Reynolds, and G.C. O'Leary, "Estimation of Handset Nonlinearity with Application to Speaker Recognition" [55]. ©2000, IEEE. Used by permission.

periodic case, tails of previous responses add to the present impulse response. Nevertheless, it has been found empirically for a range of typical pitch and multi-resonant synthetic signals that the spectral envelope of the harmonics differs negligibly from the spectrum of a single impulse response distorted by a variety of polynomials [55]. Therefore, for periodic signals of interest, phantom formants can, in practice, be predicted from the original formants of the underlying response

Handset Models and Estimation — In this section, we study a nonlinear handset model based on the three-component system model[10] consisting of a memoryless nonlinearity sandwiched between two linear filters, illustrated in Figure 14.15. The specific selections in this model are a finite-order polynomial and finite-length linear filters.

The output of the nonlinear handset model is given by

$$y[n] = Q(g[n] * x[n]) * h[n] \tag{14.11}$$

where Q is a P th-order polynomial nonlinear operator which for an input value u has an output

$$Q(u) = q_0 + q_1 u + q_2 u^2 + \ldots q_P u^P. \tag{14.12}$$

The sequence $g[n]$ denotes a J th-order FIR pre-filter and $h[n]$ a K th-order FIR post-filter. To formulate the handset model parameter estimation problem, we define a vector of model parameters $\underline{a} = [\underline{g}, \underline{q}, \underline{h}]$, where $\underline{g} = [g[0], g[1], \ldots g[J-1]]$, $\underline{q} = [q_0, q_1, \ldots q_P]$, and $\underline{h} = [h[0], h[1], \ldots h[K-1]]$ represent the linear pre-filter, nonlinearity, and linear post-filter, so that the goal is to estimate the vector $\underline{a}$.

A time-domain estimation approach is to minimize an error criterion based on waveform matching,[11] such as $\sum_{n=-\infty}^{\infty} (s[n] - y[n; \underline{a}])^2$, where $s[n]$ is the *measurement signal*, $x[n]$ in Equation (14.11) is called the *reference signal* (assumed the input to the actual nonlinear device, as well as the model), and where we have included the parameter vector $\underline{a}$ as an argument in model output $y[n]$. However, because of sensitivity of waveform matching to phase dispersion and delay, e.g., typical alignment error between the model output and measurement, an alternate frequency-domain approach is considered based on the spectral magnitude. As with other speech processing problems we have encountered, an additional advantage of this approach is that speech phase estimation is not required.

Figure 14.15 Nonlinear handset model.

---

[10] The primary purpose of the pre-filter is to provide scaling, spectral shaping, and dispersion. The dispersion provides memory to the nonlinearity. This is in lieu of an actual handset model that might introduce a more complex process such as hysteresis. The post-filter provides some additional spectral shaping.

[11] An alternate technique for parameter estimation that exploits waveform matching is through the *Volterra series* representation of the polynomial nonlinearity; with this series, we can expand the model representation and create a linear estimation problem [72]. The properties and limitations of this approach for our particular problem are explored in Exercise 14.9.

We begin by defining an error between the spectral magnitude of the measurement and nonlinearly distorted model output. Because a speech signal is nonstationary, the error function uses the spectral magnitude over multiple frames and is given by

$$E(\underline{a}) = \sum_{p=0}^{M-1} \sum_{k=0}^{N-1} [|S(pL, \omega_k)| - |Y(pL, \omega_k; \underline{a})|]^2 \tag{14.13}$$

where $S(pL, \omega_k)$ and $Y(pL, \omega_k; \underline{a})$ are the discrete STFTs of $s[n]$ and $y[n; \underline{a}]$, respectively, over an observation interval consisting of M frames, L is the frame length, and $\omega_k = 2\pi k/N$ where N is the DFT length. Our goal is to minimize $E(\underline{a})$ with respect to the unknown model coefficients $\underline{a}$. This is a nonlinear problem with no closed-form solution. An approach to parameter estimation is solution by iteration, one in particular being the generalized Newton method [18] and is similar in style to other iterative algorithms we have encountered throughout the text such as the iterative spectral magnitude-only estimation of Chapter 7 (Section 7.5.3) and the EM algorithm described earlier in this chapter (and in Appendix 14.A).

To formulate an iterative solution, we first define the *residual vector* $\underline{f}(\underline{a})$ by

$$\underline{f}(\underline{a}) = [\underline{f}^0(\underline{a}), \underline{f}^1(\underline{a}), \dots \underline{f}^{N-1}(\underline{a})] \tag{14.14}$$

where $\underline{f}^k(\underline{a}) = \{|S(pL, \omega_k)| - |Y(pL, \omega_k; \underline{a})|\}$ with $p = 0, 1, \dots M - 1$. The error in Equation (14.13) can be interpreted as the sum of squared residuals (i.e., components of the residual vector) over all frames, i.e.,

$$E(\underline{a}) = \underline{f}^T \underline{f}(\underline{a}) \tag{14.15}$$

where T denotes matrix transpose. The gradient of $E(\underline{a})$ is given by $\nabla E = 2\mathbf{J}^T \underline{f}$ where $\mathbf{J}$ is the Jacobian matrix of first derivatives of the components of the residual vector, i.e., the elements of $\mathbf{J}$ are given by

$$J_{ij} = \frac{\partial f_i}{\partial a_j}$$

where f_i is the ith element of $\underline{f}(\underline{a})$ (Exercise 14.10). The generalized Newton iteration, motivated by the first term of a Taylor series expansion of $\underline{f}(\underline{a})$, is formulated by adding to an approximation of $\underline{a}$ at each iteration a correction term, i.e.,

$$\underline{a}_{m+1} = \underline{a}_m + \mu \Delta \underline{a}_m$$

where $\Delta \underline{a}_m = -(\mathbf{J}^T \mathbf{J})^{-1} \mathbf{J}^T \underline{f}(\underline{a}_m)$ with $\underline{f}(\underline{a})$ evaluated at the current iterate $\underline{a}_m$, and where the factor μ scales the correction term to control convergence.[12] For the residual vector definition of Equation (14.14), there is not a closed-form expression for the gradient. Nevertheless, an approximate gradient can be calculated by finite-element approximations, invoking, for example, a first-backward difference approximation to the partials needed in the Jacobian matrix [55].

Although this estimation approach is useful in obtaining handset characteristics, the ultimate goal in addressing the speaker recognition mismatch problem is not a handset model

---

[12] When the number of equations equals the number of unknowns, the generalized Newton method reduces to using a correction $\Delta \underline{a}_m = -\mathbf{J}^{-1} \underline{f}$, which is the standard Newton method.

but rather a handset *mapper* for the purpose of reducing handset mismatch between high- and
low-quality handsets. Consider a reference signal that is from a high-quality electret handset
and a measurement from a low-quality carbon-button handset. Because we assume distortion
introduced by the high-quality electret handset is linear, the model of a handset, i.e., a nonlinearity sandwiched between two linear filters, is also used for the handset mapper. We refer
to this transformation as a "forward mapper," having occasion to also invoke an "inverse mapper" from the low-quality carbon-button to the high-quality electret. We can design an inverse
carbon-to-electret mapper simply by interchanging the reference and measurement waveforms.
The following example illustrates the method:

EXAMPLE 14.6   An example of mapping an electret handset to a carbon-button handset output
is shown in Figure 14.16. The data used in estimation consist of 1.5 s of a male speaker from HTIMIT,
analyzed with a 20-ms Hamming window at a 5-ms frame interval. The pre-filter and post-filter are
both of length 5 and the polynomial nonlinearity is of order 7. In addition, three boundary constraints
are imposed on the output of the nonlinearity, given by $Q(0) = 0$, $Q(1) = 1$, and $Q(-1) = -1$.

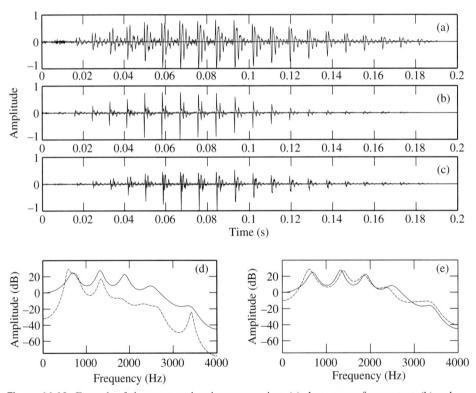

Figure 14.16 Example of electret-to-carbon-button mapping: (a) electret waveform output; (b) carbon-
button waveform output; (c) electret-to-carbon mapped waveform; (d) comparison of all-pole spectra from
(a) (dashed) and (b) (solid); (e) comparison of all-pole spectra from (b) (solid) and (c) (dashed). All-pole
14th order spectra are estimated by the autocorrelation method of linear prediction.

SOURCE: T.F. Quatieri, D.A. Reynolds, and G.C. O'Leary, "Estimation of Handset Nonlinearity with Application
to Speaker Recognition" [55]. ©2000, IEEE. Used by permission.

Applying these conditions to Equation (14.12) yields the constraint equations

$$q_0 = 0$$
$$q_1 = 1 - (q_3 + q_5 + q_7 + \cdots + q_{N-1})$$
$$q_2 = -(q_4 + q_6 + q_8 + \cdots + q_N)$$

thus reducing the number of free variables by three.

Figures 14.16a and 14.16b show particular time slices of the electret and carbon-button handset outputs, while Figure 14.16d shows the disparity in their all-pole spectra, manifested in phantom formants, bandwidth widening, and spectral flattening. Figure 14.16c gives the corresponding waveform resulting from applying the estimated mapper to the same electret output. (The time interval was selected from a region outside of the 1.5 s interval used for estimation.) Figure 14.16e compares the carbon-button all-pole spectrum to that of mapping the electret to the carbon-button output, illustrating a close spectral match. The characteristics of the mapper estimate are shown in Figure 14.17 obtained with about 500 iterations for convergence. The post-filter takes on a bandpass characteristic, while the pre-filter (not shown) is nearly flat. The nonlinearity is convex, which is consistent with the observation (compare Figures 14.16a and 14.16b) that the carbon-button handset tends to "squelch" low-level values relative to high-level values.[13]

The inverse mapper design is superimposed on the forward design in Figures 14.17a and 14.17b. The post-filter of the forward mapper is shown with the pre-filter of the inverse mapper because we expect these filters to be inverses of each other. One notable observation is that the inverse nonlinearity is twisting in the opposite (concave) direction to that of the (convex) forward mapper, consistent with undoing the squelching imparted by the carbon-button handset. ▲

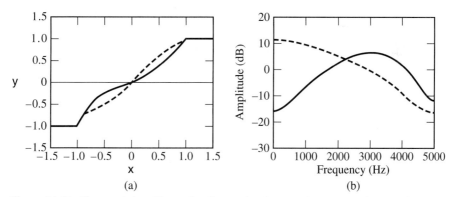

(a) (b)

Figure 14.17 Characteristics of forward and inverse handset mappings: (a) nonlinearity of forward mapper (solid) superimposed on nonlinearity of inverse mapper (dashed); (b) post-filter of forward mapper (solid) superimposed on pre-filter of inverse mapper (dashed).

SOURCE: T.F. Quatieri, D.A. Reynolds, and G.C. O'Leary, "Estimation of Handset Nonlinearity with Application to Speaker Recognition" [55]. ©2000, IEEE. Used by permission.

---

[13] The mapper is designed using a particular input energy level; with a nonlinear operator, changing this level would significantly alter the character of the output. Therefore, test signals are normalized to the energy level of the data used for estimation; this single normalization, however, may over- or under-distort the data and thus further consideration of input level is needed.

The previous examples illustrate both a forward and an inverse mapper design. Although an inverse mapper is sometimes able to reduce the spectral distortion of phantom formants, bandwidth widening, and spectral flattening, on the average the match is inferior to that achieved by the forward design. It is generally more difficult to undo nonlinear distortion than it is to introduce it.

Application to Speaker Recognition — One goal of handset mapper estimation is to eliminate handset mismatch, due to nonlinear distortion, between training and test data to improve speaker recognition over telephone channels. The strategy is to assume two handset classes: a high-quality electret and a low-quality carbon-button handset. We then design a forward electret-to-carbon-button mapper and apply the mapper according to handset detection [66] on target training and test data. In this section, we look at an example of a handset mapper designed on the HTIMIT database, and also compare the mapper approach to an alternative scheme that manipulates the log-likelihood scores of Equation (14.6) in dealing with the handset mismatch problem.

In speaker recognition, linear channel compensation is performed with both CMS and RASTA processing to reduce linear channel effects on both the training and test data. This compensation, however, does not remove a nonlinear distortion contribution nor a linear prefiltering that comes before a nonlinearity in the distortion model of Figure 14.15, i.e., only the linear postfiltering distortion can be removed. In exploiting the above handset mapper design, we apply an electret-to-carbon-button nonlinear mapper to make electret speech utterances appear to come from carbon-button handsets when a handset mismatch occurs between the target training and test data. The specific mapper design was developed in Example 14.6 using the HTIMIT database and illustrated in Figure 14.17. For each recognition test trial, the mapper is applied to either the test or target training utterances, or neither, according to the specific mismatched (or matched) condition between training and test utterances, as summarized in Table 14.4.

The second approach to account for differences in handset type across training and test data modifies the likelihood scores, rather than operating on the waveform, as does the handset mapper approach. This approach introduced by Reynolds [67] is referred to as *hnorm*, denoting handset normalization. Hnorm was motivated by the observation that, for each target model, likelihood ratio scores have different means and ranges for utterances from different handset types. The approach of hnorm is to remove these mean and range differences from the likelihood ratio scores. For each target model, we estimate the mean and standard deviation of the scores of same-sex non-target (imposter) utterances from different handset types, and then normalize out the mean and variance to obtain a zero-mean and unity-variance score distribution. For example, for the carbon-button-handset type, during testing, we use the modified score

$$\tilde{\Lambda}(\underline{x}) = \frac{\Lambda(\underline{x}) - u^{carb}}{\sigma^{carb}}$$

where the target's likelihood score $\Lambda(\underline{x})$ is normalized by the mean u^{carb} and standard deviation σ^{carb} estimates, based on the handset label assigned to the feature vector $\underline{x}$. In effect, we

Table 14.4 Mapping on testing and target training data.

		Test	
		ELEC	CARB
Train	ELEC	No map	Map train to CARB
	CARB	Map test to CARB	No map

normalize out the effect of the model (target) on the imposter score distribution for each handset type. This normalization results in improved speaker verification performance when using a single threshold for detection evaluation, as in the generation of Detection Error Tradeoff (DET) curves. More details and motivation for hnorm can be found in [62],[67].

For either hnorm or the mapper approach, we require a handset detector, i.e., we need to determine whether the utterance comes from electret or carbon-button handset type. The automatic handset detector is based on a maximum-likelihood classifier using Gaussian mixture models built to discriminate between speech originating from a carbon-button handset and speech originating from an electret handset [66] (Exercise 14.5).

EXAMPLE 14.7   We use a GMM recognizer applied to the 1997 NIST evaluation database [42] to illustrate the handset mapper and hnorm concepts [55]. Specifically, speaker verification is based on a 2048-mixture GMM-UBM classifier described earlier in Section 14.4.3 and 19 mel-cepstral coefficients (with companion delta-cepstra) over a 4000-Hz bandwidth. Training the UBM uses the 1996 NIST evaluation database [42]. Both a matched and mismatched condition were evaluated. In the matched condition, the target training and test data come from the same telephone number and the same handset type for target test trials, but different telephone numbers (and possibly different handset types) for the non-target (imposter) test trials. In the mismatched condition, training and test data come from different telephone numbers and thus possibly different handset types for both target and non-target (imposter) test trials. A performance comparison of the mapper, hnorm, and baseline (neither mapped nor score-normalized) is given in Figure 14.18 for 250 male target speakers and 2 min training and 30 s test utterances. In obtaining means and standard deviations for hnorm, 1996 NIST evaluation data is passed through the target models derived using the 1997 NIST evaluation database; 300 carbon-button and 300 electret test utterances were used.

For the mismatched condition, the handset mapper approach resulted in DET performance roughly equivalent to that using score normalization, i.e., hnorm. It is interesting that, for this example, two very different strategies, one operating directly on the waveform, as does the mapper, and the other operating on log-likelihood ratio scores, as does hnorm, provide essentially the same performance gain under the mismatched condition. On the other hand, as one might expect, using the handset mapper for the matched condition slightly hurt performance because it is applied only to non-target (imposter) test trials (for which there can be handset mismatch with target models), thus minimizing the channel mismatch in these trials and increasing the false alarm rate.   ▲

The same strategy used in the previous example has also been applied to design an optimal linear mapper, i.e., a handset mapper was designed without a nonlinear component using spectral magnitude-only estimation. The linear mapper was designed under the same conditions as the nonlinear mapper used in our previous speaker verification experiments, and provides a highpass filter characteristic. Because a linear distortion contribution is removed by CMS and RASTA, specifically the linear post-filter in our distortion model, a mapper designed with only a linear filter should not be expected to give an additional performance gain, performing inferior to the nonlinear mapper. Indeed, when this linear mapper was applied in training and testing according to the above matched and mismatched conditions, the DET performance is essentially identical to that with no mapper, and thus is inferior to that with the corresponding nonlinear mapper [55]. Finally, we note that when score normalization (hnorm) is combined with the handset mapping strategy, a small overall performance gain is obtained using the databases of Example 14.7, with some improvement over either method alone under the mismatched condition, and performance closer to that with hnorm alone under the matched condition [55]. Nevertheless, over a wide range of databases, hnorm has proven more robust than the mapping technique, indicating the need for further development of the mapper approach [55].

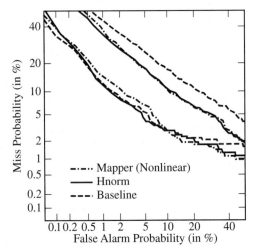

Figure 14.18 Comparison of DET performance using
the nonlinear handset mapper and handset normalization of
likelihood scores (hnorm) in speaker verification: baseline
(dashed), hnorm (solid), and nonlinear mapper (dashed-
dotted). Upper curves represent the mismatched condition,
while lower curves represent the matched condition.

SOURCE: T.F. Quatieri, D.A. Reynolds, and G.C. O'Leary, "Estimation of Handset Nonlinearity with Application to Speaker
Recognition" [55]. ©2000, IEEE. Used by permission.

14.5.3 Other Approaches

The signal processing methods (in contrast to statistical approaches such as hnorm) of CMS,
RASTA, and the nonlinear handset mapper all attempt to remove or "equalize" channel distortion
on the speech waveform to provide enhanced features for recognition. An alternate approach
is to extract features that are insensitive to channel distortion, referred to as *channel invariant*
features. The source onset times of Section 14.4.2 provide one example of features that, to some
extent, are channel invariant. These approaches to channel equalization and feature invariance
represent just a few of the many possibilities for dealing with adverse and mismatched channel
conditions. Below we briefly describe a small subset of the emerging techniques in these areas,
based on speech signal processing and modeling concepts.

Channel Equalization —
Noise reduction: With regard to additive noise, various methods of noise suppression preprocessing show evidence of helping recognition. For example, the generalized spectral subtraction
method of Section 13.5.3 applied to the mel-scale filter energies is found to enhance performance of a GMM-based recognizer under the mismatched condition of clean training data and
white-noise-corrupted test data [12]. Further performance improvement is obtained by applying
noise suppression algorithms to the noise-corrupted STFT magnitude, *prior* to determining the
mel-scale filter energies [45]. The noise suppression algorithms include generalized spectral
subtraction and the adaptive Wiener filtering described in Chapter 13.

Missing feature removal: The idea in this approach is to remove badly corrupted elements of a feature vector during testing, e.g., under the mismatched condition of training on "clean" data and testing on "dirty" data, to improve recognition performance. Motivation for this approach is the observation by Lippman [34] that speech recognition performance can improve when appropriately selecting features with known corruption by filtering and noise. A challenging aspect of this approach is feature selection when the corrupting channel is unknown. Drygajlo and El-Maliki [12] developed a missing feature detector based on the generalized spectral subtraction noise threshold rules applied to the mel-scale filter energies, and then dynamically modified the probability computations performed in the GMM recognizer based on these missing features (Exercise 14.12). This entails dividing the GMM densities into "good feature" and "bad feature" contributions based on the missing feature detector. Padilla [45] provided an alternative to this approach by developing a missing feature detector based on the STFT magnitude, rather than the mel-scale filter energies, and then mapping a frequency-based decision to the mel-scale. Because the STFT magnitude provides for improved detection resolution (over frequency) for determining good and bad spectral features, significant speaker recognition performance improvement was obtained over a detector based on mel-scale filter energies.

Channel Invariance —

Formant AM-FM: We have seen in the frequency domain that both amplitude and frequency modulation of a sinusoid add energy around the center frequency of the sinusoid as, for example, in Figure 7.8; as more modulation is added, the energy spreads more away from the center frequency. Therefore, formant AM and FM are expected to contribute only *locally* to the power spectrum of the waveform, i.e., only around the formant frequency. Motivated by this observation, Jankowski [26] has shown that as long as the frequency response of a channel is not varying too rapidly as a function of frequency, this local property of modulation is preserved, being robust not only to linear filtering but also to typical telephone-channel nonlinear distortions. This property was determined by subjecting a synthetic speech resonance with frequency and bandwidth modulation to estimated linear and nonlinear (cubic) distortion of nine different telephone channels from the NTIMIT database. A variety of modulations include increase in bandwidth associated with the *truncation effect* and the varying first formant center frequency during the glottal open phase that were described in Chapters 4 and 5. Using the Teager energy-based AM-FM estimation algorithm of Chapter 11 (Section 11.5.3), Jankowski determined both analytically and empirically the change in formant AM and FM as test signals passed through each channel. The resulting AM and FM were observed to be essentially unchanged when passed through the simulated NTIMIT channels. In other words, typical formant AM and FM barely change with linear and nonlinear distortions of the NTIMIT telephone channel. Jankowski also showed that a change in first-formant bandwidth and frequency multipliers[14] for several real vowels was negligible when the vowels were passed through the simulated channels. As such, features derived from formant modulations indicate promise for channel invariance in speaker recognition [26],[27].

Sub-cepstrum: In a series of experiments by Erzin, Cetin, and Yardimci [14] using an HMM speech recognizer, it was shown that the sub-cepstrum could outperform the mel-cepstrum features in the presence of car noise over a large range of SNR. In a variation of this approach,

---

[14] Recall from Section 4.5.2 that an effect of glottal/vocal tract interaction is to modulate formant bandwidth and frequency, represented by time-dependent multipliers.

Jabloun and Cetin [25] showed that replacing the standard energy measure by the Teager energy in deriving the sub-cepstrum can further enhance robustness to additive car noise with certain autocorrelation characteristics. This property is developed in Exercise 14.13. In yet another variation of the sub-cepstrum, changes in subband energies are accentuated using auditory-like nonlinear processing on the output of a gammachirp-based filter bank (Chapter 11, Section 11.2.2). [75]. A cepstrum derived from these enhanced energies again outperform the mel-cepstrum in a speech recognition task in noise. The demonstrated robustness of the sub-cepstrum and its extensions, however, to date, is not known to be related to the temporal resolution capability that we described in Section 14.2.3. Moreover, the sub-cepstrum has yet to be investigated for speaker recognition.

14.6 Speaker Recognition from Coded Speech

Due to the widespread use of digital speech communication systems, there has been increasing interest in the performance of automatic recognition systems from quantized speech resulting from the waveform, model-based, and hybrid speech coders described in Chapter 12. The question arises as to whether the speech quantization and encoding can affect the resulting mel-cepstral representations which are the basis for most recognition systems, including speech, speaker, and language recognition [41],[56]. There is also interest in performing recognition directly, using model-based or hybrid coder parameters rather than the synthesized coded speech [23],[57]. Possible advantages include reduced computational complexity and improved recognition performance in bypassing degradation incurred in the synthesis stage of speech coding. In this section, we investigate the effect of speech coding on the speaker verification task.

14.6.1 Synthesized Coded Speech

Three speech coders that are international standards and that were introduced in Chapter 12 are: G.729 (8 kbps), GSM (12.2 kbps), and G.723.1 (5.3 and 6.3 kbps). All three coders are based on the concept of short-time prediction corresponding to an all-pole vocal tract transfer function, followed by long-term prediction resulting in pitch (prediction delay and gain) and residual spectral contributions. The short-term prediction is typically low-order, e.g., order 10, and the residual represents the speech source as well as modeling deficiencies. The primary difference among the three coders is the manner of coding the residual.

Speaker verification has been performed on the speech output of these three coders using a GMM-UBM system [56]. The features consist of 19 appended mel-cepstra and delta cepstra (1st-order polynomial) derived from bandlimited (300–3300 Hz) mel-scale filter energies. Both CMS and RASTA filtering are performed on the mel-cepstrum features prior to training and testing. The training and testing utterances are taken from a subset of the 1998 NIST evaluation database [42]; in these experiments, both are limited to electret handsets, but possibly different phone numbers. 50 target speakers are used for each gender with 262 test utterances for males and 363 for females. Target training utterances are 30 s in duration and test utterances 1 s in duration. Background models are trained using the 1996 NIST evaluation database and 1996 NIST evaluation development database [42]. As with other recognition scenarios, we can consider both a matched condition in which target training, test, and background data all are coded, or a mismatched condition in which one subset of the three data sets is coded and another uncoded.

It was shown that, using speech synthesized from the three coders, GMM-UBM-based speaker verification generally degrades with coder bit rate, i.e., from GSM to G.729 to G.723.1, relative to an uncoded baseline whether in a matched or mismatched condition, the mismatched condition tending to be somewhat lower-performing [56].

14.6.2 Experiments with Coder Parameters

We mentioned above that in certain applications, we may want to avoid reconstructing the coded speech prior to recognition. In this section, we select one coder, G.729, and investigate speaker verification performance with its underlying parameters, exploiting both vocal tract and source modeling.

Vocal Tract Parameters — We begin with speaker verification using the mel-cepstrum corresponding to the mel-scale filter energies derived from the all-pole G.729 spectrum. Because G.729 transmits ten line spectral frequency (LSF) parameters [Chapter 12 (Section 12.6.3) introduces the LSF concept], conversion of the G.729 LSFs to mel-cepstra is first performed and is given by the following steps: (1) Convert the ten LSFs to ten all-pole predictor coefficients (Exercise 12.18) from which an all-pole frequency response is derived; (2) Sample the all-pole frequency response at DFT frequencies; (3) Apply the mel-scale filters and compute the mel-scale filter energies:

$$\hat{E}_{mel}(n, l) = \sum_{k=L_l}^{U_l} |V_l(\omega_k)\hat{H}(n, \omega_k)|^2$$

where $\hat{H}(n, \omega)$ is the all-pole transfer function estimate at frame time n and $V_l(\omega)$ is the frequency response of the lth mel-scale filter; (4) Convert mel-scale filter energies to the mel-cepstrum. In speaker verification experiments, an order 19 mel-cepstrum is used with CMS and RASTA compensation, along with the corresponding delta cepstrum.

Compared to previous speaker verification results with G.729 coded speech, performance with the above mel-cepstrum degrades significantly for both males and females. The result for the matched training and testing scenario is shown in Figure 14.19. Conditions for training and testing are identical to those used in the previous recognition experiments with synthesized coded speech. One possible explanation for the performance loss is that the mel-cepstrum derived from an all-pole spectrum is fundamentally different from the conventional mel-cepstrum: we are smoothing an already smooth all-pole envelope, rather than a high-resolution short-time Fourier transform as in the conventional scheme, thus losing spectral resolution.

An alternative is to seek a feature set that represents a more direct parameterization of the all-pole spectrum—specifically, either LSF coefficients, available within G.729, or a transformation of the LSF coefficients. In addition, we want these features to possess the property that linear channel effects are additive, as assumed in the channel compensation techniques of CMS and RASTA. It has been observed that the LSFs have this property only in the weak sense that a linear channel results in both a linear and nonlinear distortion contribution to the LSFs [57]. Cepstral coefficients, on the other hand, being a Fourier transform of a log-spectrum, have this additivity property. Moreover, it is possible to (reversibly) obtain a cepstral representation from the all-pole predictor coefficients derived from the LSFs. This is because there exists a one-to-one relation between the first p values (excluding the 0th value) of the cepstrum of the minimum-phase, all-pole spectrum, and the corresponding p all-pole predictor coefficients. This cepstral representation is distinctly different from the mel-cepstrum obtained from the mel-scale filter energies of the all-pole spectral envelope and can be computed recursively by (Exercise 6.13)

$$c[n] = a[n] + \sum_{k=1}^{n-1} \left(\frac{k}{n}\right) c[k]a[n-k], \quad n \geq 1$$

where $a[n]$ represents the linear predictor coefficients in the predictor polynomial given by $A(z) = 1 - \sum_{k=1}^{p} a[k]z^{-k}$. We refer to this recursively computed cepstrum as the *rec-cepstrum* in contrast to the mel-cepstrum derived from the mel-scale filter energies of the all-pole spectrum. Although the rec-cepstrum is, in theory, infinite in length, only the first p coefficients are needed to represent $a[n]$ (with $p = 10$ for G.729) because there exists a one-to-one relation between $a[n]$ and the first p values of $c[n]$ for $n > 0$ (Exercise 6.13).

The rec-cepstrum is seen in Figure 14.19 to give a performance improvement over the mel-cepstrum. Speaker verification is performed using a feature vector consisting of the first 10 rec-cepstral coefficients (dashed) in comparison to 19 mel-cepstral coefficients (thin-solid), along with companion delta cepstra in each case. Figure 14.19 shows some improvement for males and a significant improvement for females. The figure also shows that, in spite of improvements, we have not reached the performance of the standard mel-cepstrum of the synthesized (coded) speech, referred to as the baseline (dashed-dotted) in the caption of Figure 14.19. If we attempt to improve performance by increasing the number of rec-cepstral coefficients beyond ten, we find a decrease in performance. This is probably because the additional coefficients provide no new information while an increase in feature vector length typically requires more training data for the same recognition performance.

Source (Residual) Parameters — One approach to further recover performance of the G.729 synthesized speech is to append parameters that represent the G.729 residual to the G.729 spectral feature vector. This includes the pitch[15] (i.e., the long-term-prediction delay), gain (i.e., long-term-prediction gain), residual codebooks, and energy [24]. To determine first the importance, incrementally, of pitch,[16] we append pitch to the feature vector consisting of the rec-cepstrum and delta rec-cepstrum, yielding a 21-element vector. The resulting recognition performance is shown in Figure 14.19. We see that there is performance improvement with appending G.729 pitch (solid) relative to the use of the rec-cepstrum only (dashed), resulting in performance close to baseline. It is interesting that the extent of the relative improvement seen for female speakers is not seen for male speakers, reflecting the greater use of pitch in recognition by females, consistent with its influence on the mel-cepstrum as was illustrated in Figure 14.11. We also see in Figure 14.19 that, although we have made additional performance gains with G.729 pitch, we still have not reached the performance of using G.729 synthesized speech.[17] Nevertheless, other residual characteristics have not yet been fully exploited [57].

---

[15] For a fundamental frequency ω_o, we use $\log(\omega_o + 1)$ rather than pitch, itself, because the logarithm operation reduces dynamic range and makes the pitch probability density function (as estimated by a histogram) more Gaussian, thus more amenable to the GMM.

[16] Pitch alone has been used with some success in automatic speaker identification systems [3].

[17] The LSFs and pitch in the these experiments are unquantized. With quantization, the relative performance is intact with a negligible overall decline.

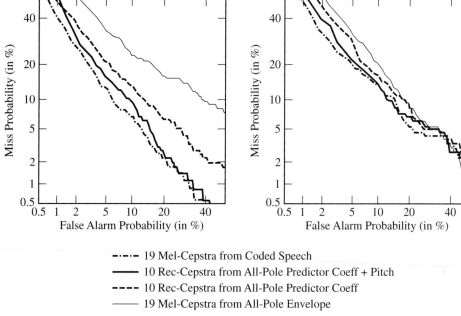

- ─·─·─ 19 Mel-Cepstra from Coded Speech
- ───── 10 Rec-Cepstra from All-Pole Predictor Coeff + Pitch
- ─ ─ ─ 10 Rec-Cepstra from All-Pole Predictor Coeff
- ───── 19 Mel-Cepstra from All-Pole Envelope

Figure 14.19 G.729 DET performance under the matched condition for baseline (dashed-dotted), rec-cepstrum + pitch (thick-solid), rec-cepstrum (dashed), and mel-cepstrum from the all-pole envelope (thin-solid). The baseline system uses the mel-cepstrum from the STFT magnitude. The rec-cepstrum is of order 10 and the mel-cepstrum is of order 19. The left panel gives the performance for female speakers and the right panel for male speakers.

14.7 Summary

In this chapter we applied principles of discrete-time speech signal processing to automatic speaker recognition. Our main focus was on signal processing, rather than recognition, leaving the multitude of statistical considerations in speaker recognition to other expositions [76]. Signal processing techniques were illustrated under the many different conditions in which speaker recognition is performed through a variety of real-world databases.

We began this chapter with the first step in speaker recognition: estimating features that characterize a speaker's voice. We introduced the spectral-based mel-ceptrum and the related sub-cepstrum, both of which use an auditory-like filter bank in deriving spectral energy measures. A comparison of the mel-cepstrum and sub-cepstrum revealed issues in representing the temporal dynamics of speech in feature selection, using the filtering interpretation of the STFT. We then introduced three different approaches to speaker recognition with the Gaussian mixture model (GMM) approach being the most widely used to date. We saw that advantages of GMM lie in its "soft" assigning of features to speech sound classes and its ability to represent any feature probability density function. We next described the use of a variety of non-spectral features in speaker recognition, specifically based on source characterization, including parameters of the glottal flow derivative and onset times of a generalized distributed source. This led to gaining insight into the relative importance of source, spectrum, and prosodic features by modifying speech chactacteristics through the sinusoidal modification techniques that we developed in

Chapter 9. This study gave a fascinating glimpse of the difference in how automatic (machine) recognizers and the human listener view speaker identifiability, and the need for automatic recognizers, currently spectral-based, to exploit non-spectral features. Non-spectral features will likely involve not only new low-level acoustic speech characteristics of a temporal nature, such as features derived from temporal envelopes of auditory-like subband filter outputs,[18] but also high-level properties that were discussed in this chapter's introduction.

In the second half of the chapter, we addressed speaker recognition under a variety of mismatched conditions in which training and test data are subjected to different channel distortions. We first applied the CMS and RASTA linear channel compensation methods of Chapter 13. We saw, however, that this compensation, although able to improve speaker recognition performance, addressed only part of the problem. In the typical mismatched condition, such as with wireline and cellular telephony or tactical communication environments, training and test data may arise not only from high-quality linear channels but also from low-quality nonlinear channels. This problem leads to the need for nonlinear channel compensation. Two such approaches were described: one that modifies the speech waveform with a nonlinearity derived from a magnitude-only spectral matching criterion, and the second that normalizes the log-likelihood scores of the classifier. Application of the noise reduction approaches of Chapter 13 and the removal of missing features, i.e., irrevocably degraded, were also described in addressing the mismatch problem. Alternatively, one can deal with this problem through channel invariant features, with source onset times, the sub-cepstrum, and parameters of the AM-FM in formants as possibilities. In the final topic of this chapter, we addressed a different class of degrading conditions in which we performed speaker recognition from coded speech. This is an increasingly important problem as the use of digital communication systems becomes more widespread.

The problems of robustness and channel mismatch in speech-related recognition remain important signal processing challenges. For example, although signal processing has led to large performance gains under the mismatched condition for telephone communications, with the onset of Internet telephony comes different and even more difficult problems of mismatch in training and testing. Indeed, new solutions are required for such environmental conditions. Such solutions will likely rely on an understanding of the human's ability to outperform the machine under these adverse conditions [73], and thus call upon new nonlinear auditory models [20],[75], as well as such models that are matched to nonlinearities in the speech production mechanism [80].

Appendix 14.A: Expectation-Maximization (EM) Estimation

In this appendix, we outline the steps required by the Expectation-Maximization (EM) algorithm for estimating GMM parameters. Further details can be found in [9],[60].

Consider the probability of occurrence of a collection of observed feature vectors, $X = \{\underline{x}_0, \underline{x}_1, \ldots \underline{x}_{M-1}\}$, from a particular speaker, as the union of probabilities of all possible states

---

[18] In Section 14.2, we refer to the mel- and sub-cepstrum as "spectral features." However, when we realize from Chapter 7 (Section 7.4.2) that a sequence can be recovered from its STFT magnitude (indeed, even possibly two samples of the STFT magnitude at each time instant), the distinction between "spectral" and "non-spectral" becomes nebulous.

(e.g., acoustic classes):

$$p(X|\lambda) = \sum_{i=1}^{I} p(X, i|\lambda)$$

$$= \sum_{i=1}^{I} p_i \prod_{n=0}^{M-1} b_i(\underline{x}_n)$$

$$= \prod_{n=0}^{M-1} \sum_{i=1}^{I} p_i b_i(\underline{x}_n)$$

where $\sum_{i=1}^{I}$ denotes the sum over all possible (hidden) acoustic classes, where p_i are the weights for each class, where the probability density function (pdf) for each state

$$b_i(\underline{x}) = \frac{1}{(2\pi)^{\frac{R}{2}}|\boldsymbol{\Sigma}_i|^{\frac{1}{2}}} e^{-\frac{1}{2}(\underline{x}-\underline{\mu}_i)^T \boldsymbol{\Sigma}_i^{-1}(\underline{x}-\underline{\mu}_i)}$$

with R the dimension of the feature vector, and where we have assumed the independence of feature vectors across the M observations (analysis frames in the speech context). The function $b_i(\underline{x})$ is an R-dimensional Gaussian pdf with a state-dependent mean vector $\underline{\mu}_i$ and covariance matrix $\boldsymbol{\Sigma}_i$. We call $p(X|\lambda)$ the Gaussian mixture model (GMM) for the feature vectors. The symbol λ represents the set of GMM mean, covariance, and weight parameters over all classes for a particular speaker, i.e.,

$$\lambda = \{p_i, \underline{\mu}_i, \boldsymbol{\Sigma}_i\}.$$

Suppose now that we have an estimate of λ denoted by λ^k. Our objective is to find a new estimate λ^{k+1} such that

$$p(X|\lambda^{k+1}) \geq p(X|\lambda^k).$$

The EM algorithm maximizes, with respect to the unknown λ^{k+1}, the expectation of the log-likelihood function $\log[p(X|\lambda^{k+1})]$, given the observed feature vector X and the current estimate (iterate) λ^k. This expectation over all possible acoustic classes is given by [9],[13],[60]

$$E(\log[p(X|\lambda^{k+1})]) = \sum_{i=1}^{I} p(X, i|\lambda^k) \log[p(X, i|\lambda^{k+1})].$$

Forming this sum is considered the *expectation* step of the EM algorithm. It can be shown, using the above formulation, that maximizing $E(\log[p(X|\lambda^{k+1})])$ over λ^{k+1} increases the kth log-likelihood, i.e., $p(X|\lambda^{k+1}) \geq p(X|\lambda^k)$. Solution to this maximization problem is obtained by differentiating $E(\log[p(X|\lambda^{k+1})])$ with respect to the unknown GMM mean, covariance, and weight parameters, i.e., $\lambda = \{p_i, \underline{\mu}_i, \mathbf{\Sigma}_i\}$, and is given by

$$p_i^{k+1} = \frac{1}{M} \sum_{n=0}^{M-1} p(i_n = i | \underline{x}_n, \lambda^k)$$

$$\underline{\mu}_i^{k+1} = \frac{\sum_{n=0}^{M-1} p(i_n = i | \underline{x}_n, \lambda^k) \underline{x}_n}{\sum_{n=0}^{M-1} p(i_n = i | \underline{x}_n, \lambda^k)}$$

$$\mathbf{\Sigma}_i^{k+1} = \frac{\sum_{n=0}^{M-1} p(i_n = i | \underline{x}_n, \lambda^k) \underline{x}_n \underline{x}_n^T}{\sum_{n=0}^{M-1} p(i_n = i | \underline{x}_n, \lambda^k)} - \underline{\mu}_i^{k+1} (\underline{\mu}_i^{k+1})^T$$

where

$$p(i_n = i | \underline{x}_n, \lambda^k) = \frac{p_i^k b_i^k(\underline{x}_n)}{\sum_{i=1}^{I} p_i^k b_i^k(\underline{x}_n)},$$

where $b_i^k(\underline{x}_n)$ is the ith pdf mixture component on the kth iteration, and where T in the covariance expression denotes matrix transpose. This is referred to as the *maximization* step of the EM algorithm. Replacement of the new model parameters is then made in the GMM to obtain the next estimate of the mixture components, $b_i^{k+1}(\underline{x}_n)$, (and weights) and the procedure is repeated.

An interesting interpretation of the EM algorithm is that it provides a "soft" *clustering* of the training feature vectors X into class components. In fact, we can think of the EM algorithm as a generalization of the k-means algorithm described in Chapter 12 (Section 12.4.2), where, rather than assigning each $\underline{x}_n$ to one centroid, each $\underline{x}_n$ is assigned probabilistically to all centroids. At each iteration in the EM algorithm, there arises a probability, $p(i_n = i | \underline{x}_n, \lambda^k)$ that an observed individual feature vector $\underline{x}_n$ comes from each of the assumed I classes. Thus we can "assign" each feature vector to a class from which the feature vector comes with the largest probability. On each iteration, the mixture component parameter estimation of the EM algorithm uses these probabilities. For example, the estimation equation for the mean vector of the ith mixture component is a weighted sum of the observed feature vectors where the weight for each feature vector is the (normalized) probability that the feature vector came from class i. Similar interpretations hold for estimation equations for the covariance matrix and the class weight associated with each Gaussian mixture component.

EXERCISES

14.1 Show that when each individual Gaussian mixture component of the GMM in Equation (14.4) integrates to unity, then the constraint $\sum_{i=1}^{I} p_i = 1$ ensures that the mixture density represents a true pdf, i.e., the mixture density itself integrates to unity. The scalars p_i are the mixture weights.

14.2 We saw in Section 14.3.3 that GMM modeling of a pdf takes place through appropriate selection of the means $\underline{\mu}_i$, covariance matrices $\mathbf{\Sigma}_i$, and probability weights p_i of the GMM.

    (a) Assuming that diagonal covariance matrices $\mathbf{\Sigma}_i$ are sufficient for good pdf approximations in GMM modeling, show that this simplification results in significantly reducing the number of unknown variables to be estimated. Assume a fixed number of mixtures in the GMM.

    (b) Argue that a diagonal covariance matrix $\mathbf{\Sigma}_i$ does not imply statistical independence of the underlying feature vectors. Under what condition on the GMM are the feature vectors statistically independent?

14.3 In this problem, we investigate the log-likelihood test associated with the GMM-based speaker verification of Section 14.3.3.

(a) Discarding the constant probability terms and applying Bayes' rule and the logarithm to Equation (14.5), show that

$$\Lambda(X) = \log[p(X|\lambda_C)] - \log[p(X|\lambda_{\bar{C}})].$$

where $X = \{\underline{x}_0, \underline{x}_1, \dots \underline{x}_{M-1}\}$, is a collection of M feature vectors obtained from a test utterance, and λ_C and $\lambda_{\bar{C}}$ denote the claimant and background speaker models, respectively.

(b) In one scenario, the probability that a test utterance does not come from the claimant speaker is determined from a set of imposter (background) speaker models $\{\lambda_{\bar{C}_1}, \lambda_{\bar{C}_2}, \dots \lambda_{\bar{C}_B}\}$. In this scenario, the single background model $\bar{C}$ is created from these individual models. Write an expression for the single (composite) background speakers' log-probability, first in terms of $p(X|\lambda_{\bar{C}_j})$ for $j = 1, 2, \dots, B$, and then as a function of the individual densities $p(\underline{x}_n|\lambda_{\bar{C}_j})$. Assume equally likely background speakers. *Hint:* Write $p(X|\lambda_{\bar{C}})$ as a joint probability density that the test utterance comes from one of the B background speakers.

(c) An alternate method for background "normalization" of the likelihood function is to first compute the claimant log-probability score and then subtract the maximum of the background scores over individual background (imposter) models, i.e.,

$$\Lambda(X) = \log[p(X|\lambda_C)] - \max_j\{\log[p(X|\lambda_{\bar{C}_j})]\}$$

where $\bar{C}_j$ denotes the individual background GMM models. Justify why this might be a reasonable approach to normalization with a variety of imposters, some close to the target and some far with respect to voice character.

(d) Observe that we could use simply the log-probability score of the claimed speaker, $\log[p(X|\lambda_C)]$, for speaker verification, without any form of background normalization. Argue why it is difficult to set a decision threshold in using this score for speaker detection (i.e., verification). Argue why, in contrast, the speaker identification task does not require a background normalization.

14.4 In this problem you consider the delta cepstrum in speaker recognition for removing a linear time-invariant channel degradation with frequency response $G(\omega)$, slowly varying in frequency, as given in Equation (14.9).

(a) Show that the difference of the mel-scale filter energy over two consecutive frames, i.e.,

$$\Delta \log[E_{mel}(pL, l)] = \log[E_{mel}(pL, l)] - \log[E_{mel}((p - 1)L, l)]$$

has no channel contribution $G(\omega)$ under the slowly varying assumption leading to Equation (14.9).

(b) Show that differencing the mel-cepstrum over two consecutive frames to form a delta cepstrum is equivalent to the log-energy difference of part (a). *Hint:* Determine the Fourier transform of $\Delta \log[E_{mel}(pL, l)]$.

14.5 Design a GMM-based handset recognition system that detects when an utterance is generated from a high-quality electret handset or a low-quality carbon-button handset. Assume you are given a training set consisting of a large set of electret and carbon-button handset utterances. *Hint:* Use the same methodology applied in the design of a GMM-based speaker recognition system.

14.6 An important problem in speaker verification is detecting the presence of a target speaker within a multi-speaker conversation. In this problem, you are asked to develop a method to determine *where* the target speaker is present within such an utterance. Assume that a GMM target model λ_C and a GMM background model $\lambda_{\bar{C}}$ have been estimated using clean, undistorted training data.

    (a) Suppose you estimate a feature vector $\underline{x}_n$ at some frame rate over a multi-speaker utterance where n here denotes the frame index. You can then compute a likelihood score for each frame as

$$\Lambda(\underline{x}_n) = \log[p(\underline{x}_n|\lambda_C)] - \log[p(\underline{x}_n|\lambda_{\bar{C}})]$$

and compare this against a threshold. Explain why this approach to speaker tracking is unreliable. Assume that there is no channel mismatch between the training and test data so that channel compensation is not needed.

    (b) An alternative approach to speaker tracking is to form a running string of feature vectors over M consecutive frames and sum all frame likelihoods above a threshold, i.e.,

$$\Lambda(I) = \sum_{\Lambda(\underline{x}_n)>\theta} \Lambda(\underline{x}_n)$$

where n represents the index of the feature vectors within the M-frame sliding "window" (denoted by I) and where $\Lambda(\underline{x}_n)$ denotes the likelihood ratio based on target and background speaker scores from part (a). When $\Lambda(I)$ falls above a fixed threshold, we say that the target speaker is present at the window center. (Note that there are many variations of this sliding-window approach.) Discuss the advantages and disadvantages of this log-likelihood statistic for tracking the presence of the target speaker. Consider the tradeoff between temporal resolution and speaker verification accuracy for various choices of the window size M, accounting for boundaries between different speakers and silence gaps.

    (c) Consider again a test utterance consisting of multi-speakers, but where the speech from each talker has traversed a different linear channel. How would you apply cepstral mean subtraction (CMS) in a meaningful way on the test utterance, knowing that more than one linear channel had affected the utterance? If you were given *a priori* the talker segmentation, i.e., the boundaries of each talker's speech, how might you modify your approach to cepstral mean subtraction?

    (d) Repeat part (c), but where RASTA is applied to the test utterance. Does RASTA provide an advantage over CMS in the scenarios of part (c)? Explain your reasoning.

    (e) You are now faced with an added complexity; not only are there present different speakers within an utterance whose speech is distorted by different linear channels, but also each speaker may talk on either an electret or a carbon-button telephone handset, and this may occur in both the training and test data. Assuming you have a handset detector that can be applied over a string of feature vectors (e.g., Exercise 14.5), design a method to account for handset differences between training and test data, as well as within the test utterance. You can use any of the approaches described in this chapter, including handset mapping and handset normalization (*hnorm*).

14.7 In speaker recognition, it is revealing to compare the importance of measured features for human and machine speaker recognition. A perception of breathiness, for example, may correspond to certain coarse- and fine-structure characteristics in the glottal flow derivative such as a large open quotient and aspiration, respectively, and its importance for the human may imply importance for machine recognition. On the other hand, certain features useful to machines may not be useful to humans. Propose some possible source and vocal tract system features for this latter case.

14.8 In Section 14.6.2, in illustrating the generation of phantom formants due to nonlinear handset distortion, we used all-pole spectral estimates of both undistorted and distorted waveforms. In such analysis, it is important, however, to ensure that the phantom formants are not artifactual effects from the analysis window length and position. This is because all-pole spectral estimates have been shown to be dependent on the window position and length [58].

Panels (a)-(c) of Figure 14.20 show the STFT magnitude and its all-pole (14th order) envelope of a time segment from a carbon-button waveform for different window positions and lengths. Pole estimation was performed using the autocorrelation method of linear prediction. Because the speech is from a low-pitched male, pole estimation suffers little from aliasing in the autocorrelation, as

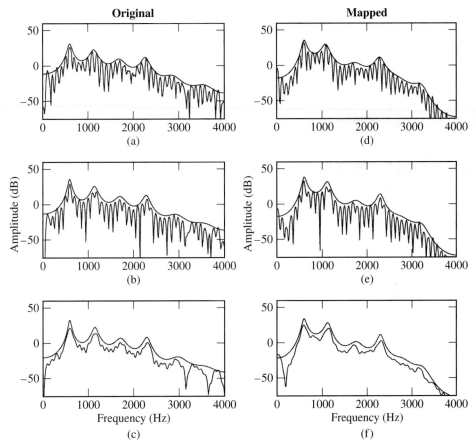

Figure 14.20 Insensitivity of phantom formants to window position and length: Panels (a) through (c) illustrate the STFT magnitude and its all-pole (14th order) envelope of a time segment of a carbon-button waveform taken from the HTIMIT database. Window specifications are (a) length of 25 ms and time zero position, i.e., center of the short-time segment, (b) length of 25 ms and displaced half a pitch period from time zero, and (c) length of 15 ms and at time zero position. Panels (d) through (f) illustrate the spectra from the corresponding electret-to-carbon-button mapped waveform segment for the same window characteristics.

SOURCE: T.F. Quatieri, D.A. Reynolds, and G.C. O'Leary, "Estimation of Handset Nonlinearity with Application to Speaker Recognition" [55]. ©2000, IEEE. Used by permission.

described in Chapter 5. Window specifications are (a) length of 25 ms and time zero position (i.e., the center of the time segment), (b) length of 25 ms and displaced half a pitch period from time zero, and (c) length of 15 ms and at time zero position. Panels (d) through (f) illustrate the spectra from the corresponding electret-to-carbon mapped waveform segment for the same window characteristics. The forward handset mapping of Figure 14.17 was applied. Explain the implication of the analysis in Figure 14.20 for the validity of the presence of phantom formants.

14.9 Let $x[n]$ denote an undistorted signal, to be referred to as the *reference signal*, and $y[n]$ the output of a nonlinear system. The pth order *Volterra series* truncated at N terms is given by [72]

$$y[n] = \sum_{m_1=0}^{N-1} h_1[m_1]x[n - m_1] + \sum_{m_1=0}^{N-1}\sum_{m_2=0}^{N-1} h_2[m_1, m_2]x[n - m_1]x[n - m_2]$$

$$+ \ldots + \sum_{m_1=0}^{N-1} \ldots \sum_{m_p=0}^{N-1} h_p[m_1, m_2, \ldots m_p]x[n - m_1]x[n - m_2] \ldots x[n - m_p] \quad (14.16)$$

which can be thought of as a generalized Taylor series with memory.

(a) Show that when the nonlinear terms are eliminated in Equation (14.16), the series reduces to the standard convolutional form of a linear system.

(b) Using a least-squared-error estimation approach, based on a quadratic error between a measurement $s[n]$ and the Volterra series model output $y[n]$, solve for the Volterra series parameters in Equation (14.16). You might consider a matrix formulation in your solution.

(c) In nonlinear system modeling and estimation, the Volterra series has the advantage of not requiring a specific form and introduces arbitrary memory into the model. It also has the advantage of converting an underlying nonlinear estimation problem into a linear one because the series in Equation (14.16) is *linear* in the unknown parameters. Nevertheless, although an arbitrary-order Volterra series in theory represents any nonlinearity, it has a number of limitations. Argue why the Volterra series has convergence problems when the nonlinearity contains discontinuities, as with a saturation operation. (Note that a Taylor series has a similar convergence problem.) Also argue that simple concatenated systems, such as a low-order FIR filter followed by a low-order polynomial nonlinearity, require a much larger-order Volterra series than the original system and thus the Volterra series can be an inefficient representation in the sense that the parameters represent a redundant expansion of the original set possibly derived from physical principles.

14.10 Show that the gradient of $E(\underline{a})$ in Equation (14.15) is given by $\nabla E = 2\mathbf{J}^T \underline{f}$, where $\mathbf{J}$ is the Jacobian matrix of first derivatives of the components of the residual vector in Equation (14.14), i.e., the elements of $\mathbf{J}$ are given by

$$J_{ij} = \frac{\partial f_i}{\partial a_j}$$

where f_i is the ith element of $\underline{f}(\underline{a})$.

14.11 In general, there will not be a unique solution in fitting the nonlinear handset representation in Equation (14.11) to a measurement, even if the measurement fits the model exactly. One problem is that the reference input $x[n]$ may be a scaled version of the actual input. Show that although the coefficients of the polynomial nonlinearity in Equation (14.11) can account for an arbitrary input scaling, estimated polynomial coefficients need not equal the underlying coefficients in spite of providing a match to the measured waveform. *Hint:* Consider $y[n] = \sum_{k=0}^{P} q_k(cx[n])^k$, where c is an arbitrary scale factor.

14.12 In this problem, you develop a missing (mel-scale filter energy) feature detector based on the generalized spectral subtraction noise threshold rules. You then dynamically modify the probability computations performed in a GMM recognizer by dividing the GMM into "good feature" and "bad feature" contributions based on the missing feature detector.

(a) Let $y[n]$ be a discrete-time noisy sequence

$$y[n] = x[n] + b[n]$$

where $x[n]$ is the desired sequence and $b[n]$ is uncorrelated background noise, with power spectra given by $S_x(\omega)$ and $S_b(\omega)$, respectively. Recall the generalized spectral subtraction suppression filter of Chapter 13 expressed in terms of relative signal level $Q(\omega)$ as

$$H_s(n, \omega) = [1 - \alpha Q(n, \omega)^{\gamma_1}]^{\gamma_2}, \quad \text{if } Q(n, \omega)^{\gamma_1} < \frac{1}{\alpha + \beta}$$

$$= [\beta Q(n, \omega)^{\gamma_1}]^{\gamma_2}, \qquad \text{otherwise}$$

where the relative signal level

$$Q(n, \omega) = \left[\frac{\hat{S}_b(\omega)}{|Y(n, \omega)|^2} \right]^{1/2}$$

with $|Y(n, \omega)|^2$ the squared STFT magnitude of the measurement $y[n]$ and with $\hat{S}_b(\omega)$ the background power spectrum estimate. Given the above threshold condition, $Q(n, \omega)^{\gamma_1} < \frac{1}{\alpha+\beta}$, decide on "good" and "bad" spectral regions along the frequency axis. Then map your decisions in frequency to the mel-scale.

(b) Suppose now that we compute mel-scale filter energies from $|Y(n, \omega)|^2$. Repeat part (a) by first designing a generalized spectral subtraction algorithm based on the mel-scale filter energies. Then design a missing feature detector based on thresholding the mel-scale energy features. Discuss the relative merits of your methods in parts (a) and (b) for detecting "good" and "bad" mel-scale filter energies.

(c) Propose a method to dynamically modify the probability computations performed in the GMM recognizer, removing missing features determined from your detector of part (a) or (b). *Hint:* Recall that for a speaker model $\lambda = \{p_i, \underline{\mu}_i, \Sigma_i\}$, the GMM pdf is given by

$$p(\underline{x}|\lambda) = \sum_{i=1}^{I} p_i b_i(\underline{x})$$

where

$$b_i(\underline{x}) = \frac{1}{(2\pi)^{\frac{R}{2}} |\Sigma_i|^{\frac{1}{2}}} e^{-\frac{1}{2}(\underline{x}-\underline{\mu}_i)^T \Sigma_i^{-1}(\underline{x}-\underline{\mu}_i)}$$

and where $\underline{\mu}_i$ is the state mean vector and Σ_i is the state covariance matrix. Then write each $b_i(\underline{x})$ as a product of pdf's of feature vector elements x_j, $j = 0, 1, \ldots R - 1$, under the assumption that each covariance matrix Σ_i is diagonal.

(d) Discuss the relation of the above missing feature strategy to the possible use of redundancy in the human auditory system, helping to enable listeners to recognize speakers in adverse conditions.

14.13 In this problem, you develop a feature set based on the Teager energy operator of Chapter 11 applied to subband-filter outputs in place of the conventional energy operation [25]. The discrete-time three-point Teager energy operator is given by

$$\Psi(x[n]) = x^2[n] - x[n-1]x[n+1]$$

and for a discrete-time AM-FM sinewave gives approximately the squared product of sinewave amplitude and frequency:

$$\Psi(x[n]) \approx A^2[n]\omega^2[n].$$

In this problem, you analyze the robustness of this energy measure to additive background noise with certain properties.

(a) Consider the sum of a desired speech signal $x[n]$ and noise background $b[n]$

$$y[n] = x[n] + b[n].$$

Show that the output of the Teager energy operator can be expressed as

$$\Psi[y] = \Psi[x] + \Psi[b] + 2\tilde{\Psi}[x, b]$$

where, for simplicity, we have removed the time index and where $\tilde{\Psi}[x, b]$, the "cross Teager energy" of $x[n]$ and $b[n]$, is given by

$$\tilde{\Psi}[x, b] = x[n]b[n] - \frac{1}{2}x[n-1]b[n+1] - \frac{1}{2}x[n+1]b[n-1].$$

(b) Show that, in a stochastic framework, with $x[n]$ and $b[n]$ uncorrelated, we have

$$E(\Psi[y]) = E(\Psi[x]) + E(\Psi[b])$$

where E denotes expected value. Then show that

$$E(\Psi[b]) = r_b[0] - r_b[2]$$

where $r_b[0]$ and $r_b[2]$ denote, respectively, the 0th and 2nd autocorrelation coefficients of the noise background, $b[n]$. Using this result, give a condition on the autocorrelation of the noise such that

$$E(\Psi[y]) = E(\Psi[x])$$

i.e., that in a statistical sense, the Teager energy is immune to the noise. Explain how the instantaneous nature of actual signal measurements, in contrast to ensemble signal averages, will affect this noise immunity in practice. How might you exploit signal ergodicity?

(c) Show that with the conventional (squared amplitude) definition of energy, we have

$$E(y^2[n]) = r_x[0] + r_b[0]$$

where $x[n]$ and $b[n]$ are uncorrelated and where $r_x[0]$ and $r_b[0]$ denote the 0th autocorrelation coefficients of $x[n]$ and $b[n]$, respectively. Then argue that, unlike with the Teager energy, one cannot find a constraint on the noise autocorrelation such that the standard energy measure is immune to noise, i.e., the "noise bias" persists. How is this difference in noise immunity of the standard and Teager energy affected when the noise is white?

(d) To obtain a feature vector, suppose the Teager energy is derived from subband signals. Therefore, we want to estimate the autocorrelation coefficients of

$$\Psi[y_l] = \Psi[y_l] + \Psi[b_l] + 2\tilde{\Psi}[x_l, b_l]$$

where $y_l[n]$, $x_l[n]$, and $b_l[n]$ are signals associated with the lth mel-scale filter $v_l[n]$ of Figure 14.2b. Repeat parts (a)-(c) for this subband signal analysis.

14.14 (MATLAB) In this problem, you investigate the time resolution properties of the mel-scale and subband filter output energy representations. You will use the speech signal *speech1_10k* in workspace *ex14M1.mat* and functions *make_mel_filters.m* and *make_sub_filters.m*, all located in companion website directory *Chap_exercises/chapter14*.

(a) Argue that the subband filters, particularly for high frequencies, are capable of greater temporal resolution of speech energy fluctuations within auditory critical bands than are the mel-scale filters. Consider the ability of the energy functions to reflect speech transitions, periodicity, and short-time events such as plosives in different spectral regions. Assume that the analysis window duration used in the STFT of the mel-scale filter bank configuration is 20 ms and is about equal to the length of the filters in the low 1000-Hz band of the subband filter bank. What constrains the temporal resolution for each filter bank?

(b) Which filter bank structure functions more similar to the wavelet transform and why?

(c) Write a MATLAB routine to compute the mel-scale filter and subband filter energies. In computing the mel-scale filter energies, use a 20-ms Hamming analysis window and the 24-component mel-scale filter bank from function *make_mel_filters.m*, assuming a 4000-Hz bandwidth. In computing the subband-filter energies, use complex zero-phase subband filters from function *make_sub_filters.m*. For each filter bank, plot different low- and high-frequency filter-bank energies in time for the voiced speech signal *speech1_10k* in workspace *ex14M1.mat*. For the subband filter bank, investigate different energy smoothing filters $p[n]$ and discuss the resulting temporal resolution differences with respect to the mel-scale filter analysis.

14.15 (MATLAB) In this problem you investigate a model for the nonlinearity of the carbon-button handset, derived from physical principles, given by [1],[43]

$$Q(u) = \beta \frac{1-u}{1-\alpha u} + \gamma \tag{14.17}$$

where typical values of α fall in the range $0 < \alpha < 1$.

(a) Show that, when we impose the constraints that the input values of 0 and 1 map to themselves and that $Q(u) = 1$ for $u > 1$, this results in the simplified model

$$Q(u) = \frac{(1-\alpha)u}{1-\alpha u}, \quad u \le 1$$
$$= 1, \quad u > 1 \tag{14.18}$$

thus containing one free variable and a fixed upper saturation level of $+1$.

(b) The nonlinearity in Equation (14.18) can take on a variety of asymmetric shapes, reflecting the physical phenomenon that the resistance of carbon-button microphone carbon granules reacts differently to positive pressure variation than to negative variation. Write a MATLAB routine to plot the nonlinearity of Equation (14.18) for a variety of values of α. Then apply your nonlinear operation to a synthetic voiced speech waveform of your choice, illustrating the ability of the handset nonlinearity model to create phantom formants. Explain why, in spite of the nonlinearity of Equation (14.17) being based on physical principles and able to create phantom formants, a polynomial model may be more effective when considering differences and age of handsets.

BIBLIOGRAPHY

[1] M.H. Abuelma'atti, "Harmonic and Intermodulation Distortion of Carbon Microphones," *Applied Acoustics*, vol. 31, pp. 233–243, 1990.

[2] T.V. Ananthapadmanabha and G. Fant, "Calculation of True Glottal Flow and its Components," *Speech Communications*, vol. 1, pp. 167–184, 1982.

[3] B.S. Atal, "Automatic Recognition of Speakers from Their Voices," *IEEE Proc.*, vol. 64, no. 4, pp. 460–475, 1976.

[4] B.S. Atal, "Effectiveness of Linear Prediction Characteristics of the Speech Wave for Automatic Speaker Identification and Verification," *J. Acoustical Society of America*, vol. 55, no. 6, pp. 1304–1312, June 1974.

[5] J.P. Campbell, "Speaker Recognition: A Tutorial," *Proc. IEEE*, vol. 85, no. 9, pp. 1437–1462, Sept. 1997.

[6] F.R. Clarke and R.W. Becker, "Comparison of Techniques for Discriminating Talkers," *J. Speech and Hearing Research*, vol. 12, pp. 747–761, 1969.

[7] S.B. Davies and P. Mermelstein, "Comparison of Parametric Representations for Monosyllabic Word Recognition in Continuously Spoken Sentences," *IEEE Trans. Acoustics, Speech, and Signal Processing*, vol. ASSP–28, no. 4, pp. 357–366, Aug. 1980.

[8] J.R. Deller, J.G. Proakis, and J.H.L. Hansen, *Discrete-Time Processing of Speech*, Macmillan Publishing Co., New York, NY, 1993.

[9] A. Dempster, N. Laird, and D. Rubin, "Maximum Likelihood from Incomplete Data via the EM Algorithm," *J. Royal Statistical Society*, vol. 39, pp. 1–38, 1977.

[10] G.R. Doddington, "Speaker Recognition—Identifying People by Their Voices," *Proc. IEEE*, vol. 73, pp. 1651–1664, 1985.

[11] A.W. Drake, *Fundamentals of Applied Probability Theory*, McGraw-Hill, New York, NY, 1967.

[12] A. Drygajlo and M. El-Maliki, "Use of Generalized Spectral Subtraction and Missing Feature Compensation for Robust Speaker Verification," *Speaker Recognition and Its Commercial and Forensic Applications (RLA2C)*, Avignon, France, April 1998.

[13] R.O. Duda, P.E. Hart, and D.G. Stork, *Pattern Classification*, 2nd Edition, John Wiley and Sons, New York, NY, 2001.

[14] E. Erzin, A. Cetin, and Y. Yardimci, "Subband Analysis for Robust Speech Recognition in the Presence of Car Noise," *Proc. IEEE Int. Conf. Acoustics, Speech, and Signal Processing*, vol. 1, pp. 417–420, Detroit, MI, May 1995.

[15] European Telecommunication Standards Institute, "European Digital Telecommunications System (Phase 2); Full Rate Speech Processing Functions (GSM 06.01)," ETSI, 1994.

[16] G. Fant, "Glottal Flow: Models and Interaction," *J. Phonetics*, vol. 14, pp. 393–399, 1986.

[17] W.M. Fisher, G.R. Doddington, and K.M. Goudie-Marshall, "The DARPA Speech Recognition Research Database: Specifications and Status," *Proc. DARPA Speech Recognition Workshop*, pp. 93–99, Palo Alto, CA, 1986.

[18] R. Fletcher, "Generalized Inverses for Nonlinear Equations and Optimization," in *Numerical Methods for Nonlinear Algebraic Equations*, P. Rabinowitz, Gordon, and Breach Science Publishers, eds., New York, NY, 1970.

[19] S. Furui, "Cepstral Analysis Technique for Automatic Speaker Verification," *IEEE Trans. Acoustics, Speech, and Signal Processing*, vol. 29, no. 2, pp. 254–272, April 1981.

[20] O. Ghitza, "Auditory Models and Human Performance in Tasks Related to Speech Coding and Speech Recognition," *IEEE Trans. Speech and Audio Process*, vol. 2, no. 1, pp. 115–132, Jan. 1994.

[21] H. Gish and M. Schmidt, "Text-Independent Speaker Identification," *IEEE Signal Processing Magazine*, vol. 11, no. 4, pp. 18–32, Oct. 1994.

[22] B. Gold and N. Morgan, *Speech and Audio Signal Processing*, John Wiley and Sons, New York, NY, 2000.

[23] J.M. Huerta and R.M. Stern, "Speech Recognition from GSM Coder Parameters," *Proc. 5th Int. Conf. on Spoken Language Processing*, vol. 4, pp. 1463–1466, Sydney, Australia, 1998.

[24] ITU-T Recommendation G.729, "Coding of Speech at 8 kbps Using Conjugate-Structure Algebraic-Code-Excited Linear Prediction," June 1995.

[25] F. Jabloun and A.E. Cetin, "The Teager Energy Based Feature Parameters for Robust Speech Recognition in Noise," *Proc. IEEE Int. Conf. Acoustics, Speech, and Signal Processing*, vol. 1, pp. 273–276, Phoenix, AZ, March 1999.

[26] C.R. Jankowski, *Fine Structure Features for Speaker Identification*, Ph.D. Thesis, Massachusetts Institute of Technology, Dept. Electrical Engineering and Computer Science, June 1996.

[27] C.R. Jankowski, T.F. Quatieri, and D.A. Reynolds, "Measuring Fine Structure in Speech: Application to Speaker Identification," *Proc. IEEE Int. Conf. Acoustics, Speech, and Signal Processing*, vol. 1, pp. 325–328, Detroit, MI, May 1995.

[28] C.R. Jankowski, A. Kalyanswamy, S. Basson, and J. Spitz, "NTIMIT: A Phonetically Balanced, Continuous Speech, Telephone Bandwidth Speech Database," *Proc. IEEE Int. Conf. Acoustics, Speech, and Signal Processing*, vol. 1, pp. 109–112, Albuquerque, NM, 1990.

[29] F. Jelinek, *Statistical Methods for Speech Recognition*, The MIT Press, Cambridge, MA, 1998.

[30] J.F. Kaiser, "On a Simple Algorithm to Calculate the 'Energy' of a Signal," in *Proc. IEEE Int. Conf. Acoustics, Speech, and Signal Processing*, vol. 1, Albuquerque, NM, pp. 381–384, April 1990.

[31] D.H. Klatt and L.C. Klatt, "Analysis, Synthesis, and Perception of Voice Quality Variations among Female and Male Talkers," *J. Acoustical Society of America*, vol. 87, no. 2, pp. 820–857, 1990.

[32] W.B. Kleijn and K.K. Paliwal, eds., *Speech Coding and Synthesis*, Elsevier, Amsterdam, the Netherlands, 1995.

[33] Linguistic Data Consortium, http://www.ldc.upenn.edu.

[34] R.P. Lippman and B.A. Carlson, "Using Missing Feature Theory to Actively Select Features for Robust Speech Recognition with Interruptions, Filtering, and Noise," *Proc. Eurospeech97*, vol. 1, pp. KN 37–40, Rhodes, Greece, Sept. 1997.

[35] J. He, L. Liu, and G. Palm, "On the Use of Features from Prediction Residual Signals in Speaker Identification," *Proc. Eurospeech95*, vol. 1, pp. 313–316, Madrid, Spain, 1995.

[36] P. Maragos, J. F. Kaiser, and T. F. Quatieri, "On Amplitude and Frequency Demodulation Using Energy Operators," *IEEE Trans. Signal Processing*, vol. 41, no. 4, pp. 1532–1550, April 1993.

[37] P. Maragos, J.F. Kaiser, and T.F. Quatieri, "Energy Separation in Signal Modulations with Application to Speech Analysis," *IEEE Trans. Signal Processing*, vol. 41, no. 10, pp. 3025–3051, Oct. 1993.

[38] A. Martin, G. Doddington, T. Kamm, M. Ordowski, and M. Przybocki, "The DET Curve in Assessment of Detection Task Performance," *Proc. Eurospeech97*, vol. 4, pp. 1895–1898, Rhodes, Greece, 1997.

[39] T. Masuko, T. Hitotsumatsu, K. Tokuda, and T. Kobayashi, "On the Security of HMM-Based Speaker Verification Systems against Imposture Using Synthetic Speech," *Proc. Eurospeech99*, vol. 3, pp. 1223–1226, Budapest, Hungary, Sept. 1999.

[40] G. McLachlan, *Mixture Models*, Macmillan Publishing Co., New York, NY, 1971.

[41] C. Mokbel, L. Mauuary, D. Jouvet, J. Monne, C. Sorin, J. Simonin, and K. Bartkova, "Towards Improving ASR Robustness for PSN and GSM Telephone Applications," *2nd IEEE Workshop on Interactive Voice Technology for Telecommunications Applications*, vol. 1, pp. 73–76, 1996.

[42] National Institute of Standards and Technology (NIST), "NIST Speaker Recognition Workshop Notebook," NIST Administered Speaker Recognition Evaluation on the Switchboard Corpus, Spring 1996–2001.

[43] H.F. Olson, *Elements of Acoustical Engineering*, Chapman and Hall, London, England, 1940.

[44] D. O'Shaughnessy, *Speech Communication: Human and Machine*, Addison-Wesley, Reading, MA, 1987.

[45] M.T. Padilla, *Applications of Missing Feature Theory to Speaker Recognition*, Masters Thesis, Massachusetts Institute of Technology, Dept. Electrical Engineering and Computer Science, Feb. 2000.

[46] A. Papoulis, *Probability, Random Variables, and Stochastic Processes*, McGraw-Hill, New York, NY, 1965.

[47] A. Pickles, *An Introduction to Auditory Physiology*, Academic Press, 2nd Edition, New York, NY, 1988.

[48] M.D. Plumpe, T.F. Quatieri, and D.A. Reynolds, "Modeling of the Glottal Flow Derivative Waveform with Application to Speaker Identification," *IEEE Trans. Speech and Audio Processing*, vol. 1, no. 5, pp. 569–586, Sept. 1999.

[49] M.D. Plumpe, *Modeling of the Glottal Flow Derivative Waveform with Application to Speaker Identification*, Masters Thesis, Massachusetts Institute of Technology, Dept. Electrical Engineering and Computer Science, Feb. 1997.

[50] M. Pryzybocki and A. Martin, "NIST Speaker Recognition Evaluation—1997," *Speaker Recognition and Its Commercial and Forensic Applications (RLA2C)*, Avignon, France, pp. 120–123, April 1998.

[51] T.F. Quatieri and R.J. McAulay, "Shape-Invariant Time-Scale and Pitch Modification of Speech," *IEEE Trans. Acoustics, Speech, and Signal Processing*, vol. 40, no. 3, pp. 497–510, March 1992.

[52] T.F. Quatieri, C.R. Jankowski, and D.A. Reynolds, "Energy Onset Times for Speaker Identification," *IEEE Signal Processing Letters*, vol. 1, no. 11, pp. 160–162, Nov. 1994.

[53] T.F. Quatieri, R.B. Dunn, and D.A. Reynolds, "On the Influence of Rate, Pitch, and Spectrum on Automatic Speaker Recognition Performance," *Proc. Int. Conf. on Spoken Language Processing*, Beijing, China, Oct. 2000.

[54] T.F. Quatieri, D.A. Reynolds, and G.C. O'Leary, "Handset Nonlinearity Estimation with Application to Speaker Recognition," in NIST Speaker Recognition Notebook, NIST Administered Speaker Recognition Evaluation on the Switchboard Corpus, June 1997.

[55] T.F. Quatieri, D.A. Reynolds, and G.C. O'Leary, "Estimation of Handset Nonlinearity with Application to Speaker Recognition," *IEEE Trans. Speech and Audio Processing*, vol. 8, no. 5, pp. 567–584, Sept. 2000.

[56] T.F. Quatieri, E. Singer, R.B. Dunn, D.A. Reynolds, and J.P. Campbell, "Speaker and Language Recognition using Speech Codec Parameters," *Proc. Eurospeech99*, vol. 2, pp. 787–790, Budapest, Hungary, Sept. 1999.

[57] T.F. Quatieri, R.B. Dunn, D.A. Reynolds, J.P. Campbell, and E. Singer, "Speaker Recognition using G.729 Speech Codec Parameters," *Proc. IEEE Int. Conf. Acoustics, Speech, and Signal Processing*, vol. 2., pp. 1089–1093, Istanbul, Turkey, June 2000.

[58] L.R. Rabiner and R.W. Schafer, *Digital Processing of Speech Signals*, Prentice Hall, Englewood Cliffs, NJ, 1978.

[59] L.R. Rabiner and B. Juang, *Fundamentals of Speech Recognition*, Prentice Hall, Englewood Cliffs, NJ, 1993.

[60] D.A. Reynolds, *A Gaussian Mixture Modeling Approach to Text-Independent Speaker Identification*, Ph.D. Thesis, Georgia Institute of Technology, Atlanta, GA, 1992.

[61] D.A. Reynolds, "Speaker Identification and Verification Using Gaussian Mixture Speaker Models," *Speech Communication*, vol. 17, pp. 91–108, Aug. 1995.

[62] D.A. Reynolds, T.F. Quatieri, and R.B. Dunn, "Speaker Verification Using Adapted Gaussian Mixture Models," *Digital Signal Processing*, Special Issue: NIST 1999 Speaker Recognition Workshop, J. Schroeder and J.P. Campbell, eds., Academic Press, vol. 10, no. 1–3, pp. 19–41, Jan./April/July 2000.

[63] D.A. Reynolds, "Effects of Population Size and Telephone Degradations on Speaker Identification Performance," *Proc. SPIE Conference on Automatic Systems for the Identification and Inspection of Humans*, 1994.

[64] D.A. Reynolds, "Speaker Identification and Verification Using Gaussian Mixture Speaker Models," *Proc. ESCA Workshop on Automatic Speaker Recognition*, pp. 27–30, Martigny, Switzerland, 1994.

[65] D.A. Reynolds, M.A. Zissman, T.F. Quatieri, G.C. O'Leary, and B.A. Carlson, "The Effects of Telephone Transmission Degradations on Speaker Recognition Performance," *Proc. IEEE Int. Conf. Acoustics, Speech, and Signal Processing*, Detroit, MI, May 1995.

[66] D.A. Reynolds, "HTIMIT and LLHDB: Speech Corpora for the Study of Handset Transducer Effects," *Proc. IEEE Int. Conf. Acoustics, Speech, and Signal Processing*, vol. 2, pp. 1535–1538, Munich, Germany, April 1997.

[67] D.A. Reynolds, "Comparison of Background Normalization Methods for Text-Independent Speaker Verification," *Proc. Eurospeech97*, vol. 1, pp. 963–967, Rhodes, Greece, Sept. 1997.

[68] D.A. Reynolds, "Automatic Speaker Recognition Using Gaussian Mixture Speaker Models," *MIT Lincoln Laboratory Journal*, vol. 8, no. 2, pp. 173–191, Fall 1995.

[69] M. Sambur, "Selection of Acoustical Features for Speaker Identification," *IEEE Trans. Acoustics, Speech, and Signal Processing*, vol. ASSP–23, no. 2, pp. 176–182, April 1975.

[70] R. Salami, C. Laflamme, J.P. Adoul, A. Kataoka, S. Hayashi, T. Moriya, C. Lamblin, D. Massaloux, S. Proust, P. Kroon, and Y. Shoham, "Design and Description of CS-ACELP: A Toll Quality 8-kbps Speech Coder," *IEEE Trans. Speech and Audio Processing*, vol. 6, no. 2, pp. 116–130, March 1998.

[71] M. Savic and S.K. Gupta, "Variable Parameter Speaker Verification System Based on Hidden Markov Modeling," *Proc. Int. Conf. Acoustics, Speech, and Signal Processing*, vol. 1, pp. 281–284, Albuquerque, NM, 1990.

[72] M. Schetzen, *The Volterra and Wiener Theories of Nonlinear Systems*, John Wiley and Sons, New York, NY, 1980.

[73] A. Schmidt-Nielsen and T.H. Crystal, "Speaker Verification by Human Listeners: Experiments Comparing Human and Machine Performance Using the NIST 1998 Speaker Evaluation Data," *Digital Signal Processing*, Special Issue: NIST 1999 Speaker Recognition Workshop, J. Schroeder and J.P. Campbell, eds., Academic Press, vol. 10, no. 1–3, pp. 249–266, Jan./April/July 2000.

[74] J. Schroeter and M.M. Sondhi, "Speech Coding Based on Physiological Models of Speech Production," chapter in *Advances in Speech Signal Processing*, S. Furui and M.M. Sondhi, eds., Marcel Dekker, New York, NY, 1992.

[75] S. Seneff, "A Joint Synchrony/Mean-Rate Model of Auditory Speech Processing," *J. Phonetics*, vol. 16, no. 1, pp. 55–76, Jan. 1988.

[76] J. Shroeder and J.P. Campbell, eds., *Digital Signal Processing*, Special Issue: NIST 1999 Speaker Recognition Workshop, Academic Press, vol. 10, no. 1–3, Jan./April/July 2000.

[77] F. Soong, A. Rosenberg, L. Rabiner, and B. Juang, "A Vector Quantization Approach to Speaker Recognition," *Proc. Int. Conf. Acoustics, Speech, and Signal Processing*, vol. 1, pp. 387–390, Tampa, FL, 1985.

[78] F. Soong and A. Rosenberg, "On the Use of Instantaneous and Transitional Spectral Information in Speaker Recognition," *IEEE Trans. Acoustics, Speech, and Signal Processing*, vol. ASSP–36, no. 6, pp. 871–879, June 1988.

[79] J. Tchorz and B. Kollmeier, "A Model of Auditory Perception as Front End for Automatic Speech Recognition," *J. Acoustical Society of America*, vol. 106, no. 4, Oct. 1999.

[80] H.M. Teager and S.M. Teager, "A Phenomenological Model for Vowel Production in the Vocal Tract," chapter in *Speech Science: Recent Advances*, R.G. Daniloff, ed., College-Hill Press, pp. 73–109, San Diego, CA, 1985.

[81] P. Thévenaz and H. Hügli, "Usefulness of the LPC-Residue in Text-Independent Speaker Verification," *Speech Communication*, vol. 17, no. 1–2, pp. 145–157, Aug. 1995.

[82] W.D. Voiers, "Perceptual Bases of Speaker Identity," *J. Acoustical Society of America*, vol. 36, pp. 1065–1073, 1964.

[83] S. van Vuuren and H. Hermansky, "On the Importance of Components of the Modulation Spectrum for Speaker Verification," *Proc. Int. Conf. on Spoken Language Processing*, Sydney, Australia, Nov. 1998.

[84] R. Zelinski and P. Noll, "Adaptive Transform Coding of Speech Signals," *IEEE Trans. Acoustics, Speech, and Signal Processing*, vol. ASSP–25, no. 4, pp. 299–309, Aug. 1977.

[85] M.A. Zissman, "Comparison of Four Approaches to Automatic Language Identification of Telephone Speech," *IEEE Trans. on Speech and Audio Processing*, vol. 4, no. 1, pp. 31–44, Jan. 1996.

Glossary

Speech Signal Processing

1-D	One-Dimensional	FBS	Filter Bank Summation
2-D	Two-Dimensional	FFT	Fast Fourier Transform
3-D	Three-Dimensional	FIR	Finite Impulse Response
A/D	Analog-to-Digital	FM	Frequency Modulation
ADPCM	Adaptive Differential Pulse Code Modulation	GFD	Glottal Flow Derivative
AFR	Average Firing Rate	GMM	Gaussian Mixture Model
AM	Amplitude Modulation	GMM-UBM	Gaussian Mixture Model-Universal Background Model
ANF	Auditory Nerve Fiber	HMM	Hidden Markov Model
AR	AutoRegressive	hnorm	handset normalization
AWT	Auditory Wavelet Transform	ICWT	Inverse Continuous Wavelet Transform
BW	BandWidth		
C/D	Continuous-to-Discrete	IDFT	Inverse Discrete Fourier Transform
CELP	Code Excited Linear Prediction	IHC	Inner Hair Cell
CF	Characteristic Frequency	IIR	Infinite Impulse Response
CMS	Cepstral Mean Subtraction	KFH	Key, Fowle, and Haggarty (phase function)
CWT	Continuous Wavelet Transform		
D/A	Digital-to-Analog	LF	Liljencrants-Fant (glottal flow model)
DAM	Diagnostic Acceptability Measure	LMAP	Linearized Maximum A Posteriori
D/C	Discrete-to-Continuous		
DCC	Delta Cepstral Coefficients	LPC	Linear Prediction Coding
DET	Detection Error Tradeoff	LSE	Least-Squared Error
DFT	Discrete Fourier Transform	LSF	Line Spectral Frequencies
DMM	Dynamical Mechanical Model	LTI	Linear Time-Invariant
DPCM	Differential Pulse Code Modulation	MAP	Maximum A Posteriori
DRT	Diagnostic Rhyme Test	MBE	Multi-Band Excitation
EER	Equal Error Rate	MELP	Mixed Excitation Linear Prediction
EM	Expectation-Maximization	ML	Maximum Likelihood
		MMSE	Minimum-Mean-Squared Error

MOS	Mean Opinion Score	RPELPC	Regular Pulse-Excited Linear Prediction Coding
MSE	Mean-Squared Error		
NL2SOL	NonLinear secant approximation To the Second Order part of the Least squared Hessian	SEEVOC	Spectral Envelope Estimation VOCoder
		SNR	Signal-to-Noise Ratio
OLA	OverLap and Add	SPL	Sound Pressure Level (with respect to ambient pressure)
PARCOR	PARtial CORrelation		
PCM	Pulse Code Modulation	STFT	Short-Time Fourier Transform
pdf	probability density function	STFTM	Short-Time Fourier Transform Magnitude
PM	Phase Modulation	STU	Secure Terminal Unit
PV	Principle phase Value	TSM	Time Scale Modification
RASTA	RelAtive SpecTrAl processing	UBM	Universal Background Model
rms	root mean square (square root of mean energy)	VOT	Voice Onset Time
		VQ	Vector Quantization
ROC	Region Of Convergence		

Units

bps	bits per second	kHz	kilohertz
cm	centimeter	m	meter
dB	deciBels	mm	millimeter
gm	gram	ms	milliseconds
Hz	Hertz	μs	microseconds
kbps	kilobits per second	Rad	Radians
kg	kilogram	s	seconds

Databases

(see http://www.ldc.upenn.edu)

HTIMIT	TIMIT passed through electret and carbon-button telephone handsets	NTIMIT	TIMIT passed through a telephone channel
		TIMIT	High-quality corpus of read speech
KING	High-quality and telephone-channel corpus of conversational speech		
		TSID	Tactical Speaker Identification corpus of read and conversational (giving directions) speech from radio transmitters and receivers
LDC	Linguistic Data Consortium		
NIST	National Institute of Standards and Technology		

Index

About the Author

Thomas F. Quatieri received the B.S. degree (summa cum laude) from Tufts University, Medford, Massachusetts, in 1973, and the SM, EE, and ScD degrees from the Massachusetts Institute of Technology (MIT), Cambridge, Massachusetts, in 1975, 1977, and 1979, respectively. He is currently a Senior Member of the Technical Staff at MIT Lincoln Laboratory, Lexington, Massachusetts.

In 1980, Dr. Quatieri joined the Sensor Processing Technology Group of MIT Lincoln Laboratory, where he worked on problems in multi-dimensional digital signal processing and image processing. Since 1983 he has been a member of the Information Systems Technology Group at Lincoln Laboratory, where he has been involved in digital signal processing for speech and audio applications, underwater sound enhancement, and data communications. His current focus is in the areas of speech and audio modification, enhancement, and coding, and speaker recognition. He has many publications in journals and conference proceedings, holds several patents, and co-authored chapters in numerous edited books including: Advanced Topics in Signal Processing (Prentice Hall, 1987), Advances in Speech Signal Processing (Marcel Dekker, 1991), and Speech Coding and Synthesis (Elsevier, 1995). He holds the position of Lecturer at MIT, where he has developed the graduate course Digital Speech Processing, and is active in advising graduate students on the MIT campus.

Dr. Quatieri is the recipient of the 1982 Paper Award of the IEEE Acoustics, Speech and Signal Processing Society for the paper, "Implementation of 2-D Digital Filters by Iterative Methods." His publications also include "Speech Analysis/Synthesis Based on a Sinusoidal Representation," winner of the 1990 IEEE Signal Processing Society's Senior Award; and "Energy Separation in Signal Modulations with Application to Speech Analysis," winner of the 1994 IEEE Signal Processing Society's Senior Award and the 1995 IEEE W.R.G. Baker Prize Award. He was a member of the IEEE Digital Signal Processing Technical Committee, from 1983 to 1992 served on the steering committee for the bi-annual Digital Signal Processing Workshop, and was Associate Editor for the IEEE Transactions on Signal Processing in the area of nonlinear systems. He currently serves on the IEEE Speech Processing Technical Committee. Dr. Quatieri is a Fellow of the IEEE and is also a member of Tau Beta Pi, Eta Kappa Nu, Sigma Xi, and the Acoustical Society of America.

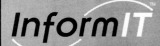